REFLECTIONS ON THE FOUNDATIONS OF MATHEMATICS

Lecture Notes in Logic

A Publication of

The Association for Symbolic Logic

LECTURE NOTES IN LOGIC 15

REFLECTIONS ON THE FOUNDATIONS OF MATHEMATICS

Essays in honor of Solomon Feferman

Edited by

Wilfried Sieg
Department of Philosophy
Carnegie Mellon University

Richard Sommer
Center for the Study of Language and Information
Stanford University

Carolyn Talcott
Computer Science Department
Stanford University

ASL
ASSOCIATION FOR SYMBOLIC LOGIC

CRC Press
Taylor & Francis Group
Boca Raton London New York

CRC Press is an imprint of the
Taylor & Francis Group, an informa business
AN A K PETERS BOOK

First published 2002 by A K Peters, Ltd.

Published 2018 by CRC Press
Taylor & Francis Group
6000 Broken Sound Parkway NW, Suite 300
Boca Raton, FL 33487-2742

ISBN 13: 978-1-56881-169-7 (hbk)

Visit the Taylor & Francis Web site at
http://www.taylorandfrancis.com

and the CRC Press Web site at
http://www.crcpress.com

Library of Congress Cataloging-in-Publication Data

Reflections on the foundations of mathematics : essays in honor of Solomon Feferman / edited by Wilfried Sieg, Richard Sommer, Carolyn Talcott.
p. cm. – (Lecture notes in logic ; 15)
Includes bibliographical references.
ISBN 1-56881-169-1 – ISBN 1-56881-1-170-5 (pbk.)
1. Logic, Symbolic and mathematical–Congresses. 2. Mathematics–Philosophy–Congresses. I. Feferman, Solomon. II. Sieg, Wilfried, 1945- III. Sommer, Richard, 1961- IV. Talcott, Carolyn L. V. Series.

QA9.A1 R44 2002
511.3–dc21 2002025778

Publisher's note: This book was typeset in LaTeX, by the ASL Typesetting Office, from electronic files produced by the authors, using the ASL documentclass `asl.cls`. The fonts are Monotype Times Roman. The cover design is by Richard Hannus, Hannus Design Associates, Boston, Massachusetts.

Association for Symbolic Logic: C. Ward Henson, Publisher Mathematics Department University of Illinois 1409 West Green Street Urbana, Illinois 61801

PREFACE

Reflections, a symposium on the foundations of mathematics, was held at Stanford University on December 11–13, 1998. The symposium was organized to honor Solomon Feferman who has played an enormous role in shaping the field over the last 40 years. It was timed so that its last day would coincide with Feferman's 70-th birthday; this provided a very special occasion to celebrate him and his career-long dedication to foundational research.

Jon Barwise and Wilfried Sieg, both doctoral students of Feferman, took the initiative in early 1996 of planning what became playfully called the *Feferfest*; Carolyn Talcott and Rick Sommer soon joined as the local Stanford organizers. Jon was instrumental in our subsequent venture to shape a program; he opened the symposium and gave a lecture on his latest work; he helped with the initial steps towards this volume, even after he had been diagnosed with cancer. We miss him.

The symposium was structured around proof-theoretically inspired themes. True to their origin in the work of David Hilbert and Paul Bernays, proof-theoretic investigations have sustained a special emphasis on or, at least a genuine connection to, broad philosophical issues. Stanford University has had an important role in fostering such work through actively engaged faculty, doctoral students, and visitors. Feferman has been at the very center of these activities.

This was an opportune moment to reflect broadly on such investigations, but also to connect them systematically with topics in Feferman's work. His primary contributions have been to proof theory, recursion theory and, in more recent years, to an analysis of the development of mathematical logic in the twentieth century. Indeed, all of these matters are of intense interest in the current discussion concerning modern mathematical thought.

The symposium had six sessions. The details of the program - with the names of contributors and chairs - can be found at the very back of the book. The papers in this volume were submitted by symposium participants, as well as by some of Feferman's students and former collaborators, as a tribute to Feferman. They are grouped, somewhat differently from the symposium program, into four parts: Proof-theoretic analysis, Logic and computation, Applicative and self-applicative theories, and Philosophy of modern mathematical and logical thought.

The volume is thus dedicated to Sol, the challenging teacher, inspiring researcher, and warm friend.

We acknowledge gratefully the financial support for the symposium from the National Science Foundation and Stanford University. The greatest debt for making the symposium and this volume a success we owe, however, to the contributors of talks and essays; they were all eager to contribute in order to express their respect and admiration for Feferman. We also thank the referees for their important contributions to the preparation of this volume: Jeremy Avigad, Michael Beeson, Wilfried Buchholz, Andrea Cantini, Martin Davis, Solomon Feferman, Timothy Fernando, Dagfinn Follesdal, Robert Harper, Jeffrey Hirst, Martin Hyland, Gerhard Jaeger, Reinhard Kahle, Hans Kamp, Ulrich Kohlenbach, Grigori Mints, Yiannis Moschovakis, Charles Parsons, Wolfram Pohlers, David Pym, Michael Rathjen, Amr Sabry, Helmut Schwichtenberg, Stewart Shapiro, Wilfried Sieg, Stephen Simpson, Richard Sommer, Thomas Strahm, William Tait, Carolyn Talcott, Makoto Tatsuta, Anthony Ungar, Johan van Benthem, Stanley Wainer, Andreas Weiermann, Freek Wiedijk.

Wilfried Sieg, Rick Sommer, Carolyn Talcott

TABLE OF CONTENTS

PART I: PROOF THEORETIC ANALYSIS

PART II: LOGIC AND COMPUTATION

PART III: APPLICATIVE AND SELF-APPLICATIVE THEORIES

PART IV: PHILOSOPHY OF MODERN MATHEMATICAL AND LOGICAL THOUGHT

PART I: PROOF THEORETIC ANALYSIS

ORDINAL ANALYSIS WITHOUT PROOFS

JEREMY AVIGAD

Abstract. An approach to ordinal analysis is presented which is finitary, but highlights the semantic content of the theories under consideration, rather than the syntactic structure of their proofs. In this paper the methods are applied to the analysis of theories extending Peano arithmetic with transfinite induction and transfinite arithmetic hierarchies.

§1. Introduction. As the name implies, in the field of proof theory one tends to focus on proofs. Nowhere is this emphasis more evident than in the field of ordinal analysis, where one typically designs procedures for "unwinding" derivations in appropriate deductive systems. One might wonder, however, if this emphasis is really necessary; after all, the results of an ordinal analysis describe a relationship between a system of ordinal notations and a theory, and it is natural to think of the latter as the set of semantic consequences of some axioms. From this point of view, it may seem disappointing that we have to choose a specific deductive system before we can begin the ordinal analysis.

In fact, Hilbert's epsilon substitution method, historically the first attempt at finding a finitary consistency proof for arithmetic, has a more semantic character. With this method one uses so-called epsilon terms to reduce arithmetic to a quantifier-free calculus, and then one looks for a procedure that assigns numerical values to any finite set of closed terms, in a manner consistent with the axioms. The first ordinal analysis of arithmetic using epsilon terms is due to Ackermann [1]; for further developments see, for example, [20].

2000 *Mathematics Subject Classification.* 03F15, 03F05, 03F30.

Dedicated to Solomon Feferman on the occasion of his 70th birthday.

It is an honor to be able to contribute a paper to this volume. Although I did my graduate work under Jack Silver at Berkeley, Sol served as an informal second advisor to me, and his thoughtful advice and guidance made frequent visits to Stanford both enjoyable and well worthwhile. Mathematical logic, as a discipline, is poised between philosophy and mathematics, and so has to answer to competing standards of philosophical relevance and mathematical elegance. Throughout his career, Sol has been able to strike a harmonious balance between the two, with work that is deeply satisfying on both counts. His style sets a high standard for future generations, and one that many of us will look to as a model.

I would like to thank Lev Beklemishev for comments and suggestions, and the anonymous referees for their very careful readings.

Work partially supported by NSF Grant DMS 9614851.

Reflections on the foundations of mathematics
Essays in honor of Solomon Feferman
Edited by Wilfried Sieg, Richard Sommer, and Carolyn Talcott

More recently, investigations of nonstandard models of arithmetic due to Paris and Kirby have given rise to another approach, which incorporates Ketonen and Solovay's finitary combinatorial notion of an α-large set of natural numbers. Roughly speaking, to show that the proof-theoretic ordinal of a theory T is bounded by α, one uses an α-large interval in a nonstandard model of arithmetic to construct a model of T. These methods are surveyed and extended in [3, 4]; some of the constructions found in the second paper are derived from model-theoretic methods due to Friedman [12, 13]. Sommer [28] has shown that one can avoid references to nonstandard models and instead view the methods as providing a way of building "finite approximations" to models of arithmetic, an idea which traces its way back to Herbrand [18]. (See also the introduction to [19].)

Finally, Quinsey has shown that a notion called *fulfillment*, due to Kripke, provides yet another way of using more semantic methods to obtain traditional proof-theoretic results. His Ph.D. thesis [23] is a tour-de-force, offering a wealth of applications in wide range of areas. Similar ideas have been developed independently by Carlson [10].

Each of the approaches just described has its own advantages and disadvantages. But with their emphasis on "building models" over "unwinding proofs," the similarities between them are more striking than the differences. And the persistence with which this point of view keeps resurfacing suggests that such methods may have something to offer to the development of proof theory.

In this paper, I have tried to fashion an approach to ordinal analysis which is in concord with these themes, incorporating and adapting ideas from all the sources mentioned above. Since these ideas have appeared in so many different contexts, often arising independently, trying to sort out the proper accreditations at each step along the way would be difficult; and so I hope this broad attribution is enough to acknowledge the general debt that this work owes to that which has come before. I should mention that I have also benefited a good deal from Buss' ordinal analysis of arithmetic, using the witnessing method, in [7]; from the emphasis on ordinal recursion and its properties, in Friedman and Sheard [11]; and, of course, from the traditional Gentzen-Schütte approach to ordinal analysis, surveyed in [21, 22, 24].

One aspect of the approach developed here is that Herbrand's theorem is used in a central way. One begins by embedding a classical theory in a universal one, with symbols describing functions that are nonconstructive in the intended interpretation. By Herbrand's theorem, to extract an appropriate witness from the proof of a Σ_1 sentence, one does not need to know the *correct* interpretation of the function symbols; one only needs an interpretation that is consistent with a finite set of axioms relevant to the proof.

Another aspect of the approach is that it is cumulative: once we have analyzed a theory T^α, dependent on a parameter α, we can work "in" that theory to analyze the next nonconstructive principle. That way, as we work our way up, we can leave behind the low-level combinatorial constructions, and carry on in a more familiar mathematical and logical framework.

Despite the semantic flavor of the approach, it is entirely finitary, in a sense that will be made precise in Section 4.

In this article, I will develop semantic analogues of the traditional tools and methods of predicative proof theory. In [2], I will extend the methods to analyze Kripke-Platek set theory, $KP\omega$. To my knowledge, the latter will provide the first ordinal analysis of a theory of that strength without the use of cut-elimination.

The outline of this paper is as follows. The first few sections provide the necessary background information: Section 2 discusses some weak fragments of arithmetic; Section 3 introduces a form of ordinal recursion, which we will use to define the proof-theoretic ordinal of a theory in Section 4; Section 5 describes the systems of ordinal notations that are needed to carry out the ordinal analysis; and Section 6 reviews Herbrand's theorem for first-order logic. The rest of the paper is concerned with bounding the proof-theoretic ordinals of various theories: primitive recursive arithmetic in Section 7, theories with Π_1 transfinite induction in Section 8, theories with arithmetic transfinite induction in Section 9, and, finally, theories of transfinite arithmetic hierarchies in 10.

§2. Weak theories of arithmetic. To get us off the ground, in this section I will introduce some weak theories of arithmetic. The theories, notations, and facts discussed are fairly standard. More information on the theories discussed below, including the formal representation of sequences and syntactic objects, can be found in [8, 17, 30]. More information on the elementary and primitive recursive functions and their properties can be found in [25].

I will take the language of arithmetic to be the first-order language with symbols 0, 1, $+$, $\times$, and $<$, and if n is a natural number, I will use $\bar{n}$ to denote the corresponding numeral. Peano arithmetic, PA, consists of quantifier-free defining axioms for $+$, $\times$, and $<$, and the schema of induction,

$$\varphi(0) \wedge \forall x\ (\varphi(x) \rightarrow \varphi(x+1)) \rightarrow \forall x\ \varphi(x)$$

for arbitrary formulae φ. A formula is said to be Δ_0 if every quantifier is bounded, that is, of the form $\forall x < t$ or $\exists x < t$, where these are interpreted in the usual way. A formula is Σ_1 (resp. Π_1) if it is obtained by prefixing existential (resp. universal) quantifiers to a Δ_0 formula; more generally, with Σ_n and Π_n formulas one is allowed n alternations of quantifiers. The theory of arithmetic in which induction is restricted to formulas in a set Γ is denoted $I\Gamma$. Over a weak base theory, Σ_n and Π_n are induction are equivalent: e.g., given a Σ_n formula $\varphi(x)$ satisfying the hypotheses of the induction axiom, if there is a y satisfying $\neg\varphi(y)$, then Π_n induction on z implies $\forall z\ \neg\varphi(y \dot{-} z)$; but then $\neg\varphi(0)$, yielding a contradiction.

Theories like $I\Delta_0$ are sensitive to the choice of initial functions. The theory obtained by adding a function symbol $exp(x, y)$ for exponentiation, with the usual defining equations, is denoted $I\Delta_0^{exp}$. From a mathematical point of view,

$I\Delta_0^{exp}$ is very weak, but from a finitary, computational point of view, it is fairly strong, as the following discussion will show.

Taking constants to be functions of arity 0, the set of *elementary functions* is defined to be the smallest set of functions on the natural numbers containing 0, $+$, $\times$, and *exp*, and projections, and closed under the operations of composition and *bounded recursion*. Using $\vec{z}$ to denote a finite sequence of variables $z_0, \ldots, z_k$, closure under bounded recursion means that whenever the functions $g(\vec{z})$, $h(x, y, \vec{z})$, and $b(x, \vec{z})$ are elementary, then so is the function $f(x, \vec{z})$, defined by the equations

$$f(0, \vec{z}) = g(\vec{z})$$
$$f(x+1, \vec{z}) = \begin{cases} h(x, f(x, \vec{z}), \vec{z}) & \text{if this is less than } b(x+1, \vec{z}) \\ 0 & \text{otherwise.} \end{cases}$$

Since bounded recursion does not allow us to introduce functions that grow faster than ones that have been previously defined, a moment's reflection shows that every elementary function is bounded by some finite iteration of the function $x \mapsto 2^x$. I will say that a relation $R(\vec{x})$ is elementary if its characteristic function, $\chi_R(\vec{x})$, is elementary, and for notational convenience I will write $R(\vec{x})$ instead of $\chi_R(\vec{x}) = 1$.

One can show that the set of elementary functions is closed under bounded sums, $f(x, \vec{z}) = \sum_{y<x} h(y, \vec{z})$, and bounded products, $g(x, \vec{z}) = \prod_{y<x} h(y, \vec{z})$. The set of elementary relations is closed under boolean operations and bounded quantification, and if $R(y, \vec{z})$ is an elementary relation, then the function $f(x, \vec{z}) = \mu y < x\ R(y, \vec{z})$ is also elementary, where the right hand side is defined to be the least y less than x satisfying $R(y, \vec{z})$ if there is one and zero otherwise. We can take $\mu y \leq x\ R(y, \vec{z})$ to abbreviate $\mu y < x + 1\ R(y, \vec{z})$. One can define functions by cases: if $R(\vec{x})$, $f(\vec{x})$, and $g(\vec{x})$ are elementary, then so is the function

$$h(\vec{x}) = \begin{cases} f(\vec{x}) & \text{if } R(\vec{x}) \\ g(\vec{x}) & \text{otherwise.} \end{cases}$$

One can also code finite sequences of numbers as a single natural number in such a way that the usual operations on sequences are elementary. I will write $\langle x_0, \ldots, x_k \rangle$ to denote (the code for) the sequence with the elements shown; if s is such a sequence, $length(s)$ to denote the length of s, $last(s)$ to denote the index of the last element of s, and $(s)_i$ to denote the ith element of s (or 0 if $i > last(s)$). The concatenation of two sequences s and t will be denoted $s\hat{\ }t$. From the point of view of computational complexity, the set of elementary functions is quite large: it can be characterized alternatively as the set of functions Turing computable with time (and/or space) resources bounded by a finite iteration of an exponential function. For details, see [25].

Elementary recursive arithmetic, denoted *ERA*, is the "natural" first-order theory of the elementary functions. The language has a symbol for each such function, and its axioms include the corresponding defining equations, as well as axioms for $<$ and the usual axioms for equality. To these, one adds the schema of induction for quantifier-free formulas. *ERA* is essentially a definitional extension of $I\Delta_0^{exp}$, or Friedman's *elementary function arithmetic, EFA*.

PROPOSITION 2.1. *Every Δ_0^{exp} relation is equivalent to an elementary one, provably in ERA.*

PROOF (SKETCH). Use the facts that equality and less-than are elementary relations, and that the elementary relations are closed under boolean operations and bounded quantification, provably in *ERA*. ⊣

PROPOSITION 2.2. *ERA can be axiomatized by a set of universal sentences, and is a conservative extension of $I\Delta_0^{exp}$.*

PROOF (SKETCH). The defining equations for the function and relation symbols are universal, and one can replace the schema of induction with axioms of the form $R(x, \vec{z}) \rightarrow R(\mu y \leq x\ R(y, \vec{z}), \vec{z})$, where R is any elementary relation. (If $R(\mu y \leq x\ R(y, \vec{z}), \vec{z})$, the defining equations guarantee that either $\mu y \leq x\ R(y, \vec{z})$ is 0 or $R(\cdot, \vec{z})$ does not hold of its predecessor.)

Proposition 2.1 guarantees that *ERA includes* $I\Delta_0^{exp}$. To show that *ERA* is a conservative extension, one shows that every elementary function is definable by a Δ_0^{exp} formula, provably in $I\Delta_0^{exp}$. See, for example, [8, 17]. ⊣

Moving on, the primitive recursive functions are obtained by dropping the bound requirement in the recursion schema; and primitive recursive arithmetic, or *PRA*, is the corresponding theory. Here we can omit the special treatment of $+$, $\times$, *exp*, and $<$, since these can defined using primitive recursion. Since bounded recursion can be seen as a special case of primitive recursion, we can view the language of *PRA* as including that of *ERA*. *PRA* is properly stronger: in *PRA* one can define iterated exponential functions, as well as a function which evaluates closed terms of *ERA*.

One can relativize the definitions of the elementary recursive and primitive recursive functions by adding additional functions to the initial set. On the axiomatic level, we will consider extensions of *ERA* and *PRA*, denoted by $ERA(f_0, \ldots, f_k)$ and $PRA(f_0, \ldots, f_k)$ respectively, obtained by adding new unary function symbols $f_0, \ldots, f_k$ to the underlying language. These are taken as initial functions in the inductively defined set of function symbols, and so may "appear" in the definitions of other functions; so, for example, there are symbols for functions defined using composition and bounded recursion (resp. primitive recursion) from these. Otherwise, however, there are no axioms governing their behavior. If f is such an uninterpreted function symbol, I will write $t(\vec{x}, f)$ to indicate that the term t depends on f, when f occurs in t or in the definition of one of the function symbols occurring in t. If $g(\vec{x}, f)$ is a k-ary function, and $h(y, \vec{z})$ is l-ary, I will use $g(\vec{x}, \lambda y\ h(y, \vec{z}))$ to denote the $k + l$-ary function obtained by replacing $f(y)$ by $h(y, \vec{z})$ everywhere in the definition of g; and I will adopt a similar convention for terms and formulas. This notation is somewhat justified by the following:

LEMMA 2.3. *Suppose ERA(f) proves $\varphi(\vec{x}, f)$, and $h(y, \vec{z})$ is a function symbol of ERA. Then ERA proves $\varphi(\vec{x}, \lambda y\ h(y, \vec{z}))$.*

PROOF. A straightforward induction on derivations. ⊣

Primitive recursive arithmetic is often taken as a reasonable representation of Hilbert and Bernays' informal notion of "finitary" mathematical reasoning. For this purpose, it is better to view *PRA* as a quantifier-free theory, obtained by dropping the universal quantifiers from the axioms and replacing quantifier rules with a substitution rule, namely, from $\varphi(x)$ conclude $\varphi(t)$ for any term t free for x in φ. Herbrand's theorem, discussed in Section 6, implies that the first-order version of *PRA* is a conservative extension of the quantifier-free one. In fact, if the first-order version of *PRA* proves a sentence of the form $\forall x\ \exists y\ R(x, y)$, where $R(x, y)$ is a primitive recursive relation, then there is a function symbol g such that the quantifier-free version proves $R(x, g(x))$; and similarly for *ERA*. In much the same way, the versions of these theories with extra function symbols may be better understood in terms of conservative second-order extensions, but developing the details for such a presentation would take us too far afield.

§3. Computations with ordinal bounds. To get a sense for the kinds of theorems we are after, consider the first ordinal analyses of arithmetic, due to Gentzen [14, 15]. Roughly speaking, Gentzen devised a means of "unwinding" proofs in arithmetic, using iterative procedures that "count down" through the ordinals below ε_0. In particular, given a proof of a Σ_1 sentence $\exists y\ \varphi(x, y, f)$ in Peano arithmetic (in a language augmented with a function symbol f), his analysis provides a procedure which, for any x and f, finds a suitable value for y.

In the next section, I will use this informal characterization to provide a formal definition of the proof-theoretic ordinal of a theory. But first, we need to make the notion of an "iterative procedure which counts down through the ordinals" more precise.

Fix an elementary relation $\prec$ that happens to be a well ordering of an elementary subset of the natural numbers. Let variables $\alpha, \beta, \gamma, \ldots$ range over the field of $\prec$; think of these as *notations* for ordinals. A *$\prec\alpha$-iterative algorithm* is given by a notation β less than α and elementary functions $start(\vec{x})$, $next(q)$, $norm(q)$, and $result(q)$. These data define a function $F(\vec{x})$, whose value at an input x is computed by starting in a state given by $start(\vec{x})$; assuming the norm of this state is less than β, iterating the *next* function until the norm of the resulting state fails to decrease; and then returning the value of *result*, applied to the final state. This algorithm can be summarized as follows:

```
clock ← β
state ← start(x⃗)
while norm(state) ≺ clock do
    clock ← norm(state)
    state ← next(state)
return result(state)
```

The fact that $\prec$ is a well ordering guarantees that the while loop always terminates.[1] To describe F more formally, say that s is a computation sequence for F at $\vec{x}$, written $CompSeq_F(s, \vec{x})$, if s is a sequence $\langle s_0, s_1, s_2, \ldots, s_k\rangle$ satisfying the following:

- $s_0 = start(\vec{x})$
- for every $i < k$, $s_{i+1} = next(s_i)$
- either

  $k = 0$ and $norm(s_0) \succeq \beta$, or

  $k > 0$, $norm(s_0) \prec \beta$, $norm(s_{i+1}) \prec norm(s_i)$ for every $i < k - 1$, and $norm(s_k) \succeq norm(s_{k-1})$

Then $F(\vec{x}) = y$ means there is a sequence s satisfying $CompSeq_F(s, \vec{x})$, and $result(s_{last(s)}) = y$. I will say that a function $F(\vec{x})$ is $\prec\alpha$-recursive if it is defined by a $\prec\alpha$-iterative algorithm.

We can refer to specific $\prec\alpha$-recursive functions within *ERA*, by identifying such a function with the numeral $\bar{\beta}$ and function symbols *start*, *norm*, *next*, and *result* that define it. We have to be careful, since in *ERA* the relation $\prec$ may not be provably well ordered. In that case, we can use the notation $F(\vec{x}) \downarrow$ to abbreviate $\exists s\ CompSeq_F(s, \vec{x})$, indicating that F is "defined" at $\vec{x}$. One can show, in *ERA*, that there is at most one computation sequence for F at $\vec{x}$, so $F(\vec{x}) = y$ is equivalent to saying that F is defined at $\vec{x}$ and y is the result of the corresponding computation sequence. More generally, we can use the notation $F(\vec{x}) \simeq G(\vec{x})$ to abbreviate the assertion that if either side is defined, then both sides are defined and equal. It is not hard to see that, again in *ERA*, $\simeq$ has the properties of an equivalence relation. As far as composition is concerned, we can take $F(G_0(\vec{x}), \ldots, G_k(\vec{x})) \downarrow$ to denote the assertion that there are computation sequences for $G_0, \ldots, G_k$ at $\vec{x}$, and a computation sequence for F at the results of these computation sequences. Then $F(G_0(\vec{x}), \ldots, G_k(\vec{x})) = y$ means that $F(G_0(\vec{x}), \ldots, G_k(\vec{x}))$ is defined, and y is the result of the corresponding computation sequence for F. Inscriptions like $F(G_0(\vec{x}), \ldots, G_k(\vec{x})) \simeq H(\vec{x})$ are to be interpreted similarly.

We also want a notion of iterative computation relative to a function f. The easiest thing to do would be to allow *start*, *norm*, *next*, and *result* to be elementary functions relative to f, but we will need a slightly finer notion. This is obtained by requiring the basic functions to be purely elementary, as before, but allowing the computation to query a single value of f at each step. That is, a *relativized* $\prec\alpha$-*iterative algorithm* is given by a notation β less than α and elementary functions $start(\vec{x})$, $query(q)$, $next(q, z)$, $norm(q)$, and $result(q)$; values of the corresponding function $F(\vec{x}, f)$ are computed by replacing the second line of the while loop above with

$$state \leftarrow next(state, f(query(state))).$$

[1] It is not hard to extend these notions to cases where $\prec$ is, more generally, a well-founded relation; assuming $\prec$ is a linear ordering simplifies the exposition.

Let $CompSeq_F(s, \vec{x}, f)$ denote the resulting modification of $CompSeq(s, x)$, and then proceed as before. From now on I will say that a function $F(\vec{x}, f)$ is $\prec\alpha$-recursive if there is a $\prec\alpha$-iterative algorithm, relative to f, which computes it. More generally, $F(\vec{x}, f_0, \ldots, f_k)$ is $\prec\alpha$-recursive if it can be computed by an algorithm which, at each step, is allowed to pose a single query to each of the f_i.

Having set forth these definitions, I should warn the reader that there are, in fact, a number competing definitions of ordinal recursion in the literature: see, for example, the schematic presentation in [26], the $\prec\alpha$-descent recursive functions of [11], or the various characterizations in [7] and [25]. The good news is that the various definitions of the $\prec\alpha$-recursive functions usually coincide, with minimal assumptions on the system of notations and α. (See, for example, [11, Proposition 1.9], [7, Section 3.2], [25], and Lemma 7.3 below.) I have chosen the presentation above because it is easy to work with, and convenient for our applications.

§4. The proof-theoretic ordinal of a theory. Following Gentzen's lead, we would like to say that the proof theoretic ordinal of T is bounded by α when there is a finitary proof of the following:

> Whenever T proves a $\Sigma_1(f)$ formula $\exists y\ \varphi(x, y, f)$, there is a $\prec\alpha$-recursive function F, such that for any x, y, and f, if $F(x, f) = y$ then $\varphi(x, y, f)$ is true.

Call this informal statement (*). Note that (*) only makes sense for theories in a language that includes the language of arithmetic and a function symbol f, or, more generally, theories for which we can interpret a notion of provability for $\Sigma_1(f)$ formulae. Note also that (*) does not imply that $F(x, f)$ is defined at every value of x and f; only that when it *is* defined, it produces a suitable witness.

The rest of this paper is dedicated to proving (*) in a finitary way, for various theories T and notations α. This section is devoted to making the notion of "proving (*) in a finitary way" more precise, and explaining why this is a desirable goal. The reader that is already satisfied with the informal characterization should feel free to skip to Section 5.

To turn the informal statement into a mathematical one, we need a formal notion of "finitary proof," and an appropriate formalization of (*). For the former, let us take primitive recursive arithmetic relative to a function symbol f; below we will see that weaker theories will do. In $PRA(f)$, we can develop a theory of syntax, representing terms and formulae as numbers in an appropriate way; we can define the set of $\Sigma_1(f)$ formulas as well as names for the elementary functions; and we can identify $\prec\alpha$-recursive functions with the iterative algorithms that define them. In $PRA(f)$ we can also refer to the "value" of an elementary function at a given set of inputs, using an appropriate primitive recursive evaluation function for the set of elementary recursive functions, and we can refer to the truth value of a $\Delta_0(f)$ formula at a given set of parameters, using a truth predicate for $\Delta_0(f)$ formulas that is primitive recursive in f.

Expressed in greater detail, (*) asserts the following:

For every proof d of a $\Sigma_1(f)$ sentence e, there is a $\prec\alpha$-recursive function F, such that for every x, y, f, and computation sequence s for F at x and f, if the result of the computation is y, then y witnesses the truth of e at x and f.

We can get rid of the existential quantifier by requiring, more stringently, that we have a primitive recursive function $\mathcal{F}(d)$ which extracts F from the proof d; and then we can leave the universal quantification over d, e, x, y, f, and s implicit. Given a primitive recursive relation $Proof_T(d, e)$, an elementary well ordering $\prec$, and a notation α, we can then take the statement "the proof-theoretic ordinal of T is at most α" to mean that there is a function symbol $\mathcal{F}$, and a $PRA(f)$-proof of an appropriate formalization of the following assertion:

For every d, $\mathcal{F}(d)$ is a $\prec\alpha$-recursive function, and whenever

- e is a $\Sigma_1(f)$ formula with free variable x,
- $Proof_T(d, e)$,
- $CompSeq_{\mathcal{F}(d)}(s, x, f)$, and
- $result(s_{last(s)}) = y$

then y witnesses the truth of e at x and f.

There is, no doubt, much to criticize in this choice of a definition, but let us consider some of the things one can say in its favor. To start with, it is *strong*, which is to say, it implies all the usual results of an ordinal analysis. Suppose the proof-theoretic ordinal of T is at most α, according to our definition. Then we have:

1. A consistency proof for T.
2. A characterization of T's provably total computable functions.
3. A characterization of T's provably well-ordered computable relations.

For the first, ignoring x and f and taking e to be the sentence "$0 = 1$," we can conclude that the consistency of T is provable in PRA together with any principle that implies that every $\prec\alpha$-iterative procedure terminates. For the second, we can ignore f and conclude that any Π_2 statement provable in T has a $\prec\alpha$-recursive Skolem function. For the third, suppose $\prec'$ is a Σ_1-definable well ordering such that T proves $\exists y\ (f(y + 1) \not\prec' f(y))$. If the order type of $\prec'$ is greater than α, there is an order-preserving embedding g of notations less than α into the field of $\prec'$. We can use the conclusion of (*) to obtain a $\prec\alpha$-recursive function $H(g, h)$, which for any h returns a value y such that that $h(y + 1) \not\prec h(y)$. With minor assumptions on α, one can use this to define a $\prec\alpha$-recursive function $J(g, x)$ that diagonalizes the functions that are $\prec\alpha$-recursive in g, yielding a contradiction. (See [11] for additional information.) Incidentally, Rathjen [24] notes that for second-order theories that include arithmetic comprehension (or first-order theories that have such conservative extensions), 3 extends to arithmetically definable orderings; and for second-order theories that include the Σ^1_1 axiom of choice, 3 extends to the hyperarithmetically definable orderings as well.

The formalization of (*) is unpalatable, and it is tempting to take the assertion that the proof-theoretic ordinal of T is at most α to mean that (*) is simply true.

But this has the undesired consequence that adding arbitrary true Π_1 statements to T (like consistency statements) does not increase the proof-theoretic ordinal. Similarly, defining the proof-theoretic ordinal in terms of provable well orderings means that the ordinal does not change when one adds arbitrary true Σ^1_1 sentences to the theory; see [24] for a discussion. This points to a second advantage of the definition above: it is *immune* to these objections.

Our formalized version of (*) expresses a relationship between a primitive recursive *representation* of T, and a system of *notations* for ordinals. It is well known that one can always cook up representations of theories and ordinals which render the ordinal analysis trivial, or meaningless; in this respect the definition above is *honest*, since it is really the "natural" representations that we care about. Some logicians are disturbed by the absence of a formal definition of naturality, and so prefer to characterize the proof-theoretic ordinal as the least upper bound to the theory's provable well orderings; this characterization is independent of the representations, but has the drawbacks mentioned above. But the absence of such a formal definition should not concern us much. The natural representations of theories and ordinals are just those for which the provability of (*) is *interesting*; and very few mathematicians have formal criteria which tell them which theorems of their subject have this property.[2]

Bounding the proof-theoretic ordinal of a theory has two aspects: proving (*), and doing so in a finitary way. In the sections that follow, I will focus on the first; but I will proceed with the implicit understanding that once we have specified an appropriate ordinal notation system (with properties described in the next section), every definition, theorem, and proof can be formalized in $PRA(f)$. The exception is this: when I state as a theorem that "the proof-theoretic ordinal of T is at most α," I mean simply that (*) holds, again with the implicit understanding that the formalized version of (*) can be proved in $PRA(f)$.

Let me close this section with two notes. First, the choice of $PRA(f)$ is not crucial. We need a metatheory that is strong enough to formalize syntax and quantify over elementary functions, and is strong enough to prove Herbrand's theorem. For these purposes, $I\Delta_0(f)$ together with the assertion that an iterated exponential function is total will suffice. If one uses a weaker class of functions in defining the ordinal-iterative algorithms, one can get by with even weaker theories, by "pushing" the work involved in satisfying Herbrand's theorem into the computation of the $\prec\alpha$-recursive function. The possibility of using a weaker metatheory is interesting from a foundational point of view, but it does not seem to help with the analysis of weaker theories. (For evidence of this, see [29]. For a more fruitful approach to the analysis of weak theories, see the use of "dynamic ordinals" in [5].)

The second note has to do with lower bounds. One can take the statement "the proof-theoretic ordinal of T is *exactly* α" to mean that the proof-theoretic ordinal

[2]Another approach to dealing with the various definitions of "proof theoretic ordinal" in the literature is to embrace the multiplicity, and explore the relationships between them. See [6] for a development along these lines, as well as the discussion in [24].

of T is at most α, but it is not at most β for any β less than α. This takes us outside our finitary metatheory, since it requires us to show that for any such β there is *no* proof of the formalized version of (*) in $PRA(f)$. But, on the assumption that $PRA(f)$ is consistent, one obtains the desired conclusion by giving a finitary proof that for every β less than α, there is a provable $\Sigma_1(f)$ formula that is not witnessed by any $\prec\beta$-recursive function; and one typically achieves this goal by developing a theory of transfinite recursion below α in T. In this paper I will focus on the upper bounds, but, in fact, all the upper bounds I provide will be sharp in this sense. (See [21, 22] for more information on establishing the lower bounds.)

§5. Systems of ordinal notations. As noted in the last section, ordinal analysis, as understood here, involves calibrating the strength of various theories relative to an elementary recursive system of ordinal notations. In this section I will discuss the properties that our system of notations, $\prec$, needs to satisfy, provably in $ERA(f)$. For more information on ordinal notations, see, for example, [3, 21, 22, 24].

For most of the results below, we only need to assume that $\prec$ is a linear ordering, with elementary functions $+$, $\cdot$, and $\alpha \mapsto \omega^\alpha$, for which the "usual properties" hold. In other words, I will assume that $\prec$ is an elementary recursive ordinal notation system (ERONS) in the terminology of [11], such that the given functions are everywhere defined; a list of the "usual properties" they are to satisfy appears in [11, Section 1]. In particular, we will need to use the fact that any α can be written in Cantor normal form, $\alpha = \omega^{\beta_1} + \cdots + \omega^{\beta_k}$ with $\beta_1 \succeq \cdots \succeq \beta_k$. If $\alpha' = \omega^{\beta'_1} + \cdots + \omega^{\beta'_{k'}}$ is also in Cantor normal form, then the symmetric sum of α and α', written $\alpha\#\alpha'$, is equal to $\omega^{\gamma_1} + \ldots + \omega^{\gamma_{k+k'}}$, where the γ_i's list the β_i's and the β'_i's in decreasing order. Unlike ordinary ordinal addition, the symmetric sum is strictly monotone in both arguments.

Classically, ε_0 is defined to be the least fixed-point of the function $\alpha \mapsto \omega^\alpha$; equivalently, it is the limit of the sequence $\langle \omega_n \rangle_{n\in\omega}$, where $\omega_0 = 1$ and $\omega_{n+1} = \omega^{\omega_n}$. The statement of Theorem 9.7 assumes that there is a notation with these properties.

More generally, the sequence of Veblen functions on any regular cardinal is defined by letting $\varphi_0(\beta) = \omega^\beta$, and otherwise letting φ_α enumerate the simultaneous fixed points of $\{\varphi_\gamma \mid \gamma < \alpha\}$. In Section 10, we need to assume that there is a binary elementary function $\varphi(\alpha, \beta)$, defined on the system of notations, representing the Veblen functions. Writing $\varphi_\alpha(\beta)$ instead of $\varphi(\alpha, \beta)$, we will assume that for every α, β, γ, and δ, $\varphi_\alpha(\beta)$ is less than $\varphi_\gamma(\delta)$ if and only if either

- $\alpha \prec \gamma$ and $\beta \prec \varphi_\gamma(\delta)$,
- $\alpha = \gamma$ and $\beta \prec \delta$, or
- $\alpha \succ \gamma$ and $\varphi_\alpha(\beta) \prec \delta$.

I will say that an ordinal notation α is infinite if it is greater than or equal to ω. Many of the lemmata and theorems below are stated most cleanly by assuming closure properties on a notation α. For reference, here are some equivalent characterizations.

PROPOSITION 5.1. *Let α be infinite. Then*

1. *α is closed under addition if and only if it is equal to ω^γ, for some γ.*
2. *α is closed under the function $\beta \mapsto \omega \cdot \beta$ if and only if it is equal to $\omega^\omega \cdot \gamma$, for some γ.*
3. *α is closed under multiplication if and only if it is equal to $\omega^{(\omega^\gamma)}$, for some γ.*
4. *α is closed under the function $\beta \mapsto \omega^\beta$ if and only if it is equal to ε_γ (that is, $\varphi_1(\gamma)$), for some γ.*

The proof is an exercise in ordinal arithmetic (see [21]).

§6. Herbrand's theorem.

Herbrand's theorem can be stated as follows:

THEOREM 6.1. *Let L be a language with at least one constant symbol, let $\varphi(\vec{x})$ be a quantifier-free formula in L, and suppose $\exists \vec{x}\ \varphi(\vec{x})$ is provable in classical first-order logic with equality. Then there are sequences of terms $\vec{t_0}, \ldots, \vec{t_k}$, whose free variables are among those of $\exists \vec{x}\ \varphi(\vec{x})$, such that $\varphi(\vec{t_1}) \vee \varphi(\vec{t_2}) \vee \ldots \vee \varphi(\vec{t_k})$ is provable in propositional logic from substitution instances of the equality axioms.*

This theorem, which effectively enables us to extract additional information from proofs of existential sentences, will form a cornerstone to our investigations. There are model-theoretic proofs of Herbrand's theorem: if the conclusion fails, the set $\{\neg\varphi(t) \mid t \text{ is a closed term}\}$ is propositionally consistent with the set of all substitution instances of the equality axioms; and from a satisfying truth assignment, one can build a model of $\forall x\ \neg\varphi(x)$. But Herbrand's theorem is also an easy consequence of the cut-elimination theorem (see [8, 26]), and hence provable in our finitary metatheory. See also [27, 23] for alternative syntactic proofs, and [19] for Herbrand's original proof.

I will say that a theory T is *universal* if it can be axiomatized by a universal set of sentences (or, equivalently, a quantifier-free set of formulae, since $\forall \vec{y}\ \psi(\vec{y})$ follows from an axiom $\psi(\vec{y})$).

COROLLARY 6.2. *Let L and $\varphi(\vec{x})$ be as above, and let T be a universal theory in L. If T proves $\exists \vec{x}\ \varphi(\vec{x})$, then there are sequences of terms $\vec{t_0}, \ldots, \vec{t_k}$, whose free variables are among those of $\exists \vec{x}\ \varphi(\vec{x})$, such that T proves $\varphi(\vec{t_1}) \vee \varphi(\vec{t_2}) \vee \ldots \vee \varphi(\vec{t_k})$. Moreover, we can assume this formula is provable in propositional logic from substitution instances of axioms of T and equality axioms.*

PROOF. Suppose T proves $\exists \vec{x}\ \varphi(\vec{x})$. Then there are (universal closures of) axioms of T, $\psi_1, \ldots, \psi_k$, such that $\exists \vec{x}\ \varphi(\vec{x})$ is provable from $\psi_1, \ldots, \psi_k$. By the deduction theorem, $\psi_1 \wedge \ldots \wedge \psi_k \rightarrow \exists \vec{x}\ \varphi(\vec{x})$ is provable in first-order logic. Bring all the quantifiers to the front, and apply Herbrand's theorem. ⊣

In many cases (but not all the ones we will consider), the theory T will be rich enough so that for every sequence of terms $t_1(\vec{x}), \ldots, t_k(\vec{x})$ and quantifier-free formulae $\varphi_1(\vec{x}), \ldots, \varphi_{k-1}(\vec{x})$, there is a function symbol $f(\vec{x})$ such that T proves

$$f(\vec{x}) = \begin{cases} t_1(\vec{x}) & \text{if } \varphi_1(\vec{x}) \\ t_2(\vec{x}) & \text{if } \neg\varphi_1(\vec{x}) \wedge \varphi_2(\vec{x}) \\ \vdots & \\ t_k(\vec{x}) & \text{otherwise.} \end{cases}$$

In cases like this, we can replace the sequence of terms $t_1, \ldots, t_k$ in the Corollary 6.2 with a single function symbol f.

Recall that if $G(x, f)$ is a $\prec\alpha$-recursive function, we will interpret references to G in the context of *ERA*(f) as references to the elements β, *start*, *norm*, *next*, *query*, and *result* that define the iterative algorithms that computes it. As an application of Herbrand's theorem, we have the following:

THEOREM 6.3. *Suppose $\theta(x, y, f)$ is a $\Delta_0(f)$ formula with the free variables shown, and suppose there is an α-recursive function $G(x, f)$ such that ERA(f) proves $G(x, f) = y \rightarrow \theta(x, y, f)$. For any x and y, if $G(x, f)$ is defined and equal to y, then $\theta(x, y, f)$ is true.*

PROOF. The conclusion follows from the soundness of *ERA*(f), but we have to take care to make sure that our proof is finitary. The following proof uses an evaluation function for the set of functions that are elementary recursive in f, and a truth predicate for $\Delta_0(f)$ sentences.

Suppose *ERA*(f) proves $G(x, f) = y \rightarrow \theta(x, y, f)$. From the definition of $G(x, f) = y$, we see that it also proves $\mathit{CompSeq}_G(s, x, f) \wedge \mathit{result}((s)_{\mathit{last}(s)}) = y \rightarrow \theta(x, y, f)$. Now, suppose $G(x, f)$ is defined and equal to y, so that in addition, there is an s satisfying $\mathit{CompSeq}_G(s, x, f)$ and $\mathit{result}((s)_{\mathit{last}(s)}) = y$. Then for this particular s, x, and y, *ERA*(f) proves $\mathit{CompSeq}_G(\bar{s}, \bar{x}, f) \wedge \mathit{result}((\bar{s})_{\mathit{last}(\bar{s})}) = \bar{y} \rightarrow \theta(\bar{x}, \bar{y}, f)$. By Herbrand's theorem, there is a propositional proof of this fact from closed instances of equality axioms and axioms of *ERA*(f). The axioms of *ERA*(f) are true; so, by induction on the length of the proof, the conclusion is also true. As a result, we have that $\mathit{CompSeq}_G(s, x, f)$ and $\mathit{result}(s)_{\mathit{last}(s)}) = y$ imply $\theta(x, y, f)$. So $\theta(x, y, f)$ is true. ⊣

This theorem seems minor, but it will play a central role. It enables us to show that the ordinal of a theory T is less than or equal to α, by showing that whenever T proves a statement of the form $\exists y\ \theta(x, y, f)$, then there is a $\prec\alpha$-recursive function $G(x, f)$ which, provably in *ERA*(f), finds a witness. We will do this repeatedly, providing explicit translations; this is what makes the account finitary. But we will proceed in steps, successively reducing more "abstract" theories to more "concrete" ones, and working "in" the target theory as much as possible.

§7. **Primitive recursion.** In this section we will bound the proof-theoretic ordinal of primitive recursive arithmetic. To do so, we will first show that with sufficient conditions on α, the $\prec\alpha$-recursive functions have nice closure properties, provably in *ERA*(f). In particular, if α is ω^ω, we will see that one can use a single $\prec\alpha$-recursive function to assign "correct" values to a finite set of terms in $PRA(f)$, again provably in $ERA(f)$. Applying Theorem 6.3 will then yield the desired upper bound.

The first lemma states that for α greater than 1, the $\prec\alpha$-recursive functions in f include both the purely elementary functions and f itself.

LEMMA 7.1. *Suppose α is greater than 0. Then for every elementary function $g(\vec{x})$ (not involving the function f), there is a $\prec\alpha$-recursive function $G(\vec{x}, f)$ such that $ERA(f)$ proves $G(\vec{x}, f) \simeq g(\vec{x})$. Also, if α is greater than 1, there is a $\prec\alpha$-recursive function $H(x, f)$ such that $ERA(f)$ proves $H(x, f) \simeq f(x)$.*

PROOF. For the first claim, let the algorithm for H store $\vec{x}$ in the state, and then return $g(\vec{x})$ immediately. In other words, assuming g is arity k, take $\beta = 0$, $start(\vec{x}) = \langle \vec{x} \rangle$, $norm(q) = 0$, $result(q) = g((q)_0, \dots, (q)_k)$.

For the second claim, let the algorithm for H store x in the state, query f, and then return the result. That is, take $\beta = 1$, $start(x) = x$, $query(q) = q$, $next(q, z) = z$, $norm(q) = 0$, $result(q) = q$. ⊣

The next lemma gives conditions under which the $\prec\alpha$-recursive functions are closed under composition, again provably in $ERA(f)$.

LEMMA 7.2. *Suppose α is infinite and closed under addition, and suppose that $F_0(\vec{x}, f), \dots, F_k(\vec{x}, f)$ and $G(z_0, \dots, z_k)$ are $\prec\alpha$-recursive functions. Then there is a $\prec\alpha$-recursive function $H(\vec{x}, f)$ such that $ERA(f)$ proves*

$$H(\vec{x}, f) \simeq G(F_0(\vec{x}, f), \dots, F_k(\vec{x}, f)).$$

PROOF. Let the algorithm for G carry out the algorithms for F_0 through F_k on input $\vec{x}$, and then send the result to the algorithm for G.

In more detail, suppose the algorithm for each F_i is given by the data β_i, $start_i$, $query_i$, $next_i$, and $result_i$, and suppose the algorithm for G is given by β_{k+1}, $start_{k+1}$, $query_{k+1}$, $next_{k+1}$, and $result_{k+1}$. Take the states of H to code tuples of the form $\langle i, c, s, u, v \rangle$, where i indicates the current subalgorithm, c is the setting of an ordinal "clock," s is the state in the subalgorithm, u stores the original input, and v stores the results which have been computed so far. The algorithm for H then corresponds to the data β, $start$, $query$, $next$, and $result$, given as follows. First, set $\beta = \beta_{k+1} + \beta_k + \dots + \beta_0 + 1$, and

$$start(\vec{x}) = \langle 0, \beta_0, start_0(\vec{x}), \emptyset, \langle \vec{x} \rangle \rangle.$$

Assuming q is of the form $\langle i, c, s, u, v \rangle$, set

$$norm(q) = \beta_{k+1} + \dots + \beta_i + c$$

and

$$query(q) = query_i(s).$$

Then define $next(q, z)$ by cases, again assuming that q is of the form $\langle i, c, s, u, v \rangle$:

1. If $norm_i(s) \prec c$, we are in the middle of the computation of the ith algorithm. In that case, set

$$next(q, z) = \langle i, norm_i(s), next_i(s, z), u, v \rangle.$$

2. If $norm_i(s) \succeq c$, we have completed the computation of the ith algorithm.
   (a) If $i < k$, store the result and begin algorithm $i + 1$: set

   $$next(q, z) = \langle i + 1, \beta_{i+1}, start_{i+1}((u)_0, \dots, (u)_l), u, v\hat{\ }\langle result_i(s)\rangle\rangle,$$

   where l is the arity of the F_i's.
   (b) If $i = k$, begin the computation of G: set

   $$next(q, z) = \langle k + 1, \beta_{k+1}, start_{k+1}((v)_0, \dots, (v)_{k-1}, result_k(s)), \emptyset, \emptyset\rangle$$

   (c) If $i = k + 1$, we are done. Set $next(q, z) = q$ to flag this fact.

Finally, set $result(q) = result_{k+1}(s)$.

It is not hard to show, in $ERA(f)$, that from a computation sequence for H at $\vec{x}$ and f one can extract computation sequences for $F_1, \dots, F_k$ at $\vec{x}$ and f, and a computation sequence for G at the result of those computations. ⊣

From now on, I will rely on less formal descriptions of the algorithms, and leave the details of the implementation to the reader. The next lemma shows that assuming that α is closed under multiplication, the set of $\prec\alpha$-recursive functions is closed under a schema of $\prec\alpha$-recursion, in which the functions defining the algorithm are themselves $\prec\alpha$-recursive. Notice that the condition on s in the statement of the lemma is identical to $CompSeq_F(s, \vec{x})$, except that the functions defining the algorithm is no longer assumed to be elementary.

LEMMA 7.3. *Suppose α is infinite and closed under multiplication. Given β less than α and $\prec\alpha$-recursive functions $Start(\vec{x}, f)$, $Norm(q,f)$, $Next(q,f)$, and $Result(q,f)$, there is a $\prec\alpha$-recursive function $F(x, f)$, such that $ERA(f)$ proves*

$$\begin{aligned}
&F(\vec{x}, f) = y \leftrightarrow \exists s\ (\\
&\quad (s)_0 = Start(\vec{x}, f) \wedge \\
&\quad \forall i < length(s)\ ((s)_{i+1} = Next((s)_i, f)) \wedge \\
&\quad ((length(s) = 1 \wedge Norm((s)_0, f) \succeq \beta) \vee \\
&\qquad (length(s) > 1 \wedge Norm((s)_0, f) \prec \beta \wedge \\
&\qquad \forall i < (last(s) - 1)\ (Norm((s)_{i+1}, f) \prec Norm((s)_i, f)) \wedge \\
&\qquad Norm((s)_{last(s)}, f) \succeq Norm((s)_{last(s)-1}, f))) \wedge \\
&\quad Result((s)_{last(s)}) = y).
\end{aligned}$$

PROOF. The proof is similar to the preceding one. The algorithm for F first computes $Start(\vec{x}, f)$; then iteratively computes *Norm* and *Next*, until the norm of the state fails to decrease; and then computes *Result*. Assuming the algorithms for *Start*, *Norm*, *Next*, and *Result* are, respectively, γ-, δ-, ε-, ζ-recursive, the algorithm for H can be made η-recursive, where $\eta = \zeta + (\delta + \varepsilon) \cdot \beta + \gamma + 1$. ⊣

Using ω-recursion, we can simulate ordinary primitive recursion.

LEMMA 7.4. *Suppose α is infinite and closed under the function $\gamma \mapsto \gamma \cdot \omega$. Let $F_0(\vec{z}, f)$ and $F_1(x, w, \vec{z}, f)$ be $\prec\alpha$-recursive. Then there is is a $\prec\alpha$-recursive function $G(x, \vec{z}, f)$ such that $ERA(f)$ proves $G(0, \vec{z}, f) \simeq F_0(\vec{z}, f)$ and*

$$G(x + 1, \vec{z}, f) \simeq F_1(x, G(x, \vec{z}, f), \vec{z}, f).$$

Furthermore, we can define G in such a way that $ERA(f)$ proves that whenever $G(x, \vec{z}, f)$ is defined, there is a sequence of computation sequences $\langle s_0, \ldots, s_x \rangle$, such that

- *s_0 is a computation sequence for F_0 at $\vec{z}, f$.*
- *If x is greater than or equal to 1, s_1 is a computation sequence for F_1 at $(1, result_0((s_0)_{last(s_0)}), \vec{z}), f$.*
- *For each i such that $0 < i < x$, s_{i+1} is a computation sequence for F_1 at $(i, result_1((s_i)_{last(s_i)}), \vec{z}), f$.*
- *$G(x, \vec{z}, f) = result_1((s_x)_{last(s_x)})$.*

PROOF. As in the previous proof, with $\beta = \omega$, we can design an algorithm that successively computes $G(0, \vec{z}, f)$, $G(1, \vec{z}, f)$, ... $G(x, \vec{z}, f)$. ⊣

Lemmata 7.1–7.4 imply that we can assign to each function symbol $g(\vec{x}, f)$ of $PRA(f)$ a $\prec\omega^\omega$-recursive function $F_g(\vec{x}, f)$, in such a way that $ERA(f)$ proves that the axioms of $PRA(f)$ are satisfied by these functions, at least at arguments where they are defined. Recall that we can take the language of $PRA(f)$ to include that of $ERA(f)$; below we will need to know that the translation $g \mapsto F_g$ preserves elementary functions, in the following sense.

LEMMA 7.5. *Let $g(\vec{x}, f)$ be an elementary function in f. Then $ERA(f)$ proves*

$$F_g(\vec{x}, f) \downarrow \rightarrow F_g(\vec{x}, f) = g(\vec{x}, f).$$

PROOF. By induction on the definition of g, using the additional information in Lemma 7.4. ⊣

In fact, in $ERA(f)$ one can prove the existence of suitable computation sequences, and therefore show $\forall \vec{x}\, F_g(\vec{x}, f) \downarrow$. We will not, however, need this fact below.

We can now show that the proof-theoretic ordinal of $PRA(f)$ is at most ω^ω. The observations following Lemma 7.4 show that one can interpret $PRA(f)$ in $ERA(f)$ together the assumptions that each $\prec\omega^\omega$-recursive function is everywhere defined. In order to use Theorem 6.3, however, we need to show that a *single* $\prec\omega^\omega$-recursive function suffices. The idea is this: we will show that given any proof in $PRA(f)$, one can use a single $\prec\omega^\omega$-recursive function to assign correct values to all the terms appearing in the proof; and furthermore, that we can do this "within" $ERA(f)$.

Say a sequence of terms $t_0, \ldots, t_k$ in $PRA(f)$ is a *formation sequence* if each t_i is either a constant or variable, or the result of applying a function symbol of $PRA(f)$ to previous terms in the sequence. To each formation sequence S in which no variable other than x occurs, the following definition assigns a formula $Eval_S(e, x, f)$ in the language of $ERA(f)$, which asserts that the sequence e assigns the correct values to the members of S, when the symbols x and f are interpreted as x and f, respectively.

DEFINITION 7.6. *For each formation sequence S in which no variable other than x occurs, let $Eval_S(e, x, f)$ be the formula in the language of $ERA(f)$, defined inductively as follows:*

- $Eval_{\emptyset}(e, x, f)$ *is the sentence* $0 = 0$
- *If* t_k *is the variable* x, *then* $Eval_{\langle t_0, \ldots, t_k \rangle}(e, x, f)$ *is defined to be*

$$(e)_{\bar{k}} = x \wedge Eval_{\langle t_0, \ldots, t_{k-1} \rangle}(e, x, f).$$

- *If* t_k *is of the form* $g(t_{i_0}, \ldots, t_{i_l}, f)$, *where* g *is a function symbol of* $PRA(f)$, *then* $Eval_{\langle t_0, \ldots, t_k \rangle}(e, x, f)$ *is*

$$(e)_{\bar{k}} = F_g((e)_{\bar{i}_0}, \ldots, (e)_{\bar{i}_l}, f) \wedge Eval_{\langle t_0, \ldots, t_{k-1} \rangle}(e, x, f).$$

LEMMA 7.7. *Let* S *be a formation sequence of terms in the language of* $PRA(f)$ *in which at most the variable* x *is free. Then there is a* $\prec\omega^{\omega}$*-recursive function* $G(x, f)$ *such that* $ERA(f)$ *proves* $G(x, f) = e \rightarrow Eval_S(e, x, f)$.

PROOF. By induction on the length of S, using Lemmata 7.1–7.4. ⊣

Given a proof in the quantifier-free version of $PRA(f)$, we can use Lemma 7.7 to find a correct evaluation of the terms appearing in the proof.

LEMMA 7.8. *Suppose* $PRA(f)$ *proves* $\exists y\ \theta(x, y, f)$, *where* θ *is* $\Delta_0(f)$. *Then there is a* $\prec\omega^{\omega}$*-recursive function* $H(x, f)$ *such that* $ERA(f)$ *proves* $H(x, f) = y \rightarrow \theta(x, y, f)$.

PROOF. Suppose $PRA(f)$ proves $\exists y\ \theta(x, y, f)$. Then it also proves the formula $\exists y\ (\chi_\theta(x, y, f) = 1)$, where χ_θ is a an elementary recursive characteristic function representing θ. By Herbrand's theorem, there is a function symbol $g(x, f)$ of primitive recursive arithmetic, and a proof d of $\chi_\theta(x, g(x, f), f) = 1$ in propositional logic, from substitution instances of the equality axioms and axioms of $PRA(f)$. For example, we may take d to be a sequence of quantifier-free formulae in the language of $PRA(f)$ such that each line either is an instance of an axiom of $PRA(f)$, is an instance of an axiom of equality, is an instance of a propositional tautology, or follows from previous lines by modus ponens (or other valid propositional inferences).

Let S be a formation sequence that includes all the terms occurring in d. Each line of d is a boolean combination of atomic formulae of the form $t = s$, where t and s are terms occurring in S. If φ is such a formula, let φ^e denote the formula obtained by replacing each term t_i by $(e)_{\bar{i}}$. Then by induction, one can show that for each line φ of d, then $ERA(f)$ proves

$$Eval_S(e, x, f) \rightarrow \varphi^e.$$

When φ is an axiom of equality or $PRA(f)$, this follows from the definition of $Eval_S(e, x, f)$; otherwise, the propositional axioms and inferences of d can be mirrored in $ERA(f)$.

In particular, suppose $g(x, f)$ is the kth term in S and $\chi_\theta(x, g(x, f), f)$ is the lth. Since the conclusion of d is $\chi_\theta(x, g(x, f), f) = 1$, in $ERA(f)$ one can prove

$$Eval_S(e, x, f) \rightarrow (e)_{\bar{l}} = 1.$$

But if e evaluates terms correctly, $(e)_{\bar{l}}$ is equal to $F_{\chi_\theta}(x, (e)_{\bar{k}}, f)$; so $ERA(f)$ proves

$$Eval_S(e, x, f) \rightarrow (e)_{\bar{l}} = F_{\chi_\theta}(x, (e)_{\bar{k}}, f),$$

and hence $Eval_S(e, x, f) \to F_{\chi_\theta}(x, (e)_{\bar{k}}, f) = 1$. But by Lemma 7.5, $ERA(f)$ proves that $F_{\chi_\theta}(x, (e)_{\bar{k}}, f) = 1$ is equivalent to $\theta(x, (e)_{\bar{k}}, f)$.

In short, in $ERA(f)$ we can prove

$$Eval_S(e, x, f) \to \theta(x, (e)_{\bar{k}}, f).$$

Using Lemma 7.7, let $G(x, f)$ be a $\prec\alpha$-recursive function which returns an e satisfying $Eval_S(e, x, f)$. Using Lemmata 7.1 and 7.2 let $H(x, f)$ be a $\prec\alpha$-recursive function such that $ERA(f)$ proves $H(x, f) \simeq (G(x, f))_{\bar{k}}$. Putting it all together, we have that $ERA(f)$ proves $H(x, f) = y \to \theta(x, y, f)$, as desired. ⊣

By Theorem 6.3 this yields

THEOREM 7.9. *The proof-theoretic ordinal of $PRA(f)$ is at most ω^ω.*

At this point, we could extend the analysis to various forms of primitive recursion on the ordinals, and use similar methods to obtain ordinal analyses of various extensions of $PRA(f)$. But instead of pursuing that, let us turn instead to theories of transfinite induction.

§8. Π_1 Transfinite induction. In the last section, we saw that ordinal recursion can be used to simulate ordinary primitive recursion; but this should not have been very surprising. In this section we will be somewhat bolder: we will augment our basic theory with function symbols that are intended to denote *noncomputable* functions, allowing us to prove a form of transfinite induction. A judicious application of Herbrand's theorem will then enable us to extract constructive information from proofs in the augmented theory.

By Proposition 2.2, we can represent our system of notations in the language of $I\Delta_0^{exp}$. If $\varphi(x)$ is any formula in this language and β is any ordinal notation, then the *principle of transfinite induction* on β for φ is

$$\forall\gamma \prec \bar{\beta}\ (\forall\delta \prec \gamma\ \varphi(\delta) \to \varphi(\gamma)) \to \forall\gamma \prec \bar{\beta}\ \varphi(\gamma).$$

In words, this reads "if $\varphi(x)$ is progressive on β, then it holds for every ordinal less than β." Its contrapositive,

$$\exists\gamma \prec \bar{\beta}\ \neg\varphi(\gamma) \to \exists\gamma \prec \bar{\beta}\ (\neg\varphi(\gamma) \wedge \forall\delta \prec \gamma\ \varphi(\delta))$$

is the *least-element principle* on β for $\neg\varphi$. If Γ is any set of formulae, then $TI(\beta, \Gamma)$ and $LEP(\beta, \Gamma)$ denote, respectively, the principle of transfinite induction and the least-element principle on β, restricted to formulae in Γ. Similarly, $TI(\prec\alpha, \Gamma)$ and $LEP(\prec\alpha, \Gamma)$ denote these principles for arbitrary β less than α.

Our goal here is to provide ordinal analysis of the theories for the form

$$I\Delta_0^{exp}(f) + TI(\prec\alpha, \Pi_l(f)).$$

The following lemma gives some equivalent characterizations.

LEMMA 8.1. *Assume α is closed under the function $\beta \mapsto \omega \cdot \beta$. Then over $I\Delta_0^{exp}(f)$, the following schemata are equivalent:*

1. $TI(\prec\alpha, \Pi_1(f))$
2. $LEP(\prec\alpha, \Sigma_1(f))$

3. $TI(\prec\alpha, \Delta_0^{exp}(f))$
4. $LEP(\prec\alpha, \Delta_0^{exp}(f))$

PROOF. The contrapositive of any instance of 1 is equivalent to an instance of 2, and vice-versa; similarly for 3 and 4. Clearly 2 implies 4, so it suffices to show that 4 implies 2.

Let $\theta(\gamma, x)$ be Δ_0^{exp} and let β be a notation less than α. Arguing in $I\Delta_0^{exp}(f) + LEP(\prec\alpha, \Delta_0^{exp}(f))$, let us prove the least-element principle on β for $\exists x\ \theta(\gamma, x)$. Let $\theta'(\delta)$ be a formula which asserts that, if δ is written in the form $\omega \cdot \delta' + y$, then $\theta(\delta', y)$. Now suppose $\theta(\gamma, x)$. Then $\theta'(\omega \cdot \gamma + x)$. By the least-element principle on $\omega \cdot \beta$ for θ', there is a least δ satisfying $\theta'(\delta)$. But if $\delta = \omega \cdot \delta' + y$, then δ' is the least element satisfying $\exists x\ \theta(\delta', x)$. ⊣

Now let us add function symbols to $ERA(f)$ that enable us to interpret the new axioms. Using the last characterization in Lemma 8.1, it is sufficient to have, for each notation β less than α and elementary relation $R(\vec{x}, y, f)$, a function $g(\vec{x}, f)$ which returns the least γ less than β satisfying $R(\vec{x}, \gamma, f)$, whenever such a γ exists. The approach we will take is slightly more general, but not more difficult.

Given an elementary function $norm(\vec{x}, z, f)$, let "z minimizes $norm(\vec{x}, \cdot, f)$ below β" denote the following formula:

$$\forall w\ (norm(\vec{x}, w, f) \prec \beta \rightarrow norm(\vec{x}, z, f) \preceq norm(\vec{z}, w, f)).$$

In words, if anything has a norm less than β, then z has the smallest possible norm. Let

$$ERA(f) + min(\prec\alpha, \mathcal{E}(f))$$

be the theory obtained by adding, for each elementary function $norm(\vec{x}, z, f)$ and β less than α, a new function symbol, $min_{norm,\beta}(\vec{x}, f)$, to the language, and an axiom

$$\text{“}min_{norm,\beta}(\vec{x}, f) \text{ minimizes } norm(\vec{x}, \cdot, f) \text{ below } \bar{\beta}.\text{”}$$

In the name of the theory, the "$\mathcal{E}(f)$" indicates that the norm functions are required to be elementary in f; note that that the theory does not have symbols, say, for elementary functions or minimization functions defined from the ones we have just added.

Even for $\beta = 1$, a function $min_{norm,\beta}$ may be nonconstructive. For example, let $T(x, y, f)$ be an elementary relation such that $\exists y\ T(x, y, f)$ is a complete $\Sigma_1(f)$ formula; more precisely, assume T has the property that for any $\Sigma_1(f)$ formula $\varphi(\vec{w}, f)$ there is a natural number n, such that $\varphi(\vec{w}, f)$ is equivalent, in $ERA(f)$, to the formula $\exists y\ T(\langle \bar{n}, \vec{w}\rangle, y, f)$. (Think of $T(\langle \bar{n}, \vec{w}\rangle, y, f)$ as asserting that y witnesses the truth of the $\Sigma_1(f)$ formula coded by n, at the parameters $\vec{w}$; or T may be a version of Kleene's T predicate, asserting that y is a halting computation of the nth Turing machine with oracle f, on input $\vec{w}$. Below, we will also assume that n can be computed in an elementary way from a Gödel number of φ.) Let

$\beta = 1$, and let

$$norm(x, z, f) = \begin{cases} 0 & \text{if } T(x, z, f) \\ 1 & \text{otherwise.} \end{cases}$$

Then $min_{norm,\beta}(\vec{x})$ is guaranteed to return a witness y to $T(x, y, f)$, if there is one; this enables us, for example, to solve the halting problem.

LEMMA 8.2. *For each α, $ERA(f) + min(\prec\alpha, \mathcal{E}(f))$ is a universal theory containing $I\Delta_0^{exp}(f) + TI(\prec\alpha, \Pi_1(f))$.*

PROOF. The formula "$min_{norm,\beta}(\vec{x}, f)$ minimizes $norm(\vec{x}, \cdot, f)$ below $\bar{\beta}$" is universal, so it suffices to show that instances of the Σ_1 least-element principle are derivable from these axioms. Given a Δ_0^{exp} formula $\theta(y, \gamma, \vec{x}, f)$ with the free variables shown and a notation β, let

$$norm(\vec{x}, z, f) = \begin{cases} (z)_1 & \text{if } (z)_1 \prec \beta \text{ and } \theta((z)_0, (z)_1, \vec{x}, f) \\ \beta & \text{otherwise} \end{cases}$$

Arguing in $ERA(f) + min(\prec\alpha, \mathcal{E}(f))$, if there is any γ satisfying $\exists y\ \theta(y, \gamma, \vec{x}, f)$, then $(min_{norm,\beta}(\vec{x}))_1$ is a least such one. ⊣

We will carry out the ordinal analysis of $ERA(f) + min(\prec\alpha, \mathcal{E}(f))$ in two steps. First, we will show that one can reduce the problem of assigning the correct values to a set of terms in this theory to the problem of finding a value minimizing an appropriate norm, provably in $ERA(f)$. Then we will use Herbrand's theorem to replace the latter problem with a $\prec\alpha$-recursive calculation.

If $\varphi(\vec{x}, z)$ is any formula in the language of $ERA(f)$ and β is any notation, let us say that $ERA(f)$ proves that $\varphi(\vec{x}, y)$ is *solvable* (for y) *by β-minimization* if there are elementary functions $norm(\vec{x}, z, f)$ and $result(\vec{x}, z, f)$ such that $ERA(f)$ proves

$$\text{"}z \text{ minimizes } norm(\vec{x}, \cdot) \text{ below } \bar{\beta}\text{"} \rightarrow \varphi(\vec{x}, result(\vec{x}, z, f)).$$

Say that $ERA(f)$ proves that $\varphi(\vec{x}, y)$ is solvable by *$\prec\alpha$-minimization* if it proves that $\varphi(\vec{x}, y)$ is solvable by β-minimization for some β less than α. Note that the "solution" to $\varphi(\vec{x}, z)$ may not be unique. If $ERA(f)$ also proves $\varphi(\vec{x}, z) \wedge \varphi(\vec{x}, z') \rightarrow z = z'$ it makes sense to say that $ERA(f)$ proves that $\varphi(\vec{x}, z)$ defines a function that is computable by $\prec\alpha$-minimization; but for our purposes the more general notion is more useful.

The next few lemmata provide closure properties on the kinds of problems that are solvable by $\prec\alpha$-minimization.

LEMMA 8.3. *For any α, if $g(\vec{x}, f)$ is an elementary function, then $ERA(f)$ proves that the relation $g(\vec{x}, f) = y$ is solvable by $\prec\alpha$-minimization.*

PROOF. Let $norm(\vec{x}, z, f)$ be arbitrary, and let $result(\vec{x}, z, f) = g(\vec{x}, f)$. ⊣

LEMMA 8.4. *Suppose β is less than α, and $norm(\vec{x}, y, f)$ is an elementary function in f. Then $ERA(f)$ proves that the relation "z minimizes $norm(\vec{x}, \cdot, f)$ below β" is solvable by $\prec\alpha$-minimization.*

PROOF. Leave β and $norm$ alone, and let $result(\vec{x}, z, f) = z$. ⊣

LEMMA 8.5. *Let α be closed under addition. If $\varphi_0(\vec{x}, y, f), \ldots, \varphi_k(\vec{x}, y, f)$ are all solvable by $\prec\alpha$-minimization, provably in ERA(f), then so is the formula $\varphi_0(\vec{x}, (y)_0, f) \wedge \ldots \wedge \varphi_k(\vec{x}, (y)_k, f)$.*

PROOF. Suppose each φ_i is solvable by β_i, $norm_i(\vec{x}, z, f)$, and $result_i(\vec{x}, z, f)$. We can assume without loss of generality that $ERA(f)$ proves that for every $\vec{x}$ and z, $norm_i(\vec{x}, z, f)$ is less than or equal to β_i, by replacing $norm_i(\vec{x}, z, f)$ with $min(\beta_i, norm_i(\vec{x}, z, f))$ if necessary. Let $\beta = \beta_0 \# \ldots \# \beta_k$, let

$$norm(\vec{x}, z, f) = norm_0(\vec{x}, (z)_0, f) \# \ldots \# norm_k(\vec{x}, (z)_k, f),$$

and let

$$result(\vec{x}, z, f) = \langle result_0(\vec{x}, (z)_0, f), \ldots, result_k(\vec{x}, (z)_k, f)\rangle.$$

It is not hard to see that if z minimizes $norm(\vec{x}, \cdot, f)$ below β, then each $(z)_i$ minimizes $norm_i(\vec{x}, \cdot, f)$ below β_i. ⊣

LEMMA 8.6. *Let α be infinite and closed under multiplication. Suppose $\varphi_0(\vec{x}, w, f)$ and $\varphi_1(\vec{x}, w, y, f)$ are solvable by $\prec\alpha$-minimization, for w and y respectively, provably in ERA(f). Then the formula*

$$\varphi_0(\vec{x}, (y)_0) \wedge \varphi_1(\vec{x}, (y)_0, (y)_1),$$

and the formula

$$\exists w\ (\varphi_0(\vec{x}, w) \wedge \varphi_1(\vec{x}, w, y)),$$

are solvable for y by $\prec\alpha$ minimization, provably in ERA(f).

PROOF. Suppose φ_0 is solved by β_0, $norm_0(\vec{x}, z, f)$, and $result_0(\vec{x}, z, f)$, and φ_1 is solved by β_1, $norm_1(\vec{x}, w, z)$, and $result_1(\vec{x}, w, f)$. As in the previous lemma, we can assume that $norm_0$ and $norm_1$ are bounded by β_0 and β_1 respectively. Let $\beta = (\beta_1 + 1)(\beta_0 + 1)$ and let

$$\begin{aligned} norm(\vec{x}, z, f) = {} & (\beta_1 + 1) \cdot norm_0(\vec{x}, (z)_0, f) \\ & + norm_1(\vec{x}, result_0(\vec{x}, (z)_0, f), (z)_1). \end{aligned}$$

Suppose z minimizes $norm(\vec{x}, \cdot, f)$ below β. Then $(z)_0$ minimizes $norm_0(\vec{x}, \cdot, f)$ below β_0; otherwise we could change $(z)_0$ and decrease the value of sum above, independent of the behavior of the second term. Fixing $(z)_0$, we also see that $(z)_1$ minimizes $norm_1(\vec{x}, result_0(\vec{x}, (z)_0, f), \cdot, f)$ below β_1, because otherwise we could change $(z)_1$ and decrease the value of the sum above. To solve the first formula, take

$$result(\vec{x}, z, f) = \langle result_0(\vec{x}, (z)_0, f), result_1(\vec{x}, result_0(\vec{x}, (z)_0, f), (z)_1, f)\rangle.$$

To solve the second formula, take

$$result(\vec{x}, z, f) = result_1(\vec{x}, result_0(\vec{x}, (z)_0, f), (z)_1, f).$$

This completes the proof. ⊣

Taken together, the lemmata above imply that for infinite α closed under multiplication, the functions that are computable by $\prec\alpha$-minimization are closed under composition. As an exercise, the reader can try to prove that if $F(\vec{x}, f)$ is a $\prec\alpha$-recursive function, then it is computable by $\prec\alpha$-minimization. (See also the first proof of Lemma 9.2 below.)

The next step is to show that given any formation sequence S for terms in $ERA(f) + min(\prec\alpha, \mathcal{E}(f))$, the problem of finding an appropriate evaluation of the terms in S is solvable by $\prec\alpha$-minimization. We have to be careful, since, in general, there may be more than one value that we can assign to a term of the form $min_{norm,\beta}(t_1, \dots, t_k, f)$; so when more than one term of this form appears in the formation sequence, we have to make sure that the evaluation assigns values to these terms consistently.

DEFINITION 8.7. *For each formation sequence S for terms in the theory*

$$ERA(f) + min(\prec\alpha, \mathcal{E}(f))$$

with no variable other than x, let $Eval_S(e, x, f)$ be the formula in the language of $ERA(f)$, defined inductively as follows:

- *$Eval_\emptyset(e, x, f)$ is the sentence $0 = 0$*
- *If t_k is the variable x, then $Eval_{\langle t_0, \dots, t_k \rangle}(e, x, f)$ is*

$$(e)_{\bar{k}} = x \wedge Eval_{\langle t_0, \dots, t_{k-1} \rangle}(e, x, f).$$

- *If t_k is of the form $g(t_{i_0}, \dots, t_{i_l}, f)$, where g is a function symbol of $ERA(f)$, then $Eval_{\langle t_0, \dots, t_k \rangle}(e, x, f)$ is*

$$(e)_{\bar{k}} = g((e)_{\bar{i}_0}, \dots, (e)_{\bar{i}_l}, f) \wedge Eval_{\langle t_0, \dots, t_{k-1} \rangle}(e, x, f).$$

- *If t_k is of the form $min_{norm,\beta}(t_{i_0}, \dots, t_{i_l}, f)$, let $t_{j_0}, \dots, t_{j_m}$ enumerate all the terms before t_k in S that are also of this form; and for $u = 0, \dots, m$, suppose t_{j_u} is the term $min_{norm,\beta}(t_{i_{u,0}}, \dots, t_{i_{u,l}}, f)$. Then $Eval_{\langle t_0, \dots, t_k \rangle}(e, x, f)$ is the conjunction of the following*:
  - *"$(e)_{\bar{k}}$ minimizes $norm((e)_{\bar{i}_0}, \dots, (e)_{\bar{i}_l}, \cdot, f)$ below $\bar{\beta}$"*
  - *$Eval_{\langle t_0, \dots, t_{k-1} \rangle}(e, x, f)$*
  - *$\bigwedge_{u=0}^{m}((e)_{\bar{i}_0} = (e)_{\bar{i}_{u,0}} \wedge \dots \wedge (e)_{\bar{i}_l} = (e)_{\bar{i}_{u,l}}) \rightarrow (e)_{\bar{k}} = (e)_{\bar{j}_u})$.*

LEMMA 8.8. *Let α be infinite and closed under multiplication, and let S be a formation sequence of terms in the language of $ERA(f) + min(\prec\alpha, \mathcal{E}(f))$. Then $ERA(f)$ proves that the relation $Eval_S(e, x, f)$ is solvable for e by $\prec\alpha$-minimization.*

PROOF. By induction on the length of S, using Lemmata 8.3–8.6. Consider the case where S is the sequence $\langle t_0, \dots, t_k \rangle$ and t_k is of the form $min_{norm,\beta}(t_{i_0}, \dots, t_{i_l}, f)$. By the induction hypothesis, $ERA(f)$ proves that the relation $Eval_{\langle t_0, \dots, t_{k-1} \rangle}(e', x, f)$ is solvable for e' by $\prec\alpha$-minimization. By Lemma 8.4, $ERA(f)$ also proves that the problem of finding a value v minimizing $norm((e)_{\bar{i}_0}, \dots, (e)_{\bar{i}_l}, \cdot, f)$ below β is solvable by $\prec\alpha$-minimization. There is an elementary function which, given e' and v, checks the values that e' assigns to previous terms in S, decides whether to assign one of these values or v to t_k, and returns the resulting assignment. By Lemmata 8.3 and 8.6, this provides a solution to $Eval_S(e, x, f)$. ⊣

LEMMA 8.9. *Let α be infinite and closed under multiplication. If $\theta(x, y, f)$ is a $\Delta_0(f)$ formula such that $ERA(f) + min(\prec\alpha, \mathcal{E}(f))$ proves $\exists y\ \theta(x, y, f)$, then $ERA(f)$ proves that $\theta(x, y, f)$ is solvable for y by $\prec\alpha$-minimization.*

PROOF. Just as in the proof of Lemma 7.8. If $ERA(f) + min(\prec\alpha, \mathcal{E}(f))$ proves $\exists y\ \theta(x, y, f)$, then there is a sequence of terms $r_0, \ldots, r_m$ and a propositional proof of $\chi_\theta(x, r_0, f) = 1 \vee \ldots \vee \chi_\theta(x, r_m, f) = 1$ from substitution instances of equality axioms and axioms of $ERA(f) + min(\prec\alpha, \mathcal{E}(f))$. In $ERA(f)$, given x we can use $\prec\alpha$-minimization to evaluate all the terms occurring in the proof and choose one satisfying θ. ⊣

The key use of Herbrand's theorem is in the proof of the following lemma.

LEMMA 8.10. *Suppose α is infinite and closed under multiplication, and let $\theta(x, y, f)$ be a Δ_0 formula such that $ERA(f)$ proves $\theta(x, y, f)$ is solvable by $\prec\alpha$-minimization. Then there is a $\prec\alpha$-recursive function $F(x, f)$ such that $ERA(f)$ proves $F(x, f) = y \rightarrow \theta(x, y, f)$.*

PROOF. The hypothesis of the lemma means that there is a β less than α, and functions *norm* and *result*, such that $ERA(f)$ proves

$$\forall z\ (\forall w\ (norm(x, w, f) \prec \bar{\beta} \rightarrow norm(x, z, f) \preceq norm(x, w, f)) \\ \rightarrow \theta(x, result(x, z, f), f)).$$

Leaving the universal quantification over z implicit, bringing the universal quantification over w to the front (where it becomes existential), and rewriting the formula slightly, we have that $ERA(f)$ proves

$$\exists w\ (norm(x, w, f) \not\prec \beta \vee norm(x, w, f) \not\prec norm(x, z, f) \\ \rightarrow \theta(x, result(x, z, f), f)).$$

By Herbrand's theorem, there is a an elementary function $g(x, z, f)$ such that $ERA(f)$ proves

$$norm(x, g(x, z, f), f) \not\prec \beta \vee norm(x, g(x, z, f), f) \not\prec norm(x, z, f) \\ \rightarrow \theta(x, result(x, z, f), f).$$

In other words, if the norm of $g(x, z, f)$ is not less than both β and the norm of z, then $result(x, z, f)$ is the witness we are after. Our job, then, is to find a z satisfying the antecedent of this implication. An obvious iterative algorithm suggests itself: First set $z_0 = 0$. If $norm(x, g(x, z_0, f), f) \not\prec \beta$ or $norm(x, g(x, z_0, f), f) \not\prec norm(x, z_0, f)$, we are done. Otherwise, iteratively set z_{i+1} equal to $g(x, z_i, f)$, until $norm(x, z_{i+1}, f) \not\prec norm(x, z_i, f)$. Then $result(x, z_i, f)$ is the value we are after.

By Lemma 7.5, $norm(x, z, f)$ and $result(x, z, f)$ are $\prec\alpha$-recursive. The algorithm just described is a $\prec\alpha$-iterative algorithm using $\prec\alpha$-recursive functions; by

Lemma 7.3, it can be carried out by a single $\prec\alpha$-recursive function, provably in $ERA(f)$. $\dashv$

Putting the last two lemmata together yield the following

LEMMA 8.11. *Let α be infinite and closed under multiplication. If $\theta(x, y, f)$ is a $\Delta_0(f)$ formula such that $ERA(f) + min(\prec\alpha, \mathcal{E}(f))$ proves $\exists y\ \theta(x, y, f)$, then there is a $\prec\alpha$-recursive function $F(x, f)$ such that $ERA(f)$ proves $F(x, f) = y \rightarrow \theta(x, y, f)$.*

Below we will need to know that Lemma 8.11 still holds with additional function symbols, $f_0, \ldots, f_k$. To see this, note that every proof in this section can easily be generalized in this respect; alternatively, one can take f to code a finite sequence of function symbols $f_0, \ldots, f_k$, and use variant of Lemma 2.3 to reduce the more general statement to that of Lemma 8.11.

Together with Theorem 6.3, Lemma 8.11 yields

THEOREM 8.12. *Let α be infinite and closed under multiplication. Then the proof-theoretic ordinal of $I\Delta_0(f) + TI(\prec\alpha, \Pi_1(f))$ is at most α.*

Since induction on the natural numbers corresponds to transfinite induction on ω, and Π_1 and Σ_1 induction on the natural numbers are equivalent, $TI(\prec\omega^\omega, \Pi_1(f))$ includes $I\Sigma_1$, and we have

THEOREM 8.13. *The proof-theoretic ordinal of $I\Sigma_1(f)$ is at most ω^ω.*

These bounds are sharp, and, more generally, the proof-theoretic ordinal of $I\Delta_0(f) + TI(\prec\alpha, \Pi_1(f))$ is the least α' greater than or equal to α that is closed under multiplication.

§9. Arithmetic transfinite induction. Having dealt with Π_1 transfinite induction, let us now extend the analysis to theories of transfinite induction for arbitrary arithmetic formulae. Once again, the first step is to express these principles in a suitable universal theory. We will do this in a straightforward way: we will use Skolem functions to reduce any arithmetic formula to an elementary relation, and then use minimization, as in the last section.

To start with, consider Π_2 transfinite induction. Let $\exists y\ T(x, y, f)$ be the complete $\Sigma_1(f)$ formula introduced in the last section, and let wit_1 be a new function symbol with defining equation

$$(wit_1(f)) \qquad \forall y\ (T(x, y, f) \rightarrow T(x, wit_1(x), f)).$$

In words, if there is any y satisfying $T(x, y, f)$, wit_1 returns such a one.[3] Let

$$ERA(f, wit_1) + (wit_1(f)) + min(\prec\alpha, \mathcal{E}(f, wit_1))$$

denote the theory extending $ERA(f, wit_1)$ with the defining axiom for wit_1, and minimization for function symbols in the language of $ERA(f, wit_1)$.

LEMMA 9.1. *The theory $ERA(f, wit_1) + (wit_1(f)) + min(\prec\alpha, \mathcal{E}(f, wit_1))$ is a universal theory containing $I\Delta_0^{exp}(f) + TI(\prec\alpha, \Pi_2(f))$.*

[3] We could go further to fix the interpretation of wit_1 by requiring it to return the *least* such y, if there is one, and 0 otherwise; but the additional generality will be useful in [2].

PROOF. If $\varphi(\vec{z}, f)$ is a $\Sigma_1(f)$ formula of the form $\exists y\ \theta(y, \vec{z}, f)$, there is a numeral n such that $\varphi(\vec{z}, f)$ is equivalent in $ERA(f, wit_I) + (wit_I)$ to the $\mathcal{E}(f, wit_1)$ relation $T(\langle \bar{n}, \vec{z}\rangle, wit_1(\langle \bar{n}, \vec{z}\rangle), f)$. $\Pi_2(f)$ formulae are then equivalent to formulae that are $\Pi_1(f, wit_1)$. As in the proof of Lemma 8.2, from $(min(\prec\alpha, \mathcal{E}(f, wit_I)))$ one can derive instances of transfinite induction for formulae of this kind. ⊣

The following lemma enables us to bound the ordinal of the theory of Lemma 9.1.

LEMMA 9.2. *Let α be infinite and closed under multiplication. If $\theta(x, y, f)$ is a $\Delta_0(f)$ formula such that $ERA(f, wit_I) + (wit_I(f)) + min(\prec\alpha, \mathcal{E}(f, wit_I))$ proves $\exists y\ \theta(x, y, f)$, then there is a $\prec\omega^\alpha$-recursive function $H(x, f)$ such that $ERA(f)$ proves $H(x, f) = y \rightarrow \theta(x, y, f)$.*

PROOF. Suppose the hypothesis of the lemma holds. By the deduction theorem, the theory $ERA(f, wit_I) + min(\prec\alpha, \mathcal{E}(f, wit_I))$ proves

$$\forall u, v\ (T(u, v, f) \rightarrow T(u, wit_1(u), f)) \rightarrow \exists y\ \theta(x, y, f).$$

Letting y code the pair $\langle u, v\rangle$ and rewriting yields

$$\exists y\ (T((y)_0, (y)_1, f) \wedge \neg T((y)_0, wit_1((y)_0), f)) \vee \exists y\ \theta(x, y, f).$$

Bringing the existential quantifiers to the front and combining them yields

$$\exists y\ ((T((y)_0, (y)_1, f) \rightarrow T((y)_0, wit_1((y)_0), f)) \rightarrow \theta(x, y, f)).$$

By Lemma 8.11, there is a $\prec\alpha$-recursive function F such that $ERA(f, wit_I)$ proves

$$F(x, f, wit_1) = y \wedge (T((y)_0, (y)_1, f) \rightarrow T((y)_0, wit_1((y)_0), f)) \rightarrow \theta(x, y, f).$$

In words, $ERA(f, wit_I)$ proves that if $F(x, f, wit_1)$ is defined, it returns either a y showing that wit_1 fails to satisfy its defining axiom at $(y)_0$, or a y satisfying $\theta(x, y, f)$.

The rest of the proof hinges on finding a finite interpretation of wit_1 that is robust enough to carry out the computation of F and pass the final test at the end. Towards that goal, note that one can code a finite *partial* function from the natural numbers to the natural numbers with a natural number. For example, one can take the number m to code the partial function

$$\tilde{m}(x) = \begin{cases} (m)_x - 1 & \text{if } x \leq last(m) \text{ and } (m)_x > 0 \\ \text{undefined} & \text{otherwise} \end{cases}$$

Let $\hat{m}(x)$ denote the extension of $\tilde{m}$ to the natural numbers, such that $\hat{m}(x) = 0$ where $\tilde{m}$ is undefined. Finally, let $eval(m, x)$ be the elementary function which returns $\hat{m}(x)$. If we now take m to be a variable in the language of $ERA(f)$, we can interpret references to $\hat{m}(x)$ as $eval(m, x)$. Returning to the conclusion of the last paragraph, using Lemma 2.3 to replace wit_1 by $\lambda x\ eval(m, x)$, we see that in $ERA(f)$ there is a proof of

$$F(x, f, \hat{m}) = y \wedge (T((y)_0, (y)_1, f) \rightarrow T((y)_0, \hat{m}((y)_0), f)) \rightarrow \theta(x, y, f).$$

Expanding the definition of $F(x, f, \hat{m}) = y$, this yields a proof of

$$(CompSeq_F(s, x, f, \hat{m}) \wedge result_F((s)_{last(s)}) = y \\ \wedge (T((y)_0, (y)_1, f) \rightarrow T((y)_0, \hat{m}((y)_0), f))) \rightarrow \theta(x, y, f).$$

We have therefore reduced the problem to finding a $\prec\omega^\alpha$-recursive function $G(x, f)$ which returns a pair $\langle s, m\rangle$ satisfying the antecedent of this last formula, provably in $ERA(f)$. In other words, we are looking for a $\prec\omega^\alpha$-recursive function that returns a finite interpretation m for wit_1 and a computation sequence s for F at x, f, and $\hat{m}$, such that if $y = result_F((s)_{last(s)})$, then $\hat{m}$ satisfies the defining equation for wit_1 at the pair coded by y. If we then let $H(x, f) \simeq result_F((G(x, f)_0)_{last(G(x,f)_0)})$, $H(x, f)$ satisfies the conclusion of the lemma.

I will now describe two ways of finding such a function G. The first is used in the proof of [7, Lemma 12]; the second is more explicit.

For the first method, note that by Lemma 8.11 it is sufficient to show that the existence of the pair s, m is provable in $ERA(f) + min(\prec\omega^\alpha, \mathcal{E}(f))$. Arguing in this theory, then, let us show how to find s and m using ordinal minimization. Without loss of generality, we can assume that if $F(x, f)$ returns y then F queries $\hat{m}$ at $(y)_0$ in the last step of its computation, since we can always replace the β-iterative algorithm for F with a $1 + \beta$-iterative algorithm which does so. Say that m is a *sound* interpretation of wit_1 if $T(x, \hat{m}(x), f)$ holds for every x in the domain of m. Let

$$norm'(m, x) = \begin{cases} 1 & \text{if } m \text{ is undefined at } x \\ 0 & \text{otherwise.} \end{cases}$$

Finally, say that s is a *partial computation sequence* for F at x, f, and $\hat{m}$ if s is a proper initial segment of a computation sequence for F at x, f, and $\hat{m}$. More explicitly, this amounts to saying that $(s)_0 = start_F(x)$, $norm_F((s)_0) \prec \beta$, and

$$\forall i < k - 1\ ((s)_{i+1} = next_F((s)_i, f(query_{F.0}((s)_i)), \hat{m}(query_{F.1}((s)_i))) \\ \wedge norm_F((s)_{i+1}) \prec norm_F((s)_i)).$$

Now define $norm(x, z, f)$ as follows:

- If z codes a pair $\langle s, m\rangle$, m is a sound interpretation of wit_1, and s is a partial computation sequence for F at x, f, and $\hat{m}$, set $norm(x, z, f)$ to

$$\omega^{norm_F((s)_0)} \cdot norm'(m, query_{F.1}((s)_0)) + \dots \\ + \omega^{norm_F((s)_{last(s)-1})} \cdot norm'(m, query_{F.1}((s)_{last(s)-1})) \\ + \omega^{norm_F((s)_{last(s)})} \cdot 2$$

(Interpret this as 0 if $last(s) = 0$.)

- If z codes a pair $\langle s, m\rangle$, m is a sound interpretation of wit_1, and s is a computation sequence for F at x, f, and $\hat{m}$, set $norm(z)$ to

$$\omega^{norm_F((s)_0)} \cdot norm'(m, query_{F,1}((s)_0)) + \dots \\ + \omega^{norm_F((s)_{last(s)-1})} \cdot norm'(m, query_{F,1}((s)_{last(s)-1}))$$

- Otherwise, set $norm(x, z, f) = \omega^\beta$.

Let z be a value minimizing $norm(x, \cdot, f)$ below ω^β. We know that for this z either the first or second case must hold, since taking s to be the sequence $\langle start_F(\vec{x})\rangle$ and m to be the partial function that is nowhere defined satisfies one of these two cases, and hence yields a norm less than ω^β. In fact, z must fall under the second case: if s is a partial computation sequence, extending it by appending $next_F((s)_{last(s)}, f(query_0((s)_{last(s)})), \hat{m}(query_1((s)_{last(s)})))$ yields either a computation sequence or partial computation sequence with a smaller norm. Finally, let us show that if $result_F((s)_{last(s)})$ is equal to y, we have $T((y)_0, (y)_1, f) \to T((y)_0, \hat{m}((y)_0), f)$. Suppose otherwise; then $T((y)_0, (y)_1, f)$ holds, but not $T((y)_0, \hat{m}((y)_0), f)$. Since m is a sound interpretation of wit_1, this means that $(y)_0$ is not in the domain of $\tilde{m}$. Let i be the least value less than $last(s)$ such that $query_1((s)_i) = (y)_0$; there is at least one such i, since we are assuming that F queries $\hat{m}$ at this value before the end of its computation. Note that we then have $norm'(m, query_{F,1}((s)_i)) = 1$. Let m' represent the partial function extending $\tilde{m}$ with $\tilde{m}'((y)_0) = (y)_1$, and let s' be the partial computation sequence which is an initial segment of s of length i. Then the norm of the pair $\langle s', m'\rangle$ is less than the norm of the pair $\langle s, m\rangle$, contradicting the fact that $\langle s, m\rangle$ is supposed to minimize $norm(x, \cdot, f)$.

For the second proof, note that we can extract an explicit description of G from the preceding argument. The iterative algorithm for G starts with the pair $\langle s, m\rangle$, where s is the sequence $\langle start_F(x)\rangle$ and m is nowhere defined, and $norm_G$ is the function $norm$ above. As long as the current state of G is a pair $\langle s, m\rangle$, where s is either a partial computation sequence or a computation sequence for F at x, $\hat{m}$, and f, and m is a sound interpretation of wit_1, the argument above provides a recipe $next_G$ for finding a pair $\langle s, m\rangle$ with a smaller norm, unless s is, in fact, a computation sequence and $\hat{m}$ satisfies the axiom for wit_1 at $result_F((s)_{last(s)})$; and in that case, $result_G$ can just return the pair $\langle s, m\rangle$. By Lemma 7.5, T is $\prec\alpha$-recursive; so $start_G$, $norm_G$, $next_G$, and $result_G$ are all $\prec\alpha$-recursive functions. By Lemma 7.3, G is be $\prec\omega^\alpha$-recursive, and the relevant properties can be proved in $ERA(f)$. ⊣

Putting this all together, we have

Corollary 9.3. *Let α be infinite and closed under multiplication. Then the proof-theoretic ordinal of $I\Delta_0^{exp}(f) + TI(\prec\alpha, \Pi_2(f))$ is at most ω^α.*

More generally, let

$$ERA(wit_1, \dots, wit_n, f) + (wit^n(f)) + min(\prec\alpha, \mathcal{E}(wit_1, \dots, wit_n, f))$$

denote the theory with n witnessing functions, axioms asserting that each wit_{i+1} returns witnesses to a complete $\Sigma_1(wit_1, \ldots, wit_i, f)$ formula, and minimization for functions that are elementary in $wit_1, \ldots, wit_n, f$. Adapting the proof of Lemma 9.2 yields

LEMMA 9.4. *Let α be infinite and closed under multiplication. Suppose*

$$\theta(x, y, wit_1, \ldots, wit_n, f)$$

is a $\Delta_0(wit_1, \ldots, wit_n, f)$ formula such that

$$ERA(wit_1, \ldots, wit_{n+1}, f) + (wit^{n+1}(f)) + min(\prec\alpha, \mathcal{E}(wit_1, \ldots, wit_{n+1}, f))$$

proves $\exists y\ \theta(x, y, wit_1, \ldots, wit_n, f)$. Then

$$ERA(wit_1, \ldots, wit_n, f) + (wit^n(f)) + min(\prec\omega^\alpha, \mathcal{E}(wit_1, \ldots, wit_n, f))$$

proves it as well.

Let $\omega_0^\alpha = \alpha$, and $\omega_{n+1}^\alpha = \omega^{\omega_n^\alpha}$. By induction, we have:

THEOREM 9.5. *Let α be infinite and closed under multiplication, and let n be greater than or equal to 0. Then the proof-theoretic ordinal of the theory*

$$I\Delta_0^{exp}(f) + TI(\prec\alpha, \Pi_{n+1}(f))$$

is at most ω_n^α.

In particular, for $\alpha = \omega^\omega$, we have:

THEOREM 9.6. *For n greater than or equal to 1, the proof-theoretic ordinal of $I\Sigma_n(f)$ is at most ω_{n+1}.*

Also, since any proof in $PA(f)$ is a proof in $I\Sigma_n(f)$, for some n, we have

THEOREM 9.7. *The proof-theoretic ordinal of $PA(f)$ is at most ε_0.*

More generally, let $PA(f) + TI(\prec\alpha)$ denote Peano arithmetic together with transfinite induction principles for arbitrary notations β less than α and arbitrary formulae in the language.

THEOREM 9.8. *Let α be infinite and closed under $\beta \mapsto \omega^\beta$. Then the proof-theoretic ordinal of $PA(f) + TI(\prec\alpha)$ is at most α.*

In all these theorems, the bounds are sharp.

§10. Transfinite arithmetic hierarchies. In this final section, we will consider theories with transfinite arithmetic hierarchies. We will continue to use functions as our basic second-order objects, and interpret references to a set X of natural numbers as references to its characteristic function, χ_X. If X is any set, the Turing jump of X, written X', is defined to be the set $\{x \mid \exists y\ T(x, y, \chi_X)\}$ where $\exists y\ T(x, y, f)$ is the complete $\Sigma_1(f)$ formula from Section 8. We can use a set X to code a sequence of such sets, by interpreting $z \in X_y$ as $\langle y, z\rangle \in X$. If β is an ordinal notation and $\langle H_\gamma\rangle_{\gamma\prec\beta}$ is a sequence of sets, then $H_{\prec\beta}$ is a set which codes this sequence (with $H_y = \emptyset$ if y is the notation for the ordinal 0, or y is not a notation less than β). Given an ordinal notation β, a jump hierarchy on β is defined to be a set H, such that for every γ less than β, $H_\gamma = (H_{\prec\gamma})'$.

Given a notation for a limit ordinal α, we can extend the language of *PA* with new symbols H^β, for β less than α. Our goal is then to bound the proof-theoretic ordinal of theories of the form

$$PA(\mathcal{H}) + H(\prec\alpha) + TI(\prec\alpha),$$

given by the following set of axioms: the axioms of *PA*, extended to the language with the new symbols; for each β less than α, an axiom asserting that H^β is a jump hierarchy on β; and transfinite induction below α, for arbitrary formulae in new language. For an ordinal analysis, we really need to consider versions of these theories in which the induction axioms and transfinite hierarchies are relativized to a function symbol f; but to simplify the exposition I will drop the references to f, with the implicit understanding that all the lemmata and theorems in this section are easily generalized in this way.

We can simplify the theories above in a number of ways. For one thing, in $I\Delta_0^{exp}(\mathcal{H})$ one can show that if H^β is a jump hierarchy on β, then for each γ less than β, $H^\beta_{\prec\gamma}$ is a transfinite hierarchy on γ. This means that for any particular proof of an arithmetic statement, it suffices to use only a single new symbol H^β, for a sufficiently large β less than α. Also, if $\varphi(H^\beta)$ is an arithmetic formula relative to one of these hierarchies, there is a natural number n such that $I\Delta_0^{exp}(\mathcal{H})$ proves that $\varphi(H^\beta)$ is equivalent to a formula that is $\Delta_0(H^{\beta+n})$. This means that nothing is lost of we restrict the transfinite induction principles to formulae that are Δ_0 in the new symbols. Finally, since the transfinite induction axioms include ordinary induction, one can replace $PA(\mathcal{H})$ by $I\Delta_0^{exp}(\mathcal{H})$.

In carrying out the ordinal analysis we will need to make use of Kleene's recursion theorem. Let $R_0, R_1, R_2, \ldots$ denote a standard enumeration of the partial computable functions (of various arities). The recursion theorem says that for any partial computable function $f(x, e)$, there is an index e for a unary partial computable function R_e, satisfying $R_e(x) \simeq f(e, x)$ for every e and x. Since there is a universal partial computable function, this allows us to define a partial computable function R_e in terms of itself, and its own index. A theory of computable functions strong enough to prove the recursion theorem can be developed in *ERA* or $I\Delta_0^{exp}$, interpreting references to such functions by references to their indices. If I say that a function f is "computable" instead of "partial computable," I mean, implicitly, that f is total.

Once again, the first step in our ordinal analysis is to embed the theories we are interested in appropriate universal theories. An obvious strategy is to extend the Σ_n witness functions of the previous section to transfinite ones. We can take a single function f to represent a sequence of functions f_x, where $f_x(y) = f(\langle x, y\rangle)$; and given a function f and a notation γ, we can take $f\restriction_\gamma$ to denote the function

$$f\restriction_\gamma(z) = \begin{cases} f(z) & \text{if } (z)_0 \prec \gamma \\ 0 & \text{otherwise} \end{cases}$$

Then a transfinite witness function *wit* on β satisfies the following:

$$\forall \gamma \prec \beta \, \forall x, y \, (T(x, y, wit \restriction_\gamma) \rightarrow T(x, wit_\gamma(x), wit \restriction_\gamma)).$$

Now, for each β less than α, we can add a function symbol wit^β, to denote a witness function on β. Then

$$ERA(\mathcal{W}) + wit(\prec\alpha) + min(\prec\alpha, \mathcal{E}(\mathcal{W}))$$

denotes the theory that extends elementary computable arithmetic (relative to the new function symbols) with axioms that assert that the new symbols denote transfinite witness functions for each β less than α, and minimization for functions in the language of $ERA(\mathcal{W})$. In $ERA(\mathcal{W})$ one can show that if *wit* is a transfinite witness function up to β and γ is less than β, then $wit \restriction_\gamma$ is a transfinite witness function up to γ; so for any proof, it suffices to have a witness function for a single β less than α that is large enough.

In the end, it makes little difference whether one has a transfinite jump hierarchy or a transfinite hierarchy of witnesses, since each one can be obtained, effectively, from the other. The following lemma makes use of this fact.

LEMMA 10.1. *Let α be a limit. Then*

$$ERA(\mathcal{W}) + wit(\prec\alpha) + min(\prec\alpha, \mathcal{E}(\mathcal{W}))$$

is a universal theory, and proves any arithmetic formula that is provable in

$$PA(\mathcal{H}) + H(\prec\alpha) + TI(\prec\alpha).$$

PROOF. It is clear that the axioms for the witness functions are universal. By the observations above, to prove the second statement, we only need to interpret references to H^β in the language of $ERA(\mathcal{W})$, for each β less than α. Fix β, write H for H^β, and write *wit* for wit^β. We will use the recursion theorem to define a total computable function $R_e(\gamma, x)$, whereby we can interpret $x \in H_\gamma$ as $T(R_e(\gamma, x), wit_\gamma(R_e(\gamma, x)), wit \restriction_\gamma)$. In other words, we aim to find an index e that allows us to interpret χ_H as

$$\chi_H(w) = \begin{cases} 1 & \text{if } (w)_0 \prec \beta \text{ and} \\ & T(R_e((w)_0, (w)_1), wit_{(w)_0}(R_e((w)_0, (w)_1)), wit \restriction_{(w)_0}) \\ 0 & \text{otherwise.} \end{cases}$$

For the moment, consider the right hand side as a function of e, as well as w; with this interpretation of H, for γ less than β, $\chi_{H \restriction_\gamma}(w)$ is a function of e, w, γ, and $wit \restriction_\gamma$, given by

$$h(w, e, \gamma, wit \restriction_\gamma) = \begin{cases} 1 & \text{if } (w)_0 \prec \gamma \text{ and } T(R_e((w)_0, (w)_1), \\ & \quad (wit \restriction_\gamma)_{(w)_0}(R_e((w)_0, (w)_1)), (wit \restriction_\gamma) \restriction_{(w)_0}) \\ 0 & \text{otherwise.} \end{cases}$$

To verify the defining axiom for H, we need to satisfy

$$T(R_e(\gamma, x), wit_\gamma(R_e(\gamma, x)), wit \restriction_\gamma) \leftrightarrow \exists y \; T(R_e(\gamma, x), y, \lambda w \, h(w, e, \gamma, wit \restriction_\gamma))$$

whenever γ is less than β. But for each e, the right hand side is Σ_1 in $wit \restriction_\gamma$; so there is a code $n(e,\gamma)$, effectively obtained from e and γ, such that this formula is equivalent to

$$\exists y\ T(\langle n(e,\gamma), x\rangle, y, wit \restriction_\gamma).$$

By the defining axiom for *wit*, this is equivalent to

$$T(\langle n(e,\gamma), x\rangle, wit_\gamma(\langle n(e,\gamma), x\rangle), wit \restriction_\gamma).$$

Using the recursion theorem to find an e such that

$$R_e(\gamma, x) = \langle n(e,\gamma), x\rangle$$

we have the equivalence we are after. ⊣

Most of the rest of this section is devoted to proving the following:

LEMMA 10.2. *Let α be infinite and closed under multiplication. If $\theta(x,y)$ is a Δ_0 formula and $ERA(\mathcal{W}) + wit(\prec\alpha) + min(\prec\alpha, \mathcal{E}(\mathcal{W}))$ proves $\exists y\ \theta(x,y)$, then $PA + TI(\prec\varphi_\alpha(0))$ proves it as well.*

From now on, fix α, and let us work in $PA + TI(\prec\varphi_\alpha(0))$. In this theory we can refer to computable functions by identifying them with their indices, and we can refer to Δ_0 formulae by identifying them with their Gödel numbers. The definitions which follow should be seen as taking place in $PA + TI(\prec\varphi_\alpha(0))$, with these conventions.

In the last section, we saw, roughly speaking, that given a $\prec\eta$-computable function that queries wit_1, and, with "adequate" responses, finds a witness to a Σ_1 formula, we can effectively obtain a $\prec\omega^\eta$-computable function that finds a witness outright. In the current setting, this will let us transform a $\prec\eta$-iterative computation that queries a witness hierarchy on $\mu + 1$, to a $\prec\omega^\eta$-iterative computation which queries a witness hierarchy on μ. To extend this reduction to the transfinite, we need to generalize the notion of an iterative computation.

A *$\prec\eta$-computable* (rather than recursive) *functional* $F(f)$ is given by a notation $\mu_F \prec\eta$, a value $start_F$, and computable (rather than elementary) functions $norm_F(q)$, $next_F(q,z)$, $query_F(q)$, and $result_F(q)$. Given f, the notions $CompSeq_F(s,f)$, $F(f)\downarrow$, $F(f) = y$, and so on, are defined as in Section 3.

Using sequences, we can express the assertion that f is a witness hierarchy up to μ with a formula $\forall y\ W(y,\mu,f)$, where $W(y,\mu,f)$ is given by

$$(y)_0 \prec \mu \wedge T((y)_1, (y)_2, f \restriction_{(y)_0}) \rightarrow T((y)_1, f((y)_1), f \restriction_{(y)_0}).$$

Think of $W(y,\mu,f)$ as asserting that f looks like a transfinite witness hierarchy on μ, "locally," at y. Let $\theta(y,f)$ be (the code of) a formula that is $\Delta_0(f)$, let η, μ, and γ be notations with γ less than μ, and let $F(f)$ be a $\prec\eta$-computable function. By way of notation, take

$$F \vdash^\eta_\mu \exists y\ \theta(y, wit^\gamma)$$

to mean

$$\forall m, y\ (F(\hat{m}) = y \wedge W(y,\mu,\hat{m}) \rightarrow Tr_{\Delta_0}(\ulcorner\theta(y, \hat{m} \restriction_\gamma)\urcorner)),$$

where Tr_{Δ_0} is a truth predicate for Δ_0 formulae. In words, $F \vdash^{\eta}_{\mu} \exists y\ \theta(y, wit_{\gamma})$ means that if $F(\hat{m})$ is equal to y and, at y, $\hat{m}$ looks like a witness hierarchy on μ at y, then $\theta(y, \hat{m} \restriction_{\gamma})$ is true. Take $\vdash^{\eta}_{\mu} \exists y\ \theta(y, wit^{\gamma})$ to assert the existence of such an F.

(Stepping outside $PA + TI(\prec\varphi_{\alpha}(0))$ for a moment, we can say that $\vdash^{\eta}_{\mu} \exists y\ \theta(y, wit^{\gamma})$ implies that $\exists y\ \theta(y, wit^{\gamma})$ holds for any transfinite witness function wit^{γ}. Reason as follows: given F satisfying $\vdash^{\eta}_{\mu} \exists y\ \theta(y, wit^{\gamma})$ and a transfinite witness function wit^{γ}, let wit^{μ} be a transfinite witness function with $wit^{\mu} \restriction_{\gamma} = wit^{\gamma}$, and let m be a finite partial function agreeing with wit^{μ}, defined at enough values to carry out the computation of F, satisfy W_{μ} at the result of the computation (call it y), and evaluate the truth of $\theta(y, wit^{\mu} \restriction_{\gamma})$. Since we have assumed $F \vdash^{\eta}_{\mu} \exists y\ \theta(y, wit^{\gamma})$, we have $\theta(y, \hat{m})$, and hence $\theta(y, wit^{\gamma})$.)

Still working in $PA + TI(\prec\varphi_{\alpha}(0))$, note that if ε is less than α, then transfinite induction up to $\varphi_{\varepsilon}(0)$ is available to us in this theory; and $\varphi_{\varepsilon}(0)$ is closed under the functions φ_{γ}, for γ less than ε. With the notation we have just introduced, the two lemmata below bear a strong resemblance to the elimination lemmata of [21, Section 18] and [22, Section 2.1.2], as well as to [4, Lemma 3.6].

Lemma 10.3. *There is a computable function $R_j(\eta, \mu, \theta, F)$, such that the following is provable in $PA + TI(\prec\varphi_{\alpha}(0))$: for any η, μ, F, and $\Delta_0(f)$ formula $\theta(x, f)$, if*

$$F \vdash^{\eta}_{\mu+1} \theta(y, wit^{\mu})$$

then

$$R_j(\eta, \mu, \theta, F) \vdash^{\omega^{\eta}}_{\mu} \theta(y, wit^{\mu}).$$

Proof. This is a straightforward adaptation of the proof of Lemma 9.2. ⊣

Lemma 10.4. *Let ε be less than α. Then there is a computable function $R_e(\rho, \eta, \mu, \theta, F)$, such that the following is provable in $PA + TI(\prec\varphi_{\alpha}(0))$: for any η and μ less than $\varphi_{\varepsilon}(0)$, any ρ less than ε, any F, and any $\Delta_0(f)$ formula $\theta(y, f)$, if*

$$F \vdash^{\eta}_{\mu+\omega^{\rho}} \exists y\ \theta(y, wit^{\mu})$$

then

$$R_e(\rho, \eta, \mu, \theta, F) \vdash^{\varphi_{\rho}(\eta)}_{\mu} \exists y\ \theta(y, wit^{\mu}).$$

Proof. We will use effective transfinite recursion to define R_e, with a primary recursion on ρ and a secondary recursion on η; and then we will use nested instances of transfinite induction in $PA + TI(\prec\varphi_{\alpha}(0))$ to prove that R_e is total and satisfies the conclusion of the lemma. In other words, we will use the recursion theorem to define $R_e(\rho, \eta, \mu, \theta, F)$ in terms of ρ, η, μ, θ, F, and the index e; and then for each ρ and η, we will verify the correctness of R_e at ρ and η, assuming it behaves correctly for arguments ρ', η', μ', θ', and F' whenever either either $\rho' \prec \rho$ or $\rho' = \rho$ and $\eta' \prec \eta$. For expository purposes, I will combine these two steps, defining R_e by cases and showing that in each case, assuming the induction hypotheses are met, we have the desired conclusion. The proof is adapted from [21, 22].

First, note that if ρ is equal to 0, then we can set $R_e(\rho, \eta, \mu, \theta, F)$ equal to $R_j(\eta, \mu, \theta, F)$, where R_j is as in the previous lemma.

Next, suppose ρ is greater than 0, and

$$F \vdash^{\eta}_{\mu+\omega^\rho} \exists y\ \theta(y, wit^\mu)$$

where F is an η'-computable function and η' is less than η. Consider the computation of F at $\hat{m}$. If $norm_F(start_F) \not\prec \eta'$, then the computation halts immediately. In that case, set $y = result_F(start_F)$; then we have

$$W(y, \mu + \omega^\rho, \hat{m}) \to \theta(y, \hat{m} \restriction_\mu).$$

But for this particular y, $W(y, \mu + \omega^\rho, \hat{m})$ only depends on values of $\hat{m}$ whose first component is at most $(y)_0$, assuming $(y)_0 \prec \mu + \omega^\rho$; so there is a $\rho' \prec \rho$ and n such that $W(\mu + \omega^\rho, y, \hat{m})$ holds if and only if $W(\mu + \omega^{\rho'} \cdot n, y, \hat{m})$ does. In other words, for this F, we have

$$F \vdash^{\eta}_{\mu+\omega^{\rho'}\cdot n} \exists y\ \theta(y, wit^\mu).$$

Applying, successively,

$$R_e(\rho', \eta, \mu, \theta, \cdot), R_e(\rho', \varphi_{\rho'}(\eta), \mu, \theta, \cdot), \dots, R_e(\rho', \varphi^{n-1}_{\rho'}(\eta), \mu, \theta, \cdot)$$

yields a G satisfying

$$G \vdash^{\varphi^n_{\rho'}(\eta)}_{\mu} \exists y\ \theta(y, wit^\mu).$$

Since $\eta \preceq \varphi_\rho(\eta)$ and $\rho' \prec \rho$ we have $\varphi^n_{\rho'}(\eta) \preceq \varphi_\rho(\eta)$, so

$$G \vdash^{\varphi_\rho(\eta)}_{\mu} \exists y\ \theta(y, wit^\mu).$$

as required.

Finally, consider the case where $norm_F(start_F) \prec \eta'$. Consider the next step in the computation of F: if the first state, q_0, is $start_F$, and $query_F(start_F)$ is equal to k, then the second state, q_1, is $next_F(start_F, \hat{m}(k))$. If $(k)_0$ is less than $\mu + \omega^\rho$, let n and ρ' be such that $(k)_0$ is less than $\mu + \omega^{\rho' \cdot n}$. (If $(k)_0$ is not less than $\mu + \omega^\rho$, then the value of $\hat{m}(k)$ is irrelevant, and we can take n and ρ' to be 0 in the argument which follows.) For each l, let G_l be the η'-computable function which continues the computation of F, assuming that $\hat{m}(k)$ returns l; in other words, $start_{G_l}$ is equal to $next_F(start_F, l)$, while $norm_{G_l}$, $next_{G_l}$, $query_{G_l}$, and $result_{G_l}$ are just $norm_F$, $next_F$, $query_F$, and $result_F$, respectively. Then it is not hard to see that for each l,

$$G_l \vdash^{\eta'}_{\mu+\omega^\rho} \exists y\ (wit^{\mu+\omega^{\rho'}\cdot n}(\bar{k}) = \bar{l} \to \theta(y, wit^{\mu+\omega^{\rho'}\cdot n} \restriction_\mu)).$$

After all, for any number m, if s is a computation sequence for G_l at $\hat{m}$, $result_{G_l}((s)_{last(s)})) = y$, $W(y, \mu + \omega^\rho, \hat{m})$, and $\hat{m} \restriction_{\mu+\omega^{\rho'}\cdot n} (k) = l$, then prepending $start_F$ to s yields a computation sequence for F at $\hat{m}$ returning the same result; and since $F \vdash^{\eta}_{\mu+\omega^\rho} \theta(y, wit^\mu)$, we have $\theta(y, \hat{m} \restriction_\mu)$. Now, note that $\mu + \omega^\rho$ is equal to $(\mu + \omega^{\rho'} \cdot n) + \omega^{\rho}$, since, in general, $\alpha + \omega^{\beta} = \omega^{\beta}$ whenever $\alpha \prec \omega^{\beta}$. For each l, let θ_l be the formula

$$f(\bar{k}) = \bar{l} \to \theta(y, f \restriction_{\mu})$$

and let

$$\hat{G}_l = R_e(\rho, \eta', \mu + \omega^{\rho'} \cdot n, \theta_l, G_l).$$

By the inductive hypothesis we have that for each l,

$$\hat{G}_l \vdash^{\varphi_\rho(\eta')}_{\mu+\omega^{\rho'} \cdot n} \exists y\ (wit^{\mu+\omega^{\rho'} \cdot n}(\bar{k}) = \bar{l} \to \theta(y, wit^{\mu+\omega^{\rho'} \cdot n} \restriction_{\mu})).$$

Let $G'(f)$ be the $\varphi_\rho(\eta')$-computable function which, in the first step, evaluates f at k; and if the result is l, continues with $\hat{G}_l$. Then we have

$$G' \vdash^{\varphi_\rho(\eta')+1}_{\mu+\omega^{\rho'} \cdot n} \exists y\ \theta(y, wit^{\mu}),$$

since if s is a computation sequence for G' at $\hat{m}$ and $l = \hat{m}(k)$, then dropping the first element of s yields a computation sequence for $\hat{G}_l$ at $\hat{m}$ with the same result. Finally, applying R_e n times with first argument ρ', we obtain a G satisfying

$$G \vdash^{\varphi^n_{\rho'}(\varphi_\rho(\eta')+1)}_{\mu} \exists y\ \theta(y, wit^{\mu}).$$

Since $\varphi^n_{\rho'}(\varphi_\rho(\eta')+1)$ is less than $\varphi_\rho(\eta)$, we have $G \vdash^{\varphi_\rho(\eta)}_{\mu} \exists y\ \theta(y, wit^{\mu})$, as desired.

To justify the use of the recursion theorem, we only need to verify that in each of the three paragraphs above, G was obtained effectively from η, μ, ρ, θ, F, and the index e. This is straightforward. Then, using transfinite induction up to ε on ρ, with a secondary transfinite induction up to $\varphi_\varepsilon(0)$ on η, we need to verify in $PA + TI(\prec\varphi_\alpha(0))$ that for every ρ, η, μ, θ, and F, R_e is defined, and if F satisfies the hypothesis of the lemma, then R_e returns a G witnessing the conclusion. But this is the argument we have just sketched. ⊣

Proof of Lemma 10.2. Suppose the theory $PA(\mathcal{H}) + H(\prec\alpha) + TI(\prec\alpha)$ proves $\exists y\ \theta(x, y)$. By Lemma 10.1, we have that $\exists y\ \theta(x, y)$ is also provable in $ERA(\mathcal{W}) + wit(\prec\alpha) + min(\prec\alpha)$. By the observations above, $\exists y\ \theta(x, y)$ is therefore provable in the theory $ERA(wit^{\alpha'}) + \forall y\ W(y, \bar{\alpha}', wit^{\alpha'}) + min(\prec\alpha)$, for an α' less than α that is large enough. By the deduction theorem and Lemma 8.11, there is a $\prec\alpha$-recursive function F such that $ERA(wit^{\alpha'})$ proves

$$F(x, wit^{\alpha'}) = y \wedge W(y, \bar{\alpha}', wit^{\alpha'}) \to \theta(x, y).$$

Suppose F is α''-recursive, where α'' is less than α. Pick ε large enough so that we have $\alpha', \alpha'' \prec \varepsilon \prec \alpha$.

Now argue in $PA + TI(\prec\varphi_\alpha(0))$. For each x, let $F_x(f)$ be α''-computable functional which, for each f, returns $F(x, f)$. Since $ERA(f)$ is sound, F_x satisfies

$$F_x \vdash^{\alpha''+1}_{\alpha'+1} \exists y\ \theta(\bar{x}, y).$$

Using Lemma 10.4 , we have that for each x there is a function $G_x(f)$ satisfying

$$G_x \vdash^{\varphi_{\alpha'+1}(\alpha''+1)}_{0} \exists y\ \theta(\bar{x}, y).$$

In other words, for any m, if $G_x(\hat{m})$ is defined then it returns a value y satisfying $\theta(\bar{x}, y)$ outright. Let m be the partial function that is nowhere defined; using transfinite induction up to $\varphi_{\alpha'+1}(\alpha''+1)$ we can show that $G_x(\hat{m})$ is defined, and so $\exists y\ \theta(\bar{x}, y)$ is true. ⊣

Relativizing this theorem to an additional function symbol f, we have

THEOREM 10.5. *Let α be infinite and closed under multiplication. Then the proof-theoretic ordinal of $PA(\mathcal{H},f) + H(\prec\alpha) + TI(\prec\alpha)$ is at most $\varphi(\prec\alpha, 0)$.*

If Γ_0 is the least fixed point of the function $\alpha \mapsto \varphi(\alpha, 0)$, we have the following:

THEOREM 10.6. *$PA(\mathcal{H},f) + H(\prec\Gamma_0) + TI(\prec\Gamma_0)$ has proof-theoretic ordinal at most Γ_0.*

Also, relativizing the proof of Theorem 10.5 to finite witness functions, and then interpreting references to these witness functions in $PA(f)$, yields

THEOREM 10.7. *Let α be infinite and closed under multiplication. Then any formula that is arithmetic in f and provable in $PA(\mathcal{H},f) + H(\prec\alpha) + TI(\prec\alpha)$ is also provable in $PA(f) + TI(\prec\varphi_\alpha(0))$.*

REFERENCES

[1] WILHELM ACKERMANN, *Zur Widerspruchsfreiheit der Zahlentheorie*, ***Mathematische Annalen***, vol. 93 (1940), pp. 1–36.

[2] JEREMY AVIGAD, *An ordinal analysis of admissible set theory using recursion on ordinal notations*, to appear in the ***Journal of Mathematical Logic***.

[3] JEREMY AVIGAD and RICHARD SOMMER, *A model-theoretic approach to ordinal analysis*, ***The Bulletin of Symbolic Logic***, vol. 3 (1997), pp. 17–52.

[4] ———, *The model-theoretic ordinal analysis of predicative theories*, ***The Journal of Symbolic Logic***, vol. 64 (1999), pp. 327–349.

[5] ARNOLD BECKMANN, *Separating fragments of bounded arithmetic*, ***Ph.D. thesis***, Westfälische Wilhelms-Universität, Münster, 1996.

[6] LEV BEKLEMISHEV, *Proof-theoretic analysis by iterated reflection*, to appear in the ***Archive for Mathematical Logic***.

[7] SAMUEL BUSS, *The witness function method and provably recursive functions of Peano arithmetic*, ***Proceedings of the ninth international congress on logic, methodology, and philosophy of science, Uppsala, Sweden, 1991*** (D. Westerstahl D. Prawitz, B. Skyrms, editor), Elsevier North-Holland, Amsterdam, 1994, pp. 29–68.

[8] ———, *First-order proof theory of arithmetic*, in Buss [9].

[9] Samuel Buss (editor), ***The handbook of proof theory***, North-Holland, Amsterdam, 1998.

[10] TIMOTHY CARLSON, *Ranked partial structures*, preprint.

[11] HARVEY FRIEDMAN and MICHAEL SHEARD, *Elementary descent recursion and proof theory*, ***Annals of Pure and Applied Logic***, vol. 71 (1995), pp. 1–45.

[12] HARVEY FRIEDMAN, *Iterated inductive definitions and $\Sigma^1_2 - AC$*, ***Intuitionism and proof theory*** (A. Kino, J. Myhill, and R. E. Vesley, editors), North-Holland, Amsterdam, 1970, pp. 435–442.

[13] HARVEY FRIEDMAN, KENNETH MCALOON, and STEPHEN SIMPSON, *A finite combinatorial principle which is equivalent to the 1-consistency of predicative analysis*, ***Patras Logic Symposion*** (G. Metakides, editor), North-Holland, Amsterdam, 1982, pp. 197–230.

[14] GERHARD GENTZEN, *Untersuchungen über das logische Schliessen*, ***Mathematische Zeitschrift***, vol. 39 (1935), pp. 176–210, 405–431, translated as "Investigations into Logical Deduction" in [16], pages 68–131.

[15] ———, *Neue Fassung des Widerspruchsfreiheit für die reine Zahlentheorie*, ***Forschungen zur Logik und zur Grundlegung der exakten Wissenschaften***, New Series 4 (1938), pp. 19–44, translated as "New Version of the Consistency Proof for Elementary Number Theory" in [16], pages 252–286.

[16] ———, ***Collected works***, North-Holland, Amsterdam, 1969, Edited by M. E. Szabo.

[17] PETR HÁJEK and PAVEL PUDLÁK, ***Metamathematics of first-order arithmetic***, Springer, Berlin, 1993.

[18] JACQUES HERBRAND, *Recherches sur la théorie de la démonstration*, ***Ph.D. thesis***, University of Paris, 1930, English translation in [19].

[19] ———, ***Collected works***, Harvard University Press, Cambridge, 1971, Warren Goldfarb, editor.

[20] GRIGORI MINTS and SERGEI TUPAILO, *Epsilon subsitution method for the ramified language and Δ^1_1-comprehension rule*, ***Logic and foundations of mathematics*** (Andrea Cantini et al., editor), Kluwer, the Netherlands, 1999, pp. 107–130.

[21] WOLFRAM POHLERS, ***Proof theory: An introduction***, Lecture Notes in Mathematics #1407, Springer, Berlin, 1989.

[22] ———, *Subsystems of set theory and second order number theory*, in Buss [9].

[23] JOSEPH QUINSEY, *Some problems in logic*, ***Ph.D. thesis***, Oxford University, 1980.

[24] MICHAEL RATHJEN, *The realm of ordinal analysis*, ***Sets and proofs (Leeds, 1997)***, Cambridge Univ. Press, Cambridge, 1999, pp. 219–279.

[25] H. E. ROSE, ***Subrecursion: Functions and hierarchies***, Clarendon Press, Oxford, 1984.

[26] HELMUT SCHWICHTENBERG, *Proof theory: Some aspects of cut-elimination*, in ***The handbook of mathematical logic*** (Jon Barwise, editor), North-Holland, Amsterdam, 1977, pp. 867–895.

[27] JOSEPH SHOENFELD , ***Mathematical logic***, Addison Wesley, Reading, Massachusetts, 1967.

[28] RICHARD SOMMER, *Transfinite induction and hierarchies generated by transfinite recursion within Peano arithmetic*, ***Ph.D. thesis***, University of California, Berkeley, 1990.

[29] ———, *Ordinal arithmetic in $I\Delta_0$*, ***Arithmetic, proof theory, and computational complexity*** (Peter Clote and Jan Krajíček, editors), Oxford University Press, Oxford, 1993, pp. 320–363.

[30] ALEX WILKIE and JEFFPARIS , *On the scheme of induction for bounded arithmetic formulas*, ***Annals of Pure and Applied Logic***, vol. 35 (1987), pp. 261–302.

DEPARTMENT OF PHILOSOPHY
CARNEGIE MELLON UNIVERSITY
PITTSBURGH, PA 15213, USA
E-mail: avigad@cmu.edu
URL: http://www.andrew.cmu.edu/~avigad

RELATING ORDINALS TO PROOFS IN A PERSPICUOUS WAY

WILFRIED BUCHHOLZ

Contemporary ordinal-theoretic proof theory (i.e., the part of proof theory concerned with ordinal analysises of strong impredicative theories) suffers from the extreme (and as it seems unavoidable) complexity and opacity of its main tool, the ordinal notation systems. This is not only a technical stumbling block which prevents most proof-theorists from a closer engagement in that field, but it also calls the achieved results into question, at least as long as these results do not have interesting consequences, such as e.g., foundational reductions or intuitively graspable combinatorial independence results. The question is, what have we gained by having shown that the proof-theoretic ordinal of a certain formal theory S equals the ordertype $\| \prec \|$ of some specific term ordering $\prec$ (disregarding for the moment possible applications of the abovementioned kind). I would say, as long as the ordering is simple and natural (as e.g., the standard orderings of order type ε_0 or Γ_0) or occurs already elsewhere in mathematics or theoretical computer science, we have obtained a result of pure mathematics which is interesting and noteworthy by itself. But I'm not sure if the latter can (already) be said about Rathjen's ([18]) and Arai's ([3], [2]) most impressive and admirable work on Π^1_2-CA and beyond. Rather I think, that extensive efforts should be made aiming at a substantial simplification and deeper understanding of that work, especially of the used ordinal notation systems (term orderings, resp.).

The purpose of the present paper is to present as an example a particularly nice piece of ordinal-theoretic, impredicative proof theory (far below Π^1_2-CA) where such a simplification and deeper understanding has already been achieved. The paper is more or less a condensation and improved presentation of parts of [6], [7], [8], [9] providing an ordinal analysis "from above" for the theories ID_ν of iterated inductive definitions by means of the so-called $\Omega_{\mu+1}$-rule. Special efforts have been made to let the assignment of ordinals to derivations not appear as an ad hoc approach but as a quite natural and perspicuous procedure.

§1. Ordinal functions and ordinal notations. Around the early seventies Feferman suggested the following definition of ordinal functions θ_α for generating large segments of recursive ordinals, which was intended to replace the extremely

Dedicated to Solomon Feferman on the occasion of his 70th birthday. I am deeply indebted to Sol for the encouragement and stimulation he gave me over many years through his interest in my work.

Reflections on the foundations of mathematics
Essays in honor of Solomon Feferman
Edited by Wilfried Sieg, Richard Sommer, and Carolyn Talcott

complex Bachmann-Pfeiffer-Isles definition procedure for hierarchies of normal functions. This definition brought a new impetus into proof theory and was the starting point of a rather successful development.

DEFINITION (Feferman 1970). Let ν_0 be a fixed countable ordinal. By transfinite recursion on α one defines sets $C(\alpha,\beta)$ and ordinal functions θ_α:

$$\begin{aligned} C(\alpha,\beta) := {} & \text{least set } X \supseteq \beta \cup \{\Omega_\sigma : 0 < \sigma < \nu_0\} \text{ such that} \\ & \forall \xi,\eta \in X(\xi + \eta \in X) \text{ and} \\ & \forall \xi,\eta \in X(\xi < \alpha \Longrightarrow \theta_\xi(\eta) \in X), \\ \theta_\alpha := {} & \text{ordering function of } \{\beta > 0 : C(\alpha,\beta) \cap \beta^+ \subseteq \beta\}, \end{aligned}$$

where $\Omega_\sigma := \aleph_\sigma$ and $\beta^+ := \min\{\Omega_{\sigma+1} : \beta < \Omega_{\sigma+1}\}$.

A simple cardinal argument shows that $C(\alpha,\beta)$ has essentially the same cardinality as β (namely $card(C(\alpha,\beta)) = \max\{\aleph_0, card(\beta)\}$), no matter how large α is. This yields $\forall \beta < \Omega_{\sigma+1}(\theta_\alpha(\beta) < \Omega_{\sigma+1})$, for any α (see Lemma 1.1 below). Hence the function $\lambda\alpha.\theta_\alpha(\Omega_\sigma)$ maps On into $\Omega_{\sigma+1}$, i.e., it is a so-called *collapsing function*.

Important contributions to the further investigation of the θ-functions were made by Weyhrauch, Aczel and (most important) Jane Bridge. Aczel generalized the definition and conjectured relationships between the θ_α's and the Pfeiffer/Isles functions. These conjectures were established by Jane Bridge in her Ph.D. thesis. She also obtained partial results on the recursiveness of the notation system associated with θ_α [4]. Starting from Bridge's thesis, recursiveness of the full θ-notation system was established in [5]. There also a variant $\overline{\theta}_\alpha$ of θ_α was introduced which has the advantage that the $<$-relation between $\overline{\theta}$-terms can be characterized in a particularly simple way. This opened the way to a successful use of these ordinal functions in proof-theory. Later it turned out that in proof-theoretic applications actually only the values of $\overline{\theta}_\alpha$ at initial ordinals Ω_σ are used, which led to the idea to define directly functions ψ_σ corresponding to $\lambda\alpha.\overline{\theta}_\alpha(\Omega_\sigma)$.

DEFINITION (Buchholz 1981). Let ν_0 be a fixed countable ordinal.

$$\begin{aligned} C(\alpha,\beta) &:= \text{closure of } \beta \text{ under } + \text{ and all } \psi_\sigma{\restriction}\alpha \quad (\sigma < \nu_0), \\ \psi_\sigma(\alpha) &:= \min\{\beta \geq \Omega^-_{\sigma+1} : C(\alpha,\beta) \cap \Omega_{\sigma+1} \subseteq \beta\} \quad (\sigma < \nu_0) \\ & \quad \text{with } \Omega^-_1 := 1 \text{ and } \Omega^-_{\sigma+1} := \Omega_\sigma \text{ for } \sigma > 0. \end{aligned}$$

Remark. $C(\alpha,\psi_\sigma(\alpha)) \cap \Omega_{\sigma+1} = \psi_\sigma(\alpha)$

Before we start to prove some basic facts about these ordinal functions ψ_σ, we establish a general Lemma which also applies to stronger systems of ordinal functions.

DEFINITION. Given a countable set $\mathcal{F}$ of ordinal functions and an ordinal β let $\mathrm{Cl}(\mathcal{F};\beta) :=$ the closure of β under $+$ and all functions in $\mathcal{F}$.

Let $\Re$ be the class of all uncountable regular cardinals.

Lemma 1.1. *If $\mathcal{F}$ is a countable set of ordinal functions, then for each $\kappa \in \mathfrak{R}$ the set $\{\beta < \kappa : \mathrm{Cl}(\mathcal{F}; \beta) \cap \kappa \subseteq \beta\}$ is closed unbounded in κ.*

Proof. *"unbounded":* Let $\beta_0 < \kappa$ be given. We set $\beta_{n+1} := \min\{\eta : \mathrm{Cl}(\mathcal{F}; \beta_n) \cap \kappa \subseteq \eta\}$, $\beta := \sup_{n<\omega} \beta_n$. Since $\forall \eta < \kappa(\mathit{card}(\mathrm{Cl}(\mathcal{F}; \eta)) < \kappa)$, we obtain (by induction on n) $\beta_n < \kappa$, and then $\beta_0 \leq \beta < \kappa$ and $\mathrm{Cl}(\mathcal{F}; \beta) \cap \kappa \subseteq \bigcup_{n<\omega} \mathrm{Cl}(\mathcal{F}; \beta_n) \cap \kappa \subseteq \bigcup_{n<\omega} \beta_{n+1} \subseteq \beta$. — *"closed":* If $\beta = \sup(X)$ and $\forall \eta \in X(\mathrm{Cl}(\mathcal{F}; \eta) \cap \kappa \subseteq \eta)$ then $\mathrm{Cl}(\mathcal{F}; \beta) \cap \kappa \subseteq \bigcup_{\eta \in X} \mathrm{Cl}(\mathcal{F}; \eta) \cap \kappa \subseteq \bigcup_{\eta \in X} \eta = \beta$. ⊣

Theorem 1.2 (Basic properties of ψ_σ).

(a) $\psi_\sigma(\alpha) < \Omega_{\sigma+1}$ *(collapsing property)*
(b) $\psi_\sigma(0) = \Omega^-_{\sigma+1}$, *and* $\psi_\sigma(\alpha)$ *is closed under* $+$
(c) $\alpha_0 < \alpha \;\&\; \alpha_0 \in C(\alpha_0, \psi_\sigma(\alpha_0)) \Longrightarrow \psi_\sigma(\alpha_0) < \psi_\sigma(\alpha)$
(d) $\psi_\sigma(\alpha_0) = \psi_\sigma(\alpha_1) \;\&\; \underbrace{\alpha_i \in C(\alpha_i, \psi_\sigma(\alpha_i))}_{\textit{normalform condition}} \;(i = 0, 1) \Longrightarrow \alpha_0 = \alpha_1$

Proof. (a) Obviously $C(\alpha, \beta) = \mathrm{Cl}(\{\psi_\sigma \restriction \alpha : \sigma < \nu_0\}; \beta)$. Hence $\{\beta < \Omega_{\sigma+1} : \beta \geq \Omega^-_{\sigma+1} \;\&\; C(\alpha, \beta) \cap \Omega_{\sigma+1} \subseteq \beta\} \neq \emptyset$ (by Lemma 1.1), and therefore $\psi_\sigma(\alpha) = \min\{\beta \geq \Omega^-_{\sigma+1} : C(\alpha, \beta) \cap \Omega_{\sigma+1} \subseteq \beta\} < \Omega_{\sigma+1}$.

(b) Obviously $C(0, \Omega^-_{\sigma+1}) = \Omega^-_{\sigma+1}$, and therefore $\psi_\sigma(0) = \Omega^-_{\sigma+1}$. $\psi_\sigma(\alpha)$ is closed under $+$, since $\psi_\sigma(\alpha) = C(\alpha, \psi_\sigma(\alpha)) \cap \Omega_{\sigma+1}$.

(c) $C(\alpha_0, \psi_\sigma(\alpha)) \cap \Omega_{\sigma+1} \subseteq C(\alpha, \psi_\sigma(\alpha)) \cap \Omega_{\sigma+1} \subseteq \psi_\sigma \alpha$ and therefore $\psi_\sigma(\alpha_0) \leq \psi_\sigma(\alpha)$ which implies $C(\alpha_0, \psi_\sigma(\alpha_0)) \subseteq C(\alpha, \psi_\sigma(\alpha))$. From $\alpha_0 < \alpha \;\&\; \alpha_0 \in C(\alpha, \psi_\sigma(\alpha))$ we conclude $\psi_\sigma(\alpha_0) \in C(\alpha, \psi_\sigma(\alpha))$. By (a) we have $\psi_\sigma(\alpha_0) < \Omega_{\sigma+1}$ and thus $\psi_\sigma(\alpha_0) \in C(\alpha, \psi_\sigma(\alpha)) \cap \Omega_{\sigma+1} \subseteq \psi_\sigma(\alpha)$.

(d) If $\alpha_0 < \alpha_1$ then the assumption $\alpha_0 \in C(\alpha_0, \psi_\sigma(\alpha_0))$ together with (c) yields $\psi_\sigma(\alpha_0) < \psi_\sigma(\alpha_1)$. ⊣

Lemma 1.3. $C(\alpha, \psi_0(\alpha)) = C(\alpha, 1)$

Proof by induction on α. Let us assume that $C(\xi, \psi_0(\xi)) = C(\xi, 1)$, for all $\xi < \alpha$ (IH). We have to prove $\psi_0(\alpha) \subseteq C(\alpha, 1)$. Let $\alpha > 0$ (otherwise $\psi_0(\alpha) = 1 \subseteq C(\alpha, 1)$). As we will show below the IH implies that $\beta := C(\alpha, 1) \cap \Omega_1$ is in fact an ordinal. Obviously $1 \leq \beta$ and $C(\alpha, \beta) \cap \Omega_1 \subseteq C(\alpha, 1) \cap \Omega_1 = \beta$ and thus $\psi_0(\alpha) \leq \beta$.

Claim: $\gamma \in C(\alpha, 1) \cap \Omega_1 \Rightarrow \gamma \subseteq C(\alpha, 1)$. Proof by side induction on the definition of $C(\alpha, 1)$:

1. $\gamma = \psi_0(\xi)$ with $\xi < \alpha \;\&\; \xi \in C(\alpha, 1)$: By the above IH we have $C(\xi, \psi_0(\xi)) = C(\xi, 1)$. Hence $\gamma = \psi_0(\xi) \subseteq C(\xi, 1) \subseteq C(\alpha, 1)$.

2. $\gamma = \gamma_0 + \gamma_1$ with $\gamma_0, \gamma_1 \in C(\alpha, 1) \cap \Omega_1$: Then by SIH $\gamma_0, \gamma_1 \subseteq C(\alpha, 1)$ which (together with $\gamma_0 \in C(\alpha, 1)$) yields $\gamma_0 + \gamma_1 \subseteq C(\alpha, 1)$. ⊣

Disgression. Of course the above definition of ψ_σ can be generalized in the same way as Aczel [1] generalized Feferman's definition, namely by incorporating a countable set $\mathcal{G}$ of cardinal valued ordinal functions into the definition of $C(\alpha, \beta)$. For the simple case $\mathcal{G} = \{\lambda x.\Omega_x\}$ this is equivalent to setting, for any $\sigma, \alpha \in On$,

$\psi_\sigma(\alpha) := \min\{\beta \geq \Omega^-_{\sigma+1} : C(\alpha,\beta) \cap \Omega_{\sigma+1} \subseteq \beta\}$ where $C(\alpha,\beta) := \mathrm{Cl}(\psi{\restriction}\alpha;\beta)$ with $\psi{\restriction}\alpha : On \times \alpha \to On, (\rho,\xi) \mapsto \psi_\rho(\xi)$.

A more substantial extension of the θ/ψ-approach was developed by Schütte (unpublished), Pohlers [14] and, most important, Jäger [12]. In addition to the ψ_σ's which are collapsing functions for successor regulars, now also collapsing functions for limit regulars, i.e., weakly inaccessible cardinals, are introduced. In order to be able to treat both kinds of collapsing functions uniformly Jäger denoted the ψ-function for a regular cardinal κ, i.e., one with values below κ, by ψ_κ. In this notation the former ψ_σ becomes $\psi_{\Omega_{\sigma+1}}$.

Definition (Jäger 1984). $C(\alpha,\beta) := \mathrm{Cl}(\lambda xy.I_x(y), \psi{\restriction}\alpha;\, \beta)$
with $\psi{\restriction}\alpha : \Re\times\alpha \to On,\ (\pi,\xi) \mapsto \psi_\pi(\xi)$
$\psi_\kappa(\alpha) := \min\{\beta > \kappa^- : C(\alpha,\beta) \cap \kappa \subseteq \beta\}$

Here I_ρ is defined by recursion on ρ as the ordering function of the topological closure of $\{\kappa \in \Re : \forall\xi < \rho(I_\xi(\kappa) = \kappa)\}$, and

$$\kappa^- := \begin{cases} 0 & \text{if } \kappa = I_\rho(0) \\ I_\rho(\sigma) & \text{if } \kappa = I_\rho(\sigma{+}1) \text{ with } \rho,\sigma < \kappa. \end{cases}$$

Later Rathjen developed further extensions up to the use of large large cardinals ([16], [17], [18]). In the first of these extensions Rathjen assumed the existence of a weakly Mahlo cardinal M and utilized the fact that the regular cardinals are stationary in M.

Definition (Rathjen 1990).

$$C(\alpha,\beta) := \mathrm{Cl}(\lambda x.\omega^{M+x}, \psi{\restriction}\alpha;\beta) \text{ with}$$
$$(\psi{\restriction}\alpha)(\pi,\xi) := \psi_\pi(\xi) \text{ for } \pi \in \Re \cap (M{+}1) \text{ and } \xi < \alpha$$

$$\psi_\kappa(\alpha) := \begin{cases} \min\{\beta \in \Re : \alpha \in C(\alpha,\beta)\ \&\ C(\alpha,\beta)\cap\kappa \subseteq \beta\} & \text{if } \kappa = M \\ \min\{\beta \in On : \kappa \in C(\alpha,\beta)\ \&\ C(\alpha,\beta)\cap\kappa \subseteq \beta\} & \text{if } \kappa < M \end{cases}$$

Let us see how the "Mahloness" of M is used for obtaining the crucial property $\forall\alpha < \varepsilon_{M+1}(\psi_M(\alpha) < M)$. According to Lemma 1.1 the set $\{\beta < M : C(\alpha,\beta) \cap M \subseteq \beta\}$ is closed unbounded in M. If $\alpha < \varepsilon_{M+1}$ then $\alpha \in C(\alpha,\beta)$ for sufficiently large $\beta < M$. Hence also $X := \{\beta < M : \alpha \in C(\alpha,\beta)\ \&\ C(\alpha,\beta) \cap M \subseteq \beta\}$ is club in M (if $\alpha < \varepsilon_{M+1}$). Since $\Re$ is stationary in M, we obtain $X \cap \Re \neq \emptyset$, and thus $\psi_M(\alpha) = \min(X \cap \Re) < M$.

The general strategy behind these extensions is that one tries to produce notations for more and more regular cardinals κ, which in turn gives more and more collapsing functions ψ_κ. In Jäger's definition the regulars are provided by the hierarchy $(I_\rho)_{\rho\in On}$. In defining this hierarchy one already assumes the existence of a weakly Mahlo cardinal M, but one does not exploit this assumption to the same extent as it is done in Rathjen's approach. Actually the limit ordinal $\psi_{\Omega_1}(\sup_n \delta_n)$ $(\delta_0 := 0,\ \delta_{n+1} := I_{\delta_n}(0))$ of [12] is "much" smaller then the limit ordinal $\psi_{\Omega_1}(\varepsilon_{M+1})$ of [16].

Now we return to the simple system of ψ-functions ψ_σ $(\sigma < \nu_0)$ defined above. In order to avoid some technical complications we will even assume $\nu_0 = \omega$. But we want to stress that no essential new difficulties would emerge when in all what follows the assumption "$\nu_0 = \omega$" would be replaced by "$\nu_0 < \omega_1^{CK}$".

From now on σ, ρ, μ, ν range over numbers $< \omega$.

Below we will introduce a system of ordinal notations based on the ordinal functions ψ_σ. The canonical way for that is to consider the set T of all terms which are generated from the constant 0 by means of function symbols $\oplus, D_0, D_1, \ldots$ for the ordinal functions $+, \psi_0, \psi_1, \ldots$. Then one looks for a (primitive) recursive characterization of the relation $<_{\mathrm{o}} := \{(a,b) \in \mathrm{T}\times\mathrm{T} : \mathrm{o}(a) < \mathrm{o}(b)\}$, where $\mathrm{o}(a) \in On$ is the canonical interpretation of $a \in \mathrm{T}$. It turns out that the relation $<_{\mathrm{o}}$ has a particularly simple characterization when it is restricted to the subset $\mathrm{OT} \subseteq \mathrm{T}$ of those terms $a \in \mathrm{T}$ which are in "normalform" (i.e., $\mathrm{o}(b) \in C(\mathrm{o}(b), \psi_\sigma(\mathrm{o}(b)))$ for each subterm $D_\sigma b$ of a, and $\mathrm{o}(a_n) \leq \ldots \leq \mathrm{o}(a_0)$ for each subterm $a_0 \oplus \ldots \oplus a_n$ of a).

DEFINITION.

$$\overline{\mathrm{o}}(a) := \text{ ordertype of } (\{x \in \mathrm{OT} : x <_{\mathrm{o}} a\}, <_{\mathrm{o}})$$
$$\mathrm{OT}_0 := \{a \in \mathrm{OT} : a <_{\mathrm{o}} D_1 0\}.$$

It can be proved that every $a \in \mathrm{T}$ has a (unique) normalform $a^* \in \mathrm{OT}$ such that $\mathrm{o}(a^*) = \mathrm{o}(a)$ which yields

$$(\Diamond) \quad \{\mathrm{o}(a) : a \in \mathrm{T}\} = \{\mathrm{o}(a) : a \in \mathrm{OT}\}.$$

Obviously $\{\mathrm{o}(a) : a \in \mathrm{T}\} = C(\Omega_\omega, 1)$, and thus by Lemma 1.3 $\{\mathrm{o}(a) : a \in \mathrm{T} \,\&\, a <_{\mathrm{o}} D_1 0\} = \psi_0(\Omega_\omega)$. So, $\{\mathrm{o}(a) : a \in \mathrm{T} \,\&\, a <_{\mathrm{o}} D_1 0\}$ is a segment of On, and the equation $(\Diamond)$ implies $\mathrm{o}(a) = \overline{\mathrm{o}}(a)$ for each $a \in \mathrm{OT}_0$. (Of course we will never have $\mathrm{o}(a) = \overline{\mathrm{o}}(a)$ for *all* $a \in \mathrm{OT}$, since T is countable and thus $\overline{\mathrm{o}}(a) < \Omega_1 \leq \mathrm{o}(a)$ for any $a \in \mathrm{OT} \setminus \mathrm{OT}_0$.) The proof of $(\Diamond)$ itself is rather tedious (cf. [10] or [19]) and we will not discuss it here. Having a closer look one will realize that actually $(\Diamond)$ is not of great importance, since in all (existing) proof theoretic applications of the ψ_σ's one can confine to terms from OT, and the only drawback when one dispenses with $(\Diamond)$ is that e.g., instead of saying "the proof theoretic ordinal of ID_ν is $\psi_0(\varepsilon_{\Omega_\nu+1})$" one has to say "the proof theoretic ordinal of ID_ν is the ordertype of $D_0 D_{\nu+1} 0$ in $(\mathrm{OT}_0, <_{\mathrm{o}})$". Usually one avoids the trouble with having to distinguish between $\mathrm{o}(a)$ and $\overline{\mathrm{o}}(a)$ (for $a \in \mathrm{OT}_0$) by following Schütte and incorporating the normalform condition into the definition of $C(\alpha, \beta)$, i.e., by closing $C(\alpha, \beta)$ only under $\widetilde{\psi}_\sigma{\restriction}\alpha$ (instead of $\psi_\sigma{\restriction}\alpha$) where

$$(\widetilde{\psi}_\sigma{\restriction}\alpha)(\xi) := \begin{cases} \psi_\sigma(\xi) & \text{if } \xi < \alpha \text{ and } \xi \in C(\xi, \psi_\sigma(\xi)) \\ \text{undefined} & \text{otherwise} \end{cases}.$$

Then $\{o(a) : a \in \mathrm{OT}\} = C(\Omega_\omega, 1)$ is easy to prove and together with Lemma 1.3 it yields $\mathrm{o}(a) = \overline{\mathrm{o}}(a)$ for all $a \in \mathrm{OT}_0$.

Now we define the set T of terms, a linear ordering $\prec$ on T, for any $a \in \mathrm{T}$ and $\sigma < \omega$ a set $G_\sigma a$ of subterms of a, and the set OT of *ordinal terms* (i.e., terms in normalform) in such a way that, for all $a, c \in \mathrm{OT}$,

(a) $c \prec a \Leftrightarrow \mathrm{o}(c) < \mathrm{o}(a)$ $[\Leftrightarrow c <_{\mathrm{o}} a]$ and

(b) $G_\sigma c \prec a \Leftrightarrow \mathrm{o}(c) \in C(\mathrm{o}(a), \psi_\sigma \mathrm{o}(a))$. (Here $G_\sigma a \prec c$ abbreviates $\forall x \in G_\sigma a (x \prec c)$.)

DEFNITION. The set T of terms is inductively defined by

1. $0 \in \mathrm{T}$.
2. If $a \in \mathrm{T}$ and $\sigma < \omega$, then $D_\sigma a \in \mathrm{T}$; we call $D_\sigma a$ a *principal term*.
3. If $a_0, \ldots, a_n$ are principal terms and $n \geq 1$, then $(a_0 \oplus \ldots \oplus a_n) \in \mathrm{T}$.

Notation. For principal terms $a_0, \ldots, a_{n-1}$ and $n \geq 0$ we set

$$a_0 \oplus \ldots \oplus a_{n-1} := \begin{cases} 0 & \text{if } n = 0 \\ a_0 & \text{if } n = 1 \\ (a_0 \oplus \ldots \oplus a_{n-1}) & \text{if } n > 1 \end{cases}.$$

So every $a \in \mathrm{T}$ can be uniquely written as $a = D_{\sigma_0} a_0 \oplus \ldots \oplus D_{\sigma_{n-1}} a_{n-1}$ with $n \geq 0$ and $a_0, \ldots, a_{n-1} \in \mathrm{T}$.

Further we define: $(a_0 \oplus \ldots \oplus a_{n-1}) \oplus (b_0 \oplus \ldots \oplus b_{m-1}) := a_0 \oplus \ldots \oplus a_{n-1} \oplus b_0 \oplus \ldots \oplus b_{m-1}$, and $a \cdot n := \underbrace{a \oplus \ldots \oplus a}_{n}$ for principal terms a_i, b_i, a.

DEFNITION. The mapping $\mathrm{o} : \mathrm{T} \longrightarrow On$ is defined by

$$\mathrm{o}(D_{\sigma_0} a_0 \oplus \ldots \oplus D_{\sigma_{n-1}} a_{n-1}) := \psi_{\sigma_0} \mathrm{o}(a_0) + \ldots + \psi_{\sigma_{n-1}} \mathrm{o}(a_{n-1}).$$

DEFNITION. The binary relation $\prec$ on T is defined by

1. $0 \prec b :\Longleftrightarrow b \neq 0$
2. $D_\sigma a \oplus \tilde{a} \prec D_\rho b \oplus \tilde{b} :\Longleftrightarrow$
$$\sigma < \rho \text{ or } (\sigma = \rho \,\&\, a \prec b) \text{ or } (\sigma = \rho \,\&\, a = b \,\&\, \tilde{a} \prec \tilde{b}).$$

Remark. $\prec$ is a linear ordering on T, but it's not a wellordering e.g.,

$$\ldots \prec D_0 D_0 D_1 0 \prec D_0 D_1 0 \prec D_1 0.$$

DEFNITION. For each $a \in \mathrm{T}$ and each σ we define a finite set $G_\sigma a$ of subterms of a by

1. $G_\sigma(a_0 \oplus \ldots \oplus a_{n-1}) := \bigcup_{i<n} G_\sigma a_i$,
2. $G_\sigma D_\mu a := \begin{cases} \{a\} \cup G_\sigma a & \text{if } \sigma \leq \mu \\ \emptyset & \text{if } \mu < \sigma \end{cases}.$

DEFNITION. The subset OT of T is inductively defined by

1. $0 \in \mathrm{OT}$.
2. $a \in \mathrm{OT} \,\&\, G_\sigma a \prec a \Longrightarrow D_\sigma a \in \mathrm{OT}$.
3. $a_0, \ldots, a_n \in \mathrm{OT}$ $(n \geq 1)$ principal terms with
$$a_n \preceq \ldots \preceq a_0 \Longrightarrow (a_0 \oplus \ldots \oplus a_n) \in \mathrm{OT}.$$

The elements of OT are called *ordinal terms*. We identify $n \in \mathbb{N}$ with the ordinal term $\underbrace{D_0 0 \oplus \ldots \oplus D_0 0}_{n}$.

Abbreviation. $\Omega_0 := \omega := D_0 1$, $\Omega_\sigma := D_\sigma 0$ for $\sigma > 0$.

THEOREM 1.4. *For $a, c \in$ OT we have*

(a) $c \prec a \Leftrightarrow \mathrm{o}(c) < \mathrm{o}(a)$, $\quad$ [*i.e.*, $c \prec a \Leftrightarrow c <_{\mathrm{o}} a$]
(b) $G_\sigma c \prec a \Leftrightarrow \mathrm{o}(c) \in C(\mathrm{o}(a), \psi_\sigma \mathrm{o}(a))$

PROOF. The proof is by induction on the length of c simultaneous for (a),(b): We only prove "$\Rightarrow$". The reverse implication of (a) follows from "$\Rightarrow$", since $\prec$ is total. The reverse direction of (b) is more difficult to obtain, but since it is not needed for the proof of (a) nor for any proof in this paper, we omit it here.

(a) Let $c = (D_\sigma c_0) \oplus c_1 \oplus \ldots \oplus c_m$ and $a = (D_\rho a_0) \oplus a_1 \oplus \ldots \oplus a_n$ with principal terms $c_1, \ldots, c_m, a_1, \ldots, a_n$.

1. $\sigma < \rho$: From $c_m \preceq \ldots \preceq c_1 \preceq D_\sigma c_0$ we get by IH $\mathrm{o}(c_m) \leq \ldots \leq \mathrm{o}(c_1) \leq \mathrm{o}(D_\sigma c_0) = \psi_\sigma \mathrm{o}(c_0) < \Omega_{\sigma+1}$ and thus $\mathrm{o}(c) < \Omega_{\sigma+1} \leq \Omega_\rho \leq \mathrm{o}(D_\rho a_0) \leq \mathrm{o}(a)$.

2. $\sigma = \rho$ and $c_0 \prec a_0$: By IH $\mathrm{o}(c_0) < \mathrm{o}(a_0)$. Since $D_\sigma c_0 \in$ OT, we have $G_\sigma c_0 \prec c_0$ and thus by IH $\mathrm{o}(c_0) \in C(\mathrm{o}(c_0), \psi_\sigma \mathrm{o}(c_0))$. Hence $\psi_\sigma \mathrm{o}(c_0) < \psi_\sigma \mathrm{o}(a_0)$ by Theorem 1.2(c). Now $\mathrm{o}(c) \prec \mathrm{o}(a)$ follows as in 1. (using that $\psi_\sigma \mathrm{o}(a)$ is additively closed).

3. $\sigma = \rho \,\&\, c_0 = a_0 \,\&\, c_1 \oplus \ldots \oplus c_m \prec a_1 \oplus \ldots \oplus a_n$: Immediate by IH.

(b) 1. $c = c_0 \oplus \ldots \oplus c_{k-1}$ with $k \neq 1$: Then $G_\sigma c_i \prec a$ and thus (by IH) $\mathrm{o}(c_i) \in C := C(\mathrm{o}(a), \psi_\sigma \mathrm{o}(a))$ for $i < k$. This yields $\mathrm{o}(c) = \mathrm{o}(c_0) + \ldots + \mathrm{o}(c_{k-1}) \in C$.

2. $c = D_\mu c_0$ with $\mu < \sigma$: Then $\mathrm{o}(c) \in \Omega_{\mu+1} \subseteq \Omega_\sigma \subseteq C$.

3. $c = D_\mu c_0$ with $\sigma \leq \mu$: Then $\{c_0\} \cup G_\sigma c_0 = G_\sigma c_0 \prec a$ and therefore by IH $\mathrm{o}(c_0) < \mathrm{o}(a) \,\&\, \mathrm{o}(c_0) \in C$ which yields $\mathrm{o}(c) = \psi_\mu \mathrm{o}(c_0) \in C$. ⊣

COROLLARY. (OT, $\prec$) *is a well ordering.*

Fundamental sequences. In order to get a better insight into the structure (T, $\prec$) and a better understanding of the collapsing functions ψ_σ we now present an assignment of (fundamental) sequences to the elements of T. For each term $a \in$ T we define its *(cofinality) type* $\mathrm{tp}(a) \in \{0, 1, \omega\} \cup \{\Omega_{\mu+1} : \mu < \omega\}$ and a family $(a[x])_{x \in |\mathrm{tp}(a)|}$ of terms, such that the following holds, where $|0| := \emptyset$, $|1| := \{0\}$, $|\omega| := \mathbb{N}$, $|\Omega_{\mu+1}| := \{D_\mu b : b \in \mathrm{T}\}$:

THEOREM 1.5. *For all $a, c, x, x' \in$ T the following holds*

(a) $x \in |\mathrm{tp}(a)| \Longrightarrow a[x] \prec a$
(b) $x, x' \in |\mathrm{tp}(a)| \,\&\, x \prec x' \Longrightarrow a[x] \prec a[x']$
(c) $\mathrm{tp}(a) = 1 \Longrightarrow a = a[0] \oplus 1$
(d) $a, c \in \mathrm{OT} \,\&\, c \prec a \,\&\, \mathrm{tp}(a) \neq 1 \Longrightarrow \exists x \in \mathrm{OT} \cap |\mathrm{tp}(a)| (c \prec a[x])$
(e) $a, x \in \mathrm{OT} \,\&\, x \in |\mathrm{tp}(a)| \Longrightarrow a[x] \in \mathrm{OT}$

Note that, according to Theorem 1.5, only for $a \in$ OT and only relative to (OT, $\prec$) is the family $(a[x])_{x \in |\mathrm{tp}(a)|}$ a fundamental sequence of a in the proper sense. But in §3 we will give a natural interpretation of the terms $a \in \mathrm{T}$ as well-founded trees (so-called *tree ordinals*) $\mathrm{t}(a)$ which harmonizes with the assignment $(a, x) \mapsto a[x]$. For example we will have $\mathrm{t}(a) = (\mathrm{t}(a[i]))_{i \in \mathbb{N}}$ if $\mathrm{tp}(a) = \omega$.

I think, one can well say that all clauses of the following definition are canonical (modulo some minor variations, such as setting $(D_\sigma a)[i] := (D_\sigma a[0]) \cdot i$ instead of $(D_\sigma a[0]) \cdot (i+1)$ in clause 4.), and therefore it seems reasonable to call $(a[x])_{x \in |\mathrm{tp}(a)|}$ *the canonical* fundamental sequence of $a \in \mathrm{OT}$. Only clause 6. requires some explanation which will be given below in the proof of Theorem 1.5.

DEFNITION. For $a \in \mathrm{T}$ we define $\mathrm{tp}(a)$ and $a[x]$ (for $x \in |\mathrm{tp}(a)|$) by the following clauses

1. $\mathrm{tp}(0) := 0$.
2. $\mathrm{tp}(D_0 0) := 1$, $(D_0 0)[0] := 0$.
3. $\mathrm{tp}(D_{\mu+1} 0) := \Omega_{\mu+1}$, $(D_{\mu+1} 0)[x] := x$.
4. If $\mathrm{tp}(a) = 1$ then $\mathrm{tp}(D_\sigma a) := \omega$ and $(D_\sigma a)[i] := (D_\sigma a[0]) \cdot (i+1)$.
5. If $\mathrm{tp}(a) \in \{\omega\} \cup \{\Omega_{\mu+1} : \mu < \sigma\}$ then $\mathrm{tp}(D_\sigma a) := \mathrm{tp}(a)$, and $(D_\sigma a)[x] := D_\sigma a[x]$.
6. If $\mathrm{tp}(a) = \Omega_{\mu+1}$ with $\mu \geq \sigma$ then $\mathrm{tp}(D_\sigma a) := \omega$, and $(D_\sigma a)[i] := D_\sigma a[x_i]$ with $x_0 := \Omega_\mu$, $x_{i+1} := D_\mu a[x_i]$.
7. $\mathrm{tp}(a_0 \oplus \ldots \oplus a_n) := \mathrm{tp}(a_n)$, $(a_0 \oplus \ldots \oplus a_n)[x] := (a_0 \oplus \ldots \oplus a_{n-1}) \oplus a_n[x]$ $(n \geq 1)$.

For technical reasons we also set $a[n] := a[0]$, if $\mathrm{tp}(a) = 1$.

PROOF OF THEOREM 1.5. (a),(b),(c) are easily verified by induction on $\ell(a)$ (length of the string a). For a proof of (e) see [10](§5) or [8](Lemma 3.3).

(d) is proved by induction on $\ell(a)$. All cases except 6. are straightforward. So let us assume $D_\sigma c \oplus \tilde{c} \prec D_\sigma a$. Then $c \prec a$, and by IH $c \prec a[x]$ for some $x \in \mathrm{OT} \cap |\mathrm{tp}(a)|$. Hence $D_\sigma c \oplus \tilde{c} \prec D_\sigma a[x]$. If $\mathrm{tp}(a) \in \{\omega\} \cup \{\Omega_{\mu+1} : \mu < \sigma\}$, it can be proved that $D_\sigma a[x] \in \mathrm{OT}$. Therefore in that case $D_\sigma a[x]$ is the canonical choice for $(D_\sigma a)[x]$.

Now let us assume that $\mathrm{tp}(a) = \Omega_{\mu+1}$ with $\mu \geq \sigma$. Then "$D_\sigma a[x] \in \mathrm{OT}$" does no longer hold for arbitrary $x \in \mathrm{OT} \cap |\mathrm{tp}(a)|$.

But by induction on $\ell(a)$ one can prove

$$\text{(1)} \quad \mathrm{tp}(a) = \Omega_{\mu+1} \,\&\, c \in \mathrm{OT} \,\&\, a[\Omega_\mu] \preceq c \prec a \Longrightarrow \exists x = D_\mu(b \oplus 1) \in \mathrm{OT}(b \in G_\mu c \,\&\, c \prec a[x]),$$

from which we conclude

$$\text{(2)} \quad \mathrm{tp}(a) = \Omega_{\mu+1} \,\&\, c \in \mathrm{OT} \,\&\, a[\Omega_\mu] \preceq c \,\&\, \{c\} \cup G_\mu c \prec a \Longrightarrow \exists b \in \mathrm{OT}(\ell(b) < \ell(c) \,\&\, \{b\} \cup G_\mu b \prec a \,\&\, c \prec a[D_\mu(b \oplus 1)]).$$

Obviously (2) suggests to define $x_0 := \Omega_\mu$, $x_{n+1} := D_\mu a[x_n]$ in order to obtain by induction on $\ell(c)$

$$\text{(3)} \quad \mathrm{tp}(a) = \Omega_{\mu+1} \,\&\, c \in \mathrm{OT} \,\&\, \{c\} \cup G_\mu c \prec a \Longrightarrow \exists n(c \prec a[x_n]).$$

(Induction step: $c \prec a[D_\mu(b\oplus 1)] \,\&\, b \prec a[x_n] \Longrightarrow$

$$c \prec a[D_\mu(b\oplus 1)] \preceq a[D_\mu a[x_n]] = a[x_{n+1}]\,)$$

Now $\exists n(D_\sigma c \oplus \tilde{c} \prec (D_\sigma a)[n])$ is obtained as follows:

$$\mathrm{OT} \ni D_\sigma c \oplus \tilde{c} \prec D_\sigma a \overset{\sigma\leq\mu}{\Longrightarrow} G_\mu c \subseteq G_\sigma c \prec c \prec a \overset{(3)}{\Longrightarrow}$$

$$\exists n(D_\sigma c \oplus \tilde{c} \prec D_\sigma a[x_n] = (D_\sigma a)[n]).$$

Proof of (1):

1. $a = \Omega_{\mu+1}$: Then $c = D_\mu c_0 \oplus \tilde{c} \prec D_\mu(c_0 \oplus 1)$. Let $b := c_0$.

2. $a = a_0 \oplus a_1$ with $\mathrm{tp}(a_1) = \Omega_{\mu+1}$: Then $c = a_0 \oplus c_1$ with $a_1[\Omega_\mu] \preceq c_1 \prec a_1$, and the claim follows immediately from the IH.

3. $a = D_\rho a_0$ with $\mu < \rho$: Then $\mathrm{tp}(a_0) = \Omega_{\mu+1}$ and $a[x] = D_\rho a_0[x]$. Further $c = D_\rho c_0 \oplus \tilde{c}$ with $a_0[\Omega_\mu] \preceq c_0 \prec a_0$. By IH we get $c_0 \prec a_0[x]$ for some $x = D_\mu(b\oplus 1) \in \mathrm{OT}$ with $b \in G_\mu c_0$. Since $\mu < \rho$, $G_\mu c_0 \subseteq G_\mu c$. From $c_0 \prec a_0[x]$ we get $c \prec D_\rho a_0[x] = a[x]$. ⊣

Lemma 1.6. *If* $a \in \mathrm{OT}_0$ *then* $\mathrm{tp}(a) \in \{0, 1, \omega\}$ *and*

$$\overline{\mathrm{o}}(a) = \begin{cases} 0 & \text{if } \mathrm{tp}(a) = 0 \\ \overline{\mathrm{o}}(a[0]) + 1 & \text{if } \mathrm{tp}(a) = 1 \\ \sup_{n\in\mathbb{N}}(\overline{\mathrm{o}}(a[n])+1) & \text{if } \mathrm{tp}(a) = \omega \end{cases}$$

Proof. We only consider the third case. Let $a \in \mathrm{OT}_0$ with $\mathrm{tp}(a) = \omega$. Then $a[n] \prec a \,\&\, a[n] \in \mathrm{OT}_0$ (by Theorem 1.5a,e) and $\overline{\mathrm{o}}(a[n]) < \overline{\mathrm{o}}(a)$ (by Theorem 1.4a). Now let $\gamma < \overline{\mathrm{o}}(a)$. Then $\gamma = \overline{\mathrm{o}}(c)$ for some $c \in \mathrm{OT}_0$ with $c \prec a$. Theorem 1.5d yields $c \prec a[n]$ for some n. Hence $\gamma < \overline{\mathrm{o}}(a[n])$. ⊣

§2. Collapsing of infinitary derivations. In this section we take a fresh start and introduce systems ID_ν^∞ of infinitary derivations together with cut-elimination $\mathcal{E}$ and collapsing operations $\mathcal{D}_\sigma$ by which every ID_ν^∞-derivation of an arithmetical formula A can be transformed into a cut-free derivation of A in ID_0^∞ (i.e., ω-arithmetic). ID_ν^∞ is so to speak an infinitary version of the formal theory ID_ν of ν-times iterated inductive definitions; every ID_ν-derivation h can be translated into an ID_ν^∞-derivation h^∞ of the same formula (or sequent). We will establish an upper bound η_ν for the proof-theoretic ordinal $|\mathrm{ID}_\nu|$ in terms of the operations $h \mapsto h^\infty, \mathcal{E}, \mathcal{D}_\sigma$, namely we will prove $|\mathrm{ID}_\nu| \leq \eta_\nu := \sup\{\|\mathcal{D}_0(\mathcal{E}^m(h^\infty))\| : m \in \mathbb{N}$ and h an ID_ν-derivation with endsequent of level $0\}$, where $\|d\|$ denotes the length (or depth) of an infinitary derivation d.

As we will see below, the operations $\mathcal{E}$ and $\mathcal{D}_\sigma$ are closely related to the assignment of fundamental sequences $(a[x])_{x\in|\mathrm{tp}(a)|}$ given in §1. In §3 we will exploit this observation and prove $\|\mathcal{D}_0(\mathcal{E}^m(h^\infty))\| \leq \psi_0\psi_\nu^{m+2}(0)$, which yields $|\mathrm{ID}_\nu| \leq \eta_\nu \leq \psi_0(\varepsilon_{\Omega_\nu+1})$.

Syntax. Let $\mathcal{L}$ be an arbitrary 1st order language (i.e., set of function and predicate symbols). Atomic $\mathcal{L}$-formulas are $Rt_1 \ldots t_n$ where R is an n-ary predicate symbol (of $\mathcal{L}$), and $t_1, \ldots, t_n$ are $\mathcal{L}$-terms. Expressions of the shape A or

$\neg A$, where A is an atomic $\mathcal{L}$-formula, are called *literals*. $\mathcal{L}$-formulas are built up from literals by means of $\wedge, \vee, \forall x, \exists x$. $\mathrm{FV}(A)$ denotes the set of free variables of A. A formula or term A is called *closed* if $\mathrm{FV}(A) = \emptyset$. The *negation* $\neg A$ of a non-atomic formula A is defined via de Morgan's laws. The *rank* $\mathrm{rk}(A)$ of a formula A is defined by: $\mathrm{rk}(A) := 0$ if A is a literal, $\mathrm{rk}(A \wedge B) := \mathrm{rk}(A \vee B) := \max\{\mathrm{rk}(A), \mathrm{rk}(B)\} + 1$, $\mathrm{rk}(\forall x A) := \mathrm{rk}(\exists x A) := \mathrm{rk}(A) + 1$. By $A(x/t)$ we denote the result of substituting t for (every free occurrence of) x in A (renaming bound variables if necessary). Expressions $\lambda x.F$ are called *predicates* and denoted by $\mathcal{F}$. For $\mathcal{F} = \lambda x.F$ we set $\mathcal{F}[t] := F(x/t)$. If $\mathcal{P}$ is a unary predicate symbol then $B(\mathcal{P}/\mathcal{F})$ denotes the result of substituting $\mathcal{F}$ for $\mathcal{P}$ in B, i.e., the formula resulting from B be replacing every atom $\mathcal{P}t$ by $\mathcal{F}[t]$.

Let X be unary predicate symbol not in $\mathcal{L}$. A *positive operator form in* $\mathcal{L}$ is an $\mathcal{L} \cup \{X\}$-formula $\mathfrak{A}$ in which X occurs only positively (i.e., $\mathfrak{A}$ has no subformula $\neg Xt$) and which has at most one free variable x. We use the following abbreviations: $\mathfrak{A}(\mathcal{F}, t) := \mathfrak{A}(X/\mathcal{F}, x/t)$, $\mathfrak{A}(\mathcal{F}) \subseteq \mathcal{F} := \forall x(\mathfrak{A}(\mathcal{F}, x) \to \mathcal{F}[x])$. For each positive operator form $\mathfrak{A}$ we introduce a unary predicate symbol $\mathcal{P}_{\mathfrak{A}}$.

Finite sets of formulas are called *sequents*. They are denoted by Γ, Γ', Δ. We mostly write $A_1, \ldots, A_n$ for $\{A_1, \ldots, A_n\}$, and A, Γ, Δ for $\{A\} \cup \Gamma \cup \Delta$, etc.

Defnition (The languages $\mathcal{L}_\sigma$ $(0 \leq \sigma < \omega)$). Let $\mathcal{L}_0$ be the language consisting of the constant 0 (*zero*), the unary function symbol S (*successor*), and predicate symbols R for primitive recursive relations.

$$\mathcal{L}_{\sigma+1} := \mathcal{L}_0 \cup \{\mathcal{P}_{\mathfrak{A}} : \mathfrak{A} \text{ positive operator form in } \mathcal{L}_\sigma\}$$
$$\mathcal{L}_{<\omega} := \bigcup_{\sigma<\omega} \mathcal{L}_\sigma.$$

The only closed $\mathcal{L}_0$-terms are the *numerals* $0, S0, SS0, \ldots$ which we identify with the corresponding natural numbers (elements of $\mathbb{N}$). Arbitrary $\mathcal{L}_0$-terms will be denoted by $t, t_1, \ldots$, and (number) variables by x, y. TRUE_0 denotes the set of all closed $\mathcal{L}_0$-literals which are true in the standard model $\mathfrak{N}$.

Definition. For each $\mathcal{L}_{<\omega}$-formula A we define its *level* $\mathrm{lev}(A)$ by

$$\begin{aligned}
\mathrm{lev}(A) &:= 0 \text{ if } A \text{ is an } \mathcal{L}_0[X]\text{-literal} \\
\mathrm{lev}(\mathcal{P}_{\mathfrak{A}} t) &:= \mathrm{lev}(\mathfrak{A}), \qquad \mathrm{lev}(\neg \mathcal{P}_{\mathfrak{A}} t) := \mathrm{lev}(\mathfrak{A}) + 1 \\
\mathrm{lev}(A \wedge B) &:= \mathrm{lev}(A \vee B) := \max\{\mathrm{lev}(A), \mathrm{lev}(B)\} \\
\mathrm{lev}(\forall x A) &:= \mathrm{lev}(\exists x A) := \mathrm{lev}(A) \\
\mathrm{lev}(\mathcal{P}_{\mathfrak{A}}) &:= \mathrm{lev}(\mathfrak{A}).
\end{aligned}$$

In addition we set $\mathrm{lev}(\Gamma) := \max\{\mathrm{lev}(A) : A \in \Gamma\}$.

Remark. $\mathrm{lev}(\mathcal{P}_{\mathfrak{A}}) < \sigma$ for each predicate symbol $\mathcal{P}_{\mathfrak{A}}$ in $\mathcal{L}_\sigma$, $\mathrm{lev}(A) \leq \sigma$ for each $\mathcal{L}_\sigma$-formula A.

From now on A, B, C always denote $\mathcal{L}_{<\omega}$-formulas.

Proof systems. We will work in a Tait-style calculus with a somewhat unusual notion of derivation which is especially useful for the purposes of this paper.

DEFINITION. A *proof system* $\mathfrak{S}$ is given by

1. a set of formal expressions called *inference symbols* (syntactic variable $\mathcal{I}$)
2. for each inference symbol $\mathcal{I}$ a set $|\mathcal{I}|$ (the *arity* of $\mathcal{I}$), a sequent $\Delta(\mathcal{I})$ and a family of sequents $(\Delta_\iota(\mathcal{I}))_{\iota\in|\mathcal{I}|}$.
   The elements of $\Delta(\mathcal{I})$ [$\bigcup_{\iota\in|\mathcal{I}|}\Delta_\iota(\mathcal{I})$, resp.] are called the *principal formulas* [*minor formulas*, resp.] of $\mathcal{I}$.
3. for each inference symbol $\mathcal{I}$ a set $\mathrm{Eig}(\mathcal{I})$ which is either empty or a singleton $\{x\}$ with x a variable not in $\mathrm{FV}(\Delta(\mathcal{I}))$; in the latter case x is called the *eigenvariable* of $\mathcal{I}$.

Notation. By writing

$$(\mathcal{I})\quad \frac{\ldots\Delta_\iota\ldots(\iota\in I)}{\Delta}[!u!]$$

we declare $\mathcal{I}$ as an inference symbol with $|\mathcal{I}| = I$, $\Delta(\mathcal{I}) = \Delta$, $\Delta_\iota(\mathcal{I}) = \Delta_\iota$, and $\mathrm{Eig}(\mathcal{I}) = \emptyset$ [$\mathrm{Eig}(\mathcal{I}) = \{u\}$, resp.]. If $I = \{0,\ldots,n-1\}$ we write

$$\frac{\Delta_0\ \ \Delta_1\ \ \ldots\ \ \Delta_{n-1}}{\Delta},\quad \text{instead of}\quad \frac{\ldots\Delta_\iota\ldots(\iota\in I)}{\Delta}.$$

DEFINITION (Inductive definition of $\mathfrak{S}$-derivations). If $\mathcal{I}$ is an inference symbol of $\mathfrak{S}$, and $(d_\iota)_{\iota\in|\mathcal{I}|}$ is a family of $\mathfrak{S}$*-derivations* such that $\mathrm{Eig}(\mathcal{I})\cap\mathrm{FV}(\Gamma)=\emptyset$ where $\Gamma := \Delta(\mathcal{I})\cup\bigcup_{\iota\in|\mathcal{I}|}(\Gamma(d_\iota)\setminus\Delta_\iota(\mathcal{I}))$, then $d := \mathcal{I}(d_\iota)_{\iota\in|\mathcal{I}|}$ is an $\mathfrak{S}$*-derivation* with $\Gamma(d) := \Gamma$ *(endsequent of d)* and $\mathrm{last}(d) := \mathcal{I}$ *(last inference of d)*.

Instead of $\mathcal{I}(d_\iota)_{\iota\in|\mathcal{I}|}$ we also write

$$\frac{\ldots d_\iota\ldots(\iota\in|\mathcal{I}|)}{\mathcal{I}}\quad\text{or}\quad \mathcal{I}d_0\ldots d_{n-1}\quad\text{or}\quad \frac{d_0\ldots d_{n-1}}{\mathcal{I}}\quad \text{if } |\mathcal{I}|=\{0,\ldots,n-1\}.$$

Abbreviation. $\mathfrak{S}\ni d\vdash\Gamma :\Longleftrightarrow d$ is an $\mathfrak{S}$-derivation with $\Gamma(d)\subseteq\Gamma$.

Remark. Our notion of derivations differs from the usual one in so far as our derivations have inferences (inference symbols) and not sequents assigned to their nodes. The sequent "belonging" to a node τ of a derivation d is not explicitly displayed, but can be computed by tree recursion from d (similarly for the free assumptions in a natural deduction style derivation).

The Tait-style inference rules in their traditional form $\dfrac{\ldots\Gamma,\Delta_\iota(\mathcal{I})\ldots}{\Gamma,\Delta(\mathcal{I})}$ are reobtained here as follows: If $\mathcal{I}\in\mathfrak{S}$ and $\mathrm{Eig}(\mathcal{I})\cap\mathrm{FV}(\Gamma)=\emptyset$, then

from $\ldots\mathfrak{S}\ni d_\iota\vdash\Gamma,\Delta_\iota(\mathcal{I})\ldots(\forall\iota\in|\mathcal{I}|)$ we conclude $\mathfrak{S}\ni\mathcal{I}(d_\iota)_{\iota\in|\mathcal{I}|}\vdash\Gamma,\Delta(\mathcal{I})$.

DEFNITION (The proof system ID_ν). The language of ID_ν is $\mathcal{L}_\nu$, and the inference symbols of ID_ν are

$$(\bigwedge_{A_0\wedge A_1})\ \frac{A_0 \qquad A_1}{A_0\wedge A_1} \qquad (\bigvee^k_{A_0\vee A_1})\ \frac{A_k}{A_0\vee A_1}\ (k\in\{0,1\})$$

$$(\bigwedge^y_{\forall xA})\ \frac{A(x/y)}{\forall xA}\ !y! \qquad (\bigvee^t_{\exists xA})\ \frac{A(x/t)}{\exists xA}$$

$$(\mathsf{Cut}_C)\ \frac{C \qquad \neg C}{\emptyset} \qquad (\mathsf{Ind}^t_{\mathcal{F}})\ \frac{}{\neg\mathcal{F}[0], \neg\forall x(\mathcal{F}[x]\to\mathcal{F}[Sx]), \mathcal{F}[t]}$$

$$(\mathsf{Cl}_{\mathcal{P}_{\mathfrak{A}}t})\ \frac{\mathfrak{A}(\mathcal{P}_{\mathfrak{A}}, t)}{\mathcal{P}_{\mathfrak{A}}t} \qquad (\mathsf{Ind}^{\mathcal{P}_{\mathfrak{A}},t}_{\mathcal{F}})\ \frac{}{\neg(\mathfrak{A}(\mathcal{F})\subseteq\mathcal{F}), \neg\mathcal{P}_{\mathfrak{A}}t, \mathcal{F}[t]}$$

$$(\mathsf{Ax}_\Delta)\ \frac{}{\Delta}\ \text{ if } \Delta\in\mathsf{Ax}(\nu)$$

where $\mathsf{Ax}(\nu)$ is a set of $\mathcal{L}_\nu$-sequents such that

(i) $\Delta\in\mathsf{Ax}(\nu)\Longrightarrow\Delta(\vec{x}/\vec{t})\in\mathsf{Ax}(\nu)$

(ii) $\Delta\in\mathsf{Ax}(\nu)\ \&\ \mathsf{FV}(\Delta)=\emptyset\Longrightarrow\begin{cases}\Delta\cap\mathsf{TRUE}_0\neq\emptyset \text{ or}\\ \{\neg P, P\}\subseteq\Delta \text{ for some } P=\mathcal{P}_{\mathfrak{A}}n\end{cases}.$

DEFNITION (The infinitary proof systems $\mathrm{ID}^\infty_\sigma$ ($\sigma<\omega$)). The language of $\mathrm{ID}^\infty_\sigma$ consists of all closed $\mathcal{L}_{<\omega}$-formulas A. We use P [P_μ, resp.] as syntactic variable for formulas of the form $\mathcal{P}_{\mathfrak{A}}n$ [with $\mathrm{lev}(\mathfrak{A})=\mu$, resp.].

The inference symbols of $\mathrm{ID}^\infty_\sigma$ are

$(\mathsf{Ax}_\Delta)]\frac{}{\Delta}$, if $\Delta=\{A\}$ with $A\in\mathsf{TRUE}_0$ or $\Delta=\{\neg P, P\}$ with $\mathrm{lev}(P)<\sigma$

$$(\bigwedge_{\forall xA})\ \frac{\ldots A(x/i)\ldots(i\in\mathbb{N})}{\forall xA} \qquad (\bigwedge_{A_0\wedge A_1})\ \frac{A_0 \qquad A_1}{A_0\wedge A_1}$$

$$(\bigvee^k_{\exists xA})\ \frac{A(x/k)}{\exists xA}\ (k\in\mathbb{N}) \qquad (\bigvee^k_{A_0\vee A_1})\ \frac{A_k}{A_0\vee A_1}(k\in\{0,1\})$$

$$(\mathsf{Cut}_C)\ \frac{C \qquad \neg C}{\emptyset}$$

$$(\mathsf{Cl}_{\mathcal{P}_{\mathfrak{A}}t})\ \frac{\mathfrak{A}(\mathcal{P}_{\mathfrak{A}}, t)}{\mathcal{P}_{\mathfrak{A}}t}\qquad(\mathrm{lev}(\mathcal{P}_{\mathfrak{A}})\leq\sigma)$$

$$(\widetilde{\Omega}_P)\ \frac{P \qquad \ldots\Delta^P_q\ldots(q\in|P|)}{\emptyset}\qquad(\mathrm{lev}(P)<\sigma)$$

$|P|:=$ set of all cutfree ID^∞_μ-derivations, where $\mu:=\mathrm{lev}(P)$

$\Delta^P_q:=\Gamma(q)\setminus\{P\}$.

An ID^∞_μ-derivation d is called *cutfree* if $\mathrm{crk}(d)=0$ where $\mathrm{crk}(d)$ is the least number greater than the ranks of all cut-formulas occurring in d, i.e.,

$$\mathrm{crk}(\mathcal{I}(d_\iota)_{\iota\in I}) := \sup(\{\mathrm{crk}(\mathcal{I})\} \cup \{\mathrm{crk}(d_\iota) : \iota \in I\}) \text{ with}$$

$$\mathrm{crk}(\mathcal{I}) := \begin{cases} \mathrm{rk}(C)+1 & \text{if } \mathcal{I} = \mathsf{Cut}_C \\ 0 & \text{otherwise} \end{cases}$$

Abbreviation. $\mathrm{ID}_\sigma^\infty \ni d \vdash_m \Gamma :\Leftrightarrow \mathrm{ID}_\sigma^\infty \ni d \vdash \Gamma \,\&\, \mathrm{crk}(d) \leq m$
$d \vdash_m \Gamma :\Leftrightarrow \mathrm{ID}_\sigma^\infty \ni d \vdash_m \Gamma$ for some $\sigma < \omega$.

The set $\mathsf{ID}_\sigma^\infty$ of all $\mathrm{ID}_\sigma^\infty$-derivations is introduced by an inductive definition (as given above for arbitrary proof systems $\mathfrak{S}$) under the assumption that the sets ID_μ^∞ for $\mu < \sigma$ are already defined.
We set $\mathsf{ID}_{<\omega}^\infty := \bigcup_{\sigma<\omega} \mathsf{ID}_\sigma^\infty$.

The $(\widetilde{\Omega}_P)$-rule can be motivated as follows (with $\mu := \mathrm{lev}(P) < \sigma$): Imitating the constructive interpretation of implication we start by saying:

"An $\mathrm{ID}_\sigma^\infty$-derivation of $P \to B$ is an operation $q \mapsto d_q$ transforming every cutfree ID_μ^∞-derivation of P into an $\mathrm{ID}_\sigma^\infty$-derivation of B".

This may be replaced by the stricter version:

"An $\mathrm{ID}_\sigma^\infty$-derivation of $P \to B$ is an operation $q \mapsto d_q$ transforming every cutfree ID_μ^∞-derivation of $A \to P$ into an $\mathrm{ID}_\sigma^\infty$-derivation of $A \to B$ (for any formula A)".

In terms of the Tait-calculus used here this amounts to the following rule:

(Ω_P) *If for each Δ and each cutfree ID_μ^∞-derivation q of Δ, P , d_q is an $\mathrm{ID}_\sigma^\infty$-derivation of Δ, Γ, then $(d_q)_{q\in|P|}$ is an $\mathrm{ID}_\sigma^\infty$-derivation of $\neg P, \Gamma$".*

Now $(\widetilde{\Omega}_P)$ is just a combination of (Ω_P) and (Cut_P).

The following definitions and Theorem 2.1 are needed for the embedding of ID_ν into ID_ν^∞, i.e., for deriving $\neg(\mathfrak{A}(\mathcal{F})\subseteq\mathcal{F}), \neg\mathcal{P}_{\mathfrak{A}}n, \mathcal{F}[n]$ by means of $(\widetilde{\Omega}_{\mathcal{P}_{\mathfrak{A}}})$.

Definition (Substitution). For each closed $\mathcal{L}_\sigma$-formula A let $\mathsf{d}_{\neg A,A}$ be the canonical cutfree $\mathrm{ID}_\sigma^\infty$-derivation of $\neg A, A$ (e.g., $\mathsf{d}_{\neg P,P} := \mathsf{Ax}_{\{\neg P,P\}}$, $\mathsf{d}_{\neg A,A} := \mathsf{Ax}_{\{A\}}$ if $A \in \mathsf{TRUE}_0$, $\mathsf{d}_{\neg\forall xA,\forall xA} := \bigwedge_{\forall xA}\left(\bigvee^i_{\neg\forall xA}\mathsf{d}_{\neg A(x/i),A(x/i)}\right)_{i\in\mathbb{N}}$).

$$\mathsf{e}^n_{\mathfrak{A},\mathcal{F}} := \bigvee^n_{\neg(\mathfrak{A}(\mathcal{F})\subseteq\mathcal{F})}\bigwedge_{\mathfrak{A}(\mathcal{F},n)\wedge\neg\mathcal{F}[n]}\mathsf{d}_{\neg\mathfrak{A}(\mathcal{F},n),\mathfrak{A}(\mathcal{F},n)}\mathsf{d}_{\neg\mathcal{F}[n],\mathcal{F}[n]}$$

$$\approx \cfrac{\cfrac{\neg\mathfrak{A}(\mathcal{F},n), \mathfrak{A}(\mathcal{F},n) \qquad \neg\mathcal{F}[n], \mathcal{F}[n]}{\mathfrak{A}(\mathcal{F},n)\wedge\neg\mathcal{F}[n], \neg\mathfrak{A}(\mathcal{F},n), \mathcal{F}[n]}}{\neg(\mathfrak{A}(\mathcal{F}) \subseteq \mathcal{F}), \neg\mathfrak{A}(\mathcal{F},n), \mathcal{F}[n]}$$

Given $\mathcal{P} = \mathcal{P}_{\mathfrak{A}}$, a predicate $\mathcal{F}$, and a sequent Π we define an operation $\mathcal{S}^\Pi_{\mathcal{P},\mathcal{F}} : \mathsf{ID}^\infty_{\mathrm{lev}(\mathcal{P})} \to \mathsf{ID}^\infty_{<\omega}$ which transforms any derivation $d \in \mathsf{ID}^\infty_{\mathrm{lev}(\mathcal{P})}$ of Γ, Π into a derivation $d^* := \mathcal{S}^\Pi_{\mathcal{P},\mathcal{F}}(d)$ of $\neg(\mathfrak{A}(\mathcal{F})\subseteq\mathcal{F}), \Gamma, \Pi(\mathcal{P}/\mathcal{F})$. Roughly speaking d^*

results from d by substituting certain occurrences of $\mathcal{P}$ by $\mathcal{F}$. In doing so, some inferences $(\mathsf{Cl}_{\mathcal{P}n})\frac{\mathfrak{A}(\mathcal{P},n)}{\mathcal{P}n}$ are turned into $\frac{\mathfrak{A}(\mathcal{F},n)}{\mathcal{F}[n]}$ which is not an inference of $\mathrm{ID}^\infty_{<\omega}$. Therefore those inferences $(\mathsf{Cl}_{\mathcal{P}n})$ are replaced by

$$\dfrac{\begin{array}{cc} d_0^* & \mathsf{e}^n_{\mathfrak{A},\mathcal{F}} \\ \mathfrak{A}(\mathcal{F},n) & \neg(\mathfrak{A}(\mathcal{F})\subseteq\mathcal{F}), \neg\mathfrak{A}(\mathcal{F},n), \mathcal{F}[n] \end{array}}{\neg(\mathfrak{A}(\mathcal{F})\subseteq\mathcal{F}), \mathcal{F}[n]}\ (Cut)$$

The precise definition of $\mathcal{S}^\Pi_{\mathcal{P},\mathcal{F}}(d)$ runs as follows

$$\mathcal{S}^\Pi_{\mathcal{P},\mathcal{F}}(\mathcal{I}(d_\iota)_{\iota\in I}) := \begin{cases} \mathsf{Cut}_{\mathfrak{A}(\mathcal{F},n)}\mathcal{S}^{\Pi\cup\Delta_0(\mathcal{I})}_{\mathcal{P},\mathcal{F}}(d_0)\mathsf{e}^n_{\mathfrak{A},\mathcal{F}} & \text{if } \mathcal{I} = \mathsf{Cl}_{\mathcal{P}n} \text{ with } \mathcal{P}n\in\Pi \\ \bigwedge_{A(\mathcal{P}/\mathcal{F})}\big(\mathcal{S}^{\Pi\cup\Delta_\iota(\mathcal{I})}_{\mathcal{P},\mathcal{F}}(d_\iota)\big)_{\iota\in I} & \text{if } \mathcal{I} = \bigwedge_A \text{ with } A\in\Pi \\ \bigvee^k_{A(\mathcal{P}/\mathcal{F})}\big(\mathcal{S}^{\Pi\cup\Delta_\iota(\mathcal{I})}_{\mathcal{P},\mathcal{F}}(d_\iota)\big)_{\iota\in I} & \text{if } \mathcal{I} = \bigvee^k_A \text{ with } A\in\Pi \\ \mathcal{I}\big(\mathcal{S}^\Pi_{\mathcal{P},\mathcal{F}}(d_\iota)\big)_{\iota\in I} & \text{otherwise} \end{cases}$$

The following theorem is easily verified. Note that the axioms $\mathsf{Ax}_{\{\neg\mathcal{P}n,\mathcal{P}n\}}$ do not belong to $\mathrm{ID}^\infty_{\mathrm{lev}(\mathcal{P})}$!

THEOREM 2.1. *If* $\mathrm{ID}^\infty_{\mathrm{lev}(\mathcal{P})} \ni d \vdash_0 \Gamma, \Pi$ *and* $\mathrm{rk}(\mathfrak{A}(\mathcal{F},x)) < m$ *then*

$$\mathcal{S}^\Pi_{\mathcal{P},\mathcal{F}}(d) \vdash_m \neg(\mathfrak{A}(\mathcal{F})\subseteq\mathcal{F}), \Gamma, \Pi(\mathcal{P}/\mathcal{F}).$$

Embedding of ID_ν into ID^∞_ν. An ID_ν-derivation h is called *closed* if every number variable occurring free in h is the eigenvariable of an inference below that occurrence. Especially $\mathrm{FV}(\Gamma(h)) = \emptyset$ for closed h.

DEFNITION. For each closed ID_ν-derivation h we define an ID^∞_ν-derivation h^∞ such that $h^\infty \vdash_m \Gamma(h)$ for some $m\in\mathbb{N}$:

0. $(\mathsf{Ax}_\Delta)^\infty := \mathsf{Ax}_{\Delta'}$ with a suitable $\Delta'\subseteq\Delta$
1. $(\bigwedge^y_{\forall xA}h_0)^\infty := \bigwedge_{\forall xA}\big(h_0(y/i)^\infty\big)_{i\in\mathbb{N}}$, where $h_0(y/i)$ is defined as expected
2. $(\mathsf{Ind}^n_{\mathcal{F}})^\infty := d_n$ with $d_0 := \mathsf{d}_{\neg\mathcal{F}[0],\mathcal{F}[0]}$,
   $d_{i+1} := \bigvee^i_{\exists x(\mathcal{F}[x]\wedge\neg\mathcal{F}[Sx])}\bigwedge_{\mathcal{F}[i]\wedge\neg\mathcal{F}[Si]}d_i\mathsf{d}_{\neg\mathcal{F}[Si],\mathcal{F}[Si]}$
3. $(\mathsf{Ind}^{\mathcal{P},n}_{\mathcal{F}})^\infty := \dfrac{\mathsf{Ax}_{\{\neg\mathcal{P}n,\mathcal{P}n\}} \quad \ldots\mathcal{S}^{\{\mathcal{P}n\}}_{\mathcal{P},\mathcal{F}}(q)\ldots(q\in|\mathcal{P}n|)}{\widetilde{\Omega}_{\mathcal{P}n}}$
4. Otherwise: $(\mathcal{I}h_0\ldots h_{n-1})^\infty := \mathcal{I}h_0^\infty\ldots h_{n-1}^\infty$

THEOREM 2.2 (Embedding). *If* $\mathrm{ID}_\nu \ni h \vdash \Gamma$ *and* h *is closed, then*

$\mathrm{ID}^\infty_\nu \ni h^\infty \vdash_m \Gamma$ *for some* $m\in\mathbb{N}$.

The proof of Theorem 2.2 is straightforward. Especially,

$$(\mathsf{Ind}^{\mathcal{P},n}_{\mathcal{F}})^\infty \vdash_m \neg(\mathfrak{A}(\mathcal{F})\subseteq\mathcal{F}), \neg\mathcal{P}n, \mathcal{F}[n]$$

(where $\mathcal{P} = \mathcal{P}_{\mathfrak{A}}$) is obtained from:

$$q \in |\mathcal{P}n| \Rightarrow$$

$$\mathrm{ID}^{\infty}_{\mathrm{lev}(\mathcal{P})} \ni q \vdash_0 \Delta^{\mathcal{P}n}_q,\ \mathcal{P}n \overset{\text{Theorem 2.1}}{\Rightarrow} \mathcal{S}^{\{\mathcal{P}n\}}_{\mathcal{P}.\mathcal{F}}(q) \vdash_m \neg(\mathfrak{A}(\mathcal{F}) \subseteq \mathcal{F}), \Delta^{\mathcal{P}n}_q, \mathcal{F}[n].$$

Abbreviation. $\bigwedge$-For := set of all formulas of the shape $A \wedge B$ or $\forall x A$.
$\bigwedge^+$-For := $\mathsf{TRUE}_0 \cup \bigwedge$-For $\cup$ set of all formulas $\mathcal{P}_{\mathfrak{A}} n$.

$$C[k] := \begin{cases} C_k & \text{if } C = C_0 \overset{\vee}{\underset{\wedge}{}} C_1 \text{ and } k \in \{0,1\} \\ A(x/k) & \text{if } C = \overset{\exists}{\underset{\forall}{}} xA \text{ and } k \in \mathbb{N} \end{cases}$$

THEOREM 2.3. *By tree recursion one can define operations* $\mathcal{J}^k_C$, $\mathcal{R}_C$, $\mathcal{E}$, $\mathcal{D}_\sigma$ *on* $\mathsf{ID}^{\infty}_{<\omega}$ *with the following properties*:

($\bigwedge$-Inversion)	$d \vdash_m \Gamma, C \;\&\; C \in \bigwedge\text{-For} \Longrightarrow \mathcal{J}^k_C(d) \vdash_m C[k]$
(Reduction)	$\left.\begin{matrix} e \vdash_m \Gamma, C \;\&\; d \vdash_m \Gamma, \neg C \;\& \\ C \in \bigwedge^+\text{-For} \;\&\; \mathrm{rk}(C) \le m \end{matrix}\right\} \Longrightarrow \mathcal{R}_C(e,d) \vdash_m \Gamma$
(Elimination)	$d \vdash_{m+1} \Gamma \Longrightarrow \mathcal{E}(d) \vdash_m \Gamma$
(Collapsing)	$d \vdash_0 \Gamma \;\&\; \mathrm{lev}(\Gamma) \le \sigma \Longrightarrow \mathrm{ID}^{\infty}_{\sigma} \ni \mathcal{D}_\sigma(d) \vdash_0 \Gamma$

PROOF. For $d = \mathcal{I}(d_\iota)_{\iota \in I} \in \mathsf{ID}^{\infty}_{<\omega}$ and $e \in \mathsf{ID}^{\infty}_{<\omega}$ we define

$$\mathcal{J}^k_C(d) := \begin{cases} \mathcal{J}^k_C(d_k) & \text{if } \mathcal{I} = \bigwedge_C \\ \mathcal{I}(\mathcal{J}^k_C(d_\iota))_{\iota \in I} & \text{otherwise} \end{cases} \qquad (C \in \bigwedge\text{-For})$$

$$\mathcal{R}_C(e,d) := \begin{cases} \mathsf{Cut}_{C[k]} \mathcal{J}^k_C(e) \mathcal{R}_C(e, d_0) & \text{if } \mathcal{I} = \bigvee^k_{\neg C} \\ e & \text{if } \mathcal{I} = \mathsf{Ax}_{\{\neg C, C\}} \\ \mathcal{I}(\mathcal{R}_C(e, d_\iota))_{\iota \in I} & \text{otherwise} \end{cases} \qquad (C \in \bigwedge^+\text{-For})$$

$$\mathcal{E}(d) := \begin{cases} \mathcal{R}_C(\mathcal{E}(d_0), \mathcal{E}(d_1)) & \text{if } \mathcal{I} = \mathsf{Cut}_C \text{ with } C \in \bigwedge^+\text{-For} \\ \mathcal{R}_{\neg C}(\mathcal{E}(d_1), \mathcal{E}(d_0)) & \text{if } \mathcal{I} = \mathsf{Cut}_C \text{ with } \neg C \in \bigwedge^+\text{-For} \\ \mathcal{I}(\mathcal{E}(d_\iota))_{\iota \in I} & \text{otherwise} \end{cases}$$

$$\mathcal{D}_\sigma(d) := \begin{cases} \mathcal{D}_\sigma(d_{\mathcal{D}_\mu(d_0)}) & \text{if } \mathcal{I} = \widetilde{\Omega}_P \text{ with } \mu := \mathrm{lev}(P) \ge \sigma \\ \mathcal{I}(\mathcal{D}_\sigma(d_\iota))_{\iota \in I} & \text{otherwise} \end{cases}$$

One easily verifies that the so defined operations have the asserted properties. Let us look at $\mathcal{D}_\sigma(d)$ for $d = \widetilde{\Omega}_P(d_q)_{q \in \{0\} \cup |P|} \vdash_0 \Gamma$ with $\mathrm{lev}(\Gamma) \le \sigma \le \mu := \mathrm{lev}(P)$. Then $d_0 \vdash_0 \Gamma, P$ and $d_q \vdash_0 \Gamma, \Delta^P_q$ for all $q \in |P|$ (†). By IH $\mathrm{ID}^{\infty}_{\mu} \ni q_0 := \mathcal{D}_\mu(d_0) \vdash_0 \Gamma, P$. Hence $q_0 \in |P|$ and $\Delta^P_{q_0} \subseteq \Gamma$. Now (†) yields $d_{q_0} \vdash_0 \Gamma$, and by IH we get $\mathrm{ID}^{\infty}_{\sigma} \ni \mathcal{D}_\sigma(d_{q_0}) \vdash_0 \Gamma$. ⊣

Remark. The definition of $\mathcal{D}_\sigma(d)$ almost automatically arises if one pursues the goal to eliminate from d all $\widetilde{\Omega}_P$-inferences with $\mathrm{lev}(P) \ge \sigma$.

DEFINITION. For $\mathfrak{A}$ with $\mathrm{lev}(\mathfrak{A}) = 0$ let
$\Phi^\alpha_{\mathfrak{A}} := \{n : \mathfrak{A}(\Phi^{<\alpha}_{\mathfrak{A}}, n)\}$, where $\Phi^{<\alpha}_{\mathfrak{A}} := \bigcup_{\xi<\alpha} \Phi^\xi_{\mathfrak{A}} \quad (\alpha \in On)$.

$|n|_{\mathfrak{A}} := \min\{\alpha : n \in \Phi^\alpha_{\mathfrak{A}}\}$ (if $n \in \bigcup_{\alpha \in On} \Phi^\alpha_{\mathfrak{A}}$).
$|\mathrm{ID}_\nu| := \sup\{|n|_{\mathfrak{A}} : \mathrm{lev}(\mathfrak{A}) = 0 \,\&\, \mathrm{ID}_\nu \vdash \mathcal{P}_{\mathfrak{A}} n\}$. (*proof theoretic ordinal*)
By $(\mathfrak{N}, \Phi^{<\alpha})$ we denote the expansion of the standard model $\mathfrak{N}$ where each predicate constant $\mathcal{P}_{\mathfrak{A}}$ of level 0 is interpreted by $\Phi^{<\alpha}_{\mathfrak{A}}$.

Remark. The proof theoretic ordinal of a theory *Th* is commonly defined as the supremum of the ordertypes of primitive recursive wellorderings $\prec$ which are provably wellfounded in *Th*. In the language $\mathcal{L}_\nu$ the wellfoundedness of $\prec$ is expressed by the formula $\forall x \mathcal{P}_{\mathfrak{A}_\prec} x$ where $\mathfrak{A}_\prec(X, x) := \forall y(y \prec x \to Xy)$. Since the ordertype of $\prec$ is equal to $\sup\{|n|_{\mathfrak{A}_\prec} + 1 : n \in \Phi^{<\Omega}_{\mathfrak{A}_\prec}\}$, it easily follows that the proof theoretic ordinal of ID_ν is less or equal to the ordinal $|\mathrm{ID}_\nu|$ defined above. That actually both ordinals coincide follows from Theorem 1.3.11 in [15] where it is shown that the proof theoretic ordinal of a theory $Th \supseteq \mathrm{PA}$ is equal to its Π^1_1-ordinal.

DEFNITION. $\|\mathcal{I}(d_\iota)_{\iota \in I}\| := \sup_{\iota \in I}(\|d_\iota\| + 1)$ (*length* or *depth* of d)

THEOREM 2.4 (Boundedness). $\mathrm{ID}^\infty_0 \ni d \vdash_0 \Gamma \,\&\, \mathrm{lev}(\Gamma) = 0 \Longrightarrow (\mathfrak{N}, \Phi^{<\|d\|}) \models \Gamma$

Proof by induction on $\|d\|$.

THEOREM 2.5. *If h is a closed* ID_ν*-derivation of* Γ *with* $\mathrm{lev}(\Gamma) = 0$ *then*

$$(\mathfrak{N}, \Phi^{<\alpha}) \models \Gamma \text{ with } \alpha = \|\mathcal{D}_0(\mathcal{E}^m(h^\infty))\| \text{ for some } m \in \mathbb{N}.$$

PROOF.

$$\begin{aligned}
\mathrm{ID}_\nu \ni h \vdash \Gamma &\overset{\text{Embedding}}{\Longrightarrow} \mathrm{ID}^\infty_\nu \ni h^\infty \vdash_m \Gamma \text{ for some } m \\
&\overset{\text{Cutelim}}{\Longrightarrow} \mathrm{ID}^\infty_\nu \ni \mathcal{E}^m(h^\infty) \vdash_0 \Gamma \\
&\overset{\text{Collapsing}}{\Longrightarrow} \mathrm{ID}^\infty_0 \ni \mathcal{D}_0(\mathcal{E}^m(h^\infty)) \vdash_0 \Gamma \\
&\overset{\text{Boundedness}}{\Longrightarrow} (\mathfrak{N}, \Phi^{<\alpha}) \models \Gamma \text{ with } \alpha := \|\mathcal{D}_0(\mathcal{E}^m(h^\infty))\|
\end{aligned}$$
⊣

DEFNITION. $\eta_\nu := \sup\{\|\mathcal{D}_0(\mathcal{E}^m(h^\infty))\| : m \in \mathbb{N}$ and h a closed ID_ν-derivation with endsequent of level 0 $\}$

Then Theorem 2.5 shows that $|\mathrm{ID}_\nu| \leq \eta_\nu$. In what follows we will prove $\eta_\nu \leq \sup_{m \in \mathbb{N}} \psi_0 \psi^m_\nu(0) = \psi_0(\varepsilon_{\Omega_\nu + 1})$.

Remark. Note the similarity between
" $\mathcal{D}_\sigma(d) = \mathcal{D}_\sigma(d_{\mathcal{D}_\mu(d_0)})$ if $d = \widetilde{\Omega}_P(d_\iota)_{\iota \in \{0\} \cup |P|}$ with $\mu = \mathrm{lev}(P) \geq \sigma$ "
and
" $(D_\sigma a)[1] = D_\sigma a[D_\mu a[\Omega_\mu]]$ if $a \in \mathrm{T}$ and $\mathrm{tp}(a) = \Omega_{\mu+1}$ with $\mu \geq \sigma$ ".

This observation will be pursued in §3.

§3. Majorization of infinitary derivations by tree ordinals. We are now going to relate infinitary derivations $d \in \mathrm{ID}_\nu^\infty$ to ordinals (ordinal notations) $a \in \mathrm{OT}$. Here we heavily utilize the assignment of fundamental sequences from §1, which so to speak turns each $a \in \mathrm{OT}$ into a wellfounded tree, a so-called *tree ordinal* $\mathfrak{o}(a)$, namely $\mathfrak{o}(a) := \big(\mathfrak{o}(a[x])\big)_{x\in|\mathrm{tp}(a)|}$. On the other side, from every derivation $d \in \mathrm{ID}_\nu^\infty$ one obtains a tree ordinal $\mathfrak{o}(d)$ essentially by deleting all inference symbols (and possibly other data) assigned to the nodes of d (namely $\mathfrak{o}(\mathcal{I}(d_\iota)_{\iota\in I}) := \big(\mathfrak{o}(d_\iota)\big)_{\iota\in I}$). Now the first idea which comes into mind is that $\mathfrak{o}(d)$ should equal $\mathfrak{o}(a)$ for suitable $a \in \mathrm{OT}$ (at least if $d = h^\infty$ with $h \in \mathrm{ID}_\nu$). But this doesn't work; instead one can establish a weaker relation between $\mathfrak{o}(d)$ and $\mathfrak{o}(a)$, namely that in a certain sense $\mathfrak{o}(d)$ is "embeddable" into $\mathfrak{o}(a)$. Below we will define a relation $d \lhd \mathfrak{a}$ (d is *majorized* by $\mathfrak{a}$) between infinitary derivations d and tree ordinals $\mathfrak{a}$, which corresponds to this informal notion of embeddability. The main properties of $\lhd$ will be: (i) $d \lhd \mathfrak{a} \,\&\, d \in \mathrm{ID}_0^\infty \Rightarrow \|d\| \le \|\mathfrak{a}\|$, (ii) $d \lhd \mathfrak{a} \,\&\, d \in \mathrm{ID}_\nu^\infty \Rightarrow \mathcal{E}(d) \lhd \mathfrak{D}_\nu(\mathfrak{a})$, (iii) $d \lhd \mathfrak{a} \Rightarrow \mathcal{D}_\sigma(d) \lhd \mathfrak{D}_\sigma(\mathfrak{a})$. Here $\mathfrak{D}_\sigma$ is a collapsing function on tree ordinals defined in close analogy to $\mathcal{D}_\sigma$; at the same time $\mathfrak{D}_\sigma$ is closely related to ψ_σ, it is so to speak the "tree version" of ψ_σ, which becomes clear by comparing the definition of $\mathfrak{D}_\sigma$ (see below) with the corresponding clauses in the definition of the fundamental sequences in §1. Mainly by means of (i)-(iii) we will establish that $\|\mathcal{D}_0(\mathcal{E}^m(h^\infty))\| \le \|\mathfrak{D}_0\mathfrak{D}_\nu^{m+2}(\boldsymbol{0})\|$ and thus $\eta_\nu \le \sup_{m\in\mathbb{N}} \|\mathfrak{D}_0\mathfrak{D}_\nu^m(\boldsymbol{0})\|$. Finally we will show that $\|\mathfrak{D}_0(\mathfrak{D}_\nu^m(\boldsymbol{0}))\| = \psi_0\psi_\nu^m(0)$ which then yields $|\mathrm{ID}_\nu| \le \eta_\nu \le \sup_{m\in\mathbb{N}} \psi_0\psi_\nu^m(0) = \psi_0(\varepsilon_{\Omega_\nu+1})$.

DEFNITION (Inductive definition of classes $\mathfrak{T}_\sigma$ of tree ordinals).
1. $\boldsymbol{0} := () \in \mathfrak{T}_\sigma$
2. $\mathfrak{a} \in \mathfrak{T}_\sigma \Longrightarrow \mathfrak{a}{+}\boldsymbol{1} := (\mathfrak{a}) \in \mathfrak{T}_\sigma$
3. $\forall n \in \mathbb{N}(\mathfrak{a}_n \in \mathfrak{T}_\sigma) \Longrightarrow (\mathfrak{a}_n)_{n\in\mathbb{N}} \in \mathfrak{T}_\sigma$
4. $\mu < \sigma \,\&\, \forall\mathfrak{x} \in \mathfrak{T}_\mu(\mathfrak{a}_\mathfrak{x} \in \mathfrak{T}_\sigma) \Longrightarrow (\mathfrak{a}_\mathfrak{x})_{\mathfrak{x}\in\mathfrak{T}_\mu} \in \mathfrak{T}_\sigma$

$\mathfrak{T}_{<\omega} := \bigcup_{\sigma<\omega} \mathfrak{T}_\sigma$.

The elements of $\mathfrak{T}_{<\omega}$ are called *tree ordinals* (denoted by $\mathfrak{a}, \mathfrak{b}, \mathfrak{c}$).

Remark. Every $\mathfrak{a} \in \mathfrak{T}_\sigma$ is of the form $(\mathfrak{a}_\iota)_{\iota\in I}$ with

$I \in \{\emptyset, \{0\}, \mathbb{N}\} \cup \{\mathfrak{T}_\mu : \mu < \sigma\}$.

We define $\|(\mathfrak{a}_\iota)_{\iota\in I}\| := \sup_{\iota\in I}(\|\mathfrak{a}_\iota\| + 1)$.

Abbreviation. $\overline{0} := \boldsymbol{0}$, $\overline{n{+}1} := \overline{n}{+}\boldsymbol{1}$, $\boldsymbol{1} := \overline{1}$, $\Omega_0 := (\overline{n})_{n\in\mathbb{N}}$, $\Omega_{\mu+1} := (\mathfrak{x})_{\mathfrak{x}\in\mathfrak{T}_\mu}$

DEFNITION. For any $\mathfrak{a}, \mathfrak{b} \in \mathfrak{T}_{<\omega}$ and $n \in \mathbb{N}$ we define the tree ordinals $\mathfrak{a} + \mathfrak{b}$ and $\mathfrak{a} \cdot n$ as follows:

$\mathfrak{a} + \boldsymbol{0} := \mathfrak{a}$, $\mathfrak{a} + (\mathfrak{b}_\mathfrak{x})_{\mathfrak{x}\in I} := (\mathfrak{a} + \mathfrak{b}_\mathfrak{x})_{\mathfrak{x}\in I}$ if $I \neq \emptyset$,

$\mathfrak{a} \cdot 0 := \boldsymbol{0}$, $\mathfrak{a} \cdot (n{+}1) := (\mathfrak{a} \cdot n) + \mathfrak{a}$.

PROPOSITION. (*a*) $\mathfrak{a}, \mathfrak{b} \in \mathfrak{T}_\sigma \Rightarrow \mathfrak{a} + \mathfrak{b} \in \mathfrak{T}_\sigma$, (*b*) $\mathfrak{a} + (\mathfrak{b} + \mathfrak{c}) = (\mathfrak{a} + \mathfrak{b}) + \mathfrak{c}$.

DEFNITION. We now define functions $\mathfrak{D}_\sigma : \mathfrak{T}_{<\omega} \longrightarrow \mathfrak{T}_\sigma$. The definition of $\mathfrak{D}_\sigma(\mathfrak{a})$ proceeds by transfinite recursion on $\mathfrak{a}$ simultaneously for all $\sigma < \omega$.

$\mathfrak{D}_0(\mathit{0}) := \mathit{1}$, $\mathfrak{D}_\sigma(\mathit{0}) := \Omega_\sigma$ if $\sigma \neq 0$

$\mathfrak{D}_\sigma(\mathfrak{a}+\mathit{1}) := \big(\mathfrak{D}_\sigma(\mathfrak{a}) \cdot (n+1)\big)_{n\in\mathbb{N}}$

$$\mathfrak{D}_\sigma((\mathfrak{a}_\mathfrak{x})_{\mathfrak{x}\in I}) := \begin{cases} \big(\mathfrak{D}_\sigma(\mathfrak{a}_\mathfrak{x})\big)_{\mathfrak{x}\in I} & \text{if } I \in \{\mathbb{N}\} \cup \{\mathfrak{T}_\mu : \mu<\sigma\} \\ \big(\mathfrak{D}_\sigma(\mathfrak{a}_{\mathfrak{x}_n})\big)_{n\in\mathbb{N}} & \text{if } I = \mathfrak{T}_\mu \text{ with } \mu \geq \sigma \\ \text{with } \mathfrak{x}_0 := \Omega_\mu,\ \mathfrak{x}_{n+1} := \mathfrak{D}_\mu(\mathfrak{a}_{\mathfrak{x}_n}) & \end{cases}$$

Remark.

1. For $\mathfrak{a} = (\mathfrak{a}_\mathfrak{x})_{\mathfrak{x}\in I} \in \mathfrak{T}_\sigma \setminus \{0\}$ we have

$$\mathfrak{D}_\sigma(\mathfrak{a}) = \begin{cases} \big(\mathfrak{D}_\sigma(\mathfrak{a}_0) \cdot (n+1)\big)_{n\in\mathbb{N}} & \text{if } I = \{0\} \\ \big(\mathfrak{D}_\sigma(\mathfrak{a}_\mathfrak{x})\big)_{\mathfrak{x}\in I} & \text{otherwise} \end{cases}.$$

This means that on $\mathfrak{T}_\sigma$ the function $\mathfrak{D}_\sigma$ behaves like the ordinal function $\alpha \mapsto \omega^{\Omega_\sigma+\alpha}$ (if $\sigma > 0$) or $\alpha \mapsto \omega^\alpha$ (if $\sigma = 0$).

2. The canonical analogue to the collapsing function $\mathcal{D}_\sigma$ from §2 would be

$$\mathfrak{D}_\sigma\big((\mathfrak{a}_\mathfrak{x})_{\mathfrak{x}\in I}\big) := \begin{cases} \mathfrak{D}_\sigma \mathfrak{a}_{\mathfrak{D}_\mu(\mathfrak{a}_0)} & \text{if } I = \mathfrak{T}_\mu \text{ with } \mu \geq \sigma \\ \big(\mathfrak{D}_\sigma(\mathfrak{a}_\mathfrak{x})\big)_{\mathfrak{x}\in I} & \text{otherwise} \end{cases}$$

But we have chosen the above version of $\mathfrak{D}_\sigma$, since this precisely corresponds to ψ_σ and at the same time is not to far removed from $\mathcal{D}_\sigma$.

The following definition and lemma are auxiliary.

DEFNITION.

$$\mathfrak{a}^\ominus := \begin{cases} \mathfrak{a}_0 & \text{if } \mathfrak{a} = \mathfrak{a}_0+\mathit{1} \text{ or } \mathfrak{a} = (\mathfrak{a}_i)_{i\in\mathbb{N}} \\ \mathfrak{a}_{\Omega_\mu} & \text{if } \mathfrak{a} = (\mathfrak{a}_\mathfrak{x})_{\mathfrak{x}\in\mathfrak{T}_\mu} \end{cases}$$

$\mathfrak{b} \ll^0 \mathfrak{a} :\iff (\mathfrak{a} \neq \mathit{0} \,\&\, \mathfrak{b} = \mathfrak{a}^\ominus)$ or $(\mathfrak{a} = (\mathfrak{a}_i)_{i\in\mathbb{N}} \,\&\, \exists i\in\mathbb{N}(\mathfrak{b} = \mathfrak{a}_i))$

$\ll$ ($\lll$, resp.) is the transitive (transitive and reflexive, resp.) closure of $\ll^0$.

LEMMA 3.1.

(a) $\mathfrak{a} \neq \mathit{0} \Rightarrow (\mathfrak{c} + \mathfrak{a})^\ominus = \mathfrak{c} + \mathfrak{a}^\ominus \,\&\, \mathfrak{D}_\sigma(\mathfrak{a})^\ominus = \mathfrak{D}_\sigma(\mathfrak{a}^\ominus)$

(b) $\mathit{1} \underline{\ll} \mathfrak{a}$ *if* $\mathfrak{a} \neq \mathit{0}$

(c) $\mathfrak{b} \ll \mathfrak{a} \Rightarrow \mathfrak{c} + \mathfrak{b} \ll \mathfrak{c} + \mathfrak{a}$

(d) $\mathfrak{b} \ll \mathfrak{a} \Rightarrow \mathfrak{D}_\sigma \mathfrak{b} \ll \mathfrak{D}_\sigma \mathfrak{a}$

(e) $\overline{n} \ll \Omega_\sigma \ll \Omega_{\sigma+1}$

DEFNITION. $d \lhd \mathfrak{a}$ ($\mathfrak{a}$ *majorizes* d) if, and only if, one of the following clauses holds:

($\lhd$1) $d = \mathcal{I}(d_i)_{i\in|\mathcal{I}|}$ with $\mathcal{I} \neq \widetilde{\Omega}_P$ and $\mathfrak{a} = \mathfrak{b}+\mathit{1}$ with $d_i \lhd \mathfrak{b}$ for all $i \in |\mathcal{I}|$

($\lhd$2) $d = \widetilde{\Omega}_{P_\mu}(d_q)_{q\in\{0\}\cup|P_\mu|}$ and $\mathfrak{a} = (\mathfrak{a}_\mathfrak{x})_{\mathfrak{x}\in\mathfrak{T}_\mu}$ with $\forall q \in \{0\}\cup|P_\mu|\forall \mathfrak{x} \in \mathfrak{T}_\mu(q \lhd \mathfrak{x} \Rightarrow d_q \lhd \mathfrak{a}_\mathfrak{x})$

($\lhd$3) $d \lhd \mathfrak{b} \,\&\, \mathfrak{b} \ll \mathfrak{a}$

(By convention $0 \lhd \mathfrak{a}$ for any $\mathfrak{a}$.)

LEMMA 3.2. $d \lhd \mathfrak{a} \,\&\, \mathfrak{a} \in \mathfrak{T}_0 \Longrightarrow \|d\| \leq \|\mathfrak{a}\|$.

THEOREM 3.3.

(a) $d \lhd \mathfrak{a} \Longrightarrow \mathcal{J}^k_C(d) \lhd \mathfrak{a}$
(b) $d \lhd \mathfrak{a} \Longrightarrow \mathcal{S}^{\Pi}_{\mathcal{P},\mathcal{F}}(d) \lhd \Omega_\sigma + \mathfrak{a}$ *for each* σ
(c) $e \lhd \mathfrak{b} \,\&\, d \lhd \mathfrak{a} \Longrightarrow \mathcal{R}_C(e,d) \lhd \mathfrak{b} + \mathfrak{a}$
(d) $d \lhd \mathfrak{a} \in \mathfrak{T}_\nu \Longrightarrow \mathcal{E}(d) \lhd \mathfrak{D}_\nu(\mathfrak{a})$
(e) $d \lhd \mathfrak{a} \Longrightarrow \mathcal{D}_\sigma(d) \lhd \mathfrak{D}_\sigma(\mathfrak{a})$

PROOF BY INDUCTION ON $\mathfrak{a}$. We only carry out the essential cases of (c), (d), and (e).

(c) 1. $d = \mathsf{Ax}_{\{\neg C, C\}}$: $\mathcal{R}(e,d) = e \lhd \mathfrak{b} \underline{\ll} \mathfrak{b} + \mathfrak{a}$.
2. $d = \bigvee^k_{\neg C} d_0 \,\&\, \mathfrak{a} = \mathfrak{a}_0 + \mathbf{1} \,\&\, d_0 \lhd \mathfrak{a}_0$:
$\mathcal{R}(e,d_0) \overset{\text{IH}}{\lhd} \mathfrak{b} + \mathfrak{a}_0 \,\&\, \mathcal{J}(e) \overset{\text{(a)}}{\lhd} \mathfrak{b} \underline{\ll} \mathfrak{b} + \mathfrak{a}_0 \Longrightarrow$
$\mathcal{R}(e,d) = \mathsf{Cut}\mathcal{J}(e)\mathcal{R}(e,d_0) \lhd (\mathfrak{b} + \mathfrak{a}_0) + \mathbf{1} = \mathfrak{b} + \mathfrak{a}$.
3. $d = \widetilde{\Omega}_{P_\mu}(d_q)_{q \in I} \,\&\, \mathfrak{a} = (\mathfrak{a}_\mathfrak{x})_{\mathfrak{x} \in \mathfrak{T}_\mu} \,\&\, \forall q \in I \forall \mathfrak{x} \in \mathfrak{T}_\mu (q \lhd \mathfrak{x} \Rightarrow d_q \lhd \mathfrak{a}_\mathfrak{x})$:
IH $\Longrightarrow \forall q \in I \forall \mathfrak{x} \in \mathfrak{T}_\mu (q \lhd \mathfrak{x} \Rightarrow \mathcal{R}(e,d_q) \lhd \mathfrak{b} + \mathfrak{a}_\mathfrak{x}) \Longrightarrow$
$\mathcal{R}(e,d) = \widetilde{\Omega}_{P_\mu}\big(\mathcal{R}(e,d_q)\big)_{q \in I} \lhd (\mathfrak{b} + \mathfrak{a}_\mathfrak{x})_{\mathfrak{x} \in \mathfrak{T}_\mu} = \mathfrak{b} + \mathfrak{a}$.

(d) 1. $d = \mathsf{Cut}_C d_0 d_1$ with $C \in \bigwedge^+$-For, and $\mathfrak{a} = \mathfrak{a}_0 + \mathbf{1} \,\&\, d_0, d_1 \lhd \mathfrak{a}_0$:
IV $\Longrightarrow \mathcal{E}(d_i) \lhd \mathfrak{D}_\nu(\mathfrak{a}_0) \overset{\text{(c)}}{\Longrightarrow} \mathcal{E}(d) = \mathcal{R}_C(\mathcal{E}(d_0), \mathcal{E}(d_1)) \lhd \mathfrak{D}_\nu(\mathfrak{a}_0) + \mathfrak{D}_\nu(\mathfrak{a}_0) = \mathfrak{D}_\nu(\mathfrak{a}_0) \cdot 2 \Longrightarrow \mathcal{E}(d) \lhd \big(\mathfrak{D}_\nu(\mathfrak{a}_0) \cdot (n+1)\big)_{n \in \mathbb{N}} = \mathfrak{D}_\nu(\mathfrak{a})$.
2. $d = \widetilde{\Omega}_{P_\mu}(d_q)_{q \in \{0\} \cup |P_\mu|} \,\&\, \mathfrak{a} = (\mathfrak{a}_\mathfrak{x})_{\mathfrak{x} \in \mathfrak{T}_\mu} \,\&$
$\forall q \in \{0\} \cup |P_\mu| \forall \mathfrak{x} \in \mathfrak{T}_\mu (q \lhd \mathfrak{x} \Rightarrow d_q \lhd \mathfrak{a}_\mathfrak{x})$:
Since $\mathfrak{a} \in \mathfrak{T}_\nu$, we have $\mu < \nu$ and $\mathfrak{D}_\nu(\mathfrak{a}) = \big(\mathfrak{D}_\nu(\mathfrak{a}_\mathfrak{x})\big)_{\mathfrak{x} \in \mathfrak{T}_\mu}$.
IH $\Longrightarrow \forall q \in \{0\} \cup |P_\mu| \forall \mathfrak{x} \in \mathfrak{T}_\mu (q \lhd \mathfrak{x} \Rightarrow \mathcal{E}(d_q) \lhd \mathfrak{D}_\nu(\mathfrak{a}_\mathfrak{x})) \overset{\text{Def}}{\Longrightarrow}$
$\mathcal{E}(d) = \widetilde{\Omega}_{P_\mu}(\mathcal{E}(d_q))_{q \in \{0\} \cup |P_\mu|} \lhd \big(\mathfrak{D}_\nu(\mathfrak{a}_\mathfrak{x})\big)_{\mathfrak{x} \in \mathfrak{T}_\mu}$.

(e) 1. $d = \mathcal{I}(d_i)_{i \in I}$ with $\mathcal{I} \neq \widetilde{\Omega}_P$ and $\mathfrak{a} = \mathfrak{b} + \mathbf{1}$ with $d_i \lhd \mathfrak{b}$ for all $i \in I$:
IH $\Longrightarrow \forall i (\mathcal{D}_\sigma(d_i) \lhd \mathfrak{D}_\sigma(\mathfrak{b})) \Longrightarrow$
$\mathcal{D}_\sigma(d) = \mathcal{I}(\mathcal{D}_\sigma(d_i))_{i \in I} \lhd \mathfrak{D}_\sigma(\mathfrak{b}) + \mathbf{1} \ll \mathfrak{D}_\sigma(\mathfrak{b}) + \Omega_\sigma \underline{\ll} \mathfrak{D}_\sigma(\mathfrak{b}) + \mathfrak{D}_\sigma(\mathfrak{b}) \underline{\ll} \mathfrak{D}_\sigma(\mathfrak{b} + \mathbf{1})$.
2. $d = \widetilde{\Omega}_{P_\mu}(d_q)_{q \in I} \,\&\, \mathfrak{a} = (\mathfrak{a}_\mathfrak{x})_{\mathfrak{x} \in \mathfrak{T}_\mu} \,\&\, \forall q \in I \forall \mathfrak{x} \in \mathfrak{T}_\mu (q \lhd \mathfrak{x} \Rightarrow d_q \lhd \mathfrak{a}_\mathfrak{x})$:
2.1. $\mu < \sigma$: IH $\Longrightarrow \forall q \in I \forall \mathfrak{x} \in \mathfrak{T}_\mu (q \lhd \mathfrak{x} \Rightarrow \mathcal{D}_\sigma(d_q) \lhd \mathfrak{D}_\sigma(\mathfrak{a}_\mathfrak{x})) \Longrightarrow$
$\mathcal{D}_\sigma(d) = \widetilde{\Omega}_{P_\mu}(\mathcal{D}_\sigma(d_q))_{q \in I} \lhd (\mathfrak{D}_\sigma(\mathfrak{a}_\mathfrak{x}))_{\mathfrak{x} \in \mathfrak{T}_\mu} = \mathfrak{D}_\sigma(\mathfrak{a})$.
2.2. $\mu \geq \sigma$: Then $\mathcal{D}_\sigma(d) = \mathcal{D}_\sigma(d_{\mathcal{D}_\mu(d_0)})$ and $\mathfrak{D}_\sigma(\mathfrak{a}) = (\mathfrak{D}_\sigma(\mathfrak{a}_{\mathfrak{x}_n}))_{n \in \mathbb{N}}$ with
$\mathfrak{x}_0 = \Omega_\mu$, $\mathfrak{x}_{n+1} = \mathfrak{D}_\mu(\mathfrak{a}_{\mathfrak{x}_n})$.
$0 \lhd \mathfrak{x}_0 \Longrightarrow d_0 \lhd \mathfrak{a}_{\mathfrak{x}_0} \overset{\text{IH}}{\Longrightarrow} q := \mathcal{D}_\mu(d_0) \lhd \mathfrak{D}_\mu(\mathfrak{a}_{\mathfrak{x}_0}) = \mathfrak{x}_1 \Longrightarrow d_q \lhd \mathfrak{a}_{\mathfrak{x}_1} \overset{\text{IH}}{\Longrightarrow}$
$\mathcal{D}_\sigma(d) = \mathcal{D}_\sigma(d_q) \lhd \mathfrak{D}_\sigma(\mathfrak{a}_{\mathfrak{x}_1}) \ll \big(\mathfrak{D}_\sigma(\mathfrak{a}_{\mathfrak{x}_i})\big)_{i \in \mathbb{N}} = \mathfrak{D}_\sigma(\mathfrak{a})$. ⊣

THEOREM 3.4 (Embedding). *For each closed* ID_ν*-derivation* h *we have*

$$h^\infty \lhd \Omega_\nu{\cdot}2 + n(h),$$

where $n(\mathcal{I}h_0 \dots h_{m-1}) := \max\{0, n(h_0), \dots, n(h_{m-1})\} + 1.$

PROOF. By definition

$$(\mathsf{Ind}_{\mathcal{F}}^{\mathcal{P}.n})^\infty = \frac{\mathsf{Ax}_{\{\neg\mathcal{P}n.\mathcal{P}n\}} \quad \dots \mathcal{S}_{\mathcal{P}.\mathcal{F}}^{\{\mathcal{P}n\}}(q) \dots (q \in |\mathcal{P}n|)}{\widetilde{\Omega}_{\mathcal{P}n}}.$$

By Theorem 3.3b we have $\forall q \in |P_\mu| \forall \mathfrak{x} \in \mathfrak{T}_\mu (q \lhd \mathfrak{x} \Rightarrow \mathcal{S}_{\mathcal{P}.\mathcal{F}}^{\{\mathcal{P}n\}}(q) \lhd \Omega_\nu + \mathfrak{x})$ which together with $\forall \mathfrak{x} \in \mathfrak{T}_\mu (\mathsf{Ax}_{\{\neg\mathcal{P}n.\mathcal{P}n\}} \lhd \Omega_\nu + \mathfrak{x})$ yields

$$(\mathsf{Ind}_{\mathcal{F}}^{\mathcal{P}.n})^\infty \lhd \Omega_\nu + \Omega_{\mathrm{lev}(\mathcal{P})+1} \underline{\ll} \Omega_\nu + \Omega_\nu = \Omega_\nu{\cdot}2.$$

The other cases are easy. ⊣

THEOREM 3.5. *Let* $\nu > 0$. *If* h *is a closed* ID_ν*-derivation of* Γ *with* $\mathrm{lev}(\Gamma) = 0$ *then* $(\mathfrak{N}, \Phi^{<\alpha}) \models \Gamma$ *with* $\alpha = \|\mathfrak{D}_0(\mathfrak{D}_\nu^m(\mathit{0})\|$ *for some* $m \in \mathbb{N}$.

PROOF. Theorem 2.5 $\Rightarrow$ $(\mathfrak{N}, \Phi^{<\alpha}) \models \Gamma$ with $\alpha = \|\mathcal{D}_0(\mathcal{E}^m(h^\infty))\|$ for some $m < \omega$. $h^\infty \overset{\text{Th.3.4}}{\lhd} \Omega_\nu{\cdot}2 + n \overset{\text{L.3.1c.e}}{\ll} \Omega_\nu{\cdot}3 \overset{\text{Def}}{\ll} \mathfrak{D}_\nu(\mathit{1}) \overset{\text{L.3.1b.d}}{\underline{\ll}} \mathfrak{D}_\nu\mathfrak{D}_\nu(\mathit{0}) \overset{\text{Th.3.3d.e}}{\Longrightarrow}$ $\mathcal{D}_0(\mathcal{E}^m(h^\infty)) \lhd \mathfrak{D}_0\mathfrak{D}_\nu^{m+2}(\mathit{0}) \overset{\text{L.3.2}}{\Longrightarrow} \|\mathcal{D}_0(\mathcal{E}^m(h^\infty))\| \leq \|\mathfrak{D}_0\mathfrak{D}_\nu^{m+2}(\mathit{0})\|$. ⊣

COROLLARY. $|\mathrm{ID}_\nu| \leq \sup_{m\in\mathbb{N}} \|\mathfrak{D}_0(\mathfrak{D}_\nu^m(\mathit{0})\|$

Now we are going to prove that $\|\mathfrak{D}_0(\mathfrak{D}_\nu^m(\mathit{0})\|$ equals $\psi_0\psi_\nu^m(0)$. By comparing the definition of $\mathfrak{D}_\sigma$ with the assignment of fundamental sequences in §1 and taking Theorem 1.5 and $(\Diamond)$ into consideration this should be more or less clear. In order to obtain a rigorous proof we introduce the canonical interpretation $\mathfrak{t} : \mathrm{T} \longrightarrow \mathfrak{T}_{<\omega}$ and show that this respects the fundamental sequences $(a[x])_{x\in|\mathrm{tp}(a)}$.

DEFNITION. The mapping $\mathfrak{t} : \mathrm{T} \longrightarrow \mathfrak{T}_{<\omega}$ is defined by

$$\mathfrak{t}(D_{\sigma_0}a_0 \oplus \dots \oplus D_{\sigma_{n-1}}a_{n-1}) := \mathfrak{D}_{\sigma_0}\mathfrak{t}(a_1) + \dots + \mathfrak{D}_{\sigma_{n-1}}\mathfrak{t}(a_{n-1})$$

THEOREM 3.6. *For each* $a \in \mathrm{T}$ *we have*

(i) $\mathrm{tp}(a) = 1 \Longrightarrow \mathfrak{t}(a) = \mathfrak{t}(a[0]) + 1$,
(ii) $\mathrm{tp}(a) = \omega \Longrightarrow \mathfrak{t}(a) = (\mathfrak{t}(a[n]))_{n\in\mathbb{N}}$,
(iii) $\mathrm{tp}(a) = \Omega_{\mu+1} \Longrightarrow \mathfrak{t}(a) = (\mathfrak{a}_\mathfrak{x})_{\mathfrak{x}\in\mathfrak{T}_\mu}$ *with* $\forall x \in |\Omega_{\mu+1}| (\mathfrak{t}(a[x]) = \mathfrak{a}_{\mathfrak{t}(x)})$

PROOF. Let $\mathcal{FS}(a)$ abbreviate the claim (i)&(ii)&(iii). Then in a straightforward way one proves

(1) $\mathcal{FS}(a) \,\&\, \mathcal{FS}(b) \Longrightarrow \mathcal{FS}(b \oplus a)$, (2) $\mathcal{FS}(a) \Longrightarrow \mathcal{FS}(D_\sigma a)$,

from which one obtains $(\forall a \in \mathrm{T})\mathcal{FS}(a)$ by induction on the build up of a. ⊣

THEOREM 3.7. $a \in \mathrm{OT}_0 \Longrightarrow \overline{\mathrm{o}}(a) = \|\mathfrak{t}(a)\|$

PROOF BY INDUCTION ON $\mathfrak{t}(a)$. By L.1.6 $\mathrm{tp}(a) \in \{0, 1, \omega\}$. If $\mathrm{tp}(a) = 0$ then $a = 0$ and $\mathfrak{t}(a) = \mathit{0}$. If $\mathrm{tp}(a) = \omega$ then $\mathfrak{t}(a) \overset{3.6}{=} (\mathfrak{t}(a[n]))_{n\in\mathbb{N}}$ and therefore $\|\mathfrak{t}(a)\| = \sup_{n\in\mathbb{N}}(\|\mathfrak{t}(a[n])\| + 1) \overset{\text{IH}}{=} \sup_{n\in\mathbb{N}}(\overline{\mathrm{o}}(a[n]) + 1) \overset{\text{L.1.6}}{=} \overline{\mathrm{o}}(a)$.

The case "$\mathrm{tp}(a) = 1$" is treated in the same way. ⊣

COROLLARY.
(a) $\|\mathfrak{D}_0\mathfrak{D}_\nu^m(\mathit{0})\| = \psi_0\psi_\nu^m(0)$
(b) $|\mathrm{ID}_\nu| \leq \sup_{m\in\mathbb{N}} \psi_0\psi_\nu^m(0) = \psi_0(\varepsilon_{\Omega_\nu+1})$

PROOF. We have $|\mathrm{ID}_\nu| \leq \sup_{m\in\mathbb{N}} \|\mathfrak{D}_0(\mathfrak{D}_\nu^m(\mathit{0})\|$ and $\|\mathfrak{D}_0\mathfrak{D}_\nu^m(\mathit{0})\| \overset{\text{Def}}{=} \|\mathrm{t}(D_0D_\nu^m0)\| \overset{3.7}{=} \overline{\mathrm{o}}(D_0D_\nu^m0) \leq \mathrm{o}(D_0D_\nu^m0) \overset{\text{Def}}{=} \psi_0\psi_\nu^m(0)$. This yields (b), and (a) with $\leq$ in place of $=$. To get $=$ in (a) we have to use ($\diamondsuit$) which implies $\overline{\mathrm{o}}(a) = \mathrm{o}(a)$ for all $a \in \mathrm{OT}$ (cf. §1, pg.41). ⊣

§4. Two applications. Let $\widehat{\mathrm{T}} := \{a \in \mathrm{T} : a \text{ principal term}\}$ and $\widehat{\mathrm{OT}} := \mathrm{OT}\cap\widehat{\mathrm{T}}$. As one easily sees, the set $\widehat{\mathrm{T}}$ can be inductively generated by:

$$a_0, \ldots, a_{n-1} \in \widehat{\mathrm{T}}\ (n \geq 0)\ \&\ \sigma < \omega \Longrightarrow D_\sigma(a_0\oplus\ldots\oplus a_{n-1}) \in \widehat{\mathrm{T}}.$$

Hence $\widehat{\mathrm{T}}$ is nothing other than the set of all finite, ordered trees with labels $\sigma < \omega$, and each term $a = a_0\oplus\ldots\oplus a_{n-1} \in \mathrm{T}$ can be considered as a tree with immediate subtrees $a_0, \ldots, a_{n-1} \in \widehat{\mathrm{T}}$ and an unlabeled root. The assignment of (fundamental) sequences $(a[x])_{x\in|\mathrm{tp}(a)|}$ can then be seen as the definition of a reduction procedure (or rewriting relation) $a \hookrightarrow_x a[x]$ on T. In [9] this reduction procedure (restricted to $\mathrm{T}_0 := \{D_0a_0\oplus\ldots\oplus D_0a_{n-1} : a_0, \ldots, a_{n-1} \in \mathrm{T}\}$) had been cooked up as a so-called hydra game, where in the ith round of the game (or battle) the hydra a transforms itself into a new hydra $a[n_i]$. Using Theorem 3.6 and Theorem 3.5 one easily concludes that the hydra game terminates (i.e., $\forall a \in \mathrm{T}_0\forall(n_i)_{i\in\mathbb{N}}\exists k(a[n_0][n_1]\ldots[n_k] = 0)$), and that this fact is not provable in $\mathrm{ID}_{<\omega}$:

Let W_0 be inductively defined by the rule:

$$a \in \mathrm{T}_0\ \&[a \neq 0 \Rightarrow \forall n(a[n] \in W_0)] \Longrightarrow a \in W_0.$$

Then "$a \in W_0$" says that each $\hookrightarrow$-reduction sequence starting with a terminates. Hence "$\forall a \in \mathrm{T}_0(a \in W_0)$" expresses termination of the hydra game. Now using Theorem 3.6 by induction on $\mathrm{t}(a)$ we get $\forall a \in \mathrm{T}_0(a \in W_0\ \&\ |a|_{W_0} = \|\mathrm{t}(a)\|)$.

The unprovability result is obtained as follows:

$$\mathrm{ID}_\nu \vdash \forall x(D_0D_\nu^x0 \in W_0) \overset{\text{Th.3.5}}{\Longrightarrow} \exists m\forall n(|D_0D_\nu^n0|_{W_0} < \|\mathfrak{D}_0\mathfrak{D}_\nu^m(0)\|) \Longrightarrow$$
$$\Longrightarrow \exists m(|D_0D_\nu^m0|_{W_0} < \|\mathrm{t}(D_0D_\nu^m0)\| = |D_0D_\nu^m0|_{W_0}). \text{ Contradiction.}$$

Another interesting observation about the system $(\mathrm{OT}, \prec)$ is due to Okada [13] and provides a rather short proof of H. Friedman's result that the extended Kruskal Theorem on finite labeled trees implies the wellfoundedness of $(\mathrm{OT}, \prec)$ (provably in ACA_0). This runs as follows.

First we define a binary relation $\sqsubseteq$ on $\widehat{\mathrm{T}}$ such that $a \sqsubseteq b$ is equivalent to "there exists a homeomorphic embedding $f : a \to b$ satisfying Friedman's *gap condition* (including the gap condition for the root)".

DEFINITION. We define $a \sqsubseteq b$ for $a, b \in \widehat{\mathrm{T}}$. Let $a = D_\rho(a_0 \oplus \ldots \oplus a_{m-1})$ and $b = D_\sigma(b_0 \oplus \ldots \oplus b_{n-1})$.

$a \sqsubseteq b$ iff one of the following two clauses holds

(i) $\rho = \sigma$ and $\exists$ injective $q : \{0, \ldots, m-1\} \to \{0, \ldots, n-1\}$ such that $a_i \sqsubseteq b_{q(i)}$ for $i < m$,

(ii) $\rho \leq \sigma$ and $\exists j < n(a \sqsubseteq b_j)$.

Then we define a relation $\prec^* \subseteq \prec$ and prove $\forall a, b \in \widehat{\mathrm{OT}}(a \sqsubseteq b \Rightarrow a \preceq^* b)$.

DEFINITION. $a \prec^* b :\Leftrightarrow a \prec b \,\&\, \forall \rho(G_\rho a \preceq G_\rho b)$, where

$$X \preceq Y :\Leftrightarrow \forall x \in X \exists y \in Y(x \preceq y).$$

LEMMA 4.1. (a) $a \prec^* b \Longrightarrow D_\sigma a \prec^* D_\sigma b$

(b) $D_\rho a \preceq^* b \,\&\, \rho \leq \sigma \,\&\, G_\sigma b \prec b \Longrightarrow D_\rho a \prec^* D_\sigma b$

PROOF. (a) $a \prec^* b \,\&\, G_\rho D_\sigma a \neq \emptyset \Rightarrow G_\rho D_\sigma a = \{a\} \cup G_\rho a \preceq \{b\} \cup G_\rho b = G_\rho D_\sigma b$.

(b) 1. $D_\rho a \preceq^* b \,\&\, \mu \leq \rho \leq \sigma \Rightarrow G_\mu D_\rho a \preceq G_\mu b \subseteq G_\mu D_\sigma b$. Hence $\forall \mu(G_\mu D_\rho a \preceq G_\mu D_\sigma b)$.

2. Proof of $D_\rho a \prec D_\sigma b$: Let $\rho = \sigma$ (otherwise the claim is trivial). Then $a \in G_\sigma(D_\rho a) \preceq G_\sigma b \prec b$. ⊣

THEOREM 4.2. $a, b \in \widehat{\mathrm{OT}} \,\&\, a \sqsubseteq b \Longrightarrow a \preceq b$

PROOF. By induction on $\ell(b)$ we prove the stronger statement $a \preceq^* b$. Let

$$a = D_\rho(a_0 \oplus \ldots \oplus a_{m-1}) \text{ and } b = D_\sigma(b_0 \oplus \ldots \oplus b_{n-1}).$$

(i) $\rho = \sigma \,\&\, \forall i < m(a_i \sqsubseteq b_{q(i)}) \,\&\, \forall i, j < m(i \neq j \Rightarrow q(i) \neq q(j))$: By IH we have $a_i \preceq^* b_{q(i)}$ for $i < m$. From this we get $(a_0 \oplus \ldots \oplus a_{m-1}) \preceq^* (b_0 \oplus \ldots \oplus b_{n-1})$ and then by L.4.1a $a = D_\sigma(a_0 \oplus \ldots \oplus a_{m-1}) \preceq^* D_\sigma(b_0 \oplus \ldots \oplus b_{n-1}) = b$.

(ii) $\rho \leq \sigma$ and $\exists j < n(a \sqsubseteq b_j)$: By IH we have $a \preceq^* b_j \preceq^* (b_0 \oplus \ldots \oplus b_{n-1}) =: c$. Since $b = D_\sigma c \in \mathrm{OT}$, we also have $G_\sigma c \prec c$. By L.4.1b this yields $a = D_\rho(a_0 \oplus \ldots \oplus a_{m-1}) \prec^* D_\sigma c = b$. ⊣

REFERENCES

[1] PETER ACZEL, *A new approach to the Bachmann method for describing large countable ordinals*, unpublished.

[2] TOSHIYASU ARAI, *Proof theory for theories of ordinals II: Σ_1-stability*, Preprint (1997).

[3] ———, *Systems of ordinal diagrams*, preprint (1996).

[4] JANE BRIDGE, *A simplification of the Bachmann method for generating large countable ordinals*, ***The Journal of Symbolic Logic***, vol. 40 (1975), pp. 171–185.

[5] WILFRIED BUCHHOLZ, *Normalfunktionen und konstruktive Systeme von Ordinalzahlen*, ***Proof theory symposium, Kiel 1974*** (J. Diller and G.H. Müller, editors), Lecture Notes in Mathematics, vol. 500, Springer, 1975, pp. 4–25.

[6] ———, *The $\Omega_{\mu+1}$-rule*, ***Iterated inductive definitions and subsystems of analysis: Recent proof-theoretical studies*** (W. Buchholz, S. Feferman, W. Pohlers, and W. Sieg, editors), Lecture Notes in Mathematics, vol. 897, Springer, 1981, pp. 188–233.

[7] ———, *Ordinal analysis of* ID_ν, ***Iterated inductive definitions and subsystems of analysis: Recent proof–theoretical studies*** (W. Buchholz, S. Feferman, W. Pohlers, and W. Sieg, editors), Lecture Notes in Mathematics, vol. 897, Springer, 1981, pp. 234–260.

[8] ———, *A new system of proof-theoretic ordinal functions*, ***Annals of Pure and Applied Logic***, vol. 32 (1986), pp. 195–207.

[9] ———, *An independence result for* $(\Pi^1_1\text{-CA})+\mathrm{BI}$, ***Annals of Pure and Applied Logic***, vol. 33 (1987), pp. 131–155.

[10] Wilfried Buchholz and Kurt Schütte, ***Proof theory of impredicative subsystems of analysis***, Studies in Proof Theory Monographs, vol. 2, Bibliopolis, Napoli, 1988.

[11] Solomon Feferman, *Proof theory: A personal report*, ***Proof theory (2nd edition)*** (G. Takeuti, editor), Studies in Logic and the Foundations of Mathematics, North-Holland, 1987, pp. 447–485.

[12] Gerhard Jäger, *ρ-inaccessible ordinals, collapsing functions and a recursive notation system*, ***Archiv für mathematische Logik***, vol. 24 (1984), pp. 49–62.

[13] Mitsuhara Okada, *Note on H. Friedman's independence result on the extended Kruskal Theorem*, preprint (1988).

[14] Wolfram Pohlers, *Ordinal notations based on a hierarchy of inaccessible cardinals*, ***Annals of Pure and Applied Logic***, vol. 33 (1987), pp. 157–179.

[15] ———, *Subsystems of set theory and second order number theory*, ***Handbook of proof theory*** (S. Buss, editor), Elsevier Science B.V., 1998.

[16] Michael Rathjen, *Ordinal notations based on a weakly Mahlo cardinal*, ***Archive for Mathematical Logic***, vol. 29 (1990), pp. 249–263.

[17] ———, *Proof theory of reflection*, ***Annals of Pure and Applied Logic***, vol. 68 (1994), pp. 181–224.

[18] ———, *Recent advances in ordinal analysis:* Π^1_2-CA *and related systems*, ***The Bulletin of Symbolic Logic***, vol. 1/4 (1995), pp. 468–485.

[19] Anton Setzer, *Well-ordering proofs for Martin-Löf type theory*, ***Annals of Pure and Applied Logic***, vol. 92 (1998), pp. 113–159.

MATHEMATISCHES INSTITUT
LUDWIG-MAXIMILIANS-UNIVERSITÄT MÜNCHEN
THERESIENSTR. 39
D-80333 MÜNCHEN, GERMANY
E-mail: buchholz@rz.mathematik.uni-muenchen.de

INTERNAL FINITE TREE EMBEDDINGS

HARVEY M. FRIEDMAN

Introduction. The present author demonstrated the first example of a predicatively meaningful but predicatively unprovable theorem from the standard mathematical literature. The example was J. B. Kruskal's tree embedding theorem concerning infinite sequences of finite trees. See [Kr60] and [NW65] for proofs of this theorem, and [Si85] for the independence result (based on [Fr81], [Fr82]).

We also gave a finite form of Kruskal's theorem involving finite sequences of finite trees. See [Si85] based on [Fr81] and [Fr82], and also [Sm85] for further finite forms.

Here we present a theorem about a single sufficiently large finite tree—the Internal Finite Tree Embedding Theorem(s)—that is arguably a simpler, more elegant, and more natural theorem about finite objects, that is predicatively unprovable.

These conclusions about predicative unprovability are based on the Feferman/Schütte analysis of predicative provability, which first appeared independently in [Fe64], [Sch65]. Conceptually, Feferman's analysis has evolved over the years but has always arrived at formal systems based on the proof theoretic ordinal Γ_0. Feferman's extensive analysis can be seen in [Fe64], [Fe68a], [Fe68b], [Fe79], [Fe91], [Fe96], and [FeS00].

There is some leeway here in that the Internal Finite Tree Embedding Theorem (1,2,3,5) are independent of somewhat stronger systems than those adopted by Feferman in his analysis of predicativity. First of all, we can immediately strengthen predicative unprovability (as analyzed by Feferman/Schütte) with ATR_0 in light of our result that ATR_0 and Feferman's systems of predicative analysis prove the same arithmetic sentences (in fact, the same $\prod_1^1$ sentences). See [Si99] for a treatment of ATR_0, and [FMS82] for our conservation result.

In fact, the Internal Finite Tree Embedding Theorem (1,3,5) are all provably equivalent, over EFA, to the 1-consistency of $\prod_2^1-TI_0$ or of $ATI(< \theta\Omega^{\omega}(0))$. Here EFA is exponential function arithmetic, a convenient weak fragment of PA (Peano Arithmetic) that is very useful as a metatheory for stating metamathematical results. EFA is equivalent to the system $I\sum_0(\exp)$ of [HP93], p. 37. $\prod_2^1-TI_0$

This research was partially supported by NSF DMS-9970459.

Reflections on the foundations of mathematics
Essays in honor of Solomon Feferman
Edited by Wilfried Sieg, Richard Sommer, and Carolyn Talcott

is the subsystem of second order arithmetic given by ACA_0+ the scheme of transfinite induction for $\prod_2^1$ formulas applied to well orderings of ω (see [Si99] for background on subsystems of second order arithmetic). In [RW93], this system is referred to as $ACA_0 + \prod_2^1 - BI$, where BI stands for "bar induction." Finally, $\theta\Omega^\omega(0)$ refers to an ordinal in the Feferman Aczel notation system (see [Sch77], p. 224), and $ATI(\theta\Omega^\omega(0))$ refers to the system of PA + the scheme of transfinite induction up to each point below $\theta\Omega^\omega(0)$ for all arithmetic formulas.

These equivalences are proved here using the exact relationship between Kruskal's theorem (and certain variants) and the ordinal (notation system for) $\theta\Omega^\omega(0)$, as well as the relationship between $\prod_2^1 - TI_0$ and $\theta\Omega^\omega(0)$. It was established by the author in [Fr84] that Kruskal's theorem (without labels) is equivalent to "$\theta\Omega^\omega(0)$ is well ordered" over ACA_0, and that this is also true of Higman's theorem (even with finitely many labels). It was also shown there that left/right structure can be used without affecting these results. These results were first published in [RW93]. We also use the exact relationship between $\prod_2^1 - TI_0$ and $\theta\Omega^\omega(0)$. A detailed exposition of this result can be found in [RW93] (based on corresponding notations of Buchholz, known to be equivalent).

The reader may first wish to read Section 10 in order to see a summary of the equivalences and implications proved here.

The present versions of the Internal Finite Tree Embedding Theorem were obtained in June, 2000. They are closely related to those announced and sketched in [Fr98] and [Fr99], but are arguably more attractive than those earlier versions.

Our original independence results reported in [Fr81], [Fr82], [Si85] were based on a connection between proof theoretic ordinals and Kruskal's theorem. We relied on a certain kind of close connection—that one can derive the well foundedness of proof theoretic ordinal notation systems directly from Kruskal's theorem within very weak systems such as RCA_0.

We were not the first to see connections between proof theoretic ordinals and Kruskal's theorem. The earlier connection was through the result of [dJP77] asserting that the linear extensions of any given wpo (well partial order) are well ordered and have a maximum order type. [dJP77] computed some important examples below ϵ_0. This line of investigation was taken up by [Schm79] where ordinals $\leq \theta\Omega^\omega(0)$ appear as the maximum order type of Kruskal's theorem and its restrictions.

Apparently one needs to have our more direct connections in order to obtain the independence results.

§1. The Internal Finite Tree Embedding Theorem. A poset is a pair $(D, \leq)$ where D is a nonempty set and $\leq$ is a reflexive transitive relation obeying

$$(x \leq y \wedge y \leq x) \rightarrow x = y.$$

A tree is a poset $T = (V, \leq)$ where there is a minimum element called the root, and where for each $x \in T$, $\{y : y \leq x\}$ is linearly ordered by $\leq$.

The elements of $V = V(T)$ are called the vertices of T. A tree is said to be finite if it has finitely many vertices.

If $x \leq y$ then we call x a predecessor of y and y a successor of x. If $x < y$ then we call x a strict predecessor of y and y a strict successor of x.

If $x < y$ and there is no z such that $x < z < y$ then we call y an immediate successor of x and x an immediate predecessor of y.

For finite trees, we can define the crucial inf operation on V, where $x \inf y$ is the greatest z such that $z \leq x \wedge z \leq y$.

Let T_1 and T_2 be finite trees. We say that h is an inf preserving embedding from T_1 into T_2 if and only if

i) $h : V(T_1) \to V(T_2)$ is one-one;
ii) for all $x, y \in V(T_1), h(x \inf y) = hx \inf hy$.

The height of a vertex is the number of its strict predecessors. Thus the height of the root is 0. The height of a tree is the least cardinal number that is at least as large as the heights of all of its vertices. For finite trees, the height is the largest of the heights of its vertices.

The valence of a vertex is the number of its immediate successors. The valence of a tree is the least cardinal number that is at least as large as the valences of its vertices. For finite trees, the valence is the largest of the valences of its vertices.

A full tree is a tree where

i) all vertices of valence 0 have the same height;
ii) all vertices of valence > 0 have the same valence.

Thus there is exactly one full finite tree of a given height and valence, up to isomorphism.

We need to impose some extra structure on the trees which is entirely independent of the above definitions. The extra structure is that of a (vertex) labeling. This is a function σ whose domain is the set of vertices of the tree.

Thus a labeled tree is a triple $T = (V, \leq, \sigma)$, where $(T, \leq)$ is a tree and σ is a function with $\mathrm{dom}(\sigma) = V$.

Of particular interest are the n-labeled trees, where $n \geq 1$. These are the labeled trees where $\sigma : V \to \{1, \ldots, n\}$.

Let T_1, T_2 be labeled trees with labelings σ_1, σ_2. A function $h : V(T_1) \to V(T_2)$ is said to be label preserving if and only if for all $x \in V(T_1), \sigma_1(x) = \sigma_2(hx)$.

The following is a special case of a more general theorem of [Kr60].

THEOREM 1.1. Let $n \geq 1$ and $T_1, T_2, \ldots$ be an infinite sequence of finite n-labeled trees. There exist $i < j$ and an inf and label preserving embedding from T_i into T_j.

We actually need a somewhat sharper form of Theorem 1.1 which we easily derive as a consequence.

Let T_1, T_2 be finite trees. We say that $h : V(T_1) \to V(T_2)$ is terminal preserving if and only if for all x in $V(T_1)$ of valence 0 in T_1, $h(x)$ is of valence 0 in T_2.

This terminology makes sense because we can view the vertices of valence 0 as the terminal vertices.

THEOREM 1.2. Let $n \geq 1$ and $T_1, T_2, \ldots$ be an infinite sequence of finite n-labeled trees. There exist $i < j$ and an inf, label, and terminal preserving embedding from T_i into T_j.

PROOF. We use Theorem 1.1 for finite $2n$-labeled trees. Let $T_1, T_2, \ldots$ be as given. Let $T'_1, T'_2, \ldots$ be obtained by changing the label of each vertex of valence 0 by adding n to that label. Then any inf and label preserving embedding from T'_i into T'_j is an inf, label, and terminal preserving embedding from T_i into T_j. ⊣

In the case of full finite n-labeled trees T, there is an especially obvious notion of truncation. The truncations of T are obtained by restricting T to all vertices whose height is at most a given nonnegative integer. Thus the number of truncations of T is exactly one more than the height of T.

We now give our new finite form of Kruskal's theorem, version 1.

THEOREM 1.3. Internal Finite Tree Embedding Theorem(1). Let $k, n \geq 1$ and T be a sufficiently tall full finite n-labeled tree of valence k. There is an inf, label, and terminal preserving embedding from some truncation of T into some truncation of T of greater height.

PROOF. Let $k, n \geq 1$. Suppose this is false. Note that the full finite n-labeled trees of valence k or valence 0 themselves form an infinite tree TR under the relation of truncation. To be careful, we now require that the vertices in every full finite n-labeled tree of valence k or valence 0 used in TR be an initial segment of the positive integers. Then TR forms an infinite finitely branching tree.

Now let TR′ be the full finite n-labeled trees of valence k or valence 0 for which Theorem 1.3 is false, under the relation of truncation. It is easy to see that TR′ is also an infinite finitely branching tree. Hence TR′ has an infinite path.

We can think of this infinite path as an infinite sequence $T_0, T_1, \ldots$ of full finite n-labeled trees of valence k or valence 0 for which Theorem 1.3 fails, where each T_i is the truncation of T_{i+1} of height i.

By Theorem 1.2, let $i < j$ and h be an inf, label, and terminal preserving embedding from T_i into T_j. Then Theorem 1.3 is true for T_j. This is a contradiction. ⊣

We will also consider the following weaker second form, which is merely the case of valence 2.

THEOREM 1.4. Internal Finite Tree Embedding Theorem(2). Let $n \geq 1$ and T be a sufficiently tall full finite n-labeled tree of valence 2. There is an inf, label, and terminal preserving embedding from some truncation of T into some truncation of T of greater height.

Theorems 1.1 and 1.2 are stronger versions of the following theorem, which might be called plain vanilla Kruskal's Theorem.

THEOREM 1.5. Let $T_1, T_2, \ldots$ be an infinite sequence of finite trees. There exist $i < j$ and an inf preserving embedding from T_i into T_j.

The bounded valence case is also important.

THEOREM 1.6. Let $k \geq 1$ and $T_1, T_2, \ldots$ be an infinite sequence of finite trees of valence k. There exist $i < j$ and an inf preserving embedding from T_i into T_j.

We also put labels on Theorem 1.6 as follows.

THEOREM 1.7. Let $k, n \geq 1$ and $T_1, T_2, \ldots$ be an infinite sequence of finite n-labeled trees of valence k. There exist $i < j$ and an inf preserving embedding from T_i into T_j.

Theorems 1.6 and 1.7 are due to Higman [Hi52] in work previous to Kruskal.

Note that the proof of Theorem 1.3 only requires Theorem 1.7 and Theorem 1.4 only requires Theorem 1.7 for valence 2.

$\theta\Omega^{\omega}(0)$ refers to a particular initial segment of the Feferman Aczel ordinal notation system (see [Sch77], p. 224), which we discuss in section 5. $\prod_2^1-\mathrm{TI}_0$ is ACA_0 plus transfinite induction with respect to all $\prod_2^1$ formulas over any well ordering. This is the same as $\mathrm{ACA}_0 + \prod_2^1-\mathrm{BI}$ in [RW93].

$\prod_2^1-\mathrm{TI}$ is the same as $\prod_2^1-\mathrm{TI}_0$ except that induction on ω with respect to all formulas in the language of second order arithmetic is added.

The exact metamathematical status of Theorems 1.5–1.7 is known.

THEOREM 1.8. The following are provably equivalent in RCA_0.

i) Theorem 1.5;
ii) Theorem 1.6;
iii) Theorem 1.7;
iv) $\prod_1^1$ reflection on $\prod_2^1-\mathrm{TI}_0$;
v) $\theta\Omega^{\omega}(0)$ is well ordered.

PROOF. We proved iv $\rightarrow$ i, ii, iii in [Fr84]. See [RW93] for a published proof. We proved that any of i, ii, iii prove v in [Fr84]. See [RW93] for a published proof. iv $\rightarrow$ v was well known in the 80's or earlier. For a proof of v $\rightarrow$ iv, see Pohlers in [BFPS81] or [RW93]. ⊣

Through ordinal analysis of versions of Kruskal's theorem, it was shown in [Fr84] that Theorem 1.1 is stronger than i–v above. In particular, we proved that significantly bigger ordinals in the Feferman Aczel notation system are well ordered.

THEOREM 1.9. Theorems 1.1–1.7 are provable in $\prod_2^1-\mathrm{TI}$. Theorems 1.1–1.7 for trees of any fixed valence k (including Theorem 1.4) are provable in $\prod_2^1-\mathrm{TI}_0$. Theorems 1.1, 1.2, 1.7 for valence 2 imply Γ_0 is well ordered in RCA_0.

PROOF. The first two claims are shown in [Fr84] and (implicitly) in [RW93] for all but Theorems 1.3 and 1.4, which are, anyways, direct consequences as shown above. The third claim is from an ordinal analysis of variants of Kruskal's theorem in [Fr84], where stronger results are obtained. ⊣

We postpone consideration of further relevant metamathematical equivalences until section 10.

§2. Finite sequences of trees. In this section we derive the following finite version of the Higman/Kruskal Theorem from the Internal Finite Tree Embedding Theorem(1).

THEOREM 2.1. For all $k, n \geq 1$ there exists r such that the following holds. Let $T_0, \ldots, T_r$ be full n-labeled trees of valence k or 0, where for all $0 \leq i \leq r$, T_i has height i. There exist $0 \leq i < j \leq r$ and an inf and label preserving embedding from T_i into T_j.

PROOF. Virtually identical to the proof given of Theorem 1.3. Valence 0 is needed to accommodate height 0, which just consists of a root. ⊣

We also derive the following valence 2 version of Theorem 2.1 from the Internal Finite Tree Embedding Theorem(2):

THEOREM 2.2. For all $n \geq 1$ there exists r such that the following holds. Let $T_0, \ldots, T_r$ be full n-labeled trees of valence 2 or 0, where for all $0 \leq i \leq r$, T_i has height i. There exist $0 \leq i < j \leq r$ and an inf and label preserving embedding from T_i into T_j.

We accomplish both of these derivations by proving the following.

THEOREM 2.3. The following is provable in EFA. Let $k \geq 2$. Theorem 1.3 for k is equivalent to Theorem 2.1 for k. In particular, Theorems 1.3 and 2.1 are equivalent in EFA, and Theorems 1.4 and 2.2 are equivalent in EFA.

PROOF. We first handle the more difficult forward direction. Let $k \geq 2$ and assume Theorem 1.3. Towards deriving Theorem 2.1, let $n \geq 1$, and let $r \geq 1$ be such that for all full $4n + k + 2$-labeled trees T of valence k and height $2r + 2$, there is an inf, label, and terminal preserving embedding from some truncation of T into some truncation of T of greater height.

Now let $T_0, \ldots, T_r$ be full finite n-labeled trees of valence k or 0, where for all $0 \leq i \leq r$, T_i has height i.

We are now going to construct a full $4n+k+2$-labeled tree T^* of valence k and height $2r+2$. It will be convenient to represent the vertices in T^* by the sequences from $\{1, \ldots, k\}$ of lengths from 0 through $2r + 2$. We just have to specify what labels from $\{1, \ldots, 4n + k + 2\}$ appear on the vertices.

Some of the vertices of T^* will be used for grafting a copy of one of the T_i's. I.e., that vertex of T^* will be used as the root of that copy of T_i. The vertices of T^* used for this purpose will be called the grafting vertices.

The grafting vertices will be

i) the $r + 1$ vertices $\langle 1, 1\rangle, \langle 1, 2, 1\rangle, \ldots, \langle 1, 2, \ldots, 2, 1\rangle$, referred to as $v_0, \ldots, v_r$;
ii) the r vertices $\langle 2, 1\rangle, \langle 2, 2, 1\rangle, \ldots, \langle 2, 2, \ldots, 2, 1\rangle$, referred to as $w_1, \ldots, w_r$.

Copies of the trees $T_0, \ldots, T_r$ are grafted onto the $r+1$ grafting vertices $v_0, \ldots, v_r$, respectively. Copies of the trees $T_1, \ldots, T_r$ are grafted onto the r grafting vertices $w_1, \ldots, w_r$, respectively.

This determines the labels of the vertices that lie in the copies of the T_i's that have been grafted.

We now change the labels of copies of the T_i's that have been grafted onto the w_i's. We add n to all of their labels.

Next we add $2n$ to the labels of all of the grafting vertices.

Then we label the vertices of height 1 with distinct labels $4n+1, \ldots, 4n+k$. Next label the root with label $4n+k+1$. Finally, we label all remaining vertices of $T^\star$ with $4n+k+2$.

Note that there are these four categories of vertices, labeled in the following way.

i) the vertices in copies of the T_i's grafted onto the v_i's, other than the v_i's. Labels from $[1, n]$;
ii) the vertices in copies of the T_i's grafted onto the w_i's, other than the w_i's. Labels from $[n+1, 2n]$;
iii) the v_i's. Labels from $[2n+1, 3n]$;
iv) the w_i's. Labels from $[3n+1, 4n]$;
v) the vertices of height ≤ 1 in $T^\star$. Distinct labels from $[4n+1, 4n+k+1]$;
vi) the remaining vertices in $T^\star$. Label $4n+k+2$.

We now claim that for all $2 \leq i \leq 2r+2$, there is a grafting vertex x and $0 \leq b \leq r$ such that a copy of T_b was grafted onto x, where the terminal vertices of the copy of T_b have height i. We prove this claim by two cases.

Case 1. i is even, $i = 2b$. Then $1 \leq b \leq r+1$. Note that the tree T_{b-1} has been grafted onto v_{b-1}, which has height $b+1$, and the terminal vertices of the grafted copy of T_{b-1} have height $2b = i$.

Case 2. i is odd, $i = 2b+1$. Then $1 \leq b \leq r$. Note that the tree T_b has been grafted onto w_b, which has height $b+1$, and the terminal vertices of the grafted copy of T_b have height $2b+1 = i$.

This establishes the claim.

We also need that all labels of strict successors of grafting vertices are either $4n+k+2$ or in $[1, n]$ or in $[n+1, 2n]$, the latter two cases corresponding to the vertices in the grafted tree other than the grafting vertex itself. This second claim is clear by construction.

By the choice of r, let $0 \leq i < j \leq 2r+2$ and h be an inf, label, and terminal preserving embedding from $T^\star[\leq i]$ into $T^\star[\leq j]$. Here $T^\star[\leq i], T^\star[\leq j]$ are the truncations of $T^\star$ of height i, j. We are going to find $0 \leq i' < j' \leq r$ and an inf and label preserving embedding from T_i into T_j.

Since the vertices of $T^\star$ of height ≤ 1 have distinct labels, which are in turn distinct from the labels of all other vertices, clearly h is the identity on $T^\star[\leq 1]$ and so $2 \leq i < j \leq 2r+2$.

We now apply the first claim and let $0 \le b \le r$ and x be a grafting vertex such that a copy of T_b was grafted onto x, where the terminal vertices of the copy of T_b have height i.

Clearly $h(x)$ is a grafting vertex since grafting vertices are the only vertices with labels in $[2n+1, 4n]$. Let $0 \le c \le r$ be such that a copy of T_c was grafted onto $h(x)$ in $T^\star$.

Note that h maps the terminal vertices of the copy of T_b—which have height i—to vertices of height j.

It follows from the second claim above that h maps the copy of T_b grafted onto x into the copy of T_c grafted onto $h(x)$. Therefore h is an inf and label preserving embedding from T_b into T_c. It suffices to show that $b < c$. Since T_b has height b and T_c has height c, obviously $b \le c$.

Suppose $b = c$. Since x and $h(x)$ are both v's or both w's, we see that $h(x) = x$. Note that all successors of v of height j are labeled $4n+k+2$. Since the terminal vertices of the copy of T_b grafted onto x are mapped to vertices of height j, we have a contradiction.

So $h(x) \ne x$. Since $x, h(x)$ are both v's or both w's, we see that $b \ne c$. Hence $b < c$ as required.

For the reverse direction, let $k \ge 2$ and assume 2.1. Let $k, n \ge 1$ for 1.3. By 2.1, let r be such that the following holds. Let $T_0, \ldots, T_r$ be full $2n$-labeled trees of valence k or 0, where for all $0 \le i \le r$, T_i has height i. There exist $0 \le i < j \le r$ and an inf and label preserving embedding from T_i into T_j.

We claim that for all full finite n-labeled trees of valence k with height $\ge r$, there is an inf, label, and terminal preserving embedding from some truncation of T into some truncation of T of greater height.

To see this, let $T_0, \ldots, T_r$ be the first $r+1$ truncations of T, where n is added to the labels of all terminal vertices. By the above, let $0 \le i < j \le r$ and h be an inf and label preserving embedding from T_i into T_j. Then h is an inf, label, and terminal preserving embedding from $T[\le i]$ into $T[\le j]$. ⊣

The Internal Finite Tree Embedding Function(1) is the function $\sigma : \mathbb{N}^2 \to \mathbb{N}$ given by $\sigma(k, n) =$ the least $r \ge 1$ such that for all full n-labeled trees T of valence k and height r, there is an inf, label, and terminal preserving embedding from some truncation of T into some truncation of T of greater height.

The 2.1 Function is the function $\tau : \mathbb{N}^2 \to \mathbb{N}$ given by $\tau(k, n) =$ the least r such that the conclusion of Theorem 2.1 holds.

THEOREM 2.4. For all $n \ge 1$ and $k \ge 2$, $\sigma(k, 4n+k+2) > 2\tau(k, n)$.

PROOF. Let $n \ge 1$ and $k \ge 2$. Let r be smallest such that $2r+2 \ge \sigma(k, 4n+k+2)$. We first claim that $r \ge 1$. In other words, we claim that $\sigma(k, 4n+k+2) > 2$. This is clear because there are at least six labels to use, and we can make a tree of height 5 in which no two vertices with different heights have the same label. By the choice of r, note that $2(r-1)+2 = 2r < \sigma(k, 4n+k+2)$.

From the first paragraph above, we see that k, n, r obey the first paragraph in the proof of Theorem 2.3. According to that proof, we see that k, n, r obey the conclusion of Theorem 2.1. Hence $r \geq \tau(k, n)$, and so $\sigma(k, 4n + k + 2) > 2r \geq 2\tau(k, n)$. ⊣

§3. Elementary recursive ordinal notation systems. Here we look at what we call combinatorial well foundedness of any ordinal notation system–CWF below. This isolates the recursion theoretic part of the work from the specifics of trees and particular ordinal notation systems.

In [Fr81] and [Fr82] we considered this concept only in the context of some standard ordinal notation systems, without putting it in an informative general context.

In the exposition [Si85], a less sharp form of CWF is presented, where the sequences of notations obey the inequality $|x_i| \leq c \bullet (i + 1)$ rather than the sharper inequality $|x_i| \leq c + i$. This lead to a correspondingly stronger from of the finite forms—and thus a correspondingly weaker independence result.

Rather than rely on [Si85] here, we give a self contained treatment of CWF that is very general.

[Sm85] is an exposition of our more general approach. However, the presentation here is improved in clarity and generality. All references in [Sm85] to a forthcoming paper of Friedman and Pearce should now be construed as referring to the paper of Friedman and Sheard [FS95]. This section is a modified presentation of that work that is self contained.

Elementary recursive ordinal notation systems for ordinals were first presented in [Sm85]. These are, by definition, well founded. In the paper [FS95], the more general concept of elementary recursive ordinal notation system is introduced and developed in detail. These are not necessarily well founded. The proof theoretic results cited in [Sm85] are proved in [FS95].

In order to draw the crucial conclusions about the Internal Finite Tree Embedding Theorem(s) and Γ_0 and predicativity, we need only the relevant proof theoretic facts about the standard ordinal notation system for Γ_0, which are well known.

However, we do make a new concession to generality here. It turns out to be very convenient to require only quasi linearity on the ordering of notations rather than linearity. This is because so many notation systems are presented with quasi linearity, and one invokes a theory of normal forms in order to obtain linearity. It is best to free our development from a needless appeal to normal form theory.

Let $\mathbb{N}$ be the set of all nonnegative integers. We use $<$ for the usual ordering on $\mathbb{N}$. All other orderings are denoted by other symbols.

A quasi linear ordering $\leq'$ is a relation that is reflexive, transitive, and obeys $x \leq' y \vee y \leq' x$.

An $\in$-number is an ordinal that is closed under base ω exponentiation.

DEFNITION 3.1. An elementary recursive ordinal notation system for an $\in$-number λ is a structure $(\mathrm{A}, \leq', +, \bullet, \omega^{-})$ obeying

i) $A \subseteq \mathbb{N}$, $(A, \leq')$ is a quasi linear ordering with order type λ, $+ : A^2 \to A$, $\bullet : A^2 \to A$, and $\omega^- : A \to A$;
ii) $\leq', +, \bullet, \omega^-$ are elementary;
iii) If ord : $A \to \lambda$ is the unique ordinal assignment, then $\mathrm{ord}(x + y) = \mathrm{ord}(x) + \mathrm{ord}(y)$, $\mathrm{ord}(x \bullet y) = \mathrm{ord}(x) \bullet \mathrm{ord}(y)$, $\mathrm{ord}(\omega^x) = \omega^{\mathrm{ord}(x)}$;
iv) (Elementary Cantor normal forms). There is an elementary function $f : A \times \mathbb{N} \to A$ such that for every $x \in A$, $x = \omega^{f(x,1)} + \cdots + \omega^{f(x,n)}$, where $f(x, 1) \geq' \cdots \geq' f(x, n) <' f(x, n + 1)$.

We say that $(A, \leq', +, \bullet, \omega^-)$ is an ERONS(λ).

Here "elementary" means "elementary recursive," which refers to computation by a Turing machine with a time bound given by an exponential stack of fixed length. Here sets and relations on ω are elementary if and only if their characteristic functions are elementary, and partial functions are elementary if and only if their domains and graphs are elementary. The sup of a set refers to the least element that is at least as large as all elements of that set.

In [FS95], ERONS (without the λ) are introduced which are allowed to be non well founded (and also allowing the three operations to be partial). This notion is necessarily more complicated. It supports the general treatment of proof theory developed there.

It is not true that any ERONS(λ) in our sense is an ERONS in the sense of [FS95], because ERONS are based on a strict linear ordering. However, we still want to derive all of the proof theoretic and recursion theoretic information from [FS95] for our ERONS(λ). This is easily done by converting any ERONS(λ) to an ERONS of type λ as follows.

Let $\leq' = (A, \leq', +, \bullet, \omega^-)$ be an ERONS(λ). Let $A^\star = \{x \in A : (\forall y)(y =' x \to x \leq y\}$, where $\leq$ refers to the usual ordering on ω. For $x, y \in A^\star$, $x +' y$ is defined as the numerically least $z =' x + y$. Similarly for the other two operations. Then it is easy to verify that $\leq'^\star = (A^\star, <', +^\star, \bullet^\star, \omega^{-\star})$ is the desired ERONS of type λ. Proof theoretic and recursion theoretic conclusions about the latter immediately imply the corresponding proof theoretic and recursion theoretic conclusions about the former.

Definition 3.2. (Descent recursion). Let $\leq' = (A, \leq', +, \bullet, \omega^-)$ be an ERONS(λ). Let $g : \mathbb{N} \to \mathbb{N}$ and $h : \mathbb{N}^2 \to A$ be elementary, and $b \in A$ be such that for all $n, m \in \mathbb{N}$, $h(n, m) \leq' b$. We write $f = D(g, h)$ if for all $m \in \mathbb{N}$, $f(m) = g(\mu n[h(n, m) \leq' h(n+1, m)])$, where μn indicates the least n in the usual ordering on $\mathbb{N}$. The functions f that result from this construction are called the $\leq'$ descent recursive functions.

The $\leq'$ descent recursive functions are the same as the $<'^\star$ descent recursive functions (written DR($\leq'^\star$) as defined on page 4 of [FS95]. The verification of this is left to the reader.

All of the standard ordinal notation systems form an ERONS(λ), where standard finite sequence coding is used to make the notations elements of ω. This is true using all terms, instead of just using terms in normal form. In fact, we will have no need to consider normal forms here. In particular, the standard notation system for Γ_0 introduced and used by Feferman in [Fe64] and [Fe68a] in order to encapsulate predicativity, is an ERONS(Γ_0). We refer to this notation system as ERONS(Γ_0). We postpone any detailed discussion of ERONS(Γ_0) until section 4.

According to [FS95], the $\leq'^{\star}$ descent recursive functions are exactly the same as the functions defined by any number of schemes of ordinal recursion up to any point in $\leq'^{\star}$, including nested recursion. Also, one can consider the assignment of fundamental sequences to limit points in $\leq'^{\star}$. In [FS95], conditions are placed on an assignment of fundamental sequences in order that various versions of the Hardy and Grzegorczyk, and Wainer hierarchies also yield the same functions; i.e., the $\leq'^{\star}$ descent recursive functions. And these conditions on fundamental sequence assignments are obeyed by all standard fundamental sequence assignments. See [Ro84] for background on these hierarchies.

Analogously to [FS95], we can consider the system ATI($\leq'$). This is primitive recursive arithmetic together with transfinite induction for all arithmetic formulas up to any particular point in $\leq'$ (where $\leq'$ is an ERONS(λ)). Obviously, this is not well defined unless we give a presentation of $\leq'$. The appropriate notion of presentation is based on the obvious notion of presentation of elementary recursive data. It turns out that the provably recursive functions of ATI($\leq'$) do not depend on the choice of presentation of $\leq'$. This is a consequence of the fact that under very general circumstances, even adding all true $\prod_1^0$ sentences to a theory does not change the provably recursive functions.

This is analogous to the situation in [FS95], where the same situation is encountered. There, it is proved that the $<'$ descent recursive functions are exactly the provably recursive functions of the system ATI($<'$), which is primitive recursive arithmetic together with the scheme of arithmetic transfinite induction on $<'$, up to any particular point in $<'$ (where $\leq'$ is an ERONS).

The following is an immediate Corollary of the appropriate equivalence between ATI($\leq'$) and ATI($\leq'^{\star}$) that the $\leq'$ descent recursive functions are exactly the provably recursive functions of the system ATI($\leq'$), which is primitive recursive arithmetic together with the scheme of arithmetic transfinite induction on $\leq'$, up to any particular point in $\leq'$, and this result is independent of the presentation of the ERONS(λ), $\leq'$.

THEOREM 3.3. Let $\leq'$ be an ERONS(λ). The $\leq'$ descent recursive functions are exactly the provably recursive functions of the system ATI($\leq'$), which is primitive recursive arithmetic together with the scheme of arithmetic transfinite induction on $\leq'$, up to any particular point in $\leq'$, and this fact is independent of the presentation of the $\leq'$.

PROOF. The first claim is left to the reader. For the second claim, suppose $\leq'$ is given with a particular presentation. This lifts to a particular presentation of $\leq'^{\star}$, which is an ERONS in the sense of [FS95]. It is clear that ATI($\leq'$) and ATI($\leq'^{\star}$) are equivalent, and therefore have the same provably recursive functions. By [FS95], p. 29, these are the $\leq'^\star$ descent recursive functions. But as remarked earlier, the $\leq'$ descent recursive functions are the same as the $\leq'^\star$ descent recursive functions. $\dashv$

Standard ordinal notation systems like ERONS(Γ_0) and the others used in this paper also satisfy additional conditions. In particular, there are generating functions and an associated norm. The associated norm can be taken to be the number of operations used in a notation.

DEFINITION 3.4. A NERONS(λ) is an ERONS(λ) equipped with a norm $|\ |: A \to \mathbb{N}$ which is elementary, and where

i) There is a bound on the elements of $\{x \in A : |x| \leq n\}$ as natural numbers under the usual ordering, which is an elementary function of n;
ii) There exists $c \geq 0$ such that for all $x, y \in A$, $|x + y| \leq |x| + |y| + c$.

It is perfectly natural to put conditions corresponding to ii for the other two operations. However, we will not need such additional conditions here.

DEFINITION 3.5. Let $\leq' = (A, \leq', +, \bullet, \omega^-, |\ |)$ be a NERONS(λ). We let CWF($<'$) be the statement

$$(\forall n \geq 1)(\exists r \geq 1)(\forall x_1, \ldots, x_r \in A)((\forall i \leq r)(|x_i| \leq n+i) \to (\exists i < j)(x_i \leq' x_j)).$$

Here CWF stands for "combinatorial well foundedness."

Sublinear growth conditions have been considered in special contexts, including logarithmic growth conditions; see [LM87] and [Wei].

DEFINITION 3.6. Let $\leq' = (\mathrm{A}, \leq', +, \bullet, \omega^-, |\ |)$ be a NERONS(λ). We let $W[\leq']$: $\mathbb{N} \to \mathbb{N}$ be given by

$$W[\leq'](n) = \mu r(\forall x_1, \ldots, x_r \in \mathrm{A}((\forall i \leq r)(|x_i| \leq n + i) \to (\exists i < j)(x_i \leq' x_j)).$$

In other words, $W[\leq']$ is the function for CWF($\leq'$).

The main point in introducing norms is to establish the following connection between $W[\leq']$ and the $\leq'$ descent recursive functions.

We are aiming for Theorem 3.11, which is a sharpened form of Theorem 5, page 123, of [Sm85]. (The Cantor normal form condition on ERONS(λ) is not present in [Sm85], and it plays no role in the proof of Theorem 3.11.) Until the proof of Theorem 3.11 is complete, we fix a NERONS(λ), $\leq' = (\mathrm{A}, \leq', +, \bullet, \omega^-, |\ |)$.

We let $0'$ be any first element of A and $1'$ be any second element of A. We let $\omega' = \omega^{1'}$. Then ω' is a first limit point of $\leq'$. Let d be greater than $|0'|, |1'|, |\omega'|, c$, and 64.

For $x \in \mathrm{A}$ and $n \geq 0$ we write $x^\star n$ for $x +' \cdots +' x$, where there are n summands; $x^\star 0$ is $0'$. We write n' for $1'^\star n$.

Call a sequence $x_1, \ldots, x_r$, n-special if and only if

i) $x_1 >' x_2 >' \cdots >' x_r$;
ii) for all $1 \leq i \leq r$, $|x_i| \leq n + i$.

LEMMA 3.7. For all $n \geq 0$ and $x \in A$, $|x^\star n| \leq (n+1)(|x|+d)$ and $|n'| \leq 2d(n+1)$.

PROOF. By induction on n. ⊣

LEMMA 3.8. For all sufficiently large $n \geq 1$ there exists an n-special sequence $x_1, \ldots, x_r$ below $\omega' \bullet \omega'$, where $r > 2^n$.

PROOF. Let n be sufficiently large and let m be greatest such that $m^8 \leq n$. Consider the following strictly decreasing sequence presented in m blocks, where each block is given by $9d$ lines:

$$\omega'^\star(9dm) + m', \ldots, \omega'^\star(9dm)$$
$$\omega'^\star(9dm-1) + m', \ldots, \omega'^\star(9dm-1)$$
$$\ldots$$
$$\omega'^\star(9dm-9d+1) + m', \ldots, \omega'^\star(9dm-9d+1)$$

$$\omega'^\star(9dm-9d) + 3m', \ldots, \omega'^\star(9dm-9d)$$
$$\omega'^\star(9dm-9d-1) + 3m', \ldots, \omega'^\star(9dm-9d-1)$$
$$\ldots$$
$$\omega'^\star(9dm-18d+1) + 3m', \ldots, \omega'^\star(9dm-18d+1)$$

$$\ldots$$

$$\omega'^\star 9d + ((3^{m-1})m)', \ldots, \omega'^\star 9d$$
$$\omega'^\star(9d-1) + ((3^{m-1})m)', \ldots, \omega'^\star(9d-1)$$
$$\ldots$$
$$\omega'^\star 1 + ((3^{m-1})m)', \ldots, \omega'^\star 1$$

Each of the multiples of ω' has norm $\leq n/3$. We now examine the second summands. In the first of the $9d$ blocks, the second summands have norms $\leq 2d(m+1) < n/3$. Hence the norms of all terms in the first of the m blocks are $< n$.

Let $2 \leq i \leq m$, and α be the first term of the i-th block. Let α be the t-th term in the above strictly decreasing sequence. Then $t \geq (3^{i-2})9dm = 3^i dm$. Also the second summand of α is $((3^{i-1})m)'$. Hence $|\alpha| \leq 2d(1+(3^{i-1})m) + n/3 + d \leq t + n/3 + d \leq n + t$.

The same argument applies equally well to any term in the i-th block, $2 \leq i \leq d+1$, and also to any term in the first block. We have thus established that this sequence is n-special, and obviously its length is greater than $3^m > 2^n$ by the choice of m. ⊣

We write $E(n, m)$ for an exponential stack consisting of n 2's followed by m on top. Thus $E(0, m) = m$, $E(1, m) = 2^m$.

LEMMA 3.9. Let $\leq'$ be a NERONS(λ). For all sufficiently large $n \geq 1$ there exists an n-special sequence $x_1, \ldots, x_r$ below $\omega' \bullet \omega' \bullet \omega'$, where $r > E(n, n)$.

PROOF. Let b be a bound for Lemma 3.7 so that it holds for all $n \geq b$. Now let $n \geq b$. Let $\alpha_1, \ldots, \alpha_n$ be such that for all $1 \leq i \leq n$, α_i is an $E(i, n)$-special sequence below $\omega' \bullet \omega'$ of length $E(i+1, n)$. Then the concatenation $\alpha_1, \ldots, \alpha_n$ would be an $E(n, n)$-special sequence below $\omega' \bullet \omega'$ of length $\geq E(n, n)$ except that it may not be strictly decreasing because of the transitions between the α_i's.

Define $E'(i, m)$ by $E'(1, m) = m$, $E'(i+1, m) = 2^{(E'(i,m)-1)}$. Note that for $i, m \geq 1$, $E'(2i, m) \geq E(i, m)$.

Let n be sufficiently large. Let m be largest such that $8^m \leq n$. Choose $\beta_1, \ldots, \beta_{2m}$ so that for all $1 \leq i \leq 2m$, β_i is an $E'(i, 2m)$-special sequence below $\omega' \bullet \omega'$ of length $E'(i+1, 2m)$. Now consider the sequence

$$
\begin{aligned}
&(\omega' \bullet \omega')^\star(2m) + \beta_1 \\
&(\omega' \bullet \omega')^\star(2m-1) + \beta_2 \\
&\ldots \\
&(\omega' \bullet \omega')^\star(1) + \beta_{2m}
\end{aligned}
$$

Here the i-th line represents the sequence β_i modified by adding $(\omega' \bullet \omega')^\star(2m - i + 1)$ in front of each term.

It is easy to see that this is an n-special sequence below $\omega' \bullet \omega' \bullet \omega'$ of length at least $E'(2m, 2m) \geq E(n, n)$. ⊣

LEMMA 3.10. Let $b \in A$, $h : \mathbb{N}^2 \to A$ be elementary, and $p \in \mathbb{N}$. and assume that all values of h are $<' b$. For sufficiently large m, there is an m-special sequence of length $E(r+p, r+p)$, where $r = \mu n(h(n, m) \leq' h(n+1, m))$.

PROOF. Let h, b, p be given. Let J: $\mathbb{N}^2 \to \mathbb{N}$ be given by $J(x, y) = |h(x, y) \bullet (\omega \bullet \omega \bullet \omega)|$. Then J is elementary.

Without loss of generality, we may assume that b is a power of ω strictly above all values of h.

Now let m be sufficiently large. Let $\mu n(h(n, m) \leq' h(n+1, m)) = r$.

Using Lemma 3.9, let $\alpha_0, \ldots, \alpha_r$ be finite sequences below $\omega \bullet \omega \bullet \omega$, where

i) α_0 is $m+p$-special of length $E(m+p, m+p)$;
ii) for all $1 \leq i \leq r-1$, α_{i+1} is $[E(q, q)/2]$-special of length $E([E(q, q)/2], [E(q, q)/2])$, where q is the length of α_i.

Consider the following strictly decreasing sequence.

$$
\begin{aligned}
&h(0, m) \bullet (\omega \bullet \omega \bullet \omega) + \alpha_0 \\
&h(1, m) \bullet (\omega \bullet \omega \bullet \omega) + \alpha_1 \\
&\ldots \\
&h(r, m) \bullet (\omega \bullet \omega \bullet \omega) + \alpha_r
\end{aligned}
$$

Here the i-th line (counting from 0) represents the sequence α_i modified by adding $h(i, m) \bullet (\omega \bullet \omega \bullet \omega)$ in front of each term. This is close to the desired m-special sequence of length r that we want. The problem is the first term, whose norm is likely to be greater than $m+1$.

We fix this by creating an appropriate front end. We can do this because we have an upper bound on the first term in $\leq'$ independently of m.

By Lemma 3.9, let β be an $[m/2]$-special sequence of length $\mathrm{E}([m/2],[m/2])$. The m-special sequence that we want is

$$\begin{aligned}
&b \bullet (\omega \bullet \omega \bullet \omega) + \beta \\
&h(0,m) \bullet (\omega \bullet \omega \bullet \omega) + \alpha_0 \\
&h(1,m) \bullet (\omega \bullet \omega \bullet \omega) + \alpha_1 \\
&\dots \\
&h(r,m) \bullet (\omega \bullet \omega \bullet \omega) + \alpha_r
\end{aligned}$$

To see this, note that since m is sufficiently large, $\mathrm{J}(0,m),\dots,\mathrm{J}(r,m)$ easily get absorbed in the lengths of the α_i's. ⊣

THEOREM 3.11. Let $\le' = (\mathrm{A}, \le', +, \bullet, \omega^-, |\,|)$ be a NERONS(λ). The function $\mathrm{W}[\le']$ eventually dominates every $\le'$ descent recursive function, as well as every provably recursive function of $\mathrm{ATI}(\le')$. $\mathrm{CWF}(\le')$ is not provable in $\mathrm{ATI}(\le')$.

PROOF. The first claim is immediate from Lemma 3.10 and Theorem 3.3. The second claim follows easily, since if $\mathrm{CWF}(\le')$ is provable in $\mathrm{ATI}(\le')$ then obviously its associated function, $\mathrm{W}[\le']$, is a provably recursive function of $\mathrm{ATI}(\le')$. ⊣

§4. NERONS(Γ_0) and predicative unprovability. In the original papers [Fr81] and the exposition [Si85], no attention was paid to the issue of valence. Here we care about valence because of the formulation of the Internal Finite Tree Embedding Theorem(2), based on valence 2. Therefore, we give our own self contained treatment.

We sketch the well known NERONS(Γ_0). For this purpose, we use the following standard Gödel numbering of finite strings in a finite ordered alphabet. The finite strings are first ordered according to length and then lexicographically. The Gödel number of a finite string is its position in this ordering.

We consider all terms generated by the constant 0, and the two binary function symbols $+, \varphi$. We call these the terms.

The operations $+, \varphi$ have a standard interpretation on the countable ordinals, where $+$ is ordinal addition, and the binary function φ is defined as follows. $\varphi(0,\beta) = \omega^\beta$, and for $\alpha > 0$, $\varphi(\alpha,\beta)$ is the β-th simultaneous fixed point of all of the $\varphi_{\alpha'}$, $\alpha' < \alpha$. Γ_0 is defined to be the least ordinal such that $\varphi(\Gamma_0, 0) = \Gamma_0$.

It is well known that the ordinals represented by notations are exactly the ordinals $< \Gamma_0$. It is also well known that the ordering relation on the ordinals of the notations is elementary.

Thus we take NERONS(Γ_0) to have domain the set of terms, with the obvious operations $+, \bullet, \omega^-$, and where the norm of a term is the total number of operation symbols that appear.

It is obvious that we have defined a NERONS(Γ_0) which we will write as $< \Gamma_0 = (\mathrm{A}, \le', +, \bullet, \omega^-, |\,|)$.

A very convenient device at this point is to consider extended terms. At present, we use only $0, +, \varphi$. For the extended terms, we use $0, +_1, +_2, \varphi_1, \varphi_2$.

These four binary operations are symmetric. This reflects the fact that our trees are unstructured with no left/right.

We inductively define a map from extended terms to terms as follows.

$$\begin{aligned}
\mathrm{H}(0) &= 0;\\
\mathrm{H}(s +_1 t) &= \min(\mathrm{H}(s), \mathrm{H}(t)) + \max(\mathrm{H}(s), \mathrm{H}(t));\\
\mathrm{H}(s +_2 t) &= \max(\mathrm{H}(s), \mathrm{H}(t)) + \min(\mathrm{H}(s), \mathrm{H}(t));\\
\mathrm{H}(\varphi_1(s, t)) &= \varphi(\min(\mathrm{H}(s), \mathrm{H}(t)), \max(\mathrm{H}(s), \mathrm{H}(t)));\\
\mathrm{H}(\varphi_2(s, t)) &= \varphi(\max(\mathrm{H}(s), \mathrm{H}(t)), \min(\mathrm{H}(s), \mathrm{H}(t))).
\end{aligned}$$

Here min and max refer to $<'$, where ties are decided by crude Gödel numbers.

For extended terms x, y, we write $x \leq' y$ if and only if $\mathrm{H}(x) \leq' \mathrm{H}(y)$.

We again define $|x|$ to be the total number of operation symbols that appear in the extended term x. Note that $|x| = |\mathrm{H}(x)|$.

Lemma 4.1. There is an elementary mapping J from the extended terms into 4-labeled trees where every vertex is of valence 2 or 0, such that

i) for all extended terms x, the height of $\mathrm{J}(x)$ is at most $|x|$;
ii) for all extended terms x, y, if there is an inf and label preserving embedding from $\mathrm{J}(x)$ into $\mathrm{J}(y)$ then $x \leq' y$.

Proof. We define a mapping J′ from the extended terms into finite trees whose vertices are given two labels, the first from $\{1, 2, 3, 4\}$ and the second by an extended term, where every vertex is of valence 2 or 0.

$\mathrm{J}'(0)$ is the tree of valence 0 with labels 1 and the extended term 0. $\mathrm{J}'(s +_1 t)$ is the join of $\mathrm{J}'(s)$ and $\mathrm{J}'(t)$ by a root labeled 1 and the extended term $s +_1 t$. $\mathrm{J}'(s +_2 t)$ is the join of $\mathrm{J}'(s)$ and $\mathrm{J}'(t)$ by a root labeled 2 and the extended term $s +_2 t$. $\mathrm{J}'(\varphi_1(s, t))$ is the join of $\mathrm{J}'(s)$ and $\mathrm{J}'(t)$ by a root labeled 3 and the extended term $\varphi_1(s, t)$. $\mathrm{J}'(\varphi_2(s, t))$ is the join of $\mathrm{J}'(s)$ and $\mathrm{J}'(t)$ by a root labeled 4 and the extended term $\varphi_2(s, t)$.

Note that the height of $\mathrm{J}'(x)$ is at most $|x|$. Also note that the second label of the root of $\mathrm{J}'(x)$ is x itself.

We claim that the second label of a vertex in $\mathrm{J}'(x)$ is at least as great as the second label any of its successors. This is an easy induction that requires the laws $s, t \leq' s +_1 t, s +_2 t, \varphi_1(s, t), \varphi_2(s, t)$ for extended terms s, t.

We also need that $+_1, +_2, \varphi_1, \varphi_2$ are increasing in this sense: if $s \leq' s'$ and $t \leq' t'$, then $s +_i t \leq' s' +_i t', \varphi_i(s, t) \leq' \varphi_i(s', t')$, for all $1 \leq i \leq 2$ and extended terms s, t, s', t'.

We also need the obvious fact that $+_1, +_2, \varphi_1, \varphi_2$ are symmetric.

Using all three of these claims, it is now easy to verify that if x, y are extended terms and h is an inf and first label preserving embedding from $\mathrm{J}'(x)$ into $\mathrm{J}'(y)$, then for each vertex v of $\mathrm{J}(x)$, the second label of v in $\mathrm{J}'(x)$ is at most the second label of hv in $\mathrm{J}'(y)$. In particular, this is true for the root of $\mathrm{J}'(x)$. Then we see that the second label of the root of $\mathrm{J}'(x)$ is at most the second label of the root of $\mathrm{J}'(y)$. I.e., $x \leq' y$. Finally, we let J be the same as J′ except that the second labels of all vertices of values of J′ are suppressed. ⊣

LEMMA 4.2. There is an elementary mapping J from the terms into 4-labeled trees where every vertex is of valence 2 or 0, such that

i) for all terms x, the height of $J(x)$ is at most $|x|$;
ii) for all terms x, y, if there is an inf and label preserving embedding from $J(x)$ into $J(y)$ then $x \leq' y$.

PROOF. Let J be the mapping given by Lemma 4.1. We define the required mapping J′ as follows. Let x be a term. Then we can inductively replace the $+$ and φ in x by $+_i$ and φ_i according to the relative size of the subterms. We thus construct an extended term $x^\star$ of the same norm as x. Also note that $H(x^\star) =' x$. We then define $J'(x) = J(x^\star)$. Suppose x, y are terms and there is an inf and label preserving embedding from $J'(x)$ into $J'(y)$. Then $x^\star, y^\star$ are extended terms and there is an inf and label preserving embedding from $J(x^\star)$ into $J(y^\star)$. Hence $x^\star \leq' y^\star$. Therefore $H(x^\star) \leq' H(y^\star)$, and so $x \leq' y$. ⊣

LEMMA 4.3. There is an elementary mapping J from the terms into full 5-labeled trees of valence 2 or 0, such that

i) for all terms, the height of $J(x)$ is $|x|$;
ii) for all terms x, y, if there is an inf and label preserving embedding from $J(x)$ into $J(y)$ then $x \leq' y$.

PROOF. Let J be given by Lemma 4.2. For each term x, let $J'(x)$ be the full 5-labeled tree of valence 2 or 0 obtained by adding vertices to make $J'(x)$ full and of the same height as $J(x)$, where all new vertices are given the label 5.

Note that any inf and label preserving embedding from $J'(x)$ into $J'(y)$ must be an inf and label preserving embedding from $J(x)$ into $J(y)$, and so $x \leq' y$. Thus J′ is as required. ⊣

THEOREM 4.4. For all $n \geq 10$, $\tau(2, n) \geq W[< \Gamma_0](n)$. Hence $\tau(2, n)$, as a function of n, eventually dominates every $< \Gamma_0$ descent recursive function.

PROOF. Let $n \geq 10$ and $\tau(2, n) = r$. Then for all full n-labeled trees $T_0, \dots, T_r$ of valence 2 or 0, where for all $0 \leq i \leq r$, T_i has height i, there exists $0 \leq i < j \leq r$ and an inf and label preserving embedding from T_i into T_j.

It suffices to show that $\forall x_1, \dots, x_r \in <\Gamma_0((\forall i \leq r)(|x_i| \leq n + i) \rightarrow (\exists i < j)(x_i \leq' x_j))$. Let $x_1, \dots, x_r \in <\Gamma_0$, $(\forall i \leq r)(|x_i| \leq n + i)$.

Recall the function J from Lemma 4.3. The full trees $J(x_1), \dots, J(x_r)$ are of valence 2 or 0 and are 5-labeled, with heights at most $n+1, \dots, n+r$, respectively.

Let $T_0, \dots, T_n$ be the following full n-labeled trees of valence 2 or 0 with heights $0, \dots, n$. T_0 uses label 6. The T_i for $1 \leq i \leq n$ use exactly two labels from $\{7, \dots, n\}$, where these unordered pairs of labels must be distinct. This requires that there be at least n unordered pairs from an $n - 6$ element set. I.e., that $n \leq (n - 6)(n - 5)/2$. For this, $n \geq 10$ suffices.

Now consider $T_0, \dots, T_n, J(x_1), \dots, J(x_r)$. By the choice of r, let $0 \leq i < j \leq r$ be such that there is an inf and label preserving embedding from the i-th term of this sequence into the j-th term of this sequence. By construction, the i-th term must be past the displayed T_n. Hence we have found $1 \leq i' < j' \leq r$ and an inf and label preserving embedding from $J(x_{i'})$ into $J(x_{j'})$. Therefore $x_{i'} \leq' x_{j'}$. as required.

The second claim follows immediately from the first claim and Theorem 3.11. ⊣

THEOREM 4.5. The Internal Finite Tree Embedding Theorem(2) cannot be proved predicatively. Its associated function dominates all provably recursive functions of Feferman's IR.

PROOF. This is immediate from Theorems 4.4, 3.11, and 2.4. ⊣

§5. NERONS($\theta\Omega^k(0)$) and $\prod_2^1 - TI_0$. The standard NERONS($\theta\Omega^k(0)$), $k \geq 0$, are initial segments of the Feferman Aczel notation system (see [Sch77], p. 224). In [Fr84] we discovered a streamlined presentation of this important initial segment. We present this streamlined presentation here.

Several researchers have used this type of notation system; e.g., [Veb08], [Sch54].

DEFNITION 5.1. For $k \geq 1$, Δk is the unique n-ary function from countable ordinals into countable ordinals such that $\Delta k(\alpha_1, \dots, \alpha_k) = \lambda$ is the least power of ω with $\lambda > \alpha_1, \dots, \alpha_n$ such that for all $\lambda > \beta_1, \dots, \beta_k$ with $(\beta_1, \dots, \beta_k) <_{lex} (\alpha_1, \dots, \alpha_k)$, we have $\Delta k(\beta_1, \dots, \beta_k) < \lambda$.

Simple calculations show that $\Delta 2(1, 0) = \epsilon_0$ and $\Delta 3(1, 0, 0) = \Gamma_0$. Note that each $\Delta k(0, \dots, 0) = 1$.

The exact relationship between these functions and the more general Feferman Aczel approach is as follows. We omit the proof.

LEMMA 5.2. Let $k \geq 1$. Then $\Delta k(\alpha_1, \dots, \alpha_k) = \theta\beta(\alpha_k)$, where $\beta = \sum_{1 \leq i < k} \Omega^{k-i-1} \bullet \alpha_i$. In particular, for $k \geq 2$, $\Delta k(1, 0, \dots, 0) = \theta\Omega^{k-2}(0)$.

For each $k \geq 1$, Δk generates an ERONS($\theta\Omega^{k-1}(0)$) as follows. The Δk terms are defined by ordinary induction:

i) 0 is a Δk term;
ii) $\Delta k(t_1, \dots, t_k)$ is a Δk term if $t_1, \dots, t_k$ are Δk terms;
iii) $t_1 + \dots + t_m$ is a Δk term if $m \geq 2$ and $t_1, \dots, t_m$ are Δk terms beginning with Δk.

The ordinals represented by the Δk terms are exactly the ordinals $< \theta\Omega_{k-1}(0) = \Delta k + 1(1, 0, \dots, 0)$.

We denote the ordering relation on Δk terms by $\leq'$ (i.e., the ordinal represented by the first term is at most the ordinal represented by the second term).

By standard arguments, the ordering relation is elementary.

The norm of a normal Δk term (or any Δk term) is again defined as the number of operations that appear; i.e., the number of $+$ and Δk that appear.

Recall that we conveniently introduced extended terms in section 4. We do so here.

The extended Δk terms are the terms in $0, +_1, +_2, \Delta k\pi$, where π ranges over all permutations of $\{1, \ldots, k\}$.

We inductively define a map from extended Δk terms into Δk terms as follows.

$$\mathrm{H}(0) = 0;$$
$$\mathrm{H}(s +_1 t) = min(\mathrm{H}(s), \mathrm{H}(t)) + \max(\mathrm{H}(t), \mathrm{H}(s));$$
$$\mathrm{H}(s +_2 t) = max(\mathrm{H}(s), \mathrm{H}(t)) + \min(\mathrm{H}(s), \mathrm{H}(t));$$
$$\mathrm{H}(\Delta k\pi(s_1, \ldots, s_k)) = \Delta k(s_1, \ldots, s_k),$$

where $s_1, \ldots, s_k$ is the rearrangement of $\mathrm{H}(s_1), \ldots, \mathrm{H}(s_k)$ by first reordering $\mathrm{H}(s_1), \ldots, \mathrm{H}(s_k)$ to make it increasing under $\leq'$, and then rearranging the reordering according to the permutation π. Here the min's and max's refer to $\leq'$, and ties, which can occur there as well as in the final clause, are settled by crude Gödel numbers.

Lemma 5.3. Let $k \geq 2$. There is an elementary mapping J from the extended Δk terms into $k! + 2$-labeled trees where every vertex is of valence k or 0, such that

i) for all extended Δk terms x, the height of $\mathrm{J}(x)$ is at most $|x|$;
ii) for all extended Δk terms x, y, if there is an inf and label preserving embedding from $\mathrm{J}(x)$ into $\mathrm{J}(y)$ then $x \leq' y$.

Proof. The proof is exactly analogous to the proof of Lemma 4.1, except that we need some corresponding inequalities. These inequalities are

i) $\alpha, \beta \leq \alpha +_1 \beta, \alpha +_2 \beta$;
ii) $\alpha_1, \ldots, \alpha_k \leq \Delta k\pi(\alpha_1, \ldots, \alpha_k)$;
iii) if $1 \leq i \leq 2$, $\alpha \leq \alpha'$ and $\beta \leq \beta'$, then $\alpha +_i \beta \leq \alpha' +_i \beta'$;
iv) for all π and $\alpha_1 \leq \beta_1, \ldots, \alpha_k \leq \beta_k$, we have $\Delta k\pi(\alpha_1, \ldots, \alpha_k) \leq \Delta k\pi(\beta_1, \ldots, \beta_k)$;
v) $+_1, +_2, \Delta k\pi$ are all symmetric.

These are all immediate from the definitions except iv. But iv follows from the fact that iv is true for Δk. ⊣

Lemma 5.4. Let $k \geq 2$. There is an elementary mapping J from the Δk terms into $k! + 2$-labeled trees where every vertex is of valence k or 0, such that

i) for all Δk terms x, the height of $\mathrm{J}(x)$ is at most $|x|$;
ii) for all Δk terms x, y, if there is an inf and label preserving embedding from $\mathrm{J}(x)$ into $\mathrm{J}(y)$ then $x \leq' y$.

Proof. The proof is exactly analogous to the proof of Lemma 4.2. ⊣

Lemma 5.5. Let $k \geq 2$. There is an elementary mapping J from the Δk terms into full $k! + 3$-labeled trees where every vertex is of valence k or 0, such that

i) for all Δk terms x, the height of $\mathrm{J}(x)$ is $|x|$;
ii) for all Δk terms x, y, if there is an inf and label preserving embedding from $\mathrm{J}(x)$ into $\mathrm{J}(y)$ then $x \leq' y$.

PROOF. The proof is exactly analogous to the proof of Lemma 4.3. ⊣

LEMMA 5.6. Let $k \geq 2$. For all $n \geq 2k! + 10$, $\tau(k, n) \geq \mathrm{W}[< \theta\Omega^{k-1}(0)](n)$. Hence $\tau(k, n)$, as a function of n, eventually dominates every $< \theta\Omega^{k-1}(0)$ descent recursive function.

PROOF. Let $k \geq 2$ and n be sufficiently large. Let $\tau(k, n) = r$. Then for all full n-labeled trees $\mathrm{T}_0, \dots, \mathrm{T}_r$ of valence k or 0, where for all $0 \leq i \leq r$, T_i has height i, there exists $0 \leq i < j \leq r$ and an inf and label preserving embedding from T_i into T_j.

It suffices to show that $\forall x_1, \dots, x_r \in < \theta\Omega^{k-1}(0)((\forall i \leq r)(|x_i| \leq n + i) \rightarrow (\exists i < j)(x_i \leq' x_j))$. Let $x_1, \dots, x_r \in < \theta\Omega^{k-1}(0)$, $(\forall i \leq r)(|x_i| \leq n + i)$.

Recall the function J from Lemma 5.5. The full trees $\mathrm{J}(x_1), \dots, \mathrm{J}(x_r)$ are of valence k or 0 and are $k! + 3$-labeled, with heights at most $n + 1, \dots, n + r$, respectively.

Let $\mathrm{T}_0, \dots, \mathrm{T}_n$ be full n-labeled trees of valence k or 0 with heights $0, \dots, n$, where

i) T_0 uses label $k! + 4$;
ii) each T_i, $1 \leq i \leq n$, uses exactly two labels from $\{k! + 5, \dots, n\}$, where no two of them use the same two labels.

This is possible because if $n \geq 2k! + 10$ then the number of two element subsets of $\{k! + 5, \dots, n\}$ is at least n.

Now consider $\mathrm{T}_0, \dots, \mathrm{T}_n, \mathrm{J}(x_1), \dots, \mathrm{J}(x_r)$. By the choice of r, let $0 \leq i < j \leq r$ be such that there is an inf and label preserving embedding from the i-th term of this sequence into the j-th term of this sequence. By construction, the i-th term must be past the displayed T_n. Hence we have found $1 \leq i' < j' \leq r$ and an inf and label preserving embedding from $\mathrm{J}(x_{i'})$ into $\mathrm{J}(x_{j'})$. Therefore $x_{i'} \leq' x_{j'}$, as required.

The second claim follows immediately from the first claim and Theorem 3.11. ⊣

THEOREM 5.7. Let $k \geq 2$. The Internal Finite Tree Embedding Theorem(1) for k cannot be proved in $\mathrm{ATI}(< \theta\Omega^{k-1}(0))$. Its associated function $\tau(k, n)$ dominates all provably recursive functions of this system.

PROOF. This is immediate from Theorems 5.6, 3.11, and 2.4. ⊣

$\prod_2^1 - \mathrm{TI}_0$ is the subsystem of second order arithmetic given by ACA_0+ the scheme of transfinite induction for $\prod_2^1$ formulas applied to well orderings of ω (see [Si99] for background on subsystems of second order arithmetic). In [RW93], this system is referred to as $\mathrm{ACA}_0 + \prod_2^1 - \mathrm{BI}$, where BI stands for "bar induction."

$\prod_2^1 - \mathrm{TI}$ is the same as $\prod_2^1 - \mathrm{TI}_0$ except we add induction on ω for all formulas of second order arithmetic.

THEOREM 5.8. The Internal Finite Tree Embedding Theorem(1) cannot be proved in $\mathrm{ATI}(< \theta\Omega^{\omega}(0))$ or in $\prod_2^1 - \mathrm{TI}_0$ $(\mathrm{ACA}_0 + \prod_2^1 - \mathrm{BI})$. Its associated function $\tau(n, n)$ dominates all provably recursive functions of these systems.

PROOF. This follows immediately from Theorem 5.7 using the proof theory of $\prod_2^1-\mathrm{TI}_0$; see Pohlers in [BFPS81], and also [RW93]. ⊣

THEOREM 5.9. For each fixed $k \geq 1$, the Internal Finite Tree Embedding Theorem(1) for k can be proved in $\mathrm{ATI}(< \theta\Omega^{\omega}(0))$ or in $\prod_2^1-\mathrm{TI}_0$. The Internal Finite Tree Embedding Theorem(1) can be proved in $\prod_2^1-\mathrm{TI}$.

PROOF. From the ordinal analysis of Kruskal's theorem and its variants in [Fr84] and [RW93]. ⊣

§6. Further versions of the Internal Finite Tree Embedding Theorem. We first present a natural strengthening of the Internal Finite Tree Embedding Theorem. Here we consider n-labeled finite trees of valence $\leq k$. For such trees, the vertices of valence 0 may not have a common height, and the valences of the vertices may differ widely.

We also add left/right structure to the trees, something that was done in [Kr60] and is implicit in [Hi52]. Thus a linear ordering is assigned to every nonterminal vertex. This linear ordering is a strict linear ordering (left/right) of its immediate successors. This induces a strict linear ordering (left/right) on all vertices in the obvious way.

Structure preserving embeddings are required to have the property

if x is to the left of y then hx is to the left of hy.

Valence preserving embeddings are required to preserve the valence of each vertex.

Let T be any tree. The truncations of T are the trees resulting from restricting T to all vertices whose height is less than some given nonzero cardinal number. If T is finite, then the truncations of T are the trees resulting from restricting T to all vertices whose height is $\leq r$, for some particular nonnegative integer r.

THEOREM 6.1. Internal Finite Tree Embedding Theorem(3). Let $k, n \geq 1$ and T be a sufficiently tall finite structured n-labeled tree of valence $\leq k$. There is an inf, structure, label, and valence preserving embedding from some truncation of T into some truncation of T of greater height.

PROOF. Virtually identical to the proof of Theorem 1.3. In the original papers, Theorems 1.1 and 1.5 were proved for structured trees with structure preserving embeddings. Note that valence refers to valence in the sense of the two truncations. ⊣

The most severe direct weakening that probably puts us beyond predicativity is the restriction to 2 labels and valence 2. For instance, we know that Kruskal's Theorem (Theorem 1.1) for 2-labeled trees of valence 2 is equivalent to "Γ_0 is well ordered" over RCA_0 (see [Fr84]). This suggests that the following weakening should be predicatively unprovable.

THEOREM 6.2. Internal Finite Tree Embedding Theorem(4). Let $k \geq 1$ and T be a sufficiently tall full finite 2-labeled tree of valence 2. There is an inf, label, and terminal preserving embedding from some truncation of T of height $\geq k$ into some truncation of T of greater height.

It is already clear from the proof of Theorem 4.4 (see Lemma 4.3) that Theorem 6.2 is predicatively unprovable for full finite 10-labeled trees of valence 2.

We now consider trees without any labels. Here we must consider trees that are not necessarily full. Once again, if the height of a tree is n then there are exactly $n+1$ truncations, of heights $0, \ldots, n$.

THEOREM 6.3. Internal Finite Tree Embedding Theorem(5). Let $k \geq 1$ and T be a sufficiently tall finite tree of valence k. There is an inf, T-valence, and terminal preserving embedding from some truncation of T into some truncation of T of greater height.

THEOREM 6.4. Theorem 6.1 implies Theorem 6.3 in EFA. Theorem 1.3 implies Theorem 6.3 in EFA. Thus Theorem 6.3 cannot be proved predicatively.

PROOF. We first derive Theorem 6.3 from Theorem 6.1. Let $k \geq 1$ and T be a sufficiently tall finite tree of valence k. Let T′ result from T by labelling each vertex in T with its valence in T. By Theorem 6.1, there is an inf, label, and valence preserving embedding h from some truncation of T′ into some truncation of T′ of greater height. Note that h must be an inf, T-valence, and terminal preserving embedding from some truncation of T into some truncation of T of greater height.

Now assume Theorem 6.3. For Theorem 1.3, let $k, n \geq 1$. By Theorem 6.3, let r be such that for all finite trees T of valence $k+n$ with height $\geq r$, there is an inf, T-valence, and terminal preserving embedding from some truncation of T into some truncation of T of greater height.

For Theorem 1.3, let T be a full finite n-labeled tree of valence k with height $\geq r$. By permuting labels, we can assume without loss of generality that the root of T has label n.

We now build a corresponding finite tree T*, without labels, which is of valence $k+n$ and of height r.

We construct T* as follows. Let x be a vertex in T[$<r$] with label $1 \leq p \leq n$. We add p new immediate successors of x, each of valence 0. T* consists of T[$\leq r$] together with these new vertices. Since $r \geq 1$, T* has valence $k+n$ and height r.

By the choice of r, let h be an inf, T*-valence, and terminal preserving embedding from T*[$\leq i$] into T*[$\leq j$], $1 < i < j \leq r$.

Let x be a vertex in T[$\leq i$]. Then the valence of x in T* is greater than k, and so the valence of hx in T* is also greater than k. Hence hx is a vertex in T[$\leq j$]. Therefore $j < r$. Furthermore, the valence of x in T* is k plus the label of x, and the valence of hx in T* is k plus the label of hx. Since h is valence preserving, we see that x and hx have the same label in T. Hence h is an inf, label, and terminal preserving embedding from T[$\leq i$] into T[$\leq j$]. ⊣

§7. The Internal Finite Tree Gap Embedding Theorem. In [Fr82] we introduced and proved a strengthening of Kruskal's theorem involving an additional condition on inf and label preserving embeddings called the gap condition.

Specifically, let $n \geq 1$ and T_1 and T_2 be finite n-labeled trees with labeling functions l_1 and l_2. We say that h is a gap embedding from T_1 to T_2 if and only if

i) h is a one-one map from the vertices of T_1 to the vertices of T_2;
ii) for all vertices $ha < c < hb$ of T_2, $l_2(c) \geq l_2(hb)$.

THEOREM 7.1. Extended Kruskal's Theorem (EKT). Let $n \geq 1$ and $T_1, T_2, \ldots$ be an infinite sequence of finite n-labeled trees. There exists $i < j$ and an inf and label preserving gap embedding from T_i into T_j.

Here is an important specialization of Theorem 7.1.

THEOREM 7.2. The following are provably equivalent in RCA_0.

i) EKT (Theorem 7.1);
ii) $\theta\Omega_\omega(0)$ is well ordered;
iii) $\prod_1^1$ reflection over $\prod_1^1-CA_0$.

PROOF. EKT was proved in [Fr82] for each specific n within $\prod_1^1-CA_0$. This was done by an iterated minimal bad sequence argument. See [Si85]. From this proof, the conclusion was drawn that EKT is provable in $\prod_1^1-CA$, or even in $RCA_0 + \prod_1^1$ reflection over $\prod_1^1-CA_0$. In [Fr82] we proved i $\rightarrow$ ii by an assignment of ordinal notations to trees. Also see [Si85]. And we quoted the known equivalence ii $\leftrightarrow$ iii from Pohlers in [BFPS81], chapter VI, p. 334, Theorem 1.5. The exposition in [Si85] relies on the Buchholz system of notations, which are known to be equivalent. See [Bu87] and [BS88]. ⊣

The extended Kruskal's theorem had a role in the subsequent Robertson/Seymour graph minor theorem (GMT). It also formed the basis of a collaborative effort, resulting in the triple paper [FRS87]. We quote from [FRS87], page 240.

"A major step in the proof of the graph minor theorem consists of showing that any class L of finite graphs whose members have bounded tree-width is well-quasi-ordered by [minor inclusion]. ... Then the well-quasi-ordering can be established by applying Kruskal's theorem step by step down the hierarchy ... or, as is done in the paper "Graph Minors IV," by applying a generalized form of Friedman's [extended Kruskal's] theorem globally to the structural hierarchy. It is of metamathematical interest whether or not these theorems, Kruskal's and Friedman's, can be recovered from the resulting graph minor theorem, because of certain logical requirements known on the nature of their proofs. By showing that they can, we establish that these proof methods are also essential to the proof of the graph minor theorem."

Specifically, the following result was proved in [FRS87], p. 253.

THEOREM 7.3. The bounded GMT (GMT for graphs of bounded tree-width) is equivalent to EKT in RCA_0. In particular, the GMT, or even the bounded GMT is not provable in $\prod_1^1-CA_0$. GMT is provable in $\prod_1^1-CA + BI$, and in ID_ω.

This was proved collaboratively as follows. We came to Robertson and Seymour with the conjecture that GMT is sufficient to explicitly derive EKT. They came back with an explicit derivation of a weak form of EKT from GMT—in fact, the bounded GMT. We then derived the full EKT from the weak form of EKT. The upper bounds were established by isolating the nonconstructive parts of the proofs of GMT and bounded GMT.

In the course of showing that the relevant weak form of EKT implies EKT, we showed that still weaker forms of EKT imply EKT. For our purposes, perhaps the most natural such weakening of EKT is the following.

THEOREM 7.4. Let $n \geq 1$ and $T_1, T_2, \dots$ be an infinite sequence of full finite n-labeled trees of valence 2. There exists $i < j$ and an inf and label preserving gap embedding from T_i into T_j.

THEOREM 7.5. Theorem 7.4 and EKT are provably equivalent in RCA_0.

PROOF. Theorem 7.4 obviously implies $(\forall n)(Q_n^2)$ of [FRS87], p. 251. The result follows by Lemmas 4.1.2 and 4.1.3 of [FRS87], pp. 252–253. ⊣

Thus in the presence of gap embeddings, finite binary trees are as powerful as arbitrary finite trees. This is definitely not the case with Kruskal's theorem—with or without labels. In particular, Kruskal's theorem for binary trees without labels proof theoretically reaches to PA, whereas Kruskal's theorem for binary trees with two labels proof theoretically reaches to predicativity, and Kruskal's theorem for binary trees with n labels proof theoretically reaches beyond predicativity, but well below Kruskal's theorem for ternary trees without labels, which is in turn well below the full Kruskal's theorem. These results were proved in [Fr84], and in a somewhat different context, corresponding phenomena appear also in [Schm79], in the calculation of maximal order types.

For the sake of clarity, we now extend our list of equivalences.

THEOREM 7.6. The following are provably equivalent in RCA_0.

i) EKT (Theorem 7.1);
ii) EKT for valence 2 (Theorem 7.4);
iii) bounded GMT (GMT for bounded tree-width);
iv) $\theta\Omega_\omega(0)$ is well ordered;
v) $\prod_1^1$ reflection over $\prod_1^1-CA_0$.

We are now ready to present two forms of the Internal Finite Tree Gap Embedding Theorem.

THEOREM 7.7. Internal Finite Tree Gap Embedding Theorem(1). Let $k, n \geq 1$ and T be a sufficiently tall full finite n-labeled tree of valence k. There is an inf, label, and terminal preserving gap embedding from some truncation of T into some truncation of T of greater height.

THEOREM 7.8. Internal Finite Tree Gap Embedding Theorem(2). Let $n \geq 1$ and T be a sufficiently tall full finite n-labeled tree of valence 2. There is an inf, label, and terminal preserving gap embedding from some truncation of T into some truncation of T of greater height.

These are proved from Theorem 7.1 in the same way as Theorem 1.3 is proved from Theorem 1.1.

THEOREM 7.9. Theorems 7.1, 7.4, and 7.6, for any fixed number of labels, n, are provable in $\prod_1^1-\mathrm{CA}_0$. Theorems 7.1, 7.4, and 7.6 are provable in $\prod_1^1-\mathrm{CA}$.

PROOF. This is shown in [Fr82] and [Si85] for Theorem 7.1. It follows for Theorems 7.4 and 7.6 since they are consequences of Theorem 7.1 in RCA_0. ⊣

§8. Finite sequences of trees. In this section we derive the following finite version of the EKT (Theorem 7.1) from the Internal Finite Tree Gap Embedding Theorem(1).

THEOREM 8.1. For all $k, n \geq 1$ there exists r such that the following holds. Let $T_0, \dots, T_r$ be full n-labeled trees of valence k or 0, where for all $0 \leq i \leq r$, T_i has height i. There exists $0 \leq i < j \leq r$ and an inf, label, and terminal preserving gap embedding from T_i into T_j.

PROOF. This is proved from EKT in the same way that Theorem 2.1 is proved from Kruskal's theorem. Valence 0 is needed to accommodate height 0, which just consists of a root. ⊣

We also derive the following valence 2 version of Theorem 8.1 from the Internal Finite Tree Gap Embedding Theorem(2).

THEOREM 8.2. For all $n \geq 1$ there exists r such that the following holds. Let $T_0, \dots, T_r$ be full n-labeled trees of valence 2 or 0, where for all $0 \leq i \leq r$, T_i has height i. There exists $0 \leq i < j \leq r$ and an inf, label, and terminal preserving gap embedding from T_i into T_j.

We accomplish both of these derivations by proving the following.

THEOREM 8.3. The following is provable in EFA. Let $k \geq 2$. Theorem 7.7 for k implies Theorem 8.1 for k.

PROOF. Let $k \geq 2$ and assume Theorem 7.7. Towards deriving Theorem 8.1, let $n \geq 1$, and let $r \geq 1$ be such that for all full $4n+k+2$-labeled trees T of valence k and height $2r+2$, there is an inf, label, and terminal preserving gap embedding from some truncation of T into some truncation of T of greater height.

Now let $T_0, \dots, T_r$ be full finite n-labeled trees of valence k or 0, where for all $0 \leq i \leq r$, T_i has height i.

We let T^* be exactly the same tree as was constructed in the proof of Theorem 2.3. Let $0 \leq i < j \leq 2r+2$ and h be an inf, label, and terminal preserving gap embedding from $T^*[\leq i]$ into $T^*[\leq j]$. According to that proof, we have

i) a grafting vertex x;
ii) $0 \leq b < c \leq r$;

iii) T_b was grafted onto x with terminal vertices of height i;
iv) T_c was grafted onto hx with terminal vertices of height j;
v) h is an inf, label and terminal preserving embedding from T_b into T_c.

It remains to show that h is also a gap embedding from T_b into T_c. Let $u < v$ be vertices of T_b. Let $hu < w < hv$ in T_c. We must show that the label of w in T_c is at least the label of hv in T_c.

By the choice of h, we know that the label of w in $T^\star$ is at least the label of hv in $T^\star$. Also w and hv are vertices in T_c lying above the grafting vertex hx. It then follows that the labels of w and hv in $T^\star$ are either the labels of w and hv in T_c, or both are n more than the labels of w and hv in T_c. Hence the label of w in T_c is at least the label of hv in T_c as required. ⊣

For $n \geq 1$, let $TR^\star(k, n)$ be the set of all full n-labeled trees of valence k or 0.

LEMMA 8.4. There is an elementary function F such that the following is provable in EFA.

i) $\text{dom}(F) = \{(k, n, T) : n \geq 1 \text{ and } T \in TR^\star(k, n)\}$;
ii) for all $n \geq 1$ and $T \in TR^\star(k, n)$, $F(k, n, T) \in TR^\star(2, 2n + 2)$;
iii) for all $n \geq 1$ and $S, T \in TR^\star(k, n)$, if there is an inf and label preserving gap embedding from $F(k, n, S)$ into $F(k, n, T)$ then there is an inf and label preserving gap embedding from S into T.

PROOF. From [FRS87], pp. 252–253. The function F constructed there is much better than elementary, but this is not important for us. Neither is the expression $2n + 2$. ⊣

LEMMA 8.5. Let $c \geq 1$. The following is provable in EFA. For all $i \geq c$ there exist trees $T_i, \ldots, T_{E(c,i)} \in TR^\star(2, 4)$ of heights $i, \ldots, E(c, i)$, such that none of these trees is inf and label preserving embeddable into a later tree.

PROOF. First choose distinct trees $T_i, \ldots, T_{E(1,i)-1} \in TR^\star(2, 2)$, of height $i - 1$. Then add vertices on top of these trees to make them full trees of heights $i, i + 1, \ldots, i+E(1, i)-1$, each with exactly c new vertices labeled 3 and the rest labeled 4. Now choose distinct trees $T_{E(1,i)}, \ldots, T_{E(1,i+1)-1} \in TR^\star(2, 2)$ of height $E(1, i) - 1$. Again add vertices on top to make them full of heights $E(1, i), \ldots, E(2, i) - 1$, each with exactly $c - 1$ new vertices labeled 3 and the rest labeled 4. Continue in this way c times. ⊣

THEOREM 8.6. It is provable in EFA that Theorems 7.7, 7.8, 8.1, and 8.2 are equivalent.

PROOF. We show 7.7 → 7.8 → 8.2 → 8.1 → 7.7. The first implication is trivial. The second implication follows from Theorem 8.3 for $k = 2$. For the last implication assume 8.1. For 7.7, let $k, n \geq 1$. From 8.1, let r be such that the following holds. Let $T_0, \ldots, T_r$ be full $2n$-labeled trees of valence k or 0, where for all $0 \leq i \leq r$, T_i has height i. There exists $0 \leq i < j \leq r$ and an inf and label preserving gap embedding from T_i into T_j.

We claim the following. Let T be a full finite n-labeled tree of valence k with height $\geq r$. There is an inf, label, and terminal preserving embedding from some truncation of T into some truncation of T of greater height.

To see this, let $T_0, \ldots, T_r$ be the first $r+1$ truncations of T, where n is added to the label of every terminal vertex. Now apply the above to the sequence $T_0, \ldots, T_r$ of truncations of T.

It remains to show that Theorem 8.2 implies Theorem 8.1. Assume Theorem 8.2, and let $k, n \geq 1$.

Let F be the elementary function given by Lemma 8.4. Let $c \geq 2$ be such that for all $T \in TR^\star(k, n)$ of height i, $F(k, n, T)$ is of height $< E(c, i)$.

Define $p_0, p_1, \ldots$ inductively by $p_0 = E(c, 0)$, $p_{i+1} = E(c, p_i)$.

By Theorem 8.2, let r be such that the following holds. Let $T_0, \ldots, T_r \in TR^\star(2, 2n+8+p_0)$, where for all $0 \leq i \leq r$, T_i has height i. There exists $0 \leq i < j \leq r$ and an inf and label preserving gap embedding from T_i into T_j.

We now claim the following. Let $T_0, \ldots, T_r \in TR^\star(k, n)$, where for all $0 \leq i \leq r$, T_i has height i. There exists $0 \leq i < j \leq r$ and an inf and label preserving gap embedding from T_i into T_j.

Let $T_0, \ldots, T_r$ be as given. By Lemma 8.5, let $S_{p_0}, S_{p_0}+1, \ldots, S_{p_r} \in TR^\star(2, 4)$ be of heights $p_0, p_0+1, \ldots, p_r$ such that for all $0 \leq i \leq r$, none of the trees among $S_{p_i}, S_{p_i+1}, \ldots, S_{p_{i+1}}$ are inf and label preserving embeddable in a later one.

Now let $S_{p_0}', S_{p_0+1}', \ldots, S_{p_r}' \in TR^\star(2, 2n+3)$ also be of heights $p_0, p_0+1, \ldots, p_r$ such that

i) for all $0 \leq i \leq r$ and $p_i \leq j < p_{i+1}$, $F(k, n, T_i)$ is a truncation of S_j;
ii) all vertices of all trees $S_{p_0}', S_{p_1}', \ldots, S_{p_r}'$ other than those vertices in the truncations indicated in i) are labeled $2n+3$.

We now want to combine the trees $S_{p_0}, S_{p_0+1}, \ldots, S_{p_r}$ with the trees S_{p_0}', $S_{p_0+1}', \ldots, S_{p_r}'$ as follows. First we add $2n+3$ to all of the labels in the trees $S_{p_0}, S_{p_0+1}, \ldots, S_{p_r}$. Then for each i, we join S_i and S_i' together with a root labeled $2n+8$.

This construction results in trees $R_{p_0+1}, R_{p_0+2}, \ldots, R_{p_r+1}$, of heights $p_0+1, \ldots, p_r+1 \in T^\star(2, 2n+8)$. We now introduce trees $R_0, \ldots, R_{p_0}$ of heights $0, 1, \ldots, p_0$, where all vertices of these R_i are given the label $2n+9+i$.

We now apply the choice of r to the sequence $R_0, \ldots, R_{p_r}$. Let $p_0 < i < j \leq p_r$ and h be an inf and label preserving gap embedding from R_i into R_j. We can ignore $R_0, \ldots, R_{p_0}$ because of the way these early trees are labeled. Because of the labeling of these trees, we see that h is an inf and label preserving gap embedding from S_i into S_j and also from S_i' into S_j'.

Let b be largest such that $p_b \leq i$. Then $b < r$. By the first property of h, we see that $p_{b+1} \leq j$. Let c be largest such that $p_c \leq j$. By the second property of h, we see that h is an inf and label preserving gap embedding from $F(T_b)$ into $F(T_c)$. Hence by the choice of F, we have an inf and label preserving gap embedding from T_b into T_c as required. ⊣

§9. NERONS($\theta\Omega_\omega(0)$) and $\prod_1^1$–CA$_0$. The standard NERONS($\theta\Omega_k(0)$), $k \geq 0$, are initial segments of the Feferman Aczel notation system (see [Sch77], p. 224).

In [Fr82] we used NERONS($\theta\Omega_\omega(0)$) to establish Theorem 7.2. The published exposition in [Si85] uses the equivalent Buchholz notations, [Bu87], [BS88].

Because of the reduction to valence 2 given by Theorem 8.6, we can essentially rely on the exposition in [Si85] using Buchholz notations. Here some details are simpler than those we provided in section 5 when we were careful about valence.

For each $n \geq 1$, let ERONS(BH$_n$) be the elementary recursive ordinal notation system associated with the "countable part of the n-th level Buchholz notations," which are based on the terms generated by the constants $0, \Omega_1, \dots, \Omega_n$ and the unary functions $\psi_0, \dots, \psi_n$, addition, and ω exponentiation, provided that all outermost occurrences of the ψ_i have $i = 0$. See [Si85], p. 108, and [Bu87], [BS88].

The norm of a term is the total number of operations occurring in that term.

LEMMA 9.1. The following is provable in EFA. For each $n \geq 1$ there is an exponential time assignment J$_n$ of $n+2$-labeled trees of valence 4 to elements x of ERONS(BH$_n$) such that

i) for all x in NERONS(BH$_n$), the height of J$_n(x)$ is at most $|x|$;
ii) for all x, y in NERONS(BH$_n$), if there is an inf and label preserving gap embedding from J$_n(x)$ into J$_n(y)$ then $x \leq' y$.

PROOF. In [Si85], p. 110, for each $n \geq 1$ an assignment J$_n$ of $n+1$-labeled trees of valence ≤ 4 to elements x of NERONS(BH$_n$) is constructed, together with the corresponding assignment of elements of NERONS(BH$_n$) to each vertex. The height of each J$_n(x)$ is at most the norm of x. See [Si85], Lemma 5.10, p. 113. ⊣

LEMMA 9.2. The following is provable in EFA. For each $n \geq 1$ there is an exponential time assignment J$_n$ of full $n+2$-labeled trees of valence 4 to elements x of ERONS(BH$_n$) such that

i) for all x in NERONS(BH$_n$), the height of J$_n(x)$ is at most $|x|$;
ii) for all x, y in NERONS(BH$_n$), if there is an inf and label preserving gap embedding from J$_n(x)$ into J$_n(y)$ then $x \leq' y$.

PROOF. We can routinely fill out J$_n(x)$ by adding suitable nodes with label $n+2$. ⊣

LEMMA 9.3. The following is provable in EFA. Theorem 8.1 implies $(\forall n)$ (CWF(BH$_n$)). For all $n \geq 1$, Theorem 8.1 is not provable in ATI(BH$_n$), and hence not provable in $\prod_1^1$–CA$_0$.

PROOF. Recall that CWF(BH$_n$) asserts that

$$(\forall m \geq 1)(\exists r \geq 1)(\forall x_1, \dots, x_r \in \text{BH}_n)$$
$$((\forall i \leq r)(|x_i| \leq m + i) \rightarrow (\exists i < j)(x_i \leq' x_j)).$$

where $\leq' =$ NERONS(BH_n). Let m be given. By Theorem 8.1, let r be such that the following holds. Let $\mathrm{T}_0, \ldots, \mathrm{T}_r$ be full $n+m+1$-labeled trees of valence 4 or 0, where for all $0 \leq i \leq r$, T_i has height i. There exists $0 \leq i < j \leq r$ and an inf, label, and terminal preserving gap embedding from T_i into T_j.

Let $x_1, \ldots, x_r \in \mathrm{BH}_n$, where $(\forall i \leq r)(|x_i| \leq m+i)$. Look at the full n-labeled trees $\mathrm{J}(x_1), \ldots, \mathrm{J}(x_r)$ of valence 4 or 0 and of heights $m+1, \ldots, m+r$. Let $\mathrm{T}_0, \ldots, \mathrm{T}_m$ be trees of valence 4 or 0 of heights $0, \ldots, m$, where each T_i is labeled throughout with $n+i+1$.

Now apply the condition on r to $\mathrm{T}_0, \ldots, \mathrm{T}_m, \mathrm{J}(x_1), \ldots, \mathrm{J}(x_r)$. We cannot use the T′s, and so we find $1 \leq i < j \leq r$ such that there is an inf, label, and terminal preserving gap embedding from $\mathrm{J}(x_i)$ into $\mathrm{J}(x_j)$. Hence $x_i \leq' x_j$ by Lemma 9.2 ii.

That Theorem 8.1 is not provable in ATI(BH_n) follows from the first claim by Theorem 3.11. That Theorem 8.1 is not provable in $\prod_1^1-\mathrm{CA}_0$ follows from the fact that every $\prod_2^0$ (in fact arithmetical) theorem of $\prod_1^1-\mathrm{CA}_0$ is provable in ATI(BH_n) for some $n \geq 1$ (see Pohlers in [BFPS81]).

The second claim follows from the first claim by Theorem 3.11. ⊣

Theorem 9.4. Both forms of the Internal Finite Tree Gap Embedding Theorem are provable in $\prod_1^1-\mathrm{CA}$ but not in $\prod_1^1-\mathrm{CA}_0$.

Proof. The provability is from Theorem 7.9. The nonprovability follows immediately from Theorem 8.6 and Lemma 9.3. ⊣

In [Kriz89], EKT (Theorem 7.1) was extended to any countable well ordered set of labels. It would appear that this theorem should roughly correspond to "for all well orderings α, the α-th hyperjump exists." However, the proof seems to employ $\prod_2^1$ comprehension, which is far stronger. In addition, appropriate finite forms have not been given.

Another extension of EKT to countable well ordered sets of labels is treated in [Gor90], which is shown to roughly correspond to "for all well orderings α, the α-th hyperjump exists." But appropriate finite forms are not treated.

§10. Some metamathematical equivalences and implications. Theorems 1.8 and 7.6 provide equivalences for most of the infinite Theorems. We close with metamathematical equivalences (and implications) for most of the finite statements considered here. Most of the results are explicitly proved here (with suitable references to the literature), or are implicit (most but not all claims of provability in EFA are explicitly addressed).

Theorem 10.1. The following are provably equivalent in EFA.

i) The Internal Finite Tree Embedding Theorem(1), Theorem 1.3;
ii) The Internal Finite Tree Embedding Theorem(3), Theorem 6.1;
iii) The Internal Finite Tree Embedding Theorem(5), Theorem 6.3;
iv) Theorem 2.1;
v) CWF($< \theta\Omega^{\omega}(0)$);
vi) The 1-consistency of $\prod_2^1-\mathrm{TI}_0$;

vii) The 1-consistency of $\mathrm{ATI}(< \theta\Omega^{\omega}(0))$.

THEOREM 10.2. Each of the following provably implies the next in EFA.

i) The Internal Finite Tree Embedding Theorem(2), Theorem 1.4;
ii) Theorem 2.2;
iii) The 1-consistency of ATR_0;
iv) The 1-consistency of $\mathrm{ATI}(< \Gamma_0)$.
v) The 1-consistency of Feferman's IR.

THEOREM 10.3. The following are provably equivalent in EFA.

i) The Internal Finite Tree Gap Embedding Theorem(1), Theorem 7.7;
ii) The Internal Finite Tree Gap Embedding Theorem(2), Theorem 7.8;
iii) Theorem 8.1;
iv) Theorem 8.2;
v) The 1-consistency of $\prod_1^1-\mathrm{CA}_0$;
vi) The 1-consistency of $\mathrm{ATI}(< \theta\Omega_{\omega}(0))$.

Furthermore, the associated functions of each finite Theorem eventually dominates the provably recursive functions of the formal systems that appear alongside above.

REFERENCES

[Bu87] W. BUCHHOLZ, *An independence result for* $\prod_1^1-\mathrm{CA}+\mathrm{BI}$, ***Annals of Pure and Applied Logic***, vol. 33 (1987), pp. 131–155.

[BCW94] W. BUCHHOLZ, E.A. CICHON, and A. WEIERMANN, *A uniform approach to fundamental sequences and hierarchies*, ***Mathematical Logic Quarterly***, vol. 40 (1994), pp. 273–286.

[BFPS81] W. BUCHHOLZ, S. FEFERMAN, W. POHLERS, and W. SIEG, ***Iterated inductive definitions and subsystems of analysis: Recent proof-theoretical studies***, Springer Lecture Notes in Mathematics, vol. 897, Springer, 1981.

[BS88] W. BUCHHOLZ and K. SCHÜTTE, ***Proof theory of impredicative subsystems of analysis***, Bibliopolis, Naples, 1988.

[dJP77] D.H.J. DE JONG and R. PARIKH, *Well partial orderings and hierarchies*, ***Indigationes Mathematicae***, vol. 39 (1977), pp. 195–206.

[Fe64] S. FEFERMAN, *Systems of predicative analysis*, ***The Journal of Symbolic Logic***, vol. 29 (1964), pp. 1–30.

[Fe68b] ———, *Autonomous transfinite progressions and the extent of predicative mathematics*, ***Logic, methodology and philosophy of science III*** (B. van Rootselaar and J.F. Staal, editors), North-Holland, Amsterdam, 1968, pp. 121–135.

[Fe68a] ———, *Systems of predicative analysis, II: Representations of ordinals*, ***The Journal of Symbolic Logic***, vol. 33 (1968), pp. 193–220.

[Fe79] ———, *A more perspicuous formal system for predicativity*, ***Konstruktionen versus Positionen*** (K. Lorenz, editor), Walter de Gruyter, Berlin, 1979, pp. 68–93.

[Fe91] ———, *Reflecting on incompleteness*, ***The Journal of Symbolic Logic***, vol. 56 (1991), pp. 1–49.

[Fe96] ———, *Gödel's program for new axioms: Why, where, how and what?*, ***Gödel '96*** (P. Hajek, editor), Lecture Notes in Logic, vol. 6, Association for Symbolic Logic in conjunction with A.K. Peters, Ltd., 1996, pp. 3–22.

[FeS00] S. FEFERMAN and T. STRAHM, *The unfolding of non-finitist arithmetic*, ***Annals of Pure and Applied Logic***, vol. 104 (2000), no. 1–3, pp. 75–96.

[Fr81] H. FRIEDMAN, *Independence results in finite graph theory I–VII*, Unpublished manuscripts, Ohio State University, 76 pages, February–March 1981.

[Fr82] ———, *Beyond Kruskal's theorem I–III*, Unpublished manuscripts, Ohio State University, 48 pages, June–July 1982.

[Fr84] ———, *Notes on Kruskal's theorem*, Unpublished handwritten notes, Ohio State University, 163 pages, 1984.

[Fr98] ———, *22:finite trees/predicativity*, FOM e-mail posting,
http://www.math.psu.edu/simpson/fom/ and
http://www.math.ohio-state.edu/foundations/manuscripts.html, October 1998.

[Fr99] ———, *27:finite trees/predicativity:sketches*, FOM e-mail posting,
http://www.math.psu.edu/simpson/fom/ and
http://www.math.ohio-state.edu/foundations/manuscripts.html, January 1999.

[FMS82] H. FRIEDMAN, K. MCALOON, and S. SIMPSON, *A combinatorial principle which is equivalent to the 1-consistency of predicative analysis*, ***Patras logic symposion*** (G. Metakides, editor), Studies in Logic and the Foundations of Mathematics, North-Holland, 1982.

[FRS87] H. FRIEDMAN, N. ROBERTSON, and P. SEYMOUR, *The metamathematics of the graph minor theorem*, ***Logic and combinatorics***, Contemporary Mathematics, vol. 65, American Mathematical Society, 1987, pp. 229–261.

[FS95] H. FRIEDMAN and M. SHEARD, *Elementary descent recursion and proof theory*, ***Annals of Pure and Applied Logic***, vol. 71 (1995), pp. 1–45.

[Gor90] L. GORDEEV, *Generalizations of the Kruskal-Friedman Theorems*, ***The Journal of Symbolic Logic***, vol. 55 (1990), pp. 157–181.

[HP93] P. HAJEK and P. PUDLAK, ***Metamathematics of first-order arithmetic***, Perspectives in Mathematical Logic, Springer, 1993.

[HMSS] L. Harrington, M. Morley, A. Scedrov, and S. Simpson (editors), ***Harvey Friedman's research on the foundations of mathematics***, Studies in Logic and the Foundations of Mathematics, vol. 117, North-Holland, 1985.

[Hi52] HIGMAN, *Ordering by divisibility in abstract algebras*, ***Proceedings of the London Mathematical Society***, vol. 2 (1952), pp. 326–336.

[Kriz89] I. KRIZ, *Well-quasiordering finite trees with gap-condition*, ***Annals of Mathematics***, vol. 130 (1989), pp. 215–226.

[Kriz95] ———, *The structure of infinite Friedman trees*, ***Advances in Mathematics***, vol. 115 (1995), pp. 141–199.

[KT91] I. KRIZ and R. THOMAS, *Ordinal types in Ramsey theory and well-partial-ordering theory*, ***Mathematics of Ramsey theory*** (J. Nesetril and V. Rödl, editors), Springer, 1991.

[Kr60] J.B. KRUSKAL, *Well-quasi-ordering, the tree theorem, and Vazsonyi's conjecture*, ***Transactions of the American Mathematical Society***, vol. 95 (1960), pp. 210–225.

[LM87] M. LOEBL and J. MATOUSEK, *On undecidability of the weakened Kruskal theorem*, ***Logic and combinatorics***, Contemporary Mathematics, vol. 65, American Mathematical Society, 1987, pp. 275–280.

[NW65] C.ST.J.A. NASH-WILLIAMS, *On well-quasi-ordering finite trees*, ***Proceedings of the Cambridge Philosophical Society***, vol. 59 (1965), pp. 105–122.

[RW93] M. RATHJEN and A. WEIERMANN, *Proof-theoretic investigations on Kruskal's theorem*, ***Annals of Pure and Applied Logic***, vol. 60 (1993), pp. 49–88.

[Ro84] H.E. ROSE, ***Subrecursion: Functions and hierarchies***, Oxford Logic Guides, Oxford University Press, 1984.

[Schm79] D. SCHMIDT, ***Well-partial orderings and their maximal order types***, Habilitationsschrift, Heidelberg, 1979.

[Sch54] K. SCHÜTTE, *Kennzeichnung von Ordnungszahlen durch rekursiv erklärte Funktionen*, ***Mathematische Annalen***, vol. 127 (1954), pp. 15–32.

[Sch65] ———, *Predicative well-orderings*, ***Formal systems and recursive functions*** (Crossley and Dummett, editors), North-Holland, 1965, pp. 176–184.

[Sch77] ———, ***Proof theory***, Grundlehren der Mathematischen Wißenschaften, no. 225, Springer-Verlag, 1977.

[Si85] S.G. Simpson, *Nonprovability of certain combinatorial properties of finite trees*, ***Harvey Friedman's research on the foundations of mathematics*** (L. Harrington, M. Morley, A. Scedrov, and S. Simspon, editors), Studies in Logic and the Foundations of Mathematics, vol. 117, North-Holland, 1985, pp. 87–117.

[Si99] ———, ***Subsystems of second order arithmetic***, Perspectives in Mathematical Logic, Springer-Verlag, 1999.

[Sm85] R.L. Smith, *The consistency strengths of some finite forms of the Higman and Kruskal theorems*, ***Harvey Friedman's research on the foundations of mathematics*** (L. Harrington, M. Morley, A. Scedrov, and S. Simspon, editors), Studies in Logic and the Foundations of Mathematics, vol. 117, North-Holland, 1985, pp. 119–136.

[Veb08] O. Veblen, *Continuous increasing functions of finite and transfinite ordinals*, ***Transactions of the American Mathematical Society***, vol. 9 (1908), pp. 280–292.

[Wei] A. Weiermann, *An application of graphical enumeration to PA*, to appear in ***The Journal of Symbolic Logic***.

DEPARTMENT OF MATHEMATICS
OHIO STATE UNIVERSITY
COLUMBUS, OHIO 43210, U.S.A.
E-mail: friedman@math.ohio-state.edu
URL: www.math.ohio-state.edu/~friedman/

FOUNDATIONAL AND MATHEMATICAL USES OF HIGHER TYPES

ULRICH KOHLENBACH[†]

Dedicated to Solomon Feferman for his 70th Birthday

§1. Introduction. A central theme of proof theory is expressed by the following question:

> 'What parts of ordinary mathematics (in particular of analysis) can be carried out in certain restricted formal systems?'

The relevance of this question is twofold:

1) **Foundational relevance:** suppose a formal system $\mathcal{T}_{PA}$ allows one to formalize a great amount of mathematics but can be shown (by restricted means) to be a conservative extension of first order Peano Arithmetic **PA**, then that part of mathematics has an arithmetical foundation (partial realization of H. Weyl's program, see S. Feferman's discussion in [8]).

If we work in a system $\mathcal{T}_{PRA}$ which can be shown (finitistically) even to be conservative over Primitive Recursive Arithmetic **PRA** and identify (following [36]) **PRA** with finitism, then the parts of mathematics which can be carried out in $\mathcal{T}_{PRA}$ have a finitistic foundation (partial realization of D. Hilbert's program, see e.g., [34]).

2) **Mathematical relevance:** here the guiding question is

> 'What more do we know if we have proved a theorem by restricted means than if we merely know that it is true?' (G. Kreisel)

The aim is to get additional mathematical information out of the fact that a certain theorem S has been proved by certain restricted means. Such additional information may be the extractability of a realizing construction for an existential statement or of an algorithm or a numerical bound for a $\forall\exists$-theorem by unwinding the given proof.

Both motivations are of course closely related and research on them has mutually influenced each other: e.g., a proof of a Π^0_2-theorem carried out in a system which can be (effectively) reduced to **PRA** allows one to extract at least a primitive recursive algorithm. In the other direction, e.g., our analysis of proofs in

[†]**BRICS** Basic Research in Computer Science, funded by the Danish National Research Foundation.

Reflections on the foundations of mathematics
Essays in honor of Solomon Feferman
Edited by Wilfried Sieg, Richard Sommer, and Carolyn Talcott

approximation theory (which used the principle of the attainment of the maximum of $f \in C[0,1]$, see [20]) led us to an elimination procedure of weak König's lemma WKL over a variety of subsystems of arithmetic in all finite types thereby contributing to '1)' above (see [19]). Likewise our treatment of e.g., the Bolzano-Weierstraß principle in [26] via an elimination technique of Skolem functions yielded also new conservation results for comprehension principles ([27]).

However, there are also important differences due to the different points emphasized in 1) and 2):

Whereas there are hardly foundational (understood in the sense of Hilbert) reasons to study systems weaker than **PRA**, merely primitive recursive algorithms and bounds are in most cases much too complex to be of any mathematical value. So on the one hand further restrictions are needed to guarantee the extractability of mathematically more interesting data whereas on the other hand e.g., proofs of large classes of lemmas (having a certain logical form) can be shown not to contribute to the complexity or growth of algorithms or bounds extracted from proofs of theorems using these lemmas. Hence such lemmas can be treated simply as axioms (no matter how non-constructive their proofs might be) in the course of the analysis of a given proof. Also, for successful unwindings the complexity of the proof transformations used is critical. It has turned out that methods using functionals of finite type like appropriate versions of Gödel's functional interpretation or modified realizability combined with tools like negative translation and/or the Friedman-Dragalin translation are most useful (in particular compared to techniques which try to avoid any passage through higher types, see [28]).

Whereas we have focused on '2)' in several publications (see [21],[20],[24] among others), this paper addresses '1),' to which S. Feferman has contributed so profoundly. We study mathematical strong, but nevertheless **PRA**-reducible, systems in all finite types, emphasizing the need of third order variables already for a faithful formalization of continuous functions between Polish spaces. We investigate the relationship between the direct representation of continuous functions (which is possible in the presence of third order variables but not in systems based on the language of second order arithmetic) with their representation via certain codes used in the second order context of reverse mathematics (short: r.m.-codes). It turns out that not even a finite type extension of a second order system like RCA together with the axiom of quantifier-free choice suffices to prove that every continuous function $\Phi : \mathbb{N}^{\mathbb{N}} \to \mathbb{N}$ has an r.m.-code. So the encoding used in reverse mathematics tacitly yields a constructively enriched representation of continuous functions. More precisely, already for continuous functions $f : \mathbb{N}^{\mathbb{N}} \to \mathbb{N}$, the representation used in reverse mathematics entails the existence of a (continuous) modulus of pointwise continuity functional. In the presence of arithmetical comprehension, the difference between both representations disappears, since the existence of such a modulus of pointwise continuity can be proved using arithmetical comprehension and QF-AC^{1,0}. For the **restriction** of continuous functionals $\Phi : \mathbb{N}^{\mathbb{N}} \to \mathbb{N}^{\mathbb{N}}$ to the **Cantor space** one can show using WKL that such a code exists. This follows from a construction due to D. Normann which recently was communicated to the author. So in the presence of **WKL**, the constructively enriched encoding of continuous functions used in reverse mathematics can be shown to be faithful (i.e., not to be a genuine enrichment) for functions on the Cantor space (and more generally on **compact** Polish spaces). It remains open whether this is true also without WKL. The higher order context of our systems also allows one to investigate the relation between ε-δ-continuity and sequential continuity of functions (in reverse mathematics, the use of r.m.-codes by definition enforces ε-δ-continuity). It turns out that a form of quantifier-free choice (available in our systems) suffices to prove even the local pointwise equivalence between both concepts.

These types of results can not even be formulated in a second order setting. The use of higher order (at least third order) variables is therefore necessary to address the question of how closely various representations of anaytical objects correspond to their ordinary mathematical definitions and to develop a general theory of representations.

Let us recall very briefly some of the history of research on '1)'. As Feferman pointed out in [7], 'Hermann Weyl initiated a program for the arithmetical foundation of mathematics' in his book 'Das Kontinuum' ([40]). In this book, Weyl observed that large parts of analysis can be developed on the basis of **arithmetical** comprehension. This theme was further developed in the 50's by P. Lorenzen among others. In the late 70's Feferman [5] and G. Takeuti [37] independently designed formal systems based on arithmetical comprehension in the framework of higher order arithmetic which are conservative over **PA**. For this property it is important that the schema of induction is restricted to arithmetical formulas only.[1] Work on the program of so-called reverse mathematics by H. Friedman, S. Simpson and others has shown that almost all of the mathematics that can be developed based on arithmetical comprehension at all can also be carried out if induction is restricted in this way. This work uses a second order fragment $\mathbf{ACA}_0$ (formulated in the language of second order arithmetic) of the system from [5] (which is formulated in the language of functionals of all finite types). Via appropriate representations and codings of higher objects (like continuous functions between Polish spaces) a great deal of mathematics can be developed already in $\mathbf{ACA}_0$ (see [35] for a comprehensive treatment).

Feferman's system, however, allows a more **direct** treatment of such objects and their mathematics and also contains a strong uniform ('explicit') version of arithmetical comprehension via a non-constructive μ-operator. These features hold in an even stronger form for theories with flexible (variable) types which were developed successively by Feferman in his framework of explicit mathematics in [4],[6],[7] culminating in a formal system called **W** (where 'W' stands for 'Weyl')

[1]As was shown by Feferman in [5], the corresponding system with full induction is proof-theoretically stronger than **PA**. In [37], Takeuti considers in addition the variant where no parameters (except arithmetical parameters) are permitted in the schema of arithmetical comprehension. In this case the resulting system is conservative over **PA** even in the presence of the full schema of induction.

which was shown to be proof-theoretically reducible to and conservative over **PA** in [11]. The enormous mathematical power and flexibility of the system **W** led Feferman in [9] to the formulation of the thesis that all (or almost all) scientifically applicable mathematics can be developed in **W**.

In the late 70's, H. Friedman observed that large parts of the mathematics that can be carried out in $\mathbf{ACA}_0$ are already formalizable in a subsystem $\mathbf{WKL}_0$ which instead of the schema of arithmetical comprehension is based on the binary König's lemma (for quantifier-free trees) WKL and Σ^0_1-induction only (see again [35] for a comprehensive treatment of ordinary mathematics in $\mathbf{WKL}_0$). This fact is of foundational relevance since $\mathbf{WKL}_0$ can be proof-theoretically reduced to and is Π^0_2-conservative over **PRA** (H. Friedman (1976, unpublished) and [33]; for a historical discussion which in particular points out various errors in the literature on WKL see [23] (p. 69)).

In [19] we introduced an extension (in the spirit of Feferman's **PA**-conservative system from [5] mentioned above) of $\mathbf{WKL}_0$ to all finite types and proved among other things that this extension still can be proof-theoretically reduced to **PRA** and is Π^0_2-conservative over **PRA**.

Although this extension is already much more flexible than the system $\mathbf{WKL}_0$, the use of WKL still requires a complicated encoding of analytical objects. While working on '2)' mentioned above and investigating what parts of analysis produce only provable recursive function(al)s which can be bounded by polynomials (see [24] for a survey) we faced the problem that already the formulation of WKL involves coding devices of exponential growth. That is why we introduced a non-standard axiom F which together with some form of quantifier-free choice proves a strong principle of uniform boundedness Σ^0_1-UB which allows one to give short proofs of the usual WKL-applications in analysis relative to very weak (polynomially bounded) systems (see [23],[25]) but does not contribute to the growth of provably recursive functionals. This axiom as well as the principle of uniform boundedness is 'non-standard' in the sense that it is not true in the full set-theoretic type structure. Nevertheless all of its analytic (i.e., second order) consequences are true. In [23] we also studied a restricted version F^- of F which yields a correspondingly restricted version of uniform boundedness which is sufficient for many applications (although more complicated to use, see [25]) but which allows a very easy proof-theoretical elimination. In section 3 of this paper we show that in the presence of the axiom of extensionality and a form of quantifier-free choice, F actually is implied by F^- so that in this context (which we use throughout this paper) the F^--elimination applies to proofs based on F as well. The proof of this fact uses an argument due to Grilliot [14]. The result allows one to construct a **PRA**-reducible finite type system $\mathcal{T}^*$ which is based on Σ^0_1-UB. The relevance of this is due to fact that $\mathcal{T}^*$ allows one to develop the analysis of continuous functions between Polish spaces treating such functions directly as certain type-2-functionals and to prove all the usual WKL-consequences known from reverse mathematics without the passage through the encoding of such objects used in reverse mathematics and without formulating explicitly any continuity assumptions.

In section 5-7 we show that Σ^0_1-UB not only allows one to give shorter (coding-free) proofs for usual WKL-applications but also allows one to prove new – classically valid – third order principles which are not derivable from WKL. We develop a non-collapsing hierarchy Φ_n-WKL$_+$ of extensions of WKL. Basically, Φ_n-WKL$_+$ extends WKL from binary trees which are given by quantifier-free predicates to binary trees which are given by formulas belonging to a larger class Φ_n (see section 5 below for details). Φ_0-WKL$_+$ is equivalent to WKL, but for $n \geq 1$, Φ_n-WKL$_+$ is not even provable in E-PA$^\omega$+QF-AC^{1,0} + μ (here μ is Feferman's non-constructive μ-operator mentioned above). Nevertheless, Φ_n-WKL$_+$ is provable in $\mathcal{T}^*$ for all $n \in \mathbb{N}$ so that by the results mentioned before the whole hierarchy can be reduced proof-theoretically to **PRA**.

One might also ask for an explicit version (with flexible types) of such systems based on (extensions of) WKL. However, things are quite delicate in this case as for the uniform ('explicit') version UWKL of WKL (analogously to the uniform version of arithmetical comprehension given by μ), the strength of the resulting system crucially depends on the amount of extensionality available (see [30]).

§2. Description of the theories E-G$_n$A$^\omega$, E-PRA$^\omega$ and E-PA$^\omega$. The set $\boldsymbol{T}$ of all finite types is defined inductively by

$$(i)\ 0 \in \boldsymbol{T} \text{ and } (ii)\ \rho, \tau \in \boldsymbol{T} \Rightarrow \tau(\rho) \in \boldsymbol{T}.$$

Terms which denote a natural number have type 0. Elements of type $\tau(\rho)$ are functions which map objects of type ρ to objects of type τ.

The set $\boldsymbol{P} \subset \boldsymbol{T}$ of pure types is defined by

$$(i)\ 0 \in \boldsymbol{P} \text{ and } (ii)\ \rho \in \boldsymbol{P} \Rightarrow 0(\rho) \in \boldsymbol{P}.$$

Brackets whose occurrences are uniquely determined are often omitted, e.g., we write $0(00)$ instead of $0(0(0))$. Furthermore we write for short $\tau\rho_k \ldots \rho_1$ instead of $\tau(\rho_k)\ldots(\rho_1)$. Pure types can be represented by natural numbers: $0(n) := n+1$.

Our theories $\mathcal{T}$ used in this paper are based on many–sorted classical logic formulated in the language of functionals of all finite types plus the combinators $\Pi_{\rho,\tau}, \Sigma_{\delta,\rho,\tau}$ which allow the definition of λ–abstraction.

The systems E-G$_n$A$^\omega$ (for all $n \geq 1$) are introduced in [23] to which we refer for details. E-G$_n$A$^\omega$ has as primitive relations $=_0, \leq_0$ for objects of type 0, the constant 0^0, functions $\min_0, \max_0, S$ (successor), $A_0, \ldots, A_n$, where A_i is the i–th branch of the Ackermann function (i.e., $A_0(x,y) = y', A_1(x,y) = x+y, A_2(x,y) = x \cdot y, A_3(x,y) = x^y, \ldots$), functionals of degree 2: $\Phi_1, \ldots, \Phi_n$, where $\Phi_1 fx = \max_0(f0, \ldots, fx)$ and Φ_i is the iteration of A_{i-1} on the f–values for $i \geq 2$, i.e., $\Phi_2 fx = \sum_{i=0}^{x} fi, \Phi_3 fx = \prod_{i=0}^{x} fi, \ldots$. We also have a bounded search functional μ_b and bounded predicative recursion provided by recursor constants

$\tilde{R}_\rho$ (where 'predicative' means that recursion is possible only at the type 0 as in the case of the (unbounded) Kleene-Feferman recursors $\widehat{R}_\rho$). In this paper our systems always contain the axioms of extensionality

$$(E) : \forall x^\rho, y^\rho, z^{\tau\rho}(x =_\rho y \to zx =_\tau zy)$$

for all finite types ($x =_\rho y$ is defined as $\forall z_1^{\rho_1}, \dots, z_k^{\rho_k}(xz_1 \dots z_k =_0 yz_1 \dots z_k)$ where $\rho = 0\rho_k \dots \rho_1$).

In [23] we had in addition to the defining axioms for the constants of our theories all true sentences having the form $\forall x^\rho A_0(x)$, where A_0 is quantifier–free and $deg(\rho) \leq 2$, added as axioms.[2]

By 'true' we refer to the full set–theoretic model $\mathcal{S}^\omega$. In given proofs of course only very special universal axioms are used which can be proved in suitable extensions of our theories. Nevertheless one can include them all as axioms if one is only interested in the applied aspect '2)' discussed above, since they (more precisely their proofs) do not contribute to the provable recursive function(al)s of the system. In particular this covers all instances of the schema of quantifier-free induction. In this paper, however, we include only the schema of quantifier-free induction to E-$G_n A^\omega$ instead of taking arbitrary universal axioms, since we are interested in proof-theoretical reductions.

E-PRA$^\omega$ results if we add the functional

$$\Phi_{it}0yf =_0 y, \; \Phi_{it}x'yf =_0 f(x, \Phi_{it}xyf)$$

to E-$G_\infty A^\omega := \bigcup_{n\in\omega}$ E-$G_n A^\omega$. The system E-PRA$^\omega$ is equivalent to Feferman's system E-$\widehat{\text{PA}}^\omega\!\restriction$ from [5] since Φ_{it} allows (relative to E-$G_\infty A^\omega$) to define the predicative recursor constants $\widehat{R}_\rho$ (see [23]).

E-PA$^\omega$ is the extension of E-PRA$^\omega$ obtained by the addition of the schema of full induction and all (impredicative) primitive recursive functionals in the sense of [13].

The schema of full choice is given by

$$\text{AC}^{\rho,\tau} : \forall x^\rho \exists y^\tau A(x,y) \to \exists Y^{\tau(\rho)} \forall x^\rho A(x, Yx), \;\; \text{AC} := \bigcup_{\rho,\tau\in \boldsymbol{T}} \{\text{AC}^{\rho,\tau}\}.$$

The schema of **quantifier-free choice** QF-AC$^{\rho,\tau}$ is defined as the restriction of AC$^{\rho,\tau}$ to quantifier-free formulas A_0.[3]

REMARK 2.1. E-PRA$^\omega$+QF-AC$^{0,0} \vdash \Sigma^0_1$-IA, Δ^0_1-CA, where

Σ^0_1-IA: $\exists y^0 A_0(0,y) \wedge \forall x^0(\exists y^0 A_0(x,y) \to \exists y^0 A_0(x',y)) \to \forall x^0 \exists y^0 A_0(x,y)$,

and

Δ^0_1-CA: $\forall x^0(\exists y^0 A_0(x,y) \leftrightarrow \forall y^0 B_0(x,y)) \to \exists f^1 \forall x^0(fx = 0 \leftrightarrow \exists y^0 A_0(x,y))$,

with A_0, B_0 quantifier-free (parameters of arbitrary types allowed).

---

[2]The restriction $deg(\rho) \leq 2$ has a technical reason discussed in [23].

[3]Throughout this paper $A_0, B_0, C_0, \dots$ denote quantifier-free formulas.

Hence the system RCA_0 from reverse mathematics (see [35]) can be viewed as a subsystem of E-PRA$^\omega$+QF-AC^{0,0} by identifying sets $X \subseteq \mathbb{N}$ with their characteristic function.

The theory $\mathcal{T} + \mu$ results from $\mathcal{T}$ if we add the non-constructive μ-operator μ^2 to $\mathcal{T}$ together with the characterizing axiom

$$\mu(f) = \begin{cases} \text{the least } x \text{ such that } f(x) =_0 0, \text{ if } \exists x^0(f(x) =_0 0) \\ 0, \text{ otherwise.} \end{cases}$$

Notation: For $\rho = 0\rho_k \dots \rho_1$, we define $1^\rho := \lambda x_1^{\rho_1} \dots x_k^{\rho_k}.1^0$, where $1^0 := S0$.

Defnition 2.2. 1) Between functionals of type ρ we define the relation $\leq_\rho$:

$$\begin{cases} x_1 \leq_0 x_2 :\equiv x_1 \leq x_2, \\ x_1 \leq_{\tau\rho} x_2 :\equiv \forall y^\rho(x_1 y \leq_\tau x_2 y); \end{cases}$$

2) $\min_{\rho\tau}(x_1^{\rho\tau}, x_2^{\rho\tau}) := \lambda y^\tau. \min_\rho(x_1 y, x_2 y)$, with min_0 from above.

In the following we will need the definition of the binary ('weak') König's lemma as given in [39]:

Defnition 2.3 (Troelstra(74)).

$$\text{WKL} :\equiv \begin{cases} \forall f^1(T(f) \wedge \forall x^0 \exists n^0(lth\ n =_0 x \wedge fn =_0 0) \\ \qquad \to \exists b \leq_1 \lambda k.1 \forall x^0(f(\overline{b}x) =_0 0)), \end{cases}$$

where

$Tf :\equiv \forall n^0, m^0(f(n * m) =_0 0 \to fn =_0 0) \wedge \forall n^0, x^0(f(n * \langle x \rangle) =_0 0 \to x \leq_0 1)$ (i.e., $T(f)$ asserts that f represents a 0,1–tree).

§3. On two non-standard principles.

In this section we in particular prove a new conservation result for the non-standard axiom F which was introduced first in [23][4] (and has been applied e.g., in [25]):

$$F :\equiv \forall \Phi^{2(0)}, y^{1(0)} \exists y_0 \leq_{1(0)} y \forall k^0 \forall z \leq_1 yk(\Phi kz \leq_0 \Phi k(y_0 k)).$$

We call this axiom 'non-standard' since it does not hold in the full set-theoretic type structure $\mathcal{S}^\omega$. Nevertheless its use can be eliminated from certain proofs thereby yielding classically true results. This has been discussed extensively in [23] to which we refer for further information. In that paper we mainly made use of a weaker version F^- of F which allows a direct proof-theoretic elimination whereas the elimination of F was based on a model-theoretic argument. In this paper however we need the full version F. We show – using an argument known as Grilliot's trick in the context of recursion theory for the countable functionals (see [14])[5] – that in the fully extensional context of theories like E-PRA$^\omega$+QF-AC^{1,0},

[4]Special case of F was studied already in [21] and called also F in that paper but F_0 in [23].

[5]This argument recently has had further proof-theoretic applications in [30] and [31].

F^- actually implies F. This allows one to apply the proof-theoretic elimination of F^- to F thereby strengthening results in [23].

We apply F via one of its consequences, the following principle of uniform Σ^0_1-boundedness:

DEFINITION 3.1 ([23]). The schema[6] of **uniform Σ^0_1–boundedness** is defined as

$$\Sigma^0_1\text{–UB}: \begin{cases} \forall y^{1(0)}\big(\forall k^0 \forall x \leq_1 yk \exists z^0\, A(x,y,k,z) \\ \qquad\qquad \to \exists \chi^1 \forall k^0 \forall x \leq_1 yk \exists z \leq_0 \chi k\, A(x,y,k,z)\big), \end{cases}$$

where $A \equiv \exists \underline{l} A_0(\underline{l})$ and $\underline{l}$ is a tuple of variables of type 0 and A_0 is a quantifier–free formula (which may contain parameters of arbitrary types).

PROPOSITION 3.2 ([23]). *Let* $\mathcal{T}$:=E-G$_n$A$^\omega$ $(n \geq 2)$, E-PRA$^\omega$ *or* E-PA$^\omega$. *Then* $\mathcal{T}$+QF-AC^{1,0} + $F \vdash \Sigma^0_1$-UB.

PROPOSITION 3.3 ([23]). E-G$_3$A$^\omega$ + Σ^0_1-UB $\vdash$ WKL.

Σ^0_1-UB implies the existence of a modulus of uniform continuity for each extensional $\Phi^{1(1)}$ on $\{z^1 : z \leq_1 y\}$ (where 'continuity' refers to the usual metric on the Baire space $\mathbb{N}^{\mathbb{N}}$):

PROPOSITION 3.4 ([23]).

$$\text{E-G}_2\text{A}^\omega + \Sigma^0_1\text{-UB} \vdash$$
$$\forall \Phi^{1(1)} \forall y^1 \exists \chi^1 \forall k^0 \forall z_1, z_2 \leq_1 y\Big(\bigwedge_{i \leq_0 \chi k} (z_1 i =_0 z_2 i) \to \bigwedge_{j \leq_0 k} (\Phi z_1 j =_0 \Phi z_2 j)\Big).$$

REMARK 3.5. The argument above can actually be used to show that a sequence of functionals $\Phi_i^{1(1)}$ has a sequence of moduli of uniform continuity on a sequence of sets $\{z : z \leq_1 y_i\}$.

As mentioned above, in [23] we mainly studied a weaker version

$$F^- :\equiv$$
$$\forall \Phi^{2(0)}, y^{1(0)} \exists y_0 \leq_{1(0)} y \forall k^0, z^1, n^0 \Big(\bigwedge_{i <_0 n} (zi \leq_0 yki) \to \Phi k(\overline{z,n}) \leq_0 \Phi k(y_0 k)\Big)$$

(where, for $z^{\rho 0}$, $(\overline{z,n})(k^0) :=_\rho zk$, if $k <_0 n$ and $:= 0^\rho$, otherwise) of F and gave a proof-theoretic elimination procedure for the use of F^- which – relative to so-called weakly extensional variants WE-G$_n$A$^\omega$+QF-AC of our systems E-G$_n$A$^\omega$+QF-AC^{1,0}+QF-AC^{0,1} – applies for quite general classes of formulas. In the presence of the full extensionality axiom (E) we got corresponding results if the types involved were somewhat restricted. We now show that in the presence of (E), F is already implied by F^- and so that these results extend to F as well.

PROPOSITION 3.6. E-G$_2$A$^\omega$+QF-AC^{1,0} + $F^- \vdash F$.

[6] Σ^0_1-UB can be written as a single axiom. However the schematic version is easier to apply.

PROOF. We argue in E-G_2A^ω+QF-$AC^{1,0}$ + F^-. It is clear that F follows from F^- if

$$(1)\ \forall\Phi^2\forall f^1\exists n^0(\Phi(f) = \Phi(\overline{f,n})).$$

So let's suppose that on the contrary there exist Φ^2 and f such that

$$(2)\ \forall n^0(\Phi(f) \neq \Phi(\overline{f,n})).$$

Then for $f_i := \overline{f, i+1}$ we have

$$(3)\ \forall i^0\forall j \geq i(f_j(i) =_0 f(i))$$

and

$$(4)\ \forall i^0(\Phi(f) \neq \Phi(f_i)).$$

Define $\Psi g^1 :=_0 \begin{cases} 1, \text{ if } \Phi(g) \neq \Phi(f) \\ 0, \text{ if } \Phi(g) = \Phi(f). \end{cases}$ Then

$$(5)\ \forall i,j(\Psi(f_i) =_0 \Psi(f_j) \neq \Psi(f)).$$

Now one can apply an argument from [14], which can be formalized in E-G_2A^ω (see [30] for details on this and a further proof-theoretic application of that argument), to derive

$$(6)\ \exists\varphi^2\forall g^1(\varphi(g) = 0 \leftrightarrow \exists x(gx = 0))$$

from (3) and (5). (6), however, contradicts F^- (relative to E-G_2A^ω+QF-$AC^{1,0}$), since F^- implies that every Φ^2 is bounded on the set of all functions $\overline{g,n}$ with $g \leq_1 1, n \in \mathbb{N}$, whereas QF-$AC^{1,0}$ together with (6) yields the existence of a functional μ such that

$$(7)\ \forall g^1(\exists x^0(gx = 0) \rightarrow g(\mu(g)) = 0),$$

which obviously is unbounded on this set. □

THEOREM 3.7. *Let $\forall f^1, x^0\exists y^0 A_0(f,x,y)$ be a sentence of the language of $\mathcal{T}$ where $\mathcal{T}$:=E-G_nA^ω ($n \geq 2$), E-PRA^ω or E-PA^ω. Then the following rule holds*

$$\begin{cases} \mathcal{T} + \text{QF-AC}^{1,0}+\text{QF-AC}^{0,1} + F \vdash \forall f^1, x^0\exists y^0 A_0(f,x,y) \\ \Rightarrow \text{ one can extract a closed term } \Psi^{001} \text{ of } \mathcal{T} \text{ such that} \\ \mathcal{T}^* \vdash \forall f^1, x^0 A_0(f,x,\Psi f x), \end{cases}$$

where $\mathcal{T}^*$:=E-G_3A^ω if $\mathcal{T}$ =E-G_2A^ω and := $\mathcal{T}$, otherwise.

In particular, if $\mathcal{T}$ =E-G_2A^ω (E-G_3A^ω resp. E-PRA^ω) and if 'f' is not present, then Ψx is bounded by a polynomial in x (is an elementary recursive resp. primitive recursive function in x).

PROOF. The theorem follows from proposition 3.6 together with theorem 4.21 from [23]. □

§4. Continuous functions: direct representations versus codes. A functional $\Phi^{1(1)}$ is continuous at x^1 if

$$\forall k^0 \exists n^0 \forall y^1 \big(\bigwedge_{i=0}^{n} (xi =_0 yi) \to \bigwedge_{j=0}^{k} (\Phi xj =_0 \Phi yj) \big).$$

Φ is continuous if it is continuous at every x.

Using a suitable so-called standard representation of complete separable metric ('Polish') spaces X (which in turn relies on a representation of real numbers as Cauchy sequences of rational numbers with fixed rate of convergence), elements of X can be represented by number-theoretic functions x^1 and, moreover, every such function can be considered as a representative of a uniquely determined element of X (see [2] and [20] for details). On these representatives we have a pseudo metric d_X. The elements of X can be identified with the equivalence classes w.r.t. $x =_Y x :\equiv (d_X(x, y) =_{\mathbb{R}} 0)$. Functions $G : X \to Y$ between Polish spaces therefore are just given by functionals $\Phi_G^{1(1)}$ which respect $=_X, =_Y$, i.e.,

$$\forall x^1, y^1 (x =_X y \to \Phi_G x =_Y \Phi_G y).$$

Φ_G represents a continuous function $G : X \to Y$ if

$$\forall x^1 \forall k^0 \exists n^0 \forall y^1 \big(d_X(x, y) \leq_{\mathbb{R}} \frac{1}{n+1} \to d_Y(\Phi_G x, \Phi_G y) \leq_{\mathbb{R}} \frac{1}{k+1} \big).$$

This definition is just the usual ε-δ–definition of continuous functions. One could also consider to define continuity as sequential continuity. In the presence of QF-AC^{0,1} (which is included in all the systems we consider in this paper) both definitions are equivalent as we will show now.

As usual $G : X \to Y$ is called **sequentially continuous** in x iff

$$\forall x_{(\cdot)}^{1(0)} \big(\lim_{n \to \infty} x_n =_X x \to \lim_{n \to \infty} \Phi_G(x_n) =_Y \Phi_G(x) \big),$$

where $(\lim_{n \to \infty} x_n =_X x) :\equiv \forall k^0 \exists n^0 \forall m \geq_0 n (d_X(x_m, x) \leq \frac{1}{k+1})$.

PROPOSITION 4.1. *The theory* E-G$_3$A$^\omega$+QF-AC^{0,1} *proves that for all functions* $G : X \to Y$ *and points* $x \in X$:

G is sequentially continuous at x $\leftrightarrow$ G is ε–δ–continuous at x.

PROOF. '$\leftarrow$': Obvious!

'$\to$': Suppose that G is not ε–δ–continuous at x, i.e.,

$$(*)\ \exists k^0 \forall n^0 \exists y^1 \underbrace{\big(d_X(x, y) <_{\mathbb{R}} \frac{1}{n+1} \wedge d_Y(\Phi_G(x), \Phi_G(y)) >_{\mathbb{R}} \frac{1}{k+1} \big)}_{\equiv: A \in \Sigma_1^0}.$$

By coding pairs of natural numbers and numbers into functions one can express $\exists y^1 A$ in the form $\exists y^1 A_0$. Hence QF-AC^{0,1} applied to $(*)$ yields

$$\exists k^0, \xi^{1(0)} \forall n^0 \big(d_X(x, \xi n) <_{\mathbb{R}} \frac{1}{n+1} \wedge d_Y(\Phi_G(x), \Phi_G(\xi n)) >_{\mathbb{R}} \frac{1}{k+1} \big),$$

i.e., $(\xi n)_{n\in\mathbb{N}}$ represents a sequence of elements of X which converges to x. But $\lim_{n\to\infty} \Phi_G(\xi n) \neq_{\mathbb{R}} \Phi_G(x)$ and thus G is not sequentially continuous at the point represented by x. □

REMARK 4.2. The use of QF-AC^{0,1} in the proof of '→' in the proposition above is unavoidable already for $X = Y = \mathbb{R}$ since in this case the implication is known to be unprovable even in Zermelo–Fraenkel set theory ZF, see [16],[15] and [12].

We now discuss the indirect representation of continuous functions $G : X \to Y$ between Polish spaces X, Y via codes g as used in the context of reverse mathematics (see definition II.6.1 in [35]). Since reverse mathematics takes place in the language of second-order arithmetic (instead of a language with higher types), the direct representation of such continuous function which is available in our systems is not possible. We will show that provably in E-G$_3$A$^\omega$+QF-AC^{1,0}, for every such code g there exists a direct representation in our sense of the function coded by g, but that the reverse direction in general is not even provable in E-PA$^\omega$+QF-AC. The latter phenomenon is due to the fact that the indirect representation of continuous functions G via codes g tacitly yields a constructive enrichment of the direct representation of G by a modulus of pointwise continuity. To be more specific, let us consider the special case X = Baire space, $Y = \mathbb{N}$ (with the usual metrics). Then the existence of a code g for a continuous functional Φ^2 is (relative to E-G$_3$A$^\omega$+QF-AC^{1,0}) equivalent to the existence of a continuous modulus of pointwise continuity functional Ψ^2 for Φ^2 which in turn is equivalent to the existence of an associate of Φ in the sense of the Kleene/Kreisel countable functionals.

DEFNITION 4.3. 1) α^1 is a neighborhood function if
(a) $\forall \beta^1 \exists n^0 (\alpha(\overline{\beta}n) > 0)$ and
(b) $\forall m, n (m \sqsubseteq n \wedge \alpha(m) > 0 \to \alpha(m) = \alpha(n))$, where '$m \sqsubseteq n$' expresses the (elementary recursive) predicate that the sequence encoded by m is an initial segment of the one encoded by n.

2) α^1 is an associate of Φ^2 if
(a) $\forall \beta^1 \exists n^0 (\alpha(\overline{\beta}n) > 0)$ and
(b) $\forall \beta, n (n \text{ least s.t. } \alpha(\overline{\beta}n) > 0 \to \alpha(\overline{\beta}n) = \Phi\beta + 1)$.

Without loss of generality we may assume that an associate of Φ^2 is a neighborhood function, since otherwise we define

$$\tilde{\alpha}(n) := \begin{cases} \alpha(m), & \text{for } m \text{ shortest initial segment of } n \text{ s.t. } \alpha(m) > 0, \text{ if } \exists. \\ 0, & \text{otherwise.} \end{cases}$$

PROPOSITION 4.4. E-G$_3$A$^\omega$+QF-AC^{1,0} *proves* (*uniformly in* Φ^2) *that the following properties are pairwise equivalent*:

1) $\exists f(f$ is an r.m.-code of $\Phi)$, where 'r.m.' abbreviates 'reverse math'.[7]
2) $\exists \alpha^1(\alpha$ is an associate of $\Phi)$,
3) $\exists \omega^2_\Phi(\omega_\Phi$ is a continuous modulus of pointwise continuity for $\Phi)$.

PROOF. '1) $\rightarrow$ 3)': Let f be a r.m.-code of Φ^2. Since Φ is total, we have[8]

$$\forall \beta^1 \exists a^0, r^0, b^0, s^0 \big(d(\beta, \lambda i.(a)_i) <_{\mathbb{R}} 2^{-r} \wedge (a,r)f(b,s) \wedge 2^{-s} <_{\mathbb{Q}} 1\big)$$

and hence

$$\forall \beta^1 \exists a^0, r^0, b^0, s^0, l^0 \big(\underbrace{d(\beta, \lambda i.(a)_i) + 2^{-l} <_{\mathbb{R}} 2^{-r} \wedge (a,r)f(b,s) \wedge 2^{-s} <_{\mathbb{Q}} 1}_{\equiv : \exists v^0 A_0(f,\beta,a,r,b,s,l,v)} \big),$$

where A_0 is quantifier-free. By quantifier-free induction and QF-AC^{1,0} we obtain a functional X^2 such that

$$\forall \beta \big(X\beta \text{ minimal s.t. } A_0(f, \beta, v^6_1(X\beta), \ldots, v^6_6(X\beta))\big).$$

It is clear that X is continuous[9] and that $\Phi\beta = v^6_3(X\beta)$. With X, also

$$\omega_\Phi \beta :=_{\mathbb{Q}} 2^{-v^6_5(X\beta)}$$

is continuous. One easily verifies that ω_Φ is a modulus of pointwise continuity for Φ.

'3) $\rightarrow$ 2)': Let ω_Φ be a continuous modulus of pointwise continuity for Φ^2. Then

$$(1)\ \forall \beta, \gamma \big(\overline{\beta}(\omega_\Phi \beta) =_0 \overline{\gamma}(\omega_\Phi \beta) \rightarrow \Phi\beta =_0 \Phi\gamma\big)$$

and

$$(2)\ \forall \beta \exists n^0 \big(\omega_\Phi(\overline{\beta, n}) \leq n\big)$$

(where $\overline{\beta, n}$ is the continuation of $\overline{\beta}n$ with 0).

Define

$$\alpha(n) := \begin{cases} \Phi(\lambda i.(n)_i) + 1, & \text{if } \omega_\Phi(\lambda i.(n)_i) \leq lth(n) \\ 0, & \text{otherwise.} \end{cases}$$

(2) yields

$$\forall \beta \exists k \big(\alpha(\overline{\beta}k) > 0\big).$$

Assume that $\alpha(\overline{\beta}k) > 0$, then – by (1) and the definition of α – $\omega_\Phi(\overline{\beta, k}) \leq k \wedge \Phi(\overline{\beta, k}) = \Phi\beta$ and therefore $\alpha(\overline{\beta}k) = \Phi\beta + 1$.

---

[7] By 'r.m.-code' we here refer to definition II.6.1 in [35] specialized to $\widehat{A} := \mathbb{N}^{\mathbb{N}}$ and $\widehat{B} := \mathbb{N}$. We identify the set Φ in that definition with its characteristic function f.

[8] As in reverse mathematics we represent real numbers as Cauchy sequences with fixed rate of convergence. As a consequence of this, $<_{\mathbb{R}} \in \Sigma^0_1$. Analogously to [35](II.6.1) we write $(a,r)f(b,s)$ as abbreviation for the Σ^0_1-formula $\exists n^0(f(a,r,b,s,n) =_0 0)$.

[9] Here we use the fact that $A_0(f,\beta,a,r,b,s,l,v)$ can be written as $t_{A_0}(f,\beta,a,r,b,s,l,v) =_0 0$ for a suitable closed term t_{A_0} of E-G$_3$A$^\omega$ and that every closed term t^2 of E-G$_3$A$^\omega$ is provably pointwise continuous.

'2) $\to$ 1)': Let α be an associate for Φ. By the remark above we may assume that α is a neighborhood function. Define an r.m.-code f for Φ by

$$(a,r)f(b,s) :\equiv \alpha((\overline{\lambda i.(a)_i})r) > 0 \wedge |(\alpha((\overline{\lambda i.(a)_i})r) - 1) - b| < 2^{-s}.$$

This is a quantifier-free (and hence Σ^0_1-)predicate (which we identify with its characteristic function). It is straightforward to verify that f satisfies the properties of an r.m.-code and that f is a code for Φ. We omit the tedious details. □

REMARK 4.5. For the equivalence between 2) and 3), see also [2] (p. 143, E.8).

THEOREM 4.6. E-PA$^\omega$+QF-AC^{1,0}+QF-AC^{0,1} *does not prove that every continuous functional* Φ^2 *has an r.m.-code (i.e., that* Φ *is continuous in the sense of reverse mathematics).*

PROOF. In [32](6.4) a type-structure $A = \langle A_k \rangle_{k\in\mathbb{N}}$ over ω is constructed with the following properties:

(i) $E_2{\restriction}A_1 \notin A_2$, where $E_2(f^1) = 0 \leftrightarrow \exists x(fx = 0)$;

(ii) A is closed under computation in the sense of Kleene's schemata S1-S9.

(iii) there exists a $\Phi \in A_2$ such that Φ has no associate in A_1. By (ii), A is a model of the restriction of E-PA$^\omega$+QF-AC^{1,0} to the fragment with pure types only. Modulo the well-known reduction to pure types (see [38](1.8.5-1.8.8)), E-PA$^\omega$+QF-AC^{1,0} therefore has a model in which there exists a functional Φ^2 which has no associate and therefore – by the previous proposition – no r.m.-code f. Nevertheless, all functionals Φ^2 of type 2 are continuous: one could use here an argument due to [14] to show that the existence of a non-continuous functional in A_2 would contradict (i). However, it requires some care to verify that this argument (which usually is formulated for the full type-structure) relativises to A. We therefore use directly the construction of A which is based on a certain type-2 functional $F : \mathbb{N}^\mathbb{N} \to \mathbb{N}$ (constructed by L. Harrington using a complicated priority construction, see [32](4.21)) which has the following properties

(i) F is continuous (and therefore has an associate in $\mathbb{N}^\mathbb{N}$),

(ii) $F{\restriction}$REC is not computable (in the sense of S1-S9) and therefore has no recursive associate,

(iii) 1-sc(F) =REC, where 1-sc(F) is the set of functions computable in F.

A_1 :=REC, $A_{k+1} := \{\Phi : A_k \to \mathbb{N} : \Phi$ computable in $F{\restriction}$REC.$\}$

It is clear that every $\Phi \in A_2$ is continuous.

As a further consequence of this, QF-AC^{0,1} reduces in A to QF-AC^{0,0} since $\forall x^0 \exists f^1 A_0(x,f) \to \forall x^0 \exists y^0 A_0(\lambda i.(y)_i)$. So $A \models$ QF-AC^{0,1}. □

The next proposition shows that in the presence of arithmetical comprehension Π^0_∞-CA : $\exists f^1 \forall x^0 (f(x) = 0 \leftrightarrow A(x))$, where A is arithmetical with arbitrary parameters, every continuous function on $\mathbb{N}^\mathbb{N}$ has a code in the sense of reverse mathematics:

PROPOSITION 4.7. E-PRA$^\omega$+QF-AC^{1,0} + Π^0_∞-CA *proves that every continuous functional* $\Phi^{1(1)}$ *has an r.m.-code (i.e., that* Φ *is continuous in the sense of reverse mathematics).*

PROOF. Let $\Phi^{1(1)}$ be pointwise continuous. Then

$$(+)\forall f^1, x^0 \exists y^0 \forall i^0, j^0 \big(\Phi(x, \overline{f}y * \lambda k.(i)_k) =_0 \Phi(x, \overline{f}y * \lambda k.(j)_k)\big),$$

where $(a^0 * f^1)(k) := (a)_k$ for $k < lth(a)$ and $:= f(k - lth(a))$ otherwise.

By Π^0_∞-CA there exists a function χ such that

$$\forall x^0, a^0 \big(\chi(x,a) =_0 0 \leftrightarrow \forall i, j \big(\Phi(x, a * \lambda k.(i)_k) =_0 \Phi(x, a * \lambda k.(j)_k)\big)\big).$$

Hence $(+)$ can be written as

$$\forall f, x \exists y \big(\chi(x, \overline{f}y) =_0 0\big).$$

By QF-AC^{1,0} and QF-IA we obtain a functional $\tilde{\omega}(f,x)$ such that

$$\forall f, x \big(\chi(x, \overline{f}(\tilde{\omega}(f,x)) =_0 0 \wedge \forall z < \tilde{\omega}(f,x)(\chi(x, \overline{f}z) \neq 0)\big).$$

By the pointwise continuity of Φ one easily verifies that

$$\omega(f,x) := \max_{i \leq x}(\tilde{\omega}(f,i))$$

is a (pointwise continuous) modulus of (pointwise) continuity of Φ. The claim now follows with proposition 4.4. □

PROPOSITION 4.8. E-PA$^\omega$+QF-AC+'all continuous functionals $\Phi^{1(1)}$ have an r.m.-code' *is* Π^1_∞*-conservative over* E-PA$^\omega$+QF-AC^{0,0}.

PROOF. Formalizing the fact that the extensional continuous functionals ECF form a model of the first theory and the proof for the faithfulness of this model for the analytical fragment (see [38](2.6.5-2.6.12)), a proof of $A \in \Pi^1_\infty$ in the first theory translates into a proof of $[A]_{\text{ECF}}$ (and hence of A) in E-PA$^\omega$+QF-AC^{0,0}. □

COROLLARY 4.9. E-PA$^\omega$+QF-AC+'all continuous functionals $\Phi^{1(1)}$ have an r.m.-code' *proves neither* Π^0_∞-CA *nor* WKL.

PROOF. The corollary follows from proposition 4.8 and the fact that the hereditarily effective operations **HEO** form a model of E-PA$^\omega$+QF-AC^{0,0} (see [38](2.4.11, 2.6.13,2.6.20)) but not of WKL (and hence a-fortiori not of Π^0_∞-CA). □

The fact that the representation of continuous functions in reverse mathematics via codes goes together with a constructive enrichment is used in many proofs of basic properties of continuous functions in the system WKL$_0$. So the question arises whether WKL is sufficient to prove the same results for our direct representation. We discuss this for simplicity again for the case of continuous functions $\Phi : \mathbb{N}^\mathbb{N} \to \mathbb{N}$. As we have seen above, reverse mathematics treats Φ via an associate α^1. This representation allows one to prove the uniform continuity of Φ on the Cantor space of all 0-1-functions by WKL as follows. Define a binary tree by

$$f(n) := \begin{cases} 1, \text{ if } \forall i < lth(n)\big((n)_i \leq 1\big) \wedge \alpha(n) > 0 \\ 0, \text{ otherwise.} \end{cases}$$

Since we may assume that α is a neighborhood function, f satisfies $T(f)$. The contraposition of WKL applied to f yields

$$\forall\beta \leq_1 1\exists x^0(\alpha(\overline{\beta}x) > 0) \to \exists x\forall\beta \leq_1 1(\alpha(\overline{\beta}x) > 0),$$

i.e., $\Phi\beta = \alpha(\overline{\beta}\min n[\alpha(\overline{\beta}n) > 0]) - 1$ is uniformly continuous on $\{\beta : \beta \leq_1 1\}$.

This argument can be adopted to real functions encoded as in reverse mathematics to show in that context that e.g., every continuous function $f : [0,1] \to \mathbb{R}$ is uniformly continuous. Together with QF-AC^{0,0} one even gets a modulus of uniform continuity (see proposition 4.10).

In our direct type-2-treatment of continuous functions $\Phi : \mathbb{N}^{\mathbb{N}} \to \mathbb{N}$ as functionals Φ^2 satisfying

$$\forall f^1\exists n^0\forall g^1(\overline{f}n = \overline{g}n \to \Phi f = \Phi g),$$

the binary tree to which we have to apply König's lemma in order to prove the uniform continuity of Φ on $\{f : f \leq_1 1\}$ is given by

$$Tree(n) :\equiv \exists g, h \leq_1 1\Big(\bigwedge_{i<lth(n)} (g(i) = (n)_i = h(i)) \wedge \Phi g \neq \Phi h\Big)$$

which no longer is quantifier-free and apparently does not possess a characteristic function in E-PA$^\omega$+QF-AC^{1,0} which would be necessary to apply WKL right away. However, a construction due to D. Normann[10], which easily can be formalized in our setting, shows that WKL allows one to prove the existence of a characteristic function χ_T for $Tree(n)$ (for pointwise continuous functionals Φ). Applying then WKL to '$\chi_T(n) = 0$' yields the uniform continuity (and – together with QF-AC^{0,0} – even a modulus of uniform continuity) of the restriction of $\Phi^{1(1)}$ to the Cantor space. In particular this implies that this restriction has an r.m.-code, but note that even the proof of this corollary, which doesn't mention the uniform continuity of Φ, uses WKL.

Proposition 4.10. E-PRA$^\omega$+QF-AC^{0,0}+WKL *proves that if the restriction of a functional* $\Phi^{1(1)}$ *to the Cantor space* 2^ω *is pointwise continuous then it has a modulus of uniform continuity.*

Proof. We prove the proposition in two steps. We first show (using QF-AC^{0,0} and WKL) that the modulus of uniform continuity can be constructed if we have a characteristic function χ_T for

$$(1)\ T(k^0, n^0) :\equiv \exists g, h \leq_1 1\Big(\bigwedge_{i<lth(n)} (g(i) = (n)_i = h(i)) \wedge \Phi(g,k) \neq \Phi(h,k)\Big)$$

and then show the existence of χ_T using again WKL.

Let χ_T be such that

$$(2)\ \forall k, n\big(\chi_T(k,n) = 0 \leftrightarrow T(k,n)\big).$$

By WKL and (2), we have

$$(3)\ \exists k\forall n\exists f \leq_1 1\, T(k, \overline{f}n) \to \exists k\exists f \leq_1 1\forall n\, T(k, \overline{f}n).$$

[10]We are grateful to Professor Dag Normann for communicating this construction to us.

In E-PRA$^\omega$, '$\exists f \leq_1 1(\chi_T(k, \overline{f}n) = 0)$' can be written as a quantifier-free formula $A_0(k, n)$. Hence using QF-AC^{0,0} and (2), (3) implies

$$(4)\ \forall h^1 \exists k \exists f \leq_1 1\, T(k, \overline{f}(hk)) \to \exists k \exists f \leq_1 1 \forall n\, T(k, \overline{f}n).$$

Contraposition of (4) together with the assumption of the pointwise continuity of $\Phi|_{2^\omega}$ yields

$$(5)\ \exists h^1 \forall k \forall f \leq_1 1 \neg T(k, \overline{f}(hk)).$$

Hence $\tilde{h}(k) := \max_{i \leq k}(hi)$ is a modulus of uniform continuity of the restriction of Φ to the Cantor space.

We now show (applying again WKL) the existence of χ satisfying (2) using a construction due to D. Normann. We first notice that because of the pointwise continuity of $\Phi|_{2^\omega}$ it suffices to show the existence of χ such that[11]

$$(6)\ \forall k \forall n \in 2^{<\omega} (\chi(k, n) = 0 \leftrightarrow \forall i \in 2^{<\omega} (\Phi(n * \lambda j.(i)_j, k) = \Phi(n * 0^1, k))),$$

where '$n \in 2^{<\omega}$' denotes the primitive recursive predicate 'n is the code of a finite 0-1-sequence'. Primitive recursively in k, n, Φ we define a tree as follows:

$$(7)\ T_{k,n}(q) := \begin{cases} 0, \text{ if } q \in 2^{<\omega} \wedge \\ \quad (\forall i \in 2^{<\omega}(lth(i) \leq lth(q) \to \Phi(n * q * 0^1, k) = \Phi(n * i * 0^1, k)) \\ \quad \vee \exists \tilde{q} \in 2^{<\omega} \exists l \leq lth(q)[q = \tilde{q} * \overline{0}(l) \text{ with } lth(\tilde{q}) \text{ minimal s.t.} \\ \qquad \Phi(n * \tilde{q} * 0^1, k) \neq \Phi(n * 0^1, k)]) \\ 1, \text{ otherwise.} \end{cases}$$

For every $k \in \mathbb{N}, n \in 2^{<\omega}$, $T_{k,n}$ is an infinite binary tree. So WKL yields the existence of an infinite path $f \leq_1 1$ in that tree. One easily verifies (using again the pointwise continuity of $\Phi|_{2^\omega}$) that

$$(8)\ \forall i \in 2^{<\omega}(\Phi(n * \lambda j.(i)_j, k) = \Phi(n * 0^1, k)) \leftrightarrow \Phi(n * f, k) = \Phi(n * 0^1, k).$$

In E-PRA$^\omega$ one can code the sequence of trees $T_{k,n}$ into a single infinite binary tree T such that any infinite path g of that tree (which by WKL exists) yields a whole sequence $\lambda k, n.f_{k,n}$ of infinite paths in $T_{k,n}$ (for $n \in 2^{<\omega}$; see [30] for details) and hence – in view of (8) – the characteristic function χ satisfying (6) which concludes the proof. □

COROLLARY 4.11. E-PRA$^\omega$+QF-AC^{0,0}+WKL *proves that the restriction of every continuous functional* $\Phi^{1(1)}$ *to the Cantor space has an r.m.-code.*

Open problems:

1) Does corollary 4.11 hold without WKL?
2) Does theorem 4.6 hold with WKL added?

[11] Here $n * f$ is defined as in the proof of proposition 4.7 above.

§5. Generalization of WKL to more complex trees: Φ_∞-WKL$_+$. The discussion before proposition 4.10 above as well as the proof of that proposition, showed that a direct application of a König's lemma-based argument to prove the uniform continuity of a pointwise continuous functional $\Phi^{1(1)}$ on 2^ω would require a form of the binary König's lemma for trees given by predicates of the form $T(n) \equiv \exists g \leq_1 1 A_0(g, n)$ with quantifier-free A_0 (allowing arbitrary parameters of higher type). In the particular application in 4.10, WKL plus QF-AC was sufficient to reduce this to a quantifier-free tree predicate (so that WKL could then again be applied to that tree). But this was possible only because $A_0(g, n)$ was pointwise continuous in g. However, without this continuity assumption, the corresponding extension of WKL to trees of the form above still results in a classically valid form of König's lemma. Since Σ^0_1-UB allows one to prove the (uniform) continuity of every functional $\Phi^{1(1)}$ on 2^ω, this more general form of WKL can be derived from Σ^0_1-UB and hence is **PRA**-reducible. This obervation can be largely extended to develop a whole strict hierarchy of extensions of WKL which are not derivable from WKL but which all can be shown to be proof-theoretically reducible to **PRA**.

Definition 5.1. 1) $A \in \Phi_n$ if

$$A \equiv \forall f_1 \leq_1 s_1[\underline{a}]\exists f_2 \leq_1 s_2[\underline{a}] \dots \forall^{(d)} f_n \leq_1 s_n[\underline{a}]\forall x^0 A_0(\underline{a}, f_1, \dots, f_n, x),$$

where A_0 is quantifier-free and $\underline{a}$ contains all free variables of A and s_i (which may have arbitrary types). The f_i must not occur in $\underline{a}$.[12]

2) $A \in \Psi_n$ if

$$A \equiv \exists f_1 \leq_1 s_1[\underline{a}]\forall f_2 \leq_1 s_2[\underline{a}] \dots \exists^{(d)} f_n \leq_1 s_n[\underline{a}]\forall x^0 A_0(\underline{a}, f_1, \dots, f_n, x),$$

where A_0 and s_i as above.

3) The classes Φ_n^- and Ψ_n^- result if we restrict ourselves to parameters $\underline{a}$ of type level ≤ 1 in A_0 and s_i.

Remark 5.2. One could also allow further universal number quantifiers $\forall x^0$ (but no existential quantifiers) to occur in between the bounded function quantifiers in the definition of Φ_n. The results of this paper easily extend to this slightly generalized case. However, for notational simplicity we restrict ourselves to the definition of Φ_n as stated above.

Defnition 5.3. The generalization of WKL to Φ_n-trees is given by

$$\Phi_n\text{-WKL}: \forall n^0 \exists f \leq_1 1 \forall \tilde{n} \leq n\, A(\overline{f}\tilde{n}) \to \exists f \leq_1 1 \forall n^0 A(\overline{f}n),$$

where $A(k^0) \in \Phi_n$ (with arbitrary further parameters of arbitrary types). Ψ_n-WKL is defined analogously. Φ_∞-WKL$:= \bigcup_{n\in\omega}\{\Phi_n\text{-WKL}\}$.

The next proposition shows that in the absence of parameters of types ≥ 2 (and so in particular in a second-order context) there is no point in considering Φ_n-WKL instead of WKL.[13] For its proof we need the following

[12] Here $\forall^d = \exists$, $\exists^d = \forall$.

[13] This is in sharp contrast to the case where arbitrary parameters are allowed as we will see below.

LEMMA 5.4. *Let $A_0(\underline{a}, g^1, y^0)$ be a quantifier-free formula of $\mathcal{T}$:=E-G$_n$A$^\omega$ ($n \geq 3$), E-PRA$^\omega$ or E-PA$^\omega$ containing (in addition to g, y) only parameters $\underline{a}$ of type levels ≤ 1 and let s be a term of $\mathcal{T}$ containing at most $\underline{a}$ as free variables. Then one can construct a Π^0_1-formula $B(\underline{a})$ of $\mathcal{T}$ (containing only $\underline{a}$ free) such that*

$$\mathcal{T} + \text{WKL} \vdash \forall \underline{a}\big(B(\underline{a}) \leftrightarrow \exists g \leq_1 s[\underline{a}]\forall y^0 A_0(\underline{a}, g, y)\big).$$

PROOF. For $\mathcal{T}$ =E-PRA$^\omega$ and $\mathcal{T}$ = E-PA$^\omega$ this follows from (the proofs of) proposition 4.14 and corollary 4.15 in [19]. The use of the modulus $\tilde{t}xyk$ of pointwise continuity in y used in the proof of proposition 4.14 in [19] can easily be replaced by a modulus $\widehat{t}xk$ of uniform continuity on $\{y : y \leq_1 sx\}$. For closed $t \in$E-G$_n$A$^\omega$ such a modulus $\widehat{t}$ can be constructed in E-G$_n$A$^\omega$ by the method of [18] since the majorization argument used there is available in E-G$_n$A$^\omega$ as was shown in [23]. □

PROPOSITION 5.5. *Let $m, n \geq 0$. Over $\mathcal{T}$:=E-G$_k$A$^\omega$ ($k \geq 3$), E-PRA$^\omega$ or E-PA$^\omega$ the following principles are equivalent*:
(i) WKL, (ii) Φ_0-WKL, (iii) Ψ_0-WKL, (iv) Φ^-_m-WKL, (v) Ψ^-_n-WKL.

PROOF. We first show the following
Claim: Let $A(\underline{a})$ be a Φ^-_n (or Ψ^-_n) formula containing only parameters $\underline{a}$ of type degree ≤ 1. Then one can construct a Π^0_1-formula $B(\underline{a})$ such that

$$\mathcal{T} + \text{WKL} \vdash A(\underline{a}) \leftrightarrow B(\underline{a}).$$

Proof of the claim: We proceed by meta-induction on n:
$n = 0$: In this case $A \in \Pi^0_1$ and so $B := A$ suffices.
$n \to n+1$: Case 1: $A \in \Phi_{n+1}$. Then $A(\underline{a}) \equiv \forall f \leq_1 s[\underline{a}]\, \tilde{A}(\underline{a}, f)$, where $\tilde{A} \in \Psi_n$. By the induction hypothesis there exists a formula $\tilde{B}(\underline{a}, f) \equiv \forall y^0 \tilde{B}_0(\underline{a}, f, y) \in \Pi^0_1$ with

$$\mathcal{T} + \text{WKL} \vdash A(\underline{a}) \leftrightarrow \forall f \leq_1 s[\underline{a}]\forall y^0 \tilde{B}_0(\underline{a}, f, y).$$

Let $t_{\tilde{B}_0}$ be a closed term of $\mathcal{T}$ such that

$$\mathcal{T} \vdash \forall \underline{a}, f, y\big(t_{\tilde{B}_0}(\underline{a}, f, y) =_0 0 \leftrightarrow \tilde{B}_0(\underline{a}, f, y)\big).$$

From results in [18] (using for the case of E-G$_k$A$^\omega$ also [23]) it follows that one can construct a closed term $\widehat{t}_{\tilde{B}_0}$ of $\mathcal{T}$ such that $\widehat{t}_{\tilde{B}_0}(\underline{a}, y)$ is (provably in $\mathcal{T}$) a modulus of uniform continuity for $\lambda f.t_{\tilde{B}_0}(\underline{a}, f, y)$ on $\{f : f \leq_1 s[\underline{a}]\}$. Using this modulus, $\forall f \leq_1 s[\underline{a}]\, \tilde{B}_0(\underline{a}, f, y)$ can be written as a quantifier-free formula and hence $\forall f \leq_1 s[\underline{a}]\forall y\, \tilde{B}_0(\underline{a}, f, y)$ as a Π^0_1-formula $\widehat{B}(\underline{a})$. So

$$\mathcal{T} + \text{WKL} \vdash A(\underline{a}) \leftrightarrow \widehat{B}(\underline{a}).$$

Case 2: $A(\underline{a}) \in \Psi_{n+1}$. Then $A(\underline{a}) \equiv \exists f \leq_1 s[\underline{a}]\, \tilde{A}(\underline{a}, f)$ with $\tilde{A}(\underline{a}, f) \in \Phi_n$. By I.H. there exists a formula $\tilde{B}(\underline{a}, f) \equiv \forall y^0 \tilde{B}_0(\underline{a}, f, y) \in \Pi^0_1$ with

$$\mathcal{T} + \text{WKL} \vdash A(\underline{a}) \leftrightarrow \exists f \leq_1 s[\underline{a}]\forall y^0 \tilde{B}_0(\underline{a}, f, y).$$

By the lemma, there exists a Π^0_1-formula $\widehat{B}(\underline{a})$ such that

$$\mathcal{T} + \text{WKL} \vdash \widehat{B}(\underline{a}) \leftrightarrow \exists f \leq_1 s[\underline{a}]\forall y^0 \tilde{B}_0(\underline{a}, f, y).$$

So again

$$\mathcal{T} + \text{WKL} \vdash A(\underline{a}) \leftrightarrow \widehat{B}(\underline{a})$$

with $\widehat{B} \in \Pi^0_1$. This finishes the proof of the claim.

The claim implies that

$$\mathcal{T} + \text{WKL} \vdash \Phi^-_m\text{-WKL} \leftrightarrow \Psi^-_n\text{-WKL}$$

for all $m, n \geq 0$. Since trivially Φ^-_0-WKL $\leftrightarrow$ Φ_0-WKL, it therefore remains to show that

$$\mathcal{T} \vdash \Phi_0\text{-WKL} \leftrightarrow \Psi_0\text{-WKL} \leftrightarrow \text{WKL}.$$

Φ_0-WKL$\equiv$ Ψ_0-WKL holds by definition. It is an easy exercise to show that WKL$\leftrightarrow$ Φ_0-WKL which we leave to the reader. □

In the presence of higher type parameters, however, we get non-collapsing hierarchies of principles Φ_n-WKL and Ψ_n-WKL, as we will show now.

DEFINITION 5.6. We define the classes of formulas $\Pi^{1,b}_n$ and $\Psi^{1,b}_n$ simultaneously by induction on n:

(i) $A \in \Pi^{1,b}_0 = \Sigma^{1,b}_0$, if A is quantifier-free;
(ii) if $A(f) \in \Pi^{1,b}_n$, then $\exists f \leq_1 1\, A(f) \in \Sigma^{1,b}_{n+1}$;
(iii) if $A(f) \in \Sigma^{1,b}_n$, then $\forall f \leq_1 1\, A(f) \in \Pi^{1,b}_{n+1}$.

A may contain arbitrary parameters (of arbitrary types).

REMARK 5.7. $\Pi^{1,b}_n \subseteq \Phi_n$ and $\Sigma^{1,b}_n \subseteq \Psi_n$.

DEFINITION 5.8. 1) The schema of $\Pi^{1,b}_n$-comprehension is given by

$$\Pi^{1,b}_n\text{-CA} : \exists g^1 \forall x^0 (gx = 0 \leftrightarrow A(x)),$$

where $A(x) \in \Pi^{1,b}_n$ and may contain arbitrary parameters (of arbitrary types) in addition to x. $\Sigma^{1,b}_n$-CA is defined analogously but with $\Sigma^{1,b}_n$ instead of $\Pi^{1,b}_n$.

2) The schema of $\Pi^{1,b}_n$-choice for numbers is given by

$$\Pi^{1,b}_n\text{-AC}^{0,0} : \forall x^0 \exists y \leq_0 1\, A(x, y) \to \exists g \leq_1 1 \forall x\, A(x, gx),$$

where $A(x, y) \in \Pi^{1,b}_n$ and may contain arbitrary parameters.

PROPOSITION 5.9. *Let* $\mathcal{T}$:=E-PA$^\omega$. *Then*

$$\mathcal{T} + \Phi_{n+1}\text{-WKL} \vdash \Pi^{1,b}_n\text{-CA}$$

(*Likewise for* Ψ_{n+1}-WKL).

PROOF. We use the following tree-predicate from [39]:

$$\tilde{A}(k) :\equiv \begin{cases} (k)_{lth(k)\dot{-}1} \leq 1 \wedge ((k)_{lth(k)\dot{-}1} = 0 \to A(lth(k)\dot{-}1)) \wedge \\ \qquad ((k)_{lth(k)\dot{-}1} = 1 \to \neg A(lth(k)\dot{-}1))), \text{ if } lth(k) > 0 \\ true, \text{ otherwise.} \end{cases}$$

For $A \in \Pi^{1,b}_n$, $\tilde{A}(k)$ can be written as a Φ_{n+1}-formula. By induction on n we can prove in E-PA$^\omega$ that

$$\forall n^0 \exists f \leq_1 1 \forall \tilde{n} \leq n \tilde{A}(\overline{f}\tilde{n}).$$

Φ_{n+1}-WKL therefore yields the characteristic function for $A(n)$. □

PROPOSITION 5.10. E-PA$^\omega$ + $\Pi^{1,b}_n$-CA+μ *contains (modulo a canonical embedding which doesn't change the first order part) the second order system* (Π^1_n-CA) *known from reverse mathematics.*[14]

PROOF. Systems formulated in the language of second-order arithmetic with set variables like (Π^1_n-CA) can be embedded in (suitable) systems formulated in the language of functionals of all finite types by representing sets X by their characteristic functions χ_X and replacing formulas '$t \in X$' by '$\chi_X(t) =_0 0$'. In doing so and using the fact that the presence of μ allows one to absorb an arbitrary arithmetical quantifier-prefix in front of a quantifier-free formula with arbitrary parameters uniformly in these parameters, the comprehension schema of (Π^1_n-CA) reduces to $\Pi^{1,b}_n$-CA above. □

The two propositions above show that the systems E-PA$^\omega$+QF-AC^{1,0}+QF-AC^{0,1}+μ+Φ_n-WKL (and similar with Ψ_n-WKL) form a non-collapsing hierarchy which as n increases eventually exhausts full second-order arithmetic.

Together with the result due to Feferman that E-PA$^\omega$+QF-AC^{1,0}+QF-AC^{0,1} + μ can be reduced proof-theoretically to $(\Pi^0_1\text{-CA})_{<\varepsilon_0}$[15] and hence is proof-theoretically much weaker than (Π^1_1-CA), it in particular follows that for $n \geq 2$, Φ_n-WKL and Ψ_n-WKL are underivable in E-PA$^\omega$+QF-AC^{1,0}+QF-AC^{0,1} + μ. The next proposition sharpens this further:

PROPOSITION 5.11. E-PA$^\omega$+QF-AC^{1,0}+QF-AC^{0,1} + $\mu \nvdash \Phi_1$-WKL.

PROOF. One easily verifies that E-PA$^\omega$ + Φ_1-WKL proves $\Pi^{1,b}_1$-AC^{0,0} which in the presence of μ yields the so-called Σ^1_1-separation principle (see [35]). Hence (again by [35]) the subsystem ATR of second order arithmetic, whose proof-theoretic strength is much higher than that of $(\Pi^0_1\text{-CA})_{<\varepsilon_0}$, is contained in E-PA$^\omega$+QF-AC^{1,0}+QF-AC^{0,1} + μ + Φ_1-WKL. □

In '(4)' in the proof of proposition 4.10 we used a slightly more general form of WKL which, however, could be derived from WKL by the use of QF-AC^{0,0}. For our extensions Φ_n-WKL ($n > 0$), quantifier-free choice does not seem to be sufficient to do the same trick. This suggests to generalize Φ_n-WKL as follows:

[14]In the notation of [35], (Π^1_n-CA) is the system Π^1_n-CA$_0$+full induction.

[15]This follows from [5] together with elimination of extensionality (see also [1]).

DEFINITION 5.12. Let $A(a^0, k^0) \in \Phi_n$ (with arbitrary parameters).

Φ_n-WKL$_+$: $\forall h^1 \exists a^0 \exists f \leq_1 1 \forall \tilde{n} \leq h(a)\, A(a, \overline{f\tilde{n}}) \rightarrow \exists a \exists f \leq_1 1 \forall n^0 A(a, \overline{f}n)$

(Ψ_n-WKL$_+$ is defined analogously with $A \in \Psi_n$.)

It is an easy exercise to show

PROPOSITION 5.13. E-G$_3$A$^\omega$+QF-AC^{0,0} $\vdash$ Φ_0-WKL$\leftrightarrow$ Φ_0-WKL$_+$.

§6. PRA-reducible theories.

We now show that F (and in fact Σ^0_1-UB) suffices to prove the whole hierarchy Φ_∞-WKL$_+$:

PROPOSITION 6.1. *Let* $\mathcal{T}$:=E-G$_k$A$^\omega$ ($k \geq 3$), E-PRA$^\omega$ *or* E-PA$^\omega$. *Then*

$$\mathcal{T} + \text{QF-AC}^{1,0} + F^- \vdash \Phi_\infty\text{-WKL}_+.$$

PROOF. Because of proposition 3.6 it suffices to show that

$$\mathcal{T} + \text{QF-AC}^{1,0} + F \vdash \Phi_\infty\text{-WKL}_+.$$

The idea of the proof is to use proposition 3.4 (together with propositions 3.2 and 3.3) to show similarly to the argument in the proof of proposition 5.5 that every $A \in \Phi_n$ (or $\in \Psi_n$) can be written as a Π^0_1-formula B. Whereas in the proof of proposition 5.5 we could use the fact that for every term $t^2[\underline{a}]$ of $\mathcal{T}$ containing only variables $\underline{a}$ of type ≤ 1 one can construct a modulus of uniform continuity on $\{x : x \leq_1 b\}$ (uniformly in $\underline{a}$ and b), we have to use proposition 3.4 in the presence of arbitrary parameters. The latter provides such a modulus of uniform continuity only uniformly in number parameters but not uniformly in function parameters f unless the latter are themselves restricted to a compact set $\{f : f \leq_1 b\}$ (in which case a modulus that is independent of f does exist). However this is just the case in the situation at hand since all function variables $f_1, \ldots, f_n$ of $A \in \Phi_n$ which are not parameters are bounded. So all we need is

$$(*)\begin{cases} \forall \Phi, \underline{a} \exists \alpha^1 \forall x^0, z^0 (\lambda \underline{f}.(\Phi x z \underline{f}\, \underline{a})^0 \text{ is uniformly continuous for all} \\ f_1 \leq_1 s_1[x, \underline{a}], \ldots, f_n \leq_1 s_n[x, \underline{a}] \text{ with modulus } \alpha x z), \end{cases}$$

where $\underline{a}$ are all the remaining free variables of s_i (which may have arbitrary types).[16]

$(*)$ is implied by

$$(**)\begin{cases} \forall \Phi, \underline{a}, \underline{b}^{1(0)} \exists \alpha^1 \forall x^0, z^0 (\lambda \underline{f}.(\Phi x z \underline{f}\, \underline{a})^0 \text{ is uniformly continuous for all} \\ f_1 \leq_1 b_1 x, \ldots, f_n \leq_1 b_n x \text{ with modulus } \alpha x z). \end{cases}$$

But this follows in $\mathcal{T} + \Sigma^0_1$-UB (and therefore in $\mathcal{T}$+QF-AC^{1,0} + F by proposition 3.2) similarly to the proof of proposition 3.4. Since by proposition 3.3 also WKL is available in this theory, we can argue as in the proof of the claim in the

[16]Here 'z' is the variable from the Π^0_1-kernel of A (which of course can be merged together with x).

proof of proposition 5.5 and show that for $A(x) \in \Phi_n$ (with arbitrary additional parameters)

$$\mathcal{T} + \Sigma_1^0\text{-UB} \vdash \exists\Phi\forall x^0\big(A(x) \leftrightarrow \forall z^0(\Phi xz =_0 0)\big).$$

Hence for all $n \in \mathbb{N}$

$$(***)\mathcal{T} + \Sigma_1^0\text{-UB} \vdash \Phi_0\text{-WKL}_+ \to \Phi_n\text{-WKL}_+$$

and therefore (using propositions 3.3,5.5 and 5.13)

$$\mathcal{T} + \text{QF-AC}^{0,0} + \Sigma_1^0\text{-UB} \vdash \Phi_n\text{-WKL}_+$$

and therefore by proposition 3.2

$$\mathcal{T} + \text{QF-AC}^{1,0} + F \vdash \Phi_n\text{-WKL}_+,$$

which concludes the proof. □

Corollary to the proof of proposition 6.1:

$$\mathcal{T} + \text{QF-AC}^{0,0} + \Sigma_1^0\text{-UB} \vdash \Phi_\infty\text{-WKL}_+$$

THEOREM 6.2.

1) $\text{E-G}_3\text{A}^\omega$+QF-AC^{1,0}+QF-AC^{0,1} + Σ_1^0-UB *is* Π_2^0*-conservative over* EA,
2) E-PRA$^\omega$+QF-AC^{1,0}+QF-AC^{0,1} + Σ_1^0-UB *is* Π_2^0*-conservative over* PRA,
3) E-PA$^\omega$+QF-AC^{1,0}+QF-AC^{0,1} + Σ_1^0-UB *is conservative over* PA.

PROOF. We first prove 3): Let A be a sentence of PA which is provable in E-PA$^\omega$+QF-AC^{1,0}+QF-AC^{0,1} + Σ_1^0-UB and hence (using proposition 3.2) in E-PA$^\omega$+QF-AC^{1,0}+QF-AC^{0,1} + F. Then the Herbrand normal form $A^H \equiv \forall \underline{f}\exists \underline{y} A_0(\underline{f},\underline{y})$ of A (we may assume that A is in prenex normal form) is provable there a-fortiori. Hence by theorem 3.7

$$\text{E-PA}^\omega \vdash \forall \underline{f}\, A_0(\underline{f},\underline{\Psi}(\underline{f}))$$

for suitable closed terms $\underline{\Psi}$ of E-PA$^\omega$. Thus

$$\text{E-PA}^\omega \vdash A^H.$$

By [17](thm.4.1) we can conclude that[17]

$$\text{PA} \vdash A.$$

1) and 2): For Π_2^0-sentences A the argument above applies equally to $\text{E-G}_3\text{A}^\omega$ (resp. E-PRA$^\omega$) yielding $\text{E-G}_3\text{A}^\omega \vdash A$ (resp. E-PRA$^\omega \vdash A$). The conclusion now follows from the fact that $\text{E-G}_3\text{A}^\omega$ (resp. E-PRA$^\omega$) is Π_2^0-conservative over EA (resp. PRA). □

THEOREM 6.3.

1) $\text{E-G}_3\text{A}^\omega$+QF-AC^{1,0}+QF-AC^{0,1} + Φ_∞-WKL$_+$ *is* Π_2^0*-conservative over* EA,
2) E-PRA$^\omega$+QF-AC^{1,0}+QF-AC^{0,1}+Φ_∞-WKL$_+$ *is* Π_2^0*-conservative over* PRA,
3) E-PA$^\omega$+QF-AC^{1,0}+QF-AC^{0,1} + Φ_∞-WKL$_+$ *is conservative over* PA

[17]Warning: this argument does not apply to the subsystems E-PRA$^\omega$, PRA; see [17] for a counterexample to this.

PROOF. The theorem follows from theorem 6.2 and the corollary to the proof of proposition 6.1. □

REMARK 6.4. The purely proof-theoretic proofs of theorems 6.2 and 6.3 also yield corresponding proof-theoretic reductions.

Theorems 6.2.2) and 6.3.2) yield two new mathematically strong PRA-reducible and Π^0_2-conservative extensions of PRA. One of these systems

$$\mathcal{T}^* := \text{E-PRA}^\omega + \text{QF-AC}^{1,0} + \text{QF-AC}^{0,1} + \Sigma^0_1\text{-UB}$$

is a non-standard system in the sense that the full set-theoretic type structure $\mathcal{S}^\omega$ is not a model of $\mathcal{T}^*$.

Analysing the greater mathematical strength of $\mathcal{T}^*$ (w.r.t. to derivable consequences which **are** true in $\mathcal{S}^\omega$) in terms of generalizations of WKL to logically more complex binary trees, we developed the subsystem

$$\mathcal{T} := \text{E-PRA}^\omega + \text{QF-AC}^{1,0} + \text{QF-AC}^{0,1} + \Phi_\infty\text{-WKL}_+$$

which has $\mathcal{S}^\omega$ as a model.

There is also a different route to design PRA-reducible systems which is based on E-$G_\infty A^\omega$ instead of E-PRA$^\omega$. Although E-$G_\infty A^\omega$ contains all primitive recursive functions and primitive recursive functionals of every Grzegorczyk level n, it does not contain all ordinary Kleene-primitive recursive functionals of type 2, in particular it does not contain Φ_{it}. As a consequence of this, E-$G_\infty A^\omega$+QF-AC^{0,0} does not prove the schema of Σ^0_1-induction. As we have shown in [26],[27] and [29], one can add to E-$G_\infty A^\omega$+QF-AC^{1,0}+ QF-AC^{0,1} function parameter-free schematic forms of e.g., Π^0_1-comprehension, the Bolzano-Weierstraß principle for sequences in $[0,1]^d$, the Arzela-Ascoli lemma etc. and still obtain a PRA-reducible system (whereas the addition of any of these principles to E-PRA$^\omega$ would make the Ackermann function provably total). This result was obtained via a certain Σ^0_2-generalization of the principle Σ^0_1-UB$^-$ mentioned in the proof of proposition 3.6. Using the results of this paper we can even allow a corresponding generalization of the principle Σ^0_1-UB instead. As a consequence of this and the fact that Φ_∞-WKL$_+$ follows from Σ^0_1-UB already relative to E-$G_\infty A^\omega$, we may add Φ_∞-WKL$_+$ to the principles listed above without losing PRA-conservation. This results in a mathematically fairly strong system (note that E-$G_\infty A^\omega$+QF-AC^{0,0} contains – identifying sets $X \subseteq \mathbb{N}$ with their characteristic function – the weak base system RCA$^*_0$ from reverse mathematics and see remark X.4.3 in [35]) which is incompatible with the systems studied in this paper. A detailed treatment of this theme, however, has to be postponed for another paper.

The results of this paper and [30] suggest to propose the following extension of the program of reverse mathematics to finite types: Replace the base system RCA$_0$ by its finite type extension RCA$^\omega_0$:= E-PRA$^\omega$+QF-AC^{1,0}. This system can be shown to be conservative over (an inessential variant with function variables instead of set variable of) RCA$_0$. So for second order statements A, B (i.e., the type of statements which can be discussed in the framework of currently existing reverse mathematics) **nothing is lost** if we prove an equivalence between A and B relative to RCA_0^ω instead of RCA_0. However, the richer language allows one to consider new statements (in their direct formulation) which can not even be expressed in RCA_0 and to apply reverse mathematics to them as well. As an example, we can recast a result from [30] as a result in reverse mathematics in this extended sense:

'Relative to RCA_0^ω, the uniform weak König's lemma UWKL and the existence of Feferman's μ-operator are equivalent'.

Likewise, the equivalence between μ and strong uniform versions of analytical theorems like the attainment of the maximum of $f \in C[0, 1]$ can be obtained. This theme is devloped further in [31].

REFERENCES

[1] Avigad, J., Feferman, S., Gödel's functional ('Dialectica') interpretation. In: [3], pp. 337-405 (1998).

[2] Beeson, M.J., Foundations of Constructive Mathematics. Springer Ergebnisse der Mathematik und ihrer Grenzgebiete 3.Folge, Bd.6., Berlin Heidelberg New York Tokyo (1985).

[3] Buss, S.R. (editor), Handbook of Proof Theory. Studies in Logic and the Foundations of Mathematics Vol 137, Elsevier, vii+811 pp. (1998).

[4] Feferman, S., A language and axioms for explicit mathematics. In: Crossley, J.N. (ed.), Algebra and Logic, pp. 87-139. Springer Lecture Notes in Mathematics 450 (1975).

[5] Feferman, S., Theories of finite type related to mathematical practice. In: Barwise, J. (ed.), Handbook of Mathematical Logic, pp. 913-972, North-Holland, Amsterdam (1977).

[6] Feferman, S., Working foundations. Synthese **62**, pp. 229-254 (1985).

[7] Feferman, S., Weyl vindicated: Das Kontinuum seventy years later, in: Cellucci, C., Sambin, G. (eds.), Temi e prospettive della logica e della filosofia della scienza contemporance, vol. I, pp. 59-93 (1988), CLUEB, Bologna. Reprinted (with minor additions) in [10].

[8] Feferman, S., Infinity in mathematics: Is Cantor necessary?. In: G. Toraldo di Francia (ed.), L'infinito nella scienza, Istituto della Enciclopedia Italiana, Rome, pp. 151-209 (1987). Reprinted (with minor additions) in [10].

[9] Feferman S., Why a little bit goes a long way: Logical foundations of scientifically applicable mathematics. In PSA (1992), Vol.2 (Philosophy of Science Association, East Lansing), pp. 442-455. Reprinted (with minor additions) in [10].

[10] Feferman, S., In the Light of Logic. Oxford University Press, 340 pp. (1998).

[11] Feferman, S., Jäger, G., Systems of explicit mathematics with non-constructive μ-operator. Ann. Pure Applied Logic, Part I. **65**, pp. 243-263 (1993), Part II. **79**, pp. 37-52 (1996).

[12] Felgner, U., Models of ZF–Set Theory. Lecture Notes in Mathematics **223**, Springer, Berlin (1971).

[13] Gödel, K., Über eine bisher noch nicht benützte Erweiterung des finiten Standpunktes. Dialectica **12**, pp. 280–287 (1958).

[14] Grilliot, T.J., On effectively discontinuous type-2 objects. J. Symbolic Logic **36**, pp. 245–248 (1971).

[15] Halpern, J.D.–Levy, A., The Boolean Prime Ideal theorem does not imply the axiom of choice. In: Proc. of Symp. Pure Math. vol.XIII, pp.83–134, AMS, Providence (1971).

[16] Jägermann, M., The axiom of choice and two definitions of continuity. Bull. Acad. Polon. Sci. **13**, pp.699–704 (1965).

[17] Kohlenbach, U., Remarks on Herbrand normal forms and Herbrand realizations. Arch. Math. Logic **31**, pp. 305-317 (1992).

[18] Kohlenbach. U.. Pointwise hereditary majorization and some applications. Arch. Math. Logic **31**, pp. 227-241 (1992).

[19] Kohlenbach. U.. Effective bounds from ineffective proofs in analysis: an application of functional interpretation and majorization. J. Symbolic Logic **57**. pp. 1239–1273 (1992).

[20] Kohlenbach, U., Effective moduli from ineffective uniqueness proofs. An unwinding of de La Vallée Poussin's proof for Chebycheff approximation. Ann. Pure Appl. Logic **64**, pp. 27–94 (1993).

[21] Kohlenbach. U., Analysing proofs in analysis. In: W. Hodges. M. Hyland. C. Steinhorn. J. Truss. editors, *Logic: from Foundations to Applications. European Logic Colloquium* (Keele. 1993). pp. 225–260. Oxford University Press (1996).

[22] Kohlenbach, U., Real growth in standard parts of analysis. Habilitationsschrift. pp. xv+166, Frankfurt (1995).

[23] Kohlenbach, U., Mathematically strong subsystems of analysis with low rate of growth of provably recursive functionals. Arch. Math. Logic **36**. pp. 31–71 (1996).

[24] Kohlenbach, U., Proof theory and computational analysis. Electronic Notes in Theoretical Computer Science **13**, Elsevier (http://www.elsevier.nl/locate/entcs/volume13.html), (1998).

[25] Kohlenbach. U., The use of a logical principle of uniform boundedness in analysis. In: Cantini. A., Casari, E., Minari. P. (eds.). Logic and Foundations of Mathematics. Synthese Library 280. pp. 93–106. Kluwer Academic Publishers (1999).

[26] Kohlenbach. U.. Arithmetizing proofs in analysis. In: Larrazabal. J.M. et al. (eds.). Proceedings Logic Colloquium 96 (San Sebastian). Springer Lecture Notes in Logic **12**. pp. 115–158 (1998).

[27] Kohlenbach. U., On the arithmetical content of restricted forms of comprehension. choice and general uniform boundedness. Ann. Pure and Applied Logic **95**. pp. 257–285 (1998).

[28] Kohlenbach, U.. On the no-counterexample interpretation. J. Symbolic Logic **64**. pp. 1491–1511 (1999).

[29] Kohlenbach. U., Things that can and things that can't be done in PRA. Ann. Pure Applied Logic **102**, pp. 223–245 (2000).

[30] Kohlenbach. U., On the uniform weak König's lemma. To appear in: Annals of Pure Applied Logic.

[31] Kohlenbach, U., Higher order reverse mathematics. Preprint 14pp., submitted.

[32] Normann. D., Recursion on the Countable Functionals. Springer Lecture Notes in Mathematics **811** (1980).

[33] Sieg. W.. Fragments of arithmetic. Ann. Pure Appl. Logic **28**.pp. 33–71 (1985).

[34] Simpson. S.G.. Partial realizations of Hilbert's program. J. Symbolic Logic **53**. 349–363 (1988).

[35] Simpson. S.G.. Subsystems of Second Order Arithmetic. Perspectives in Mathematical Logic. Springer-Verlag. xiv+445 pp. (1999).

[36] Tait. W.W.. Finitism. Journal of Philosophy **78**. pp. 524–546 (1981).

[37] Takeuti. G.. A conservative extension of Peano arithmetic. Part II of 'Two applications of logic to mathematics', Publ. Math. Soc. Japan **13**. (1978).

[38] Troelstra. A.S. (ed.) Metamathematical investigation of intuitionistic arithmetic and analysis. Springer Lecture Notes in Mathematics **344** (1973).

[39] Troelstra. A.S., Note on the fan theorem. J. Symbolic Logic **39**. pp. 584–596 (1974).

[40] Weyl. H.. Das Kontinuum. Kritische Untersuchungen über die Grundlagen der Analysis. Veit. Leipzig (1918). English translation: Weyl. H., The Continuum: A critical Examination of the Foundation of Analysis. New York. Dover (1994).

DEPARTMENT OF COMPUTER SCIENCE
UNIVERSITY OF AARHUS
NY MUNKEGADE
DK-8000 AARHUS C. DENMARK
E-mail: kohlenb@brics.dk
URL: http://www.brics.dk/~kohlenb

THE EPSILON SUBSTITUTION METHOD AND CONTINUITY

GRIGORI MINTS

Dedicated to S. Feferman

Abstract. A streamlined termination proof for epsilon substitution method is presented. It uses a non-effective continuity argument.

§1. Introduction. The ε-substitution method was introduced by Hilbert (cf. [2]) as a proof-theoretic tool and used by Hilbert's school as a normalization technique: it transforms proofs of purely existential formulas in first order logic or arithmetic into explicit proofs containing realizations of existential quantifiers. The reader is referred to [4] for history and motivation. We recall necessary definitions below.

The epsilon substitution method is based on the language with epsilon terms $\varepsilon x F[x]$, read as "an x satisfying the condition $F[x]$". In number-theoretic contexts it is often interpreted as the least x satisfying $F[x]$. The main axioms of the corresponding formalism are *critical formulas*

$$F[t] \to F[\varepsilon x F[x]] \tag{1}$$

Hilbert's epsilon substitution method for a given finite system E of critical formulas generates substitutions of numerals for closed epsilon-terms. The initial substitution S_0 is identically 0. At every stage only a finite number of epsilon-terms are assigned non-zero values. If substitutions $S_0, \dots, S_i$ are already generated, and S_i is not yet a solving substitution (satisfying all critical formulas in E), then S_{i+1} is found as follows. Fix appropriate ordering of the critical formulas (1) and take the first formula in this ordering which is false under S_i, i.e., for which $S_i(F[t]) = \boldsymbol{true}$, $S_i(F[\varepsilon x F]) = \boldsymbol{false}$. Set

$$S_{i+1}(\varepsilon x F) = (\textit{the least } n \leq t)\ (S_i(F[n]) = \boldsymbol{true})$$

drop the values of higher rank and preserve remaining values.

1991 *Mathematics Subject Classification.* 03F03,03F10,03F30.

Conversation and correspondence with W. Tait influenced the content of this paper and helped to clean it up.

It is pleasant to acknowledge the influence of S. Feferman who supported the work on ε-substitution method by friendly advice, discussions, critical comments and practical help.

Reflections on the foundations of mathematics
Essays in honor of Solomon Feferman
Edited by Wilfried Sieg, Richard Sommer, and Carolyn Talcott

W. Ackermann [1] proved that the sequence $S_0, \dots, S_i$ terminates for every system E of critical formulas in first order arithmetic after a finite number of steps in a *solution* S_k satisfying all critical formulas in E. Extensions to predicatively reducible second order systems were described in [4], [3]. Extension to impredicative systems probably requires new ideas.

The main goal of this paper is to simplify a termination proof for the first order arithmetic given in the section 7 of [4] with the aim to facilitate extension to impredicative systems. We use continuity considerations. Let us recall the definition [5].

DEFINITION 1.1. Let $\alpha(\boldsymbol{x}, \boldsymbol{f})$ be a functional depending on numerical arguments $\boldsymbol{x} = x_1, \dots, x_k$ and arguments $\boldsymbol{f} = f_1, \dots, f_k$ which are functions from natural numbers to natural numbers. Then α is *continuous* if for every particular value $(\boldsymbol{n}, \beta)$ of $(\boldsymbol{x}, \boldsymbol{f})$ there are finite functions β_0 such that β_0 agree with β on the domain of β_0, and for all functions γ which agree with β_0 we have

$$\alpha(\boldsymbol{n}, \gamma) = \alpha(\boldsymbol{n}, \beta_0)$$

An important source of our approach is the following outline of a constructive termination proof for first order arithmetic presented by W. Tait in [6]. Let the system of critical formulas to be solved be written as

$$E \equiv E_n[\varepsilon^n, \varepsilon^{<n}], E_{<n}[\varepsilon^n, \varepsilon^{<n}]$$

(see section 6.2), where ε^n contains ε-matrices having maximal rank n, and E_n contains critical formulas for all such matrices. Consider a new system $E' = E_n[\varepsilon^1, \boldsymbol{f}]$ of critical formulas obtained from E_n by replacing all occurrences of matrices $\varepsilon^{<n}$ of rank $< n$ (including occurrences inside ε^n) by arbitrary new functions $\boldsymbol{f}$. Then the system E' is of rank 1 and has a solution $R(\boldsymbol{f})$ continuous in $\boldsymbol{f}$ (see Lemma 6.1). Replacement of ε^n by $R(\varepsilon^{<n})$ reduces original system E to a new system $E_{<n}[R(\varepsilon^{<n}), \varepsilon^{<n}]$of rank $< n$, which leads to a solution in n steps. It is possible that this approach works for predicative systems treated in [4], [3]. However this approach "from above" requires consideration of critical formulas where ε-matrices (and not just ε-terms) occur as arguments of operators. The approach used in [4], [3] and elaborated here works "from below" using a representation

$$E \equiv E_{\leq 1}[\varepsilon^1, \varepsilon^{>1}], E_{>1}[\varepsilon^1, \varepsilon^{>1}]$$

Assuming that the ε-substitution process for E does not terminate, ε^1 is substituted in E by a hypothetical function α of values of matrices in ε^1 resulting from this infinite process. This substitution reduces the rank of S by one.

§2. The language $L\varepsilon$.

The formulation here is very close to [4] restricted to first order ε-symbols, but now we allow free function variables, constants for functions from tuples of natural numbers to natural numbers, and no predicate constants except $=$. Syntactic identity of expressions is denoted by $\equiv$; equality by definition is $:=$.

Let us describe in detail the language of ε-calculus with constants for arbitarary functions of natural numbers.

Basic Symbols

0-variables (for natural numbers, denoted by $x, y, z, \dots$),

1-variables (for functions, denoted by $f, g, h, \dots$),

constants for functions, including the 0-ary function constant 0 (zero), the unary function constants S, pd (successor,predecessor) and the binary function constants $+, \cdot$,

predicate constant $=$ (equality),

the propositional logical constants $\neg, \wedge, \rightarrow$,

the epsilon symbol ε.

Var denotes the set of all variables.

It is understood that every constant c for functions is *interpreted* by a corresponding function α_c. In particular $+, \cdot$, S, pd are interpreted by familiar functions of natural numbers.

DEFNITION 2.1. (Terms and formulas)

1. Each ι-variable is a ι-term ($\iota = 0, 1$).
2. Every constant for a function is an 1-term.
3. If t_i $(i = 1, \dots, n)$ are 0-terms and t is an n-ary function variable or constant, then $tt_1 \dots t_n$ is a 0-term.
4. If t, s are 0-terms then $= ts$ is a formula.
5. If A, B are formulas then $\neg A$, $\wedge AB$, $\rightarrow AB$ are formulas.
6. If F is a formula and x is a 0-variable then $\varepsilon x F$ is a 0-term

To increase readability, we sometimes use infix notation for binary logical constants and $=$, insert parentheses, and use standard abbreviations like $A \vee B := \neg A \rightarrow B$.

DEFNITION 2.2. (λ-terms) If t is a 0-term and $\mathbf{z} := z_1, \dots, z_n$ are 0-variables occurring free in t, then $\lambda \mathbf{z} t$ is a λ-term (sometimes written $\lambda \mathbf{z}.t$) .

ι-terms ($\iota = 0, 1$), formulas and λ-terms are called *expressions*. Terms of the form $\varepsilon x F$ are called *ε-terms*. The 0-terms 0, S0, SS0, ... are called *numerals*. By a *constant* we mean either a numeral or a function constant.

$\mathbb{N}$ denotes the set of all numerals. For each term u we set

$$\iota(u) := \begin{cases} 0 & \text{if } u \text{ is a 0-term} \\ 1 & \text{if } u \text{ is a 1-term} \end{cases}$$

Note. Lambda-symbols are not allowed to occur inside other expressions. The result of their substitution into other expressions is understood via λ-conversion.

Syntactic variables: e, u, v, w for expressions; p for any expression and the symbols $=, \neg, \wedge, \rightarrow$; s, t for 0-terms; n for a numeral $S^n 0$; α, β, γ for numerical functions; T for λ-terms; A, B, F, G for formulas.

The set $FV(e)$ of free variables of an expression e is defined in the standard way: εx and $\lambda \mathbf{z}$ bind x and $\mathbf{z}$.

An expression e is called *closed* iff $FV(e) = \emptyset$.

We identify expressions which are equivalent modulo renaming of bound variables. If u is a $\iota(x)$-term then $e[x/u]$ denotes the result of substituting u for each free occurrence of x in e, where bound variables in e are renamed if necessary. If x is known from the context we write $e[u]$ for $e[x/u]$.

In the following we do not consider trivial ε-terms εxF where $x \notin FV(F)$.

Let $d(e)$ be the number of occurrences in e of closed subterms other than constants. Then $d(e) < \omega$, and if $d(e) = 0$ then e contains neither closed ε-subterms nor closed composite subterms other than numerals.

DEFINITION 2.3. 1. An ε-term $e \equiv \varepsilon xF$ is called canonical if it is closed and $d(e) = 1$ (i.e., $d(F) = 0$).

2. e is called simple if $d(e) = 0$ and e is closed.

3. TRUE (FALSE) denotes the set of all true (false) simple formulas. [Note that a simple formula contains no variables and is constructed from computable atomic formulas by Boolean connectives. Every simple 0-term is a numeral.]

Substitution $e[f/T]$ for an expression e, 1-variable f for n-place functions and λ-term $T \equiv \lambda z_1 \dots z_n.t$ is defined in a standard way: every subterm $fs_1 \dots s_n$ is replaced by $t[z_1/s_1, \dots, z_n/s_n]$ where it is assumed without loss of generality that no free variable of T is caught by some εx or λz in e. Notation $e[f/T]$ is used in this paper only when all occurrences of f come from subterms $fs_1 \dots s_n$.

§3. Computations with the ε-substitutions.

DEFINITION 3.1. An ε-substitution is a function S from a set $dom(S)$ (domain of S) of canonical ε-terms into $\mathbb{N} \cup \{?\}$.

An ε-substitution S is called total if $dom(S)$ is the set of all canonical ε-terms.

$\overline{S} := S \cup \{(e, ?) : e \text{ is a canonical } \varepsilon\text{-term} \notin dom(S)\}$ is called the standard extension of S.

Comment. We consider a function as a set of ordered pairs. So an ε-substitution is a set of pairs (e, u) where e is a canonical ε-term and $u \in \mathbb{N} \cup \{?\}$. Hilbert's finitist position allowed only for finite ε-substitutions S. All (canonical) ε-terms εxF not mentioned in S explicitly, have default value 0. In fact the possible value ? for a term indicates exactly this default value. The ε-substitution process is defined entirely in terms of finite substitutions: the standard extension $\overline{S}$ of S is used only to simplify notation for computations with default values.

3.1. Computation steps. Let S be an arbitrary ε-substitution. Recall that constants c are interpreted by functions α_c.

DEFNITION 3.2. (Inductive definition of $e \hookrightarrow^1_S e'$)

1.1. If $(e, u) \in S$ and $u \neq ?$, then $e \hookrightarrow^1_S u$

1.2. If $(e, ?) \in S$, then $e \hookrightarrow^1_S 0$

1.3. If $c_0, \dots, c_n$ are appropriate constants and $\alpha_c = value(\alpha_{c_0}(\alpha_{c_1}, \dots, \alpha_{c_n}))$ then $(c_0 c_1 \dots c_n) \hookrightarrow^1_S c$

2. If $0 \leq i \leq n$, $e_i \hookrightarrow^1_S e'_i$ then $(e_0e_1 \dots e_n) \hookrightarrow^1_S (e_0 \dots e_{i-1}e'_ie_{i+1} \dots e_n)$
3. If $F \hookrightarrow^1_S F'$ then $\varepsilon xF \hookrightarrow^1_S \varepsilon xF'$
4. If $G \hookrightarrow^1_S G'$ then $\lambda zG \hookrightarrow^1_S \lambda zG'$.

Comment. Since a substitution S is defined only for canonical terms which do not contain any proper closed ε-subterms, computations proceed from inside, and since canonical terms are closed, no term containing variables is immediately computed according to S: only its closed subterms can be replaced.

DEFINITION 3.3. 1. e is called S-reducible if there exists an e' with $e \hookrightarrow^1_S e'$. Otherwise e is called S-irreducible or in S-normal form.
2. $\hookrightarrow_S$ denotes the transitive and reflexive closure of $\hookrightarrow^1_S$.

LEMMA 3.4. *(a) If $e \hookrightarrow^1_S e'$ then $FV(e') = FV(e)$.*
(b) If $e \hookrightarrow^1_S e'$ then $d(e') < d(e)$ and hence $\hookrightarrow_S$ is well-founded.

PROOF. The argument by induction on the definition of $\hookrightarrow^1_S$ is standard. In the cases 1.1,1.2 one has $1 = d(e) > 0 = d(e')$. In the remaining cases $d(e)$ is strictly monotonic in d of relevant components of e. ⊣

LEMMA 3.5. *(a) If $t \hookrightarrow^1_S t'$ then $e[t] \hookrightarrow_S e[t']$.*
(b) If $e \hookrightarrow^1_S e'$ and $e \hookrightarrow^1_S e''$ then there is an expression u such that $e' \hookrightarrow_S u$ and $e'' \hookrightarrow_S u$.

PROOF. The argument by induction on e is routine. ⊣

THEOREM 3.6 (Church-Rosser Property). *For each expression e there exists a unique S-irreducible expression e^* with $e \hookrightarrow_S e^*$.*

PROOF. By Lemma 3.4 computations always terminate, and together with Lemma 3.5 this implies uniqueness. ⊣

DEFNITION 3.7 (Normal form $|e|_S$ for expression e). The unique expression e^* in the previous theorem is called the S-normalform of e and denoted by $|e|_S$.

DEFNITION 3.8. An expression e is S-computable iff $d(|e|_S) = 0$

Note. If S is total then every expression is S-computable. (Recall that all constants are interpreted by functionals total on suitable domains).
As always, $\hookrightarrow_S$ is preserved under substitution.

LEMMA 3.9. *(a) $e \hookrightarrow^1_S e'$ implies $e[\eta/u] \hookrightarrow_S e'[\eta/u]$.*
(b) $u \hookrightarrow^1_S u'$ implies $e[\eta/u] \hookrightarrow_S e[\eta/u']$.
(c) $e[\eta/u] \hookrightarrow_S |e|_S[\eta/|u|_S]$.

PROOF. The argument is routine. ⊣

§4. The rank function. In the following v denotes elements from $\mathrm{Var} \cup \{*\}$.

DEFNITION 4.1 (Definition of $\mathrm{rk}_v(e)$). 1. If $v \notin \mathrm{FV}(e) \cup \{*\}$ or e does not contain variables, then $\mathrm{rk}_v(e) := 0$.
2. For $v \in \mathrm{FV}(e) \cup \{*\}$ we define:

$$\mathrm{rk}_v(e) := \begin{cases} 0 & \text{if } e \in Var \\ \max\{\mathrm{rk}_v(p), \mathrm{rk}_v(e_1), \dots, \mathrm{rk}_v(e_n)\} & \text{if } e = pe_1 \dots e_n \\ \max\{\mathrm{rk}_v(F), \mathrm{rk}_x(F) + 1\} & \text{if } e = \varepsilon x F \\ \mathrm{rk}_v(G) & \text{if } e = \lambda z G \end{cases}$$

Note. $\mathrm{rk}_v(e) < \omega$. It is a measure of nesting of bound variables in subterms of e containing free variable v , and rk_* takes account of all epsilon subterms.

Next statements summarize the properties of the rank. Easy proofs by induction can be found in [4].

LEMMA 4.2. *(a)* $\mathrm{rk}_*(e) = \max\{\mathrm{rk}_v(F) + 1\ : \varepsilon v F$ *occur in* $e\}$.

(b) If $d(e) = 0$ *and e is not a* λ*-term then* $\mathrm{rk}_*(e) \leq \max\{\mathrm{rk}_v(e) : v \in \mathrm{FV}(e)\}$.

(c) If $\varepsilon x F$ *is canonical then* $\mathrm{rk}_*(F) \leq \mathrm{rk}_x(F)$ *and* $\mathrm{rk}_*(\varepsilon x F) = \mathrm{rk}_x(F){+}1$.

(d) The rank does not increase during computation: if $e \hookrightarrow^1_S e'$, *then* $\mathrm{rk}_v(e') \leq \mathrm{rk}_v(e)$.

(e) Substitution of a numeral for a 0*-variable does not increase the rank*:

$$\mathrm{rk}_v(e[x/n]) \leq \mathrm{rk}_v(e)$$

The next statement shows that the rank decreases when the x in the 'body' of a canonical ε-term $\varepsilon x F$ is substituted by a natural number.

LEMMA 4.3. *If* $\varepsilon x F$ *is canonical then* $\mathrm{rk}_*(F[n]) < \mathrm{rk}_*(\varepsilon x F)$.

DEFNITION 4.4. $\mathrm{rk}(e) := \mathrm{rk}_*(e)$ is called the *rank* of e.

DEFNITION 4.5. (Truncation to a given rank) For each ε-substitution S and $r \in \omega$ we set $S_{\leq r} := \{(e, u) \in S :\ \mathrm{rk}(e) \leq r\}$.

Analogously we define $S_{\geq r}$, $S_{<r}$, $S_{>r}$.

LEMMA 4.6. *If* S, S' *are* ε*-substitutions with* $S_{\leq r} = S'_{\leq r}$ *then* $|e|_S = |e|_{S'}$ *holds for all expressions e of rank* $\leq r$.

§5. The H-process.

Let us recall that critical formulas are formulas of the form

$$F[t] \to F[\varepsilon x F[x]]$$

We assume that $E \equiv\ Cr_0, \dots, Cr_N$ (with $N \in \mathbb{N}$) **is a fixed sequence of closed critical formulas.**

In this section we define a successive approximation process for finding a solution of this system according to Hilbert's approach. It is useful to recall here a Comment from Section 3. The H-process will be arranged in such a way that all non-default values of ε-terms are *correct*: $S(\varepsilon x F[x])$ is the least n satisfying $F[n]$.

DEFNITION 5.1. $F[\![x/n]\!] := F[x/n] \wedge \neg F[x/n{-}1] \wedge \dots \wedge \neg F[x/0]$,

DEFNITION 5.2. Let S be an ε-substitution:

$e \hookrightarrow_S \mathsf{TRUE}\,(\mathsf{FALSE})\ :\iff\ |e|_S \in \mathsf{TRUE}\,(\mathsf{FALSE})$.

$\mathcal{F}(S) := \{F[\![x/u]\!] :\ (\varepsilon x F, u) \in S\}$

S is *correct* iff $A \hookrightarrow_S \mathsf{TRUE}$ for all $A \in \mathcal{F}(S)$.

S is *solving* iff $Cr_I \hookrightarrow_S \mathsf{TRUE}$ for $I = 0, \dots, N$. Otherwise S is *nonsolving*.

$\overline{S} := S \cup \{(e, ?) : e$ canonical ε-term $\notin \mathrm{dom}(S)\}$ is called the *standard extension of* S, cf. Section 3.

The definition of a term whose value is to be corrected (H-term) to a new value (H-value) and of the ε-substitution $\mathrm{H}(S)$ to which S is reduced by this change of value, is given in terms of the standard extension $\overline{S}$. In fact the first term of minimal rank is chosen, and all values of higher rank are cancelled. Since ε-substitutions are defined only for canonical terms, all these operations are preceded by transforming arbitrary ε-term $\varepsilon x F$ into its *canonical form* $\varepsilon x|F|_S$.

DEFINITION 5.3. Let S be an ε-substitution such that $\overline{S}$ is nonsolving. (Then $|Cr_I|_{\overline{S}} \in \mathsf{FALSE}$ for some $I \leq N$.)

Set $r_I := \mathrm{rk}(\varepsilon x|F|_{\overline{S}})$, where $Cr_I = F_0 \to F[\varepsilon x F]$.

$\mathrm{Cr}(S) := Cr_I$, where $I \leq N$ is such that $|Cr_I|_{\overline{S}} \in \mathsf{FALSE}$ and

$$(\forall J \leq N)[\,|Cr_J|_{\overline{S}} \in \mathsf{FALSE} \Rightarrow r_I < r_J \vee (r_I = r_J \wedge I \leq J)].$$

Let $\mathrm{Cr}(S) = F_0 \to F[\varepsilon x F]$

$\varepsilon x|F|_{\overline{S}}$ is called the *H-term of* S.

The *H-value* v of S is the unique $n \in \mathbb{N}$ with $|F|_{\overline{S}}[\![n]\!] \hookrightarrow_{\overline{S}} \mathsf{TRUE}$.

The next definition is central for the substitution method.

DEFINITION 5.4. If $\overline{S}$ is nonsolving then $\mathrm{H}(S) := (S \setminus \{(e, ?)\})_{\leq \mathrm{rk}(e)} \cup \{(e, v)\}$, where e is the H-term and v the H-value of S.

LEMMA 5.5 (Correctness of $\overline{\mathrm{H}(S)}$). *Let S be an ε-substitution such that $\overline{S}$ is correct and nonsolving, and let e be the H-term, v the H-value of S. Then the following holds*:

a) $(e, ?) \in \overline{S}$,

b) $|e|_{\mathrm{H}(S)} = v \neq 0$,

c) $\overline{\mathrm{H}(S)}$ *is correct.*

PROOF. The argument is standard, cf. [2], [4]. ⊣

DEFINITION 5.6. The *H-process* (for $Cr_0, \dots, Cr_N$) is defined as follows:

$$S_0 := \emptyset, \quad S_{n+1} := \begin{cases} \mathrm{H}(S_n) & \text{if } \overline{S_n} \text{ is nonsolving} \\ S_n & \text{otherwise.} \end{cases}$$

The H-process *terminates* iff there exists an $n \in \mathbb{N}$ such that $\overline{S_n}$ is solving.

Note. According to the Definition 5.6 the first solving substitution is infinitely repeated after official termination of the H-process.

§6. Non-constructive proof of termination.

6.1. Termination for rank 1.

LEMMA 6.1. *If E is a system of closed critical formulas of rank 1, then the* H*-process for E terminates beginning with any substitution.*

PROOF. The argument goes back to Ackermann, see [2]. Let $d_\varepsilon(t)$ be the nesting number of closed ε-subterms of a closed ε-term $t \equiv \varepsilon x A$:

$$d_\varepsilon(\varepsilon x A) := \max\{d_\varepsilon(u) : u \text{ is a closed } \varepsilon\text{-subterm of } A\} + 1$$

Let $t_1, \ldots, t_m$ be the sequence of all closed ε-subterms of E such that $i > j$ if $d_\varepsilon(t_i) > d_\varepsilon(t_j)$. For every ε-substitution S define its *index*:

$$i_S := 2^{m-1}\delta_1 + \ldots + 2\delta_{m-1} + \delta_m$$

where $\delta_i := sign(|t_i|_{\overline{S}})$, i.e. $\delta_i = 0$ if $|t_i|_{\overline{S}} = 0$, and $\delta_i = 1$ otherwise. It is enough to prove that the index strictly decreases in the H-process: if S is non-solving and $S' := \mathsf{H}(S)$ then

$$i_{S'} < i_S \tag{2}$$

Indeed, consider the critical formula $F_0 \to F[\varepsilon x F]$ such that $\varepsilon x|F|_{\overline{S}}$ is the H-term of S, and let $\varepsilon x F \equiv t_I$. Then $S' := S \cup (\varepsilon x|F|_{\overline{S}}, n)$ where $n =$ the least $k(|F[x/k]|_{\overline{S}} \hookrightarrow_S \mathsf{TRUE})$,

$$|t_I|_{\overline{S}} = 0 \quad \text{and } n = |t_I|_{\overline{S'}} > 0,$$

since otherwise ***true*** $= |F[x/0]|_{\overline{S'}} = |F[x/0]|_{\overline{S}}$. Now take the least i_0 such that $|t_{i_0}|_{\overline{S}} = 0$ and $|t_{i_0}|_{\overline{S'}} > 0$. Then by induction on j one has

$$(\forall j < i_0)|t_j|_{\overline{S}} = |t_j|_{\overline{S'}}$$

and hence (2). ⊣

Note. Continuity of constants was not required in this Lemma.

Let E be a system of closed critical formulas, and the H-process for E terminates. Then α_E is the resulting system of functions corresponding to ε-matrices occurring in E.

If E contains free exactly 1-variables $\boldsymbol{f}$, then by α_E we mean the functional:

$$\alpha_E(\beta) := \alpha_{E[f/\beta]}$$

LEMMA 6.2. *If E is a system of critical formulas containing no free* 0*-variables, and α_E is total (for total arguments β) then α_E is continuous.*

PROOF. To simplify notation assume that $FV(E) \equiv \{f\}$ and let $E[\beta] := E[f/\beta]$. Since the H-process for $E[\beta]$ is finite, it uses only finite number of values of subterms, including ε-terms containing β. Find n_0 such that for all $n > n_0$ these values for β_n are the same as for β. Then the whole H-process for β_n and hence the result is the same as for β. Proceeding slightly more formally, notice the following (where syntactic expressions are supposed to be coded by natural numbers).

The following numerical values are continuous in f when both the system E and (current) ε-substitution S have continuous constants and may depend of f:

(c1) $|e|_{\overline{S}}$,

(c2) the property of $\overline{S}$ being (non)solving,

(c3) the number I of the formula $\mathrm{Cr}(S)$ in E,

(c4) the H-term and H-value of S,

(c5) the next substitution $\mathsf{H}(S)$ and the k-th next substitution $\mathsf{H}^k(S)$ beginning with $S := \mathsf{H}^0(S)$,

(c6) the whole sequence $(S, \mathsf{H}(S), \ldots, \mathsf{H}^k(S), \ldots)$ provided the H-process terminates for every f.

Proof. (c1). Induction on $d(e)$. Informally: the computation of $|e[\beta]|_{\overline{S}[\beta]}$ uses a finite number of terms $t[\beta]$, and their values are continuous by (a).

(c2),(c3). $|Cr_I|_{\overline{S}}$ is continuous by (c1).

(c4) The H-value $\mu x \leq nF$ is primitive recursive in values of terms occurring in $F[\![x/n]\!]$.

(c5),(c6). $\mathsf{H}^k(S)$ depends primitive recursively of the quantities in (c1)-(c4). ⊣

6.2. Replacing ε-matrices by functions. An occurrence of a term t in an expression e *resides* in e if no ε- or λ-symbol in e binds any free variable of this occurrence of t. In this case $e \equiv e'[y/t]$ where y has only one occurrence in e' (replacing considered occurrence of t in e) and t is substitutable for y in e'. An occurrence of t is *subordinate* in e if it does not reside in e, i.e. some of the free variables of t is bound in this occurrence by some ε, λ-symbol in e. An ε-term e is called a *matrix* if all resident occurrences of 0-terms in t are distinct free variables.

An *instance* of a matrix e with free 0-variables $\boldsymbol{z} \equiv z_1, \ldots, z_n$ is any term of the form $e[z_1/t_1, \ldots, z_n/t_n]$ or $e[\boldsymbol{z}/\boldsymbol{t}]$. Recall that the substitution $e[\boldsymbol{z}/\boldsymbol{t}]$ includes possible renaming of bound variables in e.

A *matrix* of an ε-term e is any matrix e' such that e is an *instance* of e'. A matrix of e is *unique* up to renaming free variables, since it is obtained from e by replacing maximal residing occurrences by fresh variables.

An *instance* of a λ-term $\lambda \boldsymbol{z}.s$ is any term of the form $s[\boldsymbol{z}/\boldsymbol{t}]$.

Two matrices e, e' are *unifiable* iff they have a common instance.

Let $\varepsilon x A$ be an ε-matrix, $FV(\varepsilon x A) \equiv \{z_1, \ldots, z_n\} := \boldsymbol{z}$, t be a variable or constant of type $(\overbrace{0, \ldots, 0}^{n \text{ times}} \to 0)$ and e an expression. Then $e[\varepsilon x A/t]$ is by definition the result of replacing all instances $\varepsilon x A[\boldsymbol{z}/\boldsymbol{u}]$ of $\varepsilon x A$ in e by $t\boldsymbol{u}$. More formally, for $\theta := [\varepsilon x A/t]$ one has:

$$e\theta := \begin{cases} e & \text{if } e \text{ is a variable or a constant} \\ p\theta e_1\theta \ldots e_n\theta & \text{if } e = pe_1 \ldots e_n \\ t\boldsymbol{u}\theta & \text{if } e = \varepsilon y F \equiv \varepsilon y A[x/y, \boldsymbol{z}/\boldsymbol{u}] \text{ is an instance of } \varepsilon x A \\ \varepsilon y F\theta & \text{if } e = \varepsilon y F \text{ is not an instance} \end{cases}$$

Example. 1. Let $e := \varepsilon x f(x, \varepsilon y f(y, v) = 0) = 0$, $\theta := [\varepsilon x f(x, z) = 0/g]$. Then $e\theta \equiv g(g(v))$.

2. Change the last variable: $e := \varepsilon x f(x, \varepsilon y f(y, x) = 0) = 0$. Then $e\theta \equiv \varepsilon x f(x, g(x))$.

3. Change the first ε : $e := \varepsilon v f(x, \varepsilon y f(y, v) = 0) = 0$. Then $e\theta \equiv \varepsilon v f(x, g(v))$.

For ε-matrices $e_1, \dots, e_k$ and suitable terms $t_1, \dots, t_k$ let

$$e[e_1/t_1, \dots, e_k/t_k] := e[e_1/t_1], \dots, [e_k/t_k]$$

Usually the order of e_i is chosen so that

$$\mathrm{rk}(e_i) \geq \mathrm{rk}(e_{i+1}) \tag{3}$$

for $i < k$. For a given system E of closed critical formulas we will need to partition the set of ε-matrices with instances occurring in E according to their rank:

$\varepsilon^{>1}$ is the set of all matrices of rank > 1 (listed according to (3));

ε^1 is the set of all matrices of rank 1.

Let γ, α be finite sets of function constants suitable for substitution for $\varepsilon^{>1}, \varepsilon^1$ respectively. Then we write:

$$E[\varepsilon^{>1}, \varepsilon^1] := E;\ E[\varepsilon^{>1}, \alpha] := E[\varepsilon^1/\alpha];\ E[\gamma, e^1] := E[\varepsilon^{>1}/\gamma]$$

Similarly to the partition of an ε-substitution S into the parts $S_{>1}, S_{\leq 1}$ of rank > 1 and rank 1, (end of the section 4) we partition $E \equiv (E_{>1}, E_{\leq 1})$ into critical formulas of rank > 1 and rank 1.

Let us restate the definition of the H-process in these terms including the system E of critical formulas as an additional argument : for $\mathsf{H}(S)$ of Section 5 we write $\mathsf{H}(E, S)$ or $\mathsf{H}(E[\varepsilon^{>1}, \varepsilon^1], (S_{>1}, S_{\leq 1}))$.

$$\mathsf{H}^0(E, S) := S;\ \mathsf{H}^{n+1}(E, S) := \mathsf{H}(E, H^n(E, S))$$

For the H-process beginning with a substitution (γ_0, α_0) define:

$$\mathsf{H}^n(E, (\gamma_0, \alpha_0)) := (\gamma_n, \alpha_n)$$

where $\gamma_n := (\mathsf{H}^n(E, (\gamma_0, \alpha_0)))_{>1}$; $\alpha_n := (\mathsf{H}^n(E, (\gamma_0, \alpha_0)))_{\leq 1}$.

Then for every n exactly one of the following cases holds:

The H-term of (γ_n, α_n) is of rank 1;

The H-term of (γ_n, α_n) is of rank > 1;

(γ_n, α_n) is a solution for E,

or in more detail:

$$\neg\mathrm{sol}(E_{\leq 1}[\gamma_n, \varepsilon^1], \alpha_n) \wedge (\gamma_{n+1}, \alpha_{n+1}) \equiv (0, H(E_{\leq 1}[\gamma_n, \varepsilon^1], \alpha_n)) \tag{4}$$

$$\mathrm{sol}(E_{\leq 1}[\gamma_n, \varepsilon^1], \alpha_n) \wedge (\gamma_{n+1}, \alpha_{n+1}) \equiv (H(E_{>1}[\varepsilon^{>1}, \alpha_n], \gamma_n), \alpha_n) \tag{5}$$

$$\mathrm{sol}(E[\varepsilon^{>1}, \varepsilon^1], (\gamma_n, \alpha_n)) \wedge (\gamma_{n+1}, \alpha_{n+1}) \equiv (\gamma_n, \alpha_n) \tag{6}$$

In the last case the condition is equivalent to

$$\mathrm{sol}(E_{>1}[\varepsilon^{>1}, \alpha_n], \gamma_n) \,\&\, \mathrm{sol}(E_{\leq 1}[\gamma_n, \varepsilon^1], \alpha_n)$$

6.3. Termination proof.

THEOREM 6.3. *Let E be a system of closed critical formulas with continuous constants. Then the H-process for E terminates beginning with every substitution.*

PROOF. We use induction on the maximal rank R of the critical formulas in E. Induction base is Lemma 6.1. For the induction step consider arbitrary initial substitution (γ_0, α_0) and

$$(\gamma_n, \alpha_n) := \mathsf{H}^n(E[\varepsilon^{>1}, \varepsilon^1], (\gamma_0, \alpha_0)) \tag{7}$$

Since α_{n+1} is an extension of α_n (the values of rank 1 are never deleted) there exists

$$\alpha := \lim_{n\to\infty} \alpha_n$$

Consider

$$\beta_n := \mathsf{H}^n(E[\varepsilon^{>1}, \alpha], 0)$$

The sequence β_n stabilises by IH: there exists a function β and a number N_β such that

$$(\forall n \geq N_\beta)(\beta_n = \beta \;\&\; \mathrm{sol}(E[\varepsilon^{>1}, \alpha], \beta_n)) \tag{8}$$

Let us denote by $\mathrm{sol}^?$ the Boolean function producing the truth-value of sol.

By the continuity

$$\forall k(\mathrm{sol}^?(E_{>1}[\varepsilon^{>1}, \alpha_n], \beta_k) \to_{n\to\infty} \mathrm{sol}^?(E_{>1}[\varepsilon^{>1}, \alpha], \beta_k)$$

In particular, by (8) there is an N_α such that $(\forall n \geq N_\alpha)(\forall k \geq N_\beta)$

$$\mathrm{sol}^?(E_{>1}[\varepsilon^{>1}, \alpha_n], \beta_k) = \mathrm{sol}^?(E_{>1}[\varepsilon^{>1}, \alpha], \beta_k) = \textbf{\textit{true}} = \tag{9}$$

$$= \mathrm{sol}^?(E_{>1}[\varepsilon^{>1}, \alpha_n], \beta) = \mathrm{sol}^?(E_{>1}[\varepsilon^{>1}, \alpha], \beta)$$

and

$$(\forall i \leq N_\beta)\mathsf{H}(E_{>1}[\varepsilon^{>1}, \alpha_n], \beta_i) = \mathsf{H}(E_{>1}[\varepsilon^{>1}, \alpha], \beta_i) \equiv \beta_{i+1}$$

Assume that (6) never holds, and hence (4) or (5) holds for every n.

Case 1. For some N and all $n \geq N$ (4) holds. Then for all $n \geq N$

$$\gamma_{n+1} = 0, \quad \alpha_{n+1} = \mathsf{H}(E_{\leq 1}[0, \varepsilon^1], \alpha_n)$$

and the H-process for $(E[\varepsilon^{>1}, \varepsilon^1], (\gamma_0, \alpha_0))$ terminates by Lemma 6.1 since it is cofinal with the H-process for $(E_{\leq 1}[0, \varepsilon^1], \alpha_N)$.

Case 2. For some N and all $n \geq N$ (5) holds. Then for all $n \geq N$

$$\gamma_{n+1} = \mathsf{H}(E_{>1}[\varepsilon^{>1}, \alpha_N], \gamma_n), \quad \alpha_{n+1} = \alpha_n$$

and the H-process for $(E[\varepsilon^{>1}, \varepsilon^1], (\gamma_0, \alpha_0))$ terminates by IH since it is cofinal with the H-process for $(E_{>1}[\varepsilon^{>1}, \alpha_N], \gamma_N)$.

Case 3. None of the cases 1,2 holds, i.e. both of (4), (5) occur infinitely often. In particular, for infinitely many n, m,

$$\gamma_{n+1} = \mathsf{H}(E_{>1}[\varepsilon^{>1}, \alpha_n], \gamma_n), \tag{10}$$

$\gamma_m = 0$, and moreover, for infinitely many m

$$\gamma_{m+1} = \mathsf{H}(E_{>1}[\varepsilon^{>1}, \alpha_m], \gamma_m), \text{ and } \gamma_m = 0 \tag{11}$$

Indeed, for any n such that $\gamma_n = 0$, if (10) is not the case, take the first $m \geq n$ with $\gamma_{m+1} = \mathsf{H}(E_{>1}[\varepsilon^{>1}, \alpha_m], \gamma_m)$. Then for all i with $n \leq i \leq m$ we have $\gamma_i = 0$, and hence (11).

By the continuity in α, there is an M_0 satisfying for all $i \leq N_\beta$

$$(\forall m \geq M_0)[\mathrm{sol}(E_{\leq 1}[\beta_i, \varepsilon^1], \alpha) \iff \mathrm{sol}(E_{\leq 1}[\beta_i, \varepsilon^1], \alpha_m)] \tag{12}$$

Assume that

$$M > N_\beta, M_0, \quad \gamma_M = 0, \quad \gamma_{M+1} = \mathsf{H}(E_{>1}[\varepsilon^{>1}, \alpha_N], \gamma_M)$$

and prove by induction on $i \leq N_\beta$:

$$\alpha_{M+i} = \alpha_M \ \&\ \gamma_{M+i} = \beta_i \ \&\ (\forall m \geq M)\mathrm{sol}(E_{\leq 1}[\gamma_i, \varepsilon^1], \alpha_m) \tag{13}$$

In other words, the H-process for $(E[\varepsilon^{>1}, \varepsilon^1], (\gamma_M, \alpha_M))$ is going according to the Case 2 exactly as the H-process for $(E_{>1}[\varepsilon^{>1}, \alpha], \beta_0)$.

Induction base $i = 0$. One has $\gamma_M = 0 = \beta_0$, $\mathrm{sol}(E_{\leq 1}[\gamma_M, \varepsilon^1], \alpha_M)$. By (12) $(\forall m \geq M)$ $\mathrm{sol}(E_{\leq 1}[\gamma_M, \varepsilon^1], \alpha_m)$.

Induction step. Assume (13) is satisfied up to i, and prove it for $i + 1$. Since $\mathrm{sol}(E_{\leq 1}[\beta_i, \varepsilon^1], \alpha_m)$ holds for all $m \geq M_0$, one has

$$(\gamma_{M+i+1}, \alpha_{M+i+1}) \equiv \mathsf{H}(E[\varepsilon^{>1}, \varepsilon^1], (\gamma_{M+i}, \alpha_{M+i})) \stackrel{IH}{=}$$

$$\mathsf{H}(E[\varepsilon^{>1}, \varepsilon^1], (\beta_i, \alpha_M)) \stackrel{IH}{\equiv} (\mathsf{H}(E_{>1}[\varepsilon^{>1}, \alpha_M], \beta_i), \alpha_M) \equiv (\beta_{i+1}, \alpha_M)\,,$$

as required.

Taking $k = N_\beta$ one has by (13) that $\gamma_{M+k} = \beta_k$ and hence

$$\mathrm{sol}(E[\varepsilon^{>1}, \varepsilon^1], (\gamma_{M+k}, \alpha_{M+k})) \iff \mathrm{sol}(E[\varepsilon^{>1}, \varepsilon^1], (\beta_k, \alpha_M)) \iff$$
$$\mathrm{sol}(E_{>1}[\varepsilon^{>1}, \alpha_M], \beta_k) \iff \mathrm{sol}(E_{>1}[\varepsilon^{>1}, \alpha], \beta_k) = \mathbf{true}$$

since $M > N_\beta$. ⊣

REFERENCES

[1] W. Ackermann. Zur Widerspruchsfreiheit der Zahlentheorie. *Mathematische Annalen*, 117:162–194, 1940.

[2] D. Hilbert and P. Bernays. *Grundlagen der Mathematik*, volume 2. Springer, Berlin, 2 edition, 1970.

[3] G. Mints and S. Tupailo. Epsilon Substitution Method for the Ramified Language and Δ^1_1-Comprehension Rule. In *Logic and Foundations of Mathematics*, pages 107–130. Kluwer, Amsterdam, 1999.

[4] G. Mints, S. Tupailo, and W. Buchholz. Epsilon substitution method for elementary analysis. *Archive for Mathematical Logic*, 35:103–130, 1996.

[5] D. Normann. *Recursion in the countable functionals*, volume 811. Springer, 1980.

[6] W. Tait. The substitution method. *Journal of Symbolic Logic*, 30:175–192, 1965.

DEPARTMENT OF PHILOSOPHY
STANFORD UNIVERSITY
STANFORD, CA 94305, USA
E-mail: mints@turing.stanford.edu

PREDICATIVITY: THE OUTER LIMITS

STEPHEN G. SIMPSON

Abstract. Beginning with ideas of Poincaré and Weyl. Feferman in the sixties undertook a profound analysis of the predicativist foundational program. He presented a subystem of second order arithmetic IR and argued convincingly that it represents the outer limits of what is predicatively provable. Much later, Friedman introduced another system ATR_0 which is conservative over IR for Π^1_1 sentences yet includes several well known theorems of algebra, descriptive set theory, and countable combinatorics that are not provable in IR. The proof-theoretic ordinal of both systems is Γ_0. ATR_0 has emerged as one of a handful of systems that are important for reverse mathematics. From a foundational standpoint, we may say that IR represents predicative provability while ATR_0 represents predicative reducibility. Subsequently Friedman formulated mathematically natural finite combinatorial theorems that are not only not predicatively provable but go beyond Γ_0 and therefore are not predicatively reducible.

§1. IR and ATR_0. In his first major work on systems of predicative analysis [3, 4], Feferman introduces the system IR and proposes it as an explication of predicative provability.

> "Although we strongly believe that the explications proposed in this paper for the notion of predicative provability in analysis are correct, we are not convinced that the matter has been settled conclusively by the results obtained so far. It is premature to say just what would constitute final evidence concerning this question. We expect that this will be revealed, at least in part, by further study of the theories considered here." (page 29)

In subsequent papers on predicative provability, Feferman does not back away from this proposal. The systems that he introduces in [5, 6, 7] as explications of predicative provability are conservative over IR.

In this paper I want to compare Feferman's system IR [3, 4] with another subsystem of second order arithmetic ATR_0 due to Friedman [10]. We begin by briefly reviewing the definitions of these systems.

Received by the editors October 29, 1999 and, in revised form, October 16, 2000.

1991 *Mathematics Subject Classification.* 03F35, 03D80, 03B30.

Preparation of this paper was partially supported by NSF grant DMS-0070718.

This paper is a writeup of my talk at a symposium at Stanford University, December 11–13, 1998, in honor of Solomon Feferman's 70th birthday. I would like to thank Sol for his encouragement over many years and especially when I was writing my book [26] on reverse mathematics and subsystems of second order arithmetic.

Reflections on the foundations of mathematics
Essays in honor of Solomon Feferman
Edited by Wilfried Sieg, Richard Sommer, and Carolyn Talcott

§2. Rules versus axioms. Both IR and ATR_0 are formal systems or theories in the language of second order arithmetic. While ATR_0 is easily defined by means of a finite set of axioms, IR is more conveniently described in terms of inference rules rather than axioms.

1. The Δ^1_1 Comprehension Axiom:

$$(\forall n\,((\exists X\,\alpha(n,X)) \leftrightarrow (\forall Y\,\beta(n,Y)))) \\ \rightarrow \exists Z\,\forall n\,(n \in Z \leftrightarrow \exists X\,\alpha(n,X))$$

   Here α and β are arithmetical formulas.

2. The Δ^1_1 Comprehension Rule:

$$\frac{\forall n\,((\exists X\,\alpha(n,X)) \leftrightarrow (\forall Y\,\beta(n,Y)))}{\exists Z\,\forall n\,(n \in Z \leftrightarrow \exists X\,\alpha(n,X))}$$

   Here α and β are as above.

3. The Hierarchy Axiom:

$$\forall Z\,(\mathrm{WO}(Z) \rightarrow \forall X\,\exists Y\,\mathrm{H}(Z,X,Y))$$

   Here $\mathrm{WO}(Z)$ is a Π^1_1 formula expressing that Z is a well ordering of the integers, and $\mathrm{H}(Z,X,Y)$ is an arithmetical formula expressing that Y is a Turing jump hierararchy along Z starting at X.

4. The Hierarchy Rule:

$$\frac{\mathrm{WO}(<_e)}{\forall X\,\exists Y\,\mathrm{H}(<_e,X,Y)}$$

   Here $<_e$ is a primitive recursive linear ordering of the integers, and $\mathrm{WO}(Z)$ and $\mathrm{H}(Z,X,Y)$ are as above.

5. The Transfinite Induction Axiom:

$$\forall Z\,(\mathrm{WO}(Z) \rightarrow \mathrm{TI}(Z,\gamma))$$

   Here $\mathrm{WO}(Z)$ is as above, γ is an arbitrary formula, and $\mathrm{TI}(Z,\gamma)$ expresses transfinite induction along Z with respect to γ.

6. The Transfinite Induction Rule:

$$\frac{\mathrm{WO}(<_e)}{\mathrm{TI}(<_e,\gamma)}$$

   Here $<_e$ and $\mathrm{WO}(Z)$ and $\mathrm{TI}(Z,\gamma)$ are as above.

We are now ready to define the systems IR and ATR_0.

1. IR consists of the Δ^1_1 Comprehension Rule, the Hierarchy Rule, and the Transfinite Induction Rule.
2. ATR_0 consists of the Hierarchy Axiom. It is known that ATR_0 includes the Δ^1_1 Comprehension Axiom. It is a system with restricted induction (see Friedman [10]) and so does not include the Transfinite Induction Rule.

§3. Model-theoretic properties of IR and ATR$_0$. It is known that IR and ATR$_0$ are proof-theoretically similar:

1. They have the same proof-theoretic ordinal: $|\mathsf{IR}| = |\mathsf{ATR}_0| = \Gamma_0$.
2. IR and ATR$_0$ prove the same Π^1_1 sentences. In particular, they prove the same arithmetical sentences.
3. IR and ATR$_0$ have the same proof-theoretic strength.

These results are due to Friedman [11, §4]. The main point that we would like to make here is that IR and ATR$_0$ differ greatly in some other, very significant respects. In particular:

1. IR explicates predicative provability, while ATR$_0$ explicates predicative reducibility.
2. ATR$_0$ is much stronger than IR, model-theoretically and, above all, mathematically.

The following properties of the two systems indicate how different they are from the model-theoretic point of view.

1. The minimum ω-model of IR is $\mathrm{HYP}(\Gamma_0)$, i.e., $L_{\Gamma_0} \cap P(\omega)$. This is a relatively small initial segment of
$$\mathrm{HYP} = \{X \subseteq \omega : X \text{ is hyperarithmetical}\},$$
i.e., $L_{\omega_1^{CK}} \cap P(\omega)$.
2. The minimum ω-model of the Δ^1_1 Comprehension Axiom is HYP.
3. HYP is the intersection of all ω-models of ATR$_0$.
4. ATR$_0$ has no minimal ω-model.
5. ATR$_0$ automatically holds in any β-model.
6. HYP is the intersection of all β-models.
7. There is no minimal β-model.

We can also compare IR and ATR$_0$ with the perhaps more familiar system Π^1_∞-TI$_0$ consisting of the Transfinite Induction Axiom. The latter system is sometimes known as *bar induction*. Some model-theoretic properties:

1. Π^1_∞-TI$_0$ includes both IR and ATR$_0$. The precise relationship to ATR$_0$ is that Σ^1_1-TI$_0$ = ATR$_0$ + Σ^1_1-IND (Simpson [22]).
2. Π^1_∞-TI$_0$ is proof-theoretically stronger than IR and ATR$_0$.
3. $|\Pi^1_\infty\text{-TI}_0| = \varphi_{\varepsilon_{\Omega+1}}(0)$ = the Howard ordinal.
4. Π^1_∞-TI$_0$ has no minimal ω-model.
5. Π^1_∞-TI$_0$ automatically holds in any β-model.
6. HYP is the intersection of all β-models.
7. There is no minimal β-model.

For proofs of the model-theoretic results mentioned above, see Chapters VII and VIII of Simpson [26].

REMARK 1. We see above that ATR$_0$ and Π^1_∞-TI$_0$ are model-theoretically stronger than IR, in that the ω-models are larger. One might think that the greater strength comes from the fact that ATR$_0$ and Π^1_∞-TI$_0$ deal with arbitrary well orderings of the integers, and not only primitive recursive ones. However, this is not the case. Letting ATR_0^- be ATR_0 with the Hierarchy Axiom restricted to primitive recursive linear orderings, the above model-theoretic results for ATR_0 continue to hold for ATR_0^-. Thus these results are seen to have a certain robustness.

§4. A set-theoretic version of ATR_0**.** In [5] Feferman introduced a set-theoretic version of IR. Subsequently Simpson [21] introduced a set-theoretic version of ATR_0 known as $\mathsf{ATR}_0^{\mathrm{set}}$, defined by

$$\begin{aligned}
\mathsf{ATR}_0^{\mathrm{set}} \quad = \quad & \text{Axiom of Extensionality} \\
+ \quad & \text{Axiom of Foundation:} \\
& \forall x\,(x \neq \emptyset \rightarrow \exists u \in x\,(u \cap x = \emptyset)) \\
+ \quad & \text{closure under } F_0\text{–}F_8\text{, i.e., under rudimentary functions} \\
+ \quad & \text{Axiom of Infinity} \\
+ \quad & \forall x\,(x \text{ is hereditarily countable}) \\
+ \quad & \text{Axiom Beta:} \\
& \forall r\,(\mathrm{WF}(r) \rightarrow \exists f\,(\mathrm{field}(r) \subseteq \mathrm{dom}(f) \\
& \quad \wedge \forall u \in \mathrm{dom}(f)\,(f(u) = \{f(v) : \langle v, u\rangle \in r\}))) .
\end{aligned}$$

It is shown in Simpson [21] (see also [26, §VII.3]) that $\mathsf{ATR}_0^{\mathrm{set}}$ is conservative over ATR_0. Actually, it is a definitional extension of ATR_0, where well founded trees encode hereditarily countable sets in the usual way.

§5. Reverse mathematics of ATR_0**.** As is well known, ATR_0 is one of the five basic systems of reverse mathematics. From Simpson [26, Chapter V] we have:

THEOREM 2. The following are equivalent over RCA_0.

1. ATR_0.
2. Every disjoint pair of analytic sets can be separated by a Borel set.
3. The domain of a single-valued Borel set in the plane is Borel.
4. Every uncountable closed (or analytic) set has a perfect subset.
5. Clopen (or open) determinacy.
6. The clopen (or open) Ramsey Theorem.
7. For every countable bipartite graph, there exists a *König covering*, i.e., a pair (C, M) where C is a vertex covering, M is a matching, and C consists of one vertex of each edge in M. (This is a combined result of Aharoni/Magidor/Shore 1992 [1] and Simpson 1994 [25].)
8. Comparability of countable well orderings, i.e., well orderings of the integers.
9. The Ulm theory for countable reduced Abelian p-groups.

There are some interesting open questions concerning the reverse mathematics aspect of ATR_0.

1. Is Fraïssé's conjecture for countable linear orderings provable in ATR_0? See [26, X.3.31] and Marcone 1994 [16] and Shore 1993 [19].
2. A well known consequence of the Ulm theory (see Kaplansky [14]) is:

   If each of two countable reduced Abelian p-groups is a direct summand of the other, then they are isomorphic.

   Is this statement equivalent to ATR_0? This question is due to Friedman (see [26, V.7.7]). Some recent progress on this question is in Friedman [9].

REMARK 3. The reverse mathematics investigations of [26, Chapter V] seem to indicate that, mathematically, IR is no stronger than ACA_0 (arithmetical comprehension), and Π^1_∞-TI_0 is no stronger than ATR_0. Thus ATR_0 is a much better system than IR from the viewpoint of reverse mathematics.

§6. Impredicative Π^0_2 combinatorial theorems.

In the aftermath of Paris/Harrington, it has been shown that certain mathematically appealing, finite combinatorial theorems are not provable in IR or ATR_0 or even stronger systems. In particular, such theorems are neither predicatively provable nor predicatively reducible. For a survey of this general area, see Simpson [24]. Recently Feferman [8] has cited some of these results in footnotes. We now state some of these results.

6.1. An impredicative Ramsey-type theorem. Friedman/McAloon/Simpson 1982 [11] were the first to exhibit a mathematically natural, finite combinatorial theorem which is not predicatively provable. To state their result, we need some definitions.

DEFINITION 4. Let X be a finite set of positive integers.

1. A *coloring* of X is given by $P(X) = C_1 \cup C_2$, where C_1 and C_2 are closed under initial segment. A set $Y \subseteq X$ is said to be *homogeneous* for the given coloring if either $P(Y) \subseteq C_1$ or $P(Y) \subseteq C_2$.
2. X is said to be 0-*dense* if $|X| \geq 2$ and $|X| \geq \min X$. X is said to be $(n+1)$-*dense* if for every coloring of X there exists an n-dense homogeneous set.

THEOREM 5 (Friedman/McAloon/Simpson). The following statements are pairwise equivalent over PRA.

1. $\forall n\ \exists n$-dense finite set.
2. uniform Π^0_2 reflection for IR.
3. uniform Π^0_2 reflection for ATR_0.

6.2. Friedman's work on Kruskal's Theorem. Friedman has shown that certain interesting combinatorial properties of finite trees are not provable in IR and ATR_0 and stronger systems. See the exposition in Simpson [23].

It turns out that there is a close connection between these results and recent spectacular work in finite graph theory. Namely, Friedman/Robertson/Seymour [12] have used them to show that the celebrated Robertson/Seymour Graph Minor Theorem is not provable in Π^1_1-CA_0.

6.3. A recent result of Friedman. In order to state Friedman's most recent result along these lines, we first give the necessary definitions.

DEFINITION 6.

1. A *tree* T is a finite poset with a minimum element such that the predecessors of each element are linearly ordered. The *height* of $x \in T$ is the number of predecessors of x in T. The *height of* T is the maximum height of an element of T. We say that T is *of degree* $\leq k$ if each element of T has at most k immediate successors.
2. $T(\leq i) = \{x \in T : \text{height}(x) \leq i\}$.
3. $T(i) = \{x \in T : \text{height}(x) = i\}$.
4. $T(> i) = \{x \in T : \text{height}(x) > i\}$.

Note that $T(\leq i)$ is a subtree of T.

Now consider the following combinatorial statement concerning finite trees.

> For each k there exists n so large that the following holds. If T is a tree of height n and degree $\leq k$, then there exists $1 \leq i \leq n$ and an inf-preserving embedding of $T(\leq i)$ into T which carries $T(i)$ into $T(> i)$.

Friedman has recently shown that this statement is equivalent to uniform Π^0_2 reflection for Π^1_2-TI_0. In particular, this statement of Friedman is true but not provable in ATR_0. See also Friedman's contribution to this volume.

REFERENCES

[1] RON AHARONI, MENACHEM MAGIDOR, and RICHARD A. SHORE, *On the strength of König's duality theorem for infinite bipartite graphs*, ***Journal of Combinatorial Theory, Series B***, vol. 54 (1992), pp. 257–290.

[2] J. N. Crossley, J. B. Remmel, R. A. Shore, and M. E. Sweedler (editors), ***Logical Methods***, Birkhäuser, 1993, 813 pages.

[3] SOLOMON FEFERMAN, *Systems of predicative analysis, I*, ***The Journal of Symbolic Logic***, vol. 29 (1964), pp. 1–30.

[4] ———, *Systems of predicative analysis, II*, ***The Journal of Symbolic Logic***, vol. 33 (1968), pp. 193–220.

[5] ———, *Predicatively reducible systems of set theory*, [18, part 2], 1974, pp. 11–32.

[6] ———, *A more perspicuous formal system for predicativity*, [15, volume I], 1978, pp. 68–93.

[7] ———, *Reflecting on incompleteness*, ***The Journal of Symbolic Logic***, vol. 56 (1991), pp. 1–49.

[8] ———, *Does mathematics need new axioms?*, ***American Mathematical Monthly***, vol. 106 (1999), pp. 99–111.

[9] HARVEY FRIEDMAN, *Metamathematics of Ulm theory*, June 24, 1999, unpublished, 34 pages.

[10] ———, *Systems of second order arithmetic with restricted induction, I, II* (abstracts), ***The Journal of Symbolic Logic***, vol. 41 (1976), pp. 557–559.

[11] HARVEY FRIEDMAN, KENNETH MCALOON, and STEPHEN G. SIMPSON, *A finite combinatorial principle which is equivalent to the 1-consistency of predicative analysis*, [17], 1982, pp. 197–230.

[12] HARVEY FRIEDMAN, NEIL ROBERTSON, and PAUL SEYMOUR, *The metamathematics of the graph minor theorem*, [20], 1987, pp. 229–261.

[13] L. A. Harrington, M. Morley, A. Scedrov, and S. G. Simpson (editors), ***Harvey Friedman's Research on the Foundations of Mathematics***, Studies in Logic and the Foundations of Mathematics, North-Holland, 1985, XVI + 408 pages.

[14] Irving Kaplansky, ***Infinite Abelian Groups***, revised ed., University of Michigan Press, 1969, VII + 95 pages.

[15] K. Lorenz (editor), ***Konstruktionen versus Positionen, beiträge zur Diskussion um die Konstruktive Wissenschaftstheorie***, Walter de Gruyter, Berlin, 1978, Volume I, XX + 350 pages. Volume II, X + 406 pages.

[16] Alberto Marcone, *Foundations of BQO theory*, ***Transactions of the American Mathematical Society***, vol. 345 (1994), pp. 641–660.

[17] G. Metakides (editor), ***Patras Logic Symposion***, Studies in Logic and the Foundations of Mathematics, North-Holland, 1982, IX + 391 pages.

[18] D. S. Scott and T. J. Jech (editors), ***Axiomatic Set Theory***, Proceedings of Symposia in Pure Mathematics, vol. XIII, American Mathematical Society, 1971–1974, Part 1, VI + 474 pages. Part 2, VI + 222 pages.

[19] Richard A. Shore, *On the strength of Fraïssé's conjecture*, [2], 1993, pp. 782–813.

[20] S. G. Simpson (editor), ***Logic and Combinatorics***, Contemporary Mathematics, American Mathematical Society, 1987, XI + 394 pages.

[21] Stephen G. Simpson, *Set theoretic aspects of* ATR_0, [27], 1982, pp. 255–271.

[22] ———, Σ^1_1 *and* Π^1_1 *transfinite induction*, [27], 1982, pp. 239–253.

[23] ———, *Nonprovability of certain combinatorial properties of finite trees*, [13], 1985, pp. 87–117.

[24] ———, *Unprovable theorems and fast growing functions*, [20], 1987, pp. 359–394.

[25] ———, *On the strength of König's duality theorem for countable bipartite graphs*, ***The Journal of Symbolic Logic***, vol. 59 (1994), pp. 113–123.

[26] ———, ***Subsystems of Second Order Arithmetic***, Perspectives in Mathematical Logic, Springer-Verlag, 1999, XIV + 445 pages.

[27] D. van Dalen, D. Lascar, and T. J. Smiley (editors), ***Logic Colloquium '80***, Studies in Logic and the Foundations of Mathematics, North-Holland, 1982, X + 342 pages.

DEPARTMENT OF MATHEMATICS
PENNSYLVANIA STATE UNIVERSITY
STATE COLLEGE, PA 16802, USA
E-mail: simpson@math.psu.edu
URL: http://www.math.psu.edu/simpson/

PART II: LOGIC AND COMPUTATION

A PRAGMATIC INTERPRETATION OF SUBSTRUCTURAL LOGICS

GIANLUIGI BELLIN AND CARLO DALLA POZZA

Abstract. Following work by Dalla Pozza and Garola [2, 3] on a pragmatic interpretation of intuitionistic and deontic logics, which has given evidence of their compatibility with classical semantics, we present sequent calculus system **ILP** formalizing the derivation of assertive judgements and obligations from *mixed contexts* of assertions and obligations and we prove the cut-elimination theorem for it. For the formalization of real-life *normative systems* it is essential to consider inferences from mixed contexts of assertions and obligations, and also of assertions justifiable *relatively to a given state of information* and obligations valid *in a given normative system*. In order to provide a formalization of the notion of *causal implication* and its interaction with obligations, the sequents of **ILP** have *two areas* in the antecedent, expressing the relevant and the ordinary intuitionistic consequence relation, respectively. To provide a pragmatic interpretation of reasoning with the *linear consequence relation* we consider the deductive properties of pragmatic schemes where the operators of illocutionary force are unknown (*free logic of pragmatic force*). We introduce an auxiliary system **ILLP** which formalizes such a logic, and has *substitution rules* that allow us to derive the desired mixed sequents. It is shown that in order to permit *non-uniform susbtitutions* and to preserve *Hume's law* (the underivability of obligations from truth-asserting judgements only and the potential *ineffectuality* of norms) **ILLP** must indeed be based on the *linear consequence relation*. It is also argued that the above uses of linear and relevant logics are perfectly compatible with a theory of pragmatics in a classical semantical setting and immune from popular confusions about the intuitive interpretations of substructural logics.

§1. Preface. Most forms of human reasoning can be formalized and those which deserve such a treatment also benefit from it. Not all philosophers may agree with such a statement, but never mind: the most distinguished opponent of the formalization of mathematical reasoning in the 20th century, Brouwer, was luckily contradicted by his followers and sympathisers, to whom we owe the beautifully insightful constructions of intuitionistic logic, from Heyting's axiomatization and informal semantics of proofs to Gentzen and Prawitz's natural deduction **NJ**, from Curry and Howard's correspondence between **NJ** and the typed lambda calculus to categorical logic, and so on and on. Today formal logic has been developed to represent a large variety of forms of reasoning; even such branches as *deontic logic*, which in the 1950s was the territory of a very small group of logicians, have now become popular and found applications, e.g., in Artificial Intelligence. Does this lush development threaten the idea(1) of the unity of logic?

1.1. A simple observation about the development of formal logic may give reassurances about "the unity of logic". Many mathematical systems receiving

Reflections on the foundations of mathematics
Essays in honor of Solomon Feferman
Edited by Wilfried Sieg, Richard Sommer, and Carolyn Talcott

the name "logic" *prima facie* do not deserve it. But the considerations which justify regarding such systems as logics also show their compatibility with the main body of logical theory. A formal system may be properly regarded as a logic (within the framework of a commonly accepted logical theory) only if it adequately represents pre-formal forms of reasoning which actually occur in human practice and can be shown to be correct and compatible with the standard forms of reasoning (as defined by the logical framework in question). For instance, in the predominant framework of *classical* logical theory, logic has to do with propositions, and propositions are objects which a truth-value can be assigned to: this does not mean that the only possible logic is a theory of propositions, but rather that other aspects of logical theory, e.g., a theory of judgements, must be shown to be in harmony with the logic of propositions classically understood.

There are forms of reasoning, e.g., reasoning with defaults, whose formalization is convenient for applications to Artificial Intelligence, although strictly speaking they ought to be regarded as fallacious and would indeed lead to fallacy if their range of application wasn't clearly defined. The "user manual" for such logics (implicitly or explicitly) contains a translation into the syntax and the semantics of ordinary propositional or first-order theories. In this way, "default logics" are shown to be no threat to the "unity of logic".

1.2. There are good reasons to believe that the garden of logic is not in danger of degenerating into a wild and impenetrable jungle. For instance, only those who see an irreconciliable opposition between classical logic and "deviant logics" may be tempted to mount crusades against the intuitionistic troublemakers or the classical imperialists, respectively, in the name of "the unity of logic". But there is a more interesting way of looking at this opposition, namely, to ask whether classical and intuitionistic logic are about the same aspects of reasoning and, if not, what they are about. An answer may be given by recalling Frege's distinction between *propositions* and *judgements* (or *assertions*): in [2] the second author has argued that Heyting's interpretation of intuitionistic connectives applies not to propositions but to the relations between illocutionary acts, such as assertions, commands, etc. Propositions are *true* or *false*. Illocutionary acts are *justified* or *unjustified*. But the illocutionary act of assertion can be justified only by a *proof* of the proposition expressed in it: a mathematical proof in the case of a mathematical assertion, or some kind or empirical evidence in the case of an empirical assertion. Proofs in classical logic use semantical properties of propositions given by the truth-functional semantics of connectives as a method of justification. This method of proof is not allowed intuitionistically: here the justification of an intuitionistic formula is built up according to Heyting's interpretation of intuitionistic connectives starting from proofs of atomic propositions, thus from non-logical proofs. But if we restrict the content of assertions to atomic propositions, then truth-functional interpretation of the connectives is never used as a tool for the justification of judgements; thus the restriction on the propositional content of judgements is sufficient to guarantee that restricted expressions are constructively interpreted. If this is correct, then intuitionistic logic is a special case of the logic of judgement.

Intuitionistically minded philosophers question the distinction between *truth* and *provability*. The point of view expressed in [2] is that a proof can only be a proof *of the truth* of a proposition: here "truth" is meant in the sense of Tarski's semantics as "truth-as-correspondence", and is regarded as an ontologically neutral notion.[1] This amounts to saying that the classical notion of truth is prior to the notions of proof and hence of justification and, moreover, that the logic of judgement is an extension of the classical logic of propositions.

1.3. A similar treatment can be given of other illocutionary acts: for instance, *prescriptions* are regarded as *propositions* pragmatically used in a prescriptive way. Among these, *norms* are *impersonal prescriptions* [3]. Indeed when we speak of illocutionary acts as assertions and norms, we regard them abstractly, i.e., *impersonally*: we are not considering an assertion made by Bill in a given situation with obvious physical limitations on information retrieval and on the capacity of inferring the consequences of a given body of knowledge, but the assertion of a proposition which could or could not be justifiedly made, given existing scientific information. Similarly, we are not interested in a command given in a given situation by Bill, a person in position of legitimate authority, but limited in his capacity of judgement and subject to "weakness of the will"; rather we are interested in the command that Bill should have given in the framework of the normative theory he operates in. A logic of assertions and norms along these lines is an attempt to extend the domain of logic beyond the class of expressions to which a truth-value can be assigned and to create a *formal pragmatics*, i.e., a logic of illocutionary acts.

Extending Frege's theory (later developed by Reichenbach), in Section 2 we define a formal language for pragmatics $\mathcal{L}^P$, containing not only formulas $\vdash \alpha$ ("α is assertible") and $\multimap \alpha$ ("α is obligatory") but also formulas $\vdash_k \alpha$ and $\multimap_N \alpha$, which express assertions justifiable *depending on* a system of knowledge k and norms valid *relative to* a normative system N, respectively. The main novelty of the language $\mathcal{L}^P$ is the presence of connectives representing relations between illocutionary acts, in addition to the connectives of classical logic. As indicated above, the pragmatic connectives are given Heyting's "semantics of proofs". A formula is called *pragmatically valid* (or *p-valid*) if it represents relations between illocutionary acts which hold in all circumstances. Like the notion of provability in Heyting's semantics, the notion of justifiability is informal.

In previous papers [2, 3] Hilbert-style deductive systems for the pragmatic language $\mathcal{L}^P$ have been given, in which classic, intuitionistic and deontic logics are representable. In particular the pragmatic operator of obligation has the following properties:

---

[1] The philosophical position adopted here seems to us in agreement with the point of view expressed by Professor Solomon Feferman in many occasions, in particular in his review of Prawitz [12] in the *J.S.L.* [5].

(i) It satisfies the axiom of *classical deontic necessitation*:

$$\text{if} \quad \beta_1, \ldots, \beta_n \models \alpha \quad \text{then} \quad \circ\!\!-\, \beta_1, \ldots, \circ\!\!-\, \beta_n \Rightarrow \circ\!\!-\, \alpha$$

namely, *"if α is a truth-functional consequence of $\beta_1, \ldots, \beta_n$, then $\circ\!\!-\, \alpha$ follows from $\circ\!\!-\, \beta_1, \ldots, \circ\!\!-\, \beta_n$"*.

(ii) The class of obligations is consistent:

$$\circ\!\!-\, (p \wedge \neg p) \Rightarrow\ \vdash (p \wedge \neg p)$$

namely, *"if an absurdity is an obligation then an absurdity is justifiably assertible"*.

(iii) The principle "*$\circ\!\!-\, \alpha$ or not $\circ\!\!-\, \alpha$*" is a fact about normative systems, *not a law of logic*.

Notice that (iii) is a departure from the tradition of systems such as **KD** in harmony with our Heyting-style interpretation of the pragmatic connectives.

Notice also that, unlike the case of assertions, it is not assumed that the justification of an obligation ultimately lies in the recognition of the *truth* of the proposition expressing the content of the obligation. Even in a cognitivist approach to ethics where *"bonum"* and *"verum"* ultimately meet it would be counterintuitive to postulate that the class of propositions expressing moral obligations *coincides* with the class of true propositions: the former is rather be *a proper subclass* of the latter. It is the task of ethical theory to characterize these nonlogical axioms and what counts as a justification for regarding these axioms as obligations. The task of logic is to guarantee that logical inferences may be correctly performed from the axioms.

1.4. There is a hitch in our treatment of pragmatic obligation: in reality, *legal obligations* are always expressed as conditional statements of the form: *"if β is the case, then α is obligatory"*. In coherence with the above discussion, here conditional normative expressions must have the form *"if β is assertible, then α is obligatory"*, where *"β is assertible"* means *"there is a proof of β"* rather than *"it is the case that β"*. It is not a problem to express such statements as *non-logical axioms* within the frame of the existing systems for $\mathcal{L}^P$ However, as soon as *mixed contexts* are introduced, it becomes natural to ask whether in a *sequent calculus* for $\mathcal{L}^P$ we need *bridging principles* for assertive and deontic judgements such as

$$(*0) \qquad \circ\!\!-\, \alpha,\ \vdash (\alpha \rightarrow \beta) \Rightarrow \circ\!\!-\, \beta.$$

or

$$(*\S) \qquad \circ\!\!-\, \alpha,\ (\vdash \alpha \supset\ \vdash \beta) \Rightarrow \circ\!\!-\, \beta.$$

Notice that $(*0)$ is problematic, because a proof in classical logic of $\alpha \rightarrow \beta$ may be too weak to convert a justification of $\circ\!\!-\, \alpha$ into a justification of $\circ\!\!-\, \beta$: from evidence that *"if Hitler is dead, then the Titanic has sunk"* and the assumption *"Hitler ought to have died"* we do not intuitively infer the conclusion *"the Titanic ought to have sunk"*. On the other hand, let us consider $(*\S)$: intuitionistically, $\vdash \alpha \supset\ \vdash \beta$ is justified if there is a method to convert any proof of the truth of α into a proof of the truth of β. But what kind of method is needed to transform a justification of $\circ\!\!-\, \alpha$ into a justification of $\circ\!\!-\, \beta$ according to our intuitions about justified obligations? It is not difficult to show that in $\mathcal{L}^P$ the formula

$$\vdash (\alpha \rightarrow \beta) \Rightarrow (\vdash \alpha \supset \vdash \beta)$$

is p-valid. It follows that, *if we accept classical logic*, we cannot accept $(*\S)$ without also accepting $(*0)$. (Of course, if $\alpha \rightarrow \beta$ is a tautology, then also $\vdash \alpha \supset \vdash \beta$ and $\circ\!\!-\, \alpha \supset \circ\!\!-\, \beta$ are p-valid, see [3], so the problem does not arise.)

However, there is a sense in which principle $(*\S)$ *is plausible* (at least to some logicians in this area): this is the case whenever the method that converts any proof of the truth of α into a proof of the truth of β is based upon a principle of *causality*. Suppose α causes β; then from a proof of this relation of causality and from the obligation that α we conclude that β is obligatory. Let us use the special sign "$\Supset$" to indicate this special (*causal*) case of the intuitionistic implication. The principle $(*\S)$ becomes

$$(\S) \qquad \circ\!\!-\, \alpha, (\vdash \alpha \Supset \vdash \beta) \Rightarrow \circ\!\!-\, \beta.$$

which is intuitively plausible.

In our framework we would like to have rule of deontic necessitation so as to derive sequents of the form $(\S)$. Clearly, it should not be possible to derive

$$(*1) \qquad \vdash p \Rightarrow \circ\!\!-\, p$$

because of a formal constraint which we may call *Hume's Law*: "*we cannot derive an* ought *from an* is".[2] Indeed stronger formal constraints are needed: "*we cannot derive an* ought *from an* is *and from* ought*s that are irrelevant*". Otherwise, we could derive the undesirable sequent

$$(*2) \qquad \circ\!\!-\, q, \ \vdash p \Rightarrow \circ\!\!-\, p$$

where p and q are atomic. Moreover the following sequent is counterituitive:

$$(*3) \qquad \vdash \alpha, (\circ\!\!-\, \alpha \supset \circ\!\!-\, \beta) \Rightarrow \vdash \beta.$$

as obligations may be ineffective. In general, "*we cannot derive an* is *from a relevant* ought".

We have extended the language $\mathcal{L}^P$, by introducing a new connective $\Supset$ of *causal implication*. We need to show that this extension is compatible with our framework of logical theory of propositions and judgements. An expression $\vdash \alpha \Supset \vdash \beta$ is justified if and only if $\vdash \alpha \supset \vdash \beta$ is justified as an instantiation of a first order expression $\vdash \forall x(\phi(x) \rightarrow \chi(x))$ which expresses a scientific law, where $\alpha \equiv \phi(t)$ and $\beta \equiv \chi(t)$ for some objects t. Of course, it is the task of epistemology to identify the universal statements which express scientific laws.

Now the rule

$$\frac{\Gamma \Rightarrow \vdash \alpha}{\Gamma \Rightarrow \vdash \beta \Supset \vdash \alpha}$$

[2]It should be clear from the discussion in section **1.3.** above that such a formal constraint does not rule out the possibility of expressing a cognitivist approach to ethics within $\mathcal{L}^P$.

is clearly inadmissible for *causal implication* (when $\vdash \beta$ does not occur in the contex Γ) and this show that a sequent calculus for causal implication cannot admit the unrestricted rule of Weakening.

1.5. We need a deductive system in which the logical properties of *causal implication* and of *ordinary intuitionistic implication* can be presented in the same context. This system may be regarded as a framework where new deductive principles, corresponding to intutively p-valid expressions, can be safely added. Here we present a new sequent calculus **ILP**$^-$ for the intuitionistic fragment of $\mathcal{L}^P$, inspired by Girard's **LU** [7]. Sequents have *two areas* in the antecedent:

$$\Gamma\, ;\, \Delta \Rightarrow \delta$$

In the *internal area* (where Δ lies) and in the succedent formulas are introduced according to the familiar rules of inference of Gentzen's intuitionistic sequent calculus **LJ**. In the *external area* (where Γ lies) Weakening is not allowed; formulas can be introduced here by the axiom *"ex falso quodlibet"* or by the logical rule *left causal implications*; moreover, the rule *right causal implications* is correct only with the crucial restriction that the internal area should be empty. In particular, it is incorrect to infer a causal implication through the use of the axiom *ex falso quodlibet"*:

$$\frac{\delta_1, \Gamma\, ;\, \bigwedge, \Delta \Rightarrow \delta_2}{\Gamma\, ;\, \bigwedge, \Delta \Rightarrow \delta_1 \supset \delta_2}$$

(where $\bigwedge$ is the symbol for absurdity, δ_1, δ_2 are formulas and Γ, Δ sequences of formulas of $\mathcal{L}^P$).)

In the setting of **ILP**$^-$ we can safely introduce the deductive principles characteristic of the system **ILP**:

causal implication - obligation

$$\frac{\vdash \boldsymbol{A}, \boldsymbol{Z}\, ;\, \Rightarrow\ \vdash \beta}{\multimap \boldsymbol{A}, \boldsymbol{Z}\, ;\, \Rightarrow\ \multimap \beta}$$

where $\vdash$ **A** is a sequence of assertions of atomic formulas $\vdash \alpha$ and **Z** is a sequence of formulas of the form $(\vdash \alpha_1 \supset \ldots (\vdash \alpha_n \supset \vdash \beta) \ldots)$, i.e., built from atomic assertions using *positive* causal implications only.

On one hand it is easy to see that the system **ILP** solves the problem of deriving the new deductive principles, without deriving the undesirable ones. Other extensions are possible along the same lines, as we shall see below. On the other hand the use of *relevant logic* (or any other *substructural logic*) may be regarded with suspicion by those who believe that no satisfactory account has been given so far of substructural logics in the framework of logical theory: in this case substructural logics themselves may be regarded as a threat to the "unity of logic". For this reason we shall consider reasoning with the *linear consequence relation*, where not only *Weakening* but also *Contraction* is in general not allowed, and try to provide an interpretation of such forms of reasoning that shows their compatibility with the framework of the logical theory of propositions and assertions sketched above.

1.6. Linear logic is presented as a "refinement" of both *intuitionistic* and *classical* logic. It is based on the *linear inference relation* which rules out the structural rules of *Contraction* and *Weakening*, except for formulas prefixed with special modalities, the *exponential operators*. Many features of intuitionistic logic are extended in a very interesting way to linear logic: the denotational semantics of coherent spaces, the categorical semantics of intuitionistic linear logic, the game-theoretic semantics are all important "refinements" of ideas developed for intuitionistic logic and the typed λ calculus. Moreover, classical logic can be represented within linear logic in many ways, but in such representations the proof-theory of classical logic acquires the properties of strong normalization and convergence of normalization which are notoriously lacking in Gentzen's classical sequent calculus **LK**. The list may be extended with several significant items, but all of them are *mathematical* properties which do not give an answer to the question: *what is linear logic about?*

In [8] the following explanations are given:

> "classical and intuitionistic logic deal with stable truths:
>
> if A and $A \Rightarrow B$, then B, *but A still holds.*
>
> This is perfect in mathematics, but wrong in real life, since real implication is *causal.* A causal implication cannot be iterated since the conditions are modified after its use; this process of modification of the premises (conditions) is known in physics as *reaction.*"

The suggestion is that in everyday reasoning the use of a *causal implication* modifies the deductive context in the same way as in a physical process the application of a causal law modifies the preconditions. The notion of "causal implication" is related here to the notion of assumptions as "deductive resources". Now in proof-theory we may give a technical sense to the expression "deductive resources", but it is less clear how an act of assertion may be "consumed" during an argument. What is worse, the popular examples commonly cited to suggest a natural language interpretation seem to imply a confusion between logic and its domain of application: for instance, the formula *"two atoms of hydrogen and one of oxygen yield a molecule of water"* is quoted to suggest that the meaning of "yield" in the chemical sense could help clarify the logical notion of causal implication and could give an example of reasoning where the rules of Weakening and Contraction do not apply. It is therefore a lucky circumstance that an interesting example of the role of linear logic as an auxiliary system in philosophical logic may come from the development of a formal pragmatics for *mixed assertive and deontic deductive contexts*: we shall show that in this theory it becomes natural to prohibit the unrestricted rules of Weakening and Contraction, without making appeal to the notion of "deductive resource", but preserving Girard's intuition of linear implication as causal implication. Such an example should give reassurances about the status of linear logic in the general framework of logical theory.

1.7. The main idea of this paper is that *the linear consequence relation expresses the deductive properties of pragmatic schemes where the operators of illocutionary* *force are unknown and their specification is not required to be uniform.* Let us write $\bullet\,\alpha$ for an illocutionary act whose propositional content is α and whose force is unspecified. We want to define a formal theory of illocutionary acts under such restrictions, a system which may be called the *free logic of pragmatic force.* Notice that in the popular interpretations, linguistic acts interpreting atomic expressions of linear logic are *single* events, while those interpreting the atoms of intuitionistic logic are *repeatable* events: in some sense the former are more *concrete* entities than the latter. On the contrary the atoms of linear logic are interpreted here more *abstractly* than the assertions and obligations denoted by the atoms of the intuitionistic logic of pragmatic force. Therefore in our "intended interpretation", linear logic is about more "abstract" patterns of reasoning than those of standard logical systems.

To see that Contraction is not generally valid in such a free logic, consider the rule

$$(*)\qquad \frac{\bullet\,\alpha,\ \bullet\,\alpha,\ \Gamma\ \Rightarrow\ \Delta}{\bullet\,\alpha,\ \Gamma\ \Rightarrow\ \Delta}$$

and notice that in the premise we may want to replace the first $\bullet\,\alpha$ with $\vdash\alpha$ and the second with $\circ\!-\,\alpha$. (Of course, the inference $(*)$ would be valid if the substitution $[\vdash\alpha/\bullet\,\alpha]$ or $[\circ\!-\,\alpha/\bullet\,\alpha]$ was uniform in both the sequent-premise and the sequent-conclusion!)

To see that Weakening is not generally valid, consider that from

$$\frac{\bullet\,p \Rightarrow \bullet\,p}{\bullet\,q, \bullet\,p \Rightarrow \bullet\,p}$$

through *non-uniform substitutions* we could derive $(*2)$, which is unacceptable by Hume's law. Also we could derive

$$\bullet\,p \Rightarrow \bullet\,q \multimap \bullet\,p$$

and this is inadmissible, if from the free logic by suitable substitutions we want to recover the notion of *causal implication.* But then the free logic of pragmatic force must also take into accout the relation of relevance between assumptions and conclusion, and thus it cannot allow unrestricted uses of Weakening.

1.8. Next we must show how intuitionistic forms of reasoning of the system **ILP**$^-$ can be represented in the context of the linear consequence relation and how the deductive principles characteristic of **ILP** can be implemented. This task may be achieved in several ways (for a possible alternative, see [13]). Here we take the familiar system **ILL** of *intuitionistic linear logic* and extend it with an exponential "$!_c$", loosely inspired by B. Jacob [10]: to expressions of the form $!_c\alpha$ the rule of *contraction* applies, but not the rule of *Weakening.* We define the language $\mathcal{L}_i^\ell$ and the sequent calculus **ILLP** of Intuitionistic Linear Logic for Pragmatics.

The language $\mathcal{L}_i^\ell$ is that of intuitionistic linear logic **ILL** where atoms are of the form $\bullet\,\alpha$, extended with *implicit substitution operators*, regerded as "pragmatic modalities". Such an operator, e.g., $(\lambda)^{\vdash}$, turns a linear formula λ into a *pre-sentence* ς and indicates a possible replacement of the symbol "•" with the actual sign of illocutionary force "$\vdash$" (similarly for the other operators $(\lambda)^{\vdash_K}$, $(\lambda)^{\mathcal{O}-}$ and $(\lambda)^{\mathcal{O}-_N}$). Now *given any "suitable" translation of* **IL** *into* **ILL**, (where the symbol "$!_c$" is used instead of "!" in the translation of causal implication) we consider the resulting map

$$[\ \ ]^{\bullet} : \mathcal{L}^P \to \mathcal{L}_i^{\ell}$$

and we stipulate that an *implicit substitution* indicated by our operators is realized *only for those expressions of* $\mathcal{L}_i^{\ell}$ *that are translations of formulas of* $\mathcal{L}^P$. We indicate the result of an *actual* substitution with $\overline{(\lambda^{\vdash})}$. For instance, a "suitable translation" yields

$$[\ \ ]^{\bullet} : \quad \vdash\alpha, \vdash\alpha \supset \vdash\beta \Rightarrow \vdash\beta \quad \longmapsto \quad !_c \bullet \alpha, !_c(!_c \bullet \alpha \multimap !_c \bullet \beta) \Rightarrow !_c \bullet \beta$$

and the principle (§) becomes

$$\overline{(!_c \bullet \alpha)^{\mathcal{O}-}}, \overline{(!_c(!_c \bullet \alpha \multimap !_c \bullet \beta))^{\vdash}} \Rightarrow \overline{(!_c \bullet \beta)^{\mathcal{O}-}}$$

Moreover the rules of **ILLP** are the familiar ones for **ILL**, extended with the obvious ones for "$!_c$" and also with *substitution rules*: the latter are similar to the *exponential* rules of linear logic and allow us to derive sequents $\Sigma \Rightarrow \varsigma$ where Σ is a sequence of pre-sentences, possibly containing different substitution operators. The map $[\,]^{\bullet} : \mathcal{L}^P \to \mathcal{L}_i^{\ell}$ extends to a map

$$[\,]^{\bullet} : \mathbf{ILP} \to \mathbf{ILLP},$$

thus if $\Sigma \Rightarrow \varsigma$ is the translation into **ILLP** of a sequent in the language of $\mathcal{L}^P$, then the transformation of the pre-sentences into sentences can be realized, the resulting sequent is taken as a non-logical axiom of **ILP** and the derivation in **ILLP** may be regarded as a justification for such an axiom. Therefore as represented in Figure 1, we have

$$\mathbf{ILP}^{-} \xrightarrow{[\,]^{\bullet}} \mathbf{ILLP} \xrightarrow{(\,)^{\vdash}} \mathbf{ILLP} \xrightarrow{\overline{(\,)}} \mathbf{ILP}.$$

We claim that the systems **ILP** and **ILLP** have the cut-elimination property, so that the above translations are sound.

1.9. We have not produced a formal semantics for the extended system **ILLP**, just proved its consistency through cut-elimination. But the proposed formalism well reflects the intuitive properties of the illocutionary forces under consideration and their ranking.

The interpretation of the atoms of linear logic as *unspecified illocutionary acts* provides an argument for considering the use of the *linear consequence relation* as an *extension*, rather than an *alternative* to the ordinary intuitionistic consequence relation. However, we do not have a clear and convincing interpretation of Girard's operator "!". The mapping $\overline{(\,)}$ does provide a pragmatic interpretation of linear formulas through some translation of **IL** into **ILL**, some of which make a rather unparsimonious use of the operator "!". Moreover, only a few linear formulas of $\mathcal{L}_i^{\ell}$ are translated into formulas of the pragmatic language $\mathcal{L}^P$ through

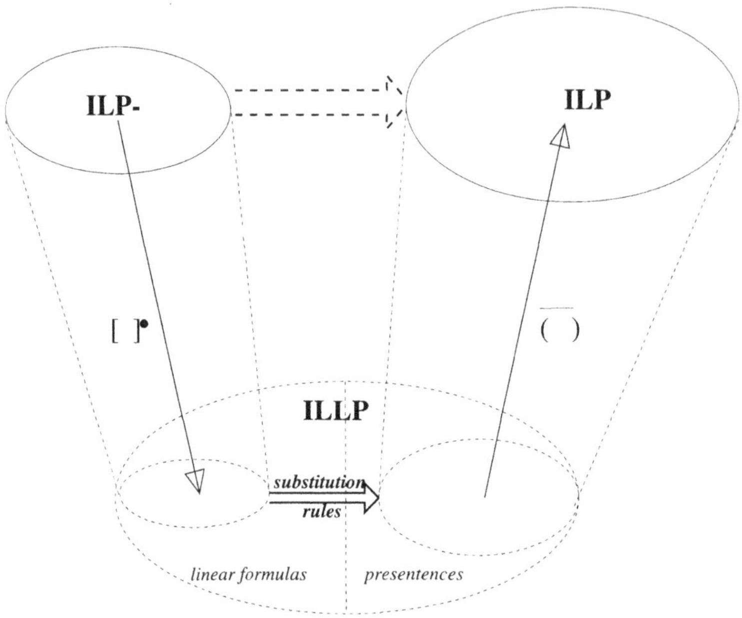

FIGURE 1

the map $\overline{(\ )}$, namely, only those which result from formulas of $\mathcal{L}^P$ *relatively to the given translation* $[\ ]^\bullet$. All other formulas of $\mathcal{L}_i^\ell$ are *uninterpreted* and play only a *syntactic* role in the calculus, similar to the role played in scientific theories by *theoretic formulas* which lack a direct observative meaning. Thus, the possibility remains open of new interpretations of the formal calculus, which may capture new forms of reasoning, different from those considered above.

A satisfactory feature of our system is that in the object language we may formalize *normative orders* as *dynamic systems*, as they are indeed, by means of a sequence of operators $\circ\!\!-_N$, $\circ\!\!-_{N'}$, ..., where N, N' ... are indices for *normative systems.*

In the same way *cognitive dynamic systems* can also be formalized by a sequence of operators $\vdash_k$, $\vdash_{k'}$, ... where k, k', ... are (indices for) *non-logical* theories. Here we may consider also situations where the language of such a theory is not included in the metalanguage, so that $\vdash_k \alpha$ may not be expressible as $K \vdash \alpha$. The logics of Gödel's provability predicate *Bew* provide interesting mathematical examples along these lines.

It is therefore clear that the cases discussed here are only instances of a large family, where the combined use of several modalities requires restrictions on the inference relation in the metalanguage according to specific properties of the modalities under consideration. The axiomatic method in logic has successfully described the properties of modal operators in isolation. New challenges now arise from the formal study of several modalities in the same context, an inevitable step towards a better understanding of the complexities of informal reasoning. It should be clear that this paper is a tentative incursion in a new territory, and that further work will certainly simplify and improve the technical development.

§2. The pragmatic language $\mathcal{L}^P$.

DEFINITION 1. (*Syntax*) (i) The language $\mathcal{L}^P$ is built from an infinite set of *propositional letters* p, p_0, p_1 ... using the *propositional connectives* $\neg$, $\wedge$, $\vee$, $\rightarrow$; these expressions are called *radical formulas*. The *elementary formulas* of the pragmatic language are obtained by prefixing a radical formula with a sign of *illocutionary force* "$\vdash$", "$\vdash_k$", "$\circ\!-$" or "$\circ\!-_N$". There is only one elementary constant for absurdity, namely $\bigwedge$. Finally, the *sentential formulas* of $\mathcal{L}^P$ are built from the elementary formulas and the constant $\bigwedge$, using the *pragmatic connectives* $\cap$, $\cup$ and $\supset$.

(ii) (*Formation Rules*) The *pragmatic* language $\mathcal{L}^P$ is the union of the sets **Rad** of *radical formulas* and **Sent** of *sentential formulas*. These sets are defined inductively as follows:

$$\alpha := p \mid \neg\alpha \mid \alpha_1 \wedge \alpha_2 \mid \alpha_1 \vee \alpha_2 \mid \alpha_1 \rightarrow \alpha_2 \mid$$
$$\eta := \bigwedge \mid \vdash \alpha \mid \vdash_k \alpha \mid \circ\!- \alpha \mid \circ\!-_N \alpha \mid$$
$$\zeta := \vdash \alpha \mid \vdash \alpha \mathbin{\dot{\supset}} \zeta \mid$$
$$\delta := \eta \mid \zeta \mid \delta_1 \supset \delta_2 \mid \delta_1 \cap \delta_2 \mid \delta_1 \cup \delta_2 \mid$$

(iii) The *intuitionistic fragment* of the language $\mathcal{L}^P$ is obtained by restricting the class of elementary sentences to those with *atomic radical* only:

$$\eta := \bigwedge \mid \vdash p \mid \vdash_k p \mid \circ\!- p \mid \circ\!-_N p \mid$$

We use the letters $\alpha, \beta, \alpha_1, \ldots$ to denote *radical* formulas, $\eta, \eta_1, \ldots \ldots$ to denote *elementary sentential* formulas, ζ, and $\delta, \delta_1, \ldots$ to denote *sentential* formulas. The letters $\zeta_1, \ldots$ denote expressions built from assertions and positive occurrences of *casual implication*. The negation of δ is defined as $\sim\delta =_{df} \delta \supset \bigwedge$.

(iv) The symbol $\bigwedge$ stands for an illocutionary act which is justifiable in no situation, e.g., asserting that "$0 = 1$". (Such an act is called *p-invalid*, see the next definition.) Let α be "$0 = 1$". We postulate

$$\bigwedge = \vdash \alpha = \circ\!- \alpha = \vdash_k \alpha = \circ\!-_N \alpha$$

We will not write down the formal axioms expressing this equivalence.

DEFINITION 2. (*Informal Interpretation*) (i) *Radical* formulas are interpreted as propositions, with the Tarskian classical semantics, as usual.

(ii) *Sentential expressions* are interpreted as follows:

1. $\vdash \alpha$ and $\vdash_k \alpha$ are interpreted as illocutionary acts of assertion; $\circ\!- \alpha$ and $\circ\!-_N \alpha$ are interpreted as illocutionary acts of prescription (or norms). All such acts are regarded as *impersonal*, i.e., making abstraction from the specific qualities of the subjects of such acts. Illocutionary acts can be *"justified"* (**J**)

or *"unjustified"* (**U**); by extension, so are also the corresponding elementary sentential expressions.

2. $\vdash \alpha$ is *justified* if and only if *there is a proof that* α *is true*; it is *unjustified* otherwise.
3. Let k denote a state of information (regarded as equal to its deductive closure). Then $\vdash_k \alpha$ is *justified* if and only if it is provable that α is true given the state of information k; $\vdash_k \alpha$ is *unjustified* otherwise.
4. Let N be a *normative system* (regarded as equal to its deductive closure). Then $\circ\!-_N \alpha$ is *justified* if and only if there is a proof that α is obligatory in the normative system N; $\circ\!-_N \alpha$ is *unjustified* otherwise.
5. $\circ\!- \alpha$ is *justified* if and only if $\circ\!-_N \alpha$ is justified for all normative systems N; $\circ\!- \alpha$ is *unjustified* otherwise.
6. $\delta_1 \cap \delta_2$ is *justified* if and only if δ_1 is justified *and* δ_2 is justified; it is *unjustified* otherwise.
7. $\delta_1 \cup \delta_2$ is *justified* if and only if δ_1 is justified *or* δ_2 is justified; it is *unjustified* otherwise.
8. $\delta_1 \supset \delta_2$ is *justified* if and only if there is a proof that a justification of δ_1 can be transformed into a justification of δ_2; it is *unjustified*, otherwise.
9. $\vdash \alpha \supset \vdash \beta$ is justified if and only if $\vdash \alpha \supset \vdash \beta$ is justified as an instantiation of a first order formula $\vdash \forall x.\phi(x) \to \chi(x)$ which expresses a scientific law, where $\alpha \equiv \phi(t)$ and $\beta \equiv \chi(t)$ for some (sequence of) objects t.

(iii) Let η be an elementary (sentential) formula of $\mathcal{L}^P$: we denote the maximal radical subformula of η by $|\alpha|$. A pragmatic interpretation π gives a *justification value* **J** (justified) or **U** (unjustified) to the sentential formulas $\mathcal{L}^P$. Let σ be a truth-value assignment to the radicals of $\mathcal{L}^P$. We say that π *depends on* σ if for every elementary formulas η, $\pi(\eta) = \mathbf{J}$ implies $\sigma(|\eta|) = T$. We write π_σ to indicate that the pragmatic interpretation π depends on the truth-value assignment σ. A sentential formula δ of $\mathcal{L}^P$ is called *pragmatically valid* (or *p-valid*) if and only if for all σ and for all π_σ we have $\pi_\sigma(\delta) = \mathbf{J}$. Similarly, given a sentential formula δ and a set of sentential formulas Γ, we say that δ is a *pragmatically valid consequence* (or a *p-consequence*) of Γ if for every σ and every π_σ, if π_σ makes all formulas in Γ justified, then π_σ makes δ justified as well. Since the notion of pragmatic interpretation is still informal, the definitions of p-validity and p-consequence are informal as well. To give a precise mathematical model for these notions goes beyond the scope of this paper: our task here is to identify some principles which seem to characterize these notions in an intuitively compelling way.

Examples. (i) A state of information k may be formalized as an axiomatic theory $\mathbf{T}_k$, say, in first-order logic. Let $\mathcal{A}_k$ be the set of axiom of $\mathbf{T}_k$ and let α^k be the translation of α in the language of $\mathbf{T}_k$. Then $\vdash_k \alpha$ is justified if and only if *there is a deduction of* α^k *from* $\mathcal{A}_k$ in first-order logic.

(ii) Extend the set of radical formulas of $\mathcal{L}^P$ with a modal operator $\Box$

$$\alpha \;:=\; p \mid \neg\alpha \mid \alpha_1 \wedge \alpha_2 \mid \alpha_1 \vee \alpha_2 \mid \alpha_1 \to \alpha_2 \mid \Box\alpha$$

and let $\mathbf{T}_{GL}$ be the classical propositional modal theory with axioms and rules of the system **GL** (Gödel-Löb) formalizing the properties of the predicate *Bew* of arithmetic provability. Let $\mathbf{T}_k$ be first-order Peano Arithmetic together with a sequence of closed sentences, one for each propositional letter of $\mathbf{T}_{GL}$. Any such $\mathbf{T}_k$ determines an interpretation $(\ )^k$ of the formulas of $\mathbf{T}_{GL}$ where $\Box$ is interpreted by the arithmetic predicate *Bew* and the interpretation commutes with classical connectives. Soloway's theorem [15] states that $\mathbf{T}_{GL} \vdash \alpha$ if and only if for all k, $\mathbf{PA} \vdash_k (\alpha)^k$. In propositional $\mathcal{L}^P$ we can express the infinite set of equivalences $\vdash_{GL} \alpha \equiv \vdash_k \alpha$, for every α and k, which we may regard as non-logical axioms.

(iii) Consider the same setting of (ii) for *intuitionistic* logic,[3] i.e., let $\mathbf{T}_{IGL}$ be the modal theory with axioms and rules of the system Gödel-Löb on *intuitionistic propositional logic* and let $\mathbf{T}_k$ be first-order Heyting Arithmetic together with a sequence of closed sentences, one for each propositional letter of $\mathbf{T}_{IGL}$. As Solovay's theorem does not hold for Heyting arithmetic, in $\mathcal{L}^P$ there will be nontrivial logical relations between $\vdash_{IGL} \alpha$ and all the expressions $\vdash_k \alpha$.

(iv) Let $\mathbf{T}_0, \ldots, \mathbf{T}_\alpha, \ldots$ be a transfinite recursive sequence of systems where $\mathbf{T}_0$ is first-order Peano Arithmetic, $\mathbf{T}_{\alpha+1}$ is $\mathbf{T}_\alpha$ extended with the Gödel sentence $\neg Bew(0 = 1)$ for $\mathbf{T}_\alpha$, as in Feferman [4].

(v) A normative system could be formalized as an axiomatic theory $\mathbf{T}_N$ possibly with mixed modalities. Let $\mathcal{A}_N$ be the set of axioms of such a theory and let α^N be the translation of α into the language of $\mathbf{T}_N$. Then $\circ\!-_N\ \alpha$ is justified if and only if α^N *is derivable from* $\mathcal{A}_N$. Let $\mathbf{T}_N, \mathbf{T}_{N'}, \ldots$ be a sequence of such theories. Then in the language of $\mathcal{L}^P$ one may hope to express the evolution of a normative system, the relative ranking of its principles, the metarules which govern the abrogation of laws when a set of laws has became inconsistent with more fundamental principles, etc.

Remark. (i) Clauses 2, 5–7 in Definition 2.(ii) are those of Heyting's semantics of proofs for intuitionistic logic, but the notion of *proof* is not restricted here to that of a mathematical proof. The claim implicit in Definition 2 is that Heyting's "semantics" belongs to the logic of judgement and is rather the beginning of a formal pragmatics: it characterizes the notion of justification of pragmatics sentences, as inductively defined relations between illocutionary acts.

(ii) Suppose the language of $\mathcal{L}^P$ is extended as in Example (*ii*) above, but with the axioms of the modal system **S4**. Then one could follow Gödel [9], McKinsey and Tarski [11] and interpret $\Box\alpha$ as "*there is a proof that* α *is true*"; thus we obtain the following modal translation of the assertive fragment of $\mathcal{L}^P$ into its radical part thus extended:

$$(\vdash \alpha)^m = \Box\alpha \qquad (\delta_1 \supset \delta_2)^m = \Box(\delta_1^m \rightarrow \delta_2^m)$$
$$(\delta_1 \cup \delta_2)^m = \delta_1^m \vee \delta_2^m \qquad (\delta_1 \cap \delta_2)^m = \delta_1^m \wedge \delta_2^m$$

Kripke's semantics for the system **S4** applies here. Notice that this is a *descriptive interpretation* of the pragmatic sign of assertive force, a *reflection* of the pragmatic

[3]We thank Giovanni Sambin for suggesting this example.

metalanguage into the set of radical modal formulas. The pragmatic and the descriptive approaches may be very closely related from a mathematical point of view, but cannot be identified: for instance, in the descriptive interpretation nested occurrences of the sign $\Box$ may occur, while as a sign of pragmatic force $\vdash$ cannot be iterated.

(iii) A similar remark applies for the sign of deontic force, which could be interpreted descriptively within some system of deontic logic, from which we would also obtain a form of Kripke's semantics. The challenge of this paper is to provide a Gentzen-style system for the language $\mathcal{L}^P$ extended with *mixed contexts*. We do not consider the descriptive interpretation of pragmatic forces here and do not provide a Kripke-style semantics for our system.

identity rules

logical axiom:

$$\delta\ ;\Rightarrow \delta$$

cut_1:

$$\frac{\Gamma\ ;\ \Delta \Rightarrow \delta \qquad \delta, \Gamma'\ ;\ \Delta \Rightarrow \delta'}{\Gamma, \Gamma'\ ;\ \Delta \Rightarrow \delta'}$$

cut_2:

$$\frac{;\ \Delta \Rightarrow \delta \qquad \Gamma\ ;\ \delta, \Delta \Rightarrow \delta'}{\Gamma\ ;\ \Delta \Rightarrow \delta'}$$

structural rules

exchange:

$$\frac{\Gamma, \delta_1, \delta_2, \Gamma'\ ;\ \Delta \Rightarrow \delta}{\Gamma, \delta_2, \delta_1, \Gamma'\ ;\ \Delta \Rightarrow \delta} \qquad \frac{\Gamma\ ;\ \Delta, \delta_1, \delta_2, \Delta' \Rightarrow \delta}{\Gamma\ ;\ \Delta, \delta_2, \delta_1, \Delta' \Rightarrow \delta}$$

contraction:

$$\frac{\delta, \delta, \Gamma\ ;\ \Delta \Rightarrow \delta'}{\delta, \Gamma\ ;\ \Delta \Rightarrow \delta'} \qquad \frac{\Gamma\ ;\ \delta, \delta, \Delta \Rightarrow \delta'}{\Gamma\ ;\ \delta, \Delta \Rightarrow \delta'}$$

permeability:

$$\frac{\Gamma, \delta\ ;\ \Delta \Rightarrow \delta'}{\Gamma\ ;\ \delta, \Delta \Rightarrow \delta'}$$

weakening:

$$\frac{\Gamma\ ;\ \Delta \Rightarrow \delta'}{\Gamma\ ;\ \delta, \Delta \Rightarrow \delta'}$$

TABLE 1. The sequent calculus $\mathbf{ILP}^-$, structural rules

2.1. Sequent calculi for intuitionistic $\mathcal{L}^P$. The sequent calculus $\mathbf{ILP}^-$ formalizes derivations of sentential formulas of $\mathcal{L}^P$. No rule is given for the derivation of elementary sentences, there is a connective expressing "causal implication" which requires a *relevant* consequence relation and the other connectives represent relations between sentential formulas informally characterized by Heyting's "semantics of proofs". The calculus therefore resembles the intuitionistic part of Girard's **LU** system (with a *relevant* rather than *linear* area for causal implication).

DEFNITION 3. Let Γ, Δ denote finite sequences of sentential formulas. The sequent calculus $\mathbf{ILP}^-$ is defined in Tables 1 and 2.

logical rules

right $\dot{\supset}$:

$$\frac{\Gamma, {\vdash}\alpha\, ; \Rightarrow \zeta}{\Gamma\, ; \Rightarrow {\vdash}\alpha \mathbin{\dot{\supset}} \zeta}$$

left $\dot{\supset}$:

$$\frac{\Gamma ; \Rightarrow {\vdash}\alpha \qquad \zeta, \Gamma' : \Delta \Rightarrow \delta}{{\vdash}\alpha \mathbin{\dot{\supset}} \zeta, \Gamma, \Gamma'\, ;\, \Delta \Rightarrow \delta}$$

right $\cap$:

$$\frac{\Gamma\, ;\, \Delta \Rightarrow \delta_1 \qquad \Gamma\, ;\, \Delta \Rightarrow \delta_2}{\Gamma\, ;\, \Delta \Rightarrow \delta_1 \cap \delta_2}$$

left $\cap$:

$$\frac{\Gamma\, ;\, \delta_i\ \Delta \Rightarrow \delta}{\Gamma\, ;\, \delta_0 \cap \delta_1, \Delta \Rightarrow \delta} \qquad \text{for } i = 0, 1.$$

right $\supset$:

$$\frac{\Gamma\, ;\, \Delta, \delta_0 \Rightarrow \delta_1}{\Gamma\, ;\, \Delta \Rightarrow \delta_0 \supset \delta_1}$$

left $\supset$:

$$\frac{;\, \Delta \Rightarrow \delta_0 \qquad \Gamma\, ; \delta_1, \Delta \Rightarrow \delta}{\Gamma\, ;\, \delta_0 \supset \delta_1, \Delta \Rightarrow \delta}$$

right $\cup$:

$$\frac{\Gamma\, ; \Delta \Rightarrow \delta_i}{\Gamma\, ; \Delta \Rightarrow \delta_0 \cup \delta_1} \qquad \text{for } i = 0, 1.$$

left $\cup$:

$$\frac{\Gamma\, ;\, \delta_0, \Delta \Rightarrow \delta \qquad \Gamma\, ;\, \delta_1, \Delta \Rightarrow \delta}{\Gamma\, ;\, \delta_0 \cup \delta_1, \Delta \Rightarrow \delta}$$

absurdity axiom:

$$\Gamma\, ;\, \bigwedge, \Delta \Rightarrow \delta$$

for any Γ, Δ and δ.

TABLE 2. The sequent calculus $\mathbf{ILP}^-$, logical rules

The proof of the following theorem is similar to that Girard's system **LU**:

THEOREM 1. *The sequent calculus* $\mathbf{ILP}^-$ *has the cut-elimination property, thus it is consistent.*

We have the following *symmetric reduction for causal implication*:

$$\dfrac{right\ \dot{\supset}\ \dfrac{\Gamma, {\vdash}\alpha\, ;\, \Rightarrow \zeta}{\Gamma\, ;\, \Rightarrow {\vdash}\alpha \mathbin{\dot{\supset}} \zeta} \qquad \dfrac{\Gamma'\, ;\, \Rightarrow {\vdash}\alpha \qquad \zeta, \Gamma''\, ;\, \Delta \Rightarrow \delta}{\alpha \mathbin{\dot{\supset}} \zeta, \Gamma', \Gamma''\, ;\, \Delta \Rightarrow \delta}\ left\ \dot{\supset}}{\Gamma, \Gamma', \Gamma''\, ;\, \Delta \Rightarrow \delta}\ cut_1$$

reduces to

$$\dfrac{\dfrac{\Gamma'\, ;\, \Rightarrow {\vdash}\alpha \qquad \Gamma, {\vdash}\alpha\, ;\, \Rightarrow \zeta}{\Gamma, \Gamma'\, ; \Rightarrow \zeta}\ cut_1 \qquad \zeta, \Gamma''\, ;\, \Delta \Rightarrow {\vdash}\delta}{\Gamma, \Gamma', \Gamma''\, ;\, \Delta \Rightarrow \delta}\ cut_1$$

The reductions strategies for the cut-elimination process are subject to constraint resulting from the restrictions on the context for the *causal implication* rules. Consider the case of a cut where the right sequent-premise is the conclusion of *right* $\dot{\supset}$:

$$\dfrac{\begin{array}{c} d_1 \\ \Gamma\,;\,\Delta \Rightarrow\ \vdash \zeta \end{array} \qquad \dfrac{\begin{array}{c} d_2 \\ \vdash \beta, \zeta, \Gamma';\ \Rightarrow \zeta' \end{array}}{\zeta, \Gamma'\,;\ \Rightarrow\ \vdash \beta \supset \zeta'}\ \textit{right} \supset}{\Gamma, \Gamma'\,;\,\Delta \Rightarrow\ \vdash \beta \supset \zeta'}\ \textit{cut}_1$$

here the cut_1 can always be permuted with the inferences of d_1. Since ζ has one of the forms $\vdash \gamma$ or $\vdash \gamma \supset \zeta'$, eventually one of the following is the case:

(i) the left premise of the cut is an *absurdity axiom*, and the cut is eliminated by taking its conclusion as a single absurdity axiom;

(ii) the left premise of the cut is a *logical axiom*, and the usual axiom reduction applies;

(iii) the left premise of the cut is also the conclusion of a *right* $\supset$ and in both sequent-premises the area to the right of ";" is empty; henceforth the cut-elimination procedure continues without further constraints in the relevant area only.

Similarly, if the right sequent-premise is the conclusion of a *left* $\supset$. Other details of the proof are left to the reader.

The above calculus can be extended in various ways, to formalize arguments that are p-valid according to the informal interpretation of the language $\mathcal{L}^P$ but cannot be represented in $\mathbf{ILP}^-$. We extend $\mathbf{ILP}^-$ to an intuitionistic system **ILP** that represent reasoning with *mixed context*. A first set of *mixed rules* connects *indexed assertions* with *assertions*, on one hand, and *indexed obligations* with *obligations*, on the other. A second kind of mixed rules connects *obligations* (indexed or not) with *causal implication*. We write $\mathbf{Z}$ for $\zeta_1, \ldots, \zeta_n$; also we write $\vdash \boldsymbol{A}$ for $\vdash \alpha_1, \ldots, \vdash \alpha_k$ and similarly $\circ\!\!- \boldsymbol{A}$.

mixed rules

k-assertability	*N-obligation*
$\vdash \delta\,; \Rightarrow\ \vdash_k \delta$	$\circ\!\!- \delta;\ \Rightarrow\ \circ\!\!-_N \delta$

for any k and N.

causal implication - obligation	*causal implication - N-obligation*
$\dfrac{\vdash \boldsymbol{A}, \boldsymbol{Z}\,;\ \Rightarrow\ \vdash \beta}{\circ\!\!- \boldsymbol{A}, \boldsymbol{Z}\,;\ \Rightarrow\ \circ\!\!- \beta}$	$\dfrac{\vdash \boldsymbol{A}, \boldsymbol{Z}\,;\ \Rightarrow\ \vdash \beta}{\circ\!\!-_N \boldsymbol{A}, \boldsymbol{Z}\,;\ \Rightarrow\ \circ\!\!-_N \beta}$

TABLE 3. The sequent calculus **ILP**, mixed rules

DEFNITION 4. Let **ILP** be $\mathbf{ILP}^-$ with the addition of the axioms *K-assertability*, *N-obligation*, *causal implication - obligation* and *causal implication - N-obligation* (see Table 3).

THEOREM 2. *The system* **ILP** *emjoys the cut-elimination property* (modulo atomic cuts with k-assertability and N-obligation axioms).

We have the following reductions:

i. causal implication - obligation / causal implication - obligation reduction:

$$\dfrac{\supset-\circ\!-\;\dfrac{\vdash \mathbf{A},\mathbf{Z}\,;\Rightarrow\,\vdash\beta}{\circ\!-\mathbf{A},\mathbf{Z}\,;\Rightarrow\,\circ\!-\beta}\qquad\dfrac{\vdash\beta,\,\vdash \mathbf{A}',\mathbf{Z}'\,;\Rightarrow\,\vdash\gamma}{\circ\!-\beta,\,\circ\!-\mathbf{A}',\mathbf{Z}'\,;\Rightarrow\,\circ\!-\gamma}\;\supset-\circ\!-}{\circ\!-\mathbf{A},\,\circ\!-\mathbf{A}',\mathbf{Z},\mathbf{Z}'\,;\Rightarrow\,\circ\!-\gamma}\;cut$$

reduces to

$$\dfrac{\dfrac{\vdash \mathbf{A},\mathbf{Z}\,;\Rightarrow\,\vdash\beta\qquad\vdash\beta,\,\vdash \mathbf{A},\mathbf{Z}'\,;\Rightarrow\,\vdash\gamma}{\vdash \mathbf{A},\,\vdash \mathbf{A}',\mathbf{Z},\mathbf{Z}'\,;\Rightarrow\,\vdash\gamma}\;cut}{\circ\!-\mathbf{A},\,\circ\!-\mathbf{A}',\mathbf{Z},\mathbf{Z}'\,;\Rightarrow\,\circ\!-\gamma}\;\supset-\circ\!-$$

ii. causal implication / causal implication - obligation reduction:

$$\dfrac{right\supset\;\dfrac{\overset{d_1}{\vdash\alpha,\Gamma;\Rightarrow\zeta}}{\Gamma;\Rightarrow\,\vdash\alpha\supset\zeta}\qquad\dfrac{\overset{d_2}{\vdash\alpha\supset\zeta,\,\vdash \mathbf{A},\mathbf{Z}\,;\Rightarrow\,\vdash\gamma}}{\vdash\alpha\supset\zeta,\,\circ\!-\mathbf{A},\mathbf{Z}\,;\Rightarrow\,\circ\!-\gamma}\;\supset-\circ\!-}{\Gamma,\,\circ\!-\mathbf{A},\mathbf{Z}\,;\Rightarrow\,\circ\!-\gamma}\;cut$$

reduces to

$$\dfrac{\dfrac{right\supset\;\dfrac{\overset{d_1}{\vdash\alpha,\Gamma;\Rightarrow\zeta}}{\Gamma;\Rightarrow\,\vdash\alpha\supset\zeta}\qquad\overset{d_2}{\vdash\alpha\supset\zeta,\,\vdash \mathbf{A},\mathbf{Z}\,;\Rightarrow\,\vdash\gamma}}{\Gamma,\,\vdash \mathbf{A},\mathbf{Z}\,;\Rightarrow\,\vdash\gamma}\;cut}{\Gamma,\,\circ\!-\mathbf{A},\mathbf{Z}\,;\Rightarrow\,\circ\!-\gamma}\;\supset-\circ\!-$$

iii. causal implication - obligation permutation:

$$\dfrac{\overset{d_1}{\Gamma\,;\,\Delta\Rightarrow\,\vdash\alpha\supset\zeta}\qquad\dfrac{\overset{d_2}{\vdash\alpha\supset\zeta,\,\vdash \mathbf{A},\mathbf{Z}\,;\Rightarrow\,\vdash\beta}}{\vdash\alpha\supset\zeta,\,\circ\!-\mathbf{A},\mathbf{Z}\,;\Rightarrow\,\circ\!-\beta}\;\supset-\circ\!-}{\Gamma,\,\circ\!-\mathbf{A},\mathbf{Z}\,;\,\Delta\Rightarrow\,\circ\!-\gamma}\;cut$$

The cut can always be permuted with the other inferences of d_1.

iv. causal implication - obligation permutation:

$$\dfrac{\overset{d_1}{\Gamma\,;\,\Delta\Rightarrow\,\circ\!-\alpha}\qquad\dfrac{\overset{d_2}{\vdash\alpha,\,\vdash \mathbf{A},\mathbf{Z}\,;\Rightarrow\,\vdash\beta}}{\circ\!-\alpha,\,\circ\!-\mathbf{A},\mathbf{Z}\,;\Rightarrow\,\circ\!-\beta}\;\supset-\circ\!-}{\Gamma,\,\circ\!-\mathbf{A},\mathbf{Z}\,;\,\Delta\Rightarrow\,\circ\!-\gamma}\;cut$$

The *cut* can always be permuted with the inferences of d_1.

v. causal implication - obligation permutation

$$\dfrac{\supset-\circ\!-\;\dfrac{\overset{d_1}{\vdash \mathbf{A},\mathbf{Z}\,;\Rightarrow\,\vdash\alpha}}{\circ\!-\mathbf{A},\mathbf{Z}\,;\Rightarrow\,\circ\!-\alpha}\qquad\overset{d_2}{\circ\!-\alpha,\Gamma\,;\,\Delta\Rightarrow\,\circ\!-\beta}}{\circ\!-\mathbf{A},\mathbf{Z},\Gamma\,;\,\Delta\Rightarrow\,\circ\!-\beta}\;cut$$

The *cut* can always be permuted with the inferences of d_2.

The formal details of the proof are left to the reader.

Another extension of $\mathbf{ILP}^-$ is obtained by allowing *classical* methods of inference: in our framework this is done by removing the restriction that the radical

$|\eta|$ of any elementary formula η should be atomic and by introducing all axioms of the form

$$\frac{\text{if } \alpha_1, \ldots, \alpha_n \models \alpha, \text{ then}}{;\ \vdash \alpha_1, \ldots, \vdash \alpha_n \Rightarrow \vdash \alpha} \quad (\textit{classical assertability}) \qquad \frac{\text{if } \alpha_1, \ldots, \alpha_n \models \alpha, \text{ then}}{;\ \multimap \alpha_1, \ldots, \multimap \alpha_n \Rightarrow \multimap \alpha} \quad (\textit{classical obligation})$$

where $\alpha_1, \ldots, \alpha_n, \alpha$ are arbitrary radical formulas.

DEFNITION 5. Consider the language $\mathcal{L}^P$ *without the connective* $\supset$ *of causal implication* and let $\mathbf{LP}^-$ be the sequent calculus obtained by adding the *classical assertability* and *classical obligation* rules to the system $\mathbf{ILP}^-$ *without the rules for* $\supset$ *and the causal implication-obligation rule*.

COROLLARY. *The system* $\mathbf{LP}^-$ *is consistent.*

Indeed, by Theorem 1 we may reduce all cuts, except those where both cut-formulas occur in *classical assertability* axioms or both in *classical obligation* axioms and similarly for indexed assertability and obligation. But these cuts can also be reduced in an obvious way, since $\alpha_1, \ldots, \alpha_m \models \alpha$ and $\alpha, \beta_1, \ldots, \beta_n \models \beta$ yield $\alpha_1, \ldots, \alpha_m, \beta_1, \ldots, \beta_n \models \beta$.

The classical extension of **ILP** seems to be more problematic. We will not pursue this extension here.

§3. The language of intuitionistic linear $\mathcal{L}_i^\ell$. We present the language of the intuitionistic linear pragmatic language $\mathcal{L}_i^\ell$. We consider translations of the intuitionistic fragment of the $\mathcal{L}^P$ into $\mathcal{L}_i^\ell$ and we define the substitution operations, giving partial translations from $\mathcal{L}_i^\ell$ back to $\mathcal{L}^P$. We define the intuitionistic linear sequent calculus **ILLP** and the bridging rules that allow us to extend the deductive principles of $\mathbf{ILP}^-$ with rules for mixed assertive and deontic contexts.

3.1. Linear expressions, translations, substitutions. $\mathcal{L}_i^\ell$ is a formal language for pragmatic sentences with *unspecified signs of illocutionary force*.

DEFNITION 6. (*Syntax*) (i) $\mathcal{L}_i^\ell$ is based on the language of *intuitionistic linear logic* **ILL**, namely, on the constants **0**, **1** and $\top$, the connectives $\multimap$, $\otimes$, & and $\oplus$ and the exponential **!** with the addition of the *relevant exponential* $!_c$. Moreover, there are operators "•", $(\ )^{\vdash}$, $(\ )^{\vdash_k}$, $(\ )^{\multimap}$ and $(\ )^{\multimap_N}$,

The elementary formulas are elementary pragmatic sentences, written $\bullet\ \alpha$, where "•" is an operator capable of being substituted by any sign among $\vdash$, $\vdash_k$, $\multimap$ or $\multimap_N$ for some k and N.

The "modal" operators $(\lambda)^{\vdash}$, $(\lambda)^{\vdash_k}$, $(\lambda)^{\multimap}$ and $(\lambda)^{\multimap_N}$, whose occurrences cannot be nested, indicate the possibility, *but only the possibility*, of a substitution of the indicated operator of pragmatic force for every occurrence of • in the given formula λ, as explained above.

(ii) (*Formation Rules*) The set **Lin** of the formulas of $\mathcal{L}_i^\ell$ is defined inductively as follows:

$$\lambda := \bullet\alpha \mid \mathbf{0} \mid \mathbf{1} \mid \top \mid \lambda_1 \multimap \lambda_2 \mid \lambda_1 \otimes \lambda_2 \mid \lambda_1 \& \lambda_2 \mid \lambda_1 \oplus \lambda_2 \mid !\lambda \mid !_c\lambda \mid$$

$$\varsigma := \lambda^{\vdash} \mid \lambda^{\vdash_k} \mid \lambda^{\circ\!-} \mid \lambda^{\circ\!-_N} \mid$$

The expressions λ are the *linear formulas*, the expressions ς will be called *pre-sentences*. We use the letter ϑ to range over linear formulas and pre-sentences.

We are interested in using linear logic as a tool to expand the deductive power of the sequent calculus $\mathbf{ILP}^-$ so as to derive *mixed deontic and assertve* sequents. For this purpose we need translations $(\ )^{\bullet} : \mathbf{IL} \to \mathbf{ILL}$ of *intuitionistic logic* **ILL** into *intuitionistic linear logic* **ILL** which are *functorial* in the sense that if d is a derivation of $\Gamma \Rightarrow \delta$ in **IL**, then $(d)^{\bullet}$ is a derivation $(\Gamma)^{\bullet} \Rightarrow (\delta)^{\bullet}$ in **ILL** and, moreover, the translation respects cut. There exist in the literature a few translations of this kind.

For example, we may use the "*boring translation*" $[\ \ ]^{+} : \mathbf{Sent} \to \mathbf{Lin}$ (see [6], p.81) which is related to the **S4** interpretation of intuitionistic logic:

i. $[\vdash \alpha]^{+} = [\vdash_k \alpha]^{+} = [\circ\!- \alpha]^{+} = [\circ\!-_N \alpha]^{+} =_{df} \ ! \bullet \alpha;$
ii. $[\delta_1 \cap \delta_2]^{+} = ![\delta_1]^{+} \otimes ![\delta_2]^{+};$
iii. $[\delta_1 \cup \delta_2]^{+} = ![\delta_1]^{+} \oplus ![\delta_2]^{+};$
iv. $[\delta_1 \supset \delta_2]^{+} = !([\delta_1]^{+} \multimap [\delta_2]^{+});$
v. $[\vdash \alpha \ \mathbin{\supset\!\!\!\cdot} \vdash \beta]^{+} = !_c(!_c \bullet \alpha \multimap !_c \bullet \beta);$
$[\vdash \alpha \ \mathbin{\supset\!\!\!\cdot} \ \zeta]^{+} = !_c(!_c \bullet \alpha \multimap [\zeta]^{+});$
vi. $[\zeta \supset \delta]^{+} = !(![\zeta]^{+} \multimap [\delta]^{+});$
$[\delta \supset \zeta]^{+} = !([\delta]^{+} \multimap ![\zeta]^{+});$
vii. $[\bigwedge]^{+} =_{df} \mathbf{0}.$

We could also use the more familiar translation $[\ \ ]^{*} : \mathbf{Sent} \to \mathbf{Lin}$ as follows:

i. $[\vdash \alpha]^{*} = [\vdash_k \alpha]^{*} = [\circ\!- \alpha]^{*} = [\circ\!- \alpha]^{*} =_{df} \bullet \alpha;$
ii. $[\delta_1 \cap \delta_2]^{*} = [\delta_1]^{*} \& [\delta_2]^{*};$
iii. $[\delta_1 \cup \delta_2]^{*} = ![\delta_1]^{*} \oplus ![\delta_2]^{*};$
iv. $[\delta_1 \supset \delta_2]^{*} = ![\delta_1]^{*} \multimap [\delta_2]^{*};$
v. $[\vdash \alpha \ \mathbin{\supset\!\!\!\cdot} \vdash \beta]^{*} = !_c \bullet \alpha \multimap !_c \bullet \beta;$
$[\vdash \alpha \ \mathbin{\supset\!\!\!\cdot} \ \zeta]^{*} = !_c \bullet \alpha \multimap [\zeta]^{*};$
v. $[\delta \supset \zeta]^{\bullet} = !(![\zeta]^{\bullet} \multimap ![\delta]^{\bullet});$
vi. $[\bigwedge]^{*} =_{df} \mathbf{0}.$

In this paper we need a *fully functorial* translation $(\ )^{\bullet} : \mathbf{IL} \to \boldsymbol{ILL}$ i.e., a translation having the additional property that *for any derivation d^{ℓ} of $(\Gamma)^{\bullet} \Rightarrow (\delta)^{\bullet}$ in* **ILL** *there exists a derivation d of $\Gamma \Rightarrow \delta$ in* **IL** *such that $d^{\ell} = (d)^{\bullet}$* (modulo suitable equalities of derivations). This property can be assumed of the translation $(\ )^{*}$ (see [14], 1.3.).

First of all we must check that given one of our translations, we may indeed use them for our purpose, i.e., to substitute the signs of illocutionary force in the language of $\mathcal{L}^P$.

PROPOSITION 1. (*Partial interpretation of pre-sentences*) *Given a functorial translation* $[\ \ ]^{\bullet} : \mathbf{IL} \to \mathbf{ILL}$ *of intuitionistic logic within intuitionistic linear logic, there is a map* $\overline{(\ )} : (\mathcal{L}^P)^{\bullet} \Rightarrow \mathcal{L}^P$ *from a subset* $(\mathcal{L}^P)^{\bullet}$ *of the pre-sentences of* $\mathcal{L}_i^{\ell}$ *to the sentential formulas of* $\mathcal{L}^P$, *with the following properties*:

identity rules

logical axiom:

$$\bullet\alpha \Rightarrow^{\bullet} \alpha$$

cut :

$$\frac{\Theta \Rightarrow \vartheta \qquad \vartheta, \Theta' \Rightarrow \xi}{\Theta, \Theta' \Rightarrow \xi}$$

structural rule

exchange:

$$\frac{\Theta, \vartheta_0, \vartheta_1, \Theta' \Rightarrow \vartheta}{\Theta, \vartheta_1, \vartheta_0, \Theta' \Rightarrow \vartheta}$$

TABLE 4. The sequent calculus **ILLP**, structural part.

$$\begin{array}{rcll}
\overline{(\zeta^{\bullet})^{\vdash}} & = & \zeta & \\
\overline{(\delta^{\bullet})^{\vdash}} & = & \delta[\vdash / \vdash_k, \vdash / \circ\!-, \vdash / \circ\!-_N], & \textit{for all } k, N; \\
\overline{(\delta^{\bullet})^{\vdash_k}} & = & \delta[\vdash_k / \vdash, \vdash_k / \vdash_j, \vdash_k / \circ\!-, \vdash_k / \circ\!-_N], & \textit{for all } N, j \neq k; \\
\overline{(\delta^{\bullet})^{\circ\!-}} & = & \delta[\circ\!- / \vdash, \circ\!- / \vdash_k, \circ\!- / \circ\!-_N] & \textit{for all } k, N; \\
\overline{(\delta^{\bullet})^{\circ\!-_N}} & = & \delta[\circ\!-_N / \vdash, \circ\!-_N / \vdash_k, \circ\!-_N / \circ\!-, \circ\!-_N / \circ\!-_{N'}] & \textit{for all } k, N' \neq N.
\end{array}$$

Expressions of the form $\lambda^{\vdash}$, $\lambda^{\vdash_k}$, $\lambda^{\circ\!-}$ *and* $(\lambda)^{\circ\!-_N}$ *are* uninterpreted, *if for all* δ, *we have* $\lambda \neq \delta^{\bullet}$.

Remark. The mapping $\overline{(\,)}$ provides a partial pragmatic interpretation of linear formulas by translating them into the language $\mathcal{L}^P$, which has an intended pragmatic interpretation. But only a few formulas of $\mathcal{L}^{\ell}_i$ are translated into formulas of $\mathcal{L}^P$ through the map $\overline{(\,)}$, namely, only those which result from formulas of $\mathcal{L}^P$ *relatively to a given translation* $[\,]^{\bullet}$. All other formulas of $\mathcal{L}^{\ell}_i$ are *uninterpreted*. and play only a *syntactic* role in the calculus, similar to the role played in scientific theories by *theoretic formulas* which lack a direct observative meaning.

3.2. The sequent calculus ILLP.

DEFNITION 7. (i) (*Notation*) We use Greek capital letters Λ, Λ', ... to range over sequences of *linear formulas*, the capital letter Σ, Σ', ... to range over sequences of *pre-sentences* and lowercase and capital letters θ, Θ, Θ' to range over expressions and sequences of expressions consisting of either *linear formulas or pre-sentences*. As usual, if Λ is $\lambda_1, \ldots, \lambda_n$, then $!\Lambda$ is $!\lambda_1, \ldots, !\lambda_n$ and $\Lambda^{\vdash}$ is $\lambda_1{}^{\vdash}$, ..., $\lambda_n{}^{\vdash}$ and similarly for the other substitution operators.

(ii) The sequent calculus **ILLP** for intuitionistic linear logic for pragmatics is defined in Tables 4, 5, 6, 7 and 8.

PROPOSITION 2. *The fragment of sequent calculus* **ILLP** *without the* relevant implication/ $\circ\!-$ *rule has the cut-elimination property.*

Indeed for the fragment of **ILLP** without substitution rules, the proof is essentially the same as for the system of intuitionistic linear logic **ILL** (cfr. [1] and the bibliography in it). Also the *left* substitution rules are analogous to the *dereliction* rule for the exponential operators; the *right* rules are analogous to the *promotion*

logical rules

1-*Right* :
$$\Rightarrow \mathbf{1}$$

1-*Left* :
$$\frac{\Lambda \Rightarrow \lambda}{\Lambda, \mathbf{1} \Rightarrow \lambda}$$

$\multimap$-*Right:*
$$\frac{\lambda_0, \Lambda \Rightarrow \lambda_1}{\Lambda \Rightarrow \lambda_0 \multimap \lambda_1}$$

$\multimap$-*Left:*
$$\frac{\Lambda \Rightarrow \lambda_0 \qquad \lambda_1, \Lambda' \Rightarrow \lambda}{\lambda_0 \multimap \lambda_1, \Lambda, \Lambda' \Rightarrow \lambda}$$

$\otimes$-*Right:*
$$\frac{\Lambda \Rightarrow \lambda_0 \qquad \Lambda' \Rightarrow \lambda_1}{\Lambda, \Lambda' \Rightarrow \lambda_0 \otimes \lambda_1}$$

$\otimes$-*Left:*
$$\frac{\lambda_0, \lambda_1, \Lambda \Rightarrow \lambda}{\lambda_0 \otimes \lambda_1, \Lambda \Rightarrow \lambda}$$

$\top$-*Right:*
$$\Lambda \Rightarrow \top$$

0-*Left:*
$$\Lambda, \mathbf{0} \Rightarrow \lambda$$

&-*Right:*
$$\frac{\Lambda \Rightarrow \lambda_0 \qquad \Lambda \Rightarrow \lambda_1}{\Lambda \to \lambda_0 \& \lambda_1}$$

&-*Left:*
$$\frac{\lambda_i, \Lambda \Rightarrow \lambda}{\lambda_0 \& \lambda_1, \Lambda \Rightarrow \lambda} \quad \text{for } i = 0, 1.$$

$\oplus$-*Right:*
$$\frac{\Lambda \Rightarrow \lambda_i}{\Lambda \Rightarrow \lambda_0 \oplus \lambda_1} \quad \text{for } i = 0, 1.$$

$\oplus$-*Left:*
$$\frac{\lambda_0, \Lambda \Rightarrow \lambda \qquad \lambda_1, \Lambda \Rightarrow \lambda}{\lambda_0 \oplus \lambda_1, \Lambda \Rightarrow \lambda}$$

TABLE 5. The sequent calculus **ILLP**, logic

rule for the exponential operator. Thus the cut reductions for the substitution rules (Table 7) are those for a *promotion - dereliction* pair in **ILL**, so the same argument applies *mutatis mutandis*.

Remark. Since the *substitution operators* can neither be nested nor occur in the scope of a linear operator, we may always assume that in a *cut-free* **ILLP** derivation quasi-sentences occur only in the terminal branch of the derivation.

DEFINITION 8. The sequent calculus $\mathbf{ILP}^+$ is $\mathbf{ILP}^-$, with the addition of all rules of the *bridging rules* of Table 9.

THEOREM 3. *Let δ be a formula in the language of $\mathcal{L}^P$. The sequent* ; $\Rightarrow \delta$ *is provable in* **ILP** *if and only if it is provable in* $\mathbf{ILP}^+$.

This result is an obvious consequence of the functoriality of the given translation $[\ \ ]^\bullet : \mathbf{IL} \to \mathbf{ILL}$. The *"only if"* part of Theorem 3 is obvious. To prove the *"if"* part notice that by Proposition 2 and the following remark we may assume that every subderivation d^ℓ ending with a bridging rule is cut-free and all the substitution rules occur at the end. If the last substitution rule is $\vdash$-*Subst Right*, then the conclusion of the bridging rule has the form $(\Gamma)^\vdash \Rightarrow (\delta)^\vdash$ and by the

exponentials

dereliction:
$$\frac{\lambda, \Lambda \Rightarrow \lambda'}{!\lambda, \Lambda \Rightarrow \lambda'}$$

weakening:
$$\frac{\Lambda \Rightarrow \lambda'}{!\lambda, \Lambda \Rightarrow \lambda'}$$

contraction:
$$\frac{!\lambda, !\lambda, \Lambda \Rightarrow \lambda'}{!\lambda, \Lambda \Rightarrow \lambda'}$$

promotion:
$$\frac{!\Lambda \Rightarrow \lambda}{!\Lambda, \Sigma \Rightarrow !\lambda}$$

dereliction:
$$\frac{\lambda, \Lambda \Rightarrow \lambda'}{!_c\lambda, \Lambda \Rightarrow \lambda'}$$

contraction:
$$\frac{!_c\lambda, !_c\lambda, \Lambda \Rightarrow \lambda'}{!_c\lambda, \Lambda \Rightarrow \lambda'}$$

promotion:
$$\frac{!_c\Lambda \Rightarrow \lambda}{!_c\Lambda \Rightarrow !_c\lambda}$$

TABLE 6. The sequent calculus **ILLP**, exponentials.

$\vdash$-*Subst Right:*
$$\frac{\Sigma \Rightarrow \lambda}{\Sigma \Rightarrow \lambda^{\vdash}}$$

$\vdash$-*Subst Left:*
$$\frac{\lambda, \Lambda, \Xi \Rightarrow \lambda'}{\lambda^{\vdash}, \Lambda, \Xi \Rightarrow \lambda'}$$

where all pre-sentences in Σ are of the form $(\lambda'')^{\vdash}$.

$\circ\!-$-*Subst Right:*
$$\frac{\Sigma \Rightarrow \lambda}{\Sigma \Rightarrow \lambda^{\circ-}}$$

$\circ\!-$-*Subst Left:*
$$\frac{\lambda, \Lambda, \Xi \Rightarrow \lambda'}{\lambda^{\circ-}, \Lambda, \Xi \Rightarrow \lambda'}$$

where all pre-sentences in Σ are of the form $(\lambda'')^{\circ-}$

$\vdash$-*Subst Right:*
$$\frac{\Sigma \Rightarrow \lambda}{\Sigma \Rightarrow \lambda^{\vdash_k}}$$

$\vdash$-*Subst Left:*
$$\frac{\lambda, \Lambda, \Xi \Rightarrow \lambda'}{\lambda^{\vdash_k}, \Lambda, \Xi \Rightarrow \lambda'}$$

where all pre-sentences in Σ are of the form $(\lambda'')^{\vdash_k}$.

$\circ\!-$-*Subst Right:*
$$\frac{\Sigma \Rightarrow \lambda}{\Sigma \Rightarrow \lambda^{\circ-_N}}$$

$\circ\!-$-*Subst Left:*
$$\frac{\lambda, \Lambda, \Xi \Rightarrow \lambda'}{\lambda^{\circ-_N}, \Lambda, \Xi \Rightarrow \lambda'}$$

where all pre-sentences in Σ are of the form $(\lambda'')^{\circ-_N}$

TABLE 7. Substitution rules - I

Given a fully functorial translation $[\ \ ]^\bullet$:

relevant implication / $\multimap$:

$$\frac{\Lambda, \Lambda' \Rightarrow \lambda}{(\Lambda)^{\multimap}, (\Lambda')^{\vdash} \Rightarrow \lambda^{\multimap}}$$

where all formulas in Λ and λ are of the form $!_c \bullet \alpha$ and all the formulas in Λ' are of the form $\zeta^\bullet$.

relativized assertion:

$$([\vdash \alpha]^\bullet)^{\vdash} \Rightarrow ([\vdash \alpha]^\bullet)^{\vdash_k}$$

relativized obligation:

$$([\vdash \alpha]^\bullet)^{\multimap} \Rightarrow ([\vdash \alpha]^\bullet)^{\multimap_N}$$

TABLE 8. Substitution rules - II

Given a fully functorial translation $[\ \ ]^\bullet$:

$$\frac{([\vdash \mathbf{A}]^\bullet)^{\multimap}, ([\mathbf{Z}]^\bullet)^{\vdash} \Rightarrow ([\vdash \alpha]^\bullet)^{\multimap},}{\multimap \mathbf{A}, \mathbf{Z}\,; \Rightarrow \multimap \alpha} \qquad \frac{([\vdash \mathbf{A}]^\bullet)^{\multimap_N}, ([\mathbf{Z}]^\bullet)^{\vdash} \Rightarrow ([\vdash \alpha]^\bullet)^{\multimap_N},}{\multimap \mathbf{A}, \mathbf{Z}\,; \Rightarrow \multimap \alpha}$$

$$\frac{([\Gamma]^\bullet)^{\vdash} \Rightarrow ([\delta]^\bullet)^{\vdash}}{;\ (\Gamma)^{\vdash} \Rightarrow (\delta)^{\vdash}} \qquad \frac{([\Gamma]^\bullet)^{\multimap} \Rightarrow ([\delta]^\bullet)^{\multimap}}{;\ (\Gamma)^{\multimap} \Rightarrow (\delta)^{\multimap}}$$

$$\frac{([\Gamma]^\bullet)^{\vdash_k} \Rightarrow ([\delta]^\bullet)^{\vdash_k}}{;\ (\Gamma)^{\vdash_k} \Rightarrow (\delta)^{\vdash_k}} \qquad \frac{([\Gamma]^\bullet)^{\multimap_N} \Rightarrow ([\delta]^\bullet)^{\multimap_N}}{;\ (\Gamma)^{\multimap_N} \Rightarrow (\delta)^{\multimap_N}}$$

where the upper sequents are derivable in **ILLP**.

TABLE 9. Bridging rules.

functoriality of the translation $[\ \ ]^\bullet$ we can transform d^ℓ into a derivation d' of $(\Gamma)^{\vdash} \Rightarrow (\delta)^{\vdash}$ in $\mathbf{ILP}^-$. Similarly if the last susbstitution is $\multimap$-*Subst Right*, $\vdash_k$-*Subst Right* and $\multimap_N$-*Subst Right*.

If the last substitution rule of d^ℓ is a *relevant implication* / $\multimap$ rule, then d^ℓ ends as follows:

$$\cfrac{\cfrac{[\vdash A]^\bullet, [Z]^\bullet \Rightarrow [\vdash \beta]^\bullet}{([\vdash A]^\bullet)^{\multimap}, [Z]^{\bullet\vdash} \Rightarrow ([\vdash \beta]^\bullet)^{\multimap}}\ \textit{rel. impl./}\ \multimap\ \textit{rule}}{\multimap A, Z\,; \Rightarrow \multimap \beta}\ \textit{bridging rule}$$

and by the functoriality of the translation we can transform d^ℓ into a suitable derivation d':

$$\dfrac{\begin{array}{c} d' \\ \vdash \boldsymbol{A}, \boldsymbol{Z}\,;\ \Rightarrow\ \vdash \beta \end{array}}{\circ\!\!-\ \boldsymbol{A}, \boldsymbol{Z}\,;\ \Rightarrow\ \circ\!\!-\ \beta}\ \overset{c}{\supset} /\ \circ\!\!-$$

3.3. Examples. The following derivation shows that the principle *"if p is obligatory and if the assertibility of p causally implies the assertibility of q, then q is obligatory"*, is provable in **ILP**:

$$\dfrac{\dfrac{\dfrac{\dfrac{\dfrac{\dfrac{\vdash p\,;\ \Rightarrow\ \vdash p \qquad \vdash q\,;\ \Rightarrow\ \vdash q}{\vdash p,\ \vdash p \overset{c}{\supset} \vdash q\,;\ \Rightarrow\ \vdash q}\ \overset{c}{\supset}}{\circ\!\!-\ p,\ \vdash p \overset{c}{\supset} \vdash q\,;\ \Rightarrow\ \circ\!\!-\ q}\ \overset{c}{\supset} - \circ\!\!-}{;\ \circ\!\!-\ p,\ \vdash p \overset{c}{\supset} \vdash q \Rightarrow \circ\!\!-\ q}\ \textit{derelictions}}{;\ \circ\!\!-\ p,\ \circ\!\!-\ p \cap (\vdash p \overset{c}{\supset} \vdash q) \Rightarrow \circ\!\!-\ q}\ \textit{left}\cap}{;\ \circ\!\!-\ p \cap (\vdash p \overset{c}{\supset} \vdash q) \Rightarrow \circ\!\!-\ q}\ \textit{left}\cap\textit{, contraction}}{;\ \Rightarrow (\circ\!\!-\ p \cap (\vdash p \overset{c}{\supset} \vdash q)) \supset \circ\!\!-\ q}\ \textit{right}\supset$$

In a similar way we derive $(\circ\!\!-_N\ p \cap (\vdash p \overset{c}{\supset} \vdash q)) \supset \circ\!\!-_N\ q$.

By the cut-elimination theorem we cannot derive any of the following formulas:

$$\begin{array}{ll}
(\circ\!\!-\ p \cap (\vdash p \supset \vdash q)) \supset \circ\!\!-\ q & (\vdash p \cap (\circ\!\!-\ p \supset \circ\!\!-\ q)) \supset \circ\!\!-\ q \\
(\vdash p \cap (\circ\!\!-\ p \supset \circ\!\!-\ q)) \supset \vdash q & (\vdash_K p \cap (\circ\!\!-\ p \supset \circ\!\!-\ q)) \supset \vdash_K q \\
(\vdash p \cap (\vdash_K p \supset \vdash_K q)) \supset \vdash q & (\circ\!\!-\ p \cap (\vdash_K p \supset \vdash_K q)) \supset \circ\!\!-\ q \\
(\vdash_K p \cap (\vdash_{K'} p \supset \vdash_{K'} q)) \supset \vdash_{K'} q & (\circ\!\!-\ p \cap (\circ\!\!-_N\ p \supset \circ\!\!-_N\ q)) \supset \circ\!\!-\ q
\end{array}$$

§4. Concluding remark. From a mathematical point of view the extension from $\mathbf{ILP}^-$ to $\mathbf{ILP}^+$ through the linear system **ILLP** is an immediate corollary of the existence of fully functorial translations $\mathbf{IL} \to \mathbf{ILL}$ and does not seem to add much to the direct extension from $\mathbf{ILP}^-$ to **ILP**. Its significance lies in the following facts. On one hand **ILP** formalizes reasoning with mixed contexts and with various notions of implications using a *mixed relevant and intuitionistic consequence relation*, where we can infer properties of *causal implication* by using the relevant area only and also derive standard intuitionistic formulas using the intuitionistic area of the consequence relation, but the relations between the two forms of reasoning are not fully clarified. Moreover, other uses of reasoning with mixed contexts may require other extensions of standard intuitionistic calculi: this could certainly be done on a case by case basis. On the other hand linear logic has been proposed as a general framework for such extensions, where the relations between these different forms of reasoning may be clarified. But the explanations so far given of the "intended interpretations" of linear logic and its place in the context of logical theory appear unsatisfactory to us. In our approach the language $\mathcal{L}_i^\ell$ and the calculus **ILLP** have a clear "intended interpretation" in the notion of a *free logic of pragmatic force*. It is not claimed that this is the only possible "natural reasoning interpretation" of linear logic. However, our interpretation shows that reasoning with the linear consequence relation is an *extension* of the ordinary logic of assertions and obligations rather than a *deviant logic*. The connectives and operators of intuitionistic linear logic also obtain in this way a *partial interpretation* in the pragmatic framework, although through the rather cumbersome machinery of translations and substitution operators.

REFERENCES

[1] N. Benton, G. Bierman, V. de Paiva, M. Hyland. A term calculus for intuitionistic linear logic. In *Proceedings of the Conference on Typed Lambda Calculi and Applications*. Springer LNCS 664, pp. 75-90, 1993.

[2] C. Dalla Pozza and C. Garola. A pragmatic interpretation of intuitionistic propositional logic, *Erkenntnis* **43**. 1995, pp.81-109.

[3] C. Dalla Pozza. Una logica prammatica per la concezione "espressiva" delle norme, In: A. Martino, ed. *Logica delle Norme*, Pisa, 1997

[4] S. Feferman. Transfinite recursive progressions of axiomatic theories. *The Journal of Symbolic Logic* **27**, 1962, pp. 259-316.

[5] S. Feferman. Review of Prawitz 1971. *The Journal of Symbolic Logic* **40**, 1975, pp. 232-234.

[6] J-Y. Girard. Linear Logic. *Theoretical Computer Science* **50**, 1987, pp. 1-102.

[7] J-Y. Girard. On the unity of logic. *Annals of Pure and Applied Logic* 59, 1993, pp.201-217.

[8] J-Y. Girard. Linear logic: its syntax and semantics. In: J-Y. Girard, Y Lafont and L. Regnier, eds., in *Advances in Linear Logic*, London Mathematical Society Lecture Note Series **222**, Cambridge University Press, 1995, pp. 1-42.

[9] K. Gödel. Eine Interpretation des Intuitionistischen Aussagenkalküls, *Ergebnisse eines Mathematischen Kolloquiums* IV, 1933, pp. 39-40.

[10] B. Jacobs. Semantics of Weakening and Contraction, *Annals of Pure and Applied Logic*, **69**, 1, May, 1992, pp.73-106.

[11] J.C.C. McKinsey and A. Tarski. Some theorems about the sentential calculi of Lewis and Heyting, *Journal of Symbolic Logic* **13**, 1948, pp. 1-15.

[12] D. Prawitz. Ideas and Results in Proof Theory. In *Proceedings of the Second Scandinavian Logic Symposium*, ed. Fenstad, North-Holland, 1971.

[13] P. O'Hearn and D. Pym. The Logic of Bunched Implications. *Bulletin of Symbolic Logic*, Volume 5, Number 2, June 1999, 215-243.

[14] H. Schellinx. The noble art of linear decorating, PhD Thesis, Universiteit van Amsterdam, Faculteit Wiskunde en Informatic. ILLC dissertation series, Amsterdam 1994.

[15] R. Solovay. Provability interpretations of modal logic. *Israel Journal of Mathematics*, **25**, 1976, pp. 287-304.

FACOLTÀ DI SCIENZE
UNIVERSITÀ DI VERONA
STRADA LE GRAZIE, CÀ VIGNAL 2
37134 VERONA, ITALY
and
DEPARTMENT OF COMPUTER SCIENCE
QUEEN MARY AND WESTFIELD COLLEGE
LONDON E1 4NS, ENGLAND
E-mail: bellin@sci.univr.it
URL: http://profs.sci.univr.it/~bellin/

DIPARTIMENTO DI FILOSOFIA
UNIVERSITÀ DI LECCE
VIA STAMPICCHIA, PALAZZO PARLANGELI
73100 LECCE, ITALY

COMPUTATIONAL COMPLEXITY AND INDUCTION FOR PARTIAL COMPUTABLE FUNCTIONS IN TYPE THEORY

ROBERT L. CONSTABLE AND KARL CRARY

Abstract. An adequate theory of partial computable functions should provide a basis for defining *computational complexity measures* and should justify the principle of *computational induction* for reasoning about programs on the basis of their recursive calls. There is no practical account of these notions in type theory, and consequently such concepts are not available in applications of type theory where they are greatly needed. It is also not clear how to provide a practical and adequate account in programming logics based on set theory.

This paper provides a practical theory supporting all these concepts in the setting of constructive type theories. We first introduce an extensional theory of partial computable functions in type theory. We then add support for intensional reasoning about programs by explicitly reflecting the essential properties of the underlying computation system. We use the resulting intensional reasoning tools to justify computational induction and to define computational complexity classes. Complexity classes take the form of *complexity-constrained function types.* These function types are also used in conjunction with the propositions-as-types principle to define a *resource-bounded logic* in which proofs of existence can guarantee feasibility of construction.

§1. Introduction. Over the past two decades, type theory has become the formalism of choice to support programming, verification and the logical foundations of computer science. The language of types underlies modern programming languages like Java and ML, and the theory of types drives significant efforts in compilation [29, 50, 36, 39, 42, 6, 47] and semantics [7, 23, 16]. Theorem proving systems based on type theory have been used for the verification of both hardware and software, and have also been very widely used for the formalization of mathematics [8, 10, 26, 24, 41, 46].

One of the major reasons type theory has enjoyed such wide successes is that it is a natural high-level language for computational mathematics and programming, a point that Sol Feferman has effectively made over the years [20, 22]. However, this advantage can sometimes pose a problem because the needs of mathematics and programming can diverge. In mathematics, equality is *extensional,* where only an object's value is significant. That is, if the result of $f(a)$ is b then $f(a) = b$, and functions f and g are equal (in $A \to B$, the space of functions from A to B) exactly when $f(a) = g(a)$ for every a in A. In contrast, in the analysis of programs, it is often critical to consider programs *intensionally,* reasoning about their structure as well as their result value. For example, showing partial correctness of recursive procedures requires reasoning inductively about

Reflections on the foundations of mathematics
Essays in honor of Solomon Feferman
Edited by Wilfried Sieg, Richard Sommer, and Carolyn Talcott

the computation of recursive calls. Also, computational complexity certainly depends on a program's algorithm, and not only on its result.

So a programming logic must reconcile two needs: one, the need to treat functions *extensionally* in order to interface with mathematics and to express the most general laws of substitution; and, two, the need to treat functions *intensionally* (as algorithms) in order to state the most useful rules of program reasoning and to express complexity. There have been several attempts at this reconciliation, some of which are theoretically quite interesting [11, 49, 3, 34, 35]. Nevertheless, none of these are very practical, nor have any been implemented.

We set as our goal producing a practical and adequate account of partial functions, computational induction and computational complexity in the class of constructive type theories such as those of Martin-Löf or Girard. This means that the type theory is built on an underlying *computation system* (or programming language), and reasoning about this computation system will be central to our account. The particular type theory we have chosen in which to formalize our account is summarized in Section 2.

1.1. Elements of a solution. Solving this problem is largely a matter of design informed by mathematical analysis. We know that some form of intensional analysis is required. It can be found by building internal *models* of computation, or, in constructive theories, by exposing parts of the underlying term structure (as in reflection [2]). The wrong choices lead to impractical and unusable theories. For example, building an internal model of a machine (say a random access machine) and a semantic map from machines to functions would express an obvious way to think about procedures, structure and complexity, but that formalization would be so heavy as to overwhelm the logic. Likewise, defining an internal semantics for the programming language component of a constructive type theory (a deep embedding) would be extremely complex.

Our judgement is that in the case of constructive type theories, the first key to a sufficient account is to bring into the type theory parts of the mechanism of the underlying computation system. Then the semantic rules connecting terms to objects (algorithms to functions) are the same as the typing rules of the type theory itself. The second key is to expose only those elements of the computation system that are needed for computational induction and complexity. The resulting extension to the type theory is lightweight but still quite expressive.

1.2. Salient points. The first technical step is to extend the type theory with an *extensional* notion of partial functions. We take the approach introduced by Constable and Smith [11] of adding *partial types*. An object t is said to be in the partial type $\overline{T}$ exactly when its termination implies it is in T. The partial functions from A to B are denoted by $A \to \overline{B}$. However, we must extend the theory of Constable and Smith to deal with equality, which they did not consider.

The second key technical step is to add a lightweight reflection mechanism: The metatype of terms is reflected into the type theory as the internal type *Term*.[1] This allows for intensional consideration of terms. This is another point where our work benefits from Sol Feferman's pioneering investigations [21]. A few other constructs are also introduced to reflect the computational behavior of terms. These intensional mechanisms allow us to define and axiomatize the type $[T]$, the subtype of *Term* that contains the *internal representations* of all terms in T.

Then the principle of computational induction is naturally expressible. This principle is based on the idea that if evaluation of t terminates, a well-founded order is induced on the terms through which it evaluates. This principle is *essentially* intensional; a term cannot be distinguished extensionally from a term to which it evaluates. In Section 4 we formalize this principle and present some examples of its uses.

Finally, using these tools we may define computational complexity classes. We define the *size* of a term as the size of its syntax tree, and its *time* as the length of its evaluation sequence to canonical form. These definitions may be used to refine the type $[T]$ of all terms in T to the type $[T]_s^t$ of all terms in T that compute within time t and space s. More interestingly, we may define complexity classes in all types using *complexity-constrained function types* such as the type $A \xrightarrow{\text{ptime}} B$ of polynomially-time computable functions from A to B.

One particular significant application of this complexity machinery is to define a *resource-bounded logic* using complexity-constrained function types and the propositions-as-types principle [25]. One of the key advantages of constructive logic is that when the existence of an object is proven, that object may be constructed. However, there has never been a guarantee that such a construction will be feasible to perform. By proving the existence of an object in resource-bounded logic, we guarantee that the object may be constructed feasibly [40].

§2. **Partial object type theory.** In this section we lay out the underlying type theory for the results of this paper. Our work builds on the type theory of Nuprl [8], a type theory in the style of Martin-Löf [31, 32]. We briefly summarize Nuprl in Section 2.1. This starting theory is essentially a theory of total functions only, but it may be extended to a theory of partial computable functions by adding partial types (types containing divergent terms) and a fixpoint rule for typing recursive functions [11, 12, 49, 48, 3, 17]. We use Crary's account [17] of type theory augmented by partial functions because that account is unique in providing an equality assertion within the logic, which we require in order to support extensional reasoning.

2.1. Type theory preliminaries. As data types, the theory contains integers (denoted by $\mathbb{Z}$), disjoint unions (denoted by $T_1 + T_2$), lists (denoted by $T\ list$), dependent products (denoted by $(x{:}T_1) \times T_2$) and dependent function spaces

---

[1]A similar approach was used by Allen, *et al.* [2], but our mechanism is much more lightweight because we do not attempt to reflect proofs as well.

(denoted by $(x{:}T_1) \to T_2$). As usual, when x does not appear free in T_2, we write $T_1 \times T_2$ for $(x{:}T_1) \times T_2$ and $T_1 \to T_2$ for $(x{:}T_1) \to T_2$. Types themselves are data in the theory and belong to a predicative hierarchy of universes, $\mathbb{U}_1, \mathbb{U}_2, \mathbb{U}_3$, etc. The universe $\mathbb{U}_1$ contains all types built from the base types only (*i.e.*, built without universes), and the universe $\mathbb{U}_{i+1}$ contains all types build from the base types and the universes $\mathbb{U}_1, \ldots, \mathbb{U}_i$. In particular, no universe is a member of itself.

Propositions are interpreted as types using the propositions-as-types principle [25]. This gives interpretations to the basic logical connectives and justifies using $\mathbb{U}_i$ as the type of propositions (of level i). Other assertions are interpreted using the following device: The type $n_1 \leq n_2$ is well-formed when n_1 and n_2 are integers, and either is inhabited by the term $\star$ (when, in fact, $n_1 \leq n_2$) or is empty (if $n_1 > n_2$).

Each type T comes with an intrinsic equality relation denoted by $t_1 = t_2 \in T$. Membership is also derived from this relation; we say that $t \in T$ if $t = t \in T$. The equality relation is interpreted in the logic by the type $t_1 = t_2$ *in* T using the device described above. (Note that $t_1 = t_2 \in T$ is a metatheoretical assertion whereas $t_1 = t_2$ *in* T is a type in the theory.) The type *Void* is then defined as $0 = 1$ *in* $\mathbb{Z}$ and the type *Unit* is defined as $0 = 0$ *in* $\mathbb{Z}$. Then the booleans ($\mathbb{B}$) are defined as *Unit* + *Unit* and *true*, *false* and *if-then-else* constructs are defined in the obvious manner.

Propositions may be used to construct new types using the set type and quotient type constructors. The set type $\{x : A \mid P\}$ contains all elements t of A such that $P[t/x]$ is inhabited (where $E[e/x]$ denotes the capture-avoiding substitution of e for x in E). We use the set type to define the natural numbers ($\mathbb{N}$) as $\{n : \mathbb{Z} | 0 \leq n\}$, the positive integers ($\mathbb{Z}^+$) as $\{n : \mathbb{Z} \mid 1 \leq n\}$, and the integer subranges ($[m \cdots k]$) as $\{n : \mathbb{Z} \mid m \leq n \wedge n \leq k\}$. The quotient type $x, y{:}A//P$ contains all elements of A, but t_1 and t_2 are equal in $x, y{:}A//P$ if $P[t_1, t_2/x, y]$ is inhabited. For such a type to be well-formed, the equality resulting from P must be coarser than the equality on A.

2.2. Computation. Underlying our type theory is a computation system that is summarized in Appendix B. The computation system is defined by a small-step evaluation relation (denoted by $t_1 \mapsto t_2$), and a set of canonical terms. (Of the terms introduced so far, the canonical terms are those appearing in the first and second columns of Figure 1.) Two properties of the computation system are that it is deterministic (*i.e.*, if $t \mapsto t_1$ and $t \mapsto t_2$ then $t_1 =_\alpha t_2$) and that canonical forms are terminal (*i.e.*, if $canon(t)$ then $t \not\mapsto t'$ for $t' \neq t$).

If $t \mapsto^* t'$ and $canon(t')$ then we say that t converges (abbreviated $t\downarrow$) and t converges to t' (abbreviated $t \Downarrow t'$). Note that if $t \Downarrow t_1$ and $t \Downarrow t_2$ then $t_1 =_\alpha t_2$ and that if $t \Downarrow t'$ then $t' \Downarrow t'$. We say that a term t diverges (abbreviated $t\uparrow$) if it does not converge.

The equality on types is constructed to obey an important property with respect to computation, that equality is value respecting: if $t \in T$ and $t \mapsto t'$ then $t = t' \in T$. As discussed in the introduction, this property is critical to mathematical uses of the theory, but it is this same property that makes intensional reasoning

	Type Formation	Introduction	Elimination
universe i	$\mathbb{U}_i$ (for $i \geq 1$)	type formation operators	
disjoint union	$T_1 + T_2$	$inj_1(e)$ $inj_2(e)$	$case(e, x_1.e_1, x_2.e_2)$
function space	$(x{:}T_1) \to T_2$	$\lambda x.e$	$e_1 e_2$
product space	$(x{:}T_1) \times T_2$	$\langle e_1, e_2 \rangle$	$\pi_1(e)$ $\pi_2(e)$
integers	$\mathbb{Z}$	$\ldots, -1, 0, 1, 2, \ldots$	assorted operations
equality	$t_1 = t_2 \; in \; T$	$\star$	
inequality	$t_1 \leq t_2$	$\star$	
set type	$\{x : T_1 \mid T_2\}$	operators for T_1	
quotient type	$x, y{:}T_1//T_2$	operators for T_1	

FIGURE 1. Basic type theory syntax

challenging: within the type theory, a term and its result value are equal, as are any two terms with the same result value.

2.3. Partial types. Functions (and objects in general) may be defined recursively in the basic type theory using the *fix* operator. However, all the types of the basic type theory are total (*i.e.*, contain only convergent elements), so to show that a recursively defined object is well-typed, it is necessary to show that it converges. Consequently, the basic theory is incapable of reasoning about possibly partial functions. To make the theory capable of reasoning about partial functions, we will add four new primitive types: a partial type constructor, and types asserting convergence, admissibility and totality.

The *partial type* $\overline{T}$ is like a "lifted" version of T; it contains all the members of T as well as all divergent terms. Partial functions from A to B may then be given the type $A \to \overline{B}$. For reasons that will become clear shortly, it is important that $\overline{T}$ never equate any convergent and divergent terms. Thus we say that two terms are equal in $\overline{T}$ if they both diverge, or if they both converge and are equal in T. However, we desire that T be a subtype of $\overline{T}$ (*i.e.*, if $t_1 = t_2 \in T$ then $t_1 = t_2 \in \overline{T}$), so in order for $\overline{T}$ to be allowed as a valid type, we must require that T also never equate any convergent and divergent terms. Instead of this condition, we require the stronger condition that T contain only convergent elements, which seems to work more nicely in practice.

We reason about convergence in the theory using the type $t \; in! \; T$, which is well-formed when $t \in \overline{T}$ and inhabited when $t\downarrow$. We wish to say that the types $t \; in! \; T$ and $t' \; in! \; T$ are equal when $t = t' \in \overline{T}$, and it is for this reason that we cannot permit $\overline{T}$ to equate convergent and divergent terms.

The well-formedness conditions of the partial type and the convergence type contain an apparent circularity: Well-formedness of the convergence type t *in*! T requires that t *in* $\overline{T}$, and the well-formedness of $\overline{T}$ requires that every member of T converge. To break this loop, we use a primitive totality predicate (T *total*) as the condition for partial type well-formedness within the logic.

Recursively defined objects may be typed directly (without showing they converge) using the fixpoint principle:

$$t \textit{ in } (\overline{T} \to \overline{T}) \Longrightarrow \textit{fix}(t) \textit{ in } \overline{T}$$

However, the fixpoint principle is not valid for all types (see Smith [49] for an example of such a type). Types for which the fixpoint principle is valid are called *admissible*. Admissibility is indicated within the logic by the type T *admiss*. Crary [15, 17] shows that a wide class of types are admissible, including all types used in conventional programming languages.

§3. Intensional reasoning. To lay the groundwork that makes intensional reasoning possible in our type theory, we begin by being more specific about the structure of terms in Section 3.1. This structure is quite general, allowing term constructors with arbitrarily many subterms and binding positions, and therefore makes few restrictions on the expressiveness of the theory. Moreover, the results of this paper should easily apply to different term structures, but we desire a concrete framework in which to work. In Section 3.2 we present operators and notation for intensional consideration of terms. Section 3.3 then presents the primitive type-theoretic constructs necessary for intensional reasoning.

3.1. Uniform term syntax. Terms are constructed using a set Ω of *operators*, ranged over by the metavariable ω, and a function *arity* mapping each operator to a finite (possibly empty) list of natural numbers. The arity function will be used to indicate the number of subterms of an operator and the number of bound variables for each subterm. We also take as given an infinite set of variables.

The set of terms is then defined inductively as follows:

- Every variable is a term.
- If ω is an operator then ω^* is a term, serving as the name of ω.
- If $\textit{arity}(\omega) = (k_1, \ldots, k_n)$ and $t_1, \ldots, t_n$ are terms, then $\omega(x_{11} \cdots x_{1k_1}.t_1, \ldots, x_{n1} \cdots x_{nk_n}.t_n)$ is a (compound) term, where x_{ij} is a variable for $1 \leq i \leq n$ and $1 \leq j \leq k_i$.

The variables $x_{i1}, \ldots, x_{ik_i}$ preceding each subterm are binding occurrences. Thus, in the term *lambda*$(x.\textit{add}(x, y))$, the variable x is bound and the variable y is free. As usual, terms that differ only by the alpha-variation of bound variables are considered identical.

A subset of the operators in Ω are designated as *canonical* operators. A term is considered canonical either if it is a compound term constructed with a canonical operator at the top, or if it is the name ω^* of an operator (canonical or not).

Although it will often be useful to present terms in the uniform term syntax just described, we will continue usually to rely on more conventional notation

$$\begin{aligned}
\lceil t \rceil &\stackrel{\text{def}}{=} \lceil t \rceil_{\varepsilon} \\
\lceil x_i \rceil_{x_n,\ldots,x_1} &\stackrel{\text{def}}{=} \mathit{var}_i \qquad\qquad (\text{for } 1 \leq i \leq n) \\
\lceil \omega^* \rceil_{x_n,\ldots,x_1} &\stackrel{\text{def}}{=} \mathit{opname}(\omega^*) \\
\lceil \omega(y_{k_1}\cdots y_1.t_1,\ldots,y_{k_n}\cdots y_1.t_n) \rceil_{x_m,\ldots,x_1} &\stackrel{\text{def}}{=} \langle\!\langle \omega^*, \lceil t_1 \rceil_{x_m,\ldots,x_1,y_{k_1},\ldots,y_1}, \\
&\qquad \vdots \\
&\qquad \lceil t_n \rceil_{x_m,\ldots,x_1,y_{k_n},\ldots,y_1} \rangle\!\rangle \\
&\qquad (\text{for } \mathit{arity}(\omega) = (k_1,\ldots,k_n))
\end{aligned}$$

$$\begin{aligned}
\langle\!\langle t_1,\ldots,t_n \rangle\!\rangle &\stackrel{\text{def}}{=} \mathit{inj}_1(\langle t_1, \langle t_2, \langle \cdots \langle t_n, \star \rangle \cdots \rangle\rangle\rangle) \\
\mathit{var}_i &\stackrel{\text{def}}{=} \mathit{inj}_2(\mathit{inj}_1(i)) \\
\mathit{opname}(t) &\stackrel{\text{def}}{=} \mathit{inj}_2(\mathit{inj}_2(t))
\end{aligned}$$

FIGURE 2. Term representations

such as that used in Section 2. These conventional notations should be viewed as abbreviations for formal terms in the uniform syntax above. For example, $\lambda x.t$ is shorthand for $\mathit{lambda}(x.t)$ where $\mathit{arity}(\mathit{lambda}) = (1)$.

3.2. Intensional representations. In order to analyze terms intensionally, we must have internal representations of terms. As discussed in the introduction, the representation of a term cannot be the term itself, as appealing as that might be. In particular, the representation of a term must be a canonical form, even when the term being represented is not.

The representation of any term t is denoted by $\lceil t \rceil$. If ω is an n-place operator with no binding occurrences then the term $\omega(t_1,\ldots,t_n)$ is represented by $\langle\!\langle \omega^*, \lceil t_1 \rceil,\ldots,\lceil t_n \rceil \rangle\!\rangle$. To avoid the syntactic overhead of formalizing variables, terms with binding structure are represented using de Bruijn indices [19]: for $i > 0$, var_i refers to the ith enclosing binding occurrence, counting from one. For example, $\lceil \lambda x.\lambda y.x \rceil = \langle\!\langle \mathit{lambda}^*, \langle\!\langle \mathit{lambda}^*, \mathit{var}_2 \rangle\!\rangle\rangle\!\rangle$. This is summarized formally in Figure 2.

Finally, it will prove convenient to have some notation like the quasiquote mechanism of Scheme for the representations of terms with holes to be filled by ordinary terms. We will denote these with concentric boxes: $\boxed{t}$ stands for the representation of t except that inner boxes within t denote holes and expressions within those holes are left unchanged. For example, if t is a term, $\boxed{\lambda x.\langle \boxed{t}, 0\rangle}$ denotes $\langle\!\langle \mathit{lambda}^*, \langle\!\langle \mathit{pair}^*, t, \mathit{number}_0 \rangle\!\rangle\rangle\!\rangle$.

3.3. Type theory of terms. We are now ready to introduce the principal devices for intensional reasoning in type theory. These include a type of intensional terms, an operator for converting intensional terms to their denotations, and types asserting intensional type membership, evaluation, canonicalism, and computational inducement:

- The type Term_i contains the representations of all terms with up to i free variables (that is, de Bruijn indices extending up to i binding positions out of the term). (*E.g.*, var_i is in Term_j when $i \leq j$.) Expressions are equal in Term_i if they represent alpha-equal terms. Note that the representations of all terms are canonical (since *pair* is a canonical operator), even if the terms themselves are not, and therefore Term_i is a total type. The type *Term* is defined to be the type Term_0 of closed term representations.
- Terms are related to their representations by a set of operators ref_n (for natural numbers n), which compute the meaning of intensional term representations. The operator ref_n is used when the representing term t may contain up to n free variables: $\mathit{ref}_n(t, e_1, \ldots, e_n)$ computes the denotation of t where e_i is substituted for the ith enclosing free variable of t. For example, $\mathit{ref}_2(\mathit{var}_1, t_1, t_2) \mapsto t_1$. When n is small, we will write $[\![t]\!]_{e_1,\ldots,e_n}$ for $\mathit{ref}_n(t, e_1, \ldots, e_n)$. Note that, unlike $\lceil\cdot\rceil$, ref_n is an operator and thus may be used within the theory (and indeed will be, extensively).
- Using *ref*, we may form the assertion that an intensional term t represents a member of T in the natural manner as $[\![t]\!]$ *in* T. For technical reasons,[2] the theory also requires a primitive type asserting that a term represents a member of a type. This type is t [*in*] T, asserting that $[\![t]\!] \in T$.

  Using this type, we may define an essential refinement of the type *Term*: the type $[T]$, which contains representations of all terms of type T,

  $$[T] \stackrel{\text{def}}{=} \{x : \mathit{Term} \mid x\ [\mathit{in}]\ T\}.$$

  Type $[T]$, along with the $[\![\cdot]\!]$ operator, will serve as the principal interface between the extensional and intensional aspects of our type theory. When we wish to consider an unknown term x both intensionally and extensionally and x extensionally is to have type T, we will give x the type $[T]$ and then use x bare in intensional contexts, but use it as $[\![x]\!]$ in extensional contexts.
- To reason about computation, we need types representing the assertions that one term evaluates to another and that a term is canonical. The type t_1 *evalto* t_2 is inhabited by the term $\star$ when $[\![t_1]\!] \mapsto [\![t_2]\!]$ and is empty otherwise. Similarly, the type t *canonical* is inhabited (by $\star$) exactly when $[\![t]\!]$ is canonical.
- We also need to reason about the related concept of computational inducement. If $e_1 \mapsto e_2$ then the evaluation of e_1 certainly induces the evaluation

[2] In Martin-Löf style type theories, the membership type t *in* T is well-formed only when t is a member of T. Hence, the assertion $[\![t]\!]$ *in* T is well-formed only when it is true, and thus it is useless as the antecedent of an implication. We will need to form propositions that are conditional on an intensional term's denotation belonging to a type, so we need a type that is inhabited if and only if the above type is, but that has more liberal conditions on well-formedness. From t [*in*] T we may deduce that $[\![t]\!]$ *in* T and vice versa, but t [*in*] T is well-formed whenever t is a member of *Term* (and T is a type).

of e_2. However, depending upon the structure of e_1, the evaluation of other terms may be induced as well. For example, the evaluation of $f\,a$ induces the evaluation of f. When the evaluation of e_1 to canonical form contains the evaluation of e_2 we say that e_1 *induces evaluation* of e_2. This is represented by the type t_1 *induces* t_2, which is inhabited by the term $\star$ when $[\![t_1]\!]$ induces evaluation of $[\![t_2]\!]$ and is empty otherwise. When a term t halts, inducement defines a well-founded relation on the terms induced by t; we will use that relation for computational induction in the next section.

This type theory is formalized in Appendices A and B, which present the type theory's inference rules and operational semantics. In addition to the primitive construct discussed above, we will also need a few functions operating on intensional representations:

- A substitution function $subst \in Term_I \to Term \to Term$ such that $subst\; e_1\; e_2$ substitutes e_2 for var_1 in e_1 (and shifts other variables down appropriately). We will often abbreviate $subst\; e_1\; e_2$ by $e_1[e_2/1]$.
- Effective versions of the evaluation and canonicalism assertions, that is, a function $iscanon \in Term \to \mathbb{B}$ that determines whether a term is canonical and a function $next \in Term \to (Term + Unit)$ that determines the next term that a term evaluates to, if it is not canonical or stuck.

Note that the last three functions (*subst*, *iscanon*, and *next*) can be written in the underlying type theory and need not be made primitive. Moreover, they will still be programmable in any reasonable extension of our computation system: the programmability of *subst* is inherent in the syntax of terms and their intensional representations, and the programmability of *iscanon* and *next* amount to the constructability of a universal machine, which is usually part of the definition of a reasonable computation system.

§4. Computational induction. With a type theory capable of intensional reasoning in hand, we are ready to look at some applications of intensional reasoning. The first, *computational induction,* is a useful principle of induction that is not valid in extensional type theories. Computational induction is based on the inducement order relation: if the evaluation of a term t converges, inducement is a well-founded order over those terms induced by t. (Recall that t induces t' when the evaluation of t includes the evaluation of t'.)

Computational induction has general applicability for reasoning about programs since it corresponds to the intuitive mechanism of reasoning about recursive programs by preconditions and postconditions, but it is particularly useful for proving *partial* correctness of programs. Other induction principles are inadequate for such proofs; if one could find a well-founded order on the inputs of a recursive function, one could show total correctness instead.

The principle of computational induction is as follows:

$$(\forall e{:}Term.\ (\forall e'{:}Term.\ e\ induces\ e' \Longrightarrow P[e']) \Longrightarrow P[e]) \wedge [\![t]\!]\ in!\ T \Longrightarrow P[t]$$

In words, if we may show $P[e]$ from the assumption of $P[e']$ for every e' induced by e, and if t halts, then $P[t]$.

Note that, unlike the fixpoint principle, there is no admissibility restriction on computational induction. This is because, like most induction principles but unlike fixpoint induction, computational induction is based on a well-founded order. The generality of computational induction makes it a useful tool for reasoning about partial function in circumstances where fixpoint induction is disallowed.

Note also that since inducement is an intrinsically intensional property, stating the computational induction principle demands the intensional reasoning structure developed in Section 3. Indeed, if the principle were expressed in extensional terms, it would be inconsistent: If e evaluates to e', then e also induces e', but because equality is value-respecting, e will be equal to e' is any types to which they belong. Thus, if computational induction were expressed extensionally (that is, if it respected equality), then the first condition would always be satisfied, allowing any property to be proven of any terminating term.

We will now examine some examples of the use of computational induction:

EXAMPLE 1. Consider the "$3n + 1$" function:

$$\begin{aligned} F = \mathit{fix}(\lambda f.\lambda n.\mathit{if}\ n = 1\ \mathit{then}\ 0 \\ \mathit{else\ if}\ n \bmod 2 = 0\ \mathit{then}\ 1 + f(n/2) \\ \mathit{else}\ f(3n+1)) \end{aligned}$$

This program is not known to halt for all positive integers, but it is easy to see that if $F(n)\downarrow$ then $\log_2 n \leq F(n)$. We may show this by computational induction using the proposition:

$$P[e] \;=\; \forall n{:}[\mathbb{Z}^+].\,(e = \boxed{F(\boxed{n})}\ \mathit{in\ Term}) \Longrightarrow \log_2 \llbracket n \rrbracket \leq \llbracket e \rrbracket.$$

PROOF. Let e *in Term* be arbitrary. Suppose $P[e']$ for all e' induced by e and suppose $e = \boxed{F(\boxed{n})}$ *in Term* (for some n *in* $[\mathbb{Z}^+]$). By assumption, $\llbracket n \rrbracket$ *in* $\mathbb{Z}^+$. If $\llbracket n \rrbracket = 1$ then $\llbracket e \rrbracket = 0$. Suppose $\llbracket n \rrbracket \bmod 2 = 0$. Then $\boxed{F(\boxed{n})}$ *evalto* $\boxed{1 + F(\boxed{n}/2)}$, and thus e induces $\boxed{F(\boxed{n}/2)}$. By induction $\log_2(\llbracket n \rrbracket/2) \leq F(\llbracket n \rrbracket/2)$, so $\log_2 \llbracket n \rrbracket = 1 + \log_2(\llbracket n \rrbracket/2) \leq 1 + F(\llbracket n \rrbracket/2) = F(\llbracket n \rrbracket)$.

The remaining case is similar. Suppose $\llbracket n \rrbracket \bmod 2 = 1$. Then $\boxed{F(\boxed{n})}$ *evalto* $\boxed{F(3\boxed{n}+1)}$, and thus e induces $\boxed{F(3\boxed{n}+1)}$. By induction $\log_2(3\llbracket n \rrbracket + 1) \leq F(3\llbracket n \rrbracket + 1)$, so $\log_2 \llbracket n \rrbracket \leq \log_2(3\llbracket n \rrbracket + 1) \leq F(3\llbracket n \rrbracket + 1) = F\llbracket n \rrbracket$. By computational induction we conclude $\llbracket t \rrbracket$ *in*! $\mathbb{Z} \Longrightarrow P[t]$. It follows that $\forall n{:}[\mathbb{Z}^+].\, F\llbracket n \rrbracket$ *in*! $\mathbb{Z} \Longrightarrow \log_2 \llbracket n \rrbracket \leq F\llbracket n \rrbracket$, as desired. ⊣

EXAMPLE 2. Computational induction is also useful for showing conditional termination. Consider the function *time* that computes the length of the reduction sequence of a term:

$$time \stackrel{\text{def}}{=} fix(\lambda f.\lambda t.case(next(t), t'.\,1 + f(t'), x.0))$$

We may use computational induction to show that $time(t)$ *in*! $\mathbb{N}$ whenever $[\![t]\!]$ *in*! T (for any T).[3] This is done using the proposition $P[e] = time(e)$ *in*! $\mathbb{N}$ and the observation that if $next(e) = inj_1(e')$ then e induces e' (since e *evalto* e').

§5. Computational complexity and resource-bounded logic. A second application of intensional reasoning lies in computational complexity. The evaluation relation on *Term* provides the basis for defining computational complexity measures such as time and space. These measures allow us to express traditional results about complexity classes as well as recent results concerning complexity in higher types [4, 43, 27, 13, 14, 44, 45, 28, 51]. The basic measure of time is the number of evaluation steps to canonical form. We have already defined (in the previous section) a function counting the evaluation steps of a term, so we already have a notion of running time.[4] However, it will prove more convenient to state that notion in predicate form where $Term(e, t)$ states that e runs within time t:

$$\begin{aligned} Time(e,t) \textit{ iff } \exists n{:}[0\cdots t].\,\exists f{:}[0\cdots n] \to Term.\ & f(0) = e \textit{ in } Term \wedge \\ & iscanon(f(n)) = true \textit{ in } \mathbb{B} \wedge \\ & \forall i{:}[0\cdots n-1]. \\ & \quad f(i) \textit{ evalto } f(i+1). \end{aligned}$$

We may define a notion of space in a similar manner. First, we may easily define a function *size* with type $Term \to \mathbb{N}$ which computes the number of operators in a term. Then we define the predicate $Space(e, s)$, that states that e runs in space at most s:

$$\begin{aligned} Space(e,s) \textit{ iff } \exists n{:}\mathbb{N}.\,\exists f{:}[0\cdots n] \to Term.\ & f(0) = e \textit{ in } Term \wedge \\ & iscanon(f(n)) = true \textit{ in } \mathbb{B} \wedge \\ & \forall i{:}[0\cdots n-1].\, f(i) \textit{ evalto } f(i+1) \wedge \\ & \forall i{:}[0\cdots n].\, size(f(i)) \le s. \end{aligned}$$

Using these, we may define the resource-indexed type $[T]^t_s$ of terms that evaluate (to a member of T) within time t and space s:

$$[T]^t_s \stackrel{\text{def}}{=} \{e : [T] \mid Time(e,t) \wedge Space(e,s)\}$$

---

[3] The fixpoint principle suffices to show $\forall t{:}Term.\ time(t) \in \overline{\mathbb{N}}$.

[4] Since the underlying computation system allows $(\lambda x.b)a$ to evaluate in one step to $b[a/x]$ (an action impossible in constant time on real machines), this measure of running time will not always correspond to measures derived for more realistic machine models. Fortunately, the theory we present does not depend greatly on the actual computation system used: many computation systems could validate our axioms, and we have chosen the one used in this paper only for its simplicity. A computation system based on *explicit substitutions* [1, 18] can also easily be used, and leads to a more realistic measure of running time. Still more realistic computation systems have also been proposed, such as the λ_{gc} of Morrisett *et al.* [37, 38], but we have not explored how easily our theory could be adapted to such systems.

One interesting application of the resource-indexed types is to define types like Parikh's feasible numbers [40], numbers that may be computed in a "reasonable" time. Benzinger [5] shows another application.

5.1. Complexity classes. With time complexity measures defined above, we may define complexity classes of functions. Complexity classes are expressed as function types whose members are required to fit within complexity constraints. We call such types *complexity-constrained* function types. For example, the quadratic time, polynomial time, and polynomial space computable functions may be defined in a straightforward manner:

$$\begin{aligned}
(x{:}A) \xrightarrow{\text{quad}} B &\stackrel{\text{def}}{=} \{f : [(x{:}A) \to B] \mid \\
&\qquad \exists c{:}\mathbb{N}.\, \forall a{:}[A]^0.\, \mathit{Time}(\langle\!\langle app^*, f, a\rangle\!\rangle, c \cdot \mathit{size}(a)^2)\} \\
(x{:}A) \xrightarrow{\text{ptime}} B &\stackrel{\text{def}}{=} \{f : [(x{:}A) \to B] \mid \\
&\qquad \exists c, c'{:}\mathbb{N}.\, \forall a{:}[A]^0.\, \mathit{Time}(\langle\!\langle app^*, f, a\rangle\!\rangle, c \cdot \mathit{size}(a)^{c'})\} \\
(x{:}A) \xrightarrow{\text{pspace}} B &\stackrel{\text{def}}{=} \{f : [(x{:}A) \to B] \mid \\
&\qquad \exists c, c'{:}\mathbb{N}.\, \forall a{:}[A]^0.\, \mathit{Space}(\langle\!\langle app^*, f, a\rangle\!\rangle, c \cdot \mathit{size}(a)^{c'})\}
\end{aligned}$$

While this technique provides complexity classes in a theory rich enough to express all mathematics, notions of complexity over *higher types* rely on *oracle time* complexity [33, 13, 14], where evaluation of a distinguished argument is not counted toward evaluation time, but the processing of its result is. Supporting such notions in this framework requires a mechanism for marking the distinguished argument throughout evaluation so that it may always be identified and discounted. Such a mechanism allows the definition of higher-order polynomial time computable functions [9, 33, 13, 45]. It is not difficult to produce such a mechanism in this framework, but we will not discuss it here due to space limitations.

5.2. Resource-bounded logic. One of the advantages of constructive logic is that when the existence of an object is proven, that object may be constructed. Many theorem provers based on constructive logic [8, 26, 30, 41] allow the automatic extraction of programs from proofs. However, there is no guarantee that such programs may feasibly be executed. This has been a serious problem in practice, as well as in principle.

Using the complexity-constrained functions, we may define a resource-bounded logic that solves this problem. Under the propositions-as-types principle, the universal statement $\forall x{:}A.B$ corresponds to the function space $(x{:}A) \to B$. By using the complexity-constrained function space instead, we obtain a *resource-bounded* universal quantifier. For example, let us denote the quantifier corresponding to the polynomial-time computable functions by $\forall_{(ptime)} x{:}A.B$. By proving the statement $\forall_{(ptime)} x{:}A.\, \exists y{:}B.\, P(x, y)$, we guarantee that the appropriate y may actually be feasibly computed from a given x.

§6. Conclusion. We have presented a type theory that reconciles extensional and intensional reasoning, providing the first account of computational complexity that does not incur heavy mechanisms such as reflection or deep embeddings. The constructs we have presented may be implemented in terms of the theory of reflection presented by Allen *et al.* [2], and indeed our theory meshes nicely with reflection. However, our theory is much more lightweight than the theory of reflection, which internalizes sequents, proofs and programs that construct proofs, so it can be profitable to justify the constructs of our theory as primitives instead. We argue, then, that our theory occupies an attractive middle position, expressive enough for powerful reasoning (such as complexity analysis), and consistent with richer theories such as reflection, but simple enough to be used practically.

Appendix A. Inference rules. A complete set of inference rules for the type theory with extensional partial functions is given in Crary [17]. In this appendix, we present those additional rules needed to reason intensionally:

A.1. Terms.

$$\frac{H \vdash j = j' \textit{ in } \mathbb{N}}{H \vdash \textit{Term}_j = \textit{Term}_{j'} \textit{ in } \mathbb{U}_i} \qquad \frac{}{H \vdash \textit{Op} = \textit{Op in } \mathbb{U}_i} \qquad \frac{}{H \vdash \omega^* = \omega^* \textit{ in Op}}$$

$$\frac{H \vdash t = t' \textit{ in Op}}{H \vdash \textit{ar}(t) = \textit{ar}(t') \textit{ in } \mathbb{N} \textit{ list}}$$

$$\frac{}{H \vdash \textit{ar}(\omega^*) = (k_1 :: \cdots :: k_n :: \textit{nil}) \textit{ in } \mathbb{N} \textit{ list}} \; (\textit{arity}(\omega^*) = (k_1, \ldots, k_n))$$

$$\frac{H \vdash e = e' \textit{ in } ((o{:}\textit{Op}) \times \textit{fold}(\textit{ar}(o), i\, X.\textit{Term}_{i+j} \times X, \textit{Unit})) + [1 \cdots i] + \textit{Op}}{H \vdash e = e' \textit{ in Term}_j}$$

$$\frac{H \vdash e = e' \textit{ in Term}_j}{H \vdash e = e' \textit{ in } ((o{:}\textit{Op}) \times \textit{fold}(\textit{ar}(o), i\, X.\textit{Term}_{i+j} \times X, \textit{Unit})) + [1 \cdots i] + \textit{Op}}$$

A.2. Intensional membership.

$$\frac{H \vdash t = t' \textit{ in Term} \quad H \vdash T = T' \textit{ in } \mathbb{U}_i}{H \vdash (t \,[\textit{in}]\, T) = (t' \,[\textit{in}]\, T') \textit{ in } \mathbb{U}_i} \qquad \frac{H \vdash [\![t]\!] \textit{ in } T}{H \vdash t \,[\textit{in}]\, T} \qquad \frac{H \vdash t \,[\textit{in}]\, T}{H \vdash [\![t]\!] \textit{ in } T}$$

A.3. Evaluation, canonicalism, and inducement.

$$\frac{H \vdash t_1 = t_1' \textit{ in Term} \quad H \vdash t_2 = t_2' \textit{ in Term}}{H \vdash (t_1 \textit{ evalto } t_2) = (t_1' \textit{ evalto } t_2') \textit{ in } \mathbb{U}_i}$$

$$\frac{H \vdash t = t' \textit{ in Term}}{H \vdash (t \textit{ canonical}) = (t' \textit{ canonical}) \textit{ in } \mathbb{U}_i}$$

$$\frac{H \vdash t_1 = t_1' \textit{ in Term} \quad H \vdash t_2 = t_2' \textit{ in Term}}{H \vdash (t_1 \textit{ induces } t_2) = (t_1' \textit{ induces } t_2') \textit{ in } \mathbb{U}_i}$$

$$\frac{H \vdash t_1 \textit{ induces } t_2}{H \vdash \neg(t_1 \ \textit{canonical})} \qquad \frac{H \vdash t_1 \textit{ evalto } t_2}{H \vdash t_1 \textit{ induces } t_2}$$

$$\frac{H \vdash t_1 \textit{ evalto } t_2 \quad H \vdash [\![t_1]\!] \textit{ in } T}{H \vdash [\![t_1]\!] = [\![t_2]\!] \textit{ in } T}$$

$$\frac{H \vdash t \ \textit{canonical} \quad H \vdash [\![t]\!] \textit{ in } T \quad H \vdash \overline{T} \textit{ in } \mathbb{U}_i}{H \vdash [\![t]\!] \textit{ in! } T}$$

$$\frac{H \vdash [\![t]\!] \textit{ in! } T \quad H, e{:}\textit{Term}, (\forall e'{:}\textit{Term}.\ e \textit{ induces } e' \Longrightarrow P[e'/x]) \vdash P[e/x]}{H \vdash P[t/x]}$$

A.4. Sample denotation rules.

$$\frac{H, y : [A], z : [A], w : ([\![y]\!] = [\![z]\!] \textit{ in } A) \vdash [\![b[y/1]]\!] = [\![b[z/1]]\!] \textit{ in } B[[\![y]\!]/x]}{H \vdash [\![\boxed{\lambda x.\boxed{b}}]\!] = \lambda x.[\![b]\!]_x \textit{ in } (x : A) \to B}$$

$$\frac{H \vdash e_1 \ [\textit{in}] \ (x : A) \to B \quad H \vdash e_2 \ [\textit{in}] \ A}{H \vdash [\![\boxed{\boxed{e_1}\boxed{e_2}}]\!] = [\![e_1]\!][\![e_2]\!] : B[[\![e_1]\!]/x]}$$

A.5. Sample evaluation and inducement rules.

$$\frac{H \vdash e \textit{ in Term}_l}{H \vdash \boxed{\lambda x.\boxed{e}} \ \textit{canonical}} \qquad \frac{H \vdash f \textit{ evalto } f' \quad H \vdash e \textit{ in Term}}{H \vdash \boxed{\boxed{f}\boxed{e}} \textit{ evalto } \boxed{\boxed{f'}\boxed{e}}}$$

$$\frac{H \vdash e \textit{ in Term}_l \quad H \vdash t \textit{ in Term}}{H \vdash \boxed{(\lambda x.\boxed{e})\boxed{t}} \textit{ evalto } e[t/1]} \qquad \frac{H \vdash f \textit{ in Term} \quad H \vdash e \textit{ in Term}}{H \vdash \boxed{\boxed{f}\boxed{e}} \textit{ induces } f}$$

A.6. Lists.

$$\frac{H \vdash T = T' \textit{ in } \mathbb{U}_i}{H \vdash T \ \textit{list} = T' \ \textit{list in } \mathbb{U}_i} \qquad \frac{H \vdash T \textit{ in } \mathbb{U}_i}{H \vdash \textit{nil} = \textit{nil in } T \ \textit{list}}$$

$$\frac{H \vdash h = h' \textit{ in } T \quad H \vdash t = t' \textit{ in } T \ \textit{list}}{H \vdash h :: t = h' :: t' \textit{ in } T \ \textit{list}}$$

$$\frac{H \vdash e_1 = e_1' \ \mathit{in}\ T\ \mathit{list} \quad H, x : T, y : S \vdash e_2 = e_2' \ \mathit{in}\ S \quad H \vdash e_3 = e_3' \ \mathit{in}\ S}{H \vdash \mathit{fold}(e_1, x\ y.e_2, e_3) = \mathit{fold}(e_1', x\ y.e_2', e_3') \ \mathit{in}\ S}$$

$$\frac{H \vdash e' \ \mathit{in}\ S}{H \vdash \mathit{fold}(\mathit{nil}, x\ y.e, e') = e' \ \mathit{in}\ S}$$

$$\frac{H \vdash h \ \mathit{in}\ T \quad H \vdash t \ \mathit{in}\ T\ \mathit{list} \quad H, x : T, y : S \vdash e \ \mathit{in}\ S \quad H \vdash e' \ \mathit{in}\ S}{H \vdash \mathit{fold}(h :: t, x\ y.e, e') = e[h, \mathit{fold}(t, x\ y.e, e')/x, y] \ \mathit{in}\ S}$$

Appendix B. Operational Semantics.

$$\frac{f \mapsto f'}{f\ e \mapsto f'\ e} \qquad \frac{}{(\lambda x.e)t \mapsto e[t/x]} \qquad \frac{t \mapsto t'}{\pi_i(t) \mapsto \pi_i(t')} \qquad \frac{}{\pi_i(\langle t_1, t_2 \rangle) \mapsto t_i}$$

$$\frac{t \mapsto t'}{\mathit{case}(t, x.e_1, x.e_2) \mapsto \mathit{case}(t', x.e_1, x.e_2)} \qquad \frac{}{\mathit{case}(\mathit{inj}_i(t), x.e_1, x.e_2) \mapsto e_i[t/x]}$$

$$\frac{e_1 \mapsto e_1'}{\mathit{fold}(e_1, x\ y.e_2, e_3) \mapsto \mathit{fold}(e_1', x\ y.e_2, e_3)} \qquad \frac{}{\mathit{fold}(\mathit{nil}, x\ y.e, e') \mapsto e'}$$

$$\frac{}{\mathit{fold}(h :: t, x\ y.e, e') \mapsto e[h, \mathit{fold}(t, x\ y.e, e')/x, y]} \qquad \frac{}{\mathit{fix}(f) \mapsto f\ \mathit{fix}(f)}$$

$$\frac{t \mapsto t'}{\mathit{ref}_n(t, e_1, \ldots, e_n) \mapsto \mathit{ref}_n(t', e_1, \ldots, e_n)}$$

$$\frac{t \mapsto t'}{\mathit{ref}_n(\mathit{inj}_i(t), e_1, \ldots, e_n) \mapsto \mathit{ref}_n(\mathit{inj}_i(t'), e_1, \ldots, e_n)}$$

$$\frac{t_1 \mapsto t_1'}{\mathit{ref}_n(\mathit{inj}_1(\langle t_1, t_2 \rangle), e_1, \ldots, e_n) \mapsto \mathit{ref}_n(\mathit{inj}_1(\langle t_1', t_2 \rangle), e_1, \ldots, e_n)}$$

$$\frac{\mathit{arity}(\omega) = (k_1, \ldots, k_m)}{\begin{array}{l} \mathit{ref}_n(\mathit{inj}_1(\langle \omega^*, t \rangle), e_1, \ldots, e_n) \mapsto \\ \quad \omega(x_{k_1} \cdots x_1.\mathit{ref}_{k_1+n}(\pi_1(t), x_1, \ldots, x_{k_1}, e_1, \ldots, e_n), \\ \quad \vdots \\ \quad x_{k_m} \cdots x_1.\mathit{ref}_{k_m+n}(\pi_1(\underbrace{\pi_2(\cdots(\pi_2}_{m-1 \text{ times}}(t))\cdots)), x_1, \ldots, x_{k_1}, e_1, \ldots, e_n)) \end{array}}$$

$$\frac{t \mapsto t'}{ref_n(inj_2(inj_i(t)), e_1, \ldots, e_n) \mapsto ref_n(inj_2(inj_i(t')), e_1, \ldots, e_n)}$$

$$\frac{}{ref_n(inj_2(inj_1(i)), e_1, \ldots, e_n) \mapsto e_i} \ (1 \leq i \leq n)$$

$$\frac{}{ref_n(inj_2(inj_2(\omega^*)), e_1, \ldots, e_n) \mapsto \omega^*}$$

REFERENCES

[1] MARTÍN ABADI, LUCA CARDELLI, PIERRE-LOUIS CURIEN, and JEAN-JACQUES LÉVY, *Explicit substitutions*, ***Journal of Functional Programming***, vol. 1 (1991), no. 4, pp. 375–416.

[2] STUART F. ALLEN, ROBERT L. CONSTABLE, DOUGLAS J. HOWE, and WILLIAM AITKEN, *The semantics of reflected proof*, ***Proceedings of the fifth symposium on logic in computer science***, IEEE, June 1990, pp. 95–197.

[3] P. AUDEBAUD, *Partial objects in the calculus of constructions*, ***Proc. of sixth symp. on logic in comp. sci.*** (IEEE, Vrije University, Amsterdam, The Netherlands), 1991, pp. 86–95.

[4] S. BELLANTONI and S. COOK, *A new recursion-theoretic characterization of the poly-time functions*, ***Computational Complexity***, vol. 2 (1992), pp. 97–110.

[5] RALPH BENZINGER, *Automated complexity analysis of Nuprl extracted programs*, ***Journal of Functional Programming***, (2000), no. To Appear.

[6] LARS BIRKEDAL, NICK ROTHWELL, MADS TOFTE, and DAVID N. TURNER, *The ML Kit (version 1)*, ***Technical Report 93/14***, Department of Computer Science, University of Copenhagen, 1993.

[7] L. CARDELLI and P. LONGO, *A semantic basis for Quest*, ***Journal of functional programming***, 1991, pp. 1:417–458.

[8] R.L. CONSTABLE, S.F. ALLEN, H.M. BROMLEY, W.R. CLEAVELAND, J.F. CREMER, R.W. HARPER, D.J. HOWE, T.B. KNOBLOCK, N.P. MENDLER, P. PANANGADEN, J.T. SASAKI, and S.F. SMITH, ***Implementing mathematics with the Nuprl proof development system***, Prentice-Hall, 1986.

[9] ROBERT L. CONSTABLE, *Type two computational complexity*, ***Proceedings of the 5th annual acm symposium on the theory of computing***, 1973, pp. 108–121.

[10] ———, Handbook of Proof Theory, ch. X: Types in Logic, Mathematics and Programming, Elsevier Science B.V., 1998.

[11] ROBERT L. CONSTABLE and SCOTT FRASER SMITH, *Partial objects in constructive type theory*, ***Second IEEE symposium of logic in computer science*** (Ithaca, New York), June 1987, pp. 183–193.

[12] ———, *Computational foundations of basic recursive function theory*, ***Third IEEE symposium of logic in computer science*** (Edinburgh, Scotland), July 1988, pp. 360–371.

[13] S. A. COOK and B. M. KAPRON, *Characterizations of the basic feasible functionals of finite type*, ***Proceedings of msi workshop on feasible mathematics*** (New York) (S. Buss and P.J. Scott, editors), Birkhauser-Boston, 1990, pp. 71–95.

[14] ———, *A new characterization of Mehlhorn's polynomial time functionals*, ***FOCS***, (1991).

[15] KARL CRARY, *Admissibility of fixpoint induction over partial types*, ***Fifteenth international conference on automated deduction*** (Lindau, Germany), Lecture Notes in Computer Science, vol. 1421, Springer-Verlag, July 1998, Extended version published as CMU technical report CMU-CS-98-164., pp. 270–285.

[16] ———, *Programming language semantics in foundational type theory*, ***International conference on programming concepts and methods*** (Shelter Island, New York), Chapman & Hall, 1998, Extended version published as Cornell University technical report TR98-1666.

[17] ———, *Type-theoretic methodology for practical programming languages*, ***Ph.D. thesis***, Department of Computer Science, Cornell University, Ithaca, New York, August 1998.

[18] Pierre-Louis Curien, Thérèse Hardin, and Jean-Jacques Lévy, *Confluence properties of weak and strong calculi of explicit substitutions*, ***Journal of the ACM***, vol. 43 (1996), no. 2, pp. 362–397.

[19] N. G. de Bruijn, *Lambda-calculus notation with nameless dummies: a tool for automatic formula manipulation with application to the Church-Rosser theorem*, ***Indag. Math.***, vol. 34 (1972), no. 5, pp. 381–392.

[20] Solomon Feferman, *A language and axioms for explicit mathematics*, ***Algebra and logic,*** lecture notes in mathematics, vol. 480 (J. N. Crossley, editor), Springer, Berlin, 1975, pp. 87–139.

[21] ———, *Inductively presented systems and the formalization of metamathematics*, ***Logic colloquium 80*** (D. van Dalen, D. Lascar, and J. Smiley, editors), North-Holland, 1982, pp. 95–128.

[22] ———, *Polymorphic typed lambda-calculi in a type free axiomatic framework*, ***Contemporary Mathematics***, vol. 106 (1990), pp. 101–135.

[23] Robert Harper and Chris Stone, *A type-theoretic interpretation of Standard ML*, ***Proof, language and interaction: Essays in honour of robin milner***, 2000.

[24] Jason Hickey and Aleksey Nogin, *Fast tactic-based theorem proving*, ***Proceedings of the 13th international conference on theorem proving in higher order logics***, 2000, pp. 252–266.

[25] W. Howard, *The formulas-as-types notion of construction*, ***To H.B. Curry: Essays on combinatory logic, lambda-calculus and formalism*** (J. P. Seldin and J. R. Hindley, editors), Academic Press, 1980, pp. 479–490.

[26] Gérard Huet, Gilles Kahn, and Christine Paulin-Mohring, *The coq proof assistant : A tutorial : Version 6.1*, ***Technical report***, INRIA-Rocquencourt, CNRS and ENS Lyon, August 1997.

[27] N. Immerman, *Languages which capture complexity classes*, ***SIAM Journal of Computing***, vol. 16 (1987), pp. 760–778.

[28] Daniel Leivant, *Stratified functional programs and computational complexity*, ***Twentieth annual acm sigplan-sigact symposium on principles of programming languages*** (Charleston, SC), ACM, ACM Press, January 1993, pp. 325–333.

[29] Xavier Leroy, *Unboxed objects and polymorphic typing*, ***Nineteenth ACM SIGACT-SIGPLAN symposium on principles of programming languages***, 1992, pp. 177–188.

[30] Lena Magnusson, *The implementation of alf—a proof editor based on martin-löf's monomorphic type theory with explicit substitution*, ***Ph.D. thesis***, Göteborg University, Göteborg Sweden, 1994.

[31] Per Martin-Löf, *An intuitionistic theory of types: Predicative part*, ***Proceedings of the logic colloquium, 1973***, Studies in Logic and the Foundations of Mathematics, vol. 80, North-Holland, 1975, pp. 73–118.

[32] Per Martin-Löf, *Constructive mathematics and computer programming*, ***Proceedings of the sixth international congress of logic, methodology and philosophy of science***, Studies in Logic and the Foundations of Mathematics, vol. 104, North-Holland, 1982, pp. 153–175.

[33] K. Mehlhorn, *Polynomial and abstract subrecursive classes*, ***Journal of Computer and System Sciences***, (1976), pp. 148–176.

[34] Eugenio Moggi, *Notions of computation and monads*, ***Information and Computation***, vol. 93 (1991).

[35] ———, *A general semantics for evaluation logic*, ***Proceedings of ninth annual ieee symposium on logic in computer science***, July 1994, pp. 353–362.

[36] Greg Morrisett, *Compiling with types*, ***Ph.D. thesis***, Carnegie Mellon University, School of Computer Science, Pittsburgh, Pennsylvania, December 1995.

[37] Greg Morrisett, Matthias Felleisen, and Robert Harper, *Abstract models of memory management*, ***1995 conference on functional programming languages and computer architecture***, 1995.

[38] Greg Morrisett and Robert Harper, *Semantics of memory management for polymorphic languages*, ***Technical Report CMU-CS-96-176***, Carnegie Mellon University, School of Computer Science, September 1996.

[39] Greg Morrisett, David Walker, Karl Crary, and Neal Glew, *From System F to typed assembly language*, ***Twenty-fifth ACM SIGACT-SIGPLAN symposium on principles of programming languages*** (San Diego), January 1998, To appear.

[40] R. Parikh, *Existence and feasibility in arithmetic*, ***Jour. Assoc. Symbolic Logic***, vol. 36 (1971), pp. 494–508.

[41] Christine Paulin-Mohring and Benjamin Werner, *Synthesis of ml programs in the system coq*, ***Journal of Symbolic Computations***, vol. 15 (1993), pp. 607–640.

[42] Simon L. Peyton Jones, Cordelia V. Hall, Kevin Hammond, Will Partain, and Philip Wadler, *The Glasgow Haskell compiler: a technical overview*, ***Proc. uk joint framework for information technology (jfit) technical conference***, July 1993.

[43] A. Seth, *There is no recursive axiomatization for feasible functionals of type 2*, ***Proceedings of the 7th annual ieee symposium on logic in computer science***, 1992.

[44] A. Seth, *Some desirable conditions for feasible functionals of type 2*, ***Eighth IEEE symposium of logic in computer science*** (Montreal), June 1993, pp. 320–331.

[45] ———, *Turing machine characterizations of feasible functionals of all finite types*, ***Proceedings of MSI workshop on feasible mathematics***, 1994.

[46] N. Shankar, S. Owre, and J. M. Rushby, *The PVS proof checker: A reference manual*, Computer Science Laboratory, SRI International, Menlo Park, CA, February 1993.

[47] Zhong Shao and Andrew Appel, *A type-based compiler for Standard ML*, ***1996 ACM SIGPLAN conference on programming language design and implementation*** (La Jolla), June 1995, pp. 116–129.

[48] Scott F. Smith, *Hybrid partial-total type theory*, ***International Journal of Foundations of Computer Science***, vol. 6 (1995), pp. 235–263.

[49] Scott Fraser Smith, *Partial objects in type theory*, ***Ph.D. thesis***, Department of Computer Science, Cornell University, Ithaca, New York, January 1989.

[50] D. Tarditi, G. Morrisett, P. Cheng, C. Stone, R. Harper, and P. Lee, *TIL: A type-directed optimizing compiler for ML*, ***1996 ACM SIGPLAN conference on programming language design and implementation***, May 1996, pp. 181–192.

[51] K. Weihrauch and Ch. Kreitz, *Type 2 computational complexity of functions on Cantor's space*, ***Theoretical Computer Science***, vol. 82 (1991), pp. 1–18.

DEPARTMENT OF COMPUTER SCIENCE
CORNELL UNIVERSITY
ITHACA, NY 14853, USA
E-mail: rc@cs.cornell.edu

SCHOOL OF COMPUTER SCIENCE
CARNEGIE MELLON UNIVERSITY
5000 FORBES AVENUE
PITTSBURGH, PA 15213, USA

COMPUTABILITY THEORY: STRUCTURE OR ALGORITHMS

JENS ERIK FENSTAD

To Sol Feferman for his 70th birthday

In the paper *Turing's "oracle": From absolute to relative computability - and back* S. Feferman discusses some major trends in recent work on degrees of unsolvability, theories of recursive functionals and generalized recursion theory and notes how "marching in under the banner of degree theory, these strands were to some extent woven together by the recursion theorists, but the trend has been to pull the subject of effective computability even farther away from questions of actual computation." (Feferman [1992a])

Feferman had expressed similar concerns in a paper from 1977 *Inductive schemata and recursively continuous functionals* (Feferman [1977]), and both he and Y. Moschovakis have since pursued a program of how to use "abstract recursion theory as a foundation for a theory of algorithms" (to quote the title of a paper by Moschovakis from 1984). Some key references in the development of this program are Feferman [1991], [1992b], [1996], and Moschovakis [1984], [1989], [1997]. We particularily recommend Feferman [1992b] *A new approach to abstract data types, I, Informal development*, for a clear conceptual presentation of the issues involved.

I had commented on these issues in the introductory part of the book *General Recursion Theory*, (Fenstad [1980]). The Oslo group had in the 1970's been active in many parts of generalized recursion theory, such as computation in higher types, inductive definability, continuous functionals, set recursion and other relationships between set theory and recursion theory. We attempted a unified analysis through the axiomatic notion of a *computation theory*. Our logic group shared the common "world view" of the recursion theory community. We were "inwards bound", concentrating above all on *structure* and *generality*.

It would, however, be wrong to claim that recursion theory in the late 1970's was all structure and that it paid no attention to questions of algorithmic content. But it was true that our brand of *generalized recursion theory* was an activity almost exclusively pursued by the logic community. We were running the risk, and this may even be the subtext of Feferman's remarks in 1992, that the inward bound travel toward ever more intricate structural concerns would lead to an academic profession largely irrelevant to computational praxis. In part I of this lecture I will look back – perhaps, with some nostalgia – to *the high times of generalized*

Reflections on the foundations of mathematics
Essays in honor of Solomon Feferman
Edited by Wilfried Sieg, Richard Sommer, and Carolyn Talcott

recursion theory in the late 1970's. I do this as a necessary background for the discussion in part II. In that part I will sketch some developments toward *theories for actual computational praxis*. It is in this direction that I today see the major challenges for computability theory. As the paper unfolds the reader will – I hope – be convinced of the correctness of this claim.

Remark. The tension between structure and algorithms is not an issue of recursion theory alone. In a recent lecture *Mathematical Problems for the next Century* S. Smale observes that "a reason that the Poincaré's conjecture is fundamental in the history of mathematics is that it helped give focus to a manifold as an object of study". Recall that the Poincaré's conjecture is precisely a conjecture about structure. Smale now sees a different trend: "I hold the conviction that there is a comparable phenomenon today in the notion of a "polynomial-time algorithm". Algorithms are becoming worthy of analysis in their own right, not merely as a means to solve other problems. Thus I am suggesting that as the study of the set of solutions of an equation (e.g., a manifold) played such an important role in 20th-century mathematics, the study of finding the solutions (e.g., an algorithm) may play an equally important role in the next century." (Smale [1998])

Part I: The high times of generalized recursion theory

The 1970's was a particularly good time for generalized recursion theory. The community was small and well integrated, it met frequently, and its progress was recorded through a series of conference proceedings, in particular, Manchester '69 (Gandy and Yates [1971]), Oslo '72 (Fenstad and Hinman [1974]), Oxford '76 (Gandy and Hyland [1977]), Oslo '77 (Fenstad, Gandy and Sacks [1978]), and Leeds '79 (Drake and Wainer [1980]). These were the times – to quote Feferman [1992a] – when "strands were woven together".

I shall comment on the several sources of computability theory below; see, in particular, the introduction to part II. But for "our" community the roots were clear. One knew the classics, Turing, Gödel, Kleene, Church, and Post (see the historical analysis in Kleene [1981], Gandy [1988] and Soare [1996]), but the "modern" development started with the great series of papers by S. C. Kleene in the mid-1950's. A particularily important event was the Summer Institute for Symbolic Logic at Cornell 1957, which led to new steps forwards by Spector, Kreisel, Sacks, Kripke, and Platek in the 1960's. If we add to this the highly original work of R. O. Gandy on recursive functionals of finite types (Gandy [1967]), we are well prepared for the Logic Colloquium at Manchester '69 and the progress of *generalized recursion theory* in the 1970's.

I mean this in a literal sense. The review of structure will take as its starting point the paper *Axioms for computation theories* by Y. Moschovakis, and the review of algorithms will start from the paper *Algorithmic procedures* by H. Friedman, both to be found in the proceedings of Manchester '69 (Gandy and Yates [1971]).

I.1 Axioms for caxioms omputation theories. Choosing the proper axioms is no trivial matter. General recursion theory could be - and has been - developed from many points of view. One could choose schemes and equations as a starting point for the general development, or one could abstract from the basic recursion theorems and choose an approach through inductive definability and fixed-point operators. Yet another possibility would be to base the development on some generalized model of Turing-machines, register machines, etc.

Common to all these approaches is the idea of a *computing device*, M, which when presented with an *input* sequence, $\sigma = (x_1, \ldots, x_n)$, may give an *output* z; in symbols, $M(\sigma) \simeq z$. Also common to all these approaches are various indexing or normal form theorems. In this way a "computing device" M could be replaced by its *code* or *gödel number* as possible inputs to further computations. The self-reference made possible by coding lies at the core of *undecidability results*. Coding also alows us to call on computing mechanisms in a uniform manner, thus is of importance in developing theories of *relative computability*.

When extending the classical theory to a theory of recursive functionals of finite types, Kleene [1959] adopted a coding scheme from the outset. His basic relation was

$$\{a\}(\sigma) \simeq z$$

asserting that the computing device coded or named by a and acting on the input sequence $\sigma = (x_1, \ldots, x_n)$ gives z as output. This approach was carried further by Moschovakis in his theories of prime and search computability (Moschovakis [1969]). The next step was *Axioms for computation theories*, Moschovakis [1971].

Remark. There were also other routes to axioms for recursion theory. We mention, in particular, Strong [1968], Wagner [1969], and Friedman [1971a].

The axiomatic approach starts by abstracting from the *computational process* $\{a\}(\sigma) \simeq z$ and introduces the *structure* $\Theta = \{(a, \sigma, z) \ : \ \{a\}(\sigma) \simeq z\}$ of computation tuples. Thus a computation domain is a structure $\mathcal{A} = \langle A, C\colon 0, 1\rangle$, where $C \subseteq A$ is the set of *codes* and 0 and 1 are designated elements of C. A *computation set* Θ over $\mathcal{A}$ is a set of tuples (a, σ, z), where $a \in C$, $\sigma = (x_1, \ldots, x_n)$ is an n-tuple from A, and $z \in A$. The length of (a, σ, z) is ≥ 2, i.e., we must have a code and an output, but not necessarily an input.

It is not too difficult to write down axioms which guarantee that Θ behaves as a "reasonable" computation theory over the domain $\mathcal{A}$. The model is the schemes of Kleene [1959], as further developed in the prime and search computability theories of Moschovakis [1969]. This is now well-known and can be found in Fenstad [1980].

Within this framework one can prove some of the basic "combinatorial" theorems of general recursion theory. But, as the following representation theorem shows, the framework is still not rich enough to serve as a general analysis. Let $\boldsymbol{L} = \boldsymbol{f}, \boldsymbol{F}$ be a list of partial functions and partial consistent functionals over a domain $\mathcal{A}$. Following Kleene [1959] and Moschovakis [1969] we can construct a computation theory PR[$\boldsymbol{L}$] on $\mathcal{A}$, called *the prime computation theory* generated by the finite list $\boldsymbol{L}$. It is now a simple fact that *given any theory* Θ *on* $\mathcal{A}$ *we can find a partial function* f *such that* Θ *is equivalent to* PR$[f]$, i.e. all information about Θ is coded into the "oracle" f. This is nice to know, but it is not a general analysis of computability, applicable e.g., to an analysis of hyperarithmetic theory (or how to compute in the functional ${}^2\boldsymbol{E}$).

What is missing in the above analysis is an understanding of the subcomputation structure of a give computation in Θ. An understanding of the more advanced part of recursion theory requires therefore an analysis not only of a computation tuple, but of the *whole structure of subcomputations* of a give computatin tuple. The full notion of a *computation theory* Θ is thus obtained by adding a new well-founded relation

$$(a', \sigma', z') < (a, \sigma, z),$$

as the *subcomputation relation* on the set Θ.

Remark. In a computation structure $\langle \Theta, < \rangle$ each computation tuple (a, σ, z) has an associated length $|a, \sigma, z|_\Theta$, which is a finite or infinite ordinal; this was the notion chosen by Moschovakis [1971].

Generalizing recursion theory is a game that can be played for several reasons. From our current point of view the most important pay-off is the *proper understanding of the structural content of the basic notions* such as computation/subcomputation, finiteness and selection (or search). Let *finiteness* be our prime example. We know from many areas of mathematics how important it is to find the "correct" extension of finiteness/compactness. In e.g., non-standard analysis saturation allows us to extend these notions to the concepts of hyperfinte and near-standard. In the computation-theoretic framework a subset B of the domain A is called (strongly) Θ-*finite* if the functional $\boldsymbol{E}_B$ is Θ-computable, where

$$\boldsymbol{E}_B \simeq \begin{cases} 0, & \text{if } \exists x \in B \ (f(x) \simeq 0) \\ 1, & \text{if } \forall x \in B \ (f(x) \simeq 1) \end{cases}$$

The importance of choosing the "correct" notion of finiteness in generalizing recursion theory was forcefully argued by G. Kreisel in the 1950's and 1960's; see e.g., his paper in the Manchester '69 volume, *Some reasons for generalizing recursion theory* (Kreisel [1971]). Finiteness is needed in order to lift the "hard" or combinatorial arguments of the classical theory to the generalized context. Kreisel convincingly vindicated his analysis through the example of *metarecursion theory* (Kreisel & Sacks [1965]). How this later developed into α-recursion theory, set recursion and other generalizations is by now well documented in the literature; see Sacks [1990].

We shall mention only one example of the role of finiteness in the generalized context of computation theories. It is a simple, but fundamental, insight that the set of subcomputations of a given computation tuple in Θ is "finite" in the above sense. This allows us to replace the "trivializing" representation theorem quoted above by the following *representation theorem*, which is *faithful to the notion of subcomputation*. Call a theory Θ s-*normal* if the sets $S_x = \{(a, \sigma, z) \in$

Θ; $(a, \sigma, z) < x\}$ are uniformly Θ-finite for $x \in \Theta$. Then *any s-normal Θ can be represented in the form* PR$[F]$, for some partial Θ-computable functional F.

I.2. Computations in higher types. We shall follow a few steps of the development of the axiomatic approach, chiefly to be able to exhibit some *structural concerns* which dominated the recursion theory of the 1970's. In the previous section we emphasised the notion of s-normality, i.e. the uniform finiteness of the set of subcomputations. For the general theory a related (indeed, almost equivalent) notion is more apt. Let $\langle \Theta, < \rangle$ be a computation theory. The following *prewellordering property* emerged as a key ingredient in the analysis of Gandy [1967], e.g., in his proof of *the selection principle* for recursion in a normal type-2 functional: Θ is called *p-normal* if there is a Θ-computable function p such that $p(x, y)\downarrow$ if $x \in \Theta$ or $y \in \Theta$, and

$$\begin{array}{ll} x \in \Theta \text{ and } |x|_\Theta \le |y|_\Theta & \text{implies } p(x,y) = 0 \\ |x|_\Theta > |y|_\Theta & \text{implies } p(x,y) = 1. \end{array}$$

The Gandy selection theorem states that a p-normal theory Θ always admits selection operators over $\mathbb{N}$, i.e. there is a *Θ-computable selection operator $q(a, \sigma)$* such that if $\exists n \in \mathbb{N}\ \{a\}(n, \sigma) \simeq 0$, then $q(a, \sigma)\downarrow$ and

$$\{a\}(q(a, \sigma), \sigma) \simeq 0.$$

The existence of a selection operator over the domain A implies a well-behaved theory of Θ-semicomputable sets and relations. We can prove that the Θ-semicomputable relations are closed under $\exists$-quantification and that a set $B \subseteq A$ is Θ-computable iff both B and $A - B$ are Θ-semicomputable. These insights led eventually to the notion of a Spector theory. A computation theory $\langle \Theta, < \rangle$ on a domain $\mathcal{A}$ is called a *Spector theory* if (i) the code set C is all of A; (ii) the domain A is Θ-finite; and (iii) Θ is p-normal.

Spector theories unify a number of results about the hyperarithmetic sets, recursion in a normal type-2 functional (i.e. a functional extending ${}^2\boldsymbol{E}$), inductive definability, ordinal recursion theory, and the constructible hierarchy. This is, indeed, a success story for the structuralist point of view, and it has been told in many places (for book expositions see Fenstad [1980] and Sacks [1990]).

But how to proceed? From the structural point of view higher types presented *a* (not necessarily *the*) next challenge. Kleene [1959] had laid the general framework, Gandy [1967] had obtained great success with type-2, now higher types was the next domain to explore. We highlight a few points of conceptual importance.

One insight which slowly emerged, is that Kleene recursion in higher types is really a *computation theory on two types* (see, in particular, Moldestad [1977]). Let the domain A have the structure $A = S \cup Tp(S)$, where $Tp(S) = \mathbb{N}^S$. Let $\boldsymbol{L}$ be a list $R_1, \ldots, R_k, f_1, \ldots, f_l, F_1, \ldots, F_m$, where each R_i is a relation on A, each f_i a total function on A with values in S, and each F_i a total functional acting on total functions with values in S. In the standard Kleene [1959] manner we can to each such list $\boldsymbol{L}$ construct a computation theory PR$[\boldsymbol{L}]$ on $A = S \cup T_p(S)$ which generalizes Kleene recursion in higher types.

In higher types we need to be a bit more careful with the notion of finiteness. Call the old notion *strong Θ-finiteness*; a set $B \subseteq A$ is called *weakly Θ-finite* if the following funtional is Θ-computable

$$E'_B(f) \simeq \begin{cases} 0, & \text{if } \forall x \in B\ f(x)\downarrow\ \wedge \exists x \in B\ f(x) \simeq 0 \\ 1, & \text{if } \forall x \in B\ \exists y \neq 0\ f(x) \simeq y \end{cases}$$

A list $\boldsymbol{L}$ is called *normal* if $\boldsymbol{E}'_A$ is a PR[$\boldsymbol{L}$]-computable functional and equality on S is a PR[$\boldsymbol{L}$]-computable relation. Thus the proto-type for PR[$\boldsymbol{L}$], where $\boldsymbol{L}$ is normal, is Kleene recursion in ${}^3\boldsymbol{E}$ over the integers and the reals.

Is it now possible to characterize Kleene recursion PR[$\boldsymbol{L}$] in axiomatic terms? The clue is the following characterization of Spector theories: *Let Θ be a Spector theory on* $\mathbb{N}$, *then Θ is equivalent to* PR[G] *for a normal, total type-2 functional* G *iff Θ is not Θ-Mahlo.* The notion of Θ-Mahlo is, in a straight forward way, the computation theoretic equivalent to the standard ordinal theoretic one; see Moldestad [1977].

In higher types the notion of Spector theory will be replaced by the following notion: A computation theory Θ on $A = S \cup Tp(S)$ is called *normal* if (i) the equality relation on S is Θ-computable; (ii) A is weakly and S is strongly Θ-finite; and (iii) Θ is p-normal. With this definition we can prove the following theorem due to Moldestad [1977]: *Let Θ be a normal theory on* $A = S \cup Tp(S)$, *then Θ is equivalent to* PR[$\boldsymbol{L}$], *for a normal list* $\boldsymbol{L}$ *iff Θ is not Θ-Mahlo.*

The axiomatic approach sketched so far is not the only way to attack Kleene recursion in higher types. Building on previous work on primitive and rudimentary set functions D. Norman [1978] introduced the full notion of *set recursion* as a way of studying Kleene recursion. This theory proved to be a productive frame-work for extending degree theory and priority arguments to functionals of higher types. The topic was much pursued in the 1980's; a full account of these developments is presented in Sacks [1990].

Recursion theory in higher types is strong on structure and difficult results, but is weaker on applications and links to other parts of mathematics. Recursion in ${}^2\boldsymbol{E}$, the proto-type of a Spector theory, lies at the cross-section of many mathematical theories. This is not so with recursion in ${}^3\boldsymbol{E}$ and with many other extended or generalized recursion theories.

But we should not end this part of our survey on a too critical note. As remarked above the study of recursion in higher types has given insight into phenomena of potential usefulness in other context, e.g., in the development of computation theories for numerical analysis selection principles similar to those useed in higher types may prove useful; see section II.4.

The Gandy selection theorem states that a p-normal theory Θ admits selection operators over $\mathbb{N}$. This is not true in higher types. One could try to remedy this situation by introducing some kind of multiple-valued search or selection operator, which, when presented with some non-empty semicomputable set, will return a sufficiently well-behaved non-empty subset of the given set. Analyzing the situation in more details one recognizes that it is possible to dispense with multiple-valuedness. A Θ-semicomputable subset B is given by an index, i.e. $x \in B$ iff $\{a\}(x, \sigma) \simeq 0$, for some code a and parameter sequence σ. A satisfactory *selection principle* should allow us *to compute a code for a sufficiently well-behaved non-empty subset* $B_0 \subseteq B$. "To compute" means that there should be a Θ-computable function $q(a, \sigma)$ such that if $B \neq \emptyset$, then $q(a, \sigma) \downarrow$ and is the code for some non-empty Θ-computable subset $B_0 \subseteq B$. If the domain is strongly Θ-finite, the set B_0 will even be strongly Θ-finite and thus certainly "well-behaved". This is the case with the subindividuals S in a normal theory over two types $A = S \cup Tp(S)$. In this case we have selection - not at the *extensional level* in the sense of picking an element of the given set – but at the *intensional level* in the sense of computing a code or description of a selection set.

The mathematics behind the analysis of selection principles in higher types is far from trivial and is due to the combined efforts of Grilliot, MacQueen, Harrington, Moldestad and Norman. The story is well told in Sacks [1990].

I.3. Finite algorithmic procedures. We now shift attention from *structure* to *computation*. As mentioned above our starting point in this review is the study by H. Friedman [1971b] of *computations over arbitrary algebraic structures*. Friedman based his work on the register machine model of computations introduced by Shepherdson and Sturgis [1963]. Shepherdson, in turn, has later analyzed and extended the work of Friedman; see, in particular, Shepherdson [1975], [1988]. We shall base our account on Moldestad, Stoltenberg-Hansen and Tucker [1980a, b]; see also Tucker and Zucker [1998].

In a computation theory we abstract from the computation seen as a sequence of computation steps and concentrate entirely on the *structure of the completed set of computation tuples*. In the register machine model, as well as in the more general notion of a finite algorithmic procedure as studied by Friedman, the *algorithm remains in focus*.

A *finite algebraic procedure* (fap) P is defined over an algebraic structure $\mathcal{A} = \langle A, \sigma_1, \dots, \sigma_n, R_1, \dots, R_m \rangle$. The basic structure of a register machine is kept. There is a number of registers $r_1, \dots, r_n, \dots$, each able to carry one element from A. The machine can compute the operations of $\mathcal{A}$ and decide the relations of $\mathcal{A}$. It can also relocate data among the registers. Computations of this kind is described by a finite set of instructions $I_1, \dots, I_k$ of the form

i. $r_\mu := \sigma(r_{\lambda_1}, \dots, r_{\lambda_l})$
ii. $r_\mu := r_\lambda$
iii. if $R(r_{\mu_1}, \dots, r_{\mu_l})$ then i else j
iv. if $r_\mu = r_\lambda$ then i else j.

Formally we should define an operational semantics for a program $P = I_1, \dots, I_k$ over $\mathcal{A}$. We shall, however, remain at the program level (but see section II.2 below for a more careful exposition of the general theory). We assume familiarity with the register machine model, hence, assume that the reader has the "correct" understanding of the instruction list above; e.g., an instruction of type iii is intended to express that if the result of testing the relation R on the content of registers $r_{\mu_1}, \ldots, r_{\mu_l}$ is yes, proceed to instruction I_i. else go to instruction I_j.

Any program P involves a finite list of registers $r_0, r_1, \ldots, r_{n-1}$, where some of the registers $r_1, \ldots, r_m$ are called the input registers. r_0 is the output register, and all the remaining registers $r_{m+1}, \ldots, r_{n-1}$ are called the working registers. A partial function $f : A^m \to A$ is called *fap-computable* over the algebra $\mathcal{A}$ if there exits a fap P over $\mathcal{A}$ such that whenever $a_1, \ldots, a_m$ are loaded into $r_1, \ldots, r_m$, respectively, and P applied, then $f(a_1, \ldots, a_m) = a$ iff P halts and the content of r_0 is a. We denote by FAP($\mathcal{A}$) *the class of* fap-*computable functions over* $\mathcal{A}$.

Two natural extensions of fap-computability have been studied. Already in his Manchester paper Friedman introduced the notion of fap-*computability with counting* (fap C). In this case we add to the *algebra registers* $r_0, r_1, \ldots$ a number of *counting registers* $c_0, c_1, \ldots$ which each can hold a single natural number. The machine can put 0 into a counting register, it can add or subtract one from a counting register, and it can test when two counting registers contain the same number.

In this case we can have either algebra-valued functions $f : \mathbb{N}^n \times A^m \to A$ or number-valued functions $f : \mathbb{N}^n \times A^m \to \mathbb{N}$. This is useful in an algebraic context, e.g., the order of an element in a group is, if defined, a natural number. Adding counting registers introduces an extended class of fap C-*computable* functions over $\mathcal{A}$, which we denote by FAPC($\mathcal{A}$); see Friedman [1971b].

The second extension adds a *stacking mechanism* to the notion of general fap-computability. This extension was introduced by Moldestad, Stoltenberg-Hansen and Tucker [1980 a]. It adds a new stack register and new instructions

: $s := (i, r_0, \ldots, r_m)$
: restore $(r_0, \ldots, r_{j-1}, r_{j+1}, \ldots, r_m)$
: if $s =$ empty then i else j.

The meaning of these instructions is as follows. The first one instructs the machine to store a copy of the content of all the registers on top of the stack together with instruction labeled i. The restore-instruction replaces the content of the registers $r_1, \ldots, r_{j-1}, r_{j+1}, \ldots, r_m$ by the values stored at the top of the stack. The new instructions allow us to interrupt a computation by placing the content of the registers in the stack and to start a subcomputation within the main computation by reloading some (but not necessarily all) of the registers $r_0, \ldots, r_{n-1}$. In a program P a stacking instruction will be followed by an instruction telling the machins how to start the subcomputation. If the subcomputation is successful, the result is loaded into register r_j and the content of the other registers are restored. The machine then proceeds with the main computation. Stacking can be iterated during a computation, it is therefore a device which allows unbounded algebraic memory over the structure $\mathcal{A}$. In a way similar to the introduction of the classes FAP($\mathcal{A}$) and FAPC($\mathcal{A}$), we get the class of fap S-*computable* functions, which we denote by FAPS($\mathcal{A}$).

The constructions fap C and fap S can be combined to a notion of fap CS-computability. We have the following set of inclusions

$$\begin{array}{ccccc} & \nearrow & \mathrm{FAPC}(\mathcal{A}) & \searrow & \\ \mathrm{FAP}(\mathcal{A}) & & & & \mathrm{FAPCS}(\mathcal{A}) \\ & \searrow & \mathrm{FAPS}(\mathcal{A}) & \nearrow & \end{array}$$

There existe algebras $\mathcal{A}$ where all inclusions are strict.

One noteworthy result in Moldestad, Stoltenberg-Hansen and Tucker [1980a] is the connection between stacking and inductive definability. The class $\mathrm{Ind}(\mathcal{A})$ of inductively defined functions over a structure $\mathcal{A}$ can be introduced in various, but related, ways. Moldestad [1977] followed an approach based on the equational calculus of Platek [1966]. The basic result is the following characterization theorem

$$\mathrm{Ind}(\mathcal{A}) = \mathrm{FAPS}(\mathcal{A})$$

proved by Moldestad, Stoltenberg-Hansen and Tucker. We shall briefly return to this result in connection with the multi-sorted structures introduced in part II. The notion of fap-computability was used by Tucker [1980] to study comptuability in algebraic systems. Again, this is a topic to which we will return in part II.

I.4. Algorithms and structure. The relationship between fap-computability and computation theories was investigated by Moldestad, Stoltenberg-Hansen and Tucker [1980b]. Let the algebraic structure $\mathcal{A}$ be augmented by the natural numbers, i.e. expand the structure $\mathcal{A} = \langle A, \sigma, \boldsymbol{R}\rangle$ to a structure

$$\mathcal{A}^N = \langle A \cup \mathbb{N}, \sigma, \boldsymbol{R}, s, p, 0\rangle,$$

where $s, p, 0$ are the successor function, the predecessor function, and the constant 0, respectively. It is rather immediate that $f \in \mathrm{FAP}(\mathcal{A}^N)$ iff f is fap C-computable over $\mathcal{A}$, and that $f \in \mathrm{FAPS}(\mathcal{A}^N)$ iff f is fap CS-computable over $\mathcal{A}$. When is $\mathrm{FAP}(\mathcal{A}^N)$ a computation theory?

The answer depends upon the exact properties of the term evaluation function. Over any algebra $\mathcal{A}$ we can define the syntactic notion of term. Each term t (in n indeterminates $x_1, \ldots, x_n$) defines a functin from A^n to A by substitution of algebra elements for the indeterminates. Terms in n indeterminates can be uniformly and recursively coded by numbers, i.e. there is a recursive subset $\Omega \subseteq \mathbb{N}$ such that for each $i \in \Omega$ there is a unique term $[i]$, and the correspondence is a surjection between terms and Ω. Let $E_n : \Omega \times A^n \to A$ be the *term evaluation function* defined by

$$E_n(i, a_1, \ldots, a_n) = [i](a_1, \ldots, a_n).$$

Then $\mathrm{FAP}(\mathcal{A}^N)$ is a computation theory iff E_n is uniformly in n fap C-computable (Moldestad, Stoltenberg-Hansen and Tucker [1980b]).

Adding stacking makes E_n uniformly fap CS-computable. Thus $\mathrm{FAPS}(\mathcal{A}^N)$ *is the minimal computation theory over* $(\mathcal{A}^N)$ (with code set $\mathbb{N}$).

This was the perspective which was adopted in Oslo in the late 1970's. You may look on computability from different point of view. If *structural questions* are your main concerns, there are arguments in favor of an *axiomatic approach*. This was our approach to Kleene recursion in higher types via the axiomatics of computation theories on two types. There were equally valid approaches through theories of inductive definability and set recursion. The latter was particularily well suited to study the degree structure in higher types; see Sacks [1990]. But from a purely structural point of view they were all computation theories.

If *algorithmic questions* were your main concern, various versions of *fap-computability* might be preferable. This was noted above in connection with the study of *computability in algebraic structures* (Tucker [1980]).

But whether structure or algorithms, this was an activity of a "core" recursion theory community. This core community was never all of the computational community, nor did it have a copyright on computational praxis.

Part II. Towards theories for computational praxis

If abstract computation theory is to meet actual computations, much care must be taken in choosing the proper structure for the theory. This has been a long standing concern of the computer science community; see e.g., Wirsing [1990]. This community has learned through long experience that "the proper treatment and choice of the basic data structures is an important and complex part in the process of program construction", because it is in the early phase of the program constructions that "decisions are made that crucially influence the subsequent development steps and even determine to a considerable extent the structure of the final program"; Wirsing [1990].

We shall not follow the long and complicated story how the "proper" notions emerged. And it took a long time for the different parts of the computability community to interact. As a personal experience I remember that when the computer scientists in Oslo, still a part of the Mathematics institute, were engaged in developing the *class concept* in the SIMULA programming language (see Dahl, Myrhaug and Nygaard [1970]) as a first instance of data abstraction, the logicians in the same department were busy writing *axioms for higher recursion theory*. – Two communities, both concerned with computations, but living intellectual worlds apart.

The algebraic approach to data structures, to some extent stimulated by the development of SIMULA, emerged within the computer science community around 1974/75. From our point of view a particularily important paper was Gougen, Thatcher, Wagner and Wright [1975] which presented "the first rigorous formalization of abstract data types within standard many-sorted algebra (Wirsing [1990])".

Looking back to part I we remember how many-sorted structures naturally emerged. In the study of theories on two types we worked with two domains, S and $Tp(S)$; in fap-computability we considered structures defined both over an algebra domain A and the natural numbers $\mathbb{N}$. It is an easy step to generalize to any finite set of sorts – a generalization which meets the needs of actual computational praxis.

The next question is now, which computational structure is the "correct" one for multi-sorted domains? The answer is not necessarily unique; a survey of possibilities is given in Tucker and Zucker [1998]. In Moschovakis [1984] the approach is through an extended theory of inductive definability and fixed-point operators. Similar approaches have been pursued by Feferman [1992b].

We shall follow the algebraic approach of Tucker and Zucker [1998], using a suitable version of While-computability for many-sorted structures. In many ways this is a safe choice. The While-language is a well-known tool of the computer trade. It is easy to use for writing actual alogorithms. And it allows a simple and direct development of the general theory.

II.1. Computations in many-sorted algebras. We need briefly to recall the basic definitions and constructions from the theory of many-sorted algebras; for full details see the handbook chapter by Wirsing 1990 and the survey lecture by Tucker and Zucker [1998].

A *signature* Σ for a many-sorted algebra consists of two finite lists of *sorts* and *function symbols*. With every function symbol $\mathrm{F} \in \mathbf{Func}(\Sigma)$ there is associated a *type* $s_1 \times \cdots \times s_m \to s$, where $s_1, \ldots, s_m, s \in \mathbf{Sort}(\Sigma)$. A Σ-algebra A has for each sort s a non-empty domain or *carrier set* A_s and for each function symbol F of type $s_1 \times \cdots \times s_m \to s$ a *function* $\mathrm{F}^A : A_{s_1} \times \cdots \times A_{s_m} \to A_s$. The algebra is *total* if all F^A are total functions; otherwise A is called *partial*.

In most of our examples the algebras will be built up from certain basic algebras such as the *naturals* $\boldsymbol{N}_0 = \langle \mathbb{N}; 0, \mathrm{suc} \rangle$, the ring of *reals* $\boldsymbol{R}_0 = \langle \mathbb{R}; 0, 1, +, -, \times \rangle$, or the *booleans* $\boldsymbol{B} = \langle \mathbb{B}; \mathrm{tt}, \mathrm{ff}, \mathrm{and}, \mathrm{or}, \mathrm{not} \rangle$. To be absolutely correct we should exhibit an algebra both by the signature and the algebra structure; e.g., the signature of the booleans can be displayed in the following format

signature	$\Sigma(\boldsymbol{B})$
sorts	bool
function symbols	true, false : $\to$ bool
	and, or : bool $\times$ bool $\to$ bool
	not : bool $\to$ bool
end	

Thus booleans have one sort bool and five function symbols (note that constants are specified by function symbols of an empty argument list). The corresponding algebra $\boldsymbol{B}$ has the carrier $\mathbb{B} = \{\mathrm{tt}, \mathrm{ff}\}$ of sort bool and as constants and functions the standard boolean ones, e.g., $\mathrm{true}^{\boldsymbol{B}} = \mathrm{tt}$ and $\mathrm{false}^{\boldsymbol{B}} = \mathrm{ff}$. It would for our purposes be tedious to present each example in a similar correct format. In most cases it suffices to exhibit the algebra structure, e.g., the algebra $\boldsymbol{N}_0$ of naturals has one sort nat and two function symbols, $0 : \to$ nat and suc : nat $\times$ nat.

algebra	N_0
carriers	$\mathbb{N}$
functions	$0 : \to \mathbb{N}$
	$\text{suc} : \mathbb{N} \to \mathbb{N}$
end	

The carrier $\mathbb{N}$ is of sort nat; note that we often use the same symbol for functions and function symbols.

A signature Σ is called *standard* if it includes $\Sigma(\boldsymbol{B})$, a *discriminator*

$$\text{if}_s : \text{bool} \times s^2 \to s$$

for all sorts s of Σ other than bool, and an *equality operator*

$$\text{eq}_s : s^2 \to \text{bool}$$

for certain sorts s, called *equality sorts*. A Σ-algebra A, for standard Σ, is called a *standard algebra* if it is an expansion of $\boldsymbol{B}$ and the discriminators and equality operators have their standard interpretation, e.g., for $b \in \mathbb{B}$ and $x, y \in A_s$

$$\text{if}_s(b, x, y) = \begin{cases} x & \text{if} \quad b = \text{tt} \\ y & \text{if} \quad b = \text{ff} \end{cases}$$

The standard algebra of naturals $\mathbb{N}$ is formed from the algebras N_0 and $\boldsymbol{B}$ by adding nat as an equality sort. We also add the order relation less_{nat} on $\mathbb{N}$. $\boldsymbol{N}$ can be exhibited in the following form:

algebra	$\boldsymbol{N}$
imports	$N_0, \boldsymbol{B}$
functions	$\text{if}_{nat} : \mathbb{B} \times \mathbb{N}^2 \to \mathbb{N}$
	$\text{eq}_{nat}, \text{less}_{nat} : \mathbb{N}^2 \to \mathbb{B}$
end	

We use the label **imports** to indicate that $\boldsymbol{N}$ is an extension of both N_0 and $\boldsymbol{B}$, and we exhibit in the list only the added functions. From now on we assume that all our algebras are standard.

Given any algebra A of signature Σ there are many ways of adding new data sets and operatios. We shall, in particular, consider the following extensions:

i. *Adding counters*: $A^{\boldsymbol{N}}$ of signature $\Sigma^{\boldsymbol{N}}$ is obtained from A by adding the naturals $\mathbb{N}$ and the standard operations 0, suc, eq_{nat}, less_{nat}, and if_{nat}.

ii. *Adding arrays*: A^* of signature Σ^* is obtained from A by adding for each sort s of $\Sigma(A)$ the carriers A^*_s which is defined as the set of finite arrays over A_s.

iii. *Adding streams*: $\bar{A}$ of signature $\bar{\Sigma}$ is formed from A by first expanding A to $A^{\boldsymbol{N}}$. Then for certain pre-stream sorts s add new sorts $\bar{s}\,(= \text{nat} \to s)$ and carriers $A_{\bar{s}} = A_{nat \to s} = [\mathbb{N} \to A_s]$, i.e. the set of functions from $\boldsymbol{N}$ to A_s, together with the appropriate functions, e.g. the evaluation function $\text{eval}_s(f, n) = f(n)$.

Remark. These are familiar extensions. In fap-computability, fap C was obtained by "adding counters", and adding a coding scheme amounts to the same as "adding arrays".

While-computability on standard multi-sorted algebras. . Following Tucker and Zucker [1998] we shall use an imperative programming model to introduce a theory of computability on standard multi-sorted (and possibly partial) algebras.

The syntax of the While(Σ) programming language is of a familiar type. We have the standard categories of *variables*, Var(Σ), and *terms*, Term(Σ). To introduce the categories of *statements* and *procedures* we need some notational devices. For u equal the product type $s_1 \times \cdots \times s_m$ we write $\boldsymbol{x} : u$ to mean that $\boldsymbol{x}$ is a u-tuple of distinct variables, i.e. a tuple of distinct variables of sorts $s_1, \ldots, s_m$ respectively.

iv. Stmt(Σ) is the class of Σ-*statements* generated by the rule

$$S ::= \text{skip} \mid \boldsymbol{x} := \boldsymbol{t} | S_1; S_2 | \text{ if } b \text{ then } S_1 \text{ else } S_2 \text{ fi} | \text{ while } b \text{ do } S \text{ od},$$

where b is a boolean, skip and $\boldsymbol{x} := \boldsymbol{t}$ are atomic statements. $\boldsymbol{x} := \boldsymbol{t}$ expresses concurrent assignments, where for some product type u, $\boldsymbol{x} : u$ and $\boldsymbol{t} : u$.

v. Proc(Σ) is the class of Σ-*procedures* $P, Q, \ldots$, where

$$P \equiv \text{proc in } \boldsymbol{a} \text{ out } \boldsymbol{b} \text{ aux } \boldsymbol{c} \text{ begin } S \text{ end}.$$

Here $\boldsymbol{a}$, $\boldsymbol{b}$, and $\boldsymbol{c}$ are the lists of *input*, *output*, and *auxiliary* variables, respectively. If $\boldsymbol{a} : u$ and $\boldsymbol{b} : v$, for some product types u, v, we call P a procedure of type $\text{Proc}_{u \to v}$.

We briefly turn to the semantics of While(Σ). State(A) is used to denote the set of all *states* σ *on* A, where each σ is a family of mappings $\sigma = \langle \sigma_s | s \in \text{Sort}(\Sigma) \rangle$ from variables of types s to A_s.

A state σ on A_s is a variable assignment. From this we define in the standard inductive way for each term t and sort a a (partial) function $\|t\|^A$ from states to A_s.

Turning to the *semantics of statements* we note that there is a general procedure applicable to a wide class of imperative programming languages for defining the *meaning of a statement* S *as a state transformation*

$$\|S\|^A : \text{State}(A) \to \text{State}(A).$$

The computation of $\|S\|^A(\sigma)$ will proceed in a number of steps $\sigma_0, \sigma_1, \ldots, \sigma_n, \ldots$, where $\sigma_0 = \sigma$ is the input state and

$$\sigma_n = \text{Comp}^A(S, \sigma, n)$$

is the nth step in the computation of S on A. Comp^A will be defined through a set of recursion equations. A computation sequence $\sigma_0, \sigma_1, \ldots, \sigma_n, \ldots$ may or may not terminate. It is possible to construct a function $\text{CompLength}^A(S, \sigma)$ which is defined if and only if the computation sequence terminates, and if defined, the function gives the least n for which it terminates. We can then define $\|S\|^A(\sigma) \simeq \text{Comp}^A(S, \sigma, \text{CompLength}^A(S, \sigma))$.

From this we immediately get the *semantics of a procedure P*

$$P \equiv \text{proc in } \boldsymbol{a} \text{ out } \boldsymbol{b} \text{ aux } \boldsymbol{c} \text{ begin } S \text{ end.}$$

For any $a \in A_{s_1} \times \cdots \times A_{s_n}$, where $\boldsymbol{a} : u = s_1 \times \cdots \times s_n$, let σ be any state on A such that $\sigma(\boldsymbol{a}) = a$. Then we define a function $\|P\|^A : A_u \to A_v$, where $A_u = A_{s_1} \times \cdots \times A_{s_n}$ and $A_v = A_{s'_1} \times \cdots \times A_{s'_m}$ $(\boldsymbol{b} : v = s'_1 \times \cdots \times s'_m)$, by

$$\|P\|^A(a) \simeq \begin{cases} \sigma'(\boldsymbol{b}) & \text{if } \|S\|^A(\sigma) = \sigma' \\ \uparrow & \text{if } \|S\|^A(\sigma) \uparrow \end{cases}$$

We need to verify that this is well-defined, i.e. independent of the behavior of σ outside $\boldsymbol{a}$.

While(A) is the class of functions computable on A by a While(Σ) procedure; in a similar way we introduce the classes While$^N(A)$ and While$^*(A)$. The general theory of While-computability now follows along well-known lines; see Tucker and Zucker [1998]. We note one particularily important result.

The syntax of While-computability can be arithmetised. Since computations are finite, we can for each term t assign a "gödel number" $\ulcorner t \urcorner$ in $\mathbb{N}$. Let $\boldsymbol{x} : u$ and $s \in \text{Sort}(\Sigma)$, we denote by $\text{Term}_{\boldsymbol{x},s}(\Sigma)$ the class of all Σ-terms of sort s with variables from the list $\boldsymbol{x}$ only.

The *term evaluation representing function* on A relative to $\boldsymbol{x}$ and s is the function with values in A_s defined by:

$$te^A_{\boldsymbol{x},s}(\ulcorner t \urcorner, a_1, \ldots, a_n) = \|t\|^A \sigma$$

where $t \in \text{Term}_{\boldsymbol{x},s}$ and $\ulcorner t \urcorner$ is the gödel number of t. σ is any state on A such that $\sigma(x_i) = a_i, i = 1, \ldots, n$.

The algebra A is said to have the *term evaluation property* (TEP) if for all $\boldsymbol{x}$ and s, the term evaluation representing function $te^A_{\boldsymbol{x},s}$ is While-computable on A^N.

Remark. We have met the term evaluation function before in the study of fap-computability and computation theories. We shall in the next theorem see how the previous analysis extends to the present setting.

THEOREM. *Let A be a standard Σ-algebra. The following conditions are equivalent*:

i. *A has the* TEP *property.*
ii. *For all Σ-product types u, v, there is a* While(Σ^N) *procedure* $\text{Univ}_{u\to v}$ *such that for all* $P \in \text{Proc}_{u \to v}$ *and* $a \in A^u$

$$\text{Univ}_{u\to v}(\ulcorner P \urcorner, a) \simeq \|P\|^A(a)$$

From this we can prove, as in the case with fap-computability that the *set of* While$^*(A)$ *of* While**-computable functions on an* $\boldsymbol{N}$*-standard algebra A is the smallest set of partial functions on A to satisfy the axioms of a computation theory* (Tucker and Zucker [1998], Theorem 8.4.1). – We see how the old structural patterns reappear in the new computational framework.

II.2. The algebraic theory of computation. We shall apply the theory of While-computability to the algebras $\boldsymbol{R}$ and $\boldsymbol{R}^{<}$. We recall that $\boldsymbol{R}$ is the structure

$$\begin{array}{ll} \textbf{algebra} & \boldsymbol{R} \\ \textbf{imports} & \boldsymbol{R}_0, \boldsymbol{B} \\ \textbf{functions} & \text{if}_{real} : \mathbb{B} \times \mathbb{R}^2 \to \mathbb{R} \\ & \text{eq}_{real} : \mathbb{R}^2 \to \mathbb{B} \\ \textbf{end} & \end{array}$$

where $\boldsymbol{R}_0 = \langle \mathbb{R}, 0, 1, +, \times, - \rangle$ is the ring of reals and $\boldsymbol{B}$ is the booleans; see section II.1. $\boldsymbol{R}^{<}$ is the algebra which extends $\boldsymbol{R}$ with less_{real}, i.e.

$$\begin{array}{ll} \textbf{algebra} & \boldsymbol{R}^{<} \\ \textbf{import} & \boldsymbol{R} \\ \textbf{functions} & \text{less}_{real} : \mathbb{R}^2 \to \mathbb{B} \end{array}$$

We shall, in particular, be interested in the structure of the While semicomputable subsets of $\boldsymbol{R}$ and $\boldsymbol{R}^{<}$. But first some general definitions.

The *halting set* of a procedure P on A is the set

$$\text{Halt}^A(P) = \{(a_1, \ldots, a_n) \in A^n \mid \|P\|^A(a_1, \ldots, a_n) \downarrow\}$$

A relation R on A is While *semicomputable* if it is the halting set on A of some While procedure; we have similar notions for While^N and While* semicomputability.

The key to understanding semicomputability lies in an analysis of the computation tree of a While statement. We recall that the set of *statements* is inductively defined from *concurrent assignment statements* $\boldsymbol{x} := \boldsymbol{t}$ (and skip) by means of three program construct statements *composition* $S_1; S_2$, *branching* if b then S_1 else S_2 fi and *iteration* while b do S od. The definition of the computation tree associated to a statement is by a structural induction. We shall exhibit the main points through an example and refer to Tucker and Zucker [1998] for full details. Let $S \equiv$ while $\boldsymbol{x} > 0$ do $\boldsymbol{x} := \boldsymbol{x} - 1$ od. The reader will certainly agree that S has the following subcomputation tree $\boldsymbol{T}[S, \boldsymbol{x}]$, see Figure 1.

The root of the tree $\boldsymbol{T}[S, \boldsymbol{x}]$ is labeled s (for *start*) and the leaves are labeled e (for *end*). The internal nodes are labeled with *assignment statements* and *boolean tests*. The edges are labeled with the "current state" of the program variables $\boldsymbol{x}$, assuming that the execution of S starts in the initial state given by $\boldsymbol{x}$. The construction of $\boldsymbol{T}[S, \boldsymbol{x}]$ is effective (i.e. primitive recursive) in the gödel numbers $\ulcorner S \urcorner$ and $\ulcorner \boldsymbol{x} \urcorner$. $\boldsymbol{T}[S, \boldsymbol{x}]$ is in general an infinite tree.

We observe that for each leaf λ of $\boldsymbol{T}[S, \boldsymbol{x}]$, which is a finite branch of $\boldsymbol{T}[S, \boldsymbol{x}]$ ending in a node labeled by e, there is a boolean $\boldsymbol{b}_{S,\lambda}$, with variables among $\boldsymbol{x}$, and which expresses the conjunction of all tests that $\boldsymbol{x}$ must satisfy in order for the computation S to proceed along the path from the root s to the end node e of λ. In our example the tests are of the form:

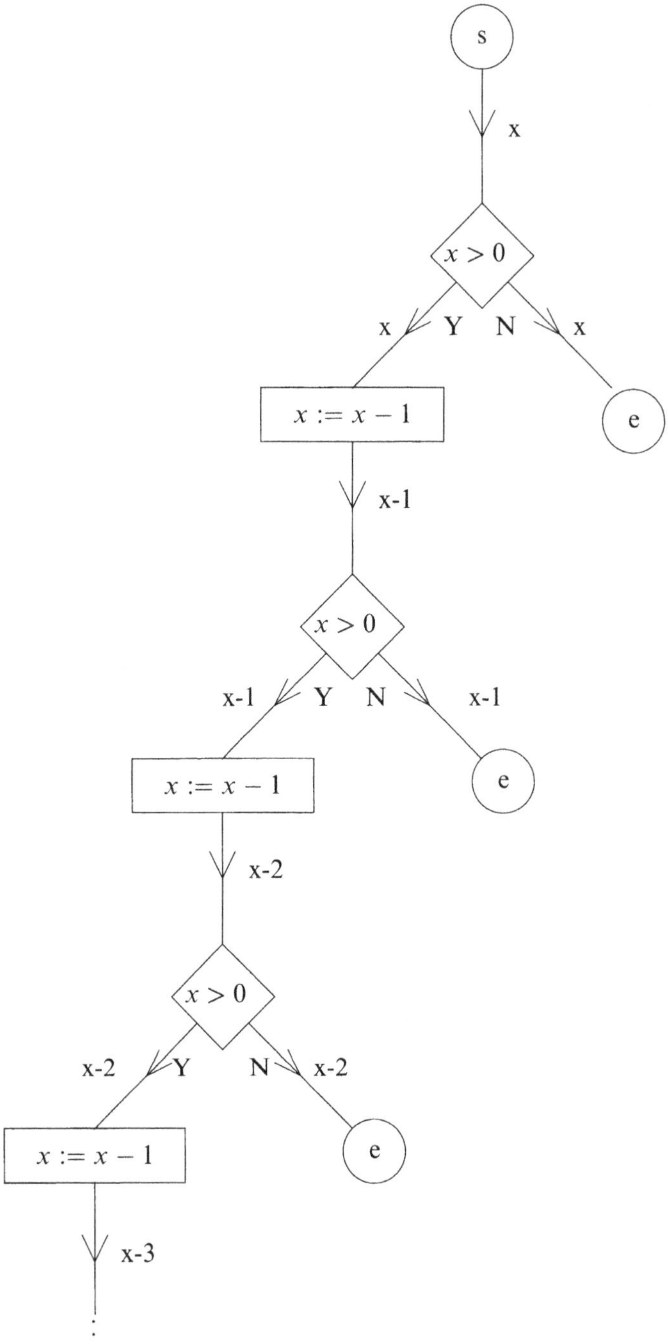

FIGURE 1

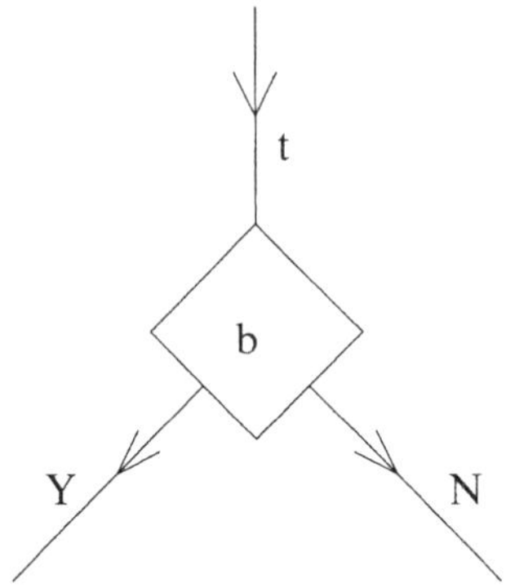

This test contributes either $\boldsymbol{b}(\boldsymbol{x}/t)$ or $\neg\boldsymbol{b}(\boldsymbol{x}/t)$ to the conjunct depending upon whether the path goes to the left or right.

Let $\lambda_0, \lambda_1, \ldots, \lambda_n, \ldots$ be some effective enumeration of the leaves of $\boldsymbol{T}[S, \boldsymbol{x}]$. We define the *halting formula of S* as the countable disjunction

$$halt_s \equiv_{def} \bigvee_{k=0}^{\infty} \boldsymbol{b}_{S.\lambda_k}$$

From this we conclude that the halting set $\mathrm{Halt}^A(P)$ can be characterized by an effective countable disjunction

$$a \in \mathrm{Halt}^A(P) \text{ iff } \bigvee_{k=0}^{\infty} b_k[a],$$

for all $a \in A^u$. (In this case the variables $\boldsymbol{x}$ of $\boldsymbol{T}[S, \boldsymbol{x}]$ is the list of input and auxiliary variables of P, and $b_k[a]$ is the evaluation of $\boldsymbol{b}_{S.\lambda_k}$ when $a \in A^u$ is assigned to the input variables $\boldsymbol{a}$.)

ENGELER'S LEMMA. *Let R be a* While-*semicomputable relation on a standard* Σ-*structure A. Then R can be expressed as an effective countable disjunction of booleans over* Σ.

This result goes back to E. Engeler [1968] and it is of crucial importance for the further development of the theory.

Assuming that A in addition has the TEP (*term evaluation property*) we have the *equivalence* of:

i. R is While semicomputable on A;
ii. R is While^N semicomputable on A;
iii. R is While* semicomputable on A;
iv. R can be expressed as an effective countable disjunction of booleans over Σ.

In the case of the algebras $\boldsymbol{R}$ and $\boldsymbol{R}^<$, which both have the term evaluation property, we first note that a *boolean over* $\boldsymbol{R}$, *with variables* $x_1, \ldots, x_n$ *of sort* real *only, is effectively equivalent to a finite conjunction of equations and negation of equations*

$$p(x) = 0 \quad \text{and} \quad q(x) \neq 0$$

where p and q are polynomials in x with coefficients in $\mathbb{Z}$. Over $\boldsymbol{R}^<$ we can reduce the booleans to finite conjunctions of equations $p(x) = 0$ and inequalities $q(x) > 0$, where p and q are polynomials in x with coefficients in $\mathbb{Z}$.

From this characterization result we can prove the following theorem (Herman and Isard [1970], Tucker and Zucker [1998]): *If $R \subset \mathbb{R}^n$ is* While**-semicomputable on* $\boldsymbol{R}$ *and contains a transcendental point* α, *then R contains some open neighborhood of* α.

The proof is simple. $R \subseteq \mathbb{R}^n$ can be expressed as an (effective) countable disjunction

$$x \in R \quad \text{iff} \quad \bigvee_i b_i(x),$$

where b_i is a finite conjunction of equations $p(x) = 0$ or negation of equations $q(x) \neq 0$. If $\alpha \in R$, then $b_i(\alpha)$ is true for some i. This $b_i(\alpha)$ cannot contain any equation since α is transcendental. Continuity properties of polynomials immediately implies the result.

This result has an interesting corollary: *Any subset of* $\mathbb{R}^n$ *which is both dense and co-dense in* $\mathbb{R}^n$ *cannot be* While**-computable on* $\boldsymbol{R}$.

For $n = 1$ we can further conclude: If $R \subseteq \mathbb{R}$ is While* semicomputable on $\boldsymbol{R}$, then R is either countable or co-finite. In particular, $R \subseteq R$ is While*-computable on $\boldsymbol{R}$ implies R is either finite or co-finite.

There are both positive and negative aspects to this analysis. While-computability on $\boldsymbol{R}$ (where $\boldsymbol{R}$ is a *total algebra with* real *as an equality sort*) leads to a satisfactory *algebraic definability theory*. This is a consequence of Engeler's lemma, which gives, in fact, a *complete structural description* of the semicomputable sets. There are interesting applications of this observation to e.g. dynamical systems; see Blum, Cucker, Shub and Smale [1998] and Tucker and Zucker [1998]. We give one simple example.

Let the system D_μ on $I = [0, 1]$ be given by the "next state" function $f_\mu(x) = \mu x (1 - x)$. It is well known that the system D_4 is chaotic. Let $\text{Orb}(D_4, x)$ denote the *orbits* of D_4. i.e. the set of points $f_4^t(x), t = 0, 1, 2, \ldots$, and $\text{Per}(D_4)$ the set of *periodic points*, i.e. the set of points x such that $f_4^t(x) = x$ for some t. Then both $\text{Orb}(D_4, x)$ and $\text{Per}(D_4)$ are While^N-semicomputable, but neither is While^N-computable. The latter follows from the fact that $\text{Orb}(D_4, x)$ and $\text{Per}(D_4)$ are both dense and co-dense in I.

If we shift attention from structure and definability to computations and algorithms,we see that the algebraic theory has limitations. In $\boldsymbol{R}$ *definability theory* prospers in the presence of totality and the definability of equality over $\mathbb{R}$. But as the resulting structural description shows, we are with this theory *far from the needs of real computer science and actual numerical analysis.*

II.3. Computing on topological partial algebras. Several severe limitations of the theory of While-computability over $\boldsymbol{R}$ were noted in the last section. But there are even further limitations. "Real" computability over $\boldsymbol{R}$ takes account of the *topological structure* of $\mathbb{R}$; in real life computations typically proceeds via *approximations*. Approximation presupposes topology and continuity, thus in a "good" theory the basic operations must be *continuous*. But the *equality* $\text{eq}_{real} : \mathbb{R}^2 \to \mathbb{B}$ is *not continuous* over $\boldsymbol{R}$. This follows from the simple fact that the only continuous functions from a connected space ($\mathbb{R}^2$) to a discrete space ($\mathbb{B}$) are the constant functions.

Feferman [1992b] also notes the *limitation of the algebraic approach*. He mentions four specific criticisms: (i) Problems with the equality sorts, e.g. in connection with streams, infinite precision reals, and more general frameworks for numerical and constructive analysis (approximations). (ii) The necessity for partial structures. (iii) The limitation of algebraic specification theory (as in the survey of Wirsing [1990]) to initial and final semantics. (iv) The theory "seems to me to be too cumbersome to support the need of practice".

In a series of papers (Feferman [1991], [1992b], and [1996]) he develops an alternative approach to overcome these weaknesses. Feferman works over a multi-sorted domain including the booleans. In addition to constants and (partial) functions he also includes type-2 partial functionals. His computability theory is based on *schemata for computational procedures on many-sorted structures*, including schemata for conditional definitions and the least fixed-point functional; see also the approach of Moschovakis [1984], [1989], and [1997].

We shall stay closer to the algebraic foundation and overcome the difficulties by stepwise extensions. In this we follow Tucker and Zucker [1999] and Stoltenberg-Hansen and Tucker [1999] and follow a path leading from a total algebraic context to a *data type of topological partial algebras over multi-sorted domains*.

A *topological partial* Σ*-algebra* A is a partial algebra with topologies on the carriers of A such that the basic functions are continuous. In a s*tandard* topological (partial) algebra we assume that the carrier $\mathbb{B}$ of $\boldsymbol{B}$ has the discrete topology. (Recall that a partial function $f : X \to Y$ is *continuous* if $f^{-1}[V] = \{x \in X | x \in \text{dom}(f) \text{ and } f(x) \in Y\}$ is open for all open $V \subseteq Y$.)

An *important example* of a topological partial algebra is $\boldsymbol{R}_p = \langle \mathbb{R}, \mathbb{B}; 0, 1, +, -, \cdot, \text{if}_{real}, \text{eq}_p, \text{less}_p \rangle$, which is formed from $\boldsymbol{R}^{<}$ by replacing the total eq_{real} and less_{real} by the partial

$$\text{eq}_p(x, y) = \begin{cases} \uparrow, & \text{if } x = y \\ \text{ff}, & \text{if } x \neq y \end{cases}$$

and

$$\text{less}_p(x, y) = \begin{cases} \text{tt}, & \text{if } x < y \\ \text{ff}, & \text{if } x > y \\ \uparrow, & \text{if } x = y \end{cases}$$

Both eq_p and less_p are continuous over $\boldsymbol{R}_p$.

As a last preparation we note that there is an almost canonical way of making A^N, A^*, and $\bar{A}$ into a topological algebra from the topologies on A.

We claimed above that computability should imply continuity. This is indeed true in the present set-up: *Let A be a standard topological partial algebra. Then any* While-*computable function on A is continuous.* (Tucker and Zucker [1998]). The result extends from While to While^N and While*.

The proof is straightforward. Any While-computable function is given by a procedure $P \equiv$ proc in $\boldsymbol{a}$ out $\boldsymbol{b}$ aux $\boldsymbol{c}$ begin S end. Recalling the definition of the state transformation $\|S\|^A$ and the function $\|P\|^A$ from section II.1. we can verify that the result follows by a simple structural induction.

We shall approach the *topic of approximation* via a study of While-computability on connected total algebras. This turns out to be no real restriction. In computing an approximating sequence we want the building blocks to be as simple as possible. In this case it turns out that they are explicitly definable in a very concrete sense.

The following is true: *Let A be a total topological algebra and assume that A^u is connected. Then any total function $f : A^u \to A^v$ is* While (While*)-*computable on A iff f is Σ-explicitly definable on A* (i.e. defined by a Σ-term on A); see Tucker and Zucker [1999]. The non-trivial part of the proof uses connectedness and totality to eliminate branching inthe computation tree of $f(a)$, $a \in A^u$. We are left with a finite and unbranching tree, from which an explicit definition immediately can be written down.

Important *examples of total connected algebras* are $\boldsymbol{R}_t^N$ and $\boldsymbol{I}_t^N$, where $\boldsymbol{R}_t^N$ is an extension of $\boldsymbol{R}_0$, $\boldsymbol{N}$, and $\boldsymbol{B}$ adding only the functions if_{real} and div_{nat} (i.e. division of reals by naturals, where division by zero is defined as zero), and where $\boldsymbol{I}_t^N$ is basically the restriction of $\boldsymbol{R}_t^N$ to the interval $I = [0, 1]$. Note that $\boldsymbol{R}_t^N$ differs from $\boldsymbol{R}$ in that basic operations are total but continuous.

These algebras are used in developing a theory of *approximable computability*. First a definition: Let $f : X \to Y$ be a total function on X with values in a metric space Y. A sequence g_n *approximates f effectively uniformly* on X if there exists a total recursive function e such that for all $n, k \in \mathbb{N}$ and all $x \in X$, if $k \geq e(n)$, then $d_Y(g_k(x), f(x)) < 2^{-n}$.

In the context of While-computability we can impose further requirements on the approximating sequence g_n. We call a function $f : A^n \to A_s$ *effectively* Σ-*Weierstrass approximable* over A if there is a total recursive function h, such that $h(n)$ is the gödel number of some term in $\text{Term}_{\boldsymbol{x},s}(\Sigma)$ and such that the sequence

$$g_n(a) =_{def} \text{te}^A_{\boldsymbol{x},s}(h(n), a)$$

approximates f effectively uniformly on A^u (see Tucker and Zucker [1999]).

To explain the word "Weierstrass" we note that a function from $\mathbb{R}^q$ (or I^q) to $\mathbb{R}$ is explicitly definable in the While formalism if and only if it is definable by a polynomial in q variables over $\mathbb{R}$ with rational coefficients; the proof of this fact can be "read off" from the computation tree as was done in the proof of the Engeler lemma and, similarily, in the proof that While-computability over a connected domain reduces to explicit definability. Thus we see that the *class of functions which are effectively Σ-Weierstrass approximable over A is exactly the class of functions which can be computed by a sequence of polynomial approximations* (Tucker and Zucker [1999]).

It remains to establish a connection to computable analysis. A total function $f : I^q \to \mathbb{R}$ is called *Grzegorcyk-Lacombe (GL) computable* if (i) f is *sequentially computable*, i.e. f maps every computable sequence of points in I^q into a computable sequence of points in $\mathbb{R}$. (ii) f is *effectively uniformly continuous*, i.e. there is a recursive function δ such that for all $x, y \in I^q$ and all $n \in \mathbb{N}$, if $|x - y| < 2^{-\delta(n)}$, then $|f(x) - f(y)| < 2^{-n}$.

The following result is one further step on the road from While-computability over multi-sorted algebras to actual numerical analysis: *Let $f : I^q \to R$ be a total function. Then f is effectively Σ-Weierstrass approximable on I_t^N if and only if f is GL-computable.*

Remark. For the notion of GL-computability see Grzegorcyk [1955]. [1957] and Lacombe [1955]. An exposition of the theory is found in Pour-El and Richard [1989]. The proof of the result above goes back to work of Pour-El and Caldwell [1975] and Shepherdson [1976]. We have followed the exposition in Tucker and Zucker [1999].

At this point we should acknowledge that GL-computability today may appear as a somewhat cumbersome and old-fashioned approach to computable analysis and that there are now several other approaches offering a better conceptual insight in the topic.

As an example of such an alternative approach to computation over topological structures we mention the recent work of Escardo [1996], [1997]. His conceptual starting point is domain theory, and he proceeds by computing directly on approximations of the objects. More precisely, he defines an extension of Plotkin's [1977] PCF to Real PCF, where the representations of the reals as the maximal elements in the continous domain of closed and bounded intervals is used as one base type. The new basic operators are essentially rational affine transformations with a "cut-off" and the boolean valued tests $x < a$, for rational a. The While-operator is replaced by a fixed-point operator in a fully typed language. In order to avoid the negative consequences of the partiallity of the boolean valued tests, Escardo introduces a non-deterministic partial if-operator. Within this framework Escardo shows that every GL-computable $f : \mathbb{R} \to \mathbb{R}$ is the restriction of a Real PCF-definable function $\hat{f}$ to the "true" reals.

We cannot pursue this theme further here. The reader is in addition invited to consult the papers *Concrete models for computation for topological algebras* (Stoltenberg-Hansen and Tucker [1999]) and *Abstract versus concrete models of computation on metric partial algebras* (Tucker and Zucker, in preparation) for a broader study of how theories of While-computability are related to a number of concrete models of computation (such as *algebraic domain representability, continuous domain representability, type-two enumeration, effective metric spaces*, and various *axiomatic approaches*).

II.4. New challenges. In the introduction I pointed out the growing importance of computations and algorithms in mathematics and modeling. I quoted - with approval - from a lecture by S. Smale *Mathematical Problems for the next Century*. Smale claimed that as "the study of the set of solutions of an equation (e.g. a manifold) played such an important role in 20-th century mathematics, the study of finding the solution (e.g. an algorithm) may play an equally important role in the next century".

This view is reinforced by a report, *Possible Trends in the Mathematics in the Coming Decades*, written by M. Gromov for the International Assessment of the U.S. Mathematical Sciences, NSF, March 1998. Gromov notes with approval "the creation of mathematical logic and then of the theoretical computer science. The latter is now coming of age. It absorbs ideas from the classical mathematics and benefits from the technological progress in the computer hardware which leads to a practical implementation of theoretically devised algorithms" (Gromov [1998]).

These are challenges that the recursion theory community must respond to. So far the community has been remarkably inward looking. "... the trend has been to pull the subject of effective computability even farther away from questions of actual computation" (Feferman [1992a]). There has been some calls for a broader engagement of the recursion theory community. Feferman is one example. Another is R. I. Soare; see his recent discussion in the survey *Computability and Recursion*, in particular, section 6 on *themes and goals of computability theory* (Soare [1996]).

I have in this paper looked at computability theory both from a structural and algorithmic point of view. One is not "more correct" than the other, and the interplay of the two has enriched our understanding of computability theory. I have traced a line of development from the structure of computation theories to the theory of effective approximations in topological algebras. But this is not the end of the story.

Typically general computability theory has been concerned with either the formal structure of computer science or the study of discrete and algebraic algorithms; see e.g. the highly regarded *Handbook of Theoretical Computer Science* [1990]. The *Handbook* has two volumes, one on *Formal Models and Semantics* and one on *Algorithms and Complexity*. There is one glaring gap, *numerical analysis* is an art which has so far been beyond the horizon of the theoretical developments in computability theory. I therefore welcome the intension expressed by S. Smale and his co-workers to use their approach as a theoretical foundation for numerical analysis; see, in particular, section 1.4 *Toward a Mathematical Foundation for Numerical Analysis* in Blum *et al* [1998].

To develop a computability theory which also includes numerical analysis is a major challenge today. There is much deep and difficult mathematics in Blum *et al* [1998], but the *foundation is not yet in place*. In our survey – from the algebra of While-computability to an approximation theory in topological partial algebras – steps have been taken toward a computability theory for numerical analysis. Continuity played an important role in this discussion. But numerical praxis does not always bow to the dictates of preconceived theory. J. V. Tucker has pointed out *how discontinuity and multivaluedness is prevalent in actual numerical analysis*; see also V. Brattka [1999]. An important example is the *pivot function*, piv : $R \to \{0, 1, \ldots, n\}$, defined as

$$\operatorname{piv}(x_1, \ldots, x_n) = \begin{cases} \text{some } i & \text{such that } x_i \neq 0 \\ \uparrow & \text{if all } x_i = 0 \end{cases}$$

This function is an essential ingredient in the Gaussian elimination algorithm. A first study of discontinuity and multivaluedness can be found in the Ph. D. thesis of K. Stewart [1998]. But it remains to see the shape of a general theory. To this end it may even happen that some of the "abstract" ideas of multiple-valued selection expounded in Part I will turn out to be of use in the "applied" numerical context.

Remark. I have in this essay been somewhat vague concerning the use of the phrases "the computability community" and "computational praxis". G. Kreisel has on several occasions called attention to the need of "informal rigour". This is a case where such an informal rigour is called for. Despite the deliberate vagueness in this essay I trust that the reader has understood my main points. But I see the urgent need of a comprehensive and professional history of computability theory as it has emerged in different branches of mathematics. The situation today is not satisfactory. Some authors collect large numbers of references and believe that this is adequate as a history of the subject; it seems to me that Blum *et al* [1998] belongs to this school. Others are admirable in their discussion but too limited in scope to serve as a general history; the survey by Soare [1996] seems to me to belong to that category. If Smale and Gromov are right, that computability and algorithms will become a major theme in the mathematics of the next century, we deserve a professional study of its roots and developments.

Acknowledgements. A number of people have helped me with this paper. I received several useful comments during the Feferfest. But above all I value the generous help and advice that I have received from Dag Normann and John V. Tucker.

REFERENCES

V. Brattka, 1999, *Recursive and computable operations over topological structures*, Ph. D. Thesis, FernUniversität Hagen.

L. Blum, F. Cucker, M. Shub and S. Smale, 1998, *Complexity and real computations*, Springer-Verlag, New York

J. Dahl, B. Myrhaug and K. Nygaard, 1970, *Common Base Language*, Norsk Regnesentral, Oslo.

F. R. Drake and S. S. Wainer, 1980, *Recursion Theory: its generalizations and applications*, Cambridge University Press, Cambridge.

E. Engeler, 1968, *Formal languages: Automata and structures*, Markham PublicationsCompany.

M. H. Escardo, 1996, *PCF extended with real numbers*, Theoretical Computer Science, 162 (1).

M. H. Escardo, 1997, *PCF extended with real numbers: a domain-theoretic approach to higher-order exact real number computation*, Ph. D. Thesis, University of London, Imperial College.

S. Feferman, 1977, *Inductive schemata and recursively continuous functionals*, in Gandy and Hyland 1977.

S. Feferman, 1991,*A new approach to abstract data types,II. Computations on ADTs as ordinary computations*, in Computer Science Logic, Lecture Notes in Computer science, 626, Springer-Verlag.

S. Feferman, 1992a, *Turing's oracle: From absolute computability - and back*, in The Space of Mathematics (ed. J. Echeverria et al), Walter de Gruyter, Berlin.

S. Feferman, 1992b, *A new approach to abstract data types I. Informal development*, Mathematical Structures in Computer Science, 2.

S. Feferman, 1996, *Computations on abstract data types. The extensional approach with an application to streams*, Annals of Pure and Applied Logic, 81.

J. E. Fenstad, 1980, *General Recursion Theory*, Springer-Verlag, Berlin-Heidelberg-New York, 1980.

J. E. Fenstad, R. O. Gandy and G. E. Sacks, 1978, *Generalized recursion theory II*, North-Holland, Amsterdam.

J. E. Fenstad and P. G. Hinman, 1974, *Generalized Recursion Theory*, North-Holland, Amsterdam.

H. Friedman, 1971a, *Axiomatic recursive function theory*, in Gandy and Yates 1971.

H. Friedman, 1971b, *Algorithmic procedures, generalized Turing algorithms, and elementary recursion theory*, in Gandy and Yates 1971.

R. O. Gandy, 1967,*General recursive functionals of finite type and hierarchies of functions*, Ann. Fac. Sci. Univ. Clermont-Ferrand, 35.

R. O. Gandy, 1988, *The confluence of ideas in 1936*, in The Universal Turing Machine. A Half-Century Survey (ed. R. Herken), Springer-Verlag, Wien-New York.

R. O. Gandy and J. E. M. Hyland, 1977,*Logic Colloquium '76*, North-Holland, Amsterdam.

R. O. Gandy and C. M. E. Yates, 1971,*Logic Colloquium '69*, North-Holland, Amsterdam.

J. A. Goguen, J. W. Thatcher, E. G. Wagner and J. B. Wright, 1975, *An initial algebra approach to the specification, correctness and implementation of abstract data types*, in Current Trends in Programming Methodology: IV Data structures (ed. R. T. Yeh), Prentice Hall, London.

A. Grzegorczyk, 1955, *Computable functions*, Fundamenta Mathematicae, 42.

A. Grzegorczyk, 1957, *On the definition of computable real continuous function*, Fundamenta Mathematicae, 44.

M. Gromov, 1998, *Possible trends in mathematics in the coming decades*, Notices AMS, 45.

G. T. Herman and S. D. Isard, 1970, *Computability over arbitrary fields*, Journal of the London Mathmatical Society, 2.

S. C. Kleene, 1959, *Recursive functionals and quantifiers of finite types, I*, Transactions of the American Mathematical Society, 91.

S. C. Kleene, 1981, *Origins of recursive function theory*, Annals of the History of Computing, 3.

G. Kreisel. 1971, *Some reasons for generalizing recursion theory*, in Gandy and Yates, 1971.

G. Kreisel and G. E. Sacks, 1965,*Metarecursive sets*, The Journal of Symbolic Logic, 30.

D. Lacombe, 1955, *Extension de la notion de fonction recursive aux fonctions d'une ou plusieurs variable reelles*, C. R. Acad. Sci. Paris, 240.

J. Moldestad, 1977, *Computations in Higher Types*, Lecture Notes in Mathematics, 574, Springer-Verlag, Berlin-Heidelberg-New York.

J. Moldestad, V. Stoltenberg-Hansen and J. V. Tucker, 1980a, *Finite algorithmic prodcedures and inductive definability*, Mathematica Scandinavia, 46.

J. Moldestad, V.Stoltenberg-Hansen and J.V.Tucker, 1980b, *Finite algorithmic procedures and computation theories*, Mathematica Scandinavia, 46.

Y. N. Moschovakis 1969, *Abstract first-order computability, I, II*, Transactions of the American Mathematical Society, 138.

Y. N. Moschovakis, 1971, *Axioms for computation theories - first draft*, in Gandy and Yates 1971.

Y. N. Moschovakis, 1984, *Abstract recursion as a foundation for the theory of algorithms*, Computation and Proof Theory, Lecture Notes in Mathematics, 1104, Springer-Verlag, Berlin-Heidelberg-New York.

Y. N. Moschovakis, 1989, *The formal language of recursion*, The Journal of Symbolic Logic, 54.

Y. N. Moschovakis, 1997, *The logic of functional recursion*, in Logic and Scientific Methods (ed M. L. Dalla Chiara et al), Kluwer Academic Publishers, Dordrecht.

D. Normann, 1978, *Set recursion*, in Fenstad, Gandy and Sacks 1978.

R. A. Platek, 1966, *Foundation of recursion theory*, Ph. D. Thesis, Stanford University.

G. Plotkin, 1977, *LCF considered as a programming language*, Theoretical Computer Science, 5 (1).

M. B. Pour-El and J. C. Caldwell, 1975, *On a simple definition of computable function of a real variable with applications to functions of a complex variable*, Zeitschr. f. math. Logik und Grundl. der Mathematik, 21.

M. B. Pour-El and J. I. Richards, 1989, *Computability in Analysis and Physics*, Springer-Verlag, New York-Heidelberg.

G. E. Sacks, 1990, *Higher Recursion Theory*, Springer-Verlag, Berlin-Heidelberg-New York.

J. C. Shepherdson, 1975,*Computations over abstract structures: serial and parallel procedures and Friedman's effective definitional schemes*, Logic Colloquium '73, North-Holland, Amsterdam.

J. C. Shepherdson. 1976, *On the definition of a computable function of a real variable*, Zeitschr. f. math. Logik und Grundl. der Mathematik, 22.

J. C. Shepherdson, 1988, *Mechanisms for computing over arbitrary structures*, in The Universal Turing Machine. A Half-Century Survey (ed. R. Haken), Springer-Verlag, Wien-New York.

J. C. Shepherdson and H. E. Sturgis, 1963, *Computability of recursive functions*, J. Ass. for Computing Machinery, 10.

S. Smale, 1998, *Mathematical problems for the next century*, The Mathematical Intelligencer, 20.

R. I. Soare, 1996, *Computability and recursion*, The Bulletin of Symbolic Logic, 2.

K. J. Stewart, 1999, *Concrete and abstract model of computation over metric algebras*, Ph. D. Thesis, University of Wales, Swansea.

V. Stoltenberg-Hansen and J. V. Tucker, 1999, *Concrete models for computation for topological algebras*, Theoretical Computer Science, 219.

H. R. Strong, 1968, *Algebraically generalized recursive function theory*, IBM J. Res. Devel., 12.

J. V. Tucker, 1980,*Computing in algebraic systems*, in Drake and Wainer 1980.

J. V. Tucker and J .I. Zucker, 1998, *Computable functions and semicomputable sets on many-sorted algebras*, in Handbook of Logic in Computer science, vol 5.

J. V. Tucker and J. I. Zucker, 1999, *Computation by While programs on topological partial algebras*, Theoretical Computer Science, 219.

J. V. Tucker and J. I. Zucker, in preparation, *Abstract versus concrete models of computation on metric partial algebras*.

J. Van Leeuwen (ed), 1990, *Handbook of Theoretical Computer Science, Vol A, Algorithms and Complexity*, Elsevier, Amsterdam.

J. Van Leeuwen (ed), 1990, *Handbook of Theoretical Computer Science, Vol B, Formal Models and Semantics*, Elsevier, Amsterdam.

E. G. Wagner, 1969, *Uniform reflexive structures*, Trans. AMS, 144.

M. Wirsing, 1990, *Algebraic specification*, in J. Van Leeuwen 1990.

INSTITUTE OF MATHEMATICS
UNIVERSITY OF OSLO
OSLO, NORWAY
E-mail: jfenstad@math.uio.no

THREE PROCESSES IN NATURAL LANGUAGE INTERPRETATION

TIM FERNANDO

Abstract. To address complications involving ambiguity, presupposition and implicature, three processes underlying natural language interpretation are isolated: translation, entailment and attunement. A meta-language integrating these processes is outlined, elaborating on a proof-theoretic approach to presupposition.

§1. Introduction. However outrageous Montague's slogan "English as a formal language" [24] may sound, the pressure to push the claim as far as it can go is, for many, irresistible. Basic to Montague's understanding of a formal language is the possibility of a model-theoretic interpretation — of obvious interest in various applications (e.g. databases) that employ models. But formulas of predicate logic (first-order or higher-order, modal or otherwise) differ significantly from English sentences marked with ambiguity and presupposition. Consider, for instance,

(s) If Sylvester gets holds of a canary, the bird will perish.

A very rough translation of (s) that falls short of predicate logic is

$$(\exists y)(canary(y) \wedge SylvesterGets(y)) \supset willPerish(\underline{\textsf{the bird}}) ,$$

from which we might expect to extract various readings of (s), depending on how we treat the ill-formed expression the bird. For now, let us focus on the reading that construes the bird anaphorically as any canary Sylvester gets hold of. The obvious rendering in predicate logic,

$$(\forall y)\ (canary(y) \wedge SylvesterGets(y)) \supset willPerish(y) ,$$

moves $\exists y$ out of the antecedent to bind the bird, turning $\exists$ into $\forall$. An alternative leaving $\exists y$ intact is provided, under a "propositions-as-types" interpretation of existential quantification $(\exists x \in A)B$ as the dependent sum/product

$$(\textstyle\sum x : A)B \;=\; \{\langle a, b\rangle \mid a \text{ in } A \text{ and } b \text{ in } B[x \mapsto a]\}$$

As wide ranging as Solomon Feferman's work on logic is, natural language is not among his driving concerns. Nevertheless, I would like to think that studying papers such as Feferman [8, 9, 10] has helped me draw certain conceptual distinctions useful in natural language interpretation. And, as a grateful student of his, I want to try to say why. I am indebted to two anonymous referees for helpful comments, and to the Feferfest committee for the opportunity of contributing to this volume. How I regret that I could not properly thank Jon Barwise.

Reflections on the foundations of mathematics
Essays in honor of Solomon Feferman
Edited by Wilfried Sieg, Richard Sommer, and Carolyn Talcott

and implication $A \supset B$ as the dependent function space

$$(\prod x : A)B = \{\text{functions mapping } a \text{ in } A \text{ to some } b \text{ in } B[x \mapsto a]\},$$

by translating (s) to

$$(1) \qquad (\prod x : (\sum y : canary)\ SylvesterGets(y))\ willPerish(lx)$$

where lx is the left projection of *any* proof x of $(\sum y : canary)\ SylvesterGets(y)$. Details are supplied in §2 below, following for the most part Martin-Löf type theory [22], where such translations have been explored extensively (Ranta [25]; see also Sundholm [27]), and where (1) is well-formed relative to a context Γ such that

$$\Gamma,\ x : (\sum y : canary)\ SylvesterGets(y) \Rightarrow willPerish(lx)\ \mathsf{type}$$

(with $\Rightarrow$ separating the context to its left from the assertion to its right, and A type saying 'A is well-formed'). To appreciate the novelty in using a (parametric) proof to construct (1), observe that in predicate logic (be it classical or intuitionistic), the notion of a well-formed formula does not depend on proofs or truth (at least not directly on that which the logic *subsequently* formalizes). By contrast, underlying the toy translation (1) of (s) is the claim that

(†) well-formedness may depend on parametric proofs — or, in symbols,

$$x_1 : A_1, \ldots, x_n : A_n \Rightarrow A\ \mathsf{type}.$$

(†) acknowledges the need for entailments to resolve problematic expressions such as the bird — the step from (s) to (1) relying (also) on all canaries being birds. The present paper elaborates on applications of (†) to natural language, addressing three questions:

(Q1) How can we keep the task of constructing logical forms managable, when proofs (implicated in that task by (†)) can get arbitrarily complicated?
(Q2) How can we interpret the parametricity in (†) model-theoretically?
(Q3) How can we bring natural language into the picture so as to constrain choices of A in (†)? (Just where is the English?)

1.1. Outline. Questions (Q1)-(Q3) are analyzed below by construing (†) as an attack on *presupposition*, and factoring that analysis into three steps specifying respectively: what presuppositions are, how they can be satisfied, and what to do if they can't. Towards that end, §2 formalizes a fragment of type theory given roughly by first-order formulas over a relational signature, with terms assembled from proofs supplied by context. §2 concludes with an answer to (Q2), on which §3 builds a reply to (Q1). Complications to the idealization assumed in §3 are exposed in §4, where (Q3) is taken up.

The attention paid to type theory is largely for the sake of concreteness. §4 applies also to other approaches (dealing say, with (s)), and can, up to a point, be read independently of the two preceding sections. Interest in the general line investigated in §§2 and 3 does, however, seem to be growing — witness, for instance, Ahn and Kolb [2], Kamareddine and Klein [15], Fox [13], Krause [21], Borghuis [6], Krahmer and Piwek [20], Cooper [7], Ranta [26] — and it is natural to relate this to *explicit mathematics* (Feferman [8]). Explicit mathematics provides in §2 the basis for an answer to (Q2) as well as a transparent treatment of subtypes, without any need for type coercion.[1]

1.2. From entailment to translation and attunement. Turning to the main part of the paper, an approach to natural language semantics that motivates a good deal of §§3 and 4 is *Discourse Representation Theory* (DRT, Kamp and Reyle [16]), addressing, as it does, the nagging question behind (Q3) of just how logical forms (or discourse representations) are to be derived from natural language text. The elusive answer is referred to in DRT as the *construction algorithm*. In an early paper, Hans Kamp writes

> it seems to me that the rules for the construction of discourse representations have at least as good a claim to being constitutive of meaning as the clauses that make up the definition of truth. ([17], page 409)

About a decade later, he reflects

> The principal challenge to DRT, experience indicates, is the exact formulation of the construction algorithm. ([18], page 37)

To meet this challenge, various proposals have been put forward, including appeals to rhetorical relations and defeasible entailments (e.g. Asher and Lascarides [3]). Abstracting over these proposals, the present paper investigates the construction algorithm in terms of a process of *translation*, conditioned by processes of *entailment* and *attunement*. Very briefly, the thrust of §§3 and 4 is to factor into the entailments mentioned in (†) the roles played by translation and attunement in natural language interpretation.

§2. An elementary system of dependent types. Let L be a (many-sorted, relational) *signature* that specifies *sorts* $(U, \dots)$ and *relation symbols* $(R, \dots)$ with associated arities. Let us agree to write $U \in L_0$ to mean U is a sort in L, and $R \in L(U_1 \cdots U_n)$ to mean that R is a relation symbol in L with n arguments of sorts U_1 to U_n. This section formalizes an elementary subsystem corresponding *roughly* to first-order L-formulas φ generated according to

$$\varphi ::= \perp \mid R(t, \dots, t) \mid \varphi \wedge \varphi \mid \varphi \supset \varphi \mid (\exists x \in U)\varphi \mid (\forall x \in U)\varphi$$

where $U \in L_0$ and $R \in L(U_1 \cdots U_n)$ for some $U_1, \dots, U_n \in L_0$. What is "rough" here is that some care must be taken in specifying how $R(t_1, \dots, t_n)$ is formed contextually from proof terms t_i.

[1]This much is implicit in the reference in Fox [13] to *Frege structures* (Aczel [1]). I hasten to add that Feferman's influence on the present work reaches beyond these applications, and into §4 (as I explain in §5.2).

2.1. Context and formation rules. As usual, contexts are finite sequences of variable typings, built from the empty sequence ε, the typings in which seep through $\Rightarrow$.

$$\frac{}{\varepsilon \text{ context}} \qquad \frac{\Gamma \Rightarrow A \text{ type}}{\Gamma, x\!:\!A \text{ context}}\ x \notin \mathsf{Var}(\Gamma) \qquad \frac{\Gamma \text{ context}}{\Gamma \Rightarrow x\!:\!A}\ \text{`}x\!:\!A\text{' in } \Gamma\,.$$

Suppressing the subscript L on $\Rightarrow_L$ for simplicity, let us pick out a subcollection wff of type corresponding to the aforementioned L-formulas φ

$$\frac{\Gamma \text{ context}}{\Gamma \Rightarrow U \text{ type}}\ U \in L_0 \qquad\qquad \frac{\Gamma \Rightarrow A \text{ wff}}{\Gamma \Rightarrow A \text{ type}}$$

with $\bot$ and atomic L-formulas from

$$\frac{\Gamma \text{ context}}{\Gamma \Rightarrow \bot \text{ wff}} \qquad \frac{\Gamma \Rightarrow t_1\!:\!U_1 \quad \cdots \quad \Gamma \Rightarrow t_n\!:\!U_n}{\Gamma \Rightarrow R(t_1,\ldots,t_n) \text{ wff}}\ R \in L(U_1 \cdots U_n)$$

$\wedge$ and $\supset$ from

$$(\wedge)\ \frac{\Gamma \Rightarrow A \text{ wff} \qquad \Gamma, x\!:\!A \Rightarrow B \text{ wff}}{\Gamma \Rightarrow (\sum x\!:\!A)B \text{ wff}} \qquad (\supset)\ \frac{\Gamma \Rightarrow A \text{ wff} \qquad \Gamma, x\!:\!A \Rightarrow B \text{ wff}}{\Gamma \Rightarrow (\prod x\!:\!A)B \text{ wff}}$$

and $\exists x \in U$ and $\forall x \in U$ from

$$\frac{\Gamma, x\!:\!U \Rightarrow A \text{ wff}}{\Gamma \Rightarrow (\sum x\!:\!U)A \text{ wff}}\ U \in L_0 \qquad \frac{\Gamma, x\!:\!U \Rightarrow A \text{ wff}}{\Gamma \Rightarrow (\prod x\!:\!U)A \text{ wff}}\ U \in L_0\,.$$

Conventions. For expressions obtained from the rules ($\wedge$) and ($\supset$), where the variable x bound does not occur in B, let us write $A \wedge B$ and $A \supset B$, respectively. Furthermore, for $R \in L(U)$, let us write $(\sum x\!:\!R)A$ instead of $(\sum x\!:\!U)(R(x) \wedge A)$, and $(\prod x\!:\!R)A$ instead of $(\prod x\!:\!U)(R(x) \supset A)$.

Example. For the translation (1) of (s) in §1, let *bird*, *willPerish*, *canary* $\in L(U)$ for some $U \in L_0$, so that $(\sum y\!:\!\mathit{canary})A$ in (1) abbreviates $(\sum y\!:\!U)(\mathit{canary}(y) \wedge A)$. To form the term lx, we need further rules for judgments $t : A$. These are, in contrast to well-formedness judgments A wff, naturally construed as semantic, although they feed into the well-formedness rule above with side condition $R \in L(U_1 \cdots U_n)$, whence the novelty noted in the introduction.

2.2. Further semantic rules. The semantic rules for $\sum$ and $\prod$ are the familiar elimination rules

$$\frac{\Gamma \Rightarrow t\!:\!(\sum x\!:\!A)B}{\Gamma \Rightarrow lt\!:\!A} \qquad \frac{\Gamma \Rightarrow t\!:\!(\sum x\!:\!A)B}{\Gamma \Rightarrow rt\!:\!B[x \mapsto lt]} \qquad \frac{\Gamma \Rightarrow t\!:\!(\prod x\!:\!A)B \qquad \Gamma \Rightarrow t'\!:\!A}{\Gamma \Rightarrow ap(t,t')\!:\!B[x \mapsto t']}$$

(with left and right projections l and r respectively, and application ap), and introduction rules

$$\frac{\Gamma \Rightarrow t\!:\!A \qquad \Gamma \Rightarrow t'\!:\!B[x \mapsto t] \qquad \Gamma, x\!:\!A \Rightarrow B \text{ type}}{\Gamma \Rightarrow \langle t, t' \rangle\!:\!(\sum x\!:\!A)B}$$

$$\frac{\Gamma, x\!:\!A \Rightarrow t\!:\!B}{\Gamma \Rightarrow \lambda x.t\!:\!(\prod x\!:\!A)B}$$

(with pairing $\langle\cdot,\cdot\rangle$ and λ-abstraction). The force of $\bot$ can be captured by

$$\frac{\Gamma \Rightarrow A \text{ wff}}{\Gamma \Rightarrow 0 : \bot \supset A}$$

(saying $\bot$ entails every well-formed formula) with double negation through

$$\frac{\Gamma \Rightarrow A \text{ wff}}{\Gamma \Rightarrow \delta_A : \neg\neg A \supset A}$$

where, as usual, $\neg A$ is $A \supset \bot$. While type subscripts on $l, r, \langle\cdot,\cdot\rangle, ap, \lambda$ and 0 can be suppressed (in accordance with a type-free formulation), double negation δ_A lacks the uniformity for such a treatment (and is, of course, intuitionistically suspect).

2.3. Formal derivations versus interpretations. Given some set $\mathcal{D}$ of rules, as in §§2.1 and 2.2, a *$\mathcal{D}$-derivation* is a derivation (in the usual sense) from the rules in $\mathcal{D}$. Let us write '$\Gamma \Rightarrow^{\mathcal{D}} \Theta$' to mean that there is a $\mathcal{D}$-derivation of '$\Gamma \Rightarrow \Theta$', and set

$$\text{context}^{\mathcal{D}} = \{\Gamma \mid \text{there is a } \mathcal{D}\text{-derivation of '}\Gamma \text{ context'}\},$$

referring to its elements as *$\mathcal{D}$-contexts*. As $\mathcal{D}$ becomes larger, so does the collection of pairs Γ, A such that

$$\text{for some term } t, \text{ '}\Gamma \Rightarrow^{\mathcal{D}} t : A\text{'},$$

an alternative approach to which (exploiting the layer of language represented by the signature L) we pursue next.

An *L-(proof-)interpretation* $[\![\cdot]\!]$ is a function mapping every $U \in L_0$ to a set $[\![U]\!]$ and every $R \in L(U_1 \cdots U_n)$ and $u_1 \in [\![U_1]\!], \ldots, u_n \in [\![U_n]\!]$ to a set $[\![R, u_1 \cdots u_n]\!]$. An L-interpretation $[\![\cdot]\!]$ induces the many-sorted L-model M that interprets $U \in L_0$ as $U_M = [\![U]\!]$ and $R \in L(U_1 \cdots U_n)$ as

$$R_M = \{u_1 \cdots u_n \in [\![U_1]\!] \times \cdots \times [\![U_n]\!] \mid [\![R, u_1 \cdots u_n]\!] \neq \emptyset\},$$

the intuition being that $[\![R, u_1 \cdots u_n]\!]$ consists of proofs of $R(u_1, \ldots, u_n)$. Combining formal derivation in $\mathcal{D}$ with an L-interpretation $[\![\cdot]\!]$, we can define for every $\mathcal{D}$-context Γ,

(i) its set $[\![\Gamma]\!]^{\mathcal{D}}$ of *$\mathcal{D}$-instantiations* ρ interpreting Γ's variables x as $\rho(x)$

simultaneously (by induction on the length of $\mathcal{D}$-derivations) with

(ii) interpretations $[\![A]\!]^{\mathcal{D}}_{\Gamma,\rho}$ of A such that '$\Gamma \Rightarrow^{\mathcal{D}} A$ type' (for $\rho \in [\![\Gamma]\!]^{\mathcal{D}}$)

and (again for $\rho \in [\![\Gamma]\!]^{\mathcal{D}}$)

(iii) interpretations $[\![t]\!]^{\mathcal{D}}_{\Gamma,\rho}$ of t such that for some A, '$\Gamma \Rightarrow^{\mathcal{D}} t : A$'.

Within (i), we set $[\![\varepsilon]\!]^{\mathcal{D}}_{\Gamma,\rho} = \{\emptyset\}$ and

$$[\![\Gamma, x : A]\!]^{\mathcal{D}}_{\Gamma,\rho} = \{\rho^x_a \mid \rho \in [\![\Gamma]\!]^{\mathcal{D}} \text{ and } a \in [\![A]\!]^{\mathcal{D}}_{\Gamma,\rho}\}$$

where ρ^x_a is $\rho \cup \{(x, a)\}$, so that, as the base clause of (iii), $[\![x]\!]^{\mathcal{D}}_{\Gamma.\rho} = \rho(x)$. Within (ii), $[\![U]\!]^{\mathcal{D}}_{\Gamma.\rho} = [\![U]\!]$, $[\![\bot]\!]^{\mathcal{D}}_{\Gamma.\rho} = \emptyset$,

$$\begin{aligned} [\![R(t_1, \ldots, t_n)]\!]^{\mathcal{D}}_{\Gamma.\rho} &= [\![R, [\![t_1]\!]^{\mathcal{D}}_{\Gamma.\rho}, \ldots, [\![t_n]\!]^{\mathcal{D}}_{\Gamma.\rho}]\!] \\ [\![(Qx : T)A]\!]^{\mathcal{D}}_{\Gamma.\rho} &= (Qt \in [\![T]\!]^{\mathcal{D}}_{\Gamma.\rho})[\![A]\!]^{\mathcal{D}}_{\Gamma.x:A.\rho^x_t} \ \text{ for } Q = \sum, \prod, \end{aligned}$$

adopting $\sum, \prod$ notation at the meta-level with : replaced by $\in$. But what meta-theory? We can adopt the framework of Feferman [10] to interpret terms t built from l, r, $\langle\cdot,\cdot\rangle$, ap, λ and 0 in a model of the type-free lambda-calculus (with products), assigning types relative to contexts such that

$(*)$ whenever $\rho \in [\![\Gamma]\!]^{\mathcal{D}}$ and '$\Gamma \Rightarrow^{\mathcal{D}} t : A$', $[\![t]\!]^{\mathcal{D}}_{\Gamma.\rho} \in [\![A]\!]^{\mathcal{D}}_{\Gamma.\rho}$.

As for double negation δ_A, let us proceed more generally and suppose we are given a set **Ax** of axioms A, with proof terms p_A introduced by

$$\frac{\Gamma \Rightarrow A \text{ wff}}{\Gamma \Rightarrow p_A : A}\ A \in \mathbf{Ax}$$

(whence δ_A amounts to $p_{\neg\neg A \supset A}$). If $[\![A]\!]^{\mathcal{D}}_{\Gamma.\rho} \neq \emptyset$, say $a \in [\![A]\!]^{\mathcal{D}}_{\Gamma.\rho}$, then we can set $[\![p_A]\!]^{\mathcal{D}}_{\Gamma.\rho} = a$. Otherwise, we might set $[\![p_A]\!]^{\mathcal{D}}_{\Gamma.\rho}$ to some arbitrary value, and classify interpretations as good or bad according to $(*)$ above. That is, let us call an L-interpretation $[\![\cdot]\!]$ *sound for* **Ax** if for every $A \in \mathbf{Ax}$,

$$[\![p_A]\!]^{\mathcal{D}}_{\Gamma.\rho} \in [\![A]\!]^{\mathcal{D}}_{\Gamma.\rho}, \ \text{whenever '}\Gamma \Rightarrow^{\mathcal{D}} A \text{ wff' and } \rho \in [\![\Gamma]\!]^{\mathcal{D}}.$$

For a sound interpretation of double negation δ_A (i.e. $p_{\neg\neg A \supset A}$) relative to a $\mathcal{D}$-instantiation ρ of a $\mathcal{D}$-context Γ such that '$\Gamma \Rightarrow^{\mathcal{D}} A$ wff', it is useful to assume the law of excluded middle for $[\![A]\!]^{\mathcal{D}}_{\Gamma.\rho}$ — if $[\![A]\!]^{\mathcal{D}}_{\Gamma.\rho} \neq \emptyset$, say $a \in [\![A]\!]^{\mathcal{D}}_{\Gamma.\rho} \neq \emptyset$, then let $[\![\delta_A]\!]^{\mathcal{D}}_{\Gamma.\rho} = \lambda x.a$; otherwise, interpret δ_A exactly as the proof 0 of $\bot \supset A$.

Returning now to the question (Q2) from §1 of interpreting proofs model-theoretically, we can turn an L-model M into an L-interpretation $[\![\cdot]\!]_M$ by setting $[\![U]\!]_M = U_M$ and

$$[\![R, u_1 \cdots u_n]\!]_M = \begin{cases} \{0\} & \text{if } R_M(u_1, \cdots, u_n) \\ \emptyset & \text{otherwise.} \end{cases}$$

While 0 can hardly be described as a *canonical* proof object of every true atomic L-sentence, it will do as *a* proof object for our purposes. Indeed, for the sequel, it will be convenient to (i) quantify away proofs t, defining

$$\Gamma \models^{\mathcal{D}}_{[\![\cdot]\!]} A \ \text{ iff '}\Gamma \Rightarrow^{\mathcal{D}} A \text{ type' and } (\forall \rho \in [\![\Gamma]\!]^{\mathcal{D}})\ [\![A]\!]^{\mathcal{D}}_{\Gamma.\rho} \neq \emptyset\,,$$

and (ii) quantify away L-interpretations $[\![\cdot]\!]$,[2] defining

$$\Gamma \models^{\mathcal{D}} A \ \text{ iff for every } \mathcal{D}\text{-sound } [\![\cdot]\!],\ \Gamma \models^{\mathcal{D}}_{[\![\cdot]\!]} A$$

[2]I drop L-interpretations $[\![\cdot]\!]$ from §3 on, in no small measure because the subscript $[\![\cdot]\!]$ on $\models^{\mathcal{D}}_{[\![\cdot]\!]}$ gets very cumbersome. Re-introducing $[\![\cdot]\!]$ would lead to different notions of presupposition and entailment in §§3 and 4 that, I suspect, could be of interest, but at the expense of notational clutter. Be that as it may, I would like to draw the reader's attention to Fernando [12], where L-interpretations $[\![\cdot]\!]$ are pushed further, for a link-up with DRT.

where $[\![\cdot]\!]$ is understood to be $\mathcal{D}$-*sound* for a rule set $\mathcal{D}$ induced by **Ax** if $[\![\cdot]\!]$ is sound for **Ax**. (**Ax** may be empty, or may contain some/all instances of double negation, as well as facts such as: canaries are birds.)

§3. Translations and presuppositions. Fix a signature L and a countable set Var of variables $(x, \dots)$. Decontextualizing the types and terms from §2, let us re-christen them *pre*types and *pre*terms, under context-free presentations. That is, let pTy be the set of *pretypes A*

$$A ::= U \mid \bot \mid R(t, \dots, t) \mid (\textstyle\sum x : A)A \mid (\textstyle\prod x : A)A$$

generated from the set pTm of *preterms t*

$$t ::= x \mid lt \mid rt \mid \langle t, t\rangle \mid ap(t, t) \mid \lambda x.t \mid 0$$

(leaving δ_A / p_A out for simplicity). Throughout this section, let $\mathcal{D}$ be a fixed set of rules that includes all rules in §2.1 (and possibly more). Our goal is to describe how example (s) from §1 obtains the reading (1), assuming a formalization of the clause <u>the bird</u> will perish as the "extended pretype"

(2) $$\{bird(z)\}\ willPerish(z)$$

where $bird(z)$ is a "restriction" on the variable z (meant in the spirit, if not letter, of situation theory; Barwise and Cooper [4], Cooper [7]). Reconceptualized as requirements on contexts Γ, restrictions are linked in §3.1 (below) to a theory of presuppositions due to Karttunen [19], with $\{\cdot\}\cdot$ construed as a binary presupposition connective (see, for instance, §4.1 of Beaver [5]). A mechanism is then provided in §3.2 that, for the example above, binds z in (2) to the preterm lx, relative to a $\mathcal{D}$-context $\Gamma, x : (\sum y : canary)\ SylvesterGets(y)$ such that

$$\Gamma \models^{\mathcal{D}} (\textstyle\prod y : canary)\ bird(y)$$

(recalling that, under the conventions mentioned in §2.1, the preceding pretype unwinds to $(\prod y : U)(canary(y) \supset bird(y)))$.

3.1. Extended pretypes and full $\mathcal{D}$-presuppositions. Let us generate the set epTy of *extended pretypes e* by adding an inductive clause for restrictions to those giving pTy

$$e ::= U \mid \bot \mid R(t, \dots, t) \mid (\textstyle\sum x : e)e \mid (\textstyle\prod x : e)e \mid \{e\}e.$$

Now, the idea is to extend the notion '$\Gamma \Rightarrow^{\mathcal{D}} A$ type' from pretypes A to extended pretypes e. Following Karttunen [19], let us define a $\mathcal{D}$-context Γ to $\mathcal{D}$-*admit e* — in symbols, $\Gamma \rhd^{\mathcal{D}} e$ — by induction on e

$$\Gamma \rhd^{\mathcal{D}} A \ \text{ iff } \ \text{`}\Gamma \Rightarrow^{\mathcal{D}} A \text{ type'} \ \text{ for } A = U, \bot, R(t_1, \dots, t_n)$$
$$\Gamma \rhd^{\mathcal{D}} (Qx : e)e' \ \text{ iff } \ \Gamma \rhd^{\mathcal{D}} e \text{ and } \Gamma, x : e^- \rhd^{\mathcal{D}} e' \ \text{ for } Q = \textstyle\sum, \prod$$
$$\Gamma \rhd^{\mathcal{D}} \{e\}e' \ \text{ iff } \ \Gamma \models^{\mathcal{D}} e^\wedge \text{ and } \Gamma \rhd^{\mathcal{D}} e'$$

where e^- and $e^\wedge$ are pretypes given by functions $\cdot^-$ and $\cdot^\wedge$ from epTy to pTy defined as follows. For $A \in$ pTy, $A^- = A^\wedge = A$, and for $Q \in \{\sum, \prod\}$,

$$((Qx : e_1)e_2)^- = ((Qx : e_1{}^-)e_2{}^-) \qquad ((Qx : e_1)e_2)^\wedge = ((Qx : e_1{}^\wedge)e_2{}^\wedge) .$$

But while $\cdot^-$ erases restrictions, $\cdot^\wedge$ incorporates them

$$(\{e_1\}e_2)^- = e_2{}^- \qquad (\{e_1\}e_2)^\wedge = e_1{}^\wedge \wedge e_2{}^\wedge$$

where, as in §2.1, $A \wedge B$ abbreviates $(\sum x : A)B$ for some fresh/dummy variable x. Clearly, for every pretype A, $\Gamma \rhd^{\mathcal{D}} A$ iff '$\Gamma \Rightarrow^{\mathcal{D}} A$ type'. Moreover, equating $\neg e$ with $(\prod x : e)\bot$ for some x not in e,

$$\Gamma \rhd^{\mathcal{D}} \neg e \text{ iff } \Gamma \rhd^{\mathcal{D}} e .$$

Now, e is said to *$\mathcal{D}$-presuppose* A if $\Gamma \models^{\mathcal{D}} A$ for every Γ that $\mathcal{D}$-admits e.

Example. Assuming $\Gamma \models^{\mathcal{D}} (\prod y : canary)\ bird(y)$, the $\mathcal{D}$-context

$$\Gamma,\ x : (\sum y : canary)\ SylvesterGets(y)$$

$\mathcal{D}$-admits the extended pretype

(3) $$\{bird(lx)\}\ willPerish(lx)$$

which, in turn, $\mathcal{D}$-presupposes $bird(lx)$. Dropping the restriction $\{bird(lx)\}$ from (3), notice that if a $\mathcal{D}$-context Γ $\mathcal{D}$-admits the pretype $willPerish(lx)$, then '$\Gamma \Rightarrow^{\mathcal{D}} lx : U$'.

Accordingly, let us say that e *$\mathcal{D}$-presupposes* $t : A$ if '$\Gamma \Rightarrow^{\mathcal{D}} t : A$' for every Γ such that $\Gamma \rhd^{\mathcal{D}} e$. A *full $\mathcal{D}$-presupposition of* an extended pretype e is a pair $\mathcal{A}, \mathcal{T}$ of finite sets $\mathcal{A}$ and $\mathcal{T}$ of pretypes A and typings $t : B$ respectively, such that for every $\mathcal{D}$-context Γ,

$$\Gamma \rhd^{\mathcal{D}} e \text{ iff } (\forall A \in \mathcal{A})\, \Gamma \models^{\mathcal{D}} A \text{ and } (\forall `t : B\text{'} \in \mathcal{T})\, `\Gamma \Rightarrow^{\mathcal{D}} t : B\text{'} .$$

Let us write this in symbols as $e \downarrow \mathcal{A}, \mathcal{T}$.

Theorem. *Suppose that $\mathcal{D}$ contains not only the rules in §2.1 but also the introduction rule for $\prod$ in §2.2, along with the usual formation rules*

$$\frac{\Gamma \Rightarrow A \text{ type} \qquad \Gamma, x : A \Rightarrow B \text{ type}}{\Gamma \Rightarrow (\sum x : A)B \text{ type}} \qquad \frac{\Gamma \Rightarrow A \text{ type} \qquad \Gamma, x : A \Rightarrow B \text{ type}}{\Gamma \Rightarrow (\prod x : A)B \text{ type}} .$$

Then there is an algorithm that given $e \in$ epTy returns a full $\mathcal{D}$-presupposition of e.

The algorithm is based on the calculations

$$e \downarrow \emptyset, \emptyset \quad \text{for } e = \bot \text{ or some } U \in L_0$$
$$R(t_1, \ldots, t_n) \downarrow \emptyset, \{t_1 : U_1, \ldots, t_n : U_n\} \quad \text{for } R \in L(U_1 \cdots U_n)$$

and if $e \downarrow \mathcal{A}, \mathcal{T}$ and $e' \downarrow \mathcal{A}', \mathcal{T}'$ then both $(\sum x : e)e'$ and $(\prod x : e)e'$ have full $\mathcal{D}$-presupposition

$$\mathcal{A} \cup \{(\prod x : e^-)A' \mid A' \in \mathcal{A}'\},\ \mathcal{T} \cup \{\lambda x.t' : (\prod x : e^-)B' \mid `t' : B'\text{'} \in \mathcal{T}'\}$$

and

$$\{e\}e' \downarrow \{e^\wedge\} \cup \mathcal{A}', \mathcal{T}'.$$

Examples. The extended pretype (3) has full $\mathcal{D}$-presupposition $\{bird(lx)\}, \{lx : U\}$ which can, in fact, be reduced to $\{bird(lx)\}, \emptyset$ since

$$\text{whenever } \Gamma \models^{\mathcal{D}} bird(lx)\ ,\ `\Gamma \Rightarrow^{\mathcal{D}} lx : U'\ .$$

An extended pretype e and its negation $\neg e$ (i.e. $(\prod x : e)\bot$) have the same full $\mathcal{D}$-presuppositions.

3.2. Presupposition binding. Missing from our account of example (s) is how to go from (2) to (3), which we presently take up, turning $z \mapsto lx$ into a substitution. Given a function θ from a subset X of Var to pTm, define the function $\hat{\theta}$: $(\mathsf{pTm} \cup \mathsf{epTy}) \to (\mathsf{pTm} \cup \mathsf{epTy})$ substituting $x \in X$ by $\theta(x)$

$$\begin{array}{rclrcl} \hat{\theta}(x) & = & \theta(x) \ \text{ for } x \in X & \hat{\theta}(ft) & = & f\hat{\theta}(t) \ \text{ for } f = l, r \\ \hat{\theta}(x) & = & x \ \text{ for } x \notin X & \hat{\theta}(ap(t, t')) & = & ap(\hat{\theta}(t), \hat{\theta}(t')) \\ \hat{\theta}(A) & = & A \ \text{ for } A = U, \bot & \hat{\theta}(\langle t, t'\rangle) & = & \langle \hat{\theta}(t), \hat{\theta}(t')\rangle \\ \hat{\theta}(R(t_1, \ldots, t_n)) & = & R(\hat{\theta}(t_1), \ldots, \hat{\theta}(t_n)) & \hat{\theta}(\{e\}e') & = & \{\hat{\theta}(e)\}\hat{\theta}(e') \end{array}$$

except where the variable occurs bound

$$\begin{array}{rcl} \hat{\theta}(\lambda x.t) & = & \lambda x.\hat{\theta_x}(t) \\ \hat{\theta}((\sum x : e)e') & = & (\sum x : \hat{\theta}(e))\hat{\theta_x}(e') \\ \hat{\theta}((\prod x : e)e') & = & (\prod x : \hat{\theta}(e))\hat{\theta_x}(e') \end{array}$$

with $\theta_x = \theta - \{(x, \theta(x))\}$ (thereby confining the substitution to free occurrences). Next, fix an extended pretype e and a set $X \subset \mathsf{Var}$ of variables that intuitively are crying out to be bound. That is, let the pair (e, X) represent an "ambiguous" extended pretype which a function θ with domain X disambiguates as $\hat{\theta}(e)$. Finding a suitable θ such that a given context Γ $\mathcal{D}$-admits $\hat{\theta}(e)$ can be non-trivial. Part of the problem is $\models^{\mathcal{D}}$, which we might calculate by extending $\mathcal{D}$ to some $\mathcal{D}' \supseteq \mathcal{D}$ such that for every A,

$$\Gamma \models^{\mathcal{D}} A \ \text{ iff for some } t,\ `\Gamma \Rightarrow^{\mathcal{D}'} t : A'.$$

The search for $\mathcal{D}'$ and t can, however, be open-ended — far harder than the calculation of full presuppositions.

Notice that we have relativized condition (†) from §1 to $\mathcal{D}$,

$$`\Gamma \Rightarrow^{\mathcal{D}} A \ \mathsf{type}',$$

and broken that into two steps. First, calculate a full $\mathcal{D}$-presupposition of A; second, show that Γ satisfies that presupposition. The first step can be carried out without Γ; the second can do without A. The approach applies to extended pretypes e (not just A), on top of which some indeterminacy can be introduced through different bindings on a set X of variables. In reply then to question (Q1) from §1, we have distinguished two tractable tasks,

the calculation of full presuppositions, and checking $\Gamma \Rightarrow t : A$,

from two other problems,

determining $\models^{\mathcal{D}}$, and resolving X through suitable bindings θ.

A handle on the indeterminacy from X is provided in the next section through a general framework for translating expressions such as (e, X) to logical forms such as pretypes, on which $\models^{\mathcal{D}}$ induces a notion of entailment.

§4. Non-determinism in translation and attunement. Translating English into a formal language is non-trivial insofar as the contextual character of ordinary discourse is taken into account. To capture discourse effects, let us apply a translation process to sequences of sentences, encoding the input/output behavior of the process by a set

$$St \subseteq \bigcup_{n \geq 0} (E^n \times \Phi^n)$$

of pairs $(\vec{e}, \vec{\varphi})$ of sequences $\vec{e} = \mathsf{e}_1 \cdots \mathsf{e}_n$ and $\vec{\varphi} = \varphi_1 \cdots \varphi_n$ from a set E of *expressions* e_i and a set Φ of *logical forms* φ_i such that the input $\vec{e}$ may translate to (the output) $\vec{\varphi}$. The input/output relation St is *not* assumed functional, allowing an input $\vec{e}$ to be ambiguous because there are distinct sequences $\vec{\varphi}$ and $\vec{\psi}$ such that both $(\vec{e}, \vec{\varphi})$ and $(\vec{e}, \vec{\psi})$ belong to St. Of course, the difference between $\vec{\varphi}$ and $\vec{\psi}$ may well be superficial — a measure of which is provided by what these sequences entail. That is, there is presumably some notion $\vdash \subseteq \Phi^* \times \Phi$ of entailment on Φ that tests just how different two outputs $\vec{\varphi}$ and $\vec{\psi}$ are. Indeed, we may well expect $\vdash$ to be the reason for mapping E to Φ.

Example (from §3, given a a fixed set $\mathcal{D}$ of rules). Let $E = \mathsf{epTy} \times Pow(\mathsf{Var})$, $\Phi = \mathsf{pTy}$, $\vdash^{\mathcal{D}} \subseteq \mathsf{pTy}^* \times \mathsf{pTy}$ be given by

$$A_1 \cdots A_n \vdash^{\mathcal{D}} A \text{ iff } x_1 : A_1, \ldots, x_n : A_n \models^{\mathcal{D}} A$$

and $St_{\mathcal{D}}$ consist, for every $n \geq 0$, of all pairs

$$((e_1, X_1)(e_2, X_2) \cdots (e_n, X_n)\ ,\ A_1 A_2 \cdots A_n) \in E^n \times \Phi^n$$

such that for some binding $\theta : (\bigcup_{i=1}^n X_i) \rightarrow \mathsf{pTm}$, whenever $1 \leq i \leq n$,

$$\mathsf{Var}(e_i) \cap dom(\theta) \subseteq X_i,\ \ A_i \text{ is } \hat{\theta}(e_i)^- \text{ and } x_1 : A_1, \ldots, x_{i-1} : A_{i-1} \rhd^{\mathcal{D}} \hat{\theta}(e_i)$$

where $\mathsf{Var}(e_i)$ is the set of all variables with free occurrences in e_i. Notice that the same binding θ disambiguates each of $e_1, \ldots, e_n$.

4.1. The language $\mathcal{L}_\circ$. To lift $\vdash$-entailments of $\vec{\varphi}$ up to $\vec{e}$ along the translation set St, it is useful to extend the set Φ of logical forms φ to a set of $\mathcal{L}_\circ(E, \Phi)$-*formulas* A generated according to

$$\mathsf{A} ::= \varphi \mid \langle \mathsf{e} \rangle \mathsf{A} \mid \tilde{\neg} \mathsf{A} \mid \mathsf{A} \tilde{\wedge} \mathsf{A}$$

with different modalities $\langle \mathsf{e} \rangle$ for each expression $\mathsf{e} \in E$, plus negation $\tilde{\neg}$ and conjunction $\tilde{\wedge}$ (not to be confused with any connectives in Φ or E). Let us shorten $\mathcal{L}_\circ(E, \Phi)$ to $\mathcal{L}_\circ$ and identify $\mathcal{L}_\circ$ with the set of $\mathcal{L}_\circ$-formulas. The semantics of $\mathcal{L}_\circ$ is specified relative to an $\mathcal{L}_\circ$-*model* $(St, \vdash)$ by a "forcing" relation $\Vdash_\circ \subseteq St \times \mathcal{L}_\circ$ as follows, where $(\vec{e}, \vec{\varphi}) \in St$:

(i) a logical form φ is forced if $\vdash$-entailed by the output

$$(\vec{e}, \vec{\varphi}) \Vdash_\circ \varphi \quad \text{iff} \quad \vec{\varphi} \vdash \varphi$$

(ii) $\langle e \rangle A$ is forced by $(\vec{e}, \vec{\varphi})$ if some extension of $(\vec{e}, \vec{\varphi})$ by e forces A

$$(\vec{e}, \vec{\varphi}) \Vdash_\circ \langle e \rangle A \quad \text{iff} \quad \text{for some } \varphi,\ (\vec{e}e, \vec{\varphi}\varphi) \Vdash_\circ A$$

(where it is understood that $(\vec{e}e, \vec{\varphi}\varphi)$ must also belong to *St*)

(iii) $\tilde{\neg}$ and $\tilde{\wedge}$ are treated classically

$$(\vec{e}, \vec{\varphi}) \Vdash_\circ \tilde{\neg} A \quad \text{iff} \quad \text{not } (\vec{e}, \vec{\varphi}) \Vdash_\circ A$$

$$(\vec{e}, \vec{\varphi}) \Vdash_\circ A \tilde{\wedge} B \quad \text{iff} \quad (\vec{e}, \vec{\varphi}) \Vdash_\circ A \text{ and } (\vec{e}, \vec{\varphi}) \Vdash_\circ B\,.$$

Observe that $\Vdash_\circ$ is just Kripke semantics, where the accessibility relation $\stackrel{e}{\rightarrow}$ underlying $\langle e \rangle$ is the binary relation on *St* such that

$$(\vec{e}, \vec{\varphi}) \stackrel{e}{\rightarrow} (\vec{e'}, \vec{\varphi'}) \quad \text{iff} \quad \vec{e'} = \vec{e}e \text{ and } \vec{\varphi'} = \vec{\varphi}\varphi \text{ for some } \varphi$$

for all $(\vec{e}, \vec{\varphi}), (\vec{e'}, \vec{\varphi'}) \in St$. The persistence of the outputs $\vec{\varphi}$ under $\stackrel{e}{\rightarrow}$ reflects a processing of the inputs which is

(i) *incremental*, translating a sequence $e_1 \cdots e_n$ one sentence at a time

$$(\varepsilon, \varepsilon) \stackrel{e_1}{\rightarrow} (e_1, \varphi_1) \stackrel{e_2}{\rightarrow} \cdots \stackrel{e_n}{\rightarrow} (e_1 \cdots e_n, \vec{\varphi})$$

(whence for all $(\vec{e}, \vec{\varphi}) \in St$, the length of $\vec{\varphi}$ is the same as that of $\vec{e}$)

and

(ii) *inertial* insofar as the incremental results are preserved: whenever $(\vec{e}, \vec{\varphi}) \stackrel{e}{\rightarrow} (\vec{e'}, \vec{\psi})$, $\vec{\varphi}$ is a prefix of $\vec{\psi}$ (not to mention $\vec{e}$ is a prefix of $\vec{e'}$).

For a translation process that is incremental and inertial (as in DRT), we can regard *St* not only as its input/output relation but also as its set of *stages*. This perspective is developed in Fernando [11]. The additional point of the present work is twofold: to tie $\mathcal{L}_\circ$ in with §§2 and 3 above, and to expand $\mathcal{L}_\circ$ so as to bring in a third process underlying interpretation.

But first, a bit more background. Ambiguity in an expression e at a stage $(\vec{e}, \vec{\varphi})$ is discerned in $\mathcal{L}_\circ$ by $\mathcal{L}_\circ$-formulas A, if any, such that

$$(\vec{e}, \vec{\varphi}) \Vdash_\circ \langle e \rangle A \ \tilde{\wedge}\ \langle e \rangle \tilde{\neg} A\,.$$

Quantifying over all stages, let us call an $\mathcal{L}_\circ$-formula A *valid in* $(St, \vdash)$ if for every $(\vec{e}, \vec{\varphi}) \in St$, $(\vec{e}, \vec{\varphi}) \Vdash_\circ A$. For example, suppose E includes the following English sentences e_1 and e_2.

e_1 Doris informed Chloe <u>she</u> aced the exam.

e_2 In fact, <u>she</u> was the only Texan who passed.

Ambiguous though e_1 may be, to say that the she's in $e_1 e_2$ co-vary in the $\mathcal{L}_\circ$-model $(St, \vdash)$ is to require that the $\mathcal{L}_\circ$-formula

$$[e_1][e_2]\ (\forall x \in \mathit{Texan})(\mathit{passed}(x) \supset \mathit{aced}(x))$$

be valid in $(St, \vdash)$, where $[e]A$ abbreviates $\tilde{\neg}\langle e \rangle \tilde{\neg} A$.

Before worrying about multiple outputs, we might check if an input e at stage $(\vec{e}, \vec{\varphi})$ has any output at all, which can be expressed as

$$(\vec{e}, \vec{\varphi}) \; \Vdash_\circ \; \langle e \rangle \tilde{\top} \quad [\text{ iff } (\vec{e}, \vec{\varphi}) \in dom(\stackrel{e}{\rightarrow}) \;]$$

where $\tilde{\top}$ is an $\mathcal{L}_\circ$-tautology (valid in every $\mathcal{L}_\circ$-model). Underlying §3.1 is the identification of the presupposition of e with its precondition for translation, $\langle e \rangle \tilde{\top}$. Under this conception, the *negation test* for presupposition, which states that the presupposition of the negation ¬e of an expression e is no different from the presupposition of e, becomes the $\mathcal{L}_\circ$-formula

$$\langle \neg e \rangle \tilde{\top} \; \tilde{\equiv} \; \langle e \rangle \tilde{\top}$$

(where $\tilde{\equiv}$ is the Boolean $\mathcal{L}_\circ$-biconditional defined from $\tilde{\neg}$ and $\tilde{\wedge}$), saying ¬e is translatable iff e is. As for a binary connective $\star$ on E (such as conjunction $\wedge$), let us agree to call $\star$ *St-sequential* if St translates the arguments of $\star$ sequentially — i.e., for all $(\vec{e}, \vec{\varphi}) \in St$ and $e, e' \in E$,

$$(\vec{e}(e \star e'), \vec{\varphi}\psi) \in St \text{ iff } \psi = \varphi \cdot \varphi' \text{ for some } \varphi \text{ and } \varphi' \text{ such that}$$
$$(\vec{e}e, \vec{\varphi}\varphi) \in St \text{ and } (\vec{e}ee', \vec{\varphi}\varphi\varphi') \in St$$

where $\cdot$ is some binary connective on Φ (possibly $\star$). If $\star$ is St-sequential then it immediately follows that the $\mathcal{L}_\circ$-scheme

$$\langle e \star e' \rangle \tilde{\top} \; \tilde{\equiv} \; \langle e \rangle \langle e' \rangle \tilde{\top}$$

is valid in $(St, \vdash)$, for any $\vdash$. The dependent type interpretations of conjunction $\wedge$ and implication $\supset$ in §§2 and 3 above are both $St_{\mathcal{D}}$-sequential (under the definition of $St_{\mathcal{D}}$ given right before this subsection). Moreover, the negation test, $\langle \neg e \rangle \tilde{\top} \tilde{\equiv} \langle e \rangle \tilde{\top}$, becomes a consequence of defining ¬e as $e \supset \perp$, provided $\langle \perp \rangle \tilde{\top}$, as is the case with our dependent type example $St_{\mathcal{D}}$.[3]

Going well beyond $\langle \perp \rangle \tilde{\top}$, an assumption about St is made in Fernando [11], motivated by the example of translating English sentences (E) into ordinary predicate logic formulas (Φ). The assumption, called $(a_\circ)$, embeds the set Φ of logical forms in the set E of inputs to translation, with the understanding that St keeps the logical forms fixed, and contains the pair $(\varepsilon, \varepsilon)$ — i.e.,

$$(a_\circ) \quad \Phi \subset E, \; (\varepsilon, \varepsilon) \in St \text{ and for every } (e_1 \cdots e_n, \varphi_1 \cdots \varphi_n) \in St,$$
$$(\forall i \in \{1, \ldots, n\}) \text{ if } e_i \in \Phi \text{ then } e_i = \varphi_i$$
$$(\forall \varphi \in \Phi) \; (e_1 \cdots e_n\varphi, \varphi_1 \cdots \varphi_n\varphi) \in St \, .$$

[3]To be more precise, we have, over $St_{\mathcal{D}}$, $\langle (\perp, \emptyset) \rangle \tilde{\top}$ rather than $\langle \perp \rangle \tilde{\top}$. We can, however, systematically confuse $(A, \emptyset)$ with A (as we see shortly). But we should not confuse presupposition satisfaction with consistency any more than translation with entailment

$$\frac{\text{translation}}{\text{presupposition satisfaction}} \approx \frac{\text{entailment}}{\text{consistency}} .$$

Inter-related though entailment and translation are, the processes are distinct, there being no shortage of English sentences that translate into what are, under ordinary notions of entailment, outright contradictions.

Under $(\mathrm{a}_\circ)$, the logical forms in Φ are extreme cases of unambiguous expressions with vacuous presuppositions. This assumption does not hold for the example of $St_{\mathcal{D}}$ above (with $\Phi = \mathsf{pTy}$), as an arbitrary context need not satisfy the presuppositions of an arbitrary pretype. Otherwise, however, notice that $(\varepsilon, \varepsilon) \in St_{\mathcal{D}}$ and that a copy of pTy in $E = \mathsf{epTy} \times Pow(\mathsf{Var})$ is given by mapping $A \in \mathsf{pTy}$ to $(A, \emptyset) \in \mathsf{epTy} \times Pow(\mathsf{Var})$. (The choice of $\emptyset$ ensures that $(A, \emptyset)$ can only be translated to A in $St_{\mathcal{D}}$.) Turning to $\vdash$, let us call $\vdash \subseteq \Phi^* \times \Phi$ $\mathcal{L}_\circ$*-transparent* if for all $\varphi_1, \dots, \varphi_n, \varphi \in \Phi$,

$$\varphi_1 \cdots \varphi_n \vdash \varphi \;\text{ iff }\; (\varphi_1 \tilde{\wedge} \cdots \tilde{\wedge} \varphi_n) \tilde{\supset} \varphi \text{ is } (St, \vdash)\text{-valid for any } St$$

(where $\tilde{\supset}$ is, like $\tilde{\equiv}$, a Boolean $\mathcal{L}_\circ$-connective defined from $\tilde{\neg}$ and $\tilde{\wedge}$).

Proposition. *$\vdash$ is $\mathcal{L}_\circ$-transparent iff $\vdash$ verifies the schemes*

$$(Weak)\ \frac{\vec{\varphi} \vdash \varphi}{\psi\vec{\varphi} \vdash \varphi}\ \ \frac{\vec{\varphi} \vdash \varphi}{\vec{\varphi}\psi \vdash \varphi} \qquad (Ref)\ \frac{}{\varphi \vdash \varphi} \qquad (Cut)\ \frac{\vec{\varphi} \vdash \psi \qquad \psi\vec{\varphi} \vdash \varphi}{\vec{\varphi} \vdash \varphi}\ .$$

Just as $(\mathrm{a}_\circ)$ does not quite hold for $St_{\mathcal{D}}$, neither does $\mathcal{L}_\circ$-transparency for $\vdash^{\mathcal{D}}$. The assumptions are, however, met by a special fragment of pTy, including all first-order L-formulas, and are of interest inasmuch as that fragment is. So continuing with our review of Fernando [11], let us add to $\mathcal{L}_\circ$-transparency the assumption that Φ has conjunction and implication (distinct from $\mathcal{L}_\circ$-connectives), recording these in

$(\mathrm{a}\vdash)$ (Weak), (Ref) and (Cut) hold in $\vdash$, and there are binary connectives $\wedge$ and $\supset$ in Φ such that for all $\varphi, \psi, \varphi' \in \Phi$ and $\vec{\varphi} \in \Phi^*$,

$$(\varphi \wedge \psi)\vec{\varphi} \vdash \varphi' \;\text{ iff }\; \varphi\psi\vec{\varphi} \vdash \varphi'$$
$$\vec{\varphi}\varphi \vdash \psi \;\text{ iff }\; \vec{\varphi} \vdash \varphi \supset \psi\ .$$

Using $\wedge$ to combine any finite number of logical forms entailed at a certain stage into one, and $\supset$ to anticipate logical forms entailed at a next stage, we can establish

Theorem. *The set of $\mathcal{L}_\circ$-formulas valid in all $\mathcal{L}_\circ$-models $(St, \vdash)$ satisfying $(\mathrm{a}_\circ)$ and $(\mathrm{a}\vdash)$ is decidable, and consists of the theorems of normal modal logic, supplemented with the $\mathcal{L}_\circ$-axiom schemes*

(A1)	$\varphi \supset \varphi$	*(A4)*	$\varphi \tilde{\supset} [\mathsf{e}]\varphi$
(A2)	$\varphi \wedge \psi \tilde{\equiv} \varphi \tilde{\wedge} \psi$	*(A5)*	$\langle\varphi\rangle\psi \tilde{\equiv} \varphi \supset \psi$
(A3)	$(\varphi \supset \psi) \tilde{\supset} (\varphi \tilde{\supset} \psi)$	*(A6)*	$\langle\varphi\rangle\mathsf{A} \tilde{\supset} [\varphi]\mathsf{A}$

Moreover, for a fixed $\vdash$ satisfying $(\mathrm{a}\vdash)$, it suffices to strenghten (A1)-(A3) to the $\mathcal{L}_\circ$-scheme

$$\varphi_1 \tilde{\wedge} \cdots \tilde{\wedge} \varphi_n \tilde{\supset} \varphi$$

for all $\varphi_1, \dots, \varphi_n, \varphi$ such that $\varphi_1 \cdots \varphi_n \vdash \varphi$.

The obvious question now is how to modify the axioms in the theorem above, when stepping beyond elementary predicate logic and into the rest of pTy.

4.2. Presuppositions and leaps between $\mathcal{L}_\circ$-models. The preceding analysis of presupposition $\langle \mathsf{e} \rangle \tilde{\top}$ in terms of *St*-sequentiality may have left the impression that presupposition is all a matter of translation *St*. This is misleading insofar as it suggests that entailment $\vdash$ is irrelevant to presupposition — which is hardly the case with our dependent type example: $St_{\mathcal{D}}$ is tightly coupled to $\vdash^{\mathcal{D}}$. More concretely, returning to example (s), let Γ_1 be the $\mathcal{D}$-context

$$x_1 : (\sum y\!:\!canary)\ SylvesterGets(y)$$

and abstracting out ambiguity for the time being, let us translate

the bird will perish

as $(\hat{e}, \emptyset)$ where $\hat{e}$ is the extended pretype $\{bird(lx_1)\}\ willPerish(lx_1)$. Assuming (as we have) that *bird*, *willPerish*$\in L(U)$ for the same $U \in L_0$, it follows that

$$\langle(\hat{e}, \emptyset)\rangle \tilde{\top}\ \hat{\equiv}\ bird(lx_1)$$

is $(St_{\mathcal{D}}, \vdash^{\mathcal{D}})$-valid, and that if $\models^{\mathcal{D}} (\prod y\!:\!canary) bird(y)$, Γ_1 $\mathcal{D}$-admits $\hat{e}$. Observe that dropping the restriction $\{bird(lx_1)\}$ on $\hat{e}$ complicates matters, as reducing the presupposition $\langle(willPerish(lx_1), \emptyset)\rangle\tilde{\top}$ to some pretype is problematic. This difficulty is not surprising, in view of well-known limitations on internal definability in explicit mathematics and in Frege structures (pp 41,42, Aczel [1]). Put positively, imposing suitable restrictions on an extended pretype e allows us to assume that e has a full $\mathcal{D}$-presupposition $\mathcal{A}, \emptyset$ where $\mathcal{A}$ consists of a single pretype A and where moreover we might arrange '$\Gamma \Rightarrow^{\mathcal{D}} A$ wff' for every $\mathcal{D}$-context Γ that $\mathcal{D}$-admits e. Re-introducing ambiguity, let us replace $(\hat{e}, \emptyset)$ by $(e', \{z\})$ where e' is the extended pretype $\{bird(z)\}\ willPerish(z)$. In this case, our reduction of presupposition stops at

$$\langle(e', \{z\})\rangle\tilde{\top}\ \hat{\equiv}\ \langle(\{bird(z)\}\bot, \{z\})\rangle\ \tilde{\top}\ .$$

Identifying the presupposition of an expression e with $\langle \mathsf{e} \rangle \tilde{\top}$ raises the question: what if $\langle \mathsf{e} \rangle \tilde{\top}$ fails? Suppose, for instance, (s) were altered ever so slightly to

(s′) If Sylvester gets holds of a canary, the poor bird will perish.

Intuitively (s′) is as interpretable as (s). But whereas the rule set $\mathcal{D}$ chosen at the outset may well contain the information that canaries are birds, it is another matter to anticipate (s′) by arranging that

(4) $$\varepsilon \models^{\mathcal{D}} (\prod y\!:\!canary)\ (SylvesterGets(y) \supset poorBird(y))\ .$$

If (4) fails, then we have presupposition failure, and interpretation of (s′) requires leaping out of the $\mathcal{L}_\circ$-model $(St_{\mathcal{D}}, \vdash^{\mathcal{D}})$ and into some other $\mathcal{L}_\circ$-model where the translation of (s′) can proceed. Such leaps, repairing defective contexts, are called *accommodation* in the presupposition literature (surveyed in Beaver [5]). They are not to be undertaken mindlessly. Consider

If Sylvester gets holds of a canary, Sylvester's son will get dinner.

Assuming we were willing to accommodate, just what assumption should we take on board: that Sylvester has a son, *Shas*, or that he does if he captures a canary

$$(\prod y : canary)\ (SylvesterGets(y) \supset Shas)\ ?$$

When and how to accommodate are notoriously tricky questions involving judgments of plausibility that may or may not be amenable to formalization. In particular, while the presuppositions calculated in §3.1 may be useful for flagging presupposition failure, they need not themselves be the appropriate assumptions to incorporate into defective contexts. And once recovery from some cases of presupposition failure is permitted, it is far from clear what intuitions about presuppositions to test. For instance, is interpreting (s′) really a matter of accommodation or not? Our somewhat slippery reply above is: that depends on whether or not interpretation begins at a rule set $\mathcal{D}$ verifying (4). The skeptical reader may inquire instead if

$$(\prod y : canary)\ (SylvesterGets(y) \supset poorBird(y))$$

is part of the presupposition, assertion or implicature of (s′)? The term presupposition has been applied as a label on so many things that perhaps any account of it is bound to be contentious.[4] What is at stake in the present work is the claim that failures in translation are interesting, as are the adjustments (to $\mathcal{L}_\circ$-models) made before, during or after a particular instance of translation.

Whatever picture of presupposition one adopts, there are many reasons (including Gricean implicatures) for considering background information α alongside the explicit discourse context $(\vec{e}, \vec{\varphi})$. The generality of logical forms (e.g. McCarthy and Buvač [23]) is brought out by keeping α separate from the translation

---

[4]A prima facie different approach pursued in (for instance) Gazdar [14] is to cancel presuppositions (initially adopted in toto, unaltered from the sentence's simple constituents) which conflict with certain implicatures derived from Grice's maxim of quantity. (As this cancellation is to be carried out after logical forms have been constructed, we may expect that some stages entail the contradiction $\perp$ — as exemplified by the stage $(\perp, \perp)$ that $(a_\circ)$ provides, embodying the saying: "garbage in, garbage out." But we can, in $\mathcal{L}_\circ$, discriminate between stages that force $\perp$ and those that force $\tilde{\neg}\perp$, leaping between $\mathcal{L}_\circ$-models on that basis.)

Instead of cancellation, the approach due to Karttunen [19] which we have been considering above takes a sentence "to be an increment to a context that satisfies its presuppositions" (page 191). It is not altogether evident, however, that increments are any simpler than decrements, as accommodation complicates the sense in which increments might be described as minimal. The effect of accommodation is to satisfy presuppositions on the basis not so much of entailment but of *compatibility*, insofar as any $\mathcal{L}_\circ$-model on which we land after a leap must (in some plausible sense) be compatible with the old $\mathcal{L}_\circ$-model (from which we take off). Compatibility here may explain why there is nothing at all odd about (s′) relative to (s). (That said, we may prefer to satisfy presuppositions without changing $\mathcal{L}_\circ$-models, resorting to accommodation only if we have to, and only if we could — difficult it may be to spell out a general rule stating when and how we could.) A wider ranging account of the coherence of (s′) might look at the interplay between (hearer) interpretation and (speaker) generation, and, with that, intentions (purposes) and planning. (How to bring all this within the present framework I can only hint at, in §5.1 below.)

$\vec{\varphi}$ of an input $\vec{\mathsf{e}}$ (rather than packing it into $\vec{\varphi^{\alpha}}$) and then applying $\vec{\varphi}$ to different α's. The process of choosing α I propose to label *attunement*, and relate to the two other processes of entailment and translation by stipulating that α determines an $\mathcal{L}_\circ$-model $(St_\alpha, \vdash_\alpha)$. Collecting α's in a set I that indexes a family $\{(St_\alpha, \vdash_\alpha)\}_{\alpha \in I}$ of $\mathcal{L}_\circ$-models, an expansion $\mathcal{L}$ of $\mathcal{L}_\circ$ is outlined in the next section, relative to which such a family of $\mathcal{L}_\circ$-models constitutes a model. Without settling what are legitimate cases of accommodation or implicature, we can nonetheless survey the possibilities in $\mathcal{L}$, passing from $\mathcal{L}_\circ$-stages $(\vec{\mathsf{e}}, \vec{\varphi})$ to $\mathcal{L}$-stages $(\vec{\mathsf{e}}, \vec{\varphi})_\alpha$, where α keeps track of the $\mathcal{L}_\circ$-model to which $(\vec{\mathsf{e}}, \vec{\varphi})$ is understood to belong. The set I of indices opens up a dimension in interpretation, beyond translation and entailment, both of which take on a monotonic character in §4.1. The expansion from $\mathcal{L}_\circ$ to $\mathcal{L}$ allows non-monotonicity to enter through leaps $(\vec{\mathsf{e}}, \vec{\varphi})_\alpha \bar{\rightarrow} (\vec{\mathsf{e}}, \vec{\varphi})_\beta$ that change α — although, in fact, already in $\mathcal{L}_\circ$, transitions $(\vec{\mathsf{e}}, \vec{\varphi}) \overset{\diamond}{\rightarrow} (\vec{\mathsf{e}}, \vec{\psi})$ exploiting ambiguity within a fixed $\mathcal{L}_\circ$-model provide a measure of non-monotonicity. Non-monotonicity aside, the mickey-mouse impression given by translations such as $SylvesterGets(y)$ points to the possibility of refining the analysis, reifying Sylvester, and beyond that, the possible events of Sylvester getting hold of a canary. Such refinements amount to cases of re-attunement, if we encode into α not only the rule set $\mathcal{D}$ but also the signature L underlying $\mathcal{D}$, allowing the notion Φ of logical form to vary with α. Having embedded Φ in E under $(\mathrm{a}_\circ)$, it is only fair to obliterate restrictions keeping expressions in E from being returned as outputs.

4.3. Reifying $\mathcal{L}_\circ$-contexts and explicit translations. Blurring the line between E and Φ, let us define an *$\mathcal{L}$-model* to be a pair $(ST, \{\vdash_\alpha\}_{\alpha \in I})$ such that $ST \subseteq \bigcup_{n \geq 0}(E^n \times E^n \times I)$ and for each $\alpha \in I$, $\vdash_\alpha \subseteq E^* \times E$. For $\alpha \in I$, let us set

$$ST_\alpha = \{(\vec{\mathsf{e}}, \vec{\varphi}) \mid (\vec{\mathsf{e}}, \vec{\varphi})_\alpha \in ST\},$$

writing $(\vec{\mathsf{e}}, \vec{\varphi})_\alpha$ for $(\vec{\mathsf{e}}, \vec{\varphi}, \alpha)$, and using $\vec{\varphi}$ and φ for expressions occurring as translation outputs. $\mathcal{L}$-formulas include all $\mathcal{L}_\circ(E, E)$-formulas, and are interpreted relative to an $\mathcal{L}$-model $(ST, \{\vdash_\alpha\}_{\alpha \in I})$ as in $\mathcal{L}_\circ$

$$
\begin{aligned}
(\vec{\mathsf{e}}, \vec{\varphi})_\alpha \Vdash \varphi \quad &\text{iff} \quad \vec{\varphi} \vdash_\alpha \varphi \\
(\vec{\mathsf{e}}, \vec{\varphi})_\alpha \Vdash \langle \mathsf{e} \rangle \mathsf{A} \quad &\text{iff} \quad (\vec{\mathsf{e}}\mathsf{e}, \vec{\varphi}\varphi)_\alpha \Vdash \mathsf{A} \text{ for some } \varphi \\
(\vec{\mathsf{e}}, \vec{\varphi})_\alpha \Vdash \bar{\neg}\mathsf{A} \quad &\text{iff} \quad \text{not } (\vec{\mathsf{e}}, \vec{\varphi})_\alpha \Vdash \mathsf{A} \\
(\vec{\mathsf{e}}, \vec{\varphi})_\alpha \Vdash \mathsf{A} \bar{\wedge} \mathsf{B} \quad &\text{iff} \quad (\vec{\mathsf{e}}, \vec{\varphi})_\alpha \Vdash \mathsf{A} \text{ and } (\vec{\mathsf{e}}, \vec{\varphi})_\alpha \Vdash \mathsf{B}
\end{aligned}
$$

for all $(\vec{\mathsf{e}}, \vec{\varphi})_\alpha \in ST$ and $\varphi, \mathsf{e} \in E$. It follows that for every $\alpha \in I$ and all $\mathcal{L}_\circ(E, E)$-formulas A, $(\vec{\mathsf{e}}, \vec{\varphi})_\alpha \Vdash \mathsf{A}$ iff $(\vec{\mathsf{e}}, \vec{\varphi}) \Vdash_\circ \mathsf{A}$ relative to $(ST_\alpha, \vdash_\alpha)$.

Stepping outside of $\mathcal{L}_\circ$, the point in expanding $\mathcal{L}_\circ$ to $\mathcal{L}$ is to allow changes in $\alpha \in I$, the simplest of which is $\bar{\rightarrow} \subseteq ST \times ST$ defined by

$$(\vec{\mathsf{e}}, \vec{\varphi})_\alpha \bar{\rightarrow} (\vec{\mathsf{e}'}, \vec{\varphi'})_{\alpha'} \quad \text{iff} \quad \vec{\mathsf{e}'} = \vec{\mathsf{e}} \text{ and } \vec{\varphi'} = \vec{\varphi},$$

keeping $\vec{e}$ and $\vec{\varphi}$ fixed, with modal operator $\langle - \rangle$

$$(\vec{\mathsf{e}}, \vec{\varphi})_\alpha \Vdash \langle - \rangle \mathsf{A} \quad \text{iff} \quad (\vec{\mathsf{e}}, \vec{\varphi})_\beta \Vdash \mathsf{A} \text{ for some } \beta \, .$$

For example, that "e's presuppositions fail at $(\vec{\mathsf{e}}, \vec{\varphi})_\alpha$ but can be accommodated consistently" can be expressed as

$$(\vec{\mathsf{e}}, \vec{\varphi})_\alpha \Vdash \tilde{\neg} \langle \mathsf{e} \rangle \tilde{\top} \, \tilde{\wedge} \, \langle - \rangle (\tilde{\neg} \bot \, \tilde{\wedge} \, \langle \mathsf{e} \rangle \tilde{\top}) \, .$$

This leaves the previous translations $\vec{\varphi}$ intact. Varying $\vec{\varphi}$ rather than α,

$$(\vec{\mathsf{e}}, \vec{\varphi})_\alpha \Vdash \Diamond \mathsf{A} \quad \text{iff} \quad (\vec{\mathsf{e}}, \vec{\psi})_\alpha \Vdash \mathsf{A} \text{ for some } \vec{\psi} \, ,$$

with accessibility relation $\overset{\Diamond}{\to} \subseteq ST \times ST$ confining the revisions to $\vec{\varphi}$

$$(\vec{\mathsf{e}}, \vec{\varphi})_\alpha \overset{\Diamond}{\to} (\vec{\mathsf{e}}', \vec{\varphi}')_{\alpha'} \quad \text{iff} \quad \vec{\mathsf{e}} = \vec{\mathsf{e}}' \text{ and } \alpha = \alpha' \, .$$

$\langle - \rangle$ can be distinguished from $\Diamond$ by introducing for every $\varphi \in E$ the atomic $\mathcal{L}$-formula $\underline{\varphi}$ saying φ is explicit:

$$(\vec{\mathsf{e}}, \vec{\varphi})_\alpha \Vdash \underline{\varphi} \quad \text{iff} \quad \varphi \text{ is listed among } \vec{\varphi} \, .$$

(In fact, $\underline{\varphi}$ can be formulated in terms of a $\bar{\to}$-leap into an $\mathcal{L}_\circ$-model where entailment amounts to look-up in the sense given by the preceding clause; but let us keep matters simple and stick to $\underline{\varphi}$.) Now, beyond the $\mathcal{L}$-schemes

$$\underline{\varphi} \; \tilde{\supset} \; [\mathsf{e}] \underline{\varphi}$$

$$\tilde{\neg} \underline{\varphi} \, \tilde{\wedge} \, \tilde{\neg} \underline{\psi} \; \tilde{\supset} \; [\mathsf{e}] \tilde{\neg} (\underline{\varphi} \, \tilde{\wedge} \, \underline{\psi}) \quad \text{for distinct } \varphi \text{ and } \psi \, ,$$

we can distinguish $\langle - \rangle$ from $\Diamond$ through

$$\underline{\varphi} \, \tilde{\supset} [-] \underline{\varphi} \qquad\qquad \tilde{\neg} \underline{\varphi} \, \tilde{\supset} [-] \tilde{\neg} \underline{\varphi}$$

(where $[-]$ is $\tilde{\neg} \langle - \rangle \tilde{\neg}$), saying previous outputs are untouched by $\bar{\to}$. With $\mathcal{L}$-formulas $\underline{\varphi}$, we can use

(5) $$\langle e \rangle \underline{e} \; \tilde{\wedge} \; [e] \underline{e}$$

to pick out expressions e from E that approximate the logical forms in an $\mathcal{L}_\circ$-model satisfying $(\mathrm{a}_\circ)$. Let us henceforth write $\mathsf{LF}(e)$ to abbreviate (5). If $\wedge$ and $\supset$ are sequential, then

$$\mathsf{LF}(\varphi) \tilde{\wedge} \langle \varphi \rangle \mathsf{LF}(\psi) \; \tilde{\cong} \; \mathsf{LF}(\varphi \wedge \psi)$$

$$\mathsf{LF}(\varphi) \tilde{\wedge} \langle \varphi \rangle \mathsf{LF}(\psi) \; \tilde{\cong} \; \mathsf{LF}(\varphi \supset \psi).$$

For the remainder of this section, let $E = \mathsf{epTy} \times Pow(\mathsf{Var})$, I be some set of rule sets $\mathcal{D}$, and, construing the pair $(A, \emptyset) \in E$ as a copy of the pretype A,

$$ST_{\mathcal{D}} = \{ (\vec{\mathsf{e}}, (A_1, \emptyset) \cdots (A_n, \emptyset)) \mid (\vec{\mathsf{e}}, A_1 \cdots A_n) \in St_{\mathcal{D}} \}$$

$$\vdash_{\mathcal{D}} = \{ (\vec{\mathsf{e}}, (A, \emptyset)) \mid (\forall (\vec{\mathsf{e}}, \vec{A}) \in St_{\mathcal{D}}) \; \vec{A} \vdash^{\mathcal{D}} A \} \, .$$

The $\mathcal{L}_\circ$-schemes (A1)-(A6) from the end of §4.1 above relativize straightforwardly to the $(ST, \{\vdash_{\mathcal{D}}\}_{\mathcal{D} \in I})$-valid $\mathcal{L}$-schemes

$$\begin{array}{rclrcl} \mathsf{LF}(\varphi) & \tilde{\supset} & \varphi \supset \varphi & \varphi & \tilde{\supset} & [\mathsf{e}]\varphi \\ \mathsf{LF}(\varphi)\tilde{\wedge}\mathsf{LF}(\psi) & \tilde{\supset} & \varphi\tilde{\wedge}\psi \,\hat{\equiv}\, \varphi \wedge \psi & \mathsf{LF}(\varphi)\tilde{\wedge}\langle\varphi\rangle\mathsf{LF}(\psi) & \tilde{\supset} & \langle\varphi\rangle\psi \,\hat{\equiv}\, \varphi \supset \psi \\ \mathsf{LF}(\psi)\tilde{\wedge}(\varphi \supset \psi) & \tilde{\supset} & \varphi\tilde{\supset}\psi & \mathsf{LF}(\varphi)\tilde{\wedge}\langle\varphi\rangle\mathsf{A} & \tilde{\supset} & [\varphi]\mathsf{A}\ . \end{array}$$

Additional $(ST, \{\vdash_{\mathcal{D}}\}_{\mathcal{D}\in I})$-valid $\mathcal{L}$-schemes include $\underline{\varphi}\,\tilde{\supset}\,\varphi$,

$$\mathsf{LF}(\varphi)\ \tilde{\supset}\ [\mathsf{e}]\mathsf{LF}(\varphi)$$

$$\tilde{\neg}\underline{\varphi}\,\tilde{\wedge}\langle\mathsf{e}\rangle(\underline{\varphi}\,\tilde{\wedge}\mathsf{LF}(\psi)\,\tilde{\wedge}\,\psi)\ \tilde{\supset}\ (\varphi \supset \psi)$$

but, given presuppositional pretypes, we should resist the $\mathcal{L}$-scheme

$$\langle\mathsf{e}\rangle\mathsf{LF}(\varphi)\ \tilde{\supset}\ \mathsf{LF}(\varphi)\ .$$

Then there are the S5 axioms for $\diamond$ and $\langle - \rangle$, and how the different modalities interact. I hope to report on these matters elsewhere.

§5. Conclusion. First, a quick summary. With an eye to interpreting a set E of expressions, we formalize in §2 an elementary system of dependent types around a signature L and rule set $\mathcal{D}$. In §3, we decontextualize the types (passing from type to pTy), introduce restrictions along with presuppositions (extending pTy to epTy), and admit ambiguity (multiplying epTy by *Pow*(Var)). These notions are analyzed in §4 through a language $\mathcal{L}_\circ$ that integrates translation, St, and entailment, $\vdash$. $\mathcal{L}_\circ$ is expanded to a language $\mathcal{L}$ for multiple translation-entailment pairs $\{(St_\alpha, \vdash_\alpha)\}_{\alpha\in I}$, carving context up between an explicit (or external) part (in $\mathcal{L}_\circ$), and an *i*mplicit (or *i*nternal) part I that incorporates the choice α of a signature L and rule set $\mathcal{D}$. $\mathcal{L}$ provides a handle on the non-determinism in natural language interpretation abstracted away by the homomorphisms invoked for "English as a formal language."

5.1. From types to processes. $\mathcal{L}$ sharpens the type-theoretic claim (†) formulated in the introduction to

(‡) whenever $(\mathsf{e}_1 \cdots \mathsf{e}_n\mathsf{e}, \varphi_1 \cdots \varphi_n\varphi) \in ST_\alpha$,

$$x_1 : \varphi_1,\ \ldots,\ x_n : \varphi_n\ \Rightarrow_\alpha\ \varphi\ \mathsf{wff}\ .$$

(‡) recognizes a place for English $\vec{e}$ and for an indexical element α determining a notion of inference $\Rightarrow_\alpha$ that conditions not only a process ST_α of translation from e's to φ's, but also the entailments $\vdash_\alpha$ between φ's. Although not explicitly linked above to indexicals and deixis, the indices α might be characterized as particular settings of deictic parameters (building on, for example, Kamp [18] and Cooper [7] for a finer analysis of attunement).

5.2. A multi-faceted approach. Rather than reducing natural language interpretation to proof theory, the present work distinguishes entailment from two other processes: translation and attunement. While the formal distinctions between the processes drawn above are perhaps crisp enough, there are intuitive differences that are somewhat harder to pin down — for instance, the differing computational pressures on the processes. Translation has a computational urgency that the open-ended process of entailment does not, as hinted in §1 by

(Q1). Although translations may depend on entailments, the entailments considered after logical forms have been translated may, in practice, be expected to be more complex than those involved in translating them. Otherwise, the price of decoding the English discourse is hardly worth the bit of information buried in it. An obvious requirement on "computationally significant" notions St, $\vdash$ of translation and entailment over a collection E of expressions is that St be decidable and $\vdash$ be semi-decidable. Or within the program of Feferman [8], we might fix some class $\mathcal{I}$ of indices and two operations $\hat{t}$, $\hat{p}$ (of translation and proof) such that

$$\hat{t} : (\mathcal{I} \times \bigcup_{n \geq 0} E^n \times E^n) \rightarrow \{0, 1\}$$
$$\hat{p} : (\mathcal{I} \times D \times E^* \times E) \rightarrow \{0, 1\}$$

for some class D of candidate derivations. An index $\alpha \in \mathcal{I}$ would then determine the $\mathcal{L}_\circ$-model $(ST_\alpha, \vdash_\alpha)$ given by

$$ST_\alpha = \bigcup_{n \geq 0} \{\langle \vec{e}, \vec{\varphi} \rangle \in E^n \times E^n \mid ap(\hat{t}, \langle \alpha, \vec{e}, \vec{\varphi} \rangle) \simeq 0\}$$
$$\vdash_\alpha = \{\langle \vec{e}, e \rangle \in E^* \times E \mid (\exists d \in D)\, ap(\hat{p}, \langle \alpha, d, \vec{e}, e \rangle) \simeq 0\}\ .$$

Failure to translate an input $\vec{e}$ may arise from incompleteness in the entailment relation $\vdash_\alpha$, inviting a leap into a different index β, as in Turing's ordinal logics (Feferman [9]). How the indices α and β are chosen is the deep mystery of attunement, which we might explore within an expansion $\mathcal{L}$ of $\mathcal{L}_\circ$. It is a considerable stretch from $\mathcal{I}$ to Church-Kleene ordinal notations $\mathcal{O}$, but just as Feferman has asked "what is a natural well-ordering?", we may ask "what is a natural subclass $I \subseteq \mathcal{I}$ for $\mathcal{L}$?" The latter question may not belong to logic (let alone proof theory), but a sense for logic (and formal systems) must surely throw some light on what it could mean.

REFERENCES

[1] PETER ACZEL, *Frege structures and the notions of proposition, truth and set*, ***The Kleene symposium*** (J. Barwise, H.J. Keisler, and K. Kunen, editors), North-Holland, Amsterdam, 1980.

[2] R. AHN and H-P. KOLB, *Discourse representation meets constructive mathematics*, ***Papers from the second symposium on logic and language*** (L. Kálmán and L. Pólos, editors), Akadémiai Kiadó, Budapest, 1990.

[3] N. ASHER and A. LASCARIDES, *Bridging*, ***Journal of Semantics***, vol. 15 (1998), pp. 83–113.

[4] J. BARWISE and R. COOPER, *Extended Kamp notation: a graphical notation for situation theory*, ***Situation theory and its applications*** (P. Aczel, D. Israel, Y. Katagiri, and S. Peters, editors), vol. 3, CSLI Lecture Notes Number 37, Stanford, 1993.

[5] DAVID IAN BEAVER, *Presupposition*, ***Handbook of logic and language*** (J. van Benthem and A. ter Meulen, editors), Elsevier, 1997.

[6] TIJN BORGHUIS, *Modal pure type systems*, ***Journal of Logic, Language and Information***, vol. 7 (1998), pp. 265–296.

[7] ROBIN COOPER, *Mixing situation theory and type theory to formalize information states in dialogue exchanges*, ***Formal semantics and pragmatics of dialogue***, Twente, the Netherlands, 1998.

[8] SOLOMON FEFERMAN, *A language and axioms for explicit mathematics*, ***Algebra and logic*** (J.N. Crossley, editor), LNM 450, Springer-Verlag, Berlin, 1975.

[9] ———, *Turing in the land of $O(z)$*, ***The universal Turing machine*** (R. Herken, editor), Oxford University Press, Oxford, 1988.

[10] ———, *Polymorphic typed lambda-calculi in a type-free axiomatic framework*, ***Logic and computation***, Contemporary Mathematics, vol. 106, A.M.S., Providence, 1990.

[11] TIM FERNANDO, *A modal logic for non-deterministic discourse processing*, ***Journal of Logic, Language and Information***, vol. 8 (1999), pp. 445–468, **Corrigendum**: the axiom scheme $(\varphi \supset \psi) \equiv (\varphi > \psi)$ in §6 (p.465) should be weakened to $(\varphi > \psi) \supset (\varphi \supset \psi)$.

[12] ———, *A type reduction from proof-conditional to dynamic semantics*, ***Journal of Philosophical Logic***, vol. 30 (2001), pp. 121–153.

[13] CHRIS FOX, *Discourse representation, type theory and property theory*, ***Proc. international workshop on computational semantics*** (H. Bunt, R. Muskens, and G. Rentier, editors), ITK, Tilburg, 1994.

[14] GERALD GAZDAR, ***Pragmatics: Implicature, presupposition and logical form***, Academic Press, New York, 1979.

[15] F. KAMAREDDINE and E. KLEIN, *Nominalization, predication and type containment*, ***Journal of Logic, Language and Information***, vol. 2 (1993), pp. 171–215.

[16] H. KAMP and U. REYLE, ***From discourse to logic***, Kluwer Academic Publishers, Dordrecht, 1993.

[17] HANS KAMP, *Events, instants and temporal reference*, ***Semantics from different points of view*** (R. Bäuerle, U. Egli, and A. von Stechow, editors), Springer, Berlin, 1979.

[18] ———, *Prolegomena to a structural account of belief and other attitudes*, ***Propositional attitudes*** (C.A. Anderson and J. Owens, editors), CSLI Lecture Notes Number 20, Stanford, 1990.

[19] LAURI KARTTUNEN, *Presupposition and linguistic context*, ***Theoretical Linguistics***, (1974), pp. 181–194.

[20] E. KRAHMER and P. PIWEK, *Presupposition projection as proof construction*, ***Computing meaning***, Kluwer Academic Publishers, Dordrecht, 1999.

[21] PETER KRAUSE, *Presupposition and abduction in type theory*, ***Computational logic and natural language processing***, South Queensferry, Scotland, 1995.

[22] PER MARTIN-LÖF, ***Intuitionistic type theory***, Bibliopolis, Napoli, 1984, Notes by Giovanni Sambin of a series of lectures given in Padua, June 1980.

[23] J. MCCARTHY and S. BUVAČ, *Formalizing context (expanded notes)*, ***Computing natural language*** (A. Aliseda, R. van Glabbeek, and D. Westerståhl, editors), CSLI Lecture Notes Number 81, Stanford, 1997.

[24] RICHARD MONTAGUE, *English as a formal language*, ***Formal philosophy*** (R. Thomason, editor), Yale University Press, New Haven, 1974.

[25] AARNE RANTA, ***Type-theoretical grammar***, Oxford University Press, Oxford, 1994.

[26] ———, *Grammatical framework tutorial*, Available at `www.xrce.xerox.com/research/mltt/gf/home.html`, 1999.

[27] GÖRAN SUNDHOLM, *Proof theory and meaning*, ***Handbook of philosophical logic*** (D. Gabbay and F. Guenthner, editors), vol. 3, Reidel, Dordrecht, 1986.

DEPARTMENT OF COMPUTER SCIENCE
TRINITY COLLEGE
DUBLIN 2, IRELAND
E-mail: Tim.Fernando@tcd.ie

INFINITARY INITIAL ALGEBRA SPECIFICATIONS FOR STREAM ALGEBRAS

J. V. TUCKER AND J. I. ZUCKER

Dedicated to Solomon Feferman on the occasion of his 70th birthday

Abstract. A *stream* is an infinite sequence of data from a set A. A wide variety of algorithms and architectures operate continuously in time, processing streams of data, for example: hardware systems, embedded systems and emergent systems. Also many models of real number computation use streams. In this paper we study the construction of an algebra $\bar{A}$ of streams over a many-sorted algebra A of data. In particular, we show how an initial algebra specification for $\bar{A}$ can be constructed from one for A. One problem is that $\bar{A}$ is uncountable even when A is finite. To handle this, we work with infinitary terms called *stream terms*, and infinitary formulae that generalise conditional equations, called *ω-conditional stream equations*. We state a Birkhoff-Mal'cev type theorem that shows conservativity for the many-sorted stream version of full infinitary first-order logic $L_{\omega_1\omega}$ over an ω-conditional equational logic, and also that certain ω-conditional stream equational theories have initial models that are complete for closed stream terms. As an application of the theory, we show that the real unit interval (or real line) can be characterised up to isomorphism by an infinitary term model, which is also an initial model of a suitable ω-conditional equational theory.

§0. Introduction. A *stream* α is an infinite sequence $\alpha(0), \alpha(1), \dots$ of data from a set A. A wide variety of algorithms process streams, operating on sequences of data indexed by discrete time $t = 0, 1, 2, \dots$. For example, parallel algorithms that typically input and output streams of data are common in the design of interactive systems such as hardware systems (*e.g.*, systolic arrays and data-flow networks), embedded systems, and emergent systems (*e.g.*, cellular automata and coupled map lattices). A survey of models for stream computation, with an emphasis on their applications in hardware and software design, is given in [24].

Streams are infinite data and constitute continuous — rather than discrete — data types. They are also used to build representations of other continuous data types, such as the sets of real and complex numbers, and the finite and infinite dimensional algebras constructed over them, such as matrix groups and function spaces. Mathematical theories of computation and reasoning with streams are needed for the theoretical foundations of hardware and software design, and scientific and engineering computation.

The research of the second author was supported by a grant from the Natural Sciences and Engineering Research Council (Canada), and by a Visiting Fellowship from the Engineering and Physical Sciences Research Council (U.K.)

Reflections on the foundations of mathematics
Essays in honor of Solomon Feferman
Edited by Wilfried Sieg, Richard Sommer, and Carolyn Talcott

Now a stream is a simply function $\alpha : \mathbb{N} \to A$, and to compute and reason with these mathematical objects requires an extension of computability theory and logic from first-order structures. The required theoretical foundations ought to be known, from the various generalisations of computability theory and logic to higher types that have accumulated, especially since the 1940s. However, theories of computation and reasoning with streams cannot be obtained as simple cases of grand theories for higher types. Indeed, *useful* theories of stream computation and stream logic must originate in the close analysis of streams and their applications. Such theories are barely developed because

(*i*) modelling, formally specifying and verifying computing systems raise new questions that are specific to streams;
(*ii*) some generalisations are too general (*e.g.*, higher-order logic over the full type structure); or
(*iii*) some generalisations are too specific (*e.g.*, recursion in higher types over the natural numbers).

In the case of generalised recursion theory — and other parts of logic and mathematics for that matter — such specific mathematical developments are inhibited by a lack of interest in the particular, an emphasis on structure rather than algorithms, and an ignorance of scientific developments in computer science, theoria and praxis: see the illuminating discussion in [5].

In computer science, the mathematical theory of data is based on many-sorted universal algebra [34, 16]. Classical computability theory and predicate logic generalise to these structures in natural ways, and some special theories have been developed to understand fundamental points of computational practice. The most important example of such a special theory is the mathematical theory of algebraic specifications for data types. This uses equations and conditional equations, their initial and final models, and term rewriting techniques for specifying, implementing and reasoning about data types.

In the case of stream computation, there are a few approaches. For example, abstract computability theory on many-sorted algebras (surveyed in [32]) has been applied to algebras containing streams in [30]. The questions addressed concern mainly primitive recursion on different algebras of streams, and reflect closely the use of models of stream computation in applications. In [4] a second-order model of stream processing is built by Feferman based on certain schemes inspired by generalised recursion theory. The model is based on many-sorted algebras, and bears comparison with models of [21] and [30].

Concrete computability theories [22, 35] also provide an important approach to computability on streams. This idea is developed for domain representations of streams in [1].

In the case of reasoning, questions of which logics to use are even more open. How does one develop axiomatisations and proof systems for stream algebras? Initial algebra specifications of *discrete* data types are important theoretically and practically [37]. Equational definability and computability are known to be equivalent for countable algebras [17, 25]. But there are problems with continuous data types such as streams.

In short, there is no shortage of either open problems or potential applications for topological data types in general, or stream data types in particular.

In this note we will study the construction of an algebra $\bar{A}$ of streams over a many-sorted algebra A of data. In particular, we will show how an *initial algebra specification* for $\bar{A}$ can be constructed from one for A in a uniform way. We will present a new class of infinitary formulae, that generalise the conditional equations, called *ω-conditional stream equations*. These have the form

$$\bigwedge\!\!\!\bigwedge_{i=0}^{\infty} P_i \;\rightarrow\; P$$

where the P_i and P are equations that may involve *stream terms* of the form

$$\langle t_0\, t_1\, t_2\, \ldots\rangle$$

where the t_i are data terms. We give a logic CondEq_ω for these formulae and prove a Birkhoff-Mal'cev type theorem (Section 5, Theorem 3) that shows conservativity for the corresponding (many-sorted) stream version of full infinitary first-order logic $L_{\omega_1\omega}$ over this logic, and also that certain ω-conditional stream equational theories have initial models that are complete for closed stream terms. Using these infinitary algebraic tools, we prove that the construction of $\bar{A}$ from A is definable by a set StrAx of ω-conditional stream equations in the following sense: If A is the initial algebra of the ω-conditional stream equational theory E then $\bar{A}$ is the initial algebra of the ω-conditional equational theory $E + \mathsf{StrAx}$.

As an application of the theory, we show that certain algebras of real numbers can be characterised up to isomorphism by infinitary term models, which are also initial algebras of suitable ω-conditional equational theories. Specifically, we show this for two algebras based on the unit interval: one with order, and another with addition and multiplication (mod 1).

This work is not directly related to our research programme in computability theory on abstract many-sorted algebras [27, 28, 29, 30, 31, 32]. It is related rather to our study of algebraic specifiability and initial algebra semantics, and their relation to computability [33], which contains more details for some of the results surveyed here.

The second author (J.Z.) has an enormous debt of gratitude to Sol Feferman, his thesis adviser. Sol's influence on his research, up to and including the work of the present paper, is twofold: first, in the specific area of proof theory, which was the subject of his thesis, and secondly and more generally, in the question of research methodology and scientific taste — developing theory so as to illuminate related subject areas, rather than as a series of recondite exercises.

We also wish to thank Martín Escardó and two anonymous referees for helpful comments, and Carolyn Talcott for technical assistance in preparing this paper for publication.

§1. Many-sorted stream algebras.

1.1. Basic definitions. A *signature* Σ (for a many-sorted partial algebra) is a pair consisting of (i) a finite set $\boldsymbol{Sort}(\Sigma)$ of *sorts*, and (ii) a finite set $\boldsymbol{Func}(\Sigma)$ of *(primitive) function symbols*, each symbol F having a *type* $s_1 \times \cdots \times s_m \to s$, where $s_1, \ldots, s_m, s \in \boldsymbol{Sort}(\Sigma)$; in which case we write $F : s_1 \times \cdots \times s_m \to s$. (The case $m = 0$ corresponds to *constant symbols*.)

A Σ*-algebra* A has, for each sort s of Σ, a non-empty *carrier set* A_s of sort s, and for each Σ-function symbol $F : s_1 \times \cdots \times s_m \to s$, a function $F^A : A_{s_1} \times \cdots \times A_{s_m} \to A_s$.

An *abstract data type* of signature Σ (Σ*-adt*) is defined to be a class $\mathbb{K}$ of Σ-algebras *closed under isomorphism*.

1.2. Adding booleans and counters: N-standard signatures and algebras. A signature Σ is *N-standard* if it includes the sorts bool and nat, and also

(i) the N-standard boolean constants and operations true, false, and, or and not, and the *discriminator* $\mathsf{if}_s : \mathsf{bool} \times s^2 \to s$ for all sorts s of Σ other than bool,

(ii) the standard operations on the naturals: 0, S, $\mathsf{eq}_{\mathsf{nat}}$, $\mathsf{less}_{\mathsf{nat}}$ for *zero*, *successor*, *equality* and *order* respectively.

Given a N-standard signature Σ, a Σ-algebra A is a *N-standard* if the carriers of sort bool and nat are (respectively) $\mathbb{B} = \{\mathtt{tt}, \mathtt{ff}\}$ and $\mathbb{N} = \{0, 1, 2, \ldots\}$, and the standard operations on these sorts listed above have their standard interpretations. [1]

Note that any many-sorted signature Σ, and Σ-algebra A, can be *N-standardised* to a signature $\Sigma^{B,N}$, and an N-standard $\Sigma^{B,N}$-algebra $A^{B,N}$, by adjoining the sorts bool and nat and the standard operations, with their standard interpretations.

We let $\boldsymbol{NStd}(\Sigma)$ be the class of all N-standard Σ-algebras.

We also consider a notion stricter than N-standardness. An N-standard signature Σ is *strictly N-standard* if the only operations of Σ with range sort nat or bool are the *standard operations* listed above. An algebra is *strictly N-standard* if its signature is.

Note that any N-standardised algebra is automatically strictly N-standard.

1.3. Adding streams: stream signatures and stream algebras. An N-standard signature Σ is called a *stream signature* if for some designated subsets $\boldsymbol{PreStr}(\Sigma)$, and $\boldsymbol{Str}(\Sigma)$ of $\boldsymbol{Sort}(\Sigma)$, the following holds:

(i) There is a bijection $s \leftrightarrow \bar{s}$ of $\boldsymbol{PreStr}(\Sigma)$ with $\boldsymbol{Str}(\Sigma)$, where we also write $\bar{s}$ as $\mathsf{nat} \to s$ for clarity;

[1] In [31, 32] there is a further condition on N-standard algebras, namely that the interpretation of equality on certain "equality sorts" should be the identity on the corresponding carrier set. These two definitions can be reconciled by assuming (in the present context) that nat is the only equality sort.

(ii) For all $\bar{s} \in \boldsymbol{Str}(\Sigma)$, the only Σ-operations with $\bar{s}$ in its domain are

$$\mathsf{eval}_s : (\mathsf{nat} \to s) \times \mathsf{nat} \to s,$$

and the discriminator $\mathsf{if}_{\bar{s}}$ on sort $\bar{s}$ (but not equality on $\bar{s}$).

The sets $\boldsymbol{PreStr}(\Sigma)$ and $\boldsymbol{Str}(\Sigma)$ are called (respectively) the sets of *pre-stream sorts* and *stream sorts* of Σ.

If Σ is a stream signature, then an N-standard Σ-algebra A is called a *Σ-stream algebra* if for each $s \in \boldsymbol{PreStr}(\Sigma)$, the carrier $\bar{A}_s$ is given by

$$\bar{A}_s = [\mathbb{N} \to A_s],$$

i.e., the set of all *streams on A_s*, or functions from $\mathbb{N}$ to A_s, and eval_s is interpreted as as the function $\mathsf{eval}_s^A : [\mathbb{N} \to A_s] \times \mathbb{N} \to A_s$ which *evaluates* a stream at an index, *i.e.*,

$$\mathsf{eval}_s^A(\alpha, n) = \alpha(n).$$

EXAMPLE. A common way to construct a stream algebra is as follows. Start with any signature Σ and Σ-algebra A. First we N-standardise them (assuming they are not already N-standard), to form $\Sigma^{B.N}$ and $A^{B.N}$. Then choose a set $\boldsymbol{S} \subseteq \boldsymbol{Sort}(\Sigma)$ of *pre-stream sorts*. We then extend $\Sigma^{B.N}$ to a *stream signature $\overline{\Sigma}^{\boldsymbol{S}}$ with respect to $\boldsymbol{S}$*, as follows. With each $s \in \boldsymbol{S}$, associate a new stream sort $\bar{s}$. Then $\boldsymbol{Sort}(\overline{\Sigma}^{\boldsymbol{S}}) = \boldsymbol{Sort}(\Sigma) \cup \overline{\boldsymbol{S}}$, where $\overline{\boldsymbol{S}} =_{df} \{\bar{s} \mid s \in \boldsymbol{S}\}$, and $\boldsymbol{Func}(\overline{\Sigma}^{\boldsymbol{S}})$ consists of $\boldsymbol{Func}(\Sigma)$, together with eval_s and $\mathsf{if}_{\bar{s}}$ for each $s \in \boldsymbol{S}$.

Now expand $A^{B.N}$ to a $\overline{\Sigma}^{\boldsymbol{S}}$-stream algebra $\bar{A}^{\boldsymbol{S}}$ by adding, for each $s \in \boldsymbol{S}$, the carrier $\bar{A}_s = [\mathbb{N} \to A_s]$. The algebra $\bar{A}^{\boldsymbol{S}}$ is the *(full) stream algebra over A with respect to $\boldsymbol{S}$*.

REMARK (on stream operations, or the absence thereof). One could, contrary to the above definition, add various primitive operations to stream algebras, such as changing stream values at particular (finitely or infinitely many) points, or mapping data elements d to streams with constant value d, or merging pairs of streams, etc. This would not have a significant effect on the theory developed in this paper. We have chosen instead a minimalist approach to stream operations, in line with our intuition of streams as arbitrary infinite sequences of data inputs to algorithms, one item at a time, rather like the oracles of classical recursion theory [23].

Note that in order for the closed terms over a signature Σ to form a *Σ-term algebra*, we should make one more assumption on Σ:

INSTANTIATION ASSUMPTION. *Each sort of Σ is instantiated, i.e., there is a closed term of each sort.*

This assumption is satisfied for all the familiar algebras used as abstract data types. We should note, however, that it is *not* satisfied by *stream signatures* at their stream sorts (with respect to the usual finite syntax), since there are no "constant" streams or, indeed, any streams, definable from the basic stream operations. Nevertheless, the assumption *is* satisfied by these stream signatures with respect to the *infinite term construction* described in the next section.

§2. Infinitary logic over stream signatures.

2.1. Infinitary terms. The problem we have is this: we want initial algebra specifications for theories based on stream signatures. Initial algebra semantics usually involve term models. However a full stream algebra $\bar{A}_s$ based on a sort s is uncountable, even if A_s is finite!

Hence, to construct a term model for a theory over a stream signature, we must work with *infinitary terms*, as follows.

Let $\boldsymbol{T} = \boldsymbol{T}(\Sigma)$ be the usual set of (finite) Σ-terms of sort s. We can define a set $\boldsymbol{T}^\omega_s = \boldsymbol{T}^\omega_s(\Sigma)$ of ω-(Σ-)terms of sort s. We omit the complete definition of $\boldsymbol{T}^\omega_s$, but note the clause:

- for a *stream sort* $\bar{s}$, if $t_0, t_1, t_2, \ldots$ is any infinite sequence of ω-terms of sort s, then $\langle t_0\, t_1\, t_2\, \ldots \rangle$ is an ω-term of sort $\bar{s}$, *i.e.*, a "stream term".

We will write $t : s$ to signify that t is a term of sort s.

2.2. Finitary and infinitary logics over Σ. We define four formal systems over Σ: based on either *finitary* or *infinitary* logic, and in either a *conditional equational* or a "*full*" form.

(1) $\mathsf{CondEq}(\Sigma)$: finitary conditional equational logic over Σ. This can conveniently be presented in the *sequent calculus*, as a subsystem of Takeuti's LJe [26]. The *Σ-atomic formulae* are *equations* between (finite) terms of the same sort, for all Σ-sorts (whether equality sorts or not). A *conditional equation* is a formula of the form

$$P_1 \wedge \ldots \wedge P_n \rightarrow P \qquad (*)$$

where $n \geq 0$ and P_i and P are equations. A *conditional equational theory* is a set of such formulae (or their universal closures). An *equational sequent* is a sequent of the form

$$P_1, \ldots, P_n \longmapsto P$$

where $n \geq 0$ and P_i and P are equations. This sequent *corresponds to* the conditional equation $(*)$.

The *initial sequents* are the Σ-equality axioms, and the *inferences* are:

- structural inferences,
- atomic cuts.

This system can be augmented in two ways:

(*a*) Adding *axioms of a conditional equation theory*, or rather all substitution instances of these, as initial sequents;

(*b*) Adding *induction* for a class $\mathcal{C}$ of formulae, in the form of the inference rule

$$\mathcal{C}\text{-}\mathsf{Ind}: \quad \frac{\Gamma \longmapsto \Delta, F(0) \qquad F(a), \Pi \longmapsto \Lambda, F(\mathsf{S}a)}{\Gamma, \Pi \longmapsto \Delta, \Lambda, F(t)}$$

where the induction variable a has sort nat, and the induction formula $F(a)$ belongs to the class $\mathcal{C}$. (We are assuming here that Σ is N-standard.) We write $\mathsf{Ind}(\Sigma)$ for full Σ-induction, *i.e.*, where $\mathcal{C}$ is the set of all first-order Σ-formulae.

Analogous augmentations can be made for the other systems considered below.

(2) $\mathsf{CondEq}_\omega(\Sigma)$**: ω-conditional equational logic over Σ.** This modifies the above system in the following ways. The atomic formulae are now *ω-(Σ-) equations*, *i.e.*, equations between ω-Σ-terms of the same sort. An *ω-conditional equation* is a formula of the form

$$\bigwedge_{i=0}^{\infty} P_i \longmapsto P \qquad (**)$$

where $n \geq 0$ and P_i and P are ω-equations. An *ω-equational sequent* is an (infinite) sequent of the form

$$P_0, P_1, \ldots, P_n, \ldots \longmapsto P$$

where P_i and P are ω-equations. This sequent *corresponds to* the conditional equation $(**)$.

The *initial sequents* are the ω-Σ-equality axioms, including the axiom

$$t_0 = t_0',\ t_1 = t_1',\ t_2 = t_2',\ \ldots \longmapsto \langle t_0\, t_1\, t_2 \ldots\rangle = \langle t_0'\, t_1'\, t_2' \ldots\rangle$$

for all sequences of terms $t_0,\ t_0',\ t_1,\ t_1', \ldots$ of some pre-stream sort; and the *inferences* are:

- structural inferences,
- atomic cuts,
- ω-cuts, *i.e.*, cuts of the form

$$\frac{\Gamma \longmapsto Q_0 \qquad \Gamma \longmapsto Q_1 \qquad \ldots \qquad Q_0, Q_1, \ldots, \Delta \longmapsto Q}{\Gamma, \Delta \longmapsto Q}$$

(3) $\mathsf{FOL}(\Sigma)$**: full finitary (first order) logic with equality over Σ.** This can be formulated in a system $\mathsf{LK_e}(\Sigma)$, which is an adaptation to the many-sorted signature Σ of the system $\mathsf{LK_e}$ of [26]. The atomic formulae are equations, as in (1).

(4) $\mathsf{FOL}_\omega(\Sigma)$**: full ω-logic with equality over Σ.** This resembles the system $L_{\omega_1\omega}$ of [9] (*i.e.*, countable conjunctions and disjunctions, plus the usual quantifiers) adapted to the signature Σ, with the additional feature (as in (2)) that the atomic formulae are ω-equations. We omit details, except to point out that the rule of induction can be subsumed under the rule of universal quantification over nat, which can be given in the form

$$\forall L : \frac{R(t), \Gamma \longmapsto \Delta}{\forall z R(z), \Gamma \longmapsto \Delta} \qquad \forall_\omega R : \frac{\ldots\ \Gamma \longmapsto \Delta, F(\bar{n})\ \ldots \quad (\text{all } n \in \mathbb{N})}{\Gamma \longmapsto \Delta, \forall z R(z)}$$

where $\bar{n}$ is the *numeral* for n, *i.e.*, the term $\mathsf{S}\ldots\mathsf{S0}$ (n times 'S'); with similar (dual) rules for existential quantification.

Note that all the infinitary systems above are assumed to satisfy the

FINITE FREE VARIABLE ASSUMPTION. *The set of free variables in each infinitary term, formula or sequent is finite.*

One reason for the importance of (finite or infinite) conditional equational logic lies in the following.

2.3. Conservativity Lemmas. First we need a definition which will be given again in context in Section 4. Let $\mathcal{F}$ be a formal system (typically $\mathsf{CondEq}(\Sigma)$ or $\mathsf{CondEq}_\omega(\Sigma)$), and let T be a theory over Σ (typically a conditional equational or ω-conditional equational theory). We say that T *determines* nat in $\mathcal{F}$ if every closed term of sort nat is, provably in $\mathcal{F}$ from T, equal to a numeral.

(1) (FOL over CondEq). Let E be a Σ-conditional equational theory, and let $\Gamma \longmapsto P$ be a Σ-equational sequent. Then $\Gamma \longmapsto P$ is provable from E in $\mathsf{FOL}(\Sigma)$ if, and only if, it is provable from E in $\mathsf{CondEq}(\Sigma)$.

(2) ($\mathsf{FOL} + \mathsf{Ind}$ over CondEq). Let E be a Σ-conditional equational theory which determines nat in $\mathsf{CondEq}(\Sigma)$, and let $\Gamma \longmapsto P$ be a *closed* Σ-equational sequent. Then $\Gamma \longmapsto P$ is provable from E in $\mathsf{FOL}(\Sigma) + \mathsf{Ind}(\Sigma)$ if, and only if, it is provable from E in $\mathsf{CondEq}(\Sigma)$.

(3) (FOL_ω over CondEq_ω). Let E be an ω-Σ-conditional equational theory over Σ which determines nat in $\mathsf{CondEq}_\omega(\Sigma)$, and let $\Gamma \longmapsto P$ be a *closed* ω-Σ-equational sequent. Then $\Gamma \longmapsto P$ is provable from E in $\mathsf{FOL}_\omega(\Sigma)$ if, and only if, it is provable from E in $\mathsf{CondEq}_\omega(\Sigma)$.

(4) (CondEq_ω over CondEq). Let E be a (finite) Σ-conditional equational theory, and let $\Gamma \longmapsto P$ be a (finite) Σ-conditional equational sequent. Then $\Gamma \longmapsto P$ is provable from E in $\mathsf{CondEq}_\omega(\Sigma)$, if, and only if, it is provable from E in $\mathsf{CondEq}(\Sigma)$.

All four lemmas can be proved by cut elimination. We omit proofs, except to note briefly that the two conditions, that E determines nat and that $\Gamma \longmapsto P$ is closed, are used in (2) to eliminate induction inferences, and in (3) to eliminate cuts of formulae universally or existentially quantified over nat.

REMARK. Conservativity lemmas 1, 2 and 3 (at least for simple equations) also follow from the Birkhoff-Mal'cev-type completeness theorems in Sections 3, 4 and 5 respectively.

§3. Initial algebras for conditional equations. In this section (only) we make no assumption about N-standardness of signatures. Let Σ be a signature and $\mathbb{K}$ a Σ-adt.

A Σ-algebra A is *pre-initial for* $\mathbb{K}$ if there is a unique Σ-homomorphism from A to every algebra in $\mathbb{K}$. An *initial algebra of* $\mathbb{K}$ is a pre-initial algebra which belongs to $\mathbb{K}$. Any two initial algebras of $\mathbb{K}$ are Σ-isomorphic. Let $\boldsymbol{Init}(\Sigma, \mathbb{K})$ denote any initial algebra of $\mathbb{K}$.

Of special interest is the case that $\mathbb{K} = \boldsymbol{Alg}(\Sigma, T)$, the class of models of a first-order theory T over Σ (typically a conditional equational theory). We write $\boldsymbol{Init}(\Sigma, T) =_{df} \boldsymbol{Init}(\Sigma, \boldsymbol{Alg}(\Sigma, T))$, the *initial model* of T (if it exists).

Let $\boldsymbol{T}(\Sigma)$ be the *closed term algebra* (in the finite language) over Σ. Consider the *quotient algebra* $\boldsymbol{T}(\Sigma, T, \mathcal{F})$ (in the finite language over Σ) formed from $\boldsymbol{T}(\Sigma)$ by identifying closed terms provably equal from T, in some formalism $\mathcal{F}$ (typically $\mathsf{CondEq}(\Sigma)$), *i.e.*,

$$\boldsymbol{T}(\Sigma, T, \mathcal{F}) =_{df} \boldsymbol{T}(\Sigma)/\approx_{T,\mathcal{F}}$$

where

$$t_1 \approx_{T,\mathcal{F}} t_2 \iff_{df} t_1 = t_2 \text{ is provable from } T \text{ in } \mathcal{F}.$$

INITIALITY LEMMA. Suppose $\mathcal{F}$ is valid for $\boldsymbol{Alg}(\Sigma, T)$. Then
(*a*) $\boldsymbol{T}(\Sigma, T, \mathcal{F})$ is pre-initial for $\boldsymbol{Alg}(\Sigma, T)$; and
(*b*) $\boldsymbol{T}(\Sigma, T, \mathcal{F}) \models T$ then $\boldsymbol{T}(\Sigma, T, \mathcal{F})$ is an initial model of T, *i.e.*,

$$\boldsymbol{T}(\Sigma, T, \mathcal{F}) \cong \boldsymbol{Init}(\Sigma, T).$$

DEFINITION. A has an *initial algebra specification* (Σ, T) if

$$A \cong \boldsymbol{Init}(\Sigma, T).$$

THEOREM 1 (MAL'CEV). *Let E be a conditional equational theory over Σ. Let $I =_{df} \boldsymbol{T}(\Sigma, E, \mathsf{CondEq}(\Sigma))$. Then I is an initial model of E. Further, for two closed Σ-terms t_1, t_2 of the same sort, the following are equivalent*:
(*i*) *t_1 and t_2 have the same value in I,*
(*ii*) *t_1 and t_2 have the same value in all models of E,*
(*iii*) *$t_1 = t_2$ is provable from E in* $\mathsf{CondEq}(\Sigma)$,
(*iv*) *$t_1 = t_2$ is provable from E in* $\mathsf{FOL}(\Sigma)$.

REMARK. Mal'cev's Theorem [12], in the form given above, can be viewed as expressing both (*a*) *completeness* of $\mathsf{CondEq}(\Sigma)$, given by the implication (*ii*) $\Longrightarrow$ (*iii*), and (*b*) *conservativity* of first order logic with equality over $\mathsf{CondEq}(\Sigma)$, given by the implication (*iv*) $\Longrightarrow$ (*iii*). (*Cf.* the remark in §2.3.)

§4. Initial N-standard algebras for conditional equations. We now repeat the theory of Section 3 for *N-standard* algebras. So assume that Σ is N-standard, and that $\mathbb{K}$ consists of *N-standard* Σ-algebras; in particular, $\mathbb{K}$ is the set of *N-standard* models of a Σ-theory T, *i.e.*, $\mathbb{K} = \boldsymbol{NStd}(\Sigma, T)$, for some Σ-theory T. Then although $\boldsymbol{T}(\Sigma, T, \mathcal{F})$, is pre-initial for $\mathbb{K}$, it might fail to be initial for $\mathbb{K}$ for two reasons: it might not satisfy T, and it might not even be N-standard! (We return to the latter point below.)

An *initial N-standard model of* T is an initial algebra of $\boldsymbol{NStd}(\Sigma, T)$. Any two initial N-standard models of T are Σ-isomorphic. We denote any such model by

$$\boldsymbol{InitNStd}(\Sigma, T) =_{df} \boldsymbol{Init}(\Sigma, \boldsymbol{NStd}(\Sigma, T)).$$

N-STANDARD INITIALITY LEMMA. *Suppose $\mathcal{F}$ is valid for $\boldsymbol{NStd}(\Sigma, T)$. If $\boldsymbol{T}(\Sigma, T, \mathcal{F}) \in \boldsymbol{NStd}(\Sigma, T)$ then it is (Σ-isomorphic to) $\boldsymbol{InitNStd}(\Sigma, T)$.*

The question arises: how can we construct (or characterise) *initial N-standard* models of T?

Let A be an N-standard Σ-algebra.

DEFINITION. A has an *initial N-standard algebra specification* (Σ, T) if

$$A \cong \boldsymbol{InitNStd}(\Sigma, T).$$

Now $\boldsymbol{T}(\Sigma, T, \mathcal{F})$ (even if it is a model of T) may fail to be *N-standard* for two reasons: namely, T proves "too little" or "too much", so to speak. The first reason is connected with non-standard interpretations of the sorts nat and bool.

Thus, there might be a Σ-function symbol f, but no corresponding axioms in T capable of "reducing" a closed term of sort nat containing f to a numeral. Similarly not all closed boolean terms (*i.e.*, terms of sort bool) may be (provably in T) equal to true or false. (In the terminology of [7], the specification (Σ, T) is not "sufficiently complete".) The second reason is that T may be *inconsistent*, in the sense that it proves 'true = false' (or, equivalently in a suitable weak background theory, '0 = 1'). This motivates the following definitions. Let $\mathcal{F}$ be a formal system and let T be a theory over Σ.

DEFINITIONS. (1) T is *consistent* in $\mathcal{F}$ if the equation 'true = false' is not provable in $\mathcal{F}$ from T.

(2) T *determines* nat in $\mathcal{F}$ if every closed term of sort nat is, provably in $\mathcal{F}$ from T, equal to a numeral; and T *determines* bool in $\mathcal{F}$ if every closed term of sort bool is, provably in $\mathcal{F}$ from T, equal to true or false.

Let $\mathsf{NStdAx}(\Sigma)$ be the following set of axioms:

$$\boxed{\begin{array}{c}
\mathsf{and}(\mathsf{true}, \mathsf{true}) = \mathsf{true}, \\
\mathsf{and}(\mathsf{true}, \mathsf{false}) = \mathsf{and}(\mathsf{false}, \mathsf{true}) = \mathsf{and}(\mathsf{false}, \mathsf{false}) = \mathsf{false}, \\
\mathsf{or}(\mathsf{false}, \mathsf{false}) = \mathsf{false}, \\
\mathsf{or}(\mathsf{true}, \mathsf{false}) = \mathsf{or}(\mathsf{false}, \mathsf{true}) = \mathsf{or}(\mathsf{true}, \mathsf{true}) = \mathsf{true}, \\
\mathsf{not}(\mathsf{true}) = \mathsf{false}, \qquad \mathsf{not}(\mathsf{false}) = \mathsf{true}, \\
\mathsf{if}_s(\mathsf{true}, x_1^s, x_2^s) = x_1^s, \qquad \mathsf{if}_s(\mathsf{false}, x_1^s, x_2^s) = x_2^s, \\
\mathsf{eq}_{\mathsf{nat}}(0, 0) = \mathsf{true}, \qquad \mathsf{eq}_{\mathsf{nat}}(\mathsf{S}z, 0) = \mathsf{eq}_{\mathsf{nat}}(0, \mathsf{S}z) = \mathsf{false}, \\
\mathsf{eq}_{\mathsf{nat}}(\mathsf{S}z_1, \mathsf{S}z_2) = \mathsf{eq}_{\mathsf{nat}}(z_1, z_2), \\
\mathsf{less}_{\mathsf{nat}}(0, \mathsf{S}z) = \mathsf{true}, \qquad \mathsf{less}_{\mathsf{nat}}(0, 0) = \mathsf{less}_{\mathsf{nat}}(\mathsf{S}z, 0) = \mathsf{false}, \\
\mathsf{less}_{\mathsf{nat}}(\mathsf{S}z_1, \mathsf{S}z_2) = \mathsf{less}_{\mathsf{nat}}(z_1, z_2)
\end{array}}$$

where s ranges over all Σ-sorts other than bool.

Note that $\mathsf{NStdAx}(\Sigma) + \mathsf{Ind}(\Sigma)$ holds in any N-standard Σ-algebra. Also, the axioms in $\mathsf{NStdAx}(\Sigma)$ are all equations.

Let $\mathsf{NStdAx}_0(\Sigma)$ be the set of all *closed Σ-substitution instances* of $\mathsf{NStdAx}(\Sigma)$.

LEMMA 1 (N-STANDARDNESS LEMMA). *Suppose that in* $\mathcal{F}$

(*i*) *T is consistent,*

(*ii*) *T determines* nat *and* bool, *and*

(*iii*) *T proves* $\mathsf{NStdAx}_0(\Sigma)$.

Then $\boldsymbol{T}(\Sigma, T, \mathcal{F})$ *is N-standard.*

LEMMA 2. *If Σ is strictly N-standard then* $\mathsf{NStdAx}_0(\Sigma)$ *determines* nat *and* bool *in* $\mathsf{CondEq}(\Sigma)$.

COROLLARY. *Suppose Σ is strictly N-standard, and E is a consistent Σ-conditional equational theory containing* $\mathsf{NStdAx}_0(\Sigma)$. *Then* $\boldsymbol{T}(\Sigma, E, \mathsf{CondEq}(\Sigma))$ *is N-standard.*

We now give the analogue of Mal'cev's Theorem (Theorem 1 in Section 3) for N-standard algebras [33].

THEOREM 2. *Let E be a conditional equational theory over Σ. Suppose that in* $\mathsf{CondEq}(\Sigma)$, *E is consistent, determines* nat *and* bool, *and proves* $\mathsf{NStdAx_0}(\Sigma)$. *Then $I =_{df} \boldsymbol{T}(\Sigma, E, \mathsf{CondEq}(\Sigma))$ is an initial N-standard model of E. Further, for two closed Σ-terms t_1, t_2 of the same sort, the following are equivalent:*

(*i*) *t_1 and t_2 have the same value in I,*
(*ii*) *t_1 and t_2 have the same value in all N-standard models of E,*
(*iii*) *$t_1 = t_2$ is provable from E in* $\mathsf{CondEq}(\Sigma)$,
(*iv*) *$t_1 = t_2$ is provable from E in* $\mathsf{FOL}(\Sigma) + \mathsf{Ind}(\Sigma)$.

REMARKS. (1) By Lemma 2, the assumption in the theorem that E determines nat and bool can be replaced by the assumption that Σ is strictly N-standard.

(2) The implication $(iv) \Longrightarrow (iii)$ can be construed as a conservativity theorem (*cf.* the remarks in §2.3 and after Theorem 1).

§5. Initial N-standard algebras for ω-conditional equations. We repeat, once again, the theory of Sections 3 and 4 for N-standard algebras with streams and infinitary logic.

Let Σ be a stream signature (§1.3). Let T be a theory in the language of $\mathsf{FOL}_\omega(\Sigma)$ (typically an ω-conditional equational theory; *cf.* §2.2), and let $\boldsymbol{T}^\omega(\Sigma)$ be the closed term algebra built from the closed ω-Σ-terms (§2.1). Consider the *quotient algebra of* $\boldsymbol{T}^\omega(\Sigma)$ with respect to T and a formalism $\mathcal{F}$ for ω-logic (typically $\mathsf{CondEq}_\omega(\Sigma)$):

$$\boldsymbol{T}^\omega(\Sigma, T, \mathcal{F}) =_{df} \boldsymbol{T}^\omega(\Sigma)/{\approx_{T,\mathcal{F}}}$$

where

$$t_1 \approx_{T,\mathcal{F}} t_2 \iff_{df} t_1 = t_2 \text{ is provable from } T \text{ in } \mathcal{F}.$$

Again (as with $\boldsymbol{T}(\Sigma, T, \mathcal{F})$ in Section 4) $\boldsymbol{T}^\omega(\Sigma, T, \mathcal{F})$ may fail to be N-standard.

Note that it *is* possible to characterise N-standard algebras in $\mathsf{FOL}_\omega(\Sigma)$ by the formulae

$$\forall z\,(\bigvee_{n=0}^{\infty} z = \bar{n}) \qquad \text{and} \qquad \forall b\,(b = \mathsf{true} \vee b = \mathsf{false})$$

(where z : nat and b : bool), but these are *not* ω-conditional equations! We therefore proceed as in the previous section with finitary logic.

We define the concepts: *T is consistent in $\mathcal{F}$*, and *T determines* nat *and* bool *in $\mathcal{F}$*, as in Section 4, but now with repect to ω-logic. Define also $\mathsf{NStdAx}(\Sigma)$ and $\mathsf{NStdAx_0}(\Sigma)$ as in Section 4.

We first present a normal form lemma for infinitary terms. A closed ω-Σ-term t is called *simple* if neither t nor any subterm of t is of a stream sort (and it is hence a finitary Σ-term). Assume, for convenience, that $\boldsymbol{PreStr}(\Sigma)$ is disjoint from $\boldsymbol{Str}(\Sigma)$. (This simplifies the formulation of the normal form lemma.)

NORMAL FORM LEMMA FOR STREAM TERMS. *Let E be an ω-conditional equational theory over Σ. Suppose that in* $\mathsf{CondEq}_\omega(\Sigma)$, *$E$ is consistent, determines* nat *and* bool, *and proves* $\mathsf{NStdAx}_0(\Sigma)$. *Then*

(*i*) *If t is a closed term not of a stream sort, then*

$$E + \mathsf{NStdAx}_0(\Sigma) \vdash t = \hat{t}$$

for some simple term $\hat{t}$.

(*ii*) *If t is a closed term of stream sort $\bar{s}$, then*

$$E + \mathsf{NStdAx}_0(\Sigma) \vdash t = \langle t_0\, t_1\, t_2 \ldots \rangle$$

for some simple terms $t_0, t_1, t_2, \ldots$ of sort s.

In both cases, '$\vdash$' refers to provability in $\mathsf{CondEq}_\omega(\Sigma)$. We give the analogue of Mal'cev's Theorem for stream signatures Σ and ω-logic.

THEOREM 3. *Let E be an ω-conditional equational theory over Σ. Suppose that in* $\mathsf{CondEq}_\omega(\Sigma)$, *$E$ is consistent, determines* nat *and* bool, *and proves* $\mathsf{NStdAx}_0(\Sigma)$. *Then $I =_{df} \boldsymbol{T}^\omega(\Sigma, E, \mathsf{CondEq}_\omega(\Sigma))$ is an initial N-standard model of E. Further, for two closed ω-Σ-terms t_1, t_2 of the same sort, the following are equivalent*:

(*i*) *t_1 and t_2 have the same value in I,*
(*ii*) *t_1 and t_2 have the same value in all N-standard models of E,*
(*iii*) *$t_1 = t_2$ is provable from E in* $\mathsf{CondEq}_\omega(\Sigma)$,
(*iv*) *$t_1 = t_2$ is provable from E in* $\mathsf{FOL}_\omega(\Sigma)$.

REMARKS. (1) By Lemma 2 in Section 4, the assumption in the normal form lemma and in Theorem 3, that E determines nat and bool, can be replaced by the assumption that Σ is strictly N-standard.

(2) The implication (*iv*) $\Longrightarrow$ (*iii*) can be construed as a conservativity theorem (*cf.* the remarks in §2.3 and after Theorems 1 and 2).

§6. Specifiability of the stream algebra operation.

6.1. Expanding operations. Assume from now on that Σ' is an N-standard signature with $\Sigma \subset \Sigma'$. Also, A is an N-standard Σ-algebra and A' is an N-standard Σ'-algebra. Also, T is a Σ-theory (in either the finite or infinitary language over Σ), and T' is a Σ'-theory.

DEFINITIONS. (1) Two Σ'-algebras A'_1 and A'_2 are *Σ'/Σ-isomorphic*, written $A'_1 \cong_{\Sigma'/\Sigma} A'_2$, if there is a Σ'-isomorphism from A'_1 to A'_2 whose restriction to Σ is the identity on $A'_1 |_\Sigma$.

(2) An operation $\Phi : \boldsymbol{NStd}(\Sigma) \to \boldsymbol{NStd}(\Sigma')$ is *expanding* (*over* Σ) if for all N-standard Σ-algebras A, $\Phi(A)$ is a Σ'-expansion of A.

EXAMPLE. For any N-standard signature Σ, fix $s \in \boldsymbol{Sort}(\Sigma)$ and let $\bar{\Sigma}$ be the extension of Σ formed by adjoining the *stream sort* $\bar{s}$ (associated with s), and the operation eval_s (which we will write below without the subscript 's'). Let $\bar{A}$ be the corresponding stream algebra (as in §1.3). Then the *stream algebra construction*

$$A \mapsto \bar{A}$$

is an expanding operation. (Note that we are working with a single pre-stream sort s for simplicity. We could just as easily take a *set* $\boldsymbol{S}$ of pre-stream sorts, as in the example in Section 1.)

Assume from now on that $\Phi : \boldsymbol{NStd}(\Sigma) \to \boldsymbol{NStd}(\Sigma')$ is an expanding operation over Σ. We will write A^{Φ} for $\Phi(A)$, and $\mathbb{K}^{\Phi}$ for $\{A^{\Phi} \mid A \in \mathbb{K}\}$ (or rather, its closure w.r.t. Σ'-isomorphism).

6.2. Specifiability of operations by theories. First we give:

DEFINITIONS. (1) Suppose A' is a Σ'-expansion of A. We say that (Σ', T') *specifies* A' *over* A if A' is the unique (up to Σ'/Σ-isomorphism) Σ'-expansion of A satisfying T'; in other words: (i) $A' \models T'$; and (ii) for all Σ'-expansions B' of A, if $B' \models T'$ then $B' \cong_{\Sigma'/\Sigma} A'$.

(2) (Σ', T') *specifies* Φ *uniformly over* Σ if for all $A \in \boldsymbol{NStd}(\Sigma)$, (Σ', T') specifies A^{Φ} over A.

PROPOSITION 1. *Suppose* (Σ', T') *specifies* Φ *uniformly over* Σ. *Then*

(i) $A \models T \iff A^{\Phi} \models T + T'$;

(ii) *If* $\mathbb{K} = \boldsymbol{NStd}(\Sigma, T)$, *then* $\mathbb{K}^{\Phi} = \boldsymbol{NStd}(\Sigma', T + T')$.

DEFINITIONS. (3) Φ has the *homomorphism extension property* (*HEP*) (w.r.t. Σ and Σ') if every homomorphism $h : A \to B$ between N-standard Σ-algebras can be *extended uniquely* to a homomorphism $h^{\Phi} : A^{\Phi} \to B^{\Phi}$ between their images under Φ.

(4) Φ is *initiality preserving* (w.r.t Σ and Σ') if for all $\mathbb{K} \subseteq \boldsymbol{NStd}(\Sigma)$ and $A \in \boldsymbol{NStd}(\Sigma)$:

$$A \text{ is initial in } \mathbb{K} \iff A^{\Phi} \text{ is initial in } \mathbb{K}^{\Phi}.$$

PROPOSITION 2. *If* Φ *has HEP, then* Φ *is initiality preserving.*

EXAMPLES. (1) The *array algebra* over any N-standard algebra A is defined in [32, 33]. Briefly, with each sort s is associated a sort s^* of *arrays* or finite sequences of data of sort s, with operations such as update, application, and length and the null array constant. The *array algebra construction* $A \mapsto A^*$ has HEP, and (hence) is initiality preserving.

(2) The *stream algebra construction* $A \mapsto \bar{A}$ also has HEP, and (hence) is also initiality preserving.

PROPOSITION 3. *Suppose* Φ *is initiality preserving. If* (Σ', T') *specifies* Φ *uniformly over* Σ, *then for any* Σ-*theory* T *and N-standard* Σ-*algebra* A,

$$A \cong \boldsymbol{InitNStd}(\Sigma, T) \iff A^{\Phi} \cong \boldsymbol{InitNStd}(\Sigma', T + T').$$

EXAMPLES. (3) In [33] a set $\mathsf{ArrAx}(\Sigma)$ of "array axioms" is defined over the signature Σ^* of A^*, such that the "array specification" $(\Sigma^*, \mathsf{ArrAx})$ specifies the array construction $A \mapsto A^*$, uniformly over Σ. Hence Proposition 3 can be applied to $T' = \mathsf{ArrAx}$.

(4) We will similary define a "stream specification" (in the ω-conditional equational language) in the next subsection.

6.3. Stream specification. Given an N-standard signature Σ and a Σ-sort s, let $\overline{\Sigma}$ be the stream signature over Σ with respect to s (as above). We are looking for a theory to specify the stream operation $A \mapsto \bar{A}$ uniformly over Σ. Note that if we had variables of stream sort, we could define this operation in $\mathsf{FOL}_\omega(\Sigma)$ by

(i) the scheme

$$\exists\alpha \bigwedge_{n=0}^{\infty}(\mathsf{eval}(\alpha,\bar{n}) = t_n)$$

for all sequences $t_0, t_1, \ldots$ of closed $\overline{\Sigma}$-terms of sort s; and

(ii) the single axiom

$$\forall\alpha, \forall\beta\big[\forall z(\mathsf{eval}(\alpha,z) = \mathsf{eval}(\beta,z)) \longrightarrow \alpha = \beta\big]$$

(where $\alpha, \beta : \bar{s}$ and z : nat); and in this way, avoid infinite terms! However neither of these is an (ω-)conditional equation. So we proceed instead as follows.

Let $\mathsf{StrAx}(\Sigma, s)$ consist of the following "stream axioms":

$$\mathsf{eval}(\langle t_0\, t_1\, t_2 \ldots\rangle, \bar{n}) = t_n,$$

$$\bigwedge_{i=0}^{\infty}(t_i = t_i') \longrightarrow \langle t_0\, t_1 \ldots\rangle = \langle t_0'\, t_1' \ldots\rangle$$

for all n, and all sequences $t_0, t_1, t_2, \ldots$ and $t_0', t_1', t_2', \ldots$ of $\overline{\Sigma}$-terms of sort s. Note that $\mathsf{StrAx}(\Sigma, s)$ is an ω-conditional equational theory.

THEOREM 4. *The specification* $(\overline{\Sigma}, \mathsf{StrAx}(\Sigma, s))$ *specifies the stream algebra construction* $A \mapsto \bar{A}$ *uniformly over* Σ.

Then from this, Proposition 3 in §6.2 and Example 2 preceding it we have:

COROLLARY 1. *For any N-standard* Σ*-algebra* A *and theory* T (*in* $\mathsf{FOL}_\omega(\Sigma)$):

$$A \cong \textit{\textbf{InitNStd}}(\Sigma, T) \iff \bar{A} \cong \textit{\textbf{InitNStd}}(\overline{\Sigma}, T + \mathsf{StrAx}(\Sigma, s)).$$

By considering the special case that T is an ω-conditional equational theory, we obtain

COROLLARY 2. If a Σ-algebra A has an initial N-standard algebra specification by a set of ω-conditional equations, then so does the stream algebra $\bar{A}$.

In particular, taking the term model of a (finite) conditional equational theory, we have from Theorems 2 and 3:

COROLLARY 3. *Let E be a conditional equational theory over Σ. Suppose that in* $\mathsf{CondEq}(\Sigma)$, *E is consistent, determines* nat *and* bool, *and proves* $\mathsf{NStdAx}_0(\Sigma)$. *Then* $I =_{df} \boldsymbol{T}(\Sigma, E, \mathsf{CondEq}(\Sigma))$ *is an initial N-standard model of E, and furthermore the stream algebra $\bar{I}$ constructed from I is (isomorphic to) the ω-term model*:

$$\bar{I} \cong \boldsymbol{T}^\omega(\overline{\Sigma}, \overline{E}, \mathsf{CondEq}_\omega(\Sigma))$$

and is also the initial N-standard model of the ω-conditional equational theory

$$\overline{E} =_{df} E + \mathsf{StrAx}(\Sigma, s).$$

§7. An initial infinite closed term model of the reals. We will now construct three ω-term models of the unit real interval $\mathbb{I} = [0, 1]$ which, by Theorem 3, will then also be initial models of certain ω-conditional equational theories. The results below could easily be adapted to the real line $\mathbb{R}$ instead of $\mathbb{I}$.

7.1. Real interval algebra. We begin with the signature $\Sigma_{\mathsf{B.N}}$ of the simplest N-standard algebra, with the sorts bool and nat, and the standard operations on these:

$$\Sigma_{\mathsf{B.N}} = \boxed{\begin{array}{ll} \mathsf{sorts} & \mathsf{bool}, \\ & \mathsf{nat} \\ \mathsf{functions} & \mathsf{true}, \mathsf{false} : \to \mathsf{bool}, \\ & \mathsf{and}, \mathsf{or} : \mathsf{bool}^2 \to \mathsf{bool}, \\ & \mathsf{not} : \mathsf{bool} \to \mathsf{bool}, \\ & 0 : \mathsf{nat}, \\ & \mathsf{S} : \mathsf{nat} \to \mathsf{nat}, \\ & \mathsf{eq}_{\mathsf{nat}}, \mathsf{less}_{\mathsf{nat}} : \mathsf{nat}^2 \to \mathsf{bool}, \\ & \mathsf{if}_{\mathsf{nat}} : \mathsf{bool} \times \mathsf{nat}^2 \to \mathsf{nat} \\ \mathsf{end} & \end{array}}$$

Next we add the sort intvl of the unit interval $\mathbb{I}$ and the corresponding discriminator:

$$\Sigma_{\mathsf{B.N.I}} = \boxed{\begin{array}{ll} \mathsf{import} & \Sigma_{\mathsf{B.N}} \\ \mathsf{sorts} & \mathsf{intvl} \\ \mathsf{functions} & \mathsf{if}_{\mathsf{intvl}} : \mathsf{bool} \times \mathsf{intvl}^2 \to \mathsf{intvl} \\ \mathsf{end} & \end{array}}$$

and then form the stream signature $\overline{\Sigma}_{\mathsf{B.N.I}}$ over $\Sigma_{\mathsf{B.N.I}}$ with respect to bool, *i.e.*, add the sort of *bitstreams*, or streams over $\mathbb{B}$:

$$\overline{\Sigma}_{\mathsf{B.N.I}} = \boxed{\begin{array}{ll} \mathsf{import} & \Sigma_{\mathsf{B.N.I}} \\ \mathsf{sorts} & \mathsf{bitstream} = \mathsf{nat} \to \mathsf{bool} \\ \mathsf{functions} & \mathsf{eval} : \mathsf{bitstream} \times \mathsf{nat} \to \mathsf{bool}, \\ & \mathsf{if}_{\mathsf{bitstream}} : \mathsf{bool} \times \mathsf{bitstream}^2 \to \mathsf{bitstream} \\ \mathsf{end} & \end{array}}$$

Finally (interpreting bitstreams as binary "decimal" expansions for reals in $\mathbb{I}$) we add the function realval which evaluates bitstreams as reals in $\mathbb{I}$:

$$\Sigma = \boxed{\begin{array}{ll} \mathsf{import} & \overline{\Sigma}_{\mathsf{B.N.I}} \\ \mathsf{functions} & \mathsf{realval} : \mathsf{bitstream} \to \mathsf{intvl} \\ \mathsf{end} & \end{array}}$$

We write truth values as 0 (for false) and 1 (for true), and refer to them as *bits*.

In order to formulate the next theorem, we need some definitions.

Let E be the theory in $\mathsf{CondEq}_\omega(\Sigma)$ consisting of:

(i) the set StrAx of stream axioms in §6.3 (taking $s = \mathsf{bool}$),

(ii) all equations of the form

$$\mathsf{realval}\,\langle \mathsf{b}_0\,\mathsf{b}_1\,\ldots\,\mathsf{b}_{n-1}\,0\,1\,1\,1\,\ldots\rangle \;=\; \mathsf{realval}\,\langle \mathsf{b}_0\,\mathsf{b}_1\,\ldots\,\mathsf{b}_{n-1}\,1\,0\,0\,0\,\ldots\rangle$$

for all $n \geq 0$ and all $\mathsf{b}_0, \mathsf{b}_1, \ldots, \mathsf{b}_{n-1}$, where each b_i is either 0 or 1.

Let $\mathcal{I}$ be the ω-Σ-term algebra

$$\mathcal{I} =_{df} \boldsymbol{T}^\omega(\Sigma,\ \mathsf{NStdAx}(\Sigma) + E,\ \mathsf{CondEq}_\omega(\Sigma)).$$

Let $\mathcal{J}$ be the N-standard Σ-algebra

$$\mathcal{J} \;=\; (\mathbb{I}, \mathbb{B}, \mathbb{N}, \overline{\mathbb{B}}, \ldots, \boldsymbol{realval})$$

over the real interval $\mathbb{I}$, where $\overline{\mathbb{B}} = [\mathbb{N} \to \mathbb{B}]$ is the set of bitstreams, and for any bitstream β, $\boldsymbol{realval}(\beta)$ is the unique real number in $\mathbb{I}$ which has β as one of its binary expansions.

THEOREM 5. *The real interval algebra $\mathcal{J}$ is isomorphic to the ω-Σ-term algebra $\mathcal{I}$, and is (therefore) the initial N-standard Σ-algebra of the ω-Σ-conditional equational theory E.*

REMARKS. (1) E is a co-semicomputable set of ω-Σ-formulae, with respect to computability on $\mathsf{CondEq}_\omega(\Sigma)$ viewed as Cantor space [35].

(2) We can thus get a specification of *equality* on $\mathbb{I}$, by a co-semicomputable set E_{eq} of formulae in $\mathsf{CondEq}_\omega(\Sigma_{\mathsf{eq}})$, where Σ_{eq} is formed from Σ by adjoining the function symbol

$$\mathsf{eq}_{\mathsf{intvl}} : \mathsf{intvl}^2 \;\to\; \mathsf{bool}$$

and E_{eq} is formed from E by adding the following two axioms:

$$\mathsf{eq}_{\mathsf{intvl}}(x, x) = \mathsf{true}, \qquad \mathsf{eq}_{\mathsf{intvl}}(x, y) = \mathsf{true} \;\to\; x = y.$$

Since equality on $\mathbb{I}$ is *not continuous*, and hence *non-computable* (in any reasonable sense), this shows that initial algebra specifiability (by a co-semicomputable set in $\mathsf{CondEq}_\omega(\Sigma_{\mathsf{eq}})$) does not imply computability. (Compare the positive results "computability $\Longrightarrow$ specifiability", and the partial converses, proved in [33].)

7.2. Real interval algebra with order. Now add a function symbol

$$\mathsf{less}_{\mathsf{intvl}} : \mathsf{intvl}^2 \;\to\; \mathsf{bool}$$

to Σ, to form the signature $\Sigma^<$.

Let F be the theory in $\mathsf{CondEq}_\omega(\Sigma^<)$ consisting of

(i) all equations of the form (writing '$t_1 < t_2$' for '$\mathsf{less}_{\mathsf{intvl}}(t_1, t_2) = \mathsf{true}$'):

$$\mathsf{realval}\,\langle \mathsf{b}_0\,\mathsf{b}_1\,\ldots\,\mathsf{b}_{n-1}\,0\,1^m\,\mathsf{b}_{n+m+1}\,\mathsf{b}_{n+m+2}\,\ldots\rangle$$
$$<\; \mathsf{realval}\,\langle \mathsf{b}_0\,\mathsf{b}_1\,\ldots\,\mathsf{b}_{n-1}\,1\,0^m\,\mathsf{b}'_{n+m+1}\,\mathsf{b}'_{n+m+2}\,\ldots\rangle$$

for all $n, m \geq 0$, and all $\mathsf{b}_0, \mathsf{b}_1, \ldots, \mathsf{b}_{n-1}, \mathsf{b}_{n+m+1}, \mathsf{b}'_{n+m+1}, \ldots,$ where 1^m denotes $11\ldots1$ (m times), and it is not the case that ($\mathsf{b}_{n+m+1} = 1$ and $\mathsf{b}'_{n+m+1} = 0$); and (*ii*) the two formulae (where variables x, y, z have sort intvl, and '$x \not< y$' stands for '$\mathsf{less}_{\mathsf{intvl}}(x, y) = \mathsf{false}$'):

$$x \not< x, \qquad x < y \wedge y < z \;\rightarrow\; x < z.$$

Let $\mathcal{I}^{<}$ be the ω-$\Sigma^{<}$-term algebra

$$\mathcal{I}^{<} =_{df} \boldsymbol{T}^{\omega}(\Sigma,\ \mathsf{NStdAx}(\Sigma) + E + F,\ \mathsf{CondEq}_{\omega}(\Sigma)).$$

LEMMA. $(\Sigma^{<},\ E + F)$ *specifies the standard ordering relation* $<$ *on* $\mathbb{I}$ *in* $\mathcal{J}$.

Let $\mathcal{J}^{<}$ be the $\Sigma^{<}$-expansion of $\mathcal{J}$ by this ordering on $\mathbb{I}$. By Theorem 5 and a modification of the argument used for Proposition 3 in §6.2, or directly, we can then show:

THEOREM 6. *The ordered real interval algebra* $\mathcal{J}^{<}$ *is isomorphic to the* ω-$\Sigma^{<}$-*term algebra* $\mathcal{I}^{<}$, *and is (therefore) the initial N-standard* $\Sigma^{<}$-*algebra of the* ω-$\Sigma^{<}$-*conditional equational theory* $E + F$.

REMARKS. (1) The axiom set F for $\mathsf{less}_{\mathsf{intvl}}$ is semicomputable, though uncountable. (It could be replaced by a countable set, by adding to the language stream variables and a 'cat' operation.)

(2) This gives us another example (*cf.* Remark 2 in §7.1) of a (discontinuous, and hence) non-computable function $\mathsf{less}_{\mathsf{intvl}}$ with an initial algebra specification $(\Sigma^{<},\ E + F)$.

7.3. Real interval algebra with arithmetic operations. For convenience we now let $\mathbb{I}$ stand for the *half-closed* real interval $[0, 1)$, and therefore add to E the axiom

$$\mathsf{realval}\,\langle 0\,0\,0\,\ldots\rangle = \mathsf{realval}\,\langle 1\,1\,1\,\ldots\rangle. \tag{$*$}$$

Now add the function symbols

$$\begin{aligned} \mathsf{plus}_{\mathsf{intvl}},\ \mathsf{times}_{\mathsf{intvl}} &: \mathsf{intvl}^2 \rightarrow \mathsf{intvl} \\ \mathsf{minus}_{\mathsf{intvl}} &: \mathsf{intvl} \rightarrow \mathsf{intvl} \end{aligned}$$

to Σ, to form the signature $\tilde{\Sigma}$.

Let G be the theory in $\mathsf{CondEq}_{\omega}(\tilde{\Sigma})$ containing, for all triples of bitstreams $\alpha = \langle \mathsf{a}_0\,\mathsf{a}_1\,\mathsf{a}_2\,\ldots\rangle$, $\beta = \langle \mathsf{b}_0\,\mathsf{b}_1\,\mathsf{b}_2\,\ldots\rangle$, $\gamma = \langle \mathsf{c}_0\,\mathsf{c}_1\,\mathsf{c}_2\,\ldots\rangle$ such that

$$\sum_{i=0}^{\infty} \frac{\mathsf{a}_i}{2^{i+1}} + \sum_{i=0}^{\infty} \frac{\mathsf{b}_i}{2^{i+1}} = \sum_{i=0}^{\infty} \frac{\mathsf{c}_i}{2^{i+1}} \pmod 1, \tag{$**$}$$

the equations

$$\mathsf{plus}_{\mathsf{intvl}}(\mathsf{realval}(\alpha),\ \mathsf{realval}(\beta)) = \mathsf{realval}\,(\gamma),$$

and similar axioms for $\mathsf{minus}_{\mathsf{intvl}}$ (additive inverse modulo 1) and $\mathsf{times}_{\mathsf{intvl}}$.

Let $\tilde{\mathcal{I}}$ be the ω-$\tilde{\Sigma}$-term algebra

$$\tilde{\mathcal{I}} =_{df} \boldsymbol{T}^{\omega}(\Sigma,\ \mathsf{NStdAx}(\Sigma) + E + G,\ \mathsf{CondEq}_{\omega}(\tilde{\Sigma})).$$

LEMMA. $(\tilde{\Sigma},\ E + G)$ *specifies the graphs of the standard arithmetic operations modulo* 1 *on* $\mathbb{I}$ *in* $\mathcal{J}$.

Let $\tilde{\mathcal{J}}$ be the $\tilde{\Sigma}$-expansion of $\mathcal{J}$ by these operations.

THEOREM 7. *The real interval algebra $\tilde{\mathcal{J}}$ is isomorphic to the ω-$\tilde{\Sigma}$-term algebra $\tilde{\mathcal{I}}$, and is (therefore) the initial N-standard $\tilde{\Sigma}$-algebra of the ω-$\tilde{\Sigma}$-conditional equational theory $E + G$.*

REMARKS. (1) The identification of (the bitstreams for) 0 and 1 (by $(*)$ above) creates a conflict with the order relation as axiomatised by F, and therefore the function symbol $\mathsf{less}_{\mathsf{intvl}}$, as well as the order axioms F, have been dropped.

(2) G is a co-semicomputable set of ω-$\tilde{\Sigma}$-formulae. The reason is that equation $(**)$ above holds if, and only if, for all n

$$\left| \sum_{i=0}^{n} \frac{\mathsf{a}_i}{2^{i+1}} + \sum_{i=0}^{n} \frac{\mathsf{b}_i}{2^{i+1}} - \sum_{i=0}^{n} \frac{\mathsf{c}_i}{2^{i+1}} \right| \leq 2^{-n}$$

(where '+' is addition modulo 1), suggesting an algorithm which, for bitstream inputs α, β and γ, can refute (but not confirm) the relation $(**)$ after finitely many steps; and similarly for the '$-$' and '$\times$' operations. Thus (*cf.* Remark 1 in §7.1) we have a *co-semicomputable specification* $(\tilde{\Sigma}, E + G)$ of the graphs of the arithmetical functions $(+, -, \times)$.

(3) We cannot define *computable* representing functions for the arithmetic functions $(+, -, \times)$ on the set $\overline{\mathbb{B}}$ of bitstreams, only *co-semicomputable* specifications for their *graphs* (as we have done above). By working, however, with *signed bitstreams* (essentially streams over a 3-point set $\{0, 1, \bar{1}\}$), we could define computable representing functions for the arithmetic functions on $\overline{\mathbb{B}}$ [36, 2, 3]. However, in that case too, specifications of their graphs would (only) be co-semicomputable.

§**8. Concluding remarks.** Continuous data types are modelled by topological many-sorted universal algebras. There is a need for a general theory of algebraic specification for classes of topological algebras — certainly, algebras with separable T_0 topologies which play a role in concrete computability theory [25, 35].

The origins of the initial algebra approach to modelling data types included the idea of modelling (certain) uncountable ordered data types. In the seminal paper [6], initial models for a class of continuous algebras are created from algebras of infinite terms. These continuous algebras are used in the fixed point semantics of recursions. However specification theory, based on equations and conditional equations, rather neglected these structures in favour of countable effective algebras. Only comparatively recently, has the need to investigate specifications and proof systems for uncountable structures re-emerged. One approach is higher-order algebra as developed by B. Möller, K. Meinke and others. A survey is given in the collection [8]. This approach began with Möller's papers [18, 19]. In [13] Meinke provided initial algebra semantics and Birkhoff-Mal'cev completeness theorems for natural type structures over a many-sorted signature. Higher order specifications have been used to handle specific stream algorithm verifications in [15]. Topological and computability questions are also examined in [11, 14]. In addition, there is infinite term rewriting as expounded in [10]. The connections between these ideas are intriguing, and probably form an important part of the needed general theory.

Stream algebras are destined to play an important role in the theory. The topology of a stream agebra $\bar{A}$ depends upon that of the data set A. In hardware, the theory of stream computation is emphasised in [20]. The computability of concrete representations of stream algebras have been considered in [1]; this includes a discussion of signals, modelled as streams indexed by continuous time.

REFERENCES

[1] J. Blanck, V. Stoltenberg-Hansen, and J.V. Tucker, *Domain representation of streams*. In Möller and Tucker [20].

[2] H.J. Boehm, R. Cartwright, M. Riggle, and M.J. O'Donnell, *Exact real arithmetic: A case study in higher programming*, ***ACM symposium on Lisp and functional programming***. 1986.

[3] M.H. Escardó, *Effective and sequential definition by cases on the reals via infinite signed-digit numerals*, ***Electronic Notes in Theoretical Computer Science***, vol. 13 (1998).

[4] S. Feferman, *Computation on abstract data types: The extensional approach, with an application to streams*, ***Annals of Pure & Applied Logic***, vol. 81 (1996), pp. 75–113.

[5] J.E. Fenstad, *Computability theory: Structure or algorithms*, In this volume.

[6] J.A. Goguen, J.W. Thatcher, E.G. Wagner, and J.B. Wright, *Initial algebra semantics and continuous algebras*, ***Journal of the Association for Computing Machinery***, vol. 24 (1977), pp. 68–95.

[7] J.V. Guttag and J.J. Horning, *The algebraic specification of abstract data types*. ***Acta Informatica***, vol. 10 (1978), pp. 27–52.

[8] J. Heering, K. Meinke, B Moller, and T Nipkow (editors), ***Hoa '93: International workshop on higher-order algebra, logic and term rewriting***, Lecture Notes in Computer Science, vol. 816, Springer-Verlag, 1994.

[9] C.R. Karp, ***Languages with expressions of infinite length***, North Holland, 1964.

[10] J.R. Kennaway, J.W. Klop, M.R. Sleep, and F.J. de Vries, *Transfinite reductions in orthogonal term rewriting systems*, ***Information & Control***, vol. 119 (1995), pp. 18–38.

[11] P. Kosiuczenko and K. Meinke, *On the power of higher-order algebraic specification methods*. ***Information & Computation***, vol. 124 (1995), pp. 85–101.

[12] A.I. Mal'cev, ***Algebraic systems***, Grundlehren der mathematischen Wissenschaften, vol. 192, Springer-Verlag, 1973.

[13] K. Meinke, *Universal algebra in higher types*, ***Theoretical Computer Science***, vol. 100 (1992), pp. 385–417.

[14] ———, *Topological methods for algebraic specification*, ***Theoretical Computer Science***, vol. 166 (1996), pp. 263–290.

[15] K. Meinke and J. Steggles, *Specification and verification in higher order algebra: a case study of convolution*, ***Higher order algebra, logic and term rewriting*** (J Heering, K Meinke, B Moller, and T Nipkow, editors), Lecture Notes in Computer Science, vol. 816, Springer-Verlag, 1994, pp. 189–222.

[16] K. Meinke and J.V. Tucker, *Universal algebra*, ***Handbook of logic in computer science*** (S. Abramsky, D. Gabbay, and T. Maibaum, editors), vol. 1, Oxford University Press, 1992, pp. 189–411.

[17] J. Meseguer and J.A. Goguen, *Initiality, induction and computability*, ***Algebraic methods in semantics*** (M. Nivat and J. Reynolds, editors), Cambridge University Press, 1985, pp. 459–541.

[18] B. Möller, *On the algebraic specification of infinite objects — ordered and continuous models of algebraic types*, ***Acta Informatica***, vol. 22 (1985), pp. 537–578.

[19] ———, *Higher-order algebraic specifications*, ***Habilitationsschrift***, Technische Universität München, 1987.

[20] B. Möller and J.V. Tucker (editors), ***Prospects for hardware foundations***, Lecture Notes in Computer Science, vol. 1546, Springer-Verlag, 1998.

[21] Y.N. MOSCHOVAKIS, *The formal language of recursion*. ***The Journal of Symbolic Logic***, vol. 54 (1989), pp. 1216–1252.

[22] M.B. POUR-EL and J.I. RICHARDS, ***Computability in analysis and physics***, Springer-Verlag, 1989.

[23] H. ROGERS, JR., ***Theory of recursive functions and effective computability***, McGraw-Hill, 1967.

[24] R. STEPHENS, *A survey of stream processing*, ***Acta Informatica***, vol. 34 (1997), pp. 491–541.

[25] V. STOLTENBERG-HANSEN and J.V. TUCKER, *Effective algebras*, ***Handbook of logic in computer science*** (S. Abramsky, D. Gabbay, and T. Maibaum, editors), vol. 4, Oxford University Press, 1995, pp. 357–526.

[26] G. TAKEUTI, ***Proof theory*** (2nd ed.), North Holland, 1987.

[27] J.V. TUCKER and J.I. ZUCKER, ***Program correctness over abstract data types, with error-state semantics***, CWI Monographs, vol. 6, North Holland, 1988.

[28] ———, *Projections of semicomputable relations on abstract data types*, ***International Journal of Foundations of Computer Science***, vol. 2 (1991), pp. 267–296.

[29] ———, *Provable computable selection functions on abstract structures*, ***Proof theory*** (P. Aczel, H. Simmons, and S.S. Wainer, editors), Cambridge University Press, 1993, pp. 277–306.

[30] ———, *Computable functions on stream algebras*, ***Proof and computation: NATO Advanced Study Institute International Summer School at Marktoberdorf, 1993*** (H. Schwichtenberg, editor), Springer-Verlag, 1994, pp. 341–382.

[31] ———, *Computation by 'while' programs on topological partial algebras*, ***Theoretical Computer Science***, vol. 219 (1999), pp. 379–420.

[32] ———, *Computable functions and semicomputable sets on many-sorted algebras*, ***Handbook of logic in computer science*** (S. Abramsky, D. Gabbay, and T. Maibaum, editors), vol. 5, Oxford University Press, 2000, pp. 317–523.

[33] ———, *Abstract computability, algebraic specification and initiality*, ***ACM Transactions on Computational Logic***, (2001), To appear.

[34] W. WECHLER, ***Universal algebra for computer scientists***, EATCS Monographs, vol. 25, Springer-Verlag, 1992.

[35] K. WEIHRAUCH, ***Computable analysis: An introduction***, Springer-Verlag, 2000.

[36] E. WIEDMER, *Computing with infinite objects*, ***Theoretical Computer Science***, vol. 10 (1980), pp. 133–155.

[37] M. WIRSING, *Algebraic specification*, ***Handbook of theoretical computer science, Vol. B: Formal methods and semantics*** (J. van Leeuwen, editor), North Holland, 1991, pp. 675–788.

DEPARTMENT OF COMPUTER SCIENCE
UNIVERSITY OF WALES
SWANSEA SA2 8PP, WALES
E-mail: J.V.Tucker@swansea.ac.uk

DEPARTMENT OF COMPUTING AND SOFTWARE
MCMASTER UNIVERSITY
HAMILTON, ONTARIO L8S 4L7, CANADA
E-mail: zucker@mcmaster.ca

PART III: APPLICATIVE AND SELF-APPLICATIVETHEORIES

ON EXTENSIONALITY, UNIFORMITY AND COMPREHENSION IN THE THEORIES OF OPERATIONS AND CLASSES

ANDREA CANTINI

For Solomon Feferman on occasion of his 70th birthday

Abstract. This note deals with theories of operations and classes, which are inspired by the investigation of the (conceptual and technical) borders of Feferman's explicit mathematics. We survey some inconsistency and consistency results.

§1. Introduction. The informal intuitions behind explicit mathematics naturally hint at a radically non-extensional ontology. In fact, there is a tension between uniform class comprehension schemata, which are typical of Feferman's theories, and class extensionality, as it is shown by a number of inconsistency results ([21], [13], [12], [1], [24], [6]).

In this note, we illustrate the issue by concentrating upon systems, which do not belong to explicit mathematics properly; nevertheless, we believe that they might be of help for discussing the problem in a broader context.

We first deal with the pure theory of explicit positive comprehension and present some negative results (refutation of power class, extensionality and others), *without involving any ground applicative structure*. The main tool is the explicit form of (positive) untyped comprehension, that is, a partial class is named by a term $\{x|\varphi(x,\vec{a})\}$, depending on the parameters $\vec{a}$. Though quite weak, the theory proves a simple version of the second recursion theorem, which gives a systematic access to standard diagonalization techniques.

Then we mainly investigate the relation between operations and classes in Quine's set theory NF, *with full extensionality*. Of course, there are serious reasons to be dissatisfied with NF (e.g., the theory does not fare well with heterogeneous collections and relations); however, once we stick to systems which involve a universal class (like Feferman's), stratification is still a resource one ought to explore, also in consideration of its logical intriguing (consistency) aspects. To this aim, we propose a theory NF_{op}[1] of partial operations and total classes, where all set existence axioms are *explicit, stratified* (in a suitably generalized form, appropriate to the new language), and *full extensionality* is admitted, together with

Research supported by "Progetto Strategico di Ateneo 1996", Università di Firenze.

[1]This system is stronger than the homonymous NF_{op} of [5].

Reflections on the foundations of mathematics
Essays in honor of Solomon Feferman
Edited by Wilfried Sieg, Richard Sommer, and Carolyn Talcott

an underlying theory of partial combinatory logic. However *not every explicitly definable* operation on classes is representable by an "internal" operation (i.e., one of of the underlying applicative structure).

As expected, the stratified framework imposes severe limitations, and the interaction between classes and operations has to be restricted: the applicative structure can only witness operations which are definable by conditions, which are *homogeneous* (i.e., arguments and values have the same implicit type) and *monotonically dependent* upon the very notion of application. So we do have explicit principles, but not all of them are *uniform* in the sense of being definable via combinators (failure of combinatory completeness).

We substantially refine the results of [5], which were mainly based on applicative extensions of NFU (New Foundations with *Urelemente*) or full NF. In particular, we study a subsystem NFI_{op} with full extensionality and operations, which proves the existence of several internal operations (like homogeneous pairing and projections, boolean operations, exponentiation, fregean successor, etc.), and yet it is interpretable into the consistent fragment NFI (see [7]) of NF. This fragment is based on loosely predicative (or better "mildly impredicative") restrictions of the stratified comprehension schema.

§2. **Inconsistencies in the theory of type-free partial classes.** Feferman ([13],[16]) has investigated theories of operations and partial classes, which strengthen in various directions Gilmore's positive comprehension axiom ([21]). We show that all the systems with type free comprehension are inconsistent with the existence of power classes, extensional singletons and extensionality, even if no underlying theory of operations is assumed; the essential point is the 'explicit' formulation of the comprehension schema and the admissibility of self-reference. The proof follows from an inseparability result, which is inspired by recursion theory.

$\mathcal{L}_s$ is the elementary set theoretic language, which comprises the binary predicate symbols $=, \in$ and the individual constant $\emptyset$. $\mathcal{L}_s$–terms are simply individual variables $(x, y, z, \dots)$ with the constant $\emptyset$, and prime formulas (atoms) have the form $t = s$, $t \in s$ (t, s terms). $\mathcal{L}_s$–formulas are inductively generated from prime formulas by means of sentential connectives and quantifiers. The elementary set theoretic language $\mathcal{L}_s^+$ is obtained by adding to $\mathcal{L}_s$ the abstraction operator $\{-|-\}$; $\mathcal{L}_s^+$–terms and formulas are then simultaneously generated. The clause for introducing class terms has the form: if φ is a formula, then $\{x|\varphi\}$ is a term where $FV(\{x|\varphi\}) = FV(\varphi - \{x\})$ ($FV(E)$ is the set of free variables occurring in the expression E).

Convention. We identify α–congruent terms (formulas); of course, two terms (formulas) are α–congruent if they only differ by renaming of bound variables.

Definition 1. POS, the collection of positive formulas, is the smallest set containing atoms and closed under $\wedge$, $\vee$, $\forall$, $\exists$. Σ^+ is the subcollection of positive existential formulas (so no universal quantifier is used), while QF^+ is the subcollection of quantifier-free positive formulas.

DEFINITION 2. If $\mathcal{F}$ is a collection of formulas, $\mathcal{F}$-CA is the theory, which comprises classical predicate calculus with equality, the axiom of non-triviality, and the explicit comprehension schema for formulas of the collection $\mathcal{F}$:

1. $\forall x(x \notin \emptyset)$;
2. $\mathcal{F}$–CA:

$$\forall u(u \in \{x|\varphi\} \leftrightarrow \varphi(u)),$$

provided $\varphi \in \mathcal{F}$ and u is free for x in φ.

$\mathcal{F}$–CA$^-$ is the schema

$$\exists y \forall u(u \in y \leftrightarrow \varphi(u))$$

provided φ is an $\mathcal{L}_s$–formula of $\mathcal{F}$, y is not free in φ.

$\mathcal{F}$–CA$^-$ will also designate the corresponding theory in the language $\mathcal{L}_s$, which is obtained from $\mathcal{F}$–CA with non-explicit comprehension principle $\mathcal{F}$–CA$^-$.

THEOREM 1. POS-CA *is consistent (provably, say, in Peano Arithmetic).*

Indeed, something stronger is true, i.e., the comprehension schema remains consistent even if the notion of positive formula is enlarged with formulas of the form $\neg t = s$. The theorem is well-known (see Gilmore [21]): a simple model-theoretic proof relies upon fixed point techniques, which are familiar from the theory of inductive definitions and have been greatly refined by Feferman in [16].

As to the possibility of relating POS–CA to an extensional theory of classes, the answer is largely negative, as shown by the following results, which are recursion-theoretic in spirit.

DEFNITION 3. We introduce some basic notions:

1. $t \subseteq s \Leftrightarrow \forall x(x \in t \rightarrow x \in s)$;
2. $t =_e s \Leftrightarrow t \subseteq s \wedge s \subseteq t$ (equiextensionality);
3. $E_x(\varphi) \Leftrightarrow \forall a, b(a =_e b \wedge \varphi[x := a] \rightarrow \varphi[x := b])$ (i.e., φ is extensional in x);
4. if $E_x(x \in t)$, where x does not occur in t, we say that t is extensional;
5. t is non-total iff for some x, $x \notin t$; t is non-empty iff for some x, $x \in t$;
6. $V = \{x|x = x\}$.
7. $\{x\} = \{u|u = x\}$; $\{x, y\} = \{u|u = x \vee u = y\}$;
8. $\langle x, y\rangle = \{\{x\}, \{x, y\}\}$.

LEMMA 1. QF$^+$–CA *proves*:

$$\forall x \forall y \forall u \forall v(\langle x, y\rangle = \langle u, v\rangle \rightarrow x = u \wedge y = v)$$

The proof is standard (apply comprehension for quantifier free positive formulas and identity logic). It is now interesting to see that, though quite weak, the positive systems provide us with a standard method[2] for solving self-referential conditions:

[2] It is a trick derived from formal semantics, [4], p.76.

LEMMA 2. (*"Second Recursion Theorem"*)
Let $fa := \{x | \langle x, a\rangle \in f\}$; *then there exists a term* I_f *such that* $FV(I_f) = \{f\}$ *and, provably in* Σ^+–CA:

$$I_f =_e fI_f$$

PROOF. Choose

$$D_f = \{z | \exists x \exists g(z = \langle x, g\rangle \wedge x \in f(gg))\}$$
$$I_f = D_f D_f$$

Then by comprehension for positive existential formulas, the pairing lemma and identity logic, we have:

$$\begin{aligned} u \in I_f &\leftrightarrow \langle u, D_f\rangle \in D_f \\ &\leftrightarrow \langle u, D_f\rangle = \langle x, g\rangle \wedge x \in f(gg) \\ &\quad \text{(for some } x, g) \\ &\leftrightarrow u \in f(D_f D_f) \equiv fI_f \end{aligned}$$ ⊣

THEOREM 2. (*"Weak Inseparability"*) *There exists a term* $f(x_1, x_2, x_3)$ *such that for any pair of formulas* φ_1, φ_2, Σ^+–CA *with intuitionistic logic proves: if* φ_1, φ_2 *are extensional,* $\varphi_1(x_1)$, $\varphi_2(x_2)$, *where* $x_1 \subseteq x_2$, *and* $\forall u(\varphi_2(u) \rightarrow u \notin x_3)$, *then*

$$\varphi_1(f(x_1, x_2, x_3)) \quad \textit{and} \quad f(x_1, x_2, x_3) \notin x_3.$$

PROOF. By the recursion lemma, there exists a term $f(x_1, x_2, x_3)$, such that

(1) $$u \in f(x_1, x_2, x_3) \leftrightarrow u \in x_1 \vee (u \in x_2 \wedge f(x_1, x_2, x_3) \in x_3)$$

provably in Σ^+–CA.

Now assume x_1, x_2, φ_1, φ_2 satisfy the hypothesis and $\forall u(\varphi_2(u) \rightarrow u \notin x_3)$.

It is enough to check $\varphi_1(f(x_1, x_2, x_3))$ and $f(x_1, x_2, x_3) \notin x_3$. Indeed, if we assume $f(x_1, x_2, x_3) \in x_3$, then $f(x_1, x_2, x_3) =_e x_2$ would follow by $x_1 \subseteq x_2$. Since $\varphi_2(x_2)$ and φ_2 is extensional, $\varphi_2(f(x_1, x_2, x_3))$ holds, whence

$$f(x_1, x_2, x_3) \notin x_3.$$

Therefore $f(x_1, x_2, x_3) =_e x_1$ and by assumption $\varphi_1(f(x_1, x_2, x_3))$. ⊣

COROLLARY 1. (*Intuitionistically provable in* Σ^+–CA *without non-triviality axiom*)

1. *Upward closure: for no* z, y, x, *we can have that* z *is extensional,* $x \in z$, $x \subseteq y$ *and* $y \notin z$;
2. *if* z *is non-total, then* $\neg\exists y \forall x(x \in y \leftrightarrow x \subseteq z)$ (*refutation of power class*);
3. *if* z *is non-total, then* $\neg\exists y \forall x(x \in y \leftrightarrow x =_e z)$ (*refutation of extensional singleton*);
4. *there exists a term* $h(x, y)$ *such that, if* x *is non-total and* $\forall v(v \in y \rightarrow v =_e x)$, *then* $h(x, y) \notin y$ *and* $h(x, y) =_e x$;
5. *there exists a term* $f(x)$ *such that, if* x *is non-total, then* $f(x) \neq x$ *and* $f(x) =_e x$ (*refutation of extensionality*);

PROOF. (1) Assume z is extensional, $x \in z$, $x \subseteq y$ and $y \notin z$. Apply the theorem with $\varphi_1(v) := (v =_e x)$, $\varphi_2(v) := (v =_e y)$, $x_1 := x$, $x_2 := y$, $x_3 := z$ (of course, φ_1, φ_2 are extensional, $\varphi_1(x)$, $\varphi_2(y)$ and by extensionality of z, $\forall v(\varphi_2(v) \to v \notin z)$).

(2) Assume that z is non-total and for some y: $\forall x(x \in y \leftrightarrow x \subseteq z)$. Then y is extensional, and moreover: $z \in y$, $V \notin y$. This contradicts item 1.

(3) Apply (1).

(4). Choose $h(x, y) := f(x, V, y)$: we apply the theorem with $\varphi_1(v) := (v =_e x)$, $\varphi_2(v) := (v =_e V)$, $x_1 := x$, $x_2 := V$, $x_3 := y$ (the assumption that x is non-total is used to check that $\forall v(\varphi_2(v) \to v \notin y)$.

(5): immediate by the preceding item. ⊣

Since the empty set axiom trivially grants the existence of a non-total class, we also have:

COROLLARY 2. Σ^+–CA *is inconsistent with each of the following statements*:

$$\forall z \exists y \forall x(x \in y \leftrightarrow x \subseteq z) \tag{2}$$

$$\forall z \exists y \forall x(x \in y \leftrightarrow x =_e z) \tag{3}$$

$$\forall x \forall y(x =_e y \to x = y) \tag{4}$$

REMARK 1. (i) Theorem 2 is a restricted version of the inseparability lemma of [6]; it can directly be proved in the weaker QF$^+$–CA. Simply choose

$$f(x_1, x_2, x_3) = q(v := s(x_1, x_2, x_3), x_1, x_2)$$

where:

$$q(v, x_1, x_2) = \{z | z \in x_1 \vee (z \in x_2 \wedge v \in v)\};$$
$$s(x_1, x_2, x_3) = \{v | q(v, x_1, x_2) \in x_3\}$$

Of the consequences stated in corollary 1, item 1 is in [4]; a special form of item 5 (for the empty class) is to be found in [21].

As to the assumption in items 2-5, Minari [26] proves that there are non-trivial models of POS–CA (actually much more), where $\forall x(x =_e V \to x = V)$.

(ii) A more direct way of regarding the inconsistency is to observe that QF$^+$–CA with each one of the principles (2), (3), (4) implies the existence of the complement, and hence the naive comprehension schema with the Russell paradox. Indeed, define $g(x, y) = \{z | z = x \wedge x \in y\}$; then trivially $x \notin y \leftrightarrow g(x, y) =_e \emptyset$, for any x. If extensionality holds or there exists w such that

$$\forall v(v \in w \leftrightarrow v \subseteq \emptyset \leftrightarrow v =_e \emptyset),$$

then by positive comprehension there exists some c such that

$$\forall x(x \in c \leftrightarrow g(x, y) \in w \leftrightarrow x \notin y)$$

(iii) The proof of theorem 2 still works within applicative theories, where sets and rules live together: for, if there exists an operation $G(h, y)$ (in the sense of partial combinatory logic) yielding the set

$$\{u | u \in x_1 \vee (u \in x_2 \wedge hy \in x_3)\},$$

the recursion theorem for operations readily provides a solution to (1). So the inconsistencies of theorem 2 easily extend to the theories of explicit mathematics ([6]), as well as to ZF with operations ([2]).

To conclude, we also mention:

Proposition 1. QF^+–CA *is essentially undecidable.*

The simplest way to see this is to observe that the minimal predicative set theory N (with the empty set and the operation $(x, y) \mapsto x \cup \{y\}$) is trivially contained in QF^+–CA. But the well-known essentially undecidable arithmetical theory Q is interpretable in N, by the main result of [27].

2.1. Consistency with positive non-explicit abstraction. The explicit formulation of comprehension (via abstraction terms that make the parameter dependence explicit) is essential for the negative results of the previous section.

Indeed, let us consider the theory which is stated in the pure set theoretic language $\mathcal{L}$, without abstraction operator and the individual constant $\emptyset$ (so that terms coincide with variables). Following [20], we define the class GPF of *generalized positive formulas*, as the smallest class which contains prime formulas of the form $x = y$, $x \in y$, is closed under disjunction, conjunction, universal and existential quantifiers and the additional clauses:

- if $\varphi \in GPF$ and y does not occur in φ, $y \neq x$, then $\forall x(x \in y \to \varphi) \in GPF$;
- if $\varphi \in GPF$ and $\theta(x)$ is an arbitrary formula with the only free variable shown, then $\forall x(\theta(x) \to \varphi) \in GPF$.

Definition 4. (Forti-Hinnion [20]) GPC is the theory, which includes classical predicate calculus with equality, and in addition:

- Empty class: $\exists y \forall x(x \notin y)$;
- Extensionality $\forall x \forall y(x =_e y \to x = y)$;
- GPC: $\exists y \forall u(u \in y \leftrightarrow \varphi(u))$, provided $\varphi \in GPF$, $y \notin FV(\varphi)$.

It is known from the work of [20] and others that GPC has models, the so-called hyperuniverses; more precisely:

Theorem 3. *If* ZF *is consistent, then* GPC+*"the existence of an infinite class" (e.g., ω) is consistent.*

On the other hand, the formulas $x \subseteq z$, $x =_e z$ are typical formulas of GPF and hence GPC proves the existence of the power class, and the singleton is obviously extensional. Hence:

Corollary 3. POS–CA$^-$ *is consistent with extensionality and the existence of power class and extensional singleton.*

Esser [10] shows how to compare extensions of GPC with the Kelley-Morse impredicative theory of classes.

§3. Operations and stratified comprehension. We introduce a theory of partial operations and total classes, where operations obey to the laws of partial combinatory logic, while classes form a model of Quine's set theory. The interaction of operations and classes is restricted and yet it comprises non-trivial instances of uniform stratified comprehension.

The language $\mathcal{L}_q$ comprises the basic symbols of $\mathcal{L}_s^+$; in addition:

- a 3–ary predicate constant Ap;
- individual constants for combinators K, S, $PAIR$, $RIGHT$, $LEFT$ (pairing and projection operators);
- a countable list c_n of individual constants ($n \in \omega$);
- a new binary function symbol H for a homogeneous pairing operation.

Terms and formulas: they are defined by simultaneous induction, as usual (variables and individual constants are terms; if φ is a formula, $\{x|\varphi\}$ is a term; if t, s are terms, $H(t,s)$ is a term; formulas are generated from atoms $t = s$, $t \in s$, $Ap(t,s,r)$ by means of boolean operations and quantifiers).

As before, we identify α–congruent terms and formulas.

Applicative terms: every term is an applicative term; if t, s are applicative terms, then (ts) is an applicative term. For applicative terms, we use the conventions adopted in combinatory logic (outermost parentheses are omitted; when parentheses are missing, they are restored by associating to the left).

DEFINITION 5. We inductively define $t \simeq y$ (t applicative term):

1. t variable or constant: $t \simeq y \Leftrightarrow t = y$;
2. $ts \simeq y \Leftrightarrow \exists z \exists v(t \simeq z \wedge s \simeq v \wedge Ap(z,v,y))$;
3. $t \downarrow \Leftrightarrow \exists y(t \simeq y)$;
4. $t \simeq s \Leftrightarrow \forall x(t \simeq x \leftrightarrow s \simeq x)$;
5. $A(t) \Leftrightarrow \exists x(t \simeq x \wedge A[x := t])$ (t applicative).

In the following, we adopt the more perspicuous notation $H(t,s) = (t,s)$, and we stipulate that $(x,y,z) = (x,(y,z))$. As usual for Quine's systems, we need stratification; we shall also define a restricted notion thereof, which is motivated by the consideration of "loosely predicative" class existence axioms.

DEFINITION 6. **(i):** φ is *stratified* iff it is possible to assign a natural number (type in short) to each term occurrence[3] of φ in such a way that

1. if $t \in s$ is a subformula of φ, the type of s is one greater than the type of t;
2. if $t = s$, $Ap(t,s,r)$ are subformulas of φ, then t, s, r receive the same type;
3. all free occurrences of the same variable in any subformula of φ have the same type;
4. if x is free in ψ and $\forall x \psi$ is a subformula of φ, then the 'x' in $\forall x$ and the free occurrences of x in ψ receive the same type;
5. if $t = \{x|\beta\}$ occurs in φ, x is free in β, then t is assigned a type one greater than the type assigned to x, and all the free occurrences of x in β receive the same type ;
6. if $t = (s,r)$, then t, r, s receive the same type;

[3]Individual constants included; these can be given any type compatible with the clauses below.

$\{x|\varphi\}$ is stratified if φ is stratified;

(ii): a term $\{x|\varphi(x,\vec{y})\}$ is *homogeneous* in $\vec{y}$, if $\varphi(x,\vec{y})$ is stratified and the parameters $\vec{y}$ are assigned the same type as $\{x|\varphi(x,\vec{y})\}$ itself ($x,\vec{y}$ comprises all the free variables of φ);

(iii): a term $\{x|\varphi(x,\vec{y})\}$ is *operative*[4] in $\vec{y}$, if the term is homogeneous in $\vec{y}$ and all the arguments of the predicate Ap occurring in φ have the same type as $\vec{y}$;

(iv): a stratified term $\{x|\varphi(x,\vec{y})\}$ is *loosely $\simeq$–predicative* iff for some type $i \in \omega$, $\{x|\varphi(x,\vec{y})\}$ has type $i+1$, no (free or bound) variable of $\varphi(x,\vec{y})$ is assigned type greater than $i+1$, and every argument of Ap in φ has type i[5]; a stratified term $\{x|\varphi(x,\vec{y})\}$ is *$\simeq$–predicative* iff $\{x|\varphi(x,\vec{y})\}$ is loosely $\simeq$–predicative and in addition no quantified variable of $\varphi(x,\vec{y})$ is assigned the same type as $\{x|\varphi(x,\vec{y})\}$ itself.

(v): If a (loosely) $\simeq$–predicative term t has no occurrence of Ap, we simply say that t is (loosely) predicative;

(vi): φ is $n+1$–stratified iff φ is stratified by means of $0,\dots,n$.

For instance, $\bigcup a = \{x|(\exists y \in a)(x \in y)\}$ is not loosely predicative, as it requires type 2, but $\bigcup a$ itself has type 1; $a \cap b = \{x|x \in a \wedge x \in b\}$ is predicative and operative in a, b. Consider

$$I(f) = \{x|\forall z(fz \subseteq z \rightarrow x \in z)\} \tag{5}$$

where

$$fz \subseteq z \Leftrightarrow \forall x(\exists y(y \simeq fz \wedge x \in y) \rightarrow x \in z)$$

Then $I(f)$ has type 1, is not loosely $\simeq$–predicative, since the arguments of Ap receive type one greater than the type of x, though it is operative in f.

$F(y) = \{x|xx \simeq \emptyset \wedge x \in y \wedge yy \simeq \emptyset\}$ is homogeneous in y, but it is not operative in y (the arguments of the second occurrence of Ap must receive a type one greater than those of the first occurrence). Of course, every term which is Ap–free and homogeneous in $\vec{y}$, is operative in $\vec{y}$.

Convention. We fix an encoding of all pairs $\{x|\beta(x,\vec{y})\}, \{x|\gamma(x,\vec{y})\}$, which are operative in $\vec{y}$ (for some list $\vec{y}$ of parameters). If k is the Gödelnumber of the pair $\{x|\beta(x,\vec{y})\}, \{x|\gamma(x,\vec{y})\}$, we only write c_β, instead of c_k; we also assume that, if c_i occurs in β then $i < k$.

DEFINITION 7. We inductively define the a–transforms φ^a, t^a (a fresh variable) by stipulating:

1. $t^a = t$, if t is a variable or a constant;
2. $\{u|\varphi(u)\}^a = \{u|\varphi^a(u)\}$;

---

[4]The notion is motivated by the transformation $a \mapsto \varphi^a$ below and the subsequent lemma: the operative terms are used to define internal operations.

[5]This is a strengthened predicativity restriction on operations; the operations involved in φ cannot act on objects of the same type $i+1$, to which the class belongs. In a sense, they are first order relative to type i.

3. $Ap(t,s,r)^a = (t^a, s^a, r^a) \in a$;
4. $a \mapsto \varphi^a$ commutes with propositional connectives and quantifiers.

Clearly $a \mapsto \varphi^a$ does not preserve stratification in general (apply the transformation to the term $F(y)$ above); however:

LEMMA 3. 1. *If a is a fresh variable and $\{x|\varphi(x,\vec{y})\}$ is operative in $\vec{y}$, then $\{x|\varphi(x,\vec{y})\}^a$ is stratified and the type of a is one greater than the type of $\vec{y}$;*
2. *if $\{x|\varphi(x,\vec{y})\}$ is (loosely) $\simeq$–predicative, then $\{x|\varphi(x,\vec{y})\}^a$ is (loosely) predicative.*

$Mon(\varphi)$ is an abbreviation for "φ is monotone in the application predicate", i.e.,

$$Mon(\varphi) \Leftrightarrow \forall a \forall b(a \subseteq b \wedge \varphi^a \to \varphi^b)$$

$CC(\beta,\gamma)$ is the conjunction of the formulas asserting completeness and consistency:

$$\forall x \forall \vec{y}(\beta(x,\vec{y}) \vee \gamma(x,\vec{y}));$$
$$\forall x \forall \vec{y}(\neg\gamma(x,\vec{y}) \vee \neg\beta(x,\vec{y}))$$

DEFNITION 8. The system NF_{op} comprises:

1. predicate logic with equality for the extended language;
2. $Ap(x,y,z) \wedge Ap(x,y,w) \to z = w$;
3. $Kxy = x \wedge Sxy\downarrow \wedge Sxyz \simeq xz(yz)$;
4. $K \neq S$.
5. $(PAIRxy \simeq (x,y) \wedge LEFT(x,y) = x \wedge RIGHT(x,y) = y$;
6. class extensionality: $\forall x \forall y(x =_e y \to x = y)$;
7. stratified explicit comprehension SCA: if φ is stratified, then
$$\forall u(u \in \{x|\varphi(x,\vec{y})\} \leftrightarrow \varphi(u,\vec{y}))$$
8. uniform monotone comprehension UMC:
$$(CC(\beta,\gamma) \wedge Mon(\beta) \wedge Mon(\gamma) \to \forall \vec{y}(c_\beta \vec{y} \simeq \{x|\beta(x,\vec{y})\});$$
Proviso: $\{x|\beta(x,\vec{y})\}$, $\{x|\gamma(x,\vec{y})\}$ are operative in $\vec{y}$.

Other systems

(i): NF is NF_{op} restricted to the set theoretic language $\mathcal{L}_s^+$[6] (so it does not comprise UMC);

(ii): NFP_{op} (NFI_{op}) is the subsystem of NF_{op}, where SCA is restricted to (loosely) $\simeq$–predicative abstracts and UMC is restricted to (loosely) predicative abstracts[7];

NFP (NFI) is the subsystem of NF, where SCA is restricted to (loosely) predicative abstracts.

---

[6]Usually NF is presented in the language without abstraction operator; however, the present formulation is conservative over the abstract–free version.

[7]So the abstracts involved in UMC must be homogeneous (in some list of parameters) and (loosely) predicative, hence Ap–free.

(iii): NF_k (NFI_k, NFP_k) is the subsystem of NF (NFI, NFP_k), where (at most) k types are allowed for stratification.

REMARK 2. If Union is the axiom " $\bigcup a$ exists, for all a ", then NFP+Union is equivalent to the full NF (cf. [7])

3.1. Elementary consequences.

PROPOSITION 2. (*Provable in* NFP_{op}). *If* $\{x|\varphi(x,\vec{a})\}$ *is predicative (hence also* Ap*–free) and homogeneous in* $\vec{a}$, *then there exists a term* c_φ *such that*

$$c_\varphi \vec{a} \simeq \{x|\varphi(x,\vec{a})\}$$

In particular, there exist INT, DOM, CO, *such that*

$$INTab = \{x|x \in a \wedge x \in b\}$$
$$COa = \{x|x \notin a\}$$
$$DOMa = \{x|\exists y((x,y) \in a)\}$$

PROOF. By assumption on φ, $\{x|\varphi(x,\vec{a})\}$, $\{x|\neg\varphi(x,\vec{a})\}$ are operative in $\vec{a}$ and monotone (as Ap does not occur); consistency and completeness are obvious. So we can apply UMC. The remaining claims are special cases. ⊣

PROPOSITION 3. (*Provable in* NFP_{op}). *There exist* APP, TOT, $EXPab$ *such that*

$$APP = \{u|\exists x\exists y\exists z(u = (x,y,z) \wedge xy \simeq z)\}$$
$$TOT = \{f|\forall x\exists y(fx \simeq y)\}$$
$$EXPab = \{f|(\forall x \in a)(\exists y \in b)(fx \simeq y)\}$$

PROOF. The first two items are obvious by $\simeq$–predicative stratified comprehension.

As to exponentiation, consider the term $G(d,a,b)$ such that

$$G(d,a,b) = \{f|(\forall x \in a)(\exists y \in b)((f,x,y) \in d)\}$$

Then we can apply the previous proposition and for some c_k, we have

$$c_k dab \simeq G(d,a,b)$$

Choose $d = APP$ and $EXP = c_k APP$ (this is a well-defined applicative term).

The definition of the image operation is similar. ⊣

Let $f : V \to V$ mean that f is everywhere defined.

PROPOSITION 4. *There exists a term* LF *such that, provably in* NF_{op}:

$$(f : V \to V) \to LFf \simeq I(f);$$
$$Mon(f) \to f(LFf) \subseteq LFf \wedge \forall x(fx \subseteq x \to LFf \subseteq x),$$

where

$$Mon(f) \Leftrightarrow \forall x\forall y\exists z\exists u(x \subseteq y \to fx \simeq z \wedge fy \simeq u \wedge z \subseteq u)$$

PROOF. First of all, the term $I(f)$ defines a set in NF_{op} (see (5) for $I(f)$). Choose

$$\beta(u, f) \Leftrightarrow \forall x(\exists y(fx \simeq y \wedge \neg y \subseteq x) \vee u \in x)$$
$$\gamma(u, f) \Leftrightarrow \exists x \exists y(fx \simeq y \wedge y \subseteq x \wedge u \notin x)$$

Observe that $\{u|\beta(u, f)\}$, $\{u|\gamma(u, f)\}$, $I(f)$ are operative in f. By uniqueness of Ap, totality of f, and since β, γ are positive in Ap, we immediately have:

$$Mon(\beta) \wedge Mon(\gamma) \wedge CC(\{u|\beta(u, f)\}, \{u|\gamma(u, f)\})$$

Hence by UIMC, there exists a constant c_β such that $c_\beta f \simeq \{x|\beta(x, f)\}$. Then by totality of f and extensionality it is easy to check that $I(f) = \{x|\beta(x, f)\}$, and we can choose $c_\beta = LF$. ⊣

REMARK 3. It would be interesting to know whether the previous proposition is already provable in NFI_{op}. In other words: is the use of instances of stratified comprehension which are not loosely predicative essential for the previous proposition?

Let $\mathsf{Ext}_{op} := \forall x(f x \simeq gx) \to f = g$, $\mathsf{Tot} := \forall x \forall y(xy\downarrow)$.

PROPOSITION 5. $\mathsf{NFP}_{op} + \mathsf{Ext}_{op}$ *and* $\mathsf{NFP}_{op} + \mathsf{Tot}$ *are inconsistent.*

PROOF. Extensionality of operations implies that every total operation has a fixed point. But CO is total by the previous propositions and hence has a fixed point: contradiction. The same holds with total application. ⊣

REMARK 4. The inseparability result of section 2 cannot be recasted in the present context: it involves either collections defined by non-stratified formulas or operations which are not homogeneous.

§4. Consistency of NF-theories with operations. Our plan is to prove that NFI_{op} (NF_{op}) is consistent, provided NFI (NF) is consistent. In the case of the theory with restricted comprehension, we can also rely upon a result of Crabbé ([7]), i.e., NFI is provably consistent in third order arithmetic. The details of the (different) consistency proofs for NFI can be found in [7] and [23]. However, let us (roughly) recall the main idea of [7], pp. 133–135. First of all, one exploits the reducibility of NFI to its fragment NFI_4, and NFI_4 is interpreted into a corresponding type theory up to level 4, TI_4, plus Amb (= the so called schema of typical ambiguity, see [31][8]). These steps are finitary and specialize well-known theorems of Grishin and Specker. Finally, TI_4+Amb is proved consistent by appealing to the Hauptsatz for second order logic (which is equivalent in primitive recursive arithmetic to the 1-consistency of full second order arithmetic[9]). Therefore, one has:

THEOREM 4. NFI *is consistent* (*in primitive recursive arithmetic plus the* 1-*consistency of second order arithmetic*).

[8] In TI_4, $\{x^i|\varphi\}$ exists, provided $i = 0, 1, 2$ and φ contains free or bound variables of type $i+1$ at most.

[9] E.g., see J.Y. Girard, *Proof Theory and Logical Complexity*, Bibliopolis, Napoli 1987, p.280.

In order to interpret NFI_{op} in NFI, it is enough to embed *each finite subsystem* of NFI_{op} into NFI: a similar argument also shows that, if NF is consistent, so is NF_{op}. To this aim, we build partial combinatory algebras on the top of NFI-models: this is carried out by inductively generating the application relation with suitable clauses, which force the truth of finitely many instances of UMC, besides the partial combinatory logic axioms and pairing. Thus we can follow Feferman[14], but then we need a pairing operation. Here is a difficulty: in order to cope with the restriction imposed by stratified comprehension, we have to rely on Quine's homogeneous pairing operation, which *does require extensionality* and the existence of a copy of the natural numbers. But it is not difficult to check that Quine's pairing is indeed well-defined already in NFI.

For the sake of clarity, it is convenient to split the proof of the theorem into two substeps: 1) interpreting NFI(pair), i.e., the fragment of NFI_{op} with homogeneous pairing but no Ap, in NFI; 2) interpreting each finite subsystem of NFI_{op} into NFI(pair).

4.1. NFI and homogeneous pairing. For technical reasons, it is convenient to identify NFI(pair) (NFP(pair), NF(pair)) with the subtheory of NFI_{op} (NFP_{op}, NF_{op} respectively), which omits the predicate Ap, but still comprises all the old individual constants and the pairing functions symbol. In addition, we fix the following abbreviations:

$K_x = (K, x)$; $S_x = (S, x)$; $S_{xy} = (S_x, y)$; $PAIR_x = (PAIR, x)$; $c_{k.\vec{y}} = (c_k, y_1, \dots y_n)$ (here k is the code of a formula φ which is operative in a sequence $\vec{z}$, $\vec{y} = y_1 \dots y_n$ is an initial subsequence of $\vec{z}$).

The axioms of NFI(pair) are simply those of NFI (in the extended language), and in addition:

1. $\forall x \forall y \forall u \forall v((x, y) = (u, v) \rightarrow x = u \wedge y = v)$;
2. "independence" axioms, stating that c_k, $c_{k.\vec{y}}$, K, S, K_x, S_x, S_{xy}, $PAIR_x$ are all pairwise different (e.g., $K_x \neq K$, $S \neq c_4 \neq S_{xy}$, etc.)

THEOREM 5. NFI(pair) *is interpretable into* NFI (*actually it is a definitional extension of* NFI).

PROOF. We first check that a homogeneous pairing operation is definable in NFI. This follows from two lemmas.

First of all, the set of Fregean natural numbers exists in NFI. In fact, define:

$$\begin{aligned}
\emptyset &= \{x | x \neq x\};\\
V &= \{x | x = x\}\\
0 &= \{\emptyset\};\\
a + 1 &= \{x \cup \{y\} | x \in a \wedge y \notin x\};\\
Cl_N(y) &\Leftrightarrow 0 \in y \wedge \forall x(x \in y \rightarrow (x + 1) \in y);\\
\mathcal{N} &= \{x | \forall y(Cl_N(y) \rightarrow x \in y)\}
\end{aligned}$$

Then NFI proves the existence of $\mathcal{N}$ and that $\mathcal{N}$ is infinite (Specker's Theorem). In fact, all the sets above are loosely predicative and moreover:

LEMMA 4. (*provable in* NFI)

$$Cl_N(\{x|\varphi(x)\}) \rightarrow \mathcal{N} \subseteq \{x|\varphi(x)\}; \tag{6}$$

$$(\forall x)(x \in \mathcal{N} \leftrightarrow x = 0 \vee (\exists y \in \mathcal{N})(x = y + 1)); \tag{7}$$

$$\emptyset \notin \mathcal{N} \wedge (\forall x \in \mathcal{N})(V \notin x); \tag{8}$$

$$(\forall x \in \mathcal{N})(x + 1 \neq 0); \tag{9}$$

$$(\forall x \in \mathcal{N})(\forall y \in \mathcal{N})(x + 1 = y + 1 \rightarrow x = y) \tag{10}$$

(*Proviso*: *in* (6) $\{x|\varphi(x)\}$ *is loosely predicative*).

For the proof, remind that (8) holds in NFI+ Union, as NFI+Union is NF, and it is a theorem of Specker that NF proves (8). On the other hand, by [7] NFI + ¬Union implies (8). The conditions (7), (6) and the Peano axioms easily follow in NFI ((10) uses the second part of (8)).

DEFNITION 9. (Homogeneous pairing; [30])

$$\phi(a) = \{y|y \in a \wedge y \notin \mathcal{N}\} \cup \{y + 1|y \in a \wedge y \in \mathcal{N}\};$$
$$\theta_1(a) = \{\phi(x)|x \in a\};$$
$$\theta_2(a) = \{\phi(x) \cup \{0\}|x \in a\};$$
$$(a, b) = \theta_1(a) \cup \theta_2(b);$$
$$Q_1(a) = \{z|\phi(z) \in a\};$$
$$Q_2(a) = \{z|\phi(z) \cup \{0\} \in a\}$$

The definitions above are (at most) loosely predicative and hence the universe of classes is closed under the corresponding (homogeneous) operations, provably in NFI. Moreover:

LEMMA 5. (*Provable in* NFI)

1. $\phi(a) = \phi(b) \rightarrow a = b$;
2. $0 \notin \phi(a)$;
3. $\theta_i(a) = \theta_i(b) \rightarrow a = b$, *where* $i = 1, 2$;
4. $Q_i((x_1, x_2)) = x_i$, *where* $i = 1, 2$.
5. $(x, y) = (u, v) \rightarrow x = u \wedge y = v$.
6. *the map* $x, y \longmapsto (x, y)$ *is* $\subseteq$*–monotone in each variable and surjective.*

The proof of the lemma is straightforward, the main ingredient being the properties of $\mathcal{N}$ with successor ([30]).

Using the properties of homogeneous pairing and of the natural numbers in NFI, it is routine to find terms $K^\diamond$, $S^\diamond$, $PAIR^\diamond$, $LEFT^\diamond$, $RIGHT^\diamond$, $c_k^\diamond$, for $k \in \omega$, such that $K^\diamond$, $S^\diamond$, $PAIR^\diamond$, $LEFT^\diamond$, $RIGHT^\diamond$, $K_x^\diamond$, $S_x^\diamond$, $PAIR_x^\diamond$, $S_{xy}^\diamond$, $c_k^\diamond$, $c_{k,\vec{y}}^\diamond$ (for $k \in \omega$) are pairwise distinct, provably in NFI. This takes care of the "independence axioms" on constants in NFI(pair).

Hence the proof of theorem 5 is completed. ⊣

We conclude the subsection with two propositions, which are (perhaps) unnoticed in the literature, though they clarify the strategic role of homogeneous pairing. In the case of NF, the proof was independently suggested by Antonelli and Holmes, as an answer to a guess by the present author about the strength of $\mathsf{NF}_3(\mathsf{pair})$:

PROPOSITION 6. $\mathsf{NF} \equiv \mathsf{NF}_3(\mathsf{pair})$

PROOF. By a theorem of Grishin [22], $\mathsf{NF} \equiv \mathsf{NF}_3 + \mathcal{E}$, where $\mathcal{E}$ is the statement $\exists y(y = E)$, and $E = \{\{\{x\}, y\} | x \in y\}$. Notice that E already exists in NFP_4. So it will suffice to show that $\mathsf{NFP}_3(\mathsf{pair})$ proves the existence of E. Let

$$\Pi = \{u | \exists x \exists y \exists z (u = (x, y) \wedge x = \{z\} \wedge z \in y)\}.$$

Clearly Π exists in $\mathsf{NFP}_3(\mathsf{pair})$. Consider the stratified formula $\psi(u, s)$:

$$\exists x \exists y \exists z (z \in u \wedge y \in u \wedge (\forall m \in u)(m = z \vee m = y) \wedge$$
$$\wedge x \in s \wedge (x, z) \in \Pi \wedge$$
$$\wedge (x, y) \in \Pi \wedge (\forall p \in s)((p, z) \in \Pi \rightarrow p = x)).$$

Then $\mathsf{NFP}_3(\mathsf{pair})$ proves

$$\forall s \exists a \forall u (u \in a \leftrightarrow \psi(u, s)).$$

If we choose $s = \{u | \exists v(\{v\} = u)\}$, then there exists a set J such that, provably in $\mathsf{NFP}_3(\mathsf{pair})$, $J = \{u | \psi(u, s)\}$. It is routine to see that $J = E$. ⊣

On the other hand, we can readily extend the proposition to the subsystems NFI, NFP.

PROPOSITION 7. $\mathsf{NFP} \equiv \mathsf{NFP}_3(\mathsf{pair})$ *and* $\mathsf{NFI} \equiv \mathsf{NFI}_3(\mathsf{pair})$

PROOF. Again by a theorem of Grishin and [7], $\mathsf{NFI} \equiv \mathsf{NFI}_3 + \mathcal{I}$, (respectively $\mathsf{NFP} \equiv \mathsf{NFP}_3 + \mathcal{I}$), where

$$\mathcal{I} \Leftrightarrow \exists y (y = \{u | \cap u \neq \emptyset\})$$

But $\mathsf{NFP}_3(\mathsf{pair})$ proves the existence of the set $M = \{(\{x\}, y) | x \in y\}$ and

$$\forall s \exists a \forall x (x \in a \leftrightarrow \exists q (q \in s \wedge (\forall z \in x)((q, z) \in M))) \tag{11}$$

Choose $s = \{\{x\} | x \in V\}$ in (11); then we obtain in $\mathsf{NFP}_3(\mathsf{pair})$ a set I such that

$$(\forall x)(x \in I \leftrightarrow \cap x \neq \emptyset)$$ ⊣

4.2. Interpreting finite subsystems of NFI_{op} and NF_{op}. In the following S is NF_{op} or NFI_{op}; accordingly, S(pair) is NF(pair) or NFI(pair). Let T be a set of finitely many instances of UMC and let S^T be the subsystem of S, with UMC restricted to instances from the set T. It is then possible to describe in S(pair) the inductive applicative clauses for pairing, projections, the combinators K and S, and, in addition, the constants c_k, representing the operations associated to the T–instances of UMC. For the sake of clarity, we explicitly consider only the special case where T contains one instance of UMC and the formulas involved β, γ are operative in a single parameter y[10]. Then we can find a formula $A^T(c, a)$, saying

[10] In general, if T contains n instances ($n \geq 2$), we ought to add n clauses of the appropriate form. The limitation to one parameter is not essentially restrictive; we simply avoid to write down many trivial clauses, corresponding to the successive applications of c_k to the parameters.

that " c is an ordered homogeneous triple and one of the following conditions on c and a holds, for some x, y, z, u, v, w ":

- $c = (K, x, K_x)$;
- $c = (K_x, y, x)$;
- $c = (S, x, S_x)$
- $c = (S_x, y, S_{xy})$;
- $c = (LEFT, (x, y), x)$
- $c = (RIGHT, (x, y), y)$;
- $c = (PAIR, x, PAIR_x)$
- $c = (PAIR_x, y, (x, y))$;
- $c = (S_{xy}, z, u)$ and $(x, z, v) \in a$ and $(y, z, w) \in a$ and $(v, w, u) \in a$;
- $c = (c_\beta, y, z)$ and $(\forall u \in z)\beta^a(u, y)$ and $\forall u(\gamma^a(u, y) \vee u \in z)$

$A^T(c, a)$ is clearly a formula in the language of S(pair). Actually, more is true. In fact, since β and γ are operative in y, u has, say, type i and y has type $i+1$; hence c has type $i+1$ and a has type $i+2$. If S(pair)=NFI(pair) and hence β and γ are loosely predicative, β and γ can contain at most variables of type $i+1$, and by inspection $A^T(c, a)$ has quantified variables of type at most $i+1$. Therefore, using also lemma 3:

LEMMA 6. 1. $\{c|A^T(c, a)\}$ *is homogeneous in a (hence stratified)*;
2. *if* $\mathsf{S}=\mathsf{NFI}_{op}$, $\{c|A^T(c, a)\}$ *is predicative.*

DEFNITION 10.

$$\Psi^T(a) := \{u : A^T(u, a)\};$$
$$Cl^T_{ap}(a) := \Psi^T(a) \subseteq a;$$
$$Ap^T := \{u : \exists x \exists y \exists z(u = (x, y, z) \wedge \forall a(Cl^T_{ap}(a) \to u \in a))\}$$

The T–translation of formulas and terms E of S is defined by composing the a–translation and the substitution operation: if a is fresh and $E[a := t]$ stands for the result of substituting t for a in E,

$$E^T = (E^a)[a := Ap^T]$$

Abbreviation: $MonCC(T)$ is the (finite) conjunction of all the formulas of the form $Mon(\beta)^T$, $Mon(\gamma)^T$, $CC(\beta, \gamma)^T$, whenever β, γ are the formulas associated to the T–instances of UMC. Observe that $Mon(\varphi)^T \equiv Mon(\varphi)$ (since Ap does not occur in $Mon(\varphi)$) and also that $CC(\beta, \gamma)^T \equiv CC(\beta^T, \gamma^T)$.

LEMMA 7. (NF(pair))

1. $\exists! a(a = Ap^T)$;
2. $MonCC(T) \to (a \subseteq b \to \Psi^T(a) \subseteq \Psi^T(a))$;
3. *if* $\{x|\varphi(x)\}$ *is a stratified term of* NF_{op}, *then* $\{x|\varphi^T(x)\}$ *is stratified and hence exists, provably in* NF(pair);
4. *if* $\{x|B(x)\}$ *is a stratified term of* NF(pair),

$$Cl^T_{ap}(\{x : B(x)\}) \to Ap^T \subseteq \{x : B(x)\};$$

5. $MonCC(T) \to Cl_{ap}(Ap^T)$;
6. $MonCC(T) \to \forall u(u \in Ap^T \to A(u, Ap^T))$;
7. *if T is a finite set of instances of* UMC *in* NFI_{op}, *the statements* 1–6 *above*[11] *are already provable in* NFI(pair), *provided $\{x|\varphi(x)\}$ is loosely $\simeq$–predicative in* 3 *and loosely predicative in* 4.

PROOF. 1. $A^T(c, a)$ is stratified and so are $Cl^T_{ap}(a)$, and Ap^T. In the case of NFI_{op}, $A^T(c, a)$ is predicative (by the previous lemma); hence Ap^T is loosely predicative and exists in NFI(pair).

2. By definition, the operator is either positive in a or monotone in a, since by assumption β, γ are monotone in the application predicate (this is trivial in the case of T, containing instances of UMC, coming from NFI_{op}).

3. If $\{x|\varphi(x)\}$ is stratified (loosely $\simeq$–predicative), then $\{x|\varphi^T(x)\}$ is stratified (loosely predicative; apply lemma 3).

4–5: by definition of Ap^T and appropriate comprehension.

6: apply item 1 and 4. $\dashv$

LEMMA 8. (NF(pair); *below, we write $xy \simeq^T z$, instead of $(x, y, z) \in Ap^T$). Assume $MonCC(T)$. Then*:

1. $\forall x \forall y(xy \simeq^T z \wedge xy \simeq^T w \to z = w)$;
2. $Kxy = x \wedge Sxy \downarrow \wedge Sxyz \simeq^T xz(yz))$;
3. $(PAIRxy \downarrow) \wedge LEFT(PAIRxy) = x \wedge RIGHT(PAIRxy) = y$;
4. $\forall y(c_\beta y \simeq^T \{x|\beta^T(x, y)\})$ (β *being an operative formula of* T).

If T is a finite set of instances of UMC *in* NFI_{op}, *the statements* 1–4 *above are provable in* NFI(pair), *provided β is also loosely predicative.*

PROOF. We only consider item 1 in the case of NFI_{op}.

Assume $MonCC(T)$, i.e., for β, γ loosely predicative :

$$\forall u(\beta^T(u, y) \to \neg\gamma^T(u, y)) \tag{12}$$

$$\forall u(\beta^T(u, y) \vee \gamma^T(u, y)) \tag{13}$$

$$Mon(\beta) \wedge Mon(\gamma) \tag{14}$$

Then consider the formula

$$C(u) \Leftrightarrow \exists x \exists y \exists z(u = (x, y, z)) \wedge u \in Ap^T$$
$$\wedge \forall v(v \in Ap^T \wedge (v)_1 = (u)_1 \wedge (u)_2 = (v)_2 \to (u)_3 = (v)_3).$$

Since C is loosely predicative, we can again apply the induction lemma 7 (items 4–7) and verify

$$\Psi^T(\{u : C(u)\}) \subseteq \{u : C(u)\} \tag{15}$$

which obviously yields uniqueness. In order to check (15), we assume

$$A^T(c, \{u : C(u)\}), \quad d \in Ap^T, \quad (c)_1 = (d)_1, \quad (c)_2 = (d)_2$$

[11] Actually, statements 2 and 5-6 can be simplified: the assumptions $MonCC(T)$ can be omitted in this case, as they are trivially satisfied for Ap–free formulas.

and we verify $(c)_3 = (d)_3$. There are several possibilities according to the form of c, but we only consider the case of UMC.

Assume $c = (c_\beta, y, z)$ and $d = (c_\beta, y, w) \in Ap^T$; then by lemma 7 (items 6–7),

$$\forall u(\gamma^T(u, y) \vee u \in w) \tag{16}$$

$$(\forall u \in w)\beta^T(u, y) \tag{17}$$

On the other hand, using the assumption $A^T(c, \{u : C(u)\})$, we have:

$$\forall u(\gamma(u, y, C) \vee u \in z), \tag{18}$$

$$(\forall u \in z)\beta(u, y, C), \tag{19}$$

where

$$\gamma(u, y, C) = (\gamma^a(u, y))[a := \{u : C(u)\}];$$
$$\beta(u, y, C) = (\beta^a(u, y))[a := \{u : C(u)\}].$$

We want $w \subseteq z$ and $z \subseteq w$.

Ad $w \subseteq z$. Then

$$\begin{aligned} u \in w &\overset{(17)}{\Rightarrow} \beta^T(u, y) \\ &\overset{(12)}{\Rightarrow} \neg\gamma^T(u.y) \\ &\Rightarrow \neg\gamma(u, y, C) \\ &\overset{(18)}{\Rightarrow} u \in z \end{aligned}$$

In the third step, we use (14) with $\{x|C(x)\} \subseteq Ap^T$.

Ad $z \subseteq w$: similar to the previous argument, applying (16), (19), besides (12), (14). ⊣

Collecting the previous lemmas, we have:

THEOREM 6. *Each finite subsystem of* NFI_{op} (NF_{op}) *is interpretable into* NFI (NF).

COROLLARY 4. *If* NFI (NF) *is consistent, so is* NFI_{op} (NF_{op}).

REMARK 5. The reader might wonder about extending ZF with a structure of partial combinatory algebra. Of course, this is a reasonable move, once we abandon the idea of a universal class and the like. An attempt in this direction is already in the literature (Beeson's system ZFR, see [2]). Let us mention that, in comparison with NF-theories, ZF-like theories can have *uniform operations representing power set and singleton*, e.g., non-homogeneous operations are allowed, too. Indeed, we have also verified that a form of monotone separation can be consistently added to ZFR (in analogy to our uniform comprehension principle UMC for NFI_{op}).

REFERENCES

[1] M.J. Beeson. ***Foundations of constructive mathematics***. Springer, Berlin, 1985.

[2] ———, *Towards a computation system based on set theory*, ***Theoretical Computer Science***, vol. 60 (1988), pp. 297–340.

[3] M. Boffa, D. van Dalen, and K. McAloon (editors), *Logic colloquium 78*, Amsterdam, North Holland, 1979.

[4] A. Cantini, ***Logical frameworks for truth and abstraction***, North Holland, Amsterdam, 1996.

[5] ———, *Relating Quine's NF to Feferman's EM*, ***Studia Logica***, vol. 62 (1999), pp. 141–162.

[6] A. Cantini and P. Minari, *Uniform inseparability in explicit mathematics*, ***The Journal of Symbolic Logic***, vol. 64 (1999), pp. 313–326.

[7] M. Crabbé, *On the consistency of an impredicative subsystem of Quine's NF*, ***The Journal of Symbolic Logic***, vol. 47 (1982), pp. 131–136.

[8] J. N. Crossley (editor), *Algebra and logic*, Lecture Notes in Mathematics 450, Berlin, Springer, 1975.

[9] J. Diller and G. H. Müller (editors), *Proof theory symposion*, Lecture Notes in Mathematics 500, Berlin, Springer, 1975.

[10] O. Esser, *On the consistency of a positive theory*, ***Mathematical Logic Quarterly***, vol. 45 (1999), to appear.

[11] S. Feferman, *Some formal systems for the unlimited theory of structures and categories*, ***The Journal of Symbolic Logic***, vol. 39 (1974), pp. 374–375, abstract.

[12] ———, *A language and axioms for explicit mathematics*, In Crossley [8], pp. 87–139.

[13] ———, *Non-extensional type-free theories of partial operations and classifications I*, In Diller and Müller [9], pp. 73–118.

[14] ———, *Recursion theory and set theory: a marriage of convenience*, In Fenstad et al. [19], pp. 55–98.

[15] ———, *Constructive theories of functions and classes*, In Boffa et al. [3], pp. 159–225.

[16] ———, *Towards useful type-free theories, I*, ***The Journal of Symbolic Logic***, vol. 49 (1984), pp. 244–259.

[17] S. Feferman and G. Jäger, *Systems of explicit mathematics with non-constructive μ-operator, I*, ***Annals of Pure and Applied Logic***, vol. 65 (1993), pp. 243–263.

[18] ———, *Systems of explicit mathematics with non-constructive μ-operator, II*, ***Annals of Pure and Applied Logic***, vol. 79 (1996), pp. 37–52.

[19] J.E. Fenstad, R.O. Gandy, and G.E. Sacks (editors), *Generalized recursion theory II*, Amsterdam, North Holland, 1978.

[20] M. Forti and R. Hinnion, *The consistency problem for positive comprehension principles*, ***The Journal of Symbolic Logic***, vol. 54 (1989), pp. 1401–1418.

[21] P. C. Gilmore, *The consistency of partial set theory without extensionality*, ***Axiomatic set theory II***, 1974, pp. 147–153.

[22] V. N. Grishin, *Consistency of a fragment of Quine's NF system*, ***Soviet Mathematics Doklady***, vol. 10 (1969), pp. 1387–1390.

[23] M. R. Holmes, *The equivalence of NF–style set theories with "tangled" type theories: the construction of ω–models of predicative NF (and more)*, ***The Journal of Symbolic Logic***, vol. 60 (1995), pp. 178–190.

[24] G. Jäger, *Power types in explicit mathematics*, ***The Journal of Symbolic Logic***, vol. 62 (1997), pp. 1142–1146.

[25] T. Jech (editor), *Axiomatic set theory II*, Proceedings of Symposia in Pure Mathematics, Providence RI, AMS, 1974.

[26] P. Minari, *Theories of types and names with positive stratified comprehension*, ***Studia Logica***, vol. 62 (1999), pp. 215–242.

[27] F. Montagna and A. Mancini, *A minimal predicative set theory*, ***Notre Dame Journal of Formal Logic***, vol. 35 (1994), pp. 186–203.

[28] E. Nagel, P. Suppes, and A. Tarski (editors), *Logic, methodology and philosophy of science*, Stanford, Stanford University Press, 1962, Proceedings of the 1960 International Congress.

[29] W. v. O. Quine, *On ordered pairs*, ***The Journal of Symbolic Logic***, vol. 10 (1945), pp. 95–96.

[30] J. B. Rosser, ***Logic for mathematicians***, Mc Graw-Hill, New York, 1953.

[31] E. Specker, *Typical ambiguity*, In Nagel et al. [28], Proceedings of the 1960 International Congress, pp. 116–123.

DIPARTIMENTO DI FILOSOFIA
UNIVERSITÀ DEGLI STUDI DI FIRENZE
FIRENZE, ITALY
E-mail: cantini@philos.unifi.it

THE PROOF-THEORETIC ANALYSIS OF THE SUSLIN OPERATOR IN APPLICATIVE THEORIES

GERHARD JÄGER AND THOMAS STRAHM

Dedicated to Sol Feferman for his 70th birthday

Abstract. In this article we introduce the Suslin quantification functional E_1 into the framework of Feferman's explicit mathematics and analyze it from the point of view of proof theory. More precisely, we work in the first order part of explicit mathematics augmented by appropriate axioms for E_1. Then we establish the exact proof-theoretic relationship between these applicative theories and (subsystems of) the second order theory (Δ^1_2-CA), depending on the induction principles permitted.

§1. Introduction. During the last decades a series of interesting axiomatic frameworks for constructive mathematics have been proposed. Depending on the philosophical position of their inventors, numerous routes have been pursued, some of which are quite different in nature. Comprehensive surveys of many of these approaches are given in the textbooks by Beeson [1] and Troelstra and Van Dalen [23, 24].

In the seventies Feferman developed his own approach towards constructive mathematics and was particularly interested in laying the logical foundations of Bishop-style constructive mathematics. In his seminal paper Feferman [4] he introduces the famous theory T_0 and more generally the program of *explicit mathematics*. Feferman [8] discusses the relationships between systems of explicit mathematics and alternative approaches towards constructivism on the one hand side and to subsystems of second order arithmetic on the other hand. It soon turned out that T_0 and its subsystems play an important role in reductive proof theory and, in particular, in reducing classical systems to constructively justifiable ones.

Feferman's development of explicit mathematics is strongly influenced by his work on generalized recursion theory and also motivated by the idea of abstract computations. It is one of the attractive features of explicit mathematics that it possesses models in the sense of classical set theory as well as models which are constructive/computational in flavor. As a consequence, we can directly relate the classical meaning of a sentence provable in explicit mathematics to its computational interpretation. This approach is brought up and studied in Feferman [7].

Reflections on the foundations of mathematics
Essays in honor of Solomon Feferman
Edited by Wilfried Sieg, Richard Sommer, and Carolyn Talcott

A further philosophically interesting approach towards mathematics goes back to work of Poincaré and Weyl and is generally referred to as the *arithmetical* foundations of mathematics. Feferman's logical analysis of this approach in Feferman [5, 9] reveals the crucial role of the so-called non-constructive minimum operator μ in this context.

Later it turned out that this work also has its natural place in explicit mathematics. Since the operations of explicit mathematics can be regarded as abstract computations, functionals of higher types can be added in a direct and perspicuous way. In explicit mathematics we basically deal with first and second order entities: *operations* which can be understood as abstract computations and *classifications* or *types* which are collections of operations. For the proof-theoretic analysis of functionals of higher types in this framework already the first-order applicative core of explicit mathematics is of significant interest. The corresponding axiomatic first order systems are often referred to as *applicative theories*. For a survey of this area see Jäger, Kahle and Strahm [18].

A first important step in the proof-theoretic treatment of functionals of higher types in the framework of applicative theories and explicit mathematics was the analysis of the non-constructive μ operator over the basic theory BON of operations and numbers in Feferman and Jäger [11]. In the subsequent articles Feferman and Jäger [12] and Glaß and Strahm [13] type structures which are closed under elementary comprehension and join have been added to $\mathsf{BON}(\mu)$.

The non-constructive μ operator is also very interesting from the point of view of type two recursion theory and corresponds to the number quantifier functional E_0. The 1-section of μ and of E_0 consists exactly of the hyperarithmetic sets and, hence, provides the least standard model of Δ^1_1 comprehension,

$$(\mathbb{N}, \text{1-sec}(\mu), \dots) \models (\Delta^1_1\text{-CA}).$$

Moreover, the least ordinal which is not recursive in E_0 is the ordinal ω_1^{ck}, the first admissible ordinal greater than ω.

It is common practice in applicative theories to represent subsets of the natural numbers as total function from natural numbers to $\{0, 1\}$. From the recursion-theoretic results we can therefore immediately read off that the sets of $\mathsf{BON}(\mu)$ in its least standard model coincide with the hyperarithmetic sets.

Proof-theoretically, however, $\mathsf{BON}(\mu)$ is comparatively weak provided that induction on the natural numbers is restricted to sets: $\mathsf{BON}(\mu)$ plus set induction $(\mathsf{S}\text{-}\mathsf{I}_\mathsf{N})$ is a conservative extension of Peano arithmetic PA and so proof-theoretically equivalent to the theory $(\Delta^1_1\text{-CA})\!\restriction$. If the schema $(\mathsf{L}\text{-}\mathsf{I}_\mathsf{N})$ of complete induction for all formulas of the underlying language L is permitted then $\mathsf{BON}(\mu)$ becomes equivalent to $(\Delta^1_1\text{-CA})$:

$$\mathsf{BON}(\mu) + (\mathsf{S}\text{-}\mathsf{I}_\mathsf{N}) \equiv (\Delta^1_1\text{-CA})\!\restriction, \qquad \mathsf{BON}(\mu) + (\mathsf{L}\text{-}\mathsf{I}_\mathsf{N}) \equiv (\Delta^1_1\text{-CA}).$$

According to the articles by Feferman, Jäger, Glaß and Strahm cited above, even elementary comprehension and join may be added to $\mathsf{BON}(\mu)$ plus set induction without going beyond PA with respect to proof-theoretic strength.

$\mathsf{BON}(\mu)$ plus set induction and with or without elementary comprehension and join provides a solid logical basis for the Poincaré-Weyl approach. It is also an interesting system in connection with a series of other formalisms introduced and studied in Feferman [6], where the problem of designing adequate higher type formalisms for mathematical practice is addressed. In this article Feferman also points out the relevance of the Suslin quantifier functional E_1.

This type two functional tests for wellfoundedness of binary relations and thus has at least the power of Π^1_1 comprehension. The 1-section of E_1 coincides with the sets of natural numbers in the constructible hierarchy up the the first recursively inaccessible ordinal ι_0. This ordinal is also the least ordinal not recursive in E_1. According to a result of Gandy, one can also show that the 1-section of E_1 builds the least standard model of Δ^1_2 comprehension,

$$(\mathbb{N}, \text{1-sec}(\mathsf{E}_1), \dots) \models (\Delta^1_2\text{-CA}).$$

In this article we will analyze the proof-theoretic strength of E_1 in the context of applicative theories. The theory $\mathsf{BON}(\mu)$ with additional axioms for E_1 is baptized SUS in the following. Then we show the proof-theoretic equivalences

$$\mathsf{SUS} + (\mathsf{S}\text{-}\mathsf{I}_\mathsf{N}) \equiv (\Delta^1_2\text{-CA})\restriction, \qquad \mathsf{SUS} + (\mathsf{L}\text{-}\mathsf{I}_\mathsf{N}) \equiv (\Delta^1_2\text{-CA}).$$

We also obtain results concerning intermediate forms of induction and their relationship to systems like $(\Delta^1_2\text{-CR})$. Obviously, there is a striking analogy concerning the relationship between proof-theoretic and recursion-theoretic results for μ and E_1.

The methods of proof, however, are rather different. For establishing the upper bounds of, say, $\mathsf{SUS} + (\mathsf{S}\text{-}\mathsf{I}_\mathsf{N})$ we make use of a very specific positive Δ^1_2 inductive definition. We interpret the application operation by a Σ definable fixed point of this inductive definition and have to exploit a delicate interplay between proper set-theoretic functions and functions defined in terms of our application relation, cf. Theorem 16 below.

The plan of the paper is a follows. In the next section we introduce the basic applicative framework, the axiomatizations of μ and E_1 and several forms of complete induction on the natural numbers. In Section 3 we establish proof-theoretic lower bounds of the various applicative theories by embedding subsystems of analysis with (iterated) Π^1_1 comprehension. Section 4 constitutes the central part of the paper. We obtain proof-theoretic upper bounds for SUS plus various forms of induction in appropriate theories for iterated admissible sets.

§2. The applicative framework and E_1. It is the purpose of this section to introduce the basic applicative framework as well as the precise axiomatizations of the non-constructive μ operator and the Suslin operator E_1. Further, we will distinguish three forms of complete induction on the natural numbers which will be relevant in the sequel.

The language of our applicative theories is a first order language L of partial terms with individual variables $a, b, c, f, g, h, u, v, w, x, y, z \dots$ (possibly with subscripts). L includes individual constants k, s (combinators), $\mathsf{p}, \mathsf{p}_0, \mathsf{p}_1$ (pairing and unpairing), 0 (zero), $\mathsf{s_N}$ (numerical successor), $\mathsf{p_N}$ (numerical predecessor), $\mathsf{d_N}$ (definition by numerical cases), $\mathsf{r_N}$ (primitive recursion), μ (non-constructive μ operator), and E_1 (Suslin operator). Further, L has a binary function symbol $\cdot$ for (partial) term application, unary relation symbols $\downarrow$ (defined) and N (natural numbers), as well as a binary relation symbol $=$ (equality).

The *individual terms* $(r, s, t, r_1, s_1, t_1, \dots)$ of L are inductively generated as follows:

1. The individual variables and individual constants are individual terms.
2. If s and t are individual terms, then so also is $(s \cdot t)$.

In the following we often abbreviate $(s \cdot t)$ simply as (st), st or sometimes also $s(t)$; the context will always ensure that no confusion arises. We further adopt the convention of association to the left so that $s_1 s_2 \dots s_n$ stands for $(\dots(s_1 s_2)\dots s_n)$. Further, we put $t' := \mathsf{s_N} t$ and $1 := 0'$. We define general n-tupling by induction on $n \geq 2$ as follows:

$$(s_1, s_2) := \mathsf{p} s_1 s_2, \qquad (s_1, \dots, s_{n+1}) := ((s_1, \dots, s_n), s_{n+1}).$$

Finally, we also use quite frequently the vector notation $\vec{\mathcal{Z}}$ for a finite string of objects $\mathcal{Z}_1, \dots, \mathcal{Z}_n$ of the same sort. Whenever we write $\vec{\mathcal{Z}}$ the length of this string is either irrelevant or given by the context.

The *formulas* $(A, B, C, A_1, B_1, C_1, \dots)$ of L are inductively generated as follows:

1. Each atomic formula $\mathsf{N}(t)$, $t\downarrow$, and $(s = t)$ is a formula.
2. If A and B are formulas, then so also are $\neg A$, $(A \vee B)$, $(A \wedge B)$, and $(A \rightarrow B)$.
3. If A is a formula, then so also are $(\exists x)A$ and $(\forall x)A$.

Our applicative theories are based on *partial* term application. Hence, it is not guaranteed that terms have a value, and $t\downarrow$ is read as "t is defined" or "t has a value". Accordingly, the *partial equality relation* $\simeq$ is introduced by

$$s \simeq t \ := \ (s\downarrow \vee\, t\downarrow) \rightarrow (s = t).$$

In addition, we write $(s \neq t)$ for $(s\downarrow \wedge\, t\downarrow \wedge\, \neg(s = t))$. Finally, we use the following abbreviations concerning the predicate N:

$$\begin{aligned}
t \in \mathsf{N} &:= \mathsf{N}(t), \\
(\exists x \in \mathsf{N})A &:= (\exists x)(x \in \mathsf{N} \wedge A), \\
(\forall x \in \mathsf{N})A &:= (\forall x)(x \in \mathsf{N} \rightarrow A), \\
t \in (\mathsf{N} \rightarrow \mathsf{N}) &:= (\forall x \in \mathsf{N})(tx \in \mathsf{N}), \\
t \in (\mathsf{N}^{m+1} \rightarrow \mathsf{N}) &:= (\forall x \in \mathsf{N})(tx \in (\mathsf{N}^m \rightarrow \mathsf{N})).
\end{aligned}$$

Now we are going to recall the basic theory BON of operations and numbers which has been introduced in Feferman and Jäger [11]. Its underlying logic is the *classical logic of partial terms* due to Beeson [1]; it is also described in Feferman [10] and corresponds to E^+ logic with strictness and equality of Troelstra and Van Dalen [23]. The non-logical axioms of BON are divided into the following five groups.

I. Partial combinatory algebra.

(1) $\mathsf{k}ab = a$,

(2) $\mathsf{s}ab\downarrow \wedge \mathsf{s}abc \simeq ac(bc)$.

II. Pairing and projection.

(3) $\mathsf{p_0}(a, b) = a \wedge \mathsf{p_1}(a, b) = b$.

III. Natural numbers.

(4) $0 \in \mathsf{N} \wedge (\forall x \in \mathsf{N})(x' \in \mathsf{N})$,

(5) $(\forall x \in \mathsf{N})(x' \neq 0 \wedge \mathsf{p_N} x' = x)$,

(6) $(\forall x \in \mathsf{N})(x \neq 0 \rightarrow \mathsf{p_N} x \in \mathsf{N} \wedge (\mathsf{p_N} x)' = x)$.

IV. Definition by numerical cases.

(7) $u \in \mathsf{N} \wedge v \in \mathsf{N} \wedge u = v \rightarrow \mathsf{d_N} abuv = a$,

(8) $u \in \mathsf{N} \wedge v \in \mathsf{N} \wedge u \neq v \rightarrow \mathsf{d_N} abuv = b$.

V. Primitive recursion on N.

(9) $f \in (\mathsf{N} \rightarrow \mathsf{N}) \wedge g \in (\mathsf{N}^3 \rightarrow \mathsf{N}) \rightarrow \mathsf{r_N} fg \in (\mathsf{N}^2 \rightarrow \mathsf{N})$,

(10) $f \in (\mathsf{N} \rightarrow \mathsf{N}) \wedge g \in (\mathsf{N}^3 \rightarrow \mathsf{N}) \wedge a \in \mathsf{N} \wedge b \in \mathsf{N} \wedge h = \mathsf{r_N} fg \rightarrow$
$ha0 = fa \wedge ha(b') = gab(hab)$.

As usual the axioms of a partial combinatory algebra allow one to define λ abstraction and to prove a recursion or fixed point theorem. For proofs of these standard results the reader is referred to [1, 4]. For the second assertion of the following lemma, which is a slight extension of the usual λ abstraction. one makes in addition use of pairing and projections.

LEMMA 1. 1. *For each* L *term* t *and all variables* x *there exists an* L *term* $(\lambda x.t)$ *whose variables are those of* t, *excluding* x, *so that* BON *proves*

$$(\lambda x.t)\downarrow \wedge (\lambda x.t)x \simeq t.$$

2. *For each* L *term* t *and all variables* $x_1, \ldots, x_n (n \geq 2)$ *there exists an* L *term* s *whose variables are those of* t, *excluding* $x_1, \ldots, x_n$, *so that* BON *proves*

$$s\downarrow \wedge s(x_1, \ldots, x_n) \simeq t.$$

3. *There exists a closed* L *term* rec *so that* BON *proves*

$$\mathsf{rec} f\downarrow \wedge \mathsf{rec} fx \simeq f(\mathsf{rec} f)x.$$

Let us now turn to the two type 2 functionals which will be relevant in the sequel. The non-constructive or unbounded μ operator is characterized by the following two axioms.

The non-constructive μ operator

(μ.1) $f \in (\mathsf{N} \rightarrow \mathsf{N}) \leftrightarrow \mu f \in \mathsf{N}$,

(μ.2) $f \in (\mathsf{N} \rightarrow \mathsf{N}) \wedge (\exists x \in \mathsf{N})(fx = 0) \rightarrow f(\mu f) = 0$.

A much stronger functional is the Suslin operator $\mathsf{E_1}$, which tests for the well-foundedness of a binary relation on N (given as a total operation from N^2 to N).

The Suslin operator E_1

$(\mathsf{E}_1.1)$ $f \in (\mathsf{N}^2 \to \mathsf{N}) \leftrightarrow \mathsf{E}_1 f \in \mathsf{N}$,

$(\mathsf{E}_1.2)$ $f \in (\mathsf{N}^2 \to \mathsf{N}) \to$
$$[(\exists g \in \mathsf{N} \to \mathsf{N})(\forall x \in \mathsf{N})(f(gx')(gx) = 0) \leftrightarrow \mathsf{E}_1 f = 0].$$

Since we want to study E_1 also in the presence of very weak induction principles, we include the non-constructive μ operator in our basic axiomatic framework for the Suslin operator. Accordingly, we let SUS denote the L theory which extends BON by the axioms about μ and E_1,

$$\mathsf{SUS} := \mathsf{BON} + (\mu.1) + (\mu.2) + (\mathsf{E}_1.1) + (\mathsf{E}_1.2).$$

In the sequel we will be interested in three forms of complete induction on the natural numbers N, namely *set induction,* N *induction* and *formula induction.*

Let us first recall the notion of a *subset of* N from [5, 11]. Sets of natural numbers are represented via their characteristic functions which are total on N. Accordingly, we define

$$f \in \mathcal{P}(\mathsf{N}) := (\forall x \in \mathsf{N})(fx = 0 \vee fx = 1)$$

with the intention that a natural number x belongs to the set $f \in \mathcal{P}(\mathsf{N})$ if and only if $(fx = 0)$. The three relevant induction principles are now given as follows.

Set induction on N $(\mathsf{S}\text{-}\mathsf{I}_\mathsf{N})$.

$$f \in \mathcal{P}(\mathsf{N}) \wedge f0 = 0 \wedge (\forall x \in \mathsf{N})(fx = 0 \to fx' = 0) \to (\forall x \in \mathsf{N})(fx = 0).$$

N induction on N $(\mathsf{N}\text{-}\mathsf{I}_\mathsf{N})$.

$$f0 \in \mathsf{N} \wedge (\forall x \in \mathsf{N})(fx \in \mathsf{N} \to fx' \in \mathsf{N}) \to (\forall x \in \mathsf{N})(fx \in \mathsf{N}).$$

Formula induction on N $(\mathsf{L}\text{-}\mathsf{I}_\mathsf{N})$. For all formulas $A(x)$ of L:

$$A(0) \wedge (\forall x \in \mathsf{N})(A(x) \to A(x')) \to (\forall x \in \mathsf{N})A(x).$$

In the rest of this paper we will be concerned with the proof-theoretic analysis of the three systems

$$\mathsf{SUS} + (\mathsf{S}\text{-}\mathsf{I}_\mathsf{N}), \qquad \mathsf{SUS} + (\mathsf{N}\text{-}\mathsf{I}_\mathsf{N}), \qquad \mathsf{SUS} + (\mathsf{L}\text{-}\mathsf{I}_\mathsf{N}).$$

In the next section we will establish lower proof-theoretic bounds by embedding suitable subsystems of analysis into these applicative systems.

§3. Lower bounds. Lower bounds for the theory SUS plus various forms of induction on the natural numbers will be established in this section by embedding suitable systems of second order arithmetic. Actually, theories with Π^1_1 comprehension and transfinitely iterated forms thereof will play a crucial role. More precisely, we will show that $\mathsf{SUS} + (\mathsf{S}\text{-}\mathsf{I}_\mathsf{N})$ contains $(\Pi^1_1\text{-}\mathsf{CA})\!\upharpoonright$, $\mathsf{SUS} + (\mathsf{N}\text{-}\mathsf{I}_\mathsf{N})$ contains $(\Pi^1_1\text{-}\mathsf{CA})_{<\omega^\omega}\!\upharpoonright$, and $\mathsf{SUS} + (\mathsf{L}\text{-}\mathsf{I}_\mathsf{N})$ contains $(\Pi^1_1\text{-}\mathsf{CA})_{<\varepsilon_0}\!\upharpoonright$.

3.1. Second order arithmetic with Π^1_1 comprehension. In the following we introduce the notions and results of second order arithmetic which will be important for our embeddings. We deviate from the usual conventions in working with a form of second order arithmetic with set and function variables. The reason for this choice is that the following Π^1_1 normal form theorem is most suitable for our purposes.

Let $\mathcal{L}_2$ denote a language of second order arithmetic with *number variables* $a, b, c, f, g, h, u, v, w, x, y, z, \dots$, *set variables* $U, V, W, X, Y, Z, \dots$ and *function variables* $F, G, H, \dots$ (all possibly with subscripts). In addition, $\mathcal{L}_2$ includes a constant 0 as well as function and relation symbols for all primitive recursive functions and relations. The *number terms* $(r, s, t, r_1, s_1, t_1, \dots)$ of $\mathcal{L}_2$ and the *formulas* $(A, B, C, A_1, B_1, C_1, \dots)$ of $\mathcal{L}_2$ are defined as usual.

An $\mathcal{L}_2$ formula is called *arithmetic*, if it does not contain bound set or function variables; let Π^1_0 denote the class of arithmetic $\mathcal{L}_2$ formulas. The Π^1_1 $[\Sigma^1_1]$ formulas of $\mathcal{L}_2$ are obtained from the arithmetic formulas by closing under universal [existential] set and function quantification.

In the following we make use of the usual primitive recursive coding machinery in $\mathcal{L}_2$: $\langle \dots \rangle$ is a standard primitive recursive function for forming n-tuples $\langle t_1, \dots, t_n \rangle$; *Seq* is the primitive recursive set of sequence numbers; $lh(t)$ denotes the length of (the sequence number coded by) t; $(t)_i$ is the ith component of (the sequence coded by) t if $i < lh(t)$, i.e. $t = \langle (t)_0, \dots, (t)_{lh(t) \dot{-} 1} \rangle$ if t is a sequence number. In addition, we write $s \in (U)_t$ for $\langle s, t \rangle \in U$.

If $\mathcal{F}$ is a collection of $\mathcal{L}_2$ formulas, then $\mathcal{F}$ comprehension ($\mathcal{F}$-CA) is the schema

$$(\exists X)(\forall x)(x \in X \leftrightarrow A(x))$$

for all formulas $A(u)$ in the collection $\mathcal{F}$. The relationship between sets and functions is given by the so-called graph principle ($\mathcal{GP}$),

$$(\forall X)[(\forall x)(\exists! y)\langle x, y \rangle \in X \rightarrow (\exists F)(\forall x)\langle x, F(x) \rangle \in X].$$

Moreover, $\mathcal{L}_2$ induction on the natural numbers ($\mathcal{L}_2$-$\mathsf{I_N}$) comprises

$$A(0) \wedge (\forall x)(A(x) \rightarrow A(x')) \rightarrow (\forall x)A(x)$$

for all formulas $A(u)$ of $\mathcal{L}_2$. Set induction (S-$\mathsf{I_N}$) with respect to $\mathcal{L}_2$, on the other hand, is the axiom

$$(\forall X)(0 \in X \wedge (\forall x)(x \in X \rightarrow x' \in X) \rightarrow (\forall x)(x \in X)).$$

(Π^1_0-CA) is the $\mathcal{L}_2$ theory which contains the usual axioms of Peano arithmetic PA, all instances of Π^1_0 comprehension, the graph principle ($\mathcal{GP}$) as well as formula induction ($\mathcal{L}_2$-$\mathsf{I_N}$). The restricted version (Π^1_0-CA)$\restriction$ of (Π^1_0-CA) is obtained if we replace ($\mathcal{L}_2$-$\mathsf{I_N}$) by (S-$\mathsf{I_N}$). The theories (Π^1_1-CA) and (Π^1_1-CA)$\restriction$ are defined accordingly.

We now state a version of the normal form theorem for Π^1_1 formulas tailored for our later purposes. Its proof is more or less folklore and can be found at many places (for example in Simpson [22]).

THEOREM 2 (Π^1_1 normal forms). *For every Π^1_1 formula A there exists an arithmetic formula $B_A(u, v)$ which contains the free variables of A plus two fresh variables u and v so that $(\Pi^1_0\text{-CA})\restriction$ proves*

$$A \leftrightarrow \neg(\exists F)(\forall x)B_A(F(x+1), F(x)).$$

Now let $\prec$ be a standard wellordering of ordertype ε_0. In addition we assume that 0 is the least element with respect to $\prec$ and the field of $\prec$ is the set of natural numbers. Furthermore, if n is a natural number, then we write $\prec_n$ for the restriction of $\prec$ to the numbers $m \prec n$.

For each Π^1_1 formula $A(U, v)$ with all its free variables in U, v, the *A-hyperjump hierarchy along $\prec_n$ starting with U* is defined by the following transfinite recursion,

$$(W)_0 := U \quad \text{and} \quad (W)_i := \{\langle m, j\rangle : j \prec i \wedge A((W)_j, m)\}$$

for all $0 \prec i \prec n$. We write $\mathsf{Hier}_A(U, W, n)$ for the $\mathcal{L}_2$ formula which formalizes this definition.

If α is an ordinal less than ε_0, then $(\Pi^1_1\text{-CA})_{<\alpha}\restriction$ is the $\mathcal{L}_2$ theory which extends $(\Pi^1_0\text{-CA})\restriction$ by the axioms

$$(\forall X)(\exists Y)\mathsf{Hier}_A(X, Y, n)$$

for each natural number n so that $\prec_n$ has order type less than α and each Π^1_1 formula $A(U, v)$ with all its free variables in U, v. Moreover, $(\Pi^1_1\text{-CA})_{<\alpha}$ denotes $(\Pi^1_1\text{-CA})_{<\alpha}\restriction$ plus $(\mathcal{L}_2\text{-I}_\mathsf{N})$.

LEMMA 3. *The theories $(\Pi^1_1\text{-CA})_{<\omega^\omega}$ and $(\Pi^1_1\text{-CA})_{<\varepsilon_0}$ are proof-theoretically equivalent to $(\Pi^1_1\text{-CA})_{<\omega^\omega}\restriction$ and $(\Pi^1_1\text{-CA})_{<\varepsilon_0}\restriction$, respectively.*

The proof of this lemma makes use of standard recursion-theoretic arguments. Actually, the lemma holds for ordinals of the form ω^λ with λ limit.

3.2. Modeling hyperjumps via E_1. Recursion-theoretic arguments indicate that E_1 can be used to deal with hyperjumps in an applicative framework. For this purpose, we work with the natural embedding of $\mathcal{L}_2$ into L so that (i) the number variables of $\mathcal{L}_2$ are interpreted as ranging over N, (ii) the set variables of $\mathcal{L}_2$ as ranging over $\mathcal{P}(\mathsf{N})$, and (iii) the function variables as ranging over $(\mathsf{N} \to \mathsf{N})$.

In the following we assume that we have a translation of the number, set and function variables of $\mathcal{L}_2$ into the variables of L so that no conflicts arise. For convenience we often simply write, for example, a, x, f for the translations of the number, set and function variable a, X, F, respectively. Furthermore, we can use the recursion operator $\mathsf{r_N}$ to associate a suitable L term to each symbol for a primitive recursive function on the natural numbers and prove the corresponding recursion equations in BON + (S-$\mathsf{I_N}$). Thus, every $\mathcal{L}_2$ term t has a canonical translation t^N in L. Similarly, each symbol for a primitive recursive relation on N can be represented by an L term which represents its characteristic function in the sense above.

Now let R be a symbol for an n-ary primitive recursive relation and t_R the corresponding L term. If $s, t_1, \ldots, t_n$ are terms of $\mathcal{L}_2$, then the atomic formulas of $\mathcal{L}_2$ are translated into L formulas as follows:

$$(s \in U)^{\mathsf{N}} := ((us^{\mathsf{N}}) = 0); \qquad (t_1 = t_2)^{\mathsf{N}} := (t_1^{\mathsf{N}} = t_2^{\mathsf{N}});$$
$$R(t_1, \dots, t_n)^{\mathsf{N}} := (t_R t_1^{\mathsf{N}} \dots t_n^{\mathsf{N}} = 0).$$

We extend this translation in the usual way and associate to each $\mathcal{L}_2$ formula $A(\vec{U}, \vec{F}, \vec{v})$ an L formula $A^{\mathsf{N}}(\vec{u}, \vec{f}, \vec{v})$ such that

$$((\exists X)A(X))^{\mathsf{N}} = (\exists x \in \mathcal{P}(\mathsf{N}))A^{\mathsf{N}}(x),$$
$$((\exists F)A(F))^{\mathsf{N}} = (\exists f \in \mathsf{N} \to \mathsf{N})A^{\mathsf{N}}(f),$$
$$((\exists y)A(y))^{\mathsf{N}} = (\exists y \in \mathsf{N})A^{\mathsf{N}}(y)$$

and similarly for universal quantifiers. A further convention is that we often identify $\mathcal{L}_2$ terms and arithmetic $\mathcal{L}_2$ formulas with their translation in L as long as no conflict arises.

Usual arguments as for example in Feferman [5] and Feferman and Jäger [11] show that with help of the unbounded minimum operator μ each arithmetic formula of $\mathcal{L}_2$ can be coded up by an L term in the sense of the following lemma.

Lemma 4. *For every arithmetic formula $A(\vec{U}, \vec{F}, \vec{v})$ of $\mathcal{L}_2$ with all its free variables in $\vec{U}, \vec{F}, \vec{v}$ there exists a closed term t_A of* L *so that* $\mathsf{BON}(\mu) + (\mathsf{S\text{-}I_N})$ *proves*

1. $(\forall \vec{x} \in \mathcal{P}(\mathsf{N}))(\forall \vec{f} \in \mathsf{N} \to \mathsf{N})(\forall \vec{y} \in \mathsf{N})(t_A(\vec{x}, \vec{f}, \vec{y}) = 0 \lor t_A(\vec{x}, \vec{f}, \vec{y}) = 1)$,
2. $(\forall \vec{x} \in \mathcal{P}(\mathsf{N}))(\forall \vec{f} \in \mathsf{N} \to \mathsf{N})(\forall \vec{y} \in \mathsf{N})(A(\vec{x}, \vec{f}, \vec{y}) \leftrightarrow t_A(\vec{x}, \vec{f}, \vec{y}) = 0)$.

This lemma is the crucial step for the interpretation of $(\Pi^1_0\text{-}\mathsf{CA})\!\restriction$ modulo the embedding described above. It is even the case that $(\Pi^1_0\text{-}\mathsf{CA})\!\restriction$ and $\mathsf{BON}(\mu) + (\mathsf{S\text{-}I_N})$ have the same proof-theoretic strength; cf. Feferman and Jäger [11]. For the following, however, we only need that $(\Pi^1_0\text{-}\mathsf{CA})\!\restriction$ is contained in $\mathsf{BON}(\mu) + (\mathsf{S\text{-}I_N})$.

Lemma 5. *Let $A(\vec{U}, \vec{F}, \vec{v})$ be an $\mathcal{L}_2$ formula with all its free variables in $\vec{U}, \vec{F}, \vec{v}$ and assume that $(\Pi^1_0\text{-}\mathsf{CA})\!\restriction$ proves $A(\vec{U}, \vec{F}, \vec{v})$. Then we have*

$$\mathsf{BON}(\mu) + (\mathsf{S\text{-}I_N}) \vdash \vec{u} \in \mathcal{P}(\mathsf{N}) \land \vec{f} \in (\mathsf{N} \to \mathsf{N}) \land \vec{v} \in \mathsf{N} \ \to\ A^{\mathsf{N}}(\vec{u}, \vec{f}, \vec{v}).$$

Proof. All axioms of Peano arithmetic PA without the schema of complete induction obviously create no problems. Set induction of $(\Pi^1_0\text{-}\mathsf{CA})\!\restriction$ directly translates into set induction in the applicative context. The schema of arithmetic comprehension follows from the lemma above. So there only remains the graph principle $(\mathcal{GP})$. Hence, let a be an element of $\mathcal{P}(\mathsf{N})$ so that

$$(\forall x \in \mathsf{N})(\exists y \in \mathsf{N})(a(\langle x, y\rangle) = 0).$$

Then we define f to be the term $(\lambda x.\mu(\lambda y.a(\langle x, y\rangle)))$. As a consequence, we have $(\lambda y.a(\langle x, y\rangle)) \in (\mathsf{N} \to \mathsf{N})$ and therefore $\mu(\lambda y.a(\langle x, y\rangle)) \in \mathsf{N}$ for all $x \in \mathsf{N}$. This implies $f \in (\mathsf{N} \to \mathsf{N})$ and, moreover,

$$(\forall x \in \mathsf{N})(a(\langle x, fx\rangle) = 0).$$

Hence, also the graph principle has been established. Straightforward induction on the length of the proof in $(\Pi^1_0\text{-}\mathsf{CA})\!\restriction$ now yields our result. $\dashv$

One important consequence of this lemma is that our Π^1_1 normal form theorem is available in $\mathsf{BON}(\mu)+(\mathsf{S}\text{-}\mathsf{I}_\mathsf{N})$. Based on these normal forms, we can now employ the Suslin operator E_1 in order to lift Lemma 4 from arithmetic to Π^1_1 formulas of $\mathcal{L}_2$.

LEMMA 6. *For every Π^1_1 formula $A(\vec{U},\vec{F},\vec{v})$ of $\mathcal{L}_2$ with all its free variables in $\vec{U},\vec{F},\vec{v}$ there exists a closed term t_A of* L *so that* $\mathsf{SUS}+(\mathsf{S}\text{-}\mathsf{I}_\mathsf{N})$ *proves*

1. $(\forall\vec{x}\in\mathcal{P}(\mathsf{N}))(\forall\vec{f}\in\mathsf{N}\to\mathsf{N})(\forall\vec{y}\in\mathsf{N})(t_A(\vec{x},\vec{f},\vec{y})=0\ \vee\ t_A(\vec{x},\vec{f},\vec{y})=1)$,
2. $(\forall\vec{x}\in\mathcal{P}(\mathsf{N}))(\forall\vec{f}\in\mathsf{N}\to\mathsf{N})(\forall\vec{y}\in\mathsf{N})(A^\mathsf{N}(\vec{x},\vec{f},\vec{y})\ \leftrightarrow\ t_A(\vec{x},\vec{f},\vec{y})=0)$.

PROOF. Let $A(\vec{U},\vec{F},\vec{v})$ be a Π^1_1 formula of $\mathcal{L}_2$ with its free variables as indicated. In view of Theorem 2 there exists an arithmetic formula B_A with the free variables of A plus the two new variables a and b so that

$$A^\mathsf{N}(\vec{u},\vec{f},\vec{v})\ \leftrightarrow\ \neg(\exists g\in\mathsf{N}\to\mathsf{N})(\forall z\in\mathsf{N})B_A(\vec{u},\vec{f},\vec{v},g(z+1),g(z))$$

for all $\vec{u}\in\mathcal{P}(\mathsf{N})$, all $\vec{f}\in(\mathsf{N}\to\mathsf{N})$ and all $\vec{v}\in\mathsf{N}$. Due to Lemma 4 there exists a closed L term t representing B_A. Then the term

$$(\lambda a,b.t(\vec{u},\vec{f},\vec{v},a,b))$$

is a total operation from N^2 to N for all $\vec{u}\in\mathcal{P}(\mathsf{N})$, all $\vec{f}\in(\mathsf{N}\to\mathsf{N})$ and all $\vec{v}\in\mathsf{N}$. If we now define s to be the term

$$1\dot{-}\mathsf{E}_1(\lambda a,b.t(\vec{u},\vec{f},\vec{v},a,b)),$$

then we can finally take t_A to be the closed L term which is associated to s according to Lemma 1(2). It is easy to check that t_A has the properties stated in the lemma. ⊣

This lemma together with Lemma 5 immediately implies our first result, namely that $(\Pi^1_1\text{-}\mathsf{CA})\!\restriction$ is contained in $\mathsf{SUS}+(\mathsf{S}\text{-}\mathsf{I}_\mathsf{N})$.

THEOREM 7. *Let $A(\vec{U},\vec{F},\vec{v})$ be an $\mathcal{L}_2$ formula with all its free variables in $\vec{U},\vec{F},\vec{v}$ and assume that $(\Pi^1_1\text{-}\mathsf{CA})\!\restriction$ proves $A(\vec{U},\vec{F},\vec{v})$. Then we have*

$$\mathsf{SUS}+(\mathsf{S}\text{-}\mathsf{I}_\mathsf{N})\vdash\vec{u}\in\mathcal{P}(\mathsf{N})\wedge\vec{f}\in(\mathsf{N}\to\mathsf{N})\wedge\vec{v}\in\mathsf{N}\ \to\ A^\mathsf{N}(\vec{u},\vec{f},\vec{v}).$$

Now we turn to transfinite hyperjump hierarchies. We first consider the system $\mathsf{SUS}+(\mathsf{L}\text{-}\mathsf{I}_\mathsf{N})$ and show that it contains $(\Pi^1_1\text{-}\mathsf{CA})_{<\varepsilon_0}$. We simply have to follow the argument in Feferman and Jäger [11] and replace ordinary jump hierarchies by hyperjump hierarchies, which is possible since E_1 is available in SUS.

LEMMA 8. *Let $A(U,v)$ be a Π^1_1 formula of $\mathcal{L}_2$ with all its free variables in U,v and assume that n is an arbitrary natural number. Then there exists a closed* L *term h_A so that* $\mathsf{SUS}+(\mathsf{L}\text{-}\mathsf{I}_\mathsf{N})$ *proves*:

1. $u\in\mathcal{P}(\mathsf{N})\ \to\ h_Au\in\mathcal{P}(\mathsf{N})$,
2. $u\in\mathcal{P}(\mathsf{N})\ \to\ \mathsf{Hier}_A(u,h_Au,n)$.

PROOF. Utilizing Lemma 6 we choose closed L terms t_A and t_B for representing the Π^1_1 formula $A(U,v)$ and the arithmetic formula $B(u,v)$,

$$B(u,v)\ :=\ (u=\langle(u)_0,(u)_1\rangle\wedge(u)_1\prec v\wedge v\prec n).$$

Then we apply the recursion theorem in order to obtain a closed term s with the property

$$suvw \simeq \begin{cases} uw, & \text{if } v = 0, \\ t_A(su(w)_1, (w)_0), & \text{if } t_B wv = 0, \\ 1, & \text{otherwise} \end{cases}$$

for all $u \in \mathcal{P}(\mathsf{N})$ and $v, w \in \mathsf{N}$. Then the following three implications are provable in $\mathsf{SUS} + (\mathsf{S\text{-}I_N})$:

(1) $u \in \mathcal{P}(\mathsf{N}) \wedge w \in \mathsf{N} \rightarrow su0w = uw$,

(2) $u \in \mathcal{P}(\mathsf{N}) \wedge v \in \mathsf{N} \wedge w \in \mathsf{N} \wedge t_B wv = 0 \rightarrow suvw \simeq t_A(su(w)_1, (w)_0)$,

(3) $u \in \mathcal{P}(\mathsf{N}) \wedge v \in \mathsf{N} \wedge w \in \mathsf{N} \wedge 0 \prec v \wedge t_B wv = 1 \rightarrow suvw = 1$.

Since full formula induction is available in $\mathsf{SUS} + (\mathsf{L\text{-}I_N})$ and the order type of $\prec_n$ is less than ε_0, we can show by transfinite induction along $\prec_n$ that

(4) $$v \in \mathsf{N} \wedge v \prec n \rightarrow suv \in \mathcal{P}(\mathsf{N})$$

for all $u \in \mathcal{P}(\mathsf{N})$. Finally, we let h_A be the term $(\lambda u, v.su(v)_1(v)_0)$. It is now easy to verify that h_A has all the desired properties. ⊣

This lemma basically states that arbitrary hyperjumps can be iterated in $\mathsf{SUS} + (\mathsf{L\text{-}I_N})$ along all ordinals less than ε_0. In view of Lemma 5 we can therefore conclude that $\mathsf{SUS} + (\mathsf{L\text{-}I_N})$ contains $(\Pi^1_1\text{-}\mathsf{CA})_{<\varepsilon_0}$.

THEOREM 9. *Let $A(\vec{U}, \vec{F}, \vec{v})$ be an $\mathcal{L}_2$ formula with all its free variables in $\vec{U}, \vec{F}, \vec{v}$ and assume that $(\Pi^1_1\text{-}\mathsf{CA})_{<\varepsilon_0}$ proves $A(\vec{U}, \vec{F}, \vec{v})$. Then we have*

$$\mathsf{SUS} + (\mathsf{L\text{-}I_N}) \vdash \vec{u} \in \mathcal{P}(\mathsf{N}) \wedge \vec{f} \in (\mathsf{N} \rightarrow \mathsf{N}) \wedge \vec{v} \in \mathsf{N} \rightarrow A^{\mathsf{N}}(\vec{u}, \vec{f}, \vec{v}).$$

This embedding of $(\Pi^1_1\text{-}\mathsf{CA})_{<\varepsilon_0}$ into $\mathsf{SUS} + (\mathsf{L\text{-}I_N})$ is rather standard. The situation is more complicated, however, in the case of the system $\mathsf{SUS} + (\mathsf{N\text{-}I_N})$, i.e. if full induction is restricted to $(\mathsf{N\text{-}I_N})$.

In $\mathsf{SUS} + (\mathsf{N\text{-}I_N})$ one can show transfinite induction with respect to formulas of the form $ta \in (\mathsf{N} \rightarrow \mathsf{N})$ up to ω^k for all k less than ω. This argument requires some technical effort, but all details are carried through in Jäger and Strahm [20], even for the theory $\mathsf{BON}(\mu) + (\mathsf{N\text{-}I_N})$.

As soon as these transfinite inductions are made available, we can step back and follow the pattern of the proof of Lemma 8 and show that hyperjump hierarchies up to each ordinal ω^k for k less than ω can be constructed in $\mathsf{SUS} + (\mathsf{N\text{-}I_N})$. Therefore, we have the following embedding theorem.

THEOREM 10. *Let $A(\vec{U}, \vec{F}, \vec{v})$ be an $\mathcal{L}_2$ formula with all its free variables in $\vec{U}, \vec{F}, \vec{v}$ and assume that $(\Pi^1_1\text{-}\mathsf{CA})_{<\omega^\omega}\!\restriction$ proves $A(\vec{U}, \vec{F}, \vec{v})$. Then we have*

$$\mathsf{SUS} + (\mathsf{N\text{-}I_N}) \vdash \vec{u} \in \mathcal{P}(\mathsf{N}) \wedge \vec{f} \in (\mathsf{N} \rightarrow \mathsf{N}) \wedge \vec{v} \in \mathsf{N} \rightarrow A^{\mathsf{N}}(\vec{u}, \vec{f}, \vec{v}).$$

§4. **Upper bounds.** A natural way for computing the proof-theoretic upper bounds of SUS with set, N and formula induction is to embed them in suitable set theories. For dealing with μ and especially E_1 we need an admissible universe which is in addition a limit of admissibles. Hence, the theory KPi or more precisely some of its subsystems, all described for example in Jäger [17], are appropriate for our purpose. By referring to known results about the relationship between these set theories and the subsystems of second order arithmetic of the previous section we will obtain exact proof-theoretic characterizations.

4.1. Set theories for a recursively inaccessible universe. Let $\mathcal{L}_1$ be the first order part of our language $\mathcal{L}_2$ from above. The theory KPi is formulated in the extension $\mathcal{L}^* = \mathcal{L}_1(\in, \mathsf{N}, \mathsf{S}, \mathsf{Ad})$ of $\mathcal{L}_1$ by the membership relation symbol $\in$, the set constant N for the set of natural numbers and the unary relation symbols S and Ad for sets and admissibles, respectively.

The terms $(\mathfrak{r}, \mathfrak{s}, \mathfrak{t}, \mathfrak{r}_1, \mathfrak{s}_1, \mathfrak{t}_1, \dots)$ of $\mathcal{L}^*$ are the terms of $\mathcal{L}_1$ plus the set constant N. The formulas $(A, B, C, A_1, B_1, C_1, \dots)$ of $\mathcal{L}^*$ as well as the Δ_0, Σ, Π, Σ_n and Π_n formulas of $\mathcal{L}^*$ are defined as usual. Equality between objects is not represented by a primitive symbol but defined by

$$(\mathfrak{s} = \mathfrak{t}) \ := \ \begin{cases} (\mathfrak{s} \in \mathsf{N} \wedge \mathfrak{t} \in \mathsf{N} \wedge (\mathfrak{s} =_{\mathsf{N}} \mathfrak{t})) \vee \\ (\mathsf{S}(\mathfrak{s}) \wedge \mathsf{S}(\mathfrak{t}) \wedge (\forall x \in \mathfrak{s})(x \in \mathfrak{t}) \wedge (\forall x \in \mathfrak{t})(x \in \mathfrak{s})) \end{cases}$$

where $=_{\mathsf{N}}$ is the symbol for the primitive recursive equality on the natural numbers. The formula $A^{\mathfrak{s}}$ is the result of replacing each unrestricted quantifier $(\exists x)(\dots)$ and $(\forall x)(\dots)$ in A by $(\exists x \in \mathfrak{s})(\dots)$ and $(\forall x \in \mathfrak{s})(\dots)$, respectively. In addition, we freely make use of all standard set-theoretic notations and write, for example $(\mathfrak{s} : \mathsf{N} \to \mathsf{N})$ to express that $\mathfrak{s}$ is a function from N to N, $\mathsf{Tran}(\mathfrak{s})$ for the Δ_0 formula saying that $\mathfrak{s}$ is a transitive set and $\mathsf{Ord}(\mathfrak{s})$ for Δ_0 formula stating that $\mathfrak{s}$ is an ordinal; in the following small Greek letters range over ordinals.

Let $\mathcal{F}$ be a collection of $\mathcal{L}^*$ formulas. Induction on the natural numbers with respect to $\mathcal{F}$ consist of all formulas

$$(\mathcal{F}\text{-}\mathsf{I}_{\mathsf{N}}) \qquad A(0) \wedge (\forall x \in \mathsf{N})(A(x) \to A(x')) \ \to \ (\forall x \in \mathsf{N})A(x),$$

so that $A(a)$ belongs to $\mathcal{F}$. Accordingly, $\in$ induction with respect to $\mathcal{F}$ comprises for all $A(a)$ from $\mathcal{F}$ the formulas

$$(\mathcal{F}\text{-}\mathsf{I}_{\in}) \qquad (\forall x)[(\forall y \in x)A(y) \to A(x)] \ \to \ (\forall x)A(x).$$

Now we introduce the theory KPi for a recursively inaccessible universe. Its logical axioms comprise the usual axioms of classical first order logic with equality. The non-logical axioms of KPi can be divided into the following five groups.

I. Ontological axioms. We have for all terms $\mathfrak{r}$, $\vec{\mathfrak{s}}$ and $\mathfrak{t}$ of $\mathcal{L}^*$, all function symbols $\mathcal{H}$ and relation symbols R of $\mathcal{L}_1$ and all axioms $A(\vec{a})$ of group III whose free variables belong to the list $\vec{a}$:

(1) $\mathfrak{r} \in \mathsf{N} \leftrightarrow \neg\mathsf{S}(\mathfrak{r})$.
(2) $\vec{\mathfrak{s}} \in \mathsf{N} \to \mathcal{H}(\vec{\mathfrak{s}}) \in \mathsf{N}$.
(3) $R(\vec{\mathfrak{s}}) \to \vec{\mathfrak{s}} \in \mathsf{N}$.

(4) $\mathfrak{r} \in \mathfrak{t} \rightarrow \mathsf{S}(\mathfrak{t})$.
(5) $\mathsf{Ad}(\mathfrak{t}) \rightarrow (\mathsf{N} \in \mathfrak{t} \wedge \mathsf{Tran}(\mathfrak{t}))$.
(6) $\mathsf{Ad}(\mathfrak{t}) \rightarrow (\forall \vec{x} \in \mathfrak{t})A^{\mathfrak{t}}(\vec{x})$.
(7) $\mathsf{Ad}(\mathfrak{r}) \wedge \mathsf{Ad}(\mathfrak{t}) \rightarrow \mathfrak{r} \in \mathfrak{t} \vee \mathfrak{r} = \mathfrak{t} \vee \mathfrak{t} \in \mathfrak{r}$.

II. Number-theoretic axioms. We have for all axioms $A(\vec{a})$ of Peano arithmetic PA which are not instances of the schema of complete induction and whose free variables belong to the list $\vec{a}$:

(Number theory): $(\forall \vec{x} \in \mathsf{N})A^{\mathsf{N}}(\vec{x})$.

III. Kripke Platek axioms. We have for all terms $\mathfrak{s}$ and $\mathfrak{t}$ and all Δ_0 formulas $A(a)$ and $B(a,b)$ of $\mathcal{L}^*$:

(Pair): $\exists x(\mathfrak{s} \in x \wedge \mathfrak{t} \in x)$.
(Tran): $\exists x(\mathfrak{s} \subset x \wedge \mathsf{Tran}(x))$.
(Δ_0-Sep): $\exists y(\mathsf{S}(y) \wedge y = \{x \in \mathfrak{s} : A(x)\})$.
(Δ_0-Coll): $(\forall x \in \mathfrak{s})\exists y B(x,y) \rightarrow \exists z(\forall x \in \mathfrak{s})(\exists y \in z)B(x,y)$.

IV. Limit axiom. It is used to formalize that each set is element of an admissible set, hence we claim:

(Lim): $(\forall x)(\exists y)(x \in y \wedge \mathsf{Ad}(y))$.

V. Induction axioms. These consist of the schemas ($\mathcal{L}^*$-$\mathsf{I_N}$) of complete induction on the natural numbers and ($\mathcal{L}^*$-$\mathsf{I}_\in$) of $\in$ induction, both for arbitrary $\mathcal{L}^*$ formulas.

The theory KPi without the limit axiom corresponds to the usual system of Kripke-Platek set theory above the natural numbers as urelements; its typical models are the structures $\mathsf{L}(\alpha)_{\mathbb{N}}$ with α an admissible ordinal greater than ω. Hence, the least standard model of KPi is the structure $\mathsf{L}(\iota)_{\mathbb{N}}$ for ι being the least recursively inaccessible ordinal.

For modeling SUS with various forms of induction, however, we do not need the full strength of KPi. So let $\mathsf{KPi^r}$ be the subsystem of KPi in which complete induction on the natural numbers and $\in$ induction are restricted to Δ_0 formulas, i.e. ($\mathcal{L}^*$-$\mathsf{I_N}$) and ($\mathcal{L}^*$-$\mathsf{I}_\in$) are replaced by (Δ_0-$\mathsf{I_N}$) and (Δ_0-$\mathsf{I}_\in$), respectively. Moreover, $\mathsf{KPi^w}$ is defined to be $\mathsf{KPi^r}$ plus ($\mathcal{L}^*$-$\mathsf{I_N}$).

The proof-theoretic analysis of these set theories has been established a long time ago and their relationship to subsystems of second order arithmetic is well-known. In the following theorem (Δ^1_2-CA) and (Δ^1_2-CR) are the usual subsystems of analysis with Δ^1_2 comprehension axiom and rule, respectively; (BI) denotes the standard schema of bar induction.

THEOREM 11. *We have the following proof-theoretic equivalences*:

1. $(\Pi^1_1\text{-CA})\!\restriction\; \equiv (\Delta^1_2\text{-CA})\!\restriction\; \equiv \mathsf{KPi^r}$,
2. $(\Pi^1_1\text{-CA})_{<\omega^\omega} \equiv (\Delta^1_2\text{-CR}) \equiv \mathsf{KPi^r} + (\Sigma_1\text{-}\mathsf{I_N})$,
3. $(\Pi^1_1\text{-CA})_{<\varepsilon_0} \equiv (\Delta^1_2\text{-CA}) \equiv \mathsf{KPi^w}$,
4. $(\Delta^1_2\text{-CA}) + (\mathsf{BI}) \equiv \mathsf{KPi}$.

For the proof of assertions (1), (3) and (4) of this theorem consult e.g. Buchholz et. al. [2], Jäger [14, 16] and Jäger and Pohlers [19]. The second assertion can be obtained by making use of similar techniques.

4.2. Modeling SUS **in** $\mathsf{KPi^r}$**.** Now we turn to the central part of this paper, namely to construct a model of SUS in $\mathsf{KPi^r}$. The crucial idea is to make use of a very specific positive Δ^1_2 inductive definition for interpreting the application relation $(rs \simeq t)$ of SUS. The comparative strength of this inductive definition is needed in order to handle E_1; additional difficulties arise since everything has to be carried through within $\mathsf{KPi^r}$. However, the syntactic structure of our Δ^1_2 inductive definition makes this possible.

An additional technical complication is created by representing the primitive recursion operator $\mathsf{r_N}$ as well as the non-constructive μ operator. Both have been treated in detail in Feferman and Jäger [11] so that we confine ourselves to repeating some basic notations.

Let $A(a,b,c)$ be an $\mathcal{L}^*$ formula with at most a, b, c free and n a natural number greater than 0. Then we define for each natural number n greater than 0 and each vector $\vec{b} = b_1, \ldots, b_n$ an $\mathcal{L}^*$ formula $\mathsf{Ap}^n_A(a, \vec{b}, c)$ by induction on n, and from that, an $\mathcal{L}^*$ formula $\mathsf{Tot}^n_A(a)$:

$$\mathsf{Ap}^1_A(a, b_1, c) := a, b_1, c \in \mathsf{N} \wedge A(a, b_1, c),$$

$$\mathsf{Ap}^{n+1}_A(a, \vec{b}, b_{n+1}, c) := (\exists x \in \mathsf{N})(\mathsf{Ap}^n_A(a, \vec{b}, x) \wedge A(x, b_{n+1}, c)),$$

$$\mathsf{Tot}^n_A(a) := (\forall \vec{x} \in \mathsf{N})(\exists y \in \mathsf{N})\mathsf{Ap}^n_A(a, \vec{x}, y).$$

Later we will choose an appropriate formula $A(a,b,c)$ as interpretation of $(ab \simeq c)$. Then $\mathsf{Ap}^n_A(a, b_1, \ldots, b_n, c)$ translates the formula $(ab_1 \ldots b_n \simeq c)$ of SUS. In this sense, $\mathsf{Tot}^n_A(a)$ means that a is (the code of) a n-ary total operation on N. Observe that $\mathsf{Ap}^n_A(a, \vec{b}, c)$ and $\mathsf{Tot}^n_A(a)$ are Σ formulas provided that $A(a,b,c)$ is a Σ formula.

Further, if a is (a code of) a unary function on N in the sense of A and b (a code of) a ternary function on N in the sense of A, then the following formula $\mathsf{Rc}_A(a, b, u, v, w)$ describes the graph of the function which is defined from a and b by primitive recursion in the sense of A:

$$\mathsf{Rc}_A(a,b,u,v,w) := \left\{ \begin{array}{l} (\exists x \in \mathsf{N})[Seq(x) \wedge lh(x) = v{+}1 \wedge A(a, u, (x)_0) \\ \quad \wedge (\forall y \in \mathsf{N})(y < v \rightarrow \mathsf{Ap}^3_A(b, u, y, (x)_y, (x)_{y+1})) \\ \quad \wedge w = (x)_v]. \end{array} \right.$$

Recall that Seq is the primitive recursive predicate for sequence numbers. We also set for all natural numbers n

$$Seq_n(\mathsf{t}) := Seq(\mathsf{t}) \wedge lh(\mathsf{t}) = n$$

and assume that our coding of sequences is so that $\neg(Seq_m(\mathsf{t}) \wedge Seq_n(\mathsf{t}))$ if the natural numbers m and n are different.

Finally, we are approaching the interpretation of L in $\mathcal{L}^*$. First we choose pairwise different numerals $\hat{\mathsf{k}}, \hat{\mathsf{s}}, \hat{\mathsf{p}}, \hat{\mathsf{p}}_0, \hat{\mathsf{p}}_1, \hat{\mathsf{s}}_\mathsf{N}, \hat{\mathsf{p}}_\mathsf{N}, \hat{\mathsf{d}}_\mathsf{N}, \hat{\mathsf{r}}_\mathsf{N}, \hat{\mu}$ and $\hat{\mathsf{E}}_1$ (the values of) which do not belong to the set $\{0\} \cup \{x \in \mathbb{N} : Seq(x)\}$; they will serve as codes of the corresponding constants of L. The treatment of compound L terms will be described below.

For the purpose of the following inductive definition we choose a fresh ternary relation symbol R and define $\mathfrak{A}(R, a, b, c)$ to be the $\mathcal{L}^*(R)$ formula which is the disjunction of the following formulas (1)–(24):

(1) $a = \hat{\mathsf{k}} \wedge c = \langle \hat{\mathsf{k}}, b \rangle$,
(2) $Seq_2(a) \wedge (a)_0 = \hat{\mathsf{k}} \wedge c = (a)_1$,
(3) $a = \hat{\mathsf{s}} \wedge c = \langle \hat{\mathsf{s}}, b \rangle$,
(4) $Seq_2(a) \wedge (a)_0 = \hat{\mathsf{s}} \wedge c = \langle \hat{\mathsf{s}}, (a)_1, b \rangle$,
(5) $Seq_3(a) \wedge (a)_0 = \hat{\mathsf{s}} \wedge$
$(\exists x, y \in \mathsf{N})[R((a)_1, b, x) \wedge R((a)_2, b, y) \wedge R(x, y, c)]$,
(6) $a = \hat{\mathsf{p}} \wedge c = \langle \hat{\mathsf{p}}, b \rangle$,
(7) $Seq_2(a) \wedge (a)_0 = \hat{\mathsf{p}} \wedge c = \langle (a)_1, b \rangle$,
(8) $a = \hat{\mathsf{p}}_0 \wedge (\exists x \in \mathsf{N})(b = \langle c, x \rangle)$,
(9) $a = \hat{\mathsf{p}}_1 \wedge (\exists x \in \mathsf{N})(b = \langle x, c \rangle)$,
(10) $a = \hat{\mathsf{s}}_{\mathsf{N}} \wedge c = b+1$,
(11) $a = \hat{\mathsf{p}}_{\mathsf{N}} \wedge b = c+1$,
(12) $a = \hat{\mathsf{d}}_{\mathsf{N}} \wedge c = \langle \hat{\mathsf{d}}_{\mathsf{N}}, b \rangle$,
(13) $Seq_2(a) \wedge (a)_0 = \hat{\mathsf{d}}_{\mathsf{N}} \wedge c = \langle \hat{\mathsf{d}}_{\mathsf{N}}, (a)_1, b \rangle$,
(14) $Seq_3(a) \wedge (a)_0 = \hat{\mathsf{d}}_{\mathsf{N}} \wedge c = \langle \hat{\mathsf{d}}_{\mathsf{N}}, (a)_1, (a)_2, b \rangle$,
(15) $Seq_4(a) \wedge (a)_0 = \hat{\mathsf{d}}_{\mathsf{N}} \wedge (a)_1 = (a)_2 \wedge c = (a)_3$,
(16) $Seq_4(a) \wedge (a)_0 = \hat{\mathsf{d}}_{\mathsf{N}} \wedge (a)_1 \neq (a)_2 \wedge c = b$,
(17) $a = \hat{\mathsf{r}}_{\mathsf{N}} \wedge c = \langle \hat{\mathsf{r}}_{\mathsf{N}}, b \rangle$,
(18) $Seq_2(a) \wedge (a)_0 = \hat{\mathsf{r}}_{\mathsf{N}} \wedge c = \langle \hat{\mathsf{r}}_{\mathsf{N}}, (a)_1, b \rangle$,
(19) $Seq_3(a) \wedge (a)_0 = \hat{\mathsf{r}}_{\mathsf{N}} \wedge c = \langle \hat{\mathsf{r}}_{\mathsf{N}}, (a)_1, (a)_2, b \rangle$,
(20) $Seq_4(a) \wedge (a)_0 = \hat{\mathsf{r}}_{\mathsf{N}} \wedge \mathsf{Rc}_R((a)_1, (a)_2, (a)_3, b, c)$,
(21) $a = \hat{\mu} \wedge c = 0 \wedge (\forall x \in \mathsf{N})(\exists y \in \mathsf{N})(y \neq 0 \wedge R(b, x, y))$,
(22) $a = \hat{\mu} \wedge R(b, c, 0) \wedge$
$(\forall x \in \mathsf{N})[x < c \rightarrow (\exists y \in \mathsf{N})(y \neq 0 \wedge R(b, x, y))]$,
(23) $a = \hat{\mathsf{E}}_1 \wedge c = 0 \wedge \mathsf{Tot}^2_R(b) \wedge$
$(\exists f : \mathsf{N} \rightarrow \mathsf{N})(\forall x \in \mathsf{N})\mathsf{Ap}^2_R(b, f(x+1), f(x), 0)$,
(24) $a = \hat{\mathsf{E}}_1 \wedge c = 1 \wedge \mathsf{Tot}^2_R(b) \wedge$
$(\forall f : \mathsf{N} \rightarrow \mathsf{N})(\exists x \in \mathsf{N})(\exists y \in \mathsf{N})(0 < y \wedge \mathsf{Ap}^2_R(b, f(x+1), f(x), y)$.

We want to remark that the clauses (1)–(22) are identical to the clauses of the inductive definition in Feferman and Jäger [11] and suffice for the treatment of $\mathsf{BON}(\mu)$. The additional clauses (23) and (24) are needed for coping with E_1. Clearly, the $\mathcal{L}^*(R)$ formula $\mathfrak{A}(R, a, b, c)$ is positive in R.

Observe that we quantify in these two clauses over proper *set-theoretic* functions from N to N, not over codes for total functions on N in the sense of R. Although formulated in the language of set theory, $\mathfrak{A}(R, a, b, c)$ is easily reformulated in the language of second order arithmetic. Then it is a very special R positive Δ^1_2 formula, more precisely arithmetical in Π^1_1.

We now employ $\mathfrak{A}(R, a, b, c)$ as a definition clause of a positive Δ^1_2 inductive definition and show several other properties needed for interpreting SUS. Since KPi^r is much too weak in order to deal with positive Δ^1_2 inductive definitions in general, some extra efforts have to be made.

In the following we freely replace the relation variable R in the formula $\mathfrak{A}(R, a, b, c)$ by sets or $\mathcal{L}^*$ formulas. So, for example, $\mathfrak{A}(u, a, b, c)$ is obtained from $\mathfrak{A}(R, a, b, c)$ by replacing all subformulas $R(\mathfrak{r}, \mathfrak{s}, \mathfrak{t})$ by $((\mathfrak{r}, \mathfrak{s}, \mathfrak{t}) \in u)$.

LEMMA 12. *There are a Σ formula $B(u, a, b, c)$ and a Π formula $C(u, a, b, c)$ of the language $\mathcal{L}^*$ with exactly the free variables u, a, b, c so that* KPi^r *proves*:

1. $u \subset \mathsf{N}^3 \wedge a, b, c \in \mathsf{N} \rightarrow (\mathfrak{A}(u, a, b, c) \leftrightarrow B(u, a, b, c))$,
2. $u \subset \mathsf{N}^3 \wedge a, b, c \in \mathsf{N} \rightarrow (\mathfrak{A}(u, a, b, c) \leftrightarrow C(u, a, b, c))$.

This lemma is an immediate consequence of the quantifier theorem in Jäger [14] (see also, e.g., [15, 17]); it says that Σ^1_2 formulas of $\mathcal{L}_2$ are provably equivalent in KPi^r to Σ_1 formulas of $\mathcal{L}^*$.

Since Σ recursion is not available in KPi^r and because of the complexity of our definition clause $\mathfrak{A}(R, a, b, c)$ we introduce hierarchies to describe the stages of the corresponding inductive definition. Accordingly, we set

$$\mathfrak{H}(f, \alpha) := \begin{cases} \mathsf{Fun}(f) \wedge \mathsf{dom}(f) = \alpha \wedge \\ (\forall \beta < \alpha)[f(\beta) = \{(x, y, z) \in \mathsf{N}^3 : \mathfrak{A}(\bigcup_{\gamma<\beta} f(\gamma), x, y, z)\}]. \end{cases}$$

Straightforward $\in$ induction for Δ_0 formulas yields that these hierarchies are unique and increasing. Following the argument in Feferman and Jäger [11] concerning the clauses (1)–(22) of the formula $\mathfrak{A}(R, a, b, c)$ and by a simple inspection of clauses (23) and (24) it is also clear that we have functionality in the third argument. I.e. we can prove in KPi^r that

(i) $\mathfrak{H}(f, \alpha) \wedge \mathfrak{H}(g, \beta) \wedge \gamma < \min(\alpha, \beta) \rightarrow f(\gamma) = g(\gamma)$,
(ii) $\mathfrak{H}(f, \alpha) \wedge \gamma < \beta < \alpha \rightarrow f(\gamma) \subset f(\beta)$,
(iii) $\mathfrak{H}(f, \alpha) \wedge \beta < \alpha \wedge (a, b, c_1) \in f(\beta) \wedge (a, b, c_2) \in f(\beta) \rightarrow c_1 = c_2$.

In view of the preceding lemma it is immediate that $\mathfrak{H}(f, \alpha)$ is provably equivalent in KPi^r to a Σ formula, with which we often identify it in the sequel. Thus the following predicate $\mathbb{S}(a, b, c)$ can be regarded as a Σ formula,

$$\mathbb{S}(a, b, c) := (\exists f)(\exists \alpha)[\mathfrak{H}(f, \alpha+1) \wedge (a, b, c) \in f(\alpha)].$$

Since only weak $\in$ induction is available in KPi^r, we cannot prove there that for all ordinals α there is a function f so that $\mathfrak{H}(f, \alpha)$. Nevertheless, we can show that $\mathbb{S}$ is a fixed point of $\mathfrak{A}(R, a, b, c)$ which is functional in its third argument.

THEOREM 13. *We can prove in* KPi^r *that*

1. $(\forall x, y, z \in \mathsf{N})[\mathbb{S}(x, y, z) \leftrightarrow \mathfrak{A}(\mathbb{S}, x, y, z)]$.
2. $(\forall x, y, z_1, z_2 \in \mathsf{N})[\mathbb{S}(x, y, z_1) \wedge \mathbb{S}(x, y, z_2) \rightarrow z_1 = z_2]$.

PROOF. Because of the properties (i)–(iii) of our hierarchies the second assertion is completely obvious. Moreover, the R positivity of $\mathfrak{A}$ immediately yields the direction from left to right of the first assertion. The crucial step thus is to show that $\mathbb{S}(a, b, c)$ follows from $\mathfrak{A}(\mathbb{S}, a, b, c)$ for all natural numbers a, b and c.

Assume $\mathfrak{A}(\mathbb{S}, a, b, c)$. Now distinction by cases according to clauses (1)–(24) of $\mathfrak{A}$ has to be carried through. Mostly the arguments are straightforward, and we confine ourselves to discussing (23) in detail. Hence, we know that

$$(1) \qquad a = \hat{\mathsf{E}}_1 \wedge c = 0,$$

$$(2) \qquad (\forall x, y \in \mathsf{N})(\exists u, v \in \mathsf{N})[\mathbb{S}(b, x, u) \wedge \mathbb{S}(u, y, v)],$$

$$(3) \qquad (\exists f : \mathsf{N} \to \mathsf{N})(\forall x \in \mathsf{N})\mathsf{Ap}^2_{\mathbb{S}}(b, f(x+1), f(x), 0).$$

Now we apply Σ reflection to (2). Because of the definition of $\mathbb{S}$ and the properties (i)–(iii) of our hierarchies we may concluded that there are a function g and an ordinal α so that $\mathfrak{H}(g, \alpha+1)$ and

$$(4) \qquad (\forall x, y \in \mathsf{N})(\exists u, v \in \mathsf{N})((b, x, u) \in g(\alpha) \wedge (u, y, v) \in g(\alpha)).$$

Hence, $g(\alpha)$ can also play the role of $\mathbb{S}$ in (3) and we obtain

$$(5) \qquad (\exists f : \mathsf{N} \to \mathsf{N})(\forall x \in \mathsf{N})\mathsf{Ap}^2_{g(\alpha)}(b, f(x+1), f(x), 0).$$

In view of (1),(4) and (5) we conclude $\mathfrak{A}(g(\alpha), a, b, c)$. Using Δ separation gives a function h so that $\mathfrak{H}(h, \alpha+2)$ and $(a, b, c) \in h(\alpha+1)$. This implies $\mathbb{S}(a, b, c)$ and concludes our proof. ⊣

For the argument in the previous proof it was crucial that we had $\mathsf{Tot}^2_{\mathbb{S}}(b)$ (i.e. line (2) above) at our disposal. Then we could apply Σ reflection and thus localize the Σ relation $\mathbb{S}$ to a specific set $g(\alpha)$. Thus it is essential that we are interpreting the functional E_1 which is applied to total operations on N; a similar argument cannot work for the partial functional $\mathsf{E}^{\#}_1$.

A last central step in showing that $\mathbb{S}$ provides an adequate application relation for interpreting SUS in KPi^r is to prove the following *inside-outside* property: given an a coding (the characteristic function of) a binary relation on N with respect to $\mathbb{S}$, we have to show that there exists a set-theoretic function which is a descending chain for a if and only if there exists a natural number coding a total function on N in the sense of $\mathbb{S}$ which is also a descending chain for a.

For the formulation (and the proof) of the following theorem some auxiliary notations are useful. First we set

$$\{a\}^{\mathbb{S}}(b_1, \dots, b_n) \simeq c \;:=\; \mathsf{Ap}^n_{\mathbb{S}}(a, b_1, \dots, b_n, c)$$

and follow the usual conventions of recursion theory when using expressions like $\{a\}^{\mathbb{S}}(\vec{b})$. Later sometimes $\mathbb{S}$ will be replaced by a subset u of N^3 with $\{a\}^u(\vec{b})$ having its obvious meaning.

Then we write $\mathsf{CDC}(a, b)$ in order to express that b is a *code of a total function* on N providing a descending chain for a, everything in the sense of $\mathbb{S}$,

$$\mathsf{CDC}_{\mathbb{S}}(a, b) \;:=\; \mathsf{Tot}^1_{\mathbb{S}}(b) \wedge (\forall x \in \mathsf{N})\mathsf{Ap}^2_{\mathbb{S}}(a, \{b\}^{\mathbb{S}}(x+1), \{b\}^{\mathbb{S}}(x), 0).$$

This is in contrast to $\mathsf{FDC}_{\mathbb{S}}(a, f)$,

$$\mathsf{FDC}_{\mathbb{S}}(a, f) \;:=\; (f : \mathsf{N} \to \mathsf{N}) \wedge (\forall x \in \mathsf{N})\mathsf{Ap}^2_{\mathbb{S}}(a, f(x+1), f(x), 0),$$

which states that f is a *set-theoretic function* providing a descending chain for a with respect to $\mathbb{S}$.

Given a hierarchy f which describes the stages of our inductive definition up to the ordinal α, i.e. $\mathfrak{H}(f, \alpha)$, it is easy to see that it can be extended up to $\alpha+k$ for each fixed natural number k. Suppose now that we have natural numbers which are known to code certain relations on the natural numbers at level α. Then the following two lemmas state that certain operations on these relations can be carried through and coded by making use of a fixed *finite* number of additional levels only.

For their formulation it is convenient to write $\mathsf{ext}^n_b(a)$ for the extension of a in the sense of b,

$$\mathsf{ext}^n_b(a) \;:=\; \{(x_1, \dots, x_n) \in \mathsf{N}^n : \mathsf{Ap}^n_b(a, x_1, \dots, x_n, 0)\}.$$

Since Ap^n_b is a Δ_0 formula, $\mathsf{ext}^n_b(a)$ defines in $\mathsf{KPi}^{\mathsf{r}}$ a subset of N^n by means of Δ_0 separation.

LEMMA 14 (Finite extension property I). *Let R be a fresh m-ary relation symbol and assume that $A(R, u, \vec{v})$ is a formula of $\mathcal{L}_1(R)$ with at most the variables u and $\vec{v} = v_1, \dots, v_n$ free. Then there exist a natural number k and primitive recursive functions $\mathcal{F}$ and $\mathcal{G}$ so that* $\mathsf{KPi}^{\mathsf{r}}$ *proves*:

1. $\mathfrak{H}(f, \alpha+k+1) \wedge a, u \in \mathsf{N} \wedge \mathsf{Tot}^m_{f(\alpha)}(a) \;\to$
   $\mathsf{Tot}^n_{f(\alpha+k)}(\mathcal{F}(a, u)) \wedge \mathsf{Tot}^1_{f(\alpha+k)}(\mathcal{G}(a))$,
2. $\mathfrak{H}(f, \alpha+k+1) \wedge a, u, \vec{v} \in \mathsf{N} \wedge \mathsf{Tot}^m_{f(\alpha)}(a) \;\to$
   $A^{\mathsf{N}}(\mathsf{ext}^m_{f(\alpha)}(a), u, \vec{v}) \leftrightarrow \{\mathcal{F}(a, u)\}^{f(\alpha+k)}(\vec{v}) \simeq 0$,
3. $\mathfrak{H}(f, \alpha+k+1) \wedge a, u \in \mathsf{N} \wedge \mathsf{Tot}^m_{f(\alpha)}(a) \;\to$
   $(\hat{\mathsf{E}}_1, \mathcal{F}(a, u), 0) \in f(\alpha+k) \leftrightarrow \{\mathcal{G}(a)\}^{f(\alpha+k)}(u) \simeq 0$.

LEMMA 15 (Finite extension property II). *Let R be a fresh m-ary and S a fresh n-ary relation symbol and assume that $B(R, S, u, v)$ is a formula of $\mathcal{L}_1(R, S)$ with at most the variables u and v free. Further assume that*

$$\mathsf{KPi}^{\mathsf{r}} \vdash (\forall x_1, x_2 \subset \mathsf{N})(\forall y \in \mathsf{N})(\exists! z \in \mathsf{N})B^{\mathsf{N}}(x_1, x_2, y, z)$$

and let $\mathcal{F}_B$ denote the (class) function defined by B^{N}. Then there exists a natural number k and a primitive recursive function $\mathcal{H}$ so that $\mathsf{KPi}^{\mathsf{r}}$ *proves*:

1. $\mathfrak{H}(f, \alpha+k+1) \wedge a, b, u \in \mathsf{N} \wedge \mathsf{Tot}^m_{f(\alpha)}(a) \wedge \mathsf{Tot}^n_{f(\alpha)}(b) \;\to$
   $\mathsf{Tot}^1_{f(\alpha+k)}(\mathcal{H}(a, b, u))$,
2. $\mathfrak{H}(f, \alpha+k+1) \wedge a, b, u, v \in \mathsf{N} \wedge \mathsf{Tot}^m_{f(\alpha)}(a) \wedge \mathsf{Tot}^n_{f(\alpha)}(b) \;\to$
   $\{\mathcal{H}(a, b, u)\}^{f(\alpha+k)}(0) \;\simeq\; u \;\wedge$
   $\{\mathcal{H}(a, b, u)\}^{f(\alpha+k)}(v+1) \;\simeq$
   $\mathcal{F}_B(\mathsf{ext}^m_{f(\alpha)}(a), \mathsf{ext}^n_{f(\alpha)}(b), \{\mathcal{H}(a, b, u)\}^{f(\alpha+k)}(v))$.

The proofs of these two finite extension lemmas are straightforward, although a bit tedious. It is essential that primitive recursion is directly built into our inductive definition. They also exploit the fact that combinatory completeness is available due to our coding of k and s. After these preparatory steps we are now ready to turn to the theorem about the *inside-outside property* mentioned above.

THEOREM 16. *We can prove in* $\mathsf{KPi^r}$ *that*

1. $(\exists b \in \mathsf{N})\mathsf{CDC}_{\mathbb{S}}(a,b) \rightarrow (\exists f)\mathsf{FDC}_{\mathbb{S}}(a,f)$,
2. $\mathsf{Tot}^2_{\mathbb{S}}(a) \wedge (\exists f)\mathsf{FDC}_{\mathbb{S}}(a,f) \rightarrow (\exists b \in \mathsf{N})\mathsf{CDC}_{\mathbb{S}}(a,b)$.

In particular, we have in $\mathsf{KPi^r}$ *the following equivalence*

$$\mathsf{Tot}^2_{\mathbb{S}}(a) \rightarrow [(\exists f)\mathsf{FDC}_{\mathbb{S}}(a,f) \leftrightarrow (\exists b \in \mathsf{N})\mathsf{CDC}_{\mathbb{S}}(a,b)].$$

PROOF. For the first assertion of our theorem choose a natural number b such that $\mathsf{CDC}_{\mathbb{S}}(a,b)$. Hence, we have $\mathsf{Tot}^1_{\mathbb{S}}(b)$, i.e.,

$$(\forall x \in \mathsf{N})(\exists y \in \mathsf{N})\mathbb{S}(b,x,y). \tag{1}$$

Because of Σ reflection and the functionality of $\mathbb{S}$ in its third argument we know that

$$f := \{(x,y) \in \mathsf{N}^2 : \mathbb{S}(b,x,y)\} \tag{2}$$

is a set-theoretic function from N to N. It follows immediately that this f satisfies $\mathsf{FDC}(a,f)$.

Throughout the proof of the second assertion, we let a be an arbitrary natural number so that $\mathsf{Tot}^2_{\mathbb{S}}(a)$, i.e.,

$$(\forall x,y \in \mathsf{N})(\exists u,v \in \mathsf{N})[\mathbb{S}(a,x,u) \wedge \mathbb{S}(u,y,v)]. \tag{3}$$

Hence, we can employ Σ reflection and find an ordinal α and a function g so that $\mathfrak{H}(g,\alpha{+}1)$ and $\mathsf{Tot}^2_{g(\alpha)}(a)$, i.e.,

$$(\forall x,y \in \mathsf{N})(\exists u,v \in \mathsf{N})((a,x,u) \in g(\alpha) \wedge (u,y,v) \in g(\alpha)). \tag{4}$$

Let us now assume that we are given a function f so that $\mathsf{FDC}_{\mathbb{S}}(a,f)$. Indeed, by (4) we have $\mathsf{FDC}_{g(\alpha)}(a,f)$, i.e., $(f : \mathsf{N} \rightarrow \mathsf{N})$ and

$$(\forall x \in \mathsf{N})\mathsf{Ap}^2_{g(\alpha)}(a,f(x{+}1),f(x),0). \tag{5}$$

We want to find a natural number b so that $\mathsf{CDC}_{\mathbb{S}}(a,b)$. Towards this aim, we have to formalize a standard *leftmost branch argument* in $\mathsf{KPi^r}$ and make substantial use of the finite extension properties I and II (Lemma 14 and 15). For definiteness, assume that k is a natural number which is big enough to carry through all the finite extensions of g needed below; accordingly, let h be a function such that $\mathfrak{H}(h,\alpha{+}2k{+}1)$ holds.

First, we set $A(R,u,v_1,v_2)$ to be the following $\mathcal{L}_1(R)$ formula

$$A(R,u,v_1,v_2) := \begin{cases} (\exists x,y)[Seq(x) \wedge lh(x) = y \wedge (x)_0 = u \wedge (x)_{y \dot{-} 1} = v_2 \\ \wedge (\forall z < y \dot{-} 1)R((x)_{z+1},(x)_z)] \wedge R(v_1,v_2). \end{cases}$$

$A(R, u, v_1, v_2)$ expresses that $R(v_1, v_2)$ and v_2 is accessible from u by means of R. We can now apply Lemma 14 and find primitive recursive functions $\mathcal{F}$ and $\mathcal{G}$ so that

$$(6) \qquad \mathsf{Tot}^2_{h(\alpha+k)}(\mathcal{F}(a,u)) \quad \text{and} \quad \mathsf{Tot}^1_{h(\alpha+k)}(\mathcal{G}(a))$$

for all natural numbers u and, moreover,

$$(7) \qquad A^{\mathsf{N}}(\mathsf{ext}^2_{h(\alpha)}(a), u, v_1, v_2) \leftrightarrow \{\mathcal{F}(a,u)\}^{h(\alpha+k)}(v_1, v_2) \simeq 0,$$

$$(8) \qquad (\hat{\mathsf{E}}_1, \mathcal{F}(a,u), 0) \in h(\alpha+k) \leftrightarrow \{\mathcal{G}(a)\}^{h(\alpha+k)}(u) \simeq 0$$

for all natural numbers u and v_1, v_2. Further, we let $B(R, S, u, v)$ denote the $\mathcal{L}_1(R, S)$ formula given by

$$B(R,S,u,v) := \begin{cases} R(v,u) \wedge S(v) \wedge (\forall w < v)(\neg R(w,u) \vee \neg S(w)) \\ \vee \neg(\exists w)(R(w,u) \wedge S(w)) \wedge v = 0. \end{cases}$$

Clearly, we have that $(\forall x_1, x_2 \subset \mathsf{N})(\forall y \in \mathsf{N})(\exists! z \in \mathsf{N})B^{\mathsf{N}}(x_1, x_2, y, z)$. Hence, we are in a position to apply Lemma 15 and obtain a primitive recursive function $\mathcal{H}$ so that we have for $b = \mathcal{H}(a, \mathcal{G}(a), f(0))$,

$$(9) \qquad \mathsf{Tot}^1_{h(\alpha+2k)}(b),$$

and, further, for all natural numbers v,

$$(10) \qquad \{b\}^{h(\alpha+2k)}(0) \simeq f(0),$$

$$(11) \qquad \{b\}^{h(\alpha+2k)}(v+1) \simeq \mathcal{F}_B(\mathsf{ext}^2_{h(\alpha)}(a), \mathsf{ext}^1_{h(\alpha+k)}(\mathcal{G}(a)), \{b\}^{h(\alpha+2k)}(v)).$$

The proof so far has been set up in such a manner that we can now apply a simple form of Δ_0 induction on N to show that indeed

$$(12) \qquad (\forall x \in \mathsf{N})\mathsf{Ap}^2_{h(\alpha)}(a, \{b\}^{h(\alpha+2k)}(x+1), \{b\}^{h(\alpha+2k)}(x), 0)$$

follows from (5) and the construction of b. Together with (9) we have thus established $\mathsf{CDC}_{h(\alpha+2k)}(a,b)$ and, hence, also $\mathsf{CDC}_{\mathbb{S}}(a,b)$. This is as desired and finishes the proof of our theorem. ⊣

Now the stage is set in order to describe a translation $*$ from L into $\mathcal{L}^*$. The central idea is to interpret the L formula $(ab \simeq c)$ by the $\mathcal{L}^*$ formula $\mathbb{S}(a,b,c)$. More precisely, let us first define an $\mathcal{L}^*$ formula $\mathbb{V}_t(u)$ for each individual term t of L so that the variable u does not occur in t. The formula $\mathbb{V}_t(u)$ expresses that u is the value of t in the sense of $\mathbb{S}$. The exact definition is by induction on the complexity of t:

1. If t is an individual variable, then $\mathbb{V}_t(u)$ is $(t = u)$.
2. If t is an individual constant, then $\mathbb{V}_t(u)$ is $(\hat{t} = u)$.
3. If t is the individual term (rs), then

$$\mathbb{V}_t(u) := (\exists x, y \in \mathsf{N})(\mathbb{V}_r(x) \wedge \mathbb{V}_s(y) \wedge \mathbb{S}(x,y,u)).$$

In a second step we define the $*$ translation of an L formula A inductively as follows:

4. If A is the formula $\mathsf{N}(t)$ or $t\downarrow$, then A^* is $(\exists x \in \mathsf{N})\mathbb{V}_t(x)$.

5. If A is the formula $(s = t)$, then A^* is $(\exists x \in \mathsf{N})(\mathbb{V}_s(x) \wedge \mathbb{V}_t(x))$.
6. If A is the formula $\neg B$, then A^* is $\neg(B^*)$.
7. If A is the formula $(B\ j\ C)$ for $j \in \{\vee, \wedge, \rightarrow\}$, then A^* is $(B^*\ j\ C^*)$.
8. If A is the formula $(Qx)B$ for $Q \in \{\exists, \forall\}$, then A^* is $(Qx \in \mathsf{N})B^*$.

Given this interpretation of L into $\mathcal{L}^*$ we can now turn to the desired embedding of SUS with set induction, N induction and formula induction on N into our set theories.

THEOREM 17. *If $A(\vec{a})$ is an* L *formula with all its free variables indicated, then we have*:

1. $\mathsf{SUS} + (\mathsf{S\text{-}I_N}) \vdash A(\vec{a}) \quad \Longrightarrow \quad \mathsf{KPi^r} \vdash \vec{a} \in \mathsf{N} \rightarrow A^*(\vec{a})$,
2. $\mathsf{SUS} + (\mathsf{N\text{-}I_N}) \vdash A(\vec{a}) \quad \Longrightarrow \quad \mathsf{KPi^r} + (\Sigma_1\text{-}\mathsf{I_N}) \vdash \vec{a} \in \mathsf{N} \rightarrow A^*(\vec{a})$,
3. $\mathsf{SUS} + (\mathsf{L\text{-}I_N}) \vdash A(\vec{a}) \quad \Longrightarrow \quad \mathsf{KPi^w} \vdash \vec{a} \in \mathsf{N} \rightarrow A^*(\vec{a})$.

PROOF. For showing the first assertion, we want to point out that all axioms of SUS plus set induction except for the axioms about $\mathsf{E_1}$ can be treated exactly as in Feferman and Jäger [11], in which the theory $\mathsf{BON}(\mu)$ plus set induction is treated.

The axiom $(\mathsf{E_1}.1)$ is an immediate consequence of the fixed point property of $\mathbb{S}$ formulated in Theorem 13. In order to obtain axiom $(\mathsf{E_1}.2)$ in $\mathsf{KPi^r}$ choose a natural number f so that $(f \in (\mathsf{N} \rightarrow \mathsf{N}))^*$. Then the fixed point property of $\mathbb{S}$ nearly yields the $*$ translation of the conclusion of $(\mathsf{E_1}.2)$. There is still one serious defect: the existential quantifier on the left hand still ranges over set-theoretic functions from N to N rather than functions in the sense of $\mathbb{S}$. By the *inside-outside property* proved in Theorem 16, however, we can now internalize this set-theoretic function into a function in the sense of $\mathbb{S}$ and vice versa.

The second assertion follows form the first assertion simply by observing that formulas of the form $(t \in \mathsf{N})$ translate into Σ formulas under our translation $*$. Hence, $(\mathsf{N\text{-}I_N})$ in the language L directly carries over to Σ_1 induction in $\mathcal{L}^*$. Clearly, the third assertion also immediately follows from the first. ⊣

Everything is available now in order to state the final proof-theoretic characterization of the theories SUS plus set induction, N induction and formula induction. Theorem 7, Theorem 9 and Theorem 10 provide their lower bounds in terms of subsystems of second order arithmetic, Theorem 17 gives their upper bounds in terms of subsystems of set theory and Theorem 11 yields the proof-theoretic equivalence of the appropriate subsystems of second order arithmetic and set theory.

COROLLARY 18. *We have the following proof-theoretic equivalences*:

1. $\mathsf{SUS} + (\mathsf{S\text{-}I_N}) \equiv (\Pi^1_1\text{-}\mathsf{CA})\!\restriction\ \equiv (\Delta^1_2\text{-}\mathsf{CA})\!\restriction\ \equiv \mathsf{KPi^r}$,
2. $\mathsf{SUS} + (\mathsf{N\text{-}I_N}) \equiv (\Pi^1_1\text{-}\mathsf{CA})_{<\omega^\omega} \equiv (\Delta^1_2\text{-}\mathsf{CR}) \equiv \mathsf{KPi^r} + (\Sigma_1\text{-}\mathsf{I_N})$,
3. $\mathsf{SUS} + (\mathsf{L\text{-}I_N}) \equiv (\Pi^1_1\text{-}\mathsf{CA})_{<\varepsilon_0} \equiv (\Delta^1_2\text{-}\mathsf{CA}) \equiv \mathsf{KPi^w}$.

Besides $(\mathsf{N\text{-}I_N})$ there exists a further form of complete induction on the natural numbers which comprises $(\mathsf{S\text{-}I_N})$ but is strictly weaker than $(\mathsf{L\text{-}I_N})$, the so-called operation induction $(\mathsf{O\text{-}I_N})$, cf. [18, 21, 20]. However, we do not have to consider

$(\mathsf{O}\text{-}\mathsf{I_N})$ in our present context since Kahle [21] shows that over $\mathsf{BON}(\mu)$ operation induction $(\mathsf{O}\text{-}\mathsf{I_N})$ and N induction $(\mathsf{N}\text{-}\mathsf{I_N})$ are equivalent. Hence, in the second assertion of this corollary we can replace $\mathsf{SUS} + (\mathsf{N}\text{-}\mathsf{I_N})$ by $\mathsf{SUS} + (\mathsf{O}\text{-}\mathsf{I_N})$.

We conclude this article with mentioning the proof-theoretic ordinals of these systems, following the notation system of Buchholz and Schütte [3], which are known since a long time for the corresponding subsystems of second order arithmetic and set theory: the proof-theoretic ordinal of $\mathsf{SUS} + (\mathsf{S}\text{-}\mathsf{I_N})$ is $\Psi 0 \Omega_{\omega}$, that of $\mathsf{SUS} + (\mathsf{N}\text{-}\mathsf{I_N})$ is $\Psi 0 \Omega_{\omega^{\omega}}$, and the one of $\mathsf{SUS} + (\mathsf{L}\text{-}\mathsf{I_N})$ is $\Psi 0 \Omega_{\varepsilon_0}$.

REFERENCES

[1] Michael J. Beeson, ***Foundations of constructive mathematics: Metamathematical studies***, Springer, Berlin, 1985.

[2] Wilfried Buchholz, Solomon Feferman, Wolfram Pohlers, and Wilfried Sieg, ***Iterated inductive definitions and subsystems of analysis: Recent proof-theoretical studies***, Lecture Notes in Mathematics, vol. 897, Springer, Berlin, 1981.

[3] Wilfried Buchholz and Kurt Schütte, ***Proof theory of impredicative subsystems of analysis***, Bibliopolis, Naples, 1988.

[4] Solomon Feferman, *A language and axioms for explicit mathematics*, ***Algebra and logic*** (J.N. Crossley, editor), Lecture Notes in Mathematics, vol. 450, Springer, Berlin, 1975, pp. 87–139.

[5] ———, *A theory of variable types*, ***Revista Colombiana de Matemáticas***, vol. 19 (1975), pp. 95–105.

[6] ———, *Theories of finite type related to mathematical practice*, ***Handbook of mathematical logic*** (J. Barwise, editor), North Holland, Amsterdam, 1977, pp. 913–971.

[7] ———, *Recursion theory and set theory: a marriage of convenience*, ***Generalized recursion theory II, Oslo 1977*** (J. E. Fenstad, R. O. Gandy, and G. E. Sacks, editors), Stud. Logic Found. Math. vol. 94, North Holland, Amsterdam, 1978, pp. 55–98.

[8] ———, *Constructive theories of functions and classes*, ***Logic colloquium '78*** (M. Boffa, D. van Dalen, and K. McAloon, editors), North Holland, Amsterdam, 1979, pp. 159–224.

[9] ———, *Weyl vindicated: "Das Kontinuum" 70 years later*, ***Temi e prospettive della logica e della filosofia della scienza contemporanee***, CLUEB, Bologna, 1988, pp. 59–93.

[10] ———, *Definedness*, ***Erkenntnis***, vol. 43 (1995), pp. 295–320.

[11] Solomon Feferman and Gerhard Jäger, *Systems of explicit mathematics with non-constructive μ-operator. Part I*, ***Annals of Pure and Applied Logic***, vol. 65 (1993), no. 3, pp. 243–263.

[12] ———, *Systems of explicit mathematics with non-constructive μ-operator. Part II*, ***Annals of Pure and Applied Logic***, vol. 79 (1996), no. 1, pp. 37–52.

[13] Thomas Glass and Thomas Strahm, *Systems of explicit mathematics with non-constructive μ-operator and join*, ***Annals of Pure and Applied Logic***, vol. 82 (1996), pp. 193–219.

[14] Gerhard Jäger, *Die konstruktible Hierarchie als Hilfsmittel zur beweistheoretischen Untersuchung von Teilsystemen der Mengenlehre und Analysis*, ***Ph.D. thesis***, Universität München, 1979.

[15] ———, *Iterating admissibility in proof theory*, ***Logic colloquium '81. Proceedings of the Herbrand symposion.***, North Holland, Amsterdam, 1982.

[16] ———, *A well-ordering proof for Feferman's theory* $\mathsf{T_0}$, ***Archiv für mathematische Logik und Grundlagenforschung***, vol. 23 (1983), pp. 65–77.

[17] ———, ***Theories for admissible sets: A unifying approach to proof theory***, Bibliopolis, Napoli, 1986.

[18] Gerhard Jäger, Reinhard Kahle, and Thomas Strahm, *On applicative theories*, ***Logic and foundations of mathematics*** (A. Cantini, E. Casari, and P. Minari, editors), Kluwer, 1999, pp. 83–92.

[19] Gerhard Jäger and Wolfram Pohlers, *Eine beweistheoretische Untersuchung von* $(\Delta^1_2\text{-}\mathsf{CA})+(\mathsf{BI})$ *und verwandter Systeme*, ***Sitzungsberichte der Bayerischen Akademie der Wissenschaften,***

Mathematisch-naturwissenschaftliche Klasse, 1982, pp. 1–28.

[20] Gerhard Jäger and Thomas Strahm, *Some theories with positive induction of ordinal strength* $\varphi\omega 0$, ***The Journal of Symbolic Logic***, vol. 61 (1996), no. 3, pp. 818–842.

[21] Reinhard Kahle, N-*strictness in applicative theories*, ***Archive for Mathematical Logic***, vol. 39 (2000), no. 2, pp. 125–144.

[22] Stephen G. Simpson, ***Subsystems of Second Order Arithmetic***, Perspectives in Mathematical Logic, Springer-Verlag, 1998.

[23] A.S. Troelstra and D. van Dalen, ***Constructivism in mathematics***, vol. I, North-Holland, Amsterdam, 1988.

[24] ———, ***Constructivism in mathematics***, vol. II, North Holland, Amsterdam, 1988.

INSTITUT FÜR INFORMATIK UND ANGEWANDTE MATHEMATIK
UNIVERSITÄT BERN
NEUBRÜCKSTRASSE 10
CH-3012 BERN, SWITZERLAND
E-mail: jaeger@iam.unibe.ch

INSTITUT FÜR INFORMATIK UND ANGEWANDTE MATHEMATIK
UNIVERSITÄT BERN
NEUBRÜCKSTRASSE 10
CH-3012 BERN, SWITZERLAND
E-mail: strahm@iam.unibe.ch

FEFERMAN-LANDIN LOGIC

IAN A. MASON AND CAROLYN L. TALCOTT

Abstract. This paper presents a logic based on Feferman's Variable Type Theories for reasoning about programs written in a class of imperative functional languages called Landinesque languages.

To Solomon Feferman on the occasion of his 70th Birthday with much gratitude.

§1. Historical background. Feferman-Landin Logic is so named because of the influence that the work of both Feferman and Landin has had on our work. Of particular relevance are two key ideas:

> *Landin's idea*: a programming language can be thought of as the lambda calculus augmented by operations on the basic data, as well as computational environment. We call such languages *Landinesque languages*.
>
> *Feferman's idea*: to formalize constructive mathematics in 2-sorted classical theories called variable type theories. In these theories both functions and data are objects of discourse in a first order setting, as are collections of such things.

Both authors were PhD students of Solomon Feferman in the early 1980's where, under his influence, we began the research program leading to the work that this paper will report on. Both the theories used by Feferman, and his approach to formalizing constructive mathematics seemed very natural and we felt that this approach should also work for developing formal systems for reasoning about programs.

The first step in such an endeavor is to develop the mathematical semantics of the programs of interest. Our focus has been on reasoning about non-functional primitives. Together we have studied lambda calculus augmented by operations for manipulating control [39], manipulating memory [29], and for communicating asynchronously in an open distributed environment [2]. In each case we developed an operational semantics based solely on syntactic entities, and studied the natural

---

The first author was partially supported by UNE URG grant 6590 881 499 and UNE small ARC grant 8282 881 499.

The second author was partially supported by NSF CCR-9633419, DARPA/Rome Labs AF F30602-96-1-0300, DARPA/NASA NAS2-98073, and ONR N00012-99-C-0198.

The authors wish to thank Jonathan Ford and the anonymous referees for their helpful comments.

Reflections on the foundations of mathematics
Essays in honor of Solomon Feferman
Edited by Wilfried Sieg, Richard Sommer, and Carolyn Talcott

operational equivalence generated by that semantics (operational equivalence meaning being equi-defined in all closing contexts). Crucial in such work is the ability to simplify the criteria for operational equivalence to some form of context lemma a la Robin Milner [31] to avoid the complexity of reasoning about arbitrary program contexts. In both the case of memory, and control, we have established such simplifications now known in the literature as *CIU* theorems (*C*losed *I*nstantiations of *U*ses). The case for communicating asynchronously in an open distributed environment remains open.

After developing this syntactic style of operational semantics, and the corresponding characterization of operational equivalence, we used this as a basis for developing logics based on Solomon Feferman's Variable typed approach. These systems are two sorted theories of operations and classes initially developed for the formalization of constructive mathematics [5, 6, 7, 8] and later applied to the study of purely functional languages [10, 11, 9].

Our work uses the syntactic operational semantics to develop expressive logics and goes well beyond traditional programming logics, such as Hoare's logic [3] and Dynamic logic [21] by treating a richer language and expressing more properties. It is close in spirit to Specification Logic [37] and to Evaluation Logic [34]. Particular logics, with corresponding reasoning principles, have been presented for the case of manipulating control [41], and for the case of manipulating memory [22], for which it is possible to establish a limited form of completeness [26].

The work discussed above studied specific Landinesque languages in a somewhat ad hoc fashion. More recently the second author has unified the ideas emerging from our ad hoc approach to syntactic operational semantics by developing a general theory of Landinesque languages [42]. The approach starts with a small step semantics in which computation state is represented using syntactic entities such as expressions and contexts. There is a single reduction rule for each operation. Care is taken so that the reduction rule for an operation is not changed when new operations or new pieces of state are added. Computation is uniform is the sense that reduction steps can be performed on states with missing parts and the missing information can be filled in later. Such a syntactic reduction system has the combined advantages of a simple transition system semantics and the symbolic reasoning of a reduction calculus. The main result is the CIU theorem for languages with uniform semantics, that simplifies reasoning about operational equivalence restricting the contexts that must be considered.

In this paper we build upon this uniform approach to present a variable typed logic for such languages, thus extending our previous work on specifying and reasoning about programs to a wide class of imperative functional languages. As such this paper subsumes the case studies presented concerning operations that manipulate control, as well as those concerning operations that manipulate memory.

The plan of this paper is as follows. In the next section we present a summary and simplification of the core results of [42]. The form and presentation of the results presented in this section has benefitted greatly from the formalization, in PVS, of this approach by Jonathan Ford at the University of New England. This formalization is reported in [15] where are much more detailed guide to our development of Landinesque languages can be found. We then present the syntax and semantics of our variable typed logic for this general framework. We do this in two stages. We first present the first-order semantics in the third section, and then in the fourth section concentrate on the notion of classes. Section five deals with the general principles that hold for these uniform languages, while section six presents a detailed example of a particular language that incorporates both control and memory manipulation. The final section describes our conclusions and further directions that this research may take.

Notation. We conclude the introduction with a summary of our notation conventions. Let X, X_0, X_1 be sets. We specify meta-variable conventions in the form: let x range over X, which should be read as: the meta-variable x and decorated variants such as x', x_0, ..., range over the set X. $\mathrm{Fmap}[X_0, X_1]$ is the set of finite maps from X_0 to X_1. We write $\mathrm{Dom}(f)$ for the domain of a function and $\mathrm{Rng}(f)$ for its range. For any function f, $f\{x \mapsto x'\}$ is the function f' such that $\mathrm{Dom}(f') = \mathrm{Dom}(f) \cup \{x\}$, $f'(x) = x'$, and $f'(z) = f(z)$ for $z \neq x, z \in \mathrm{Dom}(f)$. Also $f\lceil X$ is the restriction of f to X: the function f' such that $\mathrm{Dom}(f') = \mathrm{Dom}(f) \cap X$ and $f'(x) = f(x)$ for $x \in \mathrm{Dom}(f')$. $\boldsymbol{N} = \{0, 1, 2, \dots\}$ is the set of natural numbers and $i, j, n, n_0, \dots$ range over $\boldsymbol{N}$. In the defining equations for various syntactic classes we use two notational conventions: pointwise lifting of syntax operations to syntax classes; and the Einstein summation convention that a phrase of the form $F_n(Z^n)$ abbreviates $\bigcup_{n \in \boldsymbol{N}} F_n(Z^n)$. For example if Ω is a ranked set of operator symbols, then the terms over Ω can be defined inductively by (as the least solution to) the equation: $T_\Omega = \Omega_n(T_\Omega^n)$. Unabbreviated, this equation reads:

$$T_\Omega = \bigcup_{n \in \boldsymbol{N}} \{\omega(t_1, \dots, t_n) \mid \omega \in \Omega_n \wedge t_i \in T_\Omega \quad \text{for} \quad 1 \leq i \leq n\}.$$

§2. Uniform Landinesque languages. In this section a general framework for studying the semantics of λ-languages is set up along the lines of [42], where detailed proofs of some of the claims can be found. The syntactic entities and semantic notions of λ-languages are defined and the properties required for a uniform semantics are stated. Then several results, including the CIU theorem, valid in any λ-language with uniform semantics are presented. This is the mathematical theory that we wish to formalize following Feferman's approach. We begin with an overview of the concepts and results.

A small-step operational semantics is obtained by defining a notion of state and a single step reduction relation on states. States consist of an expression and a state context. A state context often describes dynamically created entities such as memory cells, arrays, files, etc. The form of state contexts needed depends on the choice of primitive operations. There is an empty state context, and for each state there is an associated expression representing that state. Value expressions are a subset of the set of expressions used to represent semantic values. These include variables, atoms, and lambdas. If the expression component of a state is a value, then the state is a value state and no reduction steps are possible. Otherwise, the expression decomposes uniquely into a redex placed in a reduction context. A (call-by-value) redex is a primitive operator applied to a list of values. There is one reduction rule for each primitive operator, and the single-step reduction relation on states is determined by the reduction rule for the redex operator. Of course it may happen that a redex is ill-formed (a runtime error) and no reduction step is possible. A state is defined just if it reduces (in a finite number of steps) to a value state. Using these basic notions we define the operational approximation and equivalence relations in the usual way in terms of definedness in all program contexts. This is the basic semantic framework, independent of the choice of primitive operations. Within this framework we define the notion of uniform semantics and develop tools for proving laws of approximation and equivalence in λ-languages with uniform semantics. For a particular choice of operations what remains is to define the structure of state contexts, and provide the reduction rules for each primitive operation.

The uniformity requirements are that each reduction rule hold not only for traditional expressions, but also for expressions containing parameters or meta variables. This parametric notion of computation is best presented using the idea of a context and the treatment here follows the general theory presented in [27]. Contexts can be formalized by adding meta-variables or parameters to the syntax of expressions. The novelty of our approach is that we decorate parameters with information recording pending substitutions which are to be applied when the parameter is instantiated. This has several desirable consequences. One is that alpha conversion is valid for contexts, unlike the traditional lambda-calculus contexts. The other is that symbolic execution commutes with parameter instantiation, which simplifies reasoning about program equivalence.

2.1. Expression syntax of a λ-language. Fix two disjoint countably infinite sets, $\boldsymbol{X}$, of variables, and $\boldsymbol{P}$ of parameters. The basic syntax of a λ-language is then determined by specifying three sets:

($\boldsymbol{A}$): a countable set of atoms, $\boldsymbol{A}$, disjoint from $\boldsymbol{X}$ and $\boldsymbol{P}$;

($\boldsymbol{O}$): a family of operation symbols $\boldsymbol{O} = \{\boldsymbol{O}_n \mid n \in \boldsymbol{N}\}$ ($\boldsymbol{O}_n$ is a set of n-ary operation symbols) disjoint from $\boldsymbol{X} \cup \boldsymbol{A} \cup \boldsymbol{P}$; and

($\boldsymbol{V}$): the set of value expressions, a subset of expressions, $\boldsymbol{V}$, that we will specify in more detail immediately after the definition of the syntax.

We assume that $\boldsymbol{O}$ contains at least the binary operation app (lambda application). As we shall see later by taking $\boldsymbol{A} = \{\ \}$ and $\boldsymbol{O} = \{\mathsf{app}\}$ we obtain the expressions of the pure call-by-value lambda calculus, Λ_{v}.

Definition 2.1 ($\boldsymbol{E}$, $\boldsymbol{L}$, $\boldsymbol{S}$, $\boldsymbol{F}$)**:** The set of expressions, $\boldsymbol{E}$, and the set of λ-abstractions, $\boldsymbol{L}$, the set of value substitutions, $\boldsymbol{S}$, and the set of parameter substitutions (fillings), $\boldsymbol{F}$ are defined as the least sets satisfying the following equations:

$$\boldsymbol{E} = \boldsymbol{X} \cup \boldsymbol{P}^{\boldsymbol{S}} \cup \boldsymbol{A} \cup \boldsymbol{L} \cup \boldsymbol{O}_n(\boldsymbol{E}^n)$$

$\boldsymbol{L} = \lambda \boldsymbol{X}.\boldsymbol{E}$

$\boldsymbol{S} = \text{Fmap}[\boldsymbol{X}, \boldsymbol{V}]$

$\boldsymbol{F} = \text{Fmap}[\boldsymbol{P}, \boldsymbol{E}]$

We let a range over $\boldsymbol{A}$, x, y, z range over $\boldsymbol{X}$, X, Y, Z range over $\boldsymbol{P}$, e range over $\boldsymbol{E}$, φ range over $\boldsymbol{L}$, σ range over $\boldsymbol{S}$, and Σ range over $\boldsymbol{F}$.

$\boldsymbol{P}^{\boldsymbol{S}}$ is the set of parameters, annotated or decorated by value substitutions. Value substitutions, $\boldsymbol{S}$, are finite maps from variables to value expressions. The domain of a substitution is written as $\text{Dom}(\sigma)$, and is defined in the usual way. Parameter substitutions are finite maps from parameters to expressions. We write $\{x_i \mapsto v_i \mid i < n\}$ for the value substitution, σ, with domain $\{x_i \mid i < n\}$ such that $\sigma(x_i) = v_i$ for $i < n$. Similarly we write $\{X_i \mapsto e_i \mid i < n\}$ for the parameter substitution, Σ, with domain $\{X_i \mid i < n\}$ such that $\Sigma(X_i) = e_i$ for $i < n$. Note that a parameter decorated by a value substitution is an expression. This allows us to compute parametrically with partially specified expressions, and thus our expressions generalize the usual notion of context. In this more general setting we must be somewhat more careful to define certain basic notions.

Definition 2.2 ($\text{FV}(e)$, $\text{P}(e)$)**:** The free variables, $\text{FV}(e)$, and the parameters, $\text{P}(e)$, (which are always free) of an expression e are defined inductively. The novel clauses are

$$\text{FV}(X^\sigma) = \bigcup_{x \in \text{Dom}(\sigma)} \text{FV}(\sigma(x)) \qquad \text{P}(X^\sigma) = \bigcup_{x \in \text{Dom}(\sigma)} \text{P}(\sigma(x)) \cup \{X\}$$

We extend these to substitutions is the obvious fashion:

$$\text{FV}(\Delta) = \bigcup_{\delta \in \text{Dom}(\Delta)} \text{FV}(\Delta(\delta)) \qquad \text{P}(\Delta) = \bigcup_{\delta \in \text{Dom}(\Delta)} \text{P}(\Delta(\delta)) \qquad \text{for } \Delta \in \boldsymbol{S} \cup \boldsymbol{F}.$$

Definition 2.3 ($\boldsymbol{E_0}$, term, closed)**:** We adopt the convention that an expression with no parameters is called a *term*. The set of all terms is denoted by $\boldsymbol{E_0}$. Furthermore a *term* with no free variables is *closed*. Thus being closed implies having no parameters.

Definition 2.4 (e^σ, e^Σ)**:** e^σ is the result of simultaneous substitution of free occurrences of $x \in \text{Dom}(\sigma)$ in e by $\sigma(x)$, taking care not to trap variables. Taking care not to trap variables amounts to defining simultaneous substitution into a lambda expression by the following scheme

$$(\lambda z.e)^\sigma = \lambda v.((e^{\{z \mapsto v\}})^\sigma) \qquad \text{for } v \text{ fresh, i.e., } v \notin \text{FV}(e) \cup \text{FV}(\sigma).$$

In the case of decorated parameters we define simultaneous substitution as follows, $(X^{\sigma_0})^\sigma = X^{(\sigma_0^\sigma)}$, where $\sigma_0^\sigma = \{x \mapsto \sigma_0(x)^\sigma \mid x \in \text{Dom}(\sigma_0)\}$.

e^Σ is the result of simultaneous substitution of decorated occurrences of $X \in \text{Dom}(\Sigma)$ in e by $\Sigma(X)$ instantiated by the (suitably substituted) decoration, again taking care not to trap variables (other than those in the range of the decoration). For decorated parameters it is defined as follows, $(X^\sigma)^\Sigma = \Sigma(X)^{\sigma^\Sigma}$ if

$X \in \mathrm{Dom}(\Sigma)$, and $X^{\sigma^{\Sigma}}$ otherwise, where σ^{Σ} is defined point-wise: $\sigma^{\Sigma} = \{x \mapsto \sigma(x)^{\Sigma} \mid x \in \mathrm{Dom}(\sigma)\}$. In the case of λ-abstractions, we define parameter substitution exactly as we would value substitution:

$$(\lambda x.e)^{\Sigma} = \lambda v.((e^{\{x \mapsto v\}})^{\Sigma}) \qquad \text{for } v \text{ fresh, i.e. } v \notin \mathrm{FV}(e) \cup \mathrm{FV}(\Sigma).$$

Example 2.5: Suppose that $\vartheta \in \boldsymbol{O}_2$. Let e be the expression $\lambda z.(X^{\{w \mapsto \vartheta(z,w)\}})$. Then (assuming distinct variable names name distinct variables):

(i) $e^{\{X \mapsto z\}}$ is $\lambda v.z$ since trapping is made explicit by the annotating substitution, not the surrounding context.
(ii) $e^{\{X \mapsto w\}}$ is $\lambda z.\vartheta(z, w)$ since w is trapped by the substitution annotating X.
(iii) $e^{\{X \mapsto y\}}$ is $\lambda z.y$ since y is neither trapped, nor forces us to α-convert z.
(iv) $e^{\{w \mapsto y\}}$ is $\lambda z.(X^{\{w \mapsto \vartheta(z,y)\}})$, since the only free occurrence of w is in the range of the substitution annotating X.
(v) $e^{\{w \mapsto z\}}$ is $\lambda v.(X^{\{w \mapsto \vartheta(v,z)\}})$ since substitution is defined to avoid trapping.
(vi) $e^{\{w \mapsto w\}}$ is $\lambda z.(X^{\{w \mapsto \vartheta(z,w)\}})$ since no trapping takes place.

The notion of being equivalent modulo the renaming of bound variables easily extends to this more general setting [27] by simply clarifying what can and cannot be bound: parameters are *never* bound; variables in the domain of an annotating substitution are *never* bound; variables in the range of an annotating substitution *may* be bound. Using this one can generate the α-equivalence relation, $\overset{\alpha}{\equiv}$, by a set of rules, the new one being:

$$\frac{\sigma_0(x) \overset{\alpha}{\equiv} \sigma_1(x) \qquad \text{for every } x \in \mathrm{Dom}(\sigma_i)}{X^{\sigma_0} \overset{\alpha}{\equiv} X^{\sigma_1}} \qquad \text{provided } \mathrm{Dom}(\sigma_0) = \mathrm{Dom}(\sigma_1)$$

One last piece of notation concerning annotated parameters appearing in expressions. The set of *trapped* variables, $\mathrm{Traps}(e)$, is defined to be the smallest set of variables that contains the domains of any substitution that annotates an occurrence of a parameter in e. On the other hand the *domain* of e, $\mathrm{Dom}(e)$, is defined to be the largest set of variables contained in the domain of every substitution that annotates an occurrence of a parameter in e.

Definition 2.6 ($\mathrm{Traps}(e)$, $\mathrm{Dom}(e)$)**:** These amount to a simple inductive definitions, the interesting clauses being:

$$\mathrm{Traps}(X^{\sigma}) = \bigcup_{x \in \mathrm{Dom}(\sigma)} \mathrm{Traps}(\sigma(x)) \cup \mathrm{Dom}(\sigma)$$

$$\mathrm{Dom}(X^{\sigma}) = \bigcap_{x \in \{y \in \mathrm{Dom}(\sigma) \,\mid\, \mathrm{P}(\sigma(y)) \neq \emptyset\}} \mathrm{Dom}(\sigma(x)) \cap \mathrm{Dom}(\sigma)$$

The domain of an expression is useful in expressing certain necessary closure requirements, as suggested by the following lemma.

Lemma 2.7 ($\mathrm{Dom}(e)$)**:** Suppose $\mathrm{FV}(e) = \emptyset$, and $\mathrm{FV}(\Sigma) \subseteq \mathrm{Dom}(e)$. Then $\mathrm{FV}(e^{\Sigma}) = \emptyset$

Definition 2.8 (Value Expressions ($\boldsymbol{V}$)): The set of value expressions, $\boldsymbol{V}$, contains all variables, atoms, and lambdas. It may in addition contain expressions of the form $\vartheta(v^n)$. $\boldsymbol{V}$ must also satisfy:

(triv) $\quad \boldsymbol{V}$ is closed under $\stackrel{\alpha}{\equiv}$

(vsub) $\quad v \in \boldsymbol{V} \Rightarrow v^\sigma \in \boldsymbol{V}$

(inst) $\quad v \in \boldsymbol{V} \Rightarrow v^\Sigma \in \boldsymbol{V}$

(dich) $\quad e^\Sigma \in \boldsymbol{V} \Rightarrow (e \in \boldsymbol{V}) \vee (e = X^\sigma \wedge \Sigma(X) \in \boldsymbol{V})$

We let v range over $\boldsymbol{V}$.

Operators, ϑ, that produce value expressions, are called constructors. In the languages considered here the binary pairing operation, `pr`, will serve as the prototypical constructor. The following lemma simply points out a simple consequence of the closure conditions on values.

Lemma 2.9 (inv): $e^\sigma \in \boldsymbol{V} \Rightarrow e \in \boldsymbol{V}$

Proof: Pick e_0 such that $e_0^\sigma \in \boldsymbol{V}$, and let X be a fresh parameter. Put $e = X^\sigma$, $\Sigma = \{X \mapsto e_0\}$. Then

$$e^\Sigma = (X^\sigma)^\Sigma = \Sigma(X)^{(\sigma^\Sigma)} = \Sigma(X)^\sigma = e_0^\sigma \in \boldsymbol{V}$$

since $X^\sigma \notin \boldsymbol{V}$ we can use (**dich**) to conclude that $e_0 \in \boldsymbol{V}$. □

2.2. Operational semantics. In a Landinesque language computation state is represented as a class of expressions. Each particular language will possess it's own class of state expressions, reflecting the nature of the primitive operations that it is based on. In what follows we fix a distinguished parameter $\circ$ to designate the position at which effects are to be observed in a context representing a computation state. We call this the *state* parameter.

Definition 2.10 (State expressions ($\boldsymbol{Z}$)): For a particular Landinesque language $\boldsymbol{Z}$ is assumed to be a subset of $\boldsymbol{E}$. We call $\boldsymbol{Z}$ the set of state expressions. $\boldsymbol{Z}$ is assumed to satisfy the following uniformity conditions:

(triv) $\quad \boldsymbol{Z}$ is closed under $\stackrel{\alpha}{\equiv}$

(par) $\quad \zeta \in \boldsymbol{Z} \Rightarrow \circ \in \mathrm{P}(\zeta)$

(vsub) $\quad \zeta \in \boldsymbol{Z} \Rightarrow \zeta^\sigma \in \boldsymbol{Z} \qquad$ assuming $\circ \notin \mathrm{P}(\sigma)$

(inst) $\quad \zeta \in \boldsymbol{Z} \Rightarrow \zeta^\Sigma \in \boldsymbol{Z} \qquad$ assuming $\circ \notin (\mathrm{Dom}(\Sigma) \cup \mathrm{P}(\Sigma))$

ζ ranges over $\boldsymbol{Z}$ and $\circ \in \boldsymbol{Z}$ is the empty state expression.

Definition 2.11 (Values in a State ($\boldsymbol{V}_\zeta$)): The set of *values defined in a state* ζ is $\boldsymbol{V}_\zeta$:

$$\boldsymbol{V}_\zeta = \{v \in \boldsymbol{V} \mid \mathrm{FV}(v) \subseteq \mathrm{Dom}(\zeta)\}$$

Definition 2.12 (Computation States ($\boldsymbol{CS}$)): $\boldsymbol{CS} \triangleq \boldsymbol{Z} : \boldsymbol{E}$ is the set of computation states. We let S range over $\boldsymbol{CS}$ and let $\zeta : e$ be the state with state context ζ and expression e. The computation state $\zeta : e$ is said to be a *value state* iff $e \in \boldsymbol{V}$. Given a state $\zeta : e$, we associate a corresponding expression by filling the state parameter in the context with the expression, i.e. $\zeta^{\{\circ \mapsto e\}}$. A state is *closed* just if its corresponding expression is closed, in other words if $\zeta^{\{\circ \mapsto e\}}$ has no free parameters or variables. Application of value and parameter substitutions to states is defined by in the obvious way: $(\zeta : e)^{\sigma} = \zeta^{\sigma} : e^{\sigma}$ and $(\zeta : e)^{\Sigma} = \zeta^{\Sigma} : e^{\Sigma}$. Note that by (**vsub**) (**inst**) these are only meaningful if $\circ \notin \mathrm{Dom}(\sigma)$ and $\circ \notin (\mathrm{Dom}(\Sigma) \cup \mathrm{P}(\Sigma))$.

Definition 2.13 (Reduction ($\longrightarrow$, $\longrightarrow\!\!\!\!\!\rightarrowtail$) and Definedness ($\downarrow$)): Given a reduction relation for a Landinesque language: $\zeta : e \longrightarrow \zeta' : e'$, the following definitions are standard. The *transitive* closure of $\longrightarrow$ is $\longrightarrow\!\!\!\!\!\rightarrowtail$

Definedness:

$$(\zeta : e)\downarrow \;\Leftrightarrow\; \zeta : e \longrightarrow\!\!\!\!\!\rightarrowtail \zeta' : v$$

Approximation:

$$(\zeta_0 : e_0) \preceq (\zeta_1 : e_1) \;\Leftrightarrow\; (\zeta_0 : e_0)\downarrow \;\Rightarrow\; (\zeta_1 : e_1)\downarrow$$

Equidefined:

$$(\zeta_0 : e_0) \updownarrow (\zeta_1 : e_1) \;\Leftrightarrow\; ((\zeta_0 : e_0)\downarrow \;\Leftrightarrow\; (\zeta_1 : e_1)\downarrow)$$

Equal:

$$(\zeta_0 : e_0) = (\zeta_1 : e_1) \;\Leftrightarrow\; (\exists v \in \boldsymbol{V}, \zeta \in \boldsymbol{Z})(\bigwedge_{j<2}(\zeta_i : e_i) \longrightarrow\!\!\!\!\!\rightarrowtail \zeta : v)$$

Equivalued:

$$(\zeta_0 : e_0) \sim (\zeta_1 : e_1) \;\Leftrightarrow\; (\zeta_0 : e_0) \updownarrow (\zeta_1 : e_1) \wedge (\zeta_0 : e_0)\downarrow \;\Rightarrow\; (\zeta_0 : e_0) = (\zeta_1 : e_1)$$

Length:
$|\zeta : e|$ is the least $n \in \boldsymbol{N}$ such that $\zeta : e$ reduces to a value state in n steps, if $\zeta : e\downarrow$.

To define reduction rules for general Landinesque languages, and formulate the central properties of reduction and equivalence, we introduce the notions of redex and reduction context. Since evaluation is call-by-value, a redex is simply an non-constructor operator applied to the appropriate number of value expressions. Redexes and value expressions must be disjoint, thus we must account for the fact that some expressions of the form $\vartheta(v_1, \ldots, v_n)$ may be value expressions.

Definition 2.14 (Redexes ($\boldsymbol{E}_{\mathrm{r}}$)): The set of redexes, $\boldsymbol{E}_{\mathrm{r}}$, is defined by:

$$\boldsymbol{E}_{\mathrm{r}} = \boldsymbol{O}_n(\boldsymbol{V}^n) - \boldsymbol{V}$$

Note that redexes in our framework may or may not reduce. The point is that they are simply expressions of a particular shape, in other words: candidates for reduction. The set of redexes is closed under the following uniformity conditions.

Lemma 2.15 (Redex Uniformity)**:** Redexes satisfy the following uniformity conditions:

(triv) $\boldsymbol{E}_{\mathrm{r}}$ is closed under $\stackrel{\alpha}{\equiv}$

(vsub) $r \in \boldsymbol{E}_{\mathrm{r}} \Rightarrow r^{\sigma} \in \boldsymbol{E}_{\mathrm{r}}$

(inst) $r \in \boldsymbol{E}_{\mathrm{r}} \Rightarrow r^{\Sigma} \in \boldsymbol{E}_{\mathrm{r}}$

We use the distinguished parameter • to denote the *evaluation* parameter (or hole), and we define the notion of a reduction context, R, accordingly. Reduction contexts (also called evaluation contexts in the literature) identify the subexpression of an expression in which reduction to a value must occur next. They themselves represent the remainder of the computation, i.e the *continuation*. In our approach they correspond to the left-first, call-by-value reduction strategy of [35] and were first introduced by [13].

Definition 2.16 (Reduction Contexts ($\boldsymbol{R}$))**:** The set of reduction contexts, $\boldsymbol{R}$, is the subset of $\boldsymbol{E}$ defined by

$$\boldsymbol{R} = \{\bullet\} \cup \boldsymbol{O}_{m+n+1}(\{v \in \boldsymbol{V} \mid \bullet \notin \mathrm{P}(v)\}^m, \boldsymbol{R}, \{e \in \boldsymbol{E} \mid \bullet \notin \mathrm{P}(e)\}^n)$$

We let R range over $\boldsymbol{R}$. We adopt the convention of writing $R[e]$ instead of $R^{\{\bullet \mapsto e\}}$.

Observe that both the definition of redex, and the definition of reduction contexts depend on the particular choice of values, and thus vary from one Landinesque language to another. Also note that $\boldsymbol{R}$ will satisfy a similar set of uniformity conditions as those satisfied by states (Definition 2.10):

Lemma 2.17 (Rcx Uniformity)**:** Reduction contexts satisfy the following uniformity conditions:

(triv) $\boldsymbol{R}$ is closed under $\stackrel{\alpha}{\equiv}$

(par) $R \in \boldsymbol{R} \Rightarrow \bullet \in \mathrm{P}(R)$

(vsub) $R \in \boldsymbol{R} \Rightarrow R^{\sigma} \in \boldsymbol{R}$ assuming $\bullet \notin \mathrm{P}(\sigma)$

(inst) $R \in \boldsymbol{R} \Rightarrow R^{\Sigma} \in \boldsymbol{R}$ assuming $\bullet \notin (\mathrm{Dom}(\Sigma) \cup \mathrm{P}(\Sigma))$

It is easy to check that an expression without parameters is either a value expression or decomposes uniquely into a redex placed in a reduction context (a proof can be found in [29]). This generalizes to the present situation in the following fashion.

Lemma 2.18 (Decomposition)**:** For any Landinesque language, if $e \in \boldsymbol{E}$ then either $e \in \boldsymbol{V}$ or e can be written uniquely as either

(i) $R[r]$ where R is a reduction context and $r \in \boldsymbol{E}_{\mathrm{r}}$, or else
(ii) $R[X^{\sigma}]$ where R is a reduction context, and $X^{\sigma} \in \boldsymbol{P}^{\boldsymbol{S}}$ is a decorated parameter.

In the latter case we say that the expression is *touching* the parameter, while in the former we say that the expression *may be reducible*. The requirement that the evaluation parameter does not occur in either the leading value expressions, or the trailing expressions is necessary for the uniquenss aspect of this lemma. A simple counterexample, due to Jonathan Ford, is the following:

$$R_0 = \vartheta(\bullet, \bullet) \quad R_1 = \vartheta(\bullet, r) \quad R_0\,[r] = \vartheta(r, r) = R_1\,[r].$$

For the languages considered here, the single step reduction relation is defined by giving a reduction rule for each operation. These rules are of the form:

$$\zeta_0 : R[\vartheta(v_1, \ldots, v_n)] \longrightarrow \zeta_1 : e.$$

In particular the rule for the app operation is the standard reduction rule for the call-by-value lambda calculus.

Definition 2.19 (Beta value reduction ($\beta_{\rm v}$)): The reduction rule for app is the beta value rule:

$$(\beta_{\rm v})\ \zeta : R[\mathsf{app}(\lambda x.e, v)] \longrightarrow \zeta : R[e^{\{x \mapsto v\}}].$$

Notice that this rule is schematic in the actual states, reduction contexts, and values of a particular Landinesque language.

2.3. Uniform semantics. We now specify what we mean by a λ-language having *uniform semantics*. The key requirement is that reduction steps that do not touch a parameter are uniformly independent of what the parameter might stand for. In addition, we require that: single step reduction is essentially deterministic; reduction is preserved by value substitution; a state, and its associated expression started in the empty state context, are equi-defined; and if one state reduces to another then the two states are equi-defined and the reduct has shorter computation length, if defined.

Definition 2.20 (Uniformity (U)): A λ-language is said to have *uniform semantics* if it satisfies the following:

(**i**) Functional modulo $\stackrel{\alpha}{\equiv}$ and implicit bindings:

$$(\bigwedge_{j<2}(\zeta_i : e_i) \stackrel{\alpha}{\equiv} (\zeta_i' : e_i')) \Rightarrow ((\zeta_0 : e_0 \longrightarrow \zeta_1 : e_1) \Rightarrow (\zeta_0' : e_0' \longrightarrow \zeta_1' : e_1'))$$

$$(\bigwedge_{j<2} \zeta : e \longrightarrow \zeta_i : e_i) \Rightarrow \zeta_0^{\{\circ \mapsto e_0\}} \stackrel{\alpha}{\equiv} \zeta_1^{\{\circ \mapsto e_1\}}.$$

(**ii**) Uniform in value substitutions:

$$\zeta : e \longrightarrow \zeta' : e' \Rightarrow (\zeta : e)^{\sigma} \longrightarrow (\zeta' : e')^{\sigma}$$

provided $\mathrm{Dom}(\sigma) \cap (\mathrm{Traps}(\zeta) \cup \mathrm{Traps}(\zeta')) = \emptyset$ and $\circ \notin \mathrm{P}(\sigma)$.

(**iii**) State evaluation:

$$\zeta : e \updownarrow \circ : \zeta^{\{\circ \mapsto e\}}$$

(**iv**) Well-founded:

$$(\zeta : e \longrightarrow \zeta' : e' \wedge \zeta : e\downarrow) \Rightarrow (\zeta' : e' \downarrow \wedge |\zeta' : e'| < |\zeta' : e'|)$$

(**v**) Parametric:

$\zeta_0 : e_0 \longrightarrow \zeta_1 : e_1 \Rightarrow (\zeta_0 : e_0)^\Sigma \longrightarrow (\zeta_1 : e_1)^\Sigma$

for any $\Sigma \in \boldsymbol{F}$ with $\circ \notin (\mathrm{Dom}(\Sigma) \cup \mathrm{P}(\Sigma))$.

(**vi**) Dichotomy (Either computation touches a parameter or is parametric): Assuming $\circ \notin \mathrm{Dom}(\Sigma)$, if $(\zeta : e)^\Sigma \longrightarrow \zeta' : e'$ then either

$\zeta : e$ touches a parameter in the domain of Σ, or

$\zeta : e \longrightarrow \zeta_1 : e_1$, for some $\zeta_1 : e_1$ such that $\zeta' : e' \stackrel{\alpha}{\equiv} (\zeta_1 : e_1)^\Sigma$.

(**vii**) Closure:

(i) $\quad (\zeta_0 : e_0 \longrightarrow \zeta_1 : e_1) \Rightarrow \mathrm{FV}(\zeta_1^{\{\circ \mapsto e_1\}}) \subseteq \mathrm{FV}(\zeta_0^{\{\circ \mapsto e_0\}})$

(ii) $\quad (\zeta_0 : e_0 \longrightarrow \zeta_1 : e_1) \Rightarrow \mathrm{P}(\zeta_1 : e_1) \subseteq \mathrm{P}(\zeta_0 : e_0) \wedge (\circ \notin \mathrm{P}(e_0) \Rightarrow \circ \notin \mathrm{P}(e_1))$

In the languages we consider (**U**) holds for the following reasons. (**U.i**) holds because the only non-determinism in a reduction step is the choice of names used in the state context. (**U.ii**) holds because reductions that do not depend on the values of free variables, are parametric in the values that those variables take. (**U.iii**) holds because reduction of $\circ : \zeta^{\{\circ \mapsto e\}}$ essentially recreates the state context ζ. (**U.iv**) follows since if a state is defined, then any reduction makes progress. Clearly if the reduct state is defined, then the original state is defined. (**U.v**) and (**U.vi**) formalizes the uniformity requirement for reduction steps. These are satisfied by reduction rules that treat the reduction context as an abstract entity, and that depend on the kind of construction of a redex argument, but not on any information about subparts. This is easily expressed using the parameters. Finally, (**U.vii**) holds because computation neither introduces new parameters, nor new free variables.

2.4. Approximation and equivalence. Now we define operational approximation and equivalence on terms and lay the ground work for studying properties of these relations. In what follows we fix a particular distinguished parameter, X, distinct from $\circ$ and $\bullet$. We let C range over expressions with X as the only free parameter, with the added condition that the substitutions annotating X be restricted to renamings (finite maps from variables to variables). Such expressions play the role of traditional lambda-calculus contexts, and we extend our convention, stated in definition 2.16, of sometimes writing $C[e]$ instead of $C^{\{X \mapsto e\}}$. Note however that for example the traditional context $\lambda z.\mathsf{app}(y, \lambda x.\mathsf{app}([\,], z))$ does not correspond to $\lambda z.\mathsf{app}(y, \lambda x.\mathsf{app}(X, z))$ but rather to one where the trappings have been made explicit at the occurrence of the X parameter: $\lambda z.\mathsf{app}(y, \lambda x.\mathsf{app}(X^{\{x \mapsto x, z \mapsto z\}}, z))$.

For each traditional context C there is a corresponding expression, $\hat{C}$ such that

$$C[e] = \hat{C}^{\{X \mapsto e\}}.$$

$\hat{C}$ is obtained by replacing each occurrence of $[\,]$ in C by the parameter X, and decorating each such occurrence of X with a binding substitution $\{x_i \mapsto x_i \mid x_i \in T\}$ where T is the set of lambda variables having the hole occurrence in their scope. For example, the expression corresponding to $\lambda x.[]$ is $\lambda x.X^{\{x\mapsto x\}}$. This guarantees that the following definition corresponds to the traditional notion of operational equivalence.

Definition 2.21 (Approximation $e_0 \sqsubseteq e_1$, Equivalence $e_0 \cong e_1$)**:** For terms e_0, e_1 define

$$e_0 \sqsubseteq e_1 \;\Leftrightarrow\; (\forall C \mid C[e_0], C[e_1]\,\text{closed})(\circ : C[e_0] \;\preceq\; \circ : C[e_1])$$

$$e_0 \cong e_1 \;\Leftrightarrow\; e_0 \sqsubseteq e_1 \wedge e_1 \sqsubseteq e_0$$

Note that we are restricting our attention to terms, rather than arbitrary expressions. It is easy to see that operational approximation is a congruence on terms: if $e_0 \sqsubseteq e_1$, then $C[e_0] \sqsubseteq C[e_1]$. Similarly for operational equivalence.
We may now state the main result concerning Landinesque languages with uniform semantics, the CIU theorem.

Theorem 2.22 (CIU)**:** For a λ-language with uniform semantics:

$$e_0 \sqsubseteq e_1 \;\Leftrightarrow\; (\forall \zeta, R, \sigma \mid \bigwedge_{j<2} \zeta : R[e_j^\sigma]\ \text{closed})(\zeta : R[e_0^\sigma] \;\preceq\; \zeta : R[e_1^\sigma])$$

There are several proofs of this result in the literature. The theorem first appeared in [29] and the proof (sketch) presented there used techniques similar to those developed here. A somewhat more detailed and general version of this same technique appeared in [42]. Recently this same proof has been verified using the PVS theorem prover [15]. A second, distinct, proof was presented in [22] that simply shows that the CIU relation is a congruence.

Corollary 2.23 (Substitutivity)**:** In a λ-language with uniform semantics operational approximation is preserved by substitution: if $e_0 \sqsubseteq e_1$, then $e_0^\sigma \sqsubseteq e_1^\sigma$. Similarly for operational equivalence.

Example 2.24 (Call-by-value λ-calculus (Λ_v))**:** As a first example we complete the definition of the language Λ_v. Λ_v has no atoms ($\boldsymbol{A} = \{\ \}$) and a single operation, $\boldsymbol{O} = \{\mathtt{app}\}$. There is one state, the empty state, represented by the state parameter $\circ$. In Λ_v (as in all Landinesque languages) the reduction rule for app is the β_v rule (Definition 2.19).

It is easy to see that Λ_v satisfies the uniformity conditions (Definition 2.20). The only hard part is to show that value substitution and β_v reduction commute, which is a standard result.

2.5. Context independent reduction. To state the theorems underlying computational equivalences in this general setting we introduce the notion of a context independent (CI) redex.

Definition 2.25 (Context independent reduction ($\xrightarrow{\text{ci}}\!\!\succ$))**:** A redex $\vartheta(v_1, \ldots, v_n)$ is CI if the interpretation is independent of the context: either there is no reduction possible in any computation context (state and reduction context), or the redex is replaced by the same reduct expression in any computation context. That is, exactly one of the following holds:

1. For any ζ, R there is no $\zeta' : e'$ such that $\zeta : R[\vartheta(v_1, \ldots, v_n)] \longrightarrow \zeta' : e'$.
2. There is some e such that $\zeta : R[\vartheta(v_1, \ldots, v_n)] \longrightarrow \zeta : R[e]$ for any ζ, R.

A CI redex neither examines nor modifies its context (state or reduction). For example, any redex with operator app is CI.

We write $e \xrightarrow{\text{ci}}\!\!\!\!\succ e'$ if $\zeta : R[e] \longrightarrow\!\!\!\!\succ \zeta : R[e']$ by a sequence of CI steps for any ζ and R.

The intuition that two expressions are equivalent if they have a common reduct, is justified by Theorem 2.26 below. Similarly reasoning that two expressions are equivalent if they result from placing a third expression in computationally equivalent reduction contexts is justified by Theorem 2.27.

Theorem 2.26 (Equi-reduct)**:** In a λ-language with uniform semantics, if there is some e such that $e_j \xrightarrow{\text{ci}}\!\!\!\!\succ e$ for $j < 2$ then $e_0 \cong e_1$.

Proof: Assume $e_j \xrightarrow{\text{ci}}\!\!\!\!\succ e$ for $j < 2$. By CIU (and symmetry), to show that $e_0 \cong e_1$ we need only show $\zeta : R[e_0^\sigma] \preceq \zeta : R[e_1^\sigma]$ for all closing ζ, R, and σ. Choose any such ζ, R, σ and assume $\zeta : R[e_0^\sigma] \downarrow$, furthermore we may choose ζ and σ such that $\mathrm{Dom}(\sigma) \cap \mathrm{Traps}(\zeta) = \emptyset$. By assumption pick e such that

$$\zeta : R[e_j] \longrightarrow\!\!\!\!\succ \zeta : R[e] \qquad \text{for any } \zeta \text{ and } R.$$

Thus by (**U.2**) we have

$$(\zeta : R[e_j])^\sigma \longrightarrow\!\!\!\!\succ (\zeta : R[e])^\sigma \qquad \text{for the particular } \zeta \text{ and } R \text{ chosen above.}$$

Consequently

$$\zeta : R[e_j^\sigma] \longrightarrow\!\!\!\!\succ \zeta : R[e^\sigma] \qquad \text{since } \zeta = \zeta^\sigma \text{ and } R = R^\sigma.$$

Thus, by (**U.4**), $\zeta : R[e^\sigma] \downarrow$, and hence $\zeta : R[e_1^\sigma] \downarrow$. □

Theorem 2.27 (Equi-rcx)**:** In a λ-language with uniform semantics, if for z fresh there are $e_0 \cong e_1$ such that $R_j[z] \xrightarrow{\text{ci}}\!\!\!\!\succ e_j$ for $j < 2$, then $R_0[e] \cong R_1[e]$ for any e.

To provide simple instructive examples of the use of these two principles we will introduce the ubiquitous `let` construct that will feature in the subsequent development, and establish its more well known properties.

Definition 2.28 (`let`)**:** The `let` form is very common programming language construct abbreviating λ-application:

$$\mathtt{let}\{x := v\}e \stackrel{\triangle}{=} \mathsf{app}(\lambda x.e, v).$$

Example 2.29 (`let`)**:** As mentioned above β_v reduction of $\mathsf{app}(\lambda x.e, v)$ is CI, thus an immediate consequence of Theorem 2.26 is that

(let.i) $\quad \mathtt{let}\{x := v\}e \cong e^{\{x \mapsto v\}}$.

Assume that $x \notin \mathrm{FV}(R_0)$ and $R_1 = \mathtt{let}\{x := \bullet\}R_0[x]$. Then for z fresh:

$$R_1[z] \xrightarrow{\text{ci}}\!\!\!\!\succ R_0[x]^{\{x \mapsto z\}} = R_0[z]$$

So by Theorem 2.27:

(let.ii) $R_0\,[e] \cong \texttt{let}\,\{x := e\}R_0\,[x]$

Another example is obtained by taking $R_0 = \texttt{let}\,\{x := \bullet\}\,\texttt{let}\,\{y := e_1\}e$ and $R_1 = \texttt{let}\,\{y := \texttt{let}\,\{x := \bullet\}e_1\}e$ with the added proviso that x and y are distinct, and $x \notin \mathrm{FV}(e)$. Then for z fresh:

$$R_0\,[z] = \texttt{let}\,\{x := z\}\,\texttt{let}\,\{y := e_1\}e$$

$$\xrightarrow{\text{ci}}\!\!\succ (\texttt{let}\,\{y := e_1\}e)^{\{x \mapsto z\}} = \texttt{let}\,\{y := e_1^{\{x \mapsto z\}}\}e$$

$$R_1\,[z] = \texttt{let}\,\{y := \texttt{let}\,\{x := z\}e_1\}e$$

$$\xrightarrow{\text{ci}}\!\!\succ \texttt{let}\,\{y := e_1^{\{x \mapsto z\}}\}e$$

Thus again by Theorem 2.27:

(let.iii) $\texttt{let}\,\{x := e_0\}\,\texttt{let}\,\{y := e_1\}e \simeq \texttt{let}\,\{y := \texttt{let}\,\{x := e_0\}e_1\}e$

These three principles hold in any uniform Landinesque language. They are fundamental in that they are equivalent to the let-rules of the computational Λ_c calculus [33].

An even stronger, but nevertheless useful notion, is being *ineffectual* which captures the idea of *single CI stepping to a value*. In that every reduction rule for the operation leaves its context (state and reduction) unchanged and produces a value.

Definition 2.30 (Ineffectual): An operation $\vartheta \in \boldsymbol{O}$ is said to be *ineffectual*, iff every reduction rule for the operation leaves its context (state and reduction) unchanged and produces a value, i.e is of the form:

$$\zeta : R[\vartheta(v_1, \ldots, v_n)] \longrightarrow \zeta : R[v].$$

An operation that is not ineffectual is said to have *effects*.

Of course the interesting operations are not the ineffectual ones, we simply use ineffectualness to separate out the uninteresting operations, since they have many more properties than the ones with effects. The only operation common to all Landinesque languages is app and it has effects.

§3. The syntax and semantics of formulas.

3.1. Syntax. The first order fragment of our logic is a minor generalization of classical first order logic. The atomic formulas, $\boldsymbol{W}_0$, assert the definedness, equivaluedness and operational equivalence of terms, $\boldsymbol{E}_0$, i.e. parameter-free expressions.

In addition to the usual first-order formula constructions, we add a let-assertion:

$$\texttt{let}\,\{x := e\}\Phi$$

where Φ is a formula, x a variable, and e a term. The formula $\texttt{let}\,\{x := e\}\Phi$ asserts that if execution of e terminates with value v, then Φ holds in the resulting state, with x bound to the value v. This provides a means for expressing the effects of evaluating an expression. An instructive example is the formula

(newness) $(\forall \bar{z})(\forall x)\,\mathtt{let}\,\{y := \vartheta(\bar{z})\}\neg(x \sim y)$

expresses that ϑ creates *new* values, distinct from any values existing prior to evaluation, such operations are usually called constructors, or allocators. Note that the `let`-assertion is a binding operator akin to $\forall$.

Definition 3.1 ($\boldsymbol{W}$):

$$\boldsymbol{W}_0 = (\boldsymbol{E_0} \sim \boldsymbol{E_0}) \cup (\boldsymbol{E_0} \cong \boldsymbol{E_0})$$

$$\boldsymbol{W} = \boldsymbol{W}_0 \cup (\neg \boldsymbol{W}) \cup (\boldsymbol{W} \Rightarrow \boldsymbol{W}) \cup (\mathtt{let}\,\{\boldsymbol{X} := \boldsymbol{E_0}\}\boldsymbol{W}) \cup (\forall \boldsymbol{X})(\boldsymbol{W})$$

3.2. Semantics. The meaning of formulas is given by a Tarskian satisfaction relation $\zeta \models \Phi[\sigma]$.

Definition 3.2 ($\zeta \models \Phi[\sigma]$):
Assume ζ, σ, Φ, e_j satisfy $\mathrm{FV}(\Phi^\sigma), \mathrm{FV}(e_j^\sigma) \subseteq \mathrm{Dom}(\zeta)$ and $\mathrm{P}(e_j) = \emptyset$ for $j < 2$, and $\mathrm{FV}(\zeta) = \emptyset$. Then we define the satisfaction relation $\zeta \models \Phi[\sigma]$ by induction on the structure of Φ:

$$
\begin{array}{lll}
\zeta \models (e_0 \sim e_1)[\sigma] & \text{iff} & (\forall R \in \boldsymbol{R} \mid \mathrm{FV}(R) \subseteq \mathrm{Dom}(\zeta)) \\
 & & \qquad (\zeta : R[e_0^\sigma] \sim \zeta : R[e_1^\sigma]) \\
\zeta \models (e_0 \cong e_1)[\sigma] & \text{iff} & (\forall R \in \boldsymbol{R} \mid \mathrm{FV}(R) \subseteq \mathrm{Dom}(\zeta)) \\
 & & \qquad (\zeta\,[R[e_0^\sigma]] \updownarrow \zeta\,[R[e_1^\sigma]]) \\
\zeta \models \neg\Phi[\sigma] & \text{iff} & \text{not } (\zeta \models \Phi[\sigma]) \\
\zeta \models (\Phi_0 \Rightarrow \Phi_1)[\sigma] & \text{iff} & (\zeta \models \Phi_0[\sigma]) \quad \text{implies} \quad (\zeta \models \Phi_1[\sigma]) \\
\zeta \models \mathtt{let}\,\{x := e\}\Phi[\sigma] & \text{iff} & (\zeta : e^\sigma \longrightarrow \zeta' : v) \quad \text{implies} \quad \zeta' \models \Phi[\sigma\{x := v\}]) \\
\zeta \models (\forall x)\Phi[\sigma] & \text{iff} & (\forall v \in \boldsymbol{V}_\zeta)(\zeta \models \Phi[\sigma\{x := v\}])
\end{array}
$$

As is usual in logic we define the subsidiary notions of validity and logical consequence as follows:

$$
\begin{array}{lll}
\models \Phi & \text{iff} & (\forall \zeta, \sigma \mid \mathrm{FV}(\Phi^\sigma) \subseteq \mathrm{Dom}(\zeta))(\zeta \models \Phi[\sigma]) \\
\Phi_0 \models \Phi_1 & \text{iff} & \models \Phi_0 \Rightarrow \Phi_1
\end{array}
$$

Note that conjunction, $\wedge$, and disjunction, $\vee$ and the biconditional, $\Leftrightarrow$, are all definable in the usual manner. Also, termination and non-termination are simple abbreviations. We let $\downarrow e$ (termination) abbreviate $\neg(\mathtt{let}\,\{x := e\}\mathtt{False})$ and $\uparrow e$ (non-termination) abbreviate its negation $\mathtt{let}\,\{x := e\}\mathtt{False}$ where `False` is any unsatisfiable assertion, such as $\neg(x \sim x)$. Termination is one form of definedness. In fact, there is a plethora of notions of definedness that can be expressed. A stronger notion of definedness is that of being equivalent (either via $\sim$ or via $\cong$) to a value, for example an operation that satisfies the newness principle above will never be defined in this stronger sense.

§4. Syntax and semantics of classes. Using methods of [5, 9] and [40], we extend our theory to include a general theory of classifications (classes for short). With the introduction of classes, principles such as structural induction, as well as principles accounting for the effects of an expression can easily be expressed.

4.1. Syntax of classes. We extend the syntax to include class terms. Class terms are either class variables, $\boldsymbol{X}^c$, class constants, $\boldsymbol{A}^c$, or comprehension terms, $\{x \mid \Phi\}$.

Definition 4.1 ($\boldsymbol{K}$)**:** The set $\boldsymbol{K}$ of class terms is defined by

$$\boldsymbol{K} = \boldsymbol{X}^c \cup \boldsymbol{A}^c \cup \{\boldsymbol{X} \mid \boldsymbol{W}\}$$

We extend the atomic formulas to include class membership and the set of formulas $\boldsymbol{W}$ to include quantification over class variables. We should point out that $\boldsymbol{K}$ and $\boldsymbol{W}$ form a mutual recursive definition. The definition of expressions remains unchanged.

Definition 4.2 ($\boldsymbol{W}$)**:**

$$\boldsymbol{W}_0 = (\boldsymbol{E_0} \sim \boldsymbol{E_0}) \cup (\boldsymbol{E_0} \cong \boldsymbol{E_0}) \cup (\boldsymbol{E_0} \in \boldsymbol{K})$$

$$\boldsymbol{W} = \boldsymbol{W}_0 \cup \neg\boldsymbol{W} \cup (\boldsymbol{W} \Rightarrow \boldsymbol{W}) \cup (\texttt{let}\,\{\boldsymbol{X} := \boldsymbol{E_0}\}\boldsymbol{W}) \cup (\forall \boldsymbol{X})(\boldsymbol{W}) \cup (\forall \boldsymbol{X}^c)\boldsymbol{W}$$

We let κ range over $\boldsymbol{X}^c$ and K range over $\boldsymbol{K}$. We will use identifiers beginning with an upper case letter in ***This*** font (for example ***Val***) for class constants.

4.2. Semantics of classes. To give semantics to the extended language, we extend the satisfaction relation as follows. Firstly we let $\boldsymbol{K}_\zeta$, the set of class values over ζ, be the set of subsets of $\boldsymbol{V}_\zeta$ (values over ζ) closed under $\cong$. Another possible choice for the set of class values is the set of definable sets, i.e. the set of class code extensions (cf. [7, 40]).

We extend value substitutions to map class variables to class values. This is used to define $[K]^\sigma_\zeta$, the value of a class term, K, relative to the given state, ζ, and the closing value substitution σ. In principle, the class term evaluation is relative to a valuation for class constants, but since all of our class constants are introduced by definitional extension, this can be ignored.

Definition 4.3 ($[K]^\sigma_\zeta$)**:**

$$[\kappa]^\sigma_\zeta = \sigma(\kappa)$$

$$[\{x \mid \Phi\}]^\sigma_\zeta = \{v \in \boldsymbol{V}_\zeta \mid \zeta \models \Phi[\sigma\{x := v\}]\}$$

We then extend the satisfaction relation to formulas involving class terms and quantifiers.

Definition 4.4 ($\zeta \models \Phi[\sigma]$)**:** The new clauses in the inductive definition of satisfaction are:

$$\zeta \models e \in K[\sigma] \Leftrightarrow (\exists v \in \boldsymbol{V}_{\mathrm{Dom}(\zeta)})(\zeta : e^\sigma \longrightarrow\!\!\succ \zeta : v \wedge v \in [K]^\sigma_\zeta)$$

$$\zeta \models (\forall \kappa)\Phi[\sigma] \Leftrightarrow (\forall K \in \boldsymbol{K}_{\mathrm{Dom}(\zeta)})(\zeta \models \Phi[\sigma\{\kappa := K\}])$$

It is important to note that if $\zeta \models e \in K[\sigma]$, then e evaluates (in the appropriate state) to a value without altering that state.

Example 4.5 (Class definitions)**:** We define (extensional) equality and subset relations on classes in the usual manner.

$$K_0 \subseteq K_1 \stackrel{\triangle}{=} (\forall x)(x \in K_0 \Rightarrow x \in K_1)$$

$$K_0 \equiv K_1 \stackrel{\triangle}{=} K_0 \subseteq K_1 \wedge K_1 \subseteq K_0$$

The class $K_0 \rightarrow K_1$ is the set of lambdas that define total functions that have no effect on the state context.

$$K_0 \rightarrow K_1 \stackrel{\triangle}{=} \{f \mid (\forall x \in K_0)(\exists y \in K_1)\mathsf{app}(f, x) \sim y\}$$

§5. Proof theory. Since contextual assertions are akin to modalities, we give a Hilbert style presentation. In the long run a natural deduction style system in the style of Prawitz [36] may be more desirable. Jacob Frost has developed a natural deduction presentation of this system [17, 18] for the $\lambda_{\mathtt{mk}}$ language [22], a Landinesque language with primitives for manipulating memory cells, and implemented it in the proof assistant *Isabelle*.

Definition 5.1 ($\vdash \Phi$)**:** The consequence relation, $\vdash$, is the smallest relation on $\boldsymbol{W}$ that is closed under the rules given below.

The rules are partitioned into several groups. Each group of rules is given a label, for future reference, and members of the group are numbered. For example (**E.i**) refers to the first rule in the group of equivalence and evaluation rules (the second group below). A rule has a (possibly empty) set of premises and a conclusion. In the case that the set of premises is non-empty the rule is displayed with a horizontal bar separating the premises from the conclusion.

Definition 5.2 ($\simeq$)**:** Most axioms hold true for both equivaluedness, $\sim$, and operational equivalence, $\cong$, If this is the case, then rather than write out the principle twice, we use the symbol $\simeq$ to range over these two equivalence relations.

One important reason for introducing $\sim$ is that important principles fail for $\cong$. In particular (**C.iii**) below fails as indicated in [30].

5.1. Basic equivalence and evaluation rules. The first two sets of rules concerning equivaluedness hold true also of operational equivalence. They are equivalence relations, (**E.i, E.ii, E.iii**). They satisfy a certain restricted form of substitutivity, (**E.iv**).

They are also preserved under simple forms of evaluation, (**let.i, let.ii, let.iii**) that we treated in Example 2.29.

Equivalence and Evaluation Axioms (E).

(i) $\vdash e_0 \simeq e_0$

(ii) $\vdash (e_0 \simeq e_1 \wedge e_1 \simeq e_2) \Rightarrow e_0 \simeq e_2$

(iii) $\vdash e_0 \simeq e_1 \Rightarrow e_1 \simeq e_0$

(iv) $\vdash e_0 \simeq e_1 \Rightarrow \mathtt{let}\,\{x := e_0\}e \simeq \mathtt{let}\,\{x := e_1\}e$

Evaluation Axioms (let).

(i) $\vdash \mathtt{app}(\lambda x.e, v) \simeq e^{\{x:=v\}} \simeq \mathtt{let}\{x := v\}e$

(ii) $\vdash R[e] \simeq \mathtt{let}\{x := e\}R[x]$

(iii) $\vdash \mathtt{let}\,\{x := e_0\}\,\mathtt{let}\,\{y := e_1\}e \simeq \mathtt{let}\,\{y := \mathtt{let}\,\{x := e_0\}e_1\}e$

The principle (**let.ii**) is subject to the side condition that that $x \notin \mathrm{FV}(R)$, while in (**let.iii**) we require $x \notin \mathrm{FV}(e)$. The remaining axioms and rules concerning operational equivalence are: (**≅.i**), equivaluedness implies operational equivalence; and (**≅.ii**), operational equivalence is not the same as equivaluedness on abstractions.

Operational Equivalence Rules (≅).

(i) $\vdash e_0 \sim e_1 \Rightarrow e_0 \cong e_1$

(ii) $\dfrac{\vdash e_0 \cong e_1}{\vdash \lambda x.e_0 \cong \lambda x.e_1}$ **(The ξ Rule)**

5.2. Contextual rules. A contextual assertion is a modality and as such possesses a rule akin to necessitation, (**C.i**). Note that this is a rule of proof and not an implication. The remaining axioms concerning contextual assertions are: (**C.ii**), contextual assertions distribute across the equivalences, (the converse is false); (**C.iii**), a form of contextual assertion introduction involving equivaluedness (the corresponding principle for operational equivalence is false); (**C.iv**), a principle akin to β conversion; and (**C.v**), a principle allowing for the manipulation of contexts.

Contextual Rules (C).

(i) $\dfrac{\vdash \Phi}{\vdash \mathtt{let}\,\{x := e\}\Phi}$ for any x, e, and Φ. **(Context Introduction)**

(ii) $\vdash \mathtt{let}\,\{x := e\}(e_0 \simeq e_1) \Rightarrow \mathtt{let}\,\{x := e\}e_0 \simeq \mathtt{let}\,\{x := e\}e_1$

(iii) $\vdash e_0 \sim e_1 \Rightarrow (\mathtt{let}\,\{x := e_0\}\Phi \Leftrightarrow \mathtt{let}\,\{x := e_1\}\Phi)$

(iv) $\vdash \mathtt{let}\,\{x := v\}\Phi \Leftrightarrow \Phi^{\{x:=v\}}$

(v) $\vdash \mathtt{let}\,\{x := e_0\}\,\mathtt{let}\,\{y := e_1\}\Phi \Leftrightarrow \mathtt{let}\,\{y := \mathtt{let}\,\{x := e_0\}e_1\}\Phi$

The principle (**C.v**) is subject to the condition that $x \notin \mathrm{FV}(\Phi)$.

Example 5.3: Note that using (**C.i**) in the special case that e is a value, allows us to also use (**C.iv**) to eliminate the `let` in preference for the substitution, thus the following is a derivable rule.

$\dfrac{\vdash \Phi}{\vdash \Phi^{\sigma}}$ **(Substitution)**

5.3. Logical rules. The propositional rules are, in addition to the usual Hilbert style presentation of modus ponens, (**P.iii**), and a generating set of tautologies, (**P.i**) a modal axiom corresponding to **K** and its converse, (**P.ii**).

Propositional Rules (P).

(i) $\vdash \Phi$ provided Φ is an instance of a tautology

(ii) $\vdash \mathtt{let}\,\{x := e\}(\Phi_0 \Rightarrow \Phi_1) \Leftrightarrow (\mathtt{let}\,\{x := e\}\Phi_0 \Rightarrow \mathtt{let}\,\{x := e\}\Phi_1)$

(iii) $\dfrac{\vdash \Phi_0 \qquad \vdash \Phi_0 \Rightarrow \Phi_1}{\vdash \Phi_1}$ **(Modus Ponens)**

Similarly the quantifier axioms are all standard [4] except for (**Q.iv**) which asserts that ineffectual operations have no allocation effect, that is they don't change the domain of the state context.

Quantifier Rules (Q).

(i) $\dfrac{\vdash \Phi}{\vdash (\forall x)\Phi}$ **(Generalization $\forall$I)**

(ii) $\vdash (\forall x)(\Phi_0 \Rightarrow \Phi_1) \Rightarrow (\Phi_0 \Rightarrow (\forall x)\Phi_1)$ if $x \notin \mathrm{FV}(\Phi_0)$

(iii) $\vdash (\forall x)\Phi \Rightarrow \Phi$

(iv) $\vdash (\forall x)\,\mathtt{let}\,\{z := \vartheta(\bar{y})\}\Phi \Rightarrow \mathtt{let}\,\{z := \vartheta(\bar{y})\}(\forall x)\Phi$

Where in (**Q.iv**) we require that ϑ be ineffectual (see Definition 2.30), $z \neq x$, and $x \notin \mathrm{FV}(\vartheta(\bar{y}))$.

Example 5.4: Note that by the same sort of reasoning that established the derived principle (**Substitution**) we can also establish

(i) $\vdash (\forall x)\,\mathtt{let}\,\{z := v\}\Phi \Rightarrow \mathtt{let}\,\{z := v\}(\forall x)\Phi$ for $z \neq x$

5.4. Undefinedness axiom. The most basic principle concerning undefinedness is that two undefined terms are both equivalued and operationally indistinguishable, (**U.i**).

Undefinedness Axiom (U).

(i) $\vdash \uparrow e_0 \Rightarrow (\uparrow e_1 \Leftrightarrow e_0 \simeq e_1)$

5.5. Constraint propagation axioms. An important class of axioms are those which allow assertions to be propagated into and out of assertions. In order to be succinct we write $\Phi[\pm\phi]$ to abbreviate the two formulas $\Phi[\phi]$ and $\Phi[\neg\phi]$. The following three principles require that $\vartheta \in \boldsymbol{O}$ is ineffectual, Definition 2.30, and that $x, \bar{z}$ distinct.

Constraint Propagation Axioms (S).

(i) $\vdash \mathtt{let}\,\{x := \vartheta(\bar{z})\}(x \simeq \vartheta(\bar{z}))$

(ii) $\vdash \pm(z_0 \simeq z_1) \Rightarrow \mathtt{let}\,\{x := \vartheta(\bar{y})\}(\pm(z_0 \simeq z_1))$

(iii) $\vdash \vartheta(\bar{y})\downarrow \Rightarrow (\mathtt{let}\,\{x := \vartheta(\bar{y})\}(\pm(z_0 \simeq z_1)) \Rightarrow \pm(z_0 \simeq z_1))$

5.6. Classes. As a consequence of the semantics of classifications the following are valid.

Class Axioms (Cl).

(i) $\vdash e \in K \Rightarrow \downarrow e$

(ii) $\vdash (\forall \kappa)\Phi[\kappa] \Rightarrow \Phi[K]$ where Φ contains no `let`-assertions

(iii) $\vdash (\forall x)(x \in \{x \mid \Phi\} \Leftrightarrow \Phi)$

Example 5.5: As pointed out in the earlier work [22], neither of the following two principles are valid.

(i) $e_0 \cong e_1 \wedge \mathtt{let}\,\{x := e_0\}x \in K \Rightarrow \mathtt{let}\,\{x := e_1\}x \in K$

(ii) $(\forall \kappa)\Phi[\kappa] \Rightarrow \Phi[K]$

For example (i) fails in the case of simple memory operations for class terms can observe inaccessible cells if the expressions differ only in that one generates a cell and forgets it (garbage) and the other does not. (ii) fails because evaluation of class terms (containing `let`-assertions can change the state context in which part of the formula is evaluated.

The axioms and rules above are sound in the sense that a provable formula is valid.

Theorem 5.6 (Soundness): $\vdash \Phi$ implies that $\models \Phi$.

§6. Λ_s – an example. As a particular example of a Landinesque language we define Λ_s, a language extending the call-by-value lambda calculus with the usual functional programming primitives, primitives for the manipulation of memory, and a control primitive for manipulating the current continuation (the reduction context). This combination results in a language similar to "functional" programming languages such as Scheme [38] and ML [32]. The combination of lambda abstraction, basic data structures such as numbers and pairing, memory and control primitives provides a basis for expressing a wide range of programming paradigms/styles quite naturally. Examples of the use of continuations in programming practice can be found in [14, 16].

In Λ_s the set of atoms, $\boldsymbol{A}$, contains two distinct atoms playing the role of booleans, `t` for *true* and `nil` for *false*, as well as atoms playing the role of the natural numbers, which we denote by $0, 1, \ldots$.

Definition 6.1 (Λ_s Operations):

$\boldsymbol{O} = \boldsymbol{O}^{\mathrm{f}} \cup \boldsymbol{O}^{\mathrm{m}} \cup \boldsymbol{O}^{\mathrm{c}}$

$\boldsymbol{O}^{\mathrm{f}} = \{\mathtt{app}, \mathtt{lambda?}, \mathtt{br}, \mathtt{eq?}, \mathtt{atom?}, \mathtt{pr}, \mathtt{fst}, \mathtt{snd}, \mathtt{pr?}, +\mathtt{1}, -\mathtt{1}, \mathtt{nat?}\}$

$\boldsymbol{O}^{\mathrm{m}} = \{\mathtt{mk}, \mathtt{get}, \mathtt{set}, \mathtt{cell?}\}$

$\boldsymbol{O}^{\mathrm{c}} = \{\mathtt{ncc}\}$

6.1. Informal semantics. We give a brief and informal guide to the more novel of the primitive operations.

mk is a memory allocation primitive: the evaluation of $\mathtt{mk}(v)$ results in the allocation of a *new memory cell* and initializes this cell so that it contains the value v. The value returned by this call to mk is the newly allocated cell. mk is total, it has effects.

get is the memory access primitive: the evaluation of $\mathtt{get}(v)$ is defined iff v is a memory cell. If v is a memory cell, then $\mathtt{get}(v)$ returns the value stored in that cell. Note that there is no reason why a cell cannot store itself (or some more elaborate cycle). get is both partial and ineffectual.

set is the memory modification primitive: the evaluation of $\mathtt{set}(v_0, v_1)$ is defined iff v_0 is a memory cell. If v_0 is a memory cell, then $\mathtt{set}(v_0, v_1)$ modifies that cell so that its new contents becomes v_1. The value returned by a call to set is somewhat arbitrary and somewhat irrelevant. We have chosen nil as the return value, thus if v is a cell, then $\mathtt{set}(v, v)$ will return nil, and more importantly modify v so that it contains itself. set is both partial and has effects.

cell? is the recognizer of the set of memory cells. $\mathtt{cell?}(v)$ returns t if v is a memory cell, otherwise it returns nil. It is both total and ineffectual.

br is the strict branching primitive: the evaluation of $\mathtt{br}(v_0, v_1, v_2)$ returns v_2 if v_0 is the atom nil, otherwise it returns v_1. Thus any non-nil value is considered *true*. The usual lazy branching primitive

$$\mathtt{if}(e_0, e_1, e_2) \stackrel{\triangle}{=} \mathtt{app}(\mathtt{br}(e_0, \lambda z.e_1, \lambda z.e_2), \mathtt{nil}$$

for a fresh variable z. br is total and ineffectual.

eq? is the equality primitive (solely on atoms): $\mathtt{eq?}(v_0, v_1)$ returns t if v_0 and v_1 are the same atom, otherwise it returns nil. eq? is total and ineffectual.

ncc is a primitive for capturing the current continuation (i.e the current reduction context). It gets its name from what it does, since it notes the current continuation. $\mathtt{ncc}(v)$ captures the current reduction context as a continuation, removes this reduction context from the current state, and applies v to the captured continuation at the top level. ncc, like app, is both partial and has effects.

Definition 6.2 (More notation conventions): We introduce new constant symbols by definitional extension using the notation

$$\mathtt{c} \stackrel{\triangle}{=} v$$

We also use this notation for introducing function symbols defined recursively. Thus $\mathtt{f} \stackrel{\triangle}{=} \lambda x.e$ stands for the following definition of the constant symbol f

$$\mathtt{f} \stackrel{\triangle}{=} \mathtt{Y}(\lambda \mathtt{f}.\lambda x.e)$$

where Y is some choice of call-by-value Y-combinator, see for example § 6.2.2.

We use a number of abbreviations to (hopefully) make programs more readable.

6.2.2. *Landin's Recursion Operator.* Since the Λ_s-language extends the call-by-value λ-calculus the usual call-by-value fixed point combinator is a term in the language. A somewhat different fixed point combinator, that makes use of the reference primitives, is possible:

$$\mathtt{Y} \triangleq \lambda y.\,\mathtt{let}\{z := \mathtt{mk}(\mathtt{nil})\}$$
$$\mathtt{set}(z, \lambda x.\mathrm{app}(\mathrm{app}(y, \mathtt{get}(z)), x)); \mathtt{get}(z))$$

This version of the fixed-point combinator is similar in spirit to the one suggested by Landin [25] and is operationally equivalent to the usual call-by-value Y combinator [29]. When applied to a functional F of the form $\lambda f.\lambda x.e$, Y creates a private local cell, z, with contents $G = \lambda x.\mathrm{app}(\mathrm{app}(F, \mathtt{get}(z)), x)$, and returns G. By privacy of z, G is equivalent to $F(G)$ (cf. [29]). Note that this example is typable in the simply typed lambda calculus (for provably non-empty types (cf. [22])). Thus adding operations for manipulating references to the simply typed lambda calculus causes the failure of strong normalization as well as many other of its nice mathematical properties.

6.2.3. *Integer Streams.* From an abstract point of view, a stream is simply a (possibly infinite) sequence of data [1]. In the Λ_s-language we can represent streams simply as functional objects, lambda expressions with free variables bound to cells. The sequence corresponding to a Λ_s-stream is the values returned by repeated application of the object to a fixed (and hopefully irrelevant) argument. The simplest example of a non-trivial Λ_s-stream is the stream of natural numbers.

$$\mathtt{makeStream} \triangleq \lambda m.\,\mathtt{let}\,\{z := \mathtt{mk}(m)\}$$
$$\lambda x.\,\mathtt{let}\{n := \mathtt{get}(z)\}$$
$$\mathtt{set}(z, n+1); n$$

Here makeStream applied to an integer m creates a stream of integers beginning with that integer. The so-created stream when queried (applied to any value) returns the next integer in the stream.

6.2.4. *The Sieve of Eratosthenes.* A somewhat more interesting example of a stream is the sieve of Eratosthenes [1]. We begin with the functional filter which expects an integer, n, and a stream, s and then creates a new stream. This new stream when queried repeatedly calls the stream argument, s, until an integer not divisible by the number argument, n, is returned. This number is then returned as the answer to the query.

$$\mathtt{filter} \triangleq \lambda n.\lambda s.$$
$$\lambda x.\,\mathtt{let}\{m := s(\mathtt{nil})\}$$
$$\mathtt{if}(\mathtt{divides?}(n, m)$$
$$\mathtt{filter}(n, s)(\mathtt{nil})$$
$$m)$$

sieve is an expression which when evaluated creates a new sieve of Eratosthenes. This new stream is a stream of the prime numbers. Each time the stream is queried it returns the current prime and updates its local stream to filter with this prime.

$$\texttt{sieve} \triangleq \texttt{let}\{sc := \texttt{mk}(\texttt{makeStream}(2))\}$$
$$\lambda x.\,\texttt{let}\{s := \texttt{get}(sc)\}\,\texttt{let}\,\{p := s(\texttt{nil})\}$$
$$\texttt{set}(sc, \texttt{filter}(p, s));\, p$$

6.2.5. *Co-routines.* Co-routines are a programming paradigm useful for incremental processing of streams of data. Typical examples of their use come from compiler construction, and more recently from managing data coming in segments over a network connection. Each time an additional increment of data is needed/available, the co-routine is *resumed.* The co-routine computes the next increment, remembers where it left off, and returns the incremental result. From a programming language point of view what is needed is language constructs that make it easy to remember the current state, and to continue processing when next invoked. The high-level structure of a co-routine program might look like

```
       initialize
loop:  get the next big block of data;
       process segment 1 giving result-1;
       resume with result-1;
         ...
       process segment n giving result-n;
       resume with result-n;
       repeat loop;
```

where resume implicitly saves the point in the program where the computation suspended for later resumption and returns control to the resumer of the co-routine. Thus from the point of view of the co-routine, the partner also appears to be a co-routine.

A resumption primitive can be defined using our control and memory primitives as follows.

$$\texttt{resume} \triangleq \lambda z.\lambda v.\texttt{ncc}(\lambda k.\,\texttt{let}\,\{c := \texttt{get}(z)\}\texttt{set}(z, k);\, c(v))$$

It is assumed that z is a cell where the resumee's resumption point is stored (as a continuation). $\texttt{resume}(z)(v)$ gets that state, temporarily calling it c, saves the resumer's resumption point, captured by ncc and bound to k, in z, and starts at its resumption point with argument v. The resumed program should eventually call $\texttt{resume}(z)(w)$ where w is the next increment to be returned to the resumer. Later we will show using a simple example how the Feferman–Landin logic allows us to reason about co-routines. For now we look at a sample calculation and an example co-routine. Assume that q is the resumption point of a partner co-routine that is being resumed in a context R, then

$$\zeta\{z := \texttt{mk}(q)\} : R[\texttt{resume}(z)(v)]$$

$$\twoheadrightarrow \zeta\{z := \mathtt{mk}(q)\} : R[\mathtt{ncc}(\lambda k.\,\mathtt{let}\,\{c := \mathtt{get}(z)\}\mathtt{set}(z,k); c(v))]$$

$$\twoheadrightarrow \zeta\{z := \mathtt{mk}(q)\} : \mathtt{let}\,\{c := \mathtt{get}(z)\}\mathtt{set}(z, \lambda x.\mathtt{top}(R[x])); c(v))$$

$$\twoheadrightarrow \zeta\{z := \mathtt{mk}(q)\} : \mathtt{set}(z, \lambda x.\mathtt{top}(R[x])); q(v)$$

$$\twoheadrightarrow \zeta\{z := \mathtt{mk}(\lambda x.\mathtt{top}(R[x]))\} : q(v)$$

Thus we have the derived computation rule:

$$\zeta\{z := \mathtt{mk}(q)\} : R[\mathtt{resume}(z)(v)] \twoheadrightarrow \zeta\{z := \mathtt{mk}(\lambda x.\mathtt{top}(R[x]))\} : q(v)$$

now suppose

$$\zeta\{z := \mathtt{mk}(\lambda x.\mathtt{top}(R[x]))\} : q(v)$$
$$\twoheadrightarrow \zeta'\{z := \mathtt{mk}(\lambda x.\mathtt{top}(R[x]))\} : R'[\mathtt{resume}(z)(w)]$$

then applying the above derived reduction we have

$$\zeta\{z := \mathtt{mk}(q)\} : R[\mathtt{resume}(z)(v)] \twoheadrightarrow \zeta'\{z := \mathtt{mk}(\lambda x.\mathtt{top}(R'[x]))\} : R[w]$$

with the co-routine's next resumption point saved in the cell z.

Here is a simple co-routine (schema):

$$
\begin{aligned}
\mathtt{coex} \overset{\triangle}{=} \lambda i.\lambda z.\lambda x.\\
&\mathtt{let}\,\{p := \mathtt{next}(i)\}\\
&\quad \mathtt{let}\,\{x_1 := f_1(p)\}\\
&\qquad \mathtt{resume}(z)(x_1);\\
&\qquad \mathtt{let}\,\{x_2 := f_2(p)\}\\
&\qquad\quad \mathtt{resume}(z)(x_2);\\
&\qquad\quad \mathtt{coex}(i,z)(\mathtt{nil})
\end{aligned}
$$

The initial resumption point for the co-routine is of the form $\lambda x.\mathtt{top}(\mathtt{coex}(i,z)(x))$, where i is intended to be a cell containing access to an input stream, which we leave unspecified, except that $\mathtt{next}(i)$ must produce data in the domain of f_1 and f_2, which compute functions without any use of or effect on state.

Now consider a program that does the same incremental computation but explicitly saves resumption point information.

$$
\begin{aligned}
\mathtt{stex} \overset{\triangle}{=} \lambda i\lambda z.\lambda x.\\
&\mathtt{let}\,\{s := \mathtt{get}(z)\}\\
&\quad \mathtt{if}(\mathtt{eq?}(\mathtt{fst}(s), 0),\\
&\qquad \mathtt{let}\,\{p := \mathtt{next}(i)\}\mathtt{set}(z, \mathtt{pr}(1,p)); f_1(p),\\
&\qquad \mathtt{let}\,\{p := \mathtt{snd}(s)\}\mathtt{set}(z, \mathtt{pr}(0,\mathtt{nil})); f_2(p))
\end{aligned}
$$

Then we have the following lemma. We will see later how to establish such claims in the Feferman–Landin logic.

Lemma 6.7:

$$\{z := \mathtt{mk}(\lambda x.\mathtt{top}(\mathtt{coex}(i, z)(x)))\}\mathtt{resume}(z) \cong \{z := \mathtt{mk}(\mathtt{pr}(0, \mathtt{nil}))\}\mathtt{stex}(i, z)$$

6.3. Axioms for Λ_s. The first, most basic axiom concerning operational equivalence and equivaluedness for Λ_s, is that the booleans t and nil are not equivalent.
Non-Triviality (T).

(i) $\vdash \neg(\mathtt{t} \simeq \mathtt{nil})$

To the operational equivalence axioms we can add a *garbage collection* principle, and a principle expressing the simple fact that operational equivalence and equivaluedness coincide on simple data.
Operational Equivalence Axioms ($\cong$).

(iii) $\vdash e \cong \zeta\,[e]$ provided $\mathrm{FV}(e) \cap \mathrm{Dom}(\zeta) = \emptyset$

(iv) $\vdash \tau(x) \sim \mathtt{t} \wedge \tau(y) \sim \mathtt{t} \Rightarrow (x \cong y \Leftrightarrow x \sim y)$ $\quad \tau \in \{\mathtt{atom?}, \mathtt{cell?}, \mathtt{nat?}\}$

Next we extend the quantifier and constraint propagation axioms of § 5 using the information we have about particular Λ_s operations.
Quantifier Axioms (Q).

(iv) $\vdash (\forall x)\,\mathtt{let}\,\{z := \vartheta(\bar{y})\}\Phi \Rightarrow \mathtt{let}\,\{z := \vartheta(\bar{y})\}(\forall x)\Phi$

ϑ is either ineffectual or the operation set

(v) $\vdash ((\forall x)\,\mathtt{let}\,\{z := \mathtt{mk}(y)\}\Phi(x) \wedge \mathtt{let}\,\{z := \mathtt{mk}(y)\}\Phi(z))$
$\Rightarrow \mathtt{let}\,\{z := \mathtt{mk}(y)\}(\forall x)\Phi$

Constraint Propagation Axioms (S). The following two principles require that $\vartheta_0, \vartheta_1 \in \boldsymbol{O}$ and that if $\vartheta_1 \in \{\mathtt{set}, \mathtt{app}, \mathtt{ncc}\}$, then $\vartheta_0 \in \boldsymbol{O} - \{\mathtt{get}, \mathtt{set}, \mathtt{app}, \mathtt{ncc}\}$.

(iv) $\vdash \pm(z \simeq \vartheta_0(\bar{y})) \Rightarrow \mathtt{let}\,\{x := \vartheta_1(\bar{w})\}(\pm(z \simeq \vartheta_0(\bar{y})))$

(v) $\vdash \vartheta_1(\bar{w})\!\downarrow\, \Rightarrow (\mathtt{let}\,\{x := \vartheta_1(\bar{w})\}(\pm(z \simeq \vartheta_0(\bar{y}))) \Rightarrow \pm(z \simeq \vartheta_0(\bar{y})))$

Note that in (**S.iv,v**) ncc has the same classification as app, since as noted above it applies its argument lambda to a value and thus can reduce to an arbitrary expression eventually producing arbitrary effects. Also note that taking ϑ_1 to be mk, (**S.iv**) says that allocation, does not change properties of any operations on existing values.

6.4. Definedness principles. We have already stated the most basic principle concerning undefinedness, namely that two undefined terms are both equivalued and operationally indistinguishable, (**U.i**). The rest of the principles concern the partiality of the underlying operations. Note that in the case of the memory operations mk and set, being defined is not the same as being equivalent to a value. In the other cases this is true, although we need only express the weaker form. The stronger forms are derivable.
Definedness rules (D).

(i) $\vdash \downarrow \mathtt{mk}(z)$

(ii) $\vdash \downarrow \mathtt{set}(z, x) \Leftrightarrow \mathtt{cell?}(z) \simeq \mathtt{t}$

(iii) $\vdash \downarrow \mathtt{get}(x) \Leftrightarrow \mathtt{cell?}(x) \simeq \mathtt{t} \Leftrightarrow (\exists y)(\mathtt{get}(x) \simeq y)$

(iv) $\vdash \downarrow \mathtt{ncc}(x) \Rightarrow \mathtt{lambda?}(x) \simeq \mathtt{t}$

(v) $\vdash \downarrow \mathtt{app}(x, y) \Rightarrow \mathtt{lambda?}(x) \simeq \mathtt{t}$

(vi) $\vdash \downarrow \vartheta(\bar{x})$

Where in (**D.vi**) we assume that $\vartheta \in \{\mathtt{atom?}, \mathtt{cell?}, \mathtt{lambda?}, \mathtt{nat?}, \mathtt{pr?}, \mathtt{pr}, \mathtt{eq?}, \mathtt{br}\}$ and $\bar{x}$ is the appropriate length.

6.5. Axioms for memory. The principles concerning mk are quite straightforward. (**mk.i**) describes the allocation effect of a call to mk. The remaining two principles (**mk.ii**) and (**mk.iii**) assert that the time of allocation has no discernable effect on the resulting call, however since we are in a world with control effects e_0 must be free of them for these two principles to be valid [12].

Allocation Axioms (mk).

(i) $\vdash \mathtt{let}\,\{x := \mathtt{mk}(z)\}(\neg(x \cong y) \wedge \mathtt{cell?}(x) \cong \mathtt{t} \wedge \mathtt{get}(x) \cong z)$

(ii) $\vdash \mathtt{let}\,\{y := e_0\}\,\mathtt{let}\,\{x := \mathtt{mk}(z)\}e_1 \cong \mathtt{let}\,\{x := \mathtt{mk}(z)\}\,\mathtt{let}\,\{y := e_0\}e_1$

(iii) $\vdash \mathtt{let}\,\{y := e_0\}\,\mathtt{let}\,\{x := \mathtt{mk}(z)\}\Phi \Leftrightarrow \mathtt{let}\,\{x := \mathtt{mk}(z)\}\,\mathtt{let}\,\{y := e_0\}\Phi$

The principle (**mk.i**) requires that $x \notin \{z, y\}$, while both (**mk.ii**) and (**mk.iii**) require that e_0 is closed with no control operations.

The first two contextual assertions regarding set are analogous to those of (mk.**i**). They describe what is returned and what is altered. The remaining four principles involve the commuting, cancellation, absorption, and idempotence of calls to set. For example the set absorption principle, (set.**v**), expresses that under certain simple conditions allocation followed by assignment may be replaced by a suitably altered allocation.

Modification Axioms (set).

(i) $\vdash \mathtt{cell?}(z) \Rightarrow \mathtt{let}\,\{x := \mathtt{set}(z, y)\}(\mathtt{get}(z) \cong y \wedge x \cong \mathtt{nil})$

(ii) $\vdash \mathtt{get}(x) \cong y \Rightarrow \mathtt{set}(x, y) \cong \mathtt{nil}$

(iii) $\vdash \neg(x_0 \cong x_2) \Rightarrow \mathtt{set}(x_0, x_1); \mathtt{set}(x_2, x_3) \cong \mathtt{set}(x_2, x_3); \mathtt{set}(x_0, x_1)$

(iv) $\vdash \mathtt{set}(x, y_0); \mathtt{set}(x, y_1) \cong \mathtt{set}(x, y_1)$

(v) $\vdash \mathtt{let}\,\{z := \mathtt{mk}(x)\}\mathtt{set}(z, w); e \cong \mathtt{let}\{z := \mathtt{mk}(w)\}e \qquad \text{if } z \notin \mathrm{FV}(w)$

(vi) $\vdash \mathtt{get}(x) \cong y \Rightarrow \mathtt{set}(x, y) \cong \mathtt{nil}$

Accessor Axioms (get).

(i) $\vdash \mathtt{let}\,\{x := \mathtt{get}(y)\}(x \cong \mathtt{get}(y))$

(ii) $\vdash \Phi \Rightarrow \mathtt{let}\,\{x := \mathtt{get}(y)\}\Phi \qquad x \notin \mathrm{FV}(\Phi)$

6.6. Axioms for control. In our model a continuation is a function that returns a value to the top level, thus continuations can be modeled by composition of top with an arbitrary lambda expression. The first control axiom says that within the immediate context of ncc, the current continuation is top, while the second shows how explict R and implicit (passed to the argument of ncc) continuations are related, and the third axiom expresses the fact that the argument of ncc is always applied to a continuation. The fourth axiom says that $\mathtt{ncc}(\lambda c.R[\bullet])$ behaves like a reduction context, in the sense that any expression occupying the hole is alway evaluated next, and thus can be pulled out using let analogous to (**let.ii**) as long as the continuation variable does not appear free. The fifth axiom expresses the fact that capturing the current continuation and then re-installing it is equivalent the identity operation.
Control Axioms (ncc).

(i) $\vdash \mathtt{ncc}(\lambda c.\mathtt{ncc}(e)) \cong \mathtt{ncc}(\lambda c.\mathrm{app}(e, \mathtt{top}))$

(ii) $\vdash R[\mathtt{ncc}(e)] \cong \mathtt{ncc}(\lambda c.\mathrm{app}(e, \lambda x.c(R[x])))$

if $c \notin \mathrm{FV}(e, R)$ and $x \notin \mathrm{FV}(c, R)$

(iii) $\vdash \mathtt{ncc}(\lambda c.C[c]) \cong \mathtt{ncc}(\lambda c.C[\lambda x.\mathtt{top}(c(x))])$

provided $c \notin \mathrm{Traps}(C)$.

(iv) $\vdash (\mathtt{let}\,\{x := e\}\mathtt{ncc}(\lambda k.R[x])) \cong \mathtt{ncc}(\lambda k.R[e])$

if x is fresh, and $k \notin \mathrm{FV}(e)$

(v) $\vdash \mathtt{ncc}(\lambda k.k(e)) \cong e \quad$ if $k \notin \mathrm{FV}(e)$

To illustrate contextual reasoning we derive a useful law for reasoning about ncc.

Lemma 6.8 (ncc.vi):

(vi.a) $\vdash e_0 \cong e_1 \Rightarrow (\mathtt{ncc}(\lambda k.R[e_0]) \cong \mathtt{ncc}(\lambda k.R[e_1]))$

$k \notin \mathrm{FV}(e_i)$ for $i < 2$.

(vi) $\vdash \mathtt{let}\,\{x := e\}(e_0 \cong e_1) \Rightarrow \mathtt{let}\,\{x := e\}(\mathtt{ncc}(\lambda k.R[e_0]) \cong \mathtt{ncc}(\lambda k.R[e_1]))$

if $k \notin \mathrm{FV}(e)$

Proof: To show (**ncc.vi.a**) we argue as follows.

$e_0 \cong e_1 \Rightarrow \mathtt{let}\,\{x := e_0\}\mathtt{ncc}(\lambda k.R[x]) \cong \mathtt{let}\,\{x := e_1\}\mathtt{ncc}(\lambda k.R[x])$

by (**E.iv**)

$\mathtt{ncc}(\lambda k.R[e_j]) \cong \mathtt{let}\,\{x := e_j\}\mathtt{ncc}(\lambda k.R[x])$

for $j < 2$, by (**ncc.iv**)

And by a little propositional reasoning we are done.
(**ncc.vi**) follows from (**ncc.vi.a**) using context introduction (**C.i**). □

Lemma 6.9 (top): Any reduction context surrounding an application of `top` can be discarded.

$$R[\mathtt{top}(e)] \cong \mathtt{top}(e)$$

Proof: This follows from (ncc.iv) using the `let` laws. □

6.7. Simulation principle for Λ_s. Now we introduce a simulation principle that allows us to prove equivalence of two lambdas with memory. The principle says that to show such an equivalence it is sufficient to find a correspondence between memories such that whenever started in corresponding states and applied to the same value, both lambdas

(i) return the same (first-order) value
(ii) have corresponding effects on their memory states

To simplify notation, we present only a very special case of the principle, which is adequate to illustrate the basic idea. First we must define the set of first-order values and a function that maps any value to its first-order part.

Definition 6.10 (First-orderness, $(\boldsymbol{FO}, \mathtt{fo})$): $\boldsymbol{FO}$ is the class of first-order values. This is the least class containing $\boldsymbol{At}$ and closed under pairing (essentially the well-founded S-expressions of [22]). It can be defined using the standard means of giving inductive definitions in variable typed theories as follows.

$$\boldsymbol{FO} \stackrel{\triangle}{=} \bigcap_{\kappa}(T_{FO}[\kappa] \subseteq \kappa)$$

$$\text{where} \quad T_{FO}[\kappa] \stackrel{\triangle}{=} \{x \mid \mathtt{atom?}(x) \vee (\mathtt{pr?}(x) \wedge \mathtt{fst}(x) \in \kappa \wedge \mathtt{snd}(x) \in \kappa)\}$$

$$\text{and} \quad \bigcap_{\kappa}\Phi[\kappa] \stackrel{\triangle}{=} \{x \mid (\forall\kappa)(\Phi[\kappa] \Rightarrow x \in \kappa)\}$$

The operation, `fo`, that strips away the non first order parts of its argument. It leaves atoms unchanged, commutes with pairing, and maps everything else (lambdas and cells) to `nil`.

$$\mathtt{fo} \stackrel{\triangle}{=} \lambda x.\mathtt{if}(\mathtt{atom?}(x), x, \mathtt{if}(\mathtt{pr?}(x), \mathtt{pr}(\mathtt{fo}(\mathtt{fst}(x)), \mathtt{fo}(\mathtt{snd}(x))), \mathtt{nil}))$$

Using the induction principle derived from the inductive definition of $\boldsymbol{FO}$ it is easy to show $\boldsymbol{FO} \subseteq \{x \mid \mathtt{fo}(x) \simeq x\}$. To prove the converse we need the least-fixed-point principle derived from recursive definitions using a suitable fixed-point combinator.

We state the simulation principle for the special case in which the lambda's local memory is a single cell. In the definition below we use z to name this local memory cell. Since we want the cell contents to be able to refer to the cell, the correspondence Ψ is formulated as a relation on $\boldsymbol{L}$ such that corresponding lambdas are applied to a cell to produce the actual corresponding values $(y_j(z))$. Similarly, the lambda's P_j represent abstractions over the cell variable. The assertions (**Sim.1**) and (**Sim.2**) are formal statements of the requirements (i) and (ii) given at the beginning of this subsection.

Simulation Principle (SP). If z a variable, $\Psi \subseteq \boldsymbol{L} \times \boldsymbol{L}$, and P_j is a lambda expression for $j < 2$, such that z is not free in Ψ, P_0 or P_1 and

$$(\forall (y_0, y_1) \in \Psi)(\forall x)(\exists (y_0', y_1') \in \Psi, v \in \boldsymbol{FO})$$

$$\bigwedge_{j\in\{0,1\}}$$

$$(\{z := \mathtt{mk}(y_j(z))\}(P_j(z)(x) \cong \mathtt{set}(z, y_j'(z)); v) \quad \text{(Sim.1)}$$

$$\wedge$$

$$\{z := \mathtt{mk}(y_j(z))\}(P_j(z)(x) \cong P_j(z)(\mathtt{fo}(x)))), \quad \text{(Sim.2)}$$

then $(\forall (y_0, y_1) \in \Psi)\{z := \mathtt{mk}(y_0(z))\}P_0(z) \cong \{z := \mathtt{mk}(y_1(z))\}P_1(z)$.

6.8. Using the simulation principle. Now we show how to establish equivalences such as the claim made in the co-routine lemma(6.7). We simplify the example a bit in order to be able to fill in some details in a reasonable space.

Lemma 6.11 (Simple co-routine reduction to state.): Let

$$\mathtt{inc}(f, n) \stackrel{\triangle}{=} \lambda z.\lambda x.\mathtt{resume}(z)(f(n)); \mathtt{inc}(f, n+1)(z)(\mathtt{nil})$$

$$\mathtt{ino}(f) \stackrel{\triangle}{=} \lambda z.\lambda x.\,\mathtt{let}\,\{n := \mathtt{get}(z)\}\mathtt{set}(z, n+1); f(n).$$

`inc` (c for co-routine) produces a sequence of values $f(n)$ for $n \in \boldsymbol{N}$ using resumption to remember the next element of the stream to generate. `ino` (o for object) produces the same sequence by explicitly keeping track of the next integer argument in a memory cell. Now choose $f \in \boldsymbol{FO} \to \boldsymbol{FO}$ such that $f(x) \cong f(\mathtt{fo}(x))$. Define κ_n to be the lambda term $\lambda x.\mathtt{top}(\mathtt{inc}(f, n)(z)(\mathtt{nil}))$. Then

$$\{z := \mathtt{mk}(\kappa_n)\}\mathtt{resume}(z) \cong \{z := \mathtt{mk}(n)\}\mathtt{ino}(f)(z)$$

Proof: The proof is by the simulation principle with

$$\Psi = \{(\lambda z.\kappa_n, \lambda z.n) \mid n \in \boldsymbol{N}\}$$

$$P_0 = \mathtt{resume}$$

$$P_1 = \mathtt{ino}(f)$$

The simulation requirement (**Sim.2**) follows from the assumption on f. Note that by definition 4.5 such f are total and ineffectual. Thus we only need to establish (**Sim.1**) for the two Ps. The argument for $\mathtt{ino}(f)$ is as follows. In the following we write $\zeta \vdash e_0 \cong e_1$ in place of $\vdash \zeta(e_0 \cong e_1)$. This is done to factor out the background contextual reasoning from the basic equational reasoning.

$\{z := \mathtt{mk}(n)\} \vdash$

$\quad \mathtt{ino}(f)(z)(x)$

$\quad\quad \cong \mathtt{let}\,\{n := \mathtt{get}(z)\}\mathtt{set}(z, n+1); f(n)$

$\quad\quad\quad$ by (**let.i**)

$\quad\quad \cong \mathtt{set}(z, n+1); f(n)$

$\quad\quad\quad$ by (**mk.i**) and (**let.i**)

The argument for resume is a bit more complicated. First we expand and transform $\mathtt{resume}(z)(x)$ in the given memory context.

$\{z := \mathtt{mk}(\kappa_n)\} \vdash$

$\quad \mathtt{resume}(z)(x)$

$\quad\quad \cong \mathtt{ncc}(\lambda k.\,\mathtt{let}\,\{p := \mathtt{get}(z)\}\mathtt{set}(z, k); p(x))$

$\quad\quad\quad$ by the definition of resume

$\quad\quad \cong \mathtt{ncc}(\ \lambda k.\,\mathtt{let}\,\{p := \kappa_n\}\mathtt{set}(z, k); p(x)\ )$

$\quad\quad\quad$ by (**ncc.vi**) and get laws

Now by purely equational reasoning we show that the contextually transformed expression is equivalent to an expression with directly nested nccs. First we simplify the let and unfold definitions.

$\mathtt{ncc}(\ \lambda k.\,\mathtt{let}\,\{p := \kappa_n\}\mathtt{set}(z, k); p(x)\ )$

$\quad\quad \cong \mathtt{ncc}(\ \lambda k.\mathtt{set}(z, k); \mathtt{top}(\mathtt{inc}(f, n)(z)(\mathtt{nil}))\ )$

$\quad\quad\quad$ by (**let.i**) and definition of κ_n

$\quad\quad \cong \mathtt{ncc}(\ \lambda k.(\ \mathtt{set}(z, k); \mathtt{top}(\mathtt{resume}(z)(f(n)); \mathtt{inc}(f, n+1)(z)(\mathtt{nil}))\ )\ )$

$\quad\quad\quad$ using the definition of inc

Next we unfold the resume, push the reduction context surrounding the ncc to the inside, drop the c that will be discarded by top and reduce the app.

$\mathtt{ncc}(\ \lambda k.(\ \mathtt{set}(z, k); \mathtt{top}(\mathtt{resume}(z)(f(n)); \mathtt{inc}(f, n+1)(z)(\mathtt{nil}))\ )\ )$

$\quad\quad \cong \mathtt{ncc}(\lambda k.(\ \mathtt{set}(z, k);$

$\quad\quad\quad\quad\quad \mathtt{top}(\mathtt{ncc}(\ \lambda k.\,\mathtt{let}\,\{p := \mathtt{get}(z)\}\mathtt{set}(z, k); p(f(n))\ );$

$\quad\quad\quad\quad\quad\quad \mathtt{inc}(f, n+1)(z)(\mathtt{nil}))\ )\ )$

$\quad\quad\quad$ by definition of $\mathtt{resume}(z)$

$\quad\quad \cong \mathtt{ncc}(\ \lambda k.(\ \mathtt{set}(z, k);$

$\quad\quad\quad\quad\quad \mathtt{ncc}(\ \lambda c.\mathrm{app}(\lambda k.\,\mathtt{let}\,\{p := \mathtt{get}(z)\}\mathtt{set}(z, k); p(f(n))\ );$

$\quad\quad\quad\quad\quad\quad\quad \lambda x.c(\mathtt{top}(\mathtt{inc}(f, n+1)(z)(\mathtt{nil})))\ )\ )\ )$

$\quad\quad\quad$ by (**ncc.ii**) with $R = \mathtt{top}(\bullet; \mathtt{inc}(f, n+1)(z)(\mathtt{nil}))$

$\quad\quad \cong \mathtt{ncc}(\ \lambda k.(\ \mathtt{set}(z, k);$

$\quad\quad\quad\quad\quad \mathtt{ncc}(\ \lambda c.\,\mathtt{let}\,\{p := \mathtt{get}(z)\}\mathtt{set}(z, \kappa_{n+1}); p(f(n))\ )\ )\ )$

$\quad\quad\quad$ by the (**top**) lemma and let laws

Finally, the set can be pushed inside the inner ncc and the result simplified using the memory and let laws.

$$\begin{aligned}
&\cong \mathtt{ncc}(\ \lambda k.(\ \mathtt{set}(z,k);\\
&\qquad\qquad \mathtt{ncc}(\ \lambda c.\,\mathtt{let}\,\{p := \mathtt{get}(z)\}\mathtt{set}(z,\kappa_{n+1});\, p(f(n))\ )\ )\ )\\
&\cong \mathtt{ncc}(\ \lambda k.(\ \mathtt{ncc}(\ \lambda c.\mathtt{set}(z,k);\\
&\qquad\qquad\qquad \mathtt{let}\,\{p := \mathtt{get}(z)\}\mathtt{set}(z,\kappa_{n+1});\, p(f(n))\ )\ )\ )\\
&\qquad \text{by } (\textbf{ncc.iv})\\
&\cong \mathtt{ncc}(\ \lambda k.\mathtt{ncc}(\ \lambda c.\mathtt{set}(z,\kappa_{n+1});\, k(f(n))\ )\ )\\
&\qquad \text{by the } \mathtt{set} \text{ and } \mathtt{get} \text{ laws}
\end{aligned}$$

Now we the desired 'canonical' form with directly nested nccs. This allows us to further transform the body of the inner ncc and finally to eliminate it.

$$\begin{aligned}
&\mathtt{ncc}(\ \lambda k.\mathtt{ncc}(\ \lambda c.\mathtt{set}(z,\kappa_{n+1});\, k(f(n))\ )\ )\\
&\quad\cong \mathtt{ncc}(\ \lambda k.\mathrm{app}(\lambda c.\mathtt{set}(z,\kappa_{n+1});\, k(f(n)),\mathtt{top})\ )\\
&\qquad \text{by the } (\textbf{ncc.i})\\
&\quad\cong \mathtt{ncc}(\ \lambda k.\mathtt{set}(z,\kappa_{n+1});\, k(f(n))\ )\\
&\qquad \text{by the } (\textbf{let.i})\\
&\quad\cong \mathtt{set}(z,\kappa_{n+1});\, \mathtt{ncc}(\lambda k.k(f(n)))\\
&\qquad \text{by the } (\textbf{ncc.iv})\\
&\quad\cong \mathtt{set}(z,\kappa_{n+1});\, k(f(n))\\
&\qquad \text{by the } (\textbf{ncc.v})
\end{aligned}$$

This completes the proof. □

§7. Conclusions and future directions. In this paper we have shown how to generalize our previous variable-type theory [22] for reasoning about programs of a specific language Λ_{m} to a logic for reasoning about an arbitrary Landinesque language with uniform semantics and have given an example of axiomatizing a typical language of this class. The variable-type theories have the advantage of being essentially first-order, while providing for functions (operations) as first-class objects, and providing the ability to define and reason with classifications.

This is just the foundation for putting these ideas to work in practice. There are a number of interesting directions for future work. One direction is to consider a wider class of languages, for example treating object-oriented (OO) languages such as Java [20], or treating input/output (IO) features. We believe that the sequential aspects of OO languages are not problematic. Treating IO and concurrency aspects will be more challenging, as this means removing the restriction that the semantics be deterministic. It may be that the solution here is to combine the nice features of the variable-type logics with some form of temporal logic.

Another direction for future work is implementation—building tools to aid in the process of specifying and reasoning about programs using the Feferman–Landin logic. Here the work of Frost [17] would be a good starting point.

Although complete axiomatizations of Landinesque languages are not possible, it is useful to investigate forms of relative completeness and completenss for fragments such as done in [26] for the Λ_m language. This is especially important in conjunction with implementing the logic and developing mechanized procedures for aiding the development of proofs.

When reasoning informally about programs, we use a mix of operational and equational/logical reasoning. The importance of being able to combine these forms of reasoning was first noted in [19] where operationally based principles for denotational reasoning about Lisp programs were developed. An interesting and potentially very useful direction of future research in our context is to enrich the Feferman–Landin logic by formalizing the reduction relation, and adding a reflection principle that allows meta-theorems to be reflected down to the the logic level. For example special purpose simulation principles could be added in this way. Another application would be to extend the formal meta theory with definitions of program analysis functions and derive proof principles based on program analysis results. For example analysis could characterize certain classes of effects and principles generalizing the constraint propagation axioms for Λ_s could be developed in a formal setting. Again this mix of intensional and extensional (operational and denotational) reasoning is quite natural for programmers.

It may also be fruitful to investigate the use of other methods developed for reasoning about operational approximation and equivalence. These include: general schemes for establishing equivalence; context lemmas (alternative characterizations that reduce the number of contexts to be considered); and (bi)simulation relations (alternative characterizations or approximations based on co-inductively defined relations). For example [23] develops a schema for proving congruence for a class of languages with a particular style of operational semantics. This schema succeeds in capturing many simple functional programming language features. Building on this work, Howe [24] uses an approach similar to the idea of uniform computation to define structured evaluation systems in which the form of the evaluation rules guarantees that (bi)simulation relations are congruences. The form of the rules is specified using meta variables with arities and higher-order substitutions. This syntax enrichment is very similar to the notions of place-holder and filling used here to specify uniform semantics.

A useful refinement would be to identify a form of rules that guarantees uniform semantics, generalizing the ideas of Howe [24] to functional languages with effects. Another extension would be to develop denotational tools in this setting generalizing [28].

REFERENCES

[1] H. ABELSON and G. J. SUSSMAN, ***Structure and interpretation of computer programs***, The MIT Press, McGraw-Hill Book Company, 1985.

[2] G. AGHA, I. A. MASON, S. F. SMITH, and C. L. TALCOTT, *A foundation for actor computation*, ***Journal of Functional Programming***, vol. 7 (1997), pp. 1–72.

[3] K.R. APT, *Ten years of Hoare's logic: A survey–part I*, ***ACM Transactions on Programming Languages and Systems***, vol. 4 (1981), pp. 431–483.

[4] C.C. CHANG and H.J. KEISLER, ***Model theory***, North-Holland. Amsterdam. 1973.

[5] S. FEFERMAN, *A language and axioms for explicit mathematics*, ***Algebra and logic***, Springer Lecture Notes in Mathematics, vol. 450, Springer Verlag, 1975, pp. 87–139.

[6] ———, *Non-extensional type-free theories of partial operations and classifications, i.*, ***Proof theory symposium, kiel 1974*** (J. Diller and G. H. Müller, editors), Lecture notes in mathematics, no. 500, Springer, Berlin, 1975, pp. 73–118.

[7] ———, *Constructive theories of functions and classes*, ***Logic colloquium '78***, North-Holland. 1979, pp. 159–224.

[8] ———, *A theory of variable types*, ***Revista Colombiana de Matématicas***, vol. 19 (1985), pp. 95–105.

[9] ———, *Polymorphic typed lambda-calculi in a type-free axiomatic framework*, ***Logic and computation***, Contemporary Mathematics, vol. 106, A.M.S., Providence R. I., 1990, pp. 101–136.

[10] ———, *Logics for termination and correctness of functional programs,*, ***Logic from computer science***, MSRI Publications,, vol. 21, Springer Verlag, 1992, pp. 101–136.

[11] ———, *Logics for termination and correctness of functional programs, II. logics of strength PRA.*, ***Proof theory***, Cambridge University Press, 1992.

[12] M. FELLEISEN,, 1993, Personal communication.

[13] M. FELLEISEN and D.P. FRIEDMAN, *Control operators, the SECD-machine, and the λ-calculus*, ***Formal description of programming concepts III*** (M. Wirsing, editor), North-Holland, 1986, pp. 193–217.

[14] M. FELLEISEN and A. SABRY, *Continuations in programming practice: Introduction and survey*, 1999, available at `http://www.cs.uoregon.edu/~sabry/papers/continuations.ps`.

[15] J. FORD and I. A. MASON, *Establishing a General Context Lemma in PVS*, ***2nd Australasian Workshop on Computational Logic, AWCL'01***, 2001, available at `http://mcs.une.edu.au/~pvs/`.

[16] D. P. FRIEDMAN, *Applications of continuations*, ***Technical Report 237***, Indiana Univeristy Computer Science Department, 1988, Tutorial given at POPL88.

[17] J. FROST, *Effective programming*, ***Ph.D. thesis***, Technical University of Denmark, 1996, also published as Techincal Report IT-TR 1996-001.

[18] J. FROST and I. A. MASON, *An Operational Logic of Effects*, ***Proceedings of the Australasian Theory Symposium, CATS '96***, 1996, pp. 147–156.

[19] M. J. C. GORDON, *Operational reasoning and denotational semantics*, ***Technical Report SAIL Memo AIM-264***, Artificial Intelligence Laboratory, Stanford University, 1975.

[20] J. GOSLING, B. JOY, and G. L. STEELE JR., ***The Java language specification***, Addison-Wesley, 1996.

[21] D. HAREL, *Dynamic logic*, ***Handbook of philosophical logic, vol. ii*** (D. Gabbay and G. Guenthner, editors), D. Reidel, 1984, pp. 497–604.

[22] F. HONSELL, I. A. MASON, S. F. SMITH, and C. L. TALCOTT, *A Variable Typed Logic of Effects*, ***Information and Computation***, vol. 119 (1995), no. 1, pp. 55–90.

[23] D. HOWE, *Equality in the lazy lambda calculus*, ***Fourth annual symposium on logic in computer science***, IEEE, 1989.

[24] D. J. HOWE, *Proving congruence of bisimulation in functional programming languages*, ***Information and Computation***, vol. 124 (1996), no. 2, pp. 103–112.

[25] P. J. LANDIN, *The mechanical evaluation of expressions*, ***Computer Journal***, vol. 6 (1964), pp. 308–320.

[26] I. A. Mason. *A First Order Logic of Effects*. ***Theoretical Computer Science***, vol. 185 (1997), pp. 277 – 318.

[27] ———. *Computing with contexts*. ***Higher-Order and Symbolic Computation***, vol. 12 (1999), pp. 171–201.

[28] I. A. Mason, S. F. Smith, and C. L. Talcott, *From Operational Semantics to Domain Theory*, ***Information and Computation***, vol. 128 (1996), no. 1, pp. 26–47.

[29] I. A. Mason and C. L. Talcott, *Equivalence in functional languages with effects*, ***Journal of Functional Programming***, vol. 1 (1991), pp. 287–327.

[30] ———, *Reasoning about object systems in VTLoE*, ***International Journal of Foundations of Computer Science***, vol. 6 (1995), no. 3, pp. 265–298.

[31] R. Milner, *Fully abstract models of typed λ-calculi*, ***Theoretical Computer Science***, vol. 4 (1977), pp. 1–22.

[32] R. Milner, M. Tofte, and R. Harper, ***The definition of standard ML***, MIT Press, 1990.

[33] E. Moggi, *Computational lambda-calculus and monads*, ***Fourth annual symposium on logic in computer science***, IEEE, 1989.

[34] A. M. Pitts, *Evaluation logic*, ***Ivth higher-order workshop, banff***, Workshops in Computing, vol. 283, Springer-Verlag, 1990.

[35] G. Plotkin, *Call-by-name, call-by-value and the lambda calculus*, ***Theoretical Computer Science***, vol. 1 (1975), pp. 125–159.

[36] D. Prawitz, ***Natural deduction: A proof-theoretical study***, Almquist and Wiksell, 1965.

[37] J.C. Reynolds, *Idealized ALGOL and its specification logic*, ***Tools and notions for program construction*** (D. Néel, editor), Cambridge University Press, 1982, pp. 121–161.

[38] G. L. Steele and G. J. Sussman, *Scheme: An interpreter for extended lambda calculus*, ***Higher-Order and Symbolic Computation***, vol. 11 (1999), no. 4, pp. 405–439.

[39] C. L. Talcott, *The essence of Rum: A theory of the intensional and extensional aspects of Lisp-type computation*, ***Ph.D. thesis***, Stanford University, 1985.

[40] ———, *A theory for program and data type specification*, ***Theoretical Computer Science***, vol. 104 (1992), pp. 129–159.

[41] ———, *A theory for program and data specification*, ***Theoretical Computer Science***, vol. 104 (1993), pp. 129–159.

[42] ———, *Reasoning about functions with effects*, ***Higher order operational techniques in semantics***, Cambridge University Press, 1996.

SCHOOL OF MATHEMATICS AND COMPUTER SCIENCE
UNIVERSITY OF NEW ENGLAND
ARMIDALE 2351, N.S.W. AUSTRALIA
E-mail: iam@turing.une.edu.au

DEPARTMENT OF COMPUTER SCIENCE
STANFORD UNIVERSITY
STANFORD, CA 94305, USA
E-mail: clt@cs.stanford.edu

EXPLICIT MATHEMATICS WITH MONOTONE INDUCTIVE DEFINITIONS: A SURVEY

MICHAEL RATHJEN

Abstract. This paper provides a survey of results about the strength of Feferman's theory of explicit mathematics augmented by least fixed point principles for monotone inductive definitions. The paper gathers all the results that are known up till now and also proves a heretofore unpublished result concerning an upper bound for the strength of $\boldsymbol{T_0} + \boldsymbol{MID}$.

§1. **Introduction.** Prompted by the question of constructive justification of Spector's consistency proof for analysis, Kreisel initiated in 1963 the study of formal theories featuring inductive definitions (cf. [15]). Proof-theoretic investigations (cf. [1], [5], [17]) of such theories have shown that the strength of monotone inductive definitions is not greater than that of positive or even accessibility inductive definitions, and the strength is the same regardless of whether intuitionistic or classical logic is being assumed. However, the status of monotone inductive definitions in a more general constructive setting, like Feferman's explicit mathematics $\boldsymbol{T_0}$, remained open. $\boldsymbol{T_0}$ is a formal framework serving many purposes. It is suitable for representing Bishop-style constructive mathematics as well as generalized recursion, including direct expression of structural concepts which admit self-application.

Let $\boldsymbol{MID}$ be the axiom asserting the existence of a least fixed point for any monotone operation f on classifications (the notion of set in explicit mathematics), and let $\boldsymbol{UMID}$ be its uniform rendering, where a least solution $\boldsymbol{lfp}(f)$ is presented as a function of the operation by adjoining a new constant $\boldsymbol{lfp}$ to the language of $\boldsymbol{T_0}$.

The question of the strength of systems of explicit mathematics with $\boldsymbol{MID}$ and $\boldsymbol{UMID}$ was raised by Feferman in [5]; we quote:

> *What is the strength of $\mathbf{T}_0+\boldsymbol{MID}$? [...] I have tried, but did not succeed, to extend my interpretation of $\boldsymbol{T_0}$ in $\Sigma^1_2 - AC + BI$ to include the statement $\boldsymbol{MID}$. The theory $\mathbf{T}_0 + \boldsymbol{MID}$ includes all constructive formulations of iteration of monotone inductive definitions of which I am aware, while $\boldsymbol{T_0}$ (in its IG axiom) is based squarely on the general iteration of accessibility*

Written in honor of Professor Sol Feferman's seventieth birthday.

Reflections on the foundations of mathematics
Essays in honor of Solomon Feferman
Edited by Wilfried Sieg, Richard Sommer, and Carolyn Talcott

inductive definitions. Thus it would be of great interest for the present subject to settle the relationship between these theories. (p. 88)

This paper provides a survey of results about the strength of Feferman's theory of explicit mathematics augmented by least fixed point principles for monotone inductive definitions.

Section 2 introduces the formal version of explicit mathematics used in this paper, and fixes the terminology for monotone fixed point principles. Section 3 provides a brief history of results about monotone fixed point principles. An upper bound for the strength of $\boldsymbol{T}_0 + \boldsymbol{MID}$ is proved in section 4.

§2. Some background on explicit mathematics.

2.1. Feferman's $\boldsymbol{T_0}$. The following presents the main features of $\boldsymbol{T_0}$; for full details cf. [3, 4]. The language of $\boldsymbol{T_0}$, $\mathcal{L}_{EM_0}$, is two-sorted, with individual variable $a, b, c, \dots, x, y, z, \dots$ and *classification* variables $A, B, C, \dots, X, Y, Z$, however, diverging from tradition, we do not wish these sorts to be disjoint realms of objects. The indended constructive interpretation is that individual variables range over a universe V of finite symbolic expressions, and the classification variables over the subuniverse of V which define properties of individuals.

N is a classification constant taken to define the class of natural numbers. **0**, s_N and $\boldsymbol{p_N}$ are operation constants whose intended interpretations are the natural number 0 and the successor and predecessor operations. Additional operation constants are **k**, **s**, **d**, **p**, $\boldsymbol{p_0}$ and $\boldsymbol{p_1}$ for the two basic combinators, definition by cases on **N**, pairing and the corresponding two projections. Additional classification constants are generated using the axioms and the constants **j**, **i** and $\boldsymbol{c_n}(\boldsymbol{n} < \omega)$ for *join*, *induction on well-founded parts* and *comprehension*.

There is no arity associated with the various constants. The *terms* of $\boldsymbol{T_0}$ are just the variables and constants of the two sorts. The atomic formulae of $\boldsymbol{T_0}$ are built up using the terms and three primitive relation symbols $=$, $\boldsymbol{App}$, and $\mathring{\in}$ as follows. If q, r, r_1, r_2 are terms, then $q = r$, $\boldsymbol{App}(q, r_1, r_2)$, and $q \mathring{\in} P$ (where P has to be a classification variable or constant) are atomic formulae. $\boldsymbol{App}(q, r_1, r_2)$ expresses that the operation q applied to r_1 yields the value r_2; $q \mathring{\in} P$ asserts[1] that q is in P or that q is classified under P.

We write $t_1 t_2 \simeq t_3$ for $\boldsymbol{App}(t_1, t_2, t_3)$.

Formulae are then generated from atomic formulae using the propositional connectives and the two quantifiers of each sort.

In order to facilitate the formulation of the axioms, the language of $\boldsymbol{T_0}$ is expanded definitionally with the symbol $\simeq$ and the auxiliary notion of an *application term* is introduced. The set of application terms is given by two clauses:

a) all terms of $\boldsymbol{T_0}$ are application terms; and
b) if s and t are application terms, then (st) is an application term.

---

[1] We use the symbol "$\mathring{\in}$" instead of "$\in$", the latter being reserved for the set–theoretic elementhood relation.

For s and t application terms, we have auxiliary, defined formulae of the form:

$$s \simeq t \;:=\; \forall y(s \simeq y \leftrightarrow t \simeq y),$$

if t is not a variable. Here $s \simeq a$ (for a a free variable) is inductively defined by:

$$s \simeq a \quad \text{is} \quad \begin{cases} s = a, & \text{if } s \text{ is a variable} \\ & \text{or a constant,} \\ \exists x, y[s_1 \simeq x \wedge s_2 \simeq y \wedge \boldsymbol{App}(x, y, a)] & \text{if } s \text{ is an application} \\ & \text{term } (s_1 s_2). \end{cases}$$

Some abbreviations are $t_1 t_2 \ldots t_n$ for $((\ldots(t_1 t_2)\ldots)t_n)$; $t\downarrow$ for $\exists y(t \simeq y)$ and $\phi(t)$ for $\exists y(t \simeq y \wedge \phi(y))$. If s, t are application terms, where t is not a classification constant or variable, then $s \mathrel{\mathring{\in}} t$ is short for $\exists X\,[t \simeq X \wedge s \mathrel{\mathring{\in}} X]$.

Some further conventions are useful. Systematic notation for *n-tuples* is introduced as follows: (t) is t, (s, t) is $\boldsymbol{p}st$, and $(t_1, \ldots, t_n)$ is defined by $((t_1, \ldots, t_{n-1}), t_n)$. t' is written for the term $\boldsymbol{s_N}t$, and $\bot$ is the atomic formula $\boldsymbol{0} \simeq \boldsymbol{0}'$. $s \mathrel{\mathring{\notin}} X$ and $s \neq t$ are short for $\neg(s \mathrel{\mathring{\in}} X)$ and $\neg(s = t)$, respectively. $\forall x \mathrel{\mathring{\in}} Y(\ldots)$ stands for $\forall x(x \mathrel{\mathring{\in}} Y \rightarrow \ldots)$. Similar conventions apply to $\exists$. Variables i, j, k, n, m are supposed to range over $\boldsymbol{N}$, i.e. $\forall n$ and $\exists n$ are short for $\forall n \mathrel{\mathring{\in}} \boldsymbol{N}$ and $\exists n \mathrel{\mathring{\in}} \boldsymbol{N}$, respectively.

A Gödel numbering for formulae is assumed in the axioms introducing the classification constants $\boldsymbol{c_n}$. A formula is said to be *elementary* if it contains only free occurrences of classification variables A (i.e. only as *parameters*), and even those free occurrences of A are restricted: A must occur only to the right of $\mathring{\in}$ in atomic formulae. The Gödel number $\boldsymbol{c_n}$ above is is the Gödel number of an elementary formula. We assume that a standard Gödel numbering has been chosen for $\mathcal{L}_{EM_0}$; if ϕ is an elementary formula and $a, b_1, \ldots, b_m, A_1, \ldots, A_n$ is a list of variables which includes all parameters of ϕ, then $\{x : \phi(x, b_1, \ldots, b_n, A_1, \ldots, A_n)\}$ stands for $\boldsymbol{c_n}(b_1, \ldots, b_n, A_1, \ldots, A_n)$, where $\boldsymbol{n}$ is the numeral that codes the pair of Gödel numbers

$$\langle \ulcorner\phi\urcorner, \ulcorner(a, b_1, \ldots, b_m, A_1, \ldots, A_n)\urcorner\rangle;$$

$\boldsymbol{n}$ is called the 'index' of ϕ and the list of variables.

In this paper, the **logic** of $\boldsymbol{T_0}$ is assumed to be that of the classical two-sorted predicate logic with identity. $\boldsymbol{T_0}$'s **non-logical axioms** are the following:

I. Basic Axioms.

a) $\forall X \exists x(X = x)$
b) $\mathrm{App}(a, b, c_1) \wedge \mathrm{App}(a, b, c_2) \rightarrow c_1 = c_2$

II. Applicative Axioms.

a) $(\boldsymbol{k}ab)\downarrow \wedge \boldsymbol{k}ab \simeq a$,
b) $(\boldsymbol{s}ab)\downarrow \wedge \boldsymbol{s}abc \simeq ac(bc)$,
c) $(\boldsymbol{p}a_1 a_2)\downarrow \wedge (\boldsymbol{p_1}a)\downarrow \wedge (\boldsymbol{p_2}a)\downarrow \wedge \boldsymbol{p_i}(\boldsymbol{p}a_1 a_2) \simeq a_i$ for $i = 0, 1$,

d) $c_1 \mathrel{\mathring{\in}} \boldsymbol{N} \wedge c_2 \mathrel{\mathring{\in}} \boldsymbol{N} \rightarrow \big[(\boldsymbol{d}abc_1c_2)\downarrow \wedge (c_1 = c_2 \rightarrow \boldsymbol{d}abc_1c_2 \simeq a) \wedge (c_1 \neq c_2 \rightarrow \boldsymbol{d}abc_1c_2 \simeq b)\big]$,

e) $a \mathrel{\mathring{\in}} \boldsymbol{N} \wedge b \mathrel{\mathring{\in}} \boldsymbol{N} \rightarrow \big[a' \downarrow \wedge \boldsymbol{p_N}(a') \simeq a \wedge \neg(a' = 0) \wedge (a' \simeq b' \rightarrow a \simeq b)\big]$.

III. Classification Axioms.

Elementary Comprehension Axiom (ECA)
$\exists X\big[X \simeq \{x : \psi(x)\} \wedge \forall x(x \mathrel{\mathring{\in}} X \leftrightarrow \psi(x))\big]$
for each elementary formula $\psi(a)$, which may contain additional parameters.

Natural Numbers

(N1) $\boldsymbol{0} \mathrel{\mathring{\in}} \boldsymbol{N} \wedge \forall x(x \mathrel{\mathring{\in}} \boldsymbol{N} \rightarrow x' \mathrel{\mathring{\in}} \boldsymbol{N})$

$(\boldsymbol{IND}_{\mathbb{N}})$ $\phi(\boldsymbol{0}) \wedge \forall x(\phi(x) \rightarrow \phi(x')) \rightarrow (\forall x \mathrel{\mathring{\in}} \boldsymbol{N})\phi(x)$

for each formula ϕ of $\mathcal{L}_{EM_0}$.

Join (J)
$\forall x \mathrel{\mathring{\in}} A\, \exists Y f x \simeq Y \rightarrow \exists X\big[X \simeq \boldsymbol{j}(A, f) \wedge \forall z\big(z \mathrel{\mathring{\in}} X \leftrightarrow \exists x \mathring{\in} A \exists y(z \simeq (x, y) \wedge y \mathrel{\mathring{\in}} f x)\big)\big]$

Inductive Generation (IG)

$$\exists X\big[X \simeq \boldsymbol{i}(A, B) \wedge \forall x \mathrel{\mathring{\in}} A\big(\forall y[(y, x) \mathrel{\mathring{\in}} B \rightarrow y \mathrel{\mathring{\in}} X] \rightarrow x \mathrel{\mathring{\in}} X\big)$$

$$\wedge\ \big(\forall x \mathrel{\mathring{\in}} A\,[\forall y\,((y, x) \mathrel{\mathring{\in}} B \rightarrow \phi(y)) \rightarrow \phi(x)] \rightarrow \forall x \mathrel{\mathring{\in}} X\, \phi(x)\big)\big]$$

where ϕ is an arbitrary formula of $\boldsymbol{T_0}$.

Frequently, when **IG** is invoked, it actually suffices to use a restricted form.

Restricted Inductive Generation IG $\restriction$

$$\exists X\big[X \simeq \boldsymbol{i}(A, B) \wedge \forall x \mathrel{\mathring{\in}} A\big(\forall y[(y, x) \mathrel{\mathring{\in}} B \rightarrow y \mathrel{\mathring{\in}} X] \rightarrow x \mathrel{\mathring{\in}} X\big)$$

$$\wedge\ \forall Z\big(\forall x \mathrel{\mathring{\in}} A\,[\forall y\,((y, x) \mathrel{\mathring{\in}} B \rightarrow y \mathrel{\mathring{\in}} Z) \rightarrow x \mathrel{\mathring{\in}} Z] \rightarrow \forall a \mathrel{\mathring{\in}} X\, x \mathrel{\mathring{\in}} Z)\big].$$

By $\boldsymbol{T_0}\restriction$ we denote the alteration of $\boldsymbol{T_0}$ with **IG** $\restriction$ in place of **IG** and with $\boldsymbol{N}$-induction, i.e. $\boldsymbol{IND}_{\mathbb{N}}$, replaced by the $\boldsymbol{N}$**-induction axiom**

$$\forall Z\,\big[0 \mathrel{\mathring{\in}} Z \wedge \forall x\,(x \mathrel{\mathring{\in}} Z \rightarrow x' \mathrel{\mathring{\in}} Z) \rightarrow \forall x \mathrel{\mathring{\in}} \boldsymbol{N}\,(x \mathrel{\mathring{\in}} Z)\big].$$

2.2. Formulation of monotone inductive definition principles. In this subsection we describe the monotone inductive definition principle and its uniform version. Several other principles considered in this paper will also be described.

DEFINITION 2.1. For *extensional equality* of classifications we use the shorthand "$\mathrel{\mathring{=}}$", i.e.

$$X \mathrel{\mathring{=}} Y \ :=\ \forall v(v \mathrel{\mathring{\in}} X \leftrightarrow v \mathrel{\mathring{\in}} Y).$$

Further, let

$$X \mathrel{\mathring{\subseteq}} Y \ :=\ \forall v(v \mathrel{\mathring{\in}} X \rightarrow v \mathrel{\mathring{\in}} Y).$$

To state the monotone fixed point principle for subclassifications of a given classification A we introduce the following shorthands:

$$\boldsymbol{Clop}(f, A) \quad \text{if} \quad \forall X \mathrel{\mathring{\subseteq}} A\, \exists Y \mathrel{\mathring{\subseteq}} A\; fX \simeq Y$$

$$\boldsymbol{Ext}(f, A) \quad \text{if} \quad \forall X \mathrel{\mathring{\subseteq}} A\, \forall Y \mathrel{\mathring{\subseteq}} A\, [X \mathrel{\mathring{=}} Y \to fX \mathrel{\mathring{=}} fY]$$

$$\boldsymbol{Mon}(f, A) \quad \text{if} \quad \forall X \mathrel{\mathring{\subseteq}} A\, \forall Y \mathrel{\mathring{\subseteq}} A [X \mathrel{\mathring{\subseteq}} Y \to fX \mathrel{\mathring{\subseteq}} fY].$$

$$\boldsymbol{Lfp}(Y, f, A) \quad \text{if} \quad fY \mathrel{\mathring{\subseteq}} Y \wedge Y \mathrel{\mathring{\subseteq}} A \wedge \forall X \mathrel{\mathring{\subseteq}} A \left[fX \mathrel{\mathring{\subseteq}} X \to Y \mathrel{\mathring{\subseteq}} X\right]$$

When f satisfies $\boldsymbol{Clop}(f, A)$, we call f a *classification operation on A*. When f satisfies $\boldsymbol{Clop}(f, A)$ and $\boldsymbol{Ext}(f, A)$, we call f *extensional* or an *extensional operation on A*. When f satisfies $\boldsymbol{Clop}(f, A)$ and $\boldsymbol{Mon}(f, A)$, we say that f is a *monotone operation on A*. Since monotonicity entails extensionality, a monotone operation is always extensional.

Now we state $\boldsymbol{UMID}_A$.

$\boldsymbol{MID}_A$ **(Monotone Inductive Definition on A)**

$$\forall f\, [\boldsymbol{Clop}(f, A) \wedge \boldsymbol{Mon}(f, A) \to \exists Y\, \boldsymbol{Lfp}(Y, f, A)].$$

$\boldsymbol{UMID}_A$ **(Uniform Monotone Inductive Definition on A)**

$$\forall f\, [\boldsymbol{Clop}(f, A) \wedge \boldsymbol{Mon}(f, A) \to \boldsymbol{Lfp}(\boldsymbol{lfp}(f), f, A)].$$

$\boldsymbol{UMID}_A$ states that if f is monotone on subclassifications of A, then $\boldsymbol{lfp}(f)$ is a least fixed point of f.

Let $\boldsymbol{V}$ be the universe, i.e. $\boldsymbol{V} := \{x : x = x\}$. By $\boldsymbol{MID}$ and $\boldsymbol{UMID}$ we denote the principles $\boldsymbol{MID}_{\boldsymbol{V}}$ and $\boldsymbol{UMID}_{\boldsymbol{V}}$, respectively.[2]

2.3. Subsystems of second order arithmetic. The language $\mathcal{L}_2$ of second-order arithmetic contains (free and bound) number variables $a, b, c, \ldots, x, y, z, \ldots$, (free and bound) set variables $A, B, C, \ldots, X, Y, Z, \ldots$, the constant 0, function symbols Suc, $+$, $\cdot$, and relation symbols $=, <, \in$. Suc stands for the successor function.

Terms are built up as usual. For $n \in \boldsymbol{N}$, let $\bar{n}$ be the canonical term denoting n. Formulae are built from the prime formulae $s = t$, $s < t$, and $s \in A$ using $\wedge, \vee, \neg, \forall x, \exists x, \forall X$ and $\exists X$ where s, t are terms.

We like to point out that equality in $\mathcal{L}_2$ is only a relation on numbers. However, equality of sets will be considered a defined notion, namely

$$A = B \text{ iff } \forall x[x \in A \Leftrightarrow x \in B].$$

As usual, number quantifiers are called bounded if they occur in the context $\forall x(x < s \to \ldots)$ or $\exists x(x < s \wedge \ldots)$ for a term s which does not contain x. The $\boldsymbol{\Delta}^0_0$-formulae are those formulae in which all quantifiers are bounded number quantifiers, $\boldsymbol{\Sigma}^0_k$-formulae are formulae of the form $\exists x_1 \forall x_2 \ldots Q x_k \phi$, where ϕ is

[2]The acronym for the principle $\boldsymbol{MID}$ in Feferman's paper [4], section 7 was MIG $\restriction$.

Δ_0^0, Π_k^0-formulae are those of the form $\forall x_1 \exists x_2 \ldots Qx_k\phi$. The union of all Π_k^0- and Σ_k^0-formulae for all $k \in \mathbb{N}$ is the class of *arithmetical* or Π_∞^0*-formulae*. The Σ_k^1 (Π_k^1)-formulae are the formulae $\exists X_1 \forall X_2 \ldots QX_k\phi$ (resp. $\forall X_1 \exists X_2 \ldots Qx_k\phi$) for arithmetical ϕ.

The basic axioms in all theories of second-order arithmetic are the defining axioms of $0, 1, +, \cdot, <$ and the *induction axiom*

$$\forall X(0 \in X \wedge \forall x(x \in X \rightarrow x + 1 \in X) \rightarrow \forall x(x \in X)),$$

respectively the *schema of induction*

$$\textbf{\textit{IND}} \qquad \phi(0) \wedge \forall x(\phi(x) \rightarrow \phi(x+1)) \rightarrow \forall x\phi(x),$$

where ϕ is an arbitrary $\mathcal{L}_2$-formula.

We consider the axiom schema of $\mathcal{C}$*-comprehension* for formula classes $\mathcal{C}$ which is given by

$$\mathcal{C} - \textbf{\textit{CA}} \qquad \exists X \forall x(x \in X \leftrightarrow \phi(x))$$

for all formulae $\phi \in \mathcal{C}$.

We will only consider theories containing at least $\Pi_\infty^0 - \textbf{\textit{CA}}$. For each axiom schema $\textbf{\textit{Ax}}$ we denote by $(\textbf{\textit{Ax}})$ the theory consisting of the basic arithmetical axioms, the schema $\Pi_\infty^0 - \textbf{\textit{CA}}$, the schema of induction and the schema $\textbf{\textit{Ax}}$. If we replace the schema of induction by the induction axiom, we denote the resulting theory by $(\textbf{\textit{Ax}})\restriction$.

An example for these notations is the theory $(\Pi_1^1\text{-}\textbf{\textit{CA}})$ which contains the induction schema, whereas $(\Pi_1^1\text{-}\textbf{\textit{CA}})\restriction$ only contains the induction axiom in addition to the comprehension schema for Π_1^1-formulae.

When arguing in a particular formal theory, we also say that a formula belongs to one of the aforementioned formula classes if it is equivalent to one formula of the class over this theory.

Any set R gives rise to a binary relation $\prec_R$ defined by $y \prec_R x := 2^y \cdot 3^x \in R$. We also use yRx as short for $y \prec_R x$.

Using the latter coding, one can formulate the axiom of choice for formulae ϕ in $\mathcal{C}$ by

$$\mathcal{C} - \textbf{\textit{AC}} \qquad \forall x \exists Y \phi(x, Y) \rightarrow \exists Y \forall x \phi(x, Y_x).$$

$\textbf{\textit{BI}}$ is the so-called principle of *Bar Induction*, i.e. the axiom schema

$$\forall X \left(\mathrm{WO}(<_X) \wedge \forall n \left[\forall m <_X n\, \Phi(m) \rightarrow \Phi(n)\right] \rightarrow \forall n \Phi(n)\right)$$

for all formulae Φ of the language of second order arithmetic.

2.4. Set theories. The axiom systems for set theory considered in this paper are formulated in the usual language of set theory containing $\in$ as the only non-logical symbol besides $=$. Formulae are built from prime formulae $a \in b$ and $a = b$ by use of propositional connectives and quantifiers $\forall x, \exists x$. Bounded quantifiers $\forall x \in a$, $\exists x \in a$ are defined as usual. Δ_0-formulae are the formulae wherein all quantifiers are bounded; Σ_1-formulae are those of the form $\exists x \varphi(x)$ where $\varphi(a)$ is a Δ_0-formula. For $n > 0$, Π_n-formulae (Σ_n-formulae) are the formulae with a prefix of n alternating unbounded quantifiers starting with a universal (existential) one followed by a Δ_0-formula. The class of Σ-formulae is the smallest class of formulae containing the Δ_0-formulae which is closed under $\wedge$, $\vee$, bounded quantification and unbounded existential quantification.

DEFNITION 2.2. Kripke-Platek set theory, ***KP***, consists of the axioms of Extensionality, Pairing, Union, Infinity and of the axiom schemata of Separation and Collection for Δ_0-formulae as well as the Foundation schema for arbitrary formulae.

$\boldsymbol{KP^r}$ arises from ***KP*** by replacing the axiom schema of Foundation by the Foundation axiom

$$\forall x(\exists y(y \in x) \to \exists y(y \in x \wedge \forall z \in x(z \notin y))).$$

$\boldsymbol{KP^w}$ is obtained from $\boldsymbol{KP^r}$ by adding the schema

$$\boldsymbol{IND}_\omega \quad \forall x \in \omega(\forall y \in x\phi(y) \to \phi(x)) \to \forall x \in \omega\phi(x)$$

of induction on ω to $\boldsymbol{KP^r}$ (for all formulae ϕ).

A non-empty transitive set which is a model of ***KP*** is called an *admissible set*.

DEFNITION 2.3. ***KPi*** is the theory ***KP*** plus an axiom asserting that any set is contained in an admissible set. The theories $\boldsymbol{KPi^r}$ and $\boldsymbol{KPi^w}$ are derived as in the case of ***KP***.

DEFNITION 2.4. Σ_n-Separation (abbreviated Σ_n-***Sep***) is the schema of axioms

$$\exists z\, \forall u\, \big(u{\in}z \;\leftrightarrow\; [u{\in}a \wedge \phi(u)]\big)$$

for all set-theoretic Σ_n-formulae ϕ with z not free in ϕ.

DEFNITION 2.5. We will use Gödel's constructible hierarchy $\boldsymbol{L} = (\boldsymbol{L}_\alpha)_{\alpha \in \mathrm{On}}$ in one of its usual formulations. For definiteness let

$$\boldsymbol{L}_0 = \emptyset, \quad \boldsymbol{L}_{\alpha+1} = \boldsymbol{Def}(\boldsymbol{L}_\alpha), \quad \boldsymbol{L}_\lambda = \bigcup_{\alpha<\lambda} \boldsymbol{L}_\alpha \text{ for } \lambda \in \mathrm{Lim}.$$

Here $\boldsymbol{Def}(x)$ is the set of all definable subsets of x.

DEFNITION 2.6. An ordinal κ is said to be stable if $\boldsymbol{L}_\kappa \prec_1 \boldsymbol{L}$, i.e. $\boldsymbol{L}_\kappa$ is a Σ_1-elementary substructure of $\boldsymbol{L}$.

Let $\rho > \kappa$. κ is ρ-stable if $\boldsymbol{L}_\kappa \prec_1 \boldsymbol{L}_\rho$.

LEMMA 2.7. *$\boldsymbol{KP} + \Sigma_1$-**Sep** proves that there exist recursively inaccessible ordinals $\kappa_1 < \kappa_2 < \kappa_3$ such that*

$$\boldsymbol{L}_{\kappa_1} \prec_1 \boldsymbol{L}_{\kappa_2} \prec_1 \boldsymbol{L}_{\kappa_3}.$$

*In particular, $\boldsymbol{L}_{\kappa_3}$ is a model of **KPi** plus the assertion*

$$\exists\gamma\exists\pi\,[\gamma < \pi \;\wedge\; \boldsymbol{L}_\gamma \prec_1 \boldsymbol{L}_\pi \;\wedge\; \pi \textit{ is recursively inaccessible}].$$

PROOF. See [20], Lemma 2.13. ⊣

S3. A survey. Proof-theoretic investigations of formal first order theories of positive iterated inductive definitions have shown that their strength is the same for classical and intuitionistic logic.

THEOREM 3.1. (Buchholz, Pohlers, Sieg 1978), [1]

$$\textbf{\textit{pos-ID}}^c_\alpha \equiv \textbf{\textit{pos-ID}}^i_\alpha \equiv \textbf{\textit{ID}}^i_\alpha(\mathcal{O}) \equiv \textbf{\textit{ID}}^{i,acc}_\alpha,$$

where α is an effectively presented ordinal, "c" stands for classical logic, "i" stands for intuitionistic logic, $\textbf{\textit{ID}}^i_\alpha(\mathcal{O})$ is the intuitionistic theory of α times iterated tree classes.

Feferman also showed that iterated monotone inductive definitions do not provide more strength than positive ones.

THEOREM 3.2. (Feferman 1982), [5]

$\textbf{\textit{mon-ID}}^c_\alpha$ *is conservative over* $\textbf{\textit{pos-ID}}^c_\alpha$.

The same pattern propagates when one considers theories where the iteration of inductive definitions is internalized, i.e. theories of second order arithmetic with axioms stating that first order inductive definitions (with set parameters) can be iterated along any well-ordering.

THEOREM 3.3. [17]

$$\textbf{\textit{mon-Aut-ID}}^c \equiv \textbf{\textit{pos-Aut-ID}}^c$$
$$\textbf{\textit{mon-Aut-ID}}^c \equiv \textbf{\textit{pos-Aut-ID}}^i$$

where $\equiv$ stands for proof-theoretic equivalence.

Positive inductive definitions are captured in $\textbf{\textit{T}}_0$ by the following principle:

DEFNITION 3.4. $\textbf{\textit{EID}}^+$ is the schema of *positive elementary inductive definitions*,

$$\exists X\left[\forall u\,(\phi(u,X)\to u\mathring{\in}X)\ \wedge\ \forall Y\,[\forall u\,(\phi(u,Y)\to u\mathring{\in}Y)\ \to\ X\subseteq Y]\right]$$

for each elementary formula $\phi(u, X)$ in which X occurs only positively. To be precise, $\phi(u, X)$ may have additional free variables of both kinds which are not exhibited.

THEOREM 3.5. $\textbf{\textit{T}}_0 + \textbf{\textit{EID}}^+$ *is of the same strength as* $\textbf{\textit{T}}_0$.

PROOF. It is readily seen that the recursion-theoretic model for $\textbf{\textit{T}}_0$ given by Feferman (cf. [6]) also validates the principle $\textbf{\textit{EID}}^+$. Moreover, the modelling can be carried out in the theory $(\Delta^1_2\text{-}\textbf{\textit{CA}}) + \textbf{\textit{BI}}$ and the latter theory is known to have the same strength as $\textbf{\textit{T}}_0$ by results of Jäger and Pohlers (cf. [13], [12]). ⊣

The general principle of monotone inductive definitions in explicit mathematics is captured by the principles ***MID*** and ***UMID***. First investigations in the way of strength of $\textbf{\textit{T}}_0 + \textbf{\textit{MID}}$ were begun by Takahashi (cf. [21]). It turned out that even the construction of models of $\textbf{\textit{T}}_0 + \textbf{\textit{MID}}$ was surprisingly difficult.

THEOREM 3.6. (Takahashi 1989), [21]

$\textbf{\textit{T}}_0 + \textbf{\textit{MID}}$ *is interpretable in the fragment of analysis with Π^1_2-comprehension and Bar induction.*

The question whether $\mathbf{T}_0 + \mathbf{MID}$ is stronger than $\mathbf{T}_0$ was left open by Takahashi's work. New insights into the strength of $\mathbf{MID}$ came with Rathjen's paper [18]. In [18] it was shown that $\mathbf{T}_0 + \mathbf{MID}$, when based on classical logic, also proves the existence of non-monotone inductive definitions that arise from arbitrary extensional operations on classifications. From the latter, one can deduce that $\mathbf{T}_0 + \mathbf{MID}$ is indeed a much stronger theory than $\mathbf{T}_0$.

Any extensional operation Θ gives rise to an inductively defined class Θ^∞ that is set-theoretically introduced via

$$\Theta^\infty := \bigcup_\alpha \Theta^\alpha,$$

$$\Theta^\alpha := \Theta\Big(\bigcup_{\beta<\alpha} \Theta^\beta\Big) \cup \bigcup_{\beta<\alpha} \Theta^\beta,$$

where α ranges over the ordinals.

THEOREM 3.7. [18]

(i) ($\mathbf{T}_0 \restriction + \mathbf{MID}$) *To any extensional operator Θ there can be associated a monotone operator Υ and a total operation $x \mapsto \Theta^x$, giving a classification Θ^x for all x, such that with $<_\Upsilon$ denoting the prewellordering pertaining to Υ (according to* [18], *Definition* 3.2),

(1) $$\Theta^x \mathrel{\mathring{=}} \Theta\Big(\bigcup_{y<_\Upsilon x} \Theta^y\Big) \cup \bigcup_{y<_\Upsilon x} \Theta^y,$$

and, for the classification $\mathbf{I}_\Theta$ defined by

(2) $$\mathbf{I}_\Theta := \bigcup_{x \mathring{\in} \mathbf{V}} \Theta^x,$$

it holds

$$\Theta(\mathbf{I}_\Theta) \mathrel{\mathring{\subseteq}} \mathbf{I}_\Theta.$$

Put differently, $\mathbf{I}_\Theta$ is a classification that arises by iterating Θ along $<_\Upsilon$ and is closed under Θ.

(ii) $\mathbf{T}_0 \restriction + \mathbf{MID}$ *proves that $\mathbf{T}_0$ has a model.*

Central technical results employed in [18] are that the stage comparison theorem for monotone operators can be proved in $\mathbf{T}_0 \restriction + \mathbf{MID}$ and that techniques for capturing closure ordinals of non-monotone operators via stage comparison prewellorderings of monotone operators as developed by Harrington and Kechris (cf. [10], [11]) can be recast in $\mathbf{T}_0 \restriction + \mathbf{MID}$.

Subsequently Glaß, Rathjen and Schlüter in [8] characterized the exact proof-theoretic strength of several variants of $\mathbf{T}_0 + \mathbf{MID}$. Roughly speaking, $\mathbf{MID}$ turned out to be related to lightface Π^1_2 comprehension. [8] utilized all the previous work on $\mathbf{MID}$, that is, Takahashi's subtle model constructions in [21], Glaß' thesis [7], and the crucial [18]. In addition, [8] used techniques from generalized recursion theory relating non-monotone inductive definitions to stability in set theory (cf. [2]).

conditions of [8], section 4.3, and therefore the model construction of [8], section 4.3 can be carried out in $\mathcal{K}$.

DEFINITION 4.3. For sets $X, X_1, \dots, X_k \subseteq \omega$ with $X, X_1, \dots, X_k \in \boldsymbol{L}_\pi$ let $\rho(X, X_1, \dots, X_k)$ denote the least recursively inaccessible ordinal ν with $X, X_1, \dots, X_k \in \boldsymbol{L}_\nu$. Observe that the existence of $\rho(X, X_1, \dots, X_k)$ follows from the fact that π is recursively inaccessible.

a) An operator $\Gamma : \boldsymbol{Pow}(\omega) \cap \boldsymbol{L}_\pi \to \boldsymbol{Pow}(\omega) \cap \boldsymbol{L}$ will be called a Σ_{inac+}-operator iff

$$\forall X \in \boldsymbol{Pow}(\omega) \cap \boldsymbol{L}_\pi \forall n \in \omega \left(n \in \Gamma(X) \Leftrightarrow \boldsymbol{L}_{\rho(X, X_1, \dots, X_k)} \models \phi[n, X, X_1, \dots, X_k]\right)$$

holds for some Σ_1-formula ϕ and $X_1, \dots, X_k \in \boldsymbol{Pow}(\omega) \cap \boldsymbol{L}_\pi$. The sets $X_1, \dots, X_k \subseteq \omega$ are called the parameters of Γ.

b) The iteration stages of an operator Γ are defined as

$$I_\Gamma^\alpha = \Gamma(I_\Gamma^{<\alpha}) \cup I_\Gamma^{<\alpha} \text{ where } I_\Gamma^{<\alpha} = \bigcup_{\beta<\alpha} I_\Gamma^\beta.$$

Note however that these iterates might not exist as $I_\Gamma^\alpha = \Gamma(I_\Gamma^{<\alpha})$ is not defined when $I_\Gamma^{<\alpha} \notin \boldsymbol{L}_\pi$.

c) A set $X \subseteq \omega$ is called a *sub-fixed point* of an operator Γ if $\Gamma(X) \subseteq X$.

REMARK 4.4. The definition of $Y = \Gamma(X)$ is a Σ-statement, namely it expresses that the ordinal $\rho(X, X_1, \dots, X_k)$ exists and that a certain statement holds in $\boldsymbol{L}_{\rho(X, X_1, \dots, X_k)}$.

LEMMA 4.5. *If Γ is a Σ_{inac+}-operator with parameters from L_γ, then $(I_\Gamma^\alpha)_{\alpha<\gamma}$ can be defined by Σ-recursion in $\boldsymbol{L}_\gamma$ and therefore is an element of $\boldsymbol{L}_{\gamma+1}$.*

PROOF. Note that for all $\alpha < \gamma$ there exists $\alpha < \beta < \gamma$ such that β is recursively inaccessible. The rest is standard. ⊣

LEMMA 4.6. *Let Γ be a monotone Σ_{inac+}-operator with parameters from $\boldsymbol{L}_\gamma$. Then for all $X \in \boldsymbol{L}_\gamma \cap \boldsymbol{Pow}(\omega)$ such that $\Gamma(X) \subseteq X$ and all $\alpha < \gamma$ we have $I_\Gamma^\alpha \subseteq X$.*

PROOF. Obvious induction on α. ⊣

PROPOSITION 4.7. *If Γ is a Σ_{inac+}-operator with parameters from L_γ, then $\Gamma(I_\Gamma^{<\gamma}) \subseteq I_\Gamma^{<\gamma}$.*

PROOF. Assume $n \in \Gamma(I_\Gamma^{<\gamma})$. Then

$$\boldsymbol{L}_{\rho(I_\Gamma^{<\gamma}, X_1, \dots, X_k)} \models \phi[n, I_\Gamma^{<\gamma}, X_1, \dots, X_k].$$

Hence

$$\exists \eta\, \boldsymbol{L}_{\rho(I_\Gamma^{<\eta}, X_1, \dots, X_k)} \models \phi[n, I_\Gamma^{<\eta}, X_1, \dots, X_k].$$

The latter formula is Σ_1 with parameters in $\boldsymbol{L}_\gamma$. Hence, as $\boldsymbol{L}_\gamma \prec_1 \boldsymbol{L}$, there exists $\eta < \gamma$ such that $\boldsymbol{L}_{\rho(I_\Gamma^{<\eta}, X_1, \dots, X_k)} \models \phi[n, I_\Gamma^{<\eta}, X_1, \dots, X_k]$. This implies $n \in I_\Gamma^\eta$ and hence $n \in I_\Gamma^{<\gamma}$. ⊣

COROLLARY 4.8. *Every Σ_{inac+}-operator Γ with parameters from $\boldsymbol{L}_\gamma$ has a sub-fixed point in $\boldsymbol{L}_\gamma$.*

Moreover, if Γ is monotone, then the least fixed point of Γ is in $\boldsymbol{L}_\gamma$.

PROOF. By Proposition 4.7 we have $\Gamma(I_\Gamma^{<\gamma}) \subseteq I_\Gamma^{<\gamma}$. Thus, by $\boldsymbol{L}_\gamma \prec_1 \boldsymbol{L}$, there exists $\eta < \gamma$ such that $\Gamma(I_\Gamma^{<\eta}) \subseteq I_\Gamma^{<\eta}$. Since $I_\Gamma^{<\eta} \in \boldsymbol{L}_\gamma$, this proves the claim.

The "moreover" part follows from the previous together with Lemma 4.6. ⊣

Section 4.2 of [8] defines structures

$$\mathfrak{S}_{M_0.\alpha} = (S, \boldsymbol{CL}_{M_0.\alpha}, \mathring{\in}_{M_0.\alpha}, App, N_S, \boldsymbol{k}, \boldsymbol{s}, \boldsymbol{p}, \boldsymbol{p}_0, \boldsymbol{p}_1, \boldsymbol{d}, \boldsymbol{s}_N, \boldsymbol{p}_N, 0, (\boldsymbol{c}_m)_{m\in\omega}, \boldsymbol{j}, \boldsymbol{i})$$

where M_0 is a finite subset of M.[4] $\mathfrak{S}_{M_0.\alpha}$ also depends on an assignment $j_0 : M_0 \to \boldsymbol{Pow}(S)$. Sometimes we prefer to make this dependence explicit by writing $\mathfrak{S}^{j_0}_{M_0.\alpha}$ instead of just $\mathfrak{S}_{M_0.\alpha}$.

LEMMA 4.9. *Let M_0 be finite and $j_0 : M_0 \to \boldsymbol{Pow}(S)$ satisfy $\forall b \in M_0\, j_0(b) \in \boldsymbol{L}_\alpha$ where α is recursively inaccessible, then*

(i) $\forall \beta \geq \alpha\ \mathfrak{S}_{M_0.\alpha} = \mathfrak{S}_{M_0.\beta}$.

(ii) $\mathfrak{S}_{M_0.\alpha} \models \boldsymbol{T_0}$.

PROOF. Ad (i): It suffices to show $\mathfrak{S}^{j_0}_{M_0.\alpha} = \mathfrak{S}^{j_0}_{M_0.\alpha+1}$ as the assertion follows then by induction on β. We have to show

$$\mathrm{Cl}^{j_0}_{M_0.\alpha} = \mathrm{Cl}^{j_0}_{M_0.\alpha+1}.$$

Let $c \in \mathrm{Cl}^{j_0}_{M_0.\alpha+1}$. If c is of the form $\boldsymbol{j}(a, f)$, then $a \in \mathrm{CL}^{j_0}_{M_0.\alpha}$ and $\mathfrak{S}^{j_0}_{M_0.\alpha} \models \forall x \mathring{\in} a\, \exists Y(fx \simeq Y)$. Then $a \in \mathrm{CL}^{j_0}_{M_0.\delta}$ for some $\delta < \alpha$ and

$$\forall x \in S\, \exists \beta < \alpha \left[\beta \geq \delta \wedge \mathfrak{S}^{j_0}_{M_0.\beta} \models x \mathring{\in} a \to \exists Y(fx \simeq Y)\right].$$

As the function $(\xi \mapsto \mathfrak{S}^{j_0}_{M_0.\xi})_{\xi<\alpha}$ is Σ_1- definable in $\boldsymbol{L}_\alpha$ it follows by Σ_1-collection in $\boldsymbol{L}_\alpha$ that there exists $\delta < \nu < \alpha$ such that

$$\forall x \in S\, \exists \beta < \nu \left[\beta \geq \delta \wedge \mathfrak{S}^{j_0}_{M_0.\beta} \models x \mathring{\in} a \to \exists Y(fx \simeq Y)\right].$$

The latter implies $c = \boldsymbol{j}(a, f) \in \mathrm{Cl}^{j_0}_{M_0.\nu}$, thus $c \in \mathrm{Cl}^{j_0}_{M_0.\alpha}$.

Finally, c is of any other form than the above, then the assertion follows by similar arguments.

(ii) is an immediate consequence of (i). ⊣

For the rest of this section we fix a surjection

$$\hat{} : M \to \bigcup\{X \in L_\gamma : X \in \boldsymbol{Pow}(S/F) \text{ for some finite } F \subseteq \boldsymbol{B}\}$$

as in [8], Section 4.3 and also let $\boldsymbol{CL}_{M.\alpha}$ and $\mathfrak{S}_{M.\alpha}$ be defined as in [8], Section 4.3, where $\boldsymbol{B}$ is the set of non-pairs as defined in [8], Definition 4.1.

LEMMA 4.10. *Let $\hat{}: M \to \boldsymbol{Pow}(S)$ and $\mathfrak{S}_M$ be as above. Assume that for some $f \in S$ and for all $b \in M$*

$$\mathfrak{S}_{M.\gamma} \models \exists Y(fb \simeq Y)$$

as well as for all $b_1, b_2 \in M$

$$\mathfrak{S}_{M.\gamma} \models b_1 \mathring{=} b_2 \to fb_1 \mathring{=} fb_2.$$

[4] M is a set providing names for the parameters to be coded into our model (cf. [8], Section 4.1).

Then there is a finite $F \subseteq \boldsymbol{B}$ and a Σ_{inac+}-operator $\Gamma : \boldsymbol{Pow}(\omega) \cap \boldsymbol{L}_\pi \to \boldsymbol{Pow}(\omega) \cap \boldsymbol{L}$ such that for all $b \in M$ with $\widehat{b} \in \boldsymbol{Pow}(S/F)$

$$\Gamma(\widehat{b}) \in \boldsymbol{Pow}(S/F) \text{ and } \mathfrak{S}_{M.\gamma} \models x \mathring{\in} fb \Leftrightarrow x \in \Gamma(\widehat{b}).$$

PROOF. By Lemma [8], Lemma 4.14 choose $F \supseteq \boldsymbol{supp_B}(f)$ such that $\widehat{\ }: F \cap M \to \boldsymbol{Pow}(S/F)$. Write $F \cap M = \{b_1, \dots, b_n\}$ and choose $b_0 \in M \setminus F$. Then we have $\mathfrak{S}_{M.\gamma} \models fb_0 \simeq a$ for some $a \in \boldsymbol{CL}_{M.\gamma}$. Let $M_0 := \{b_0, \dots, b_n\}$.

Then $\boldsymbol{supp_B}(a) \cap M \subseteq \boldsymbol{supp_B}(f) \cup \boldsymbol{supp_B}(b_0) \subseteq \{b_0, \dots, b_n\}$ and so by [8], Proposition 4.12 (for some $M_1 \supseteq M_0$ such that $a \in \boldsymbol{CL}_{M_1.\gamma}$) we see that $a \in \boldsymbol{CL}_{M_0.\gamma}$.

Define the operator $\Gamma : \boldsymbol{Pow}(\omega) \cap \boldsymbol{L}_\pi \to \boldsymbol{Pow}(\omega) \cap \boldsymbol{L}$ by

$$\Gamma(X) = \{x : \mathfrak{S}^{j_0}_{M_0.\alpha} \models x \mathring{\in} a, \text{ where } \alpha = \rho(X, \widehat{b_1}, \dots, \widehat{b_n}) \text{ and}$$
$$j_0(b_0) = X, j_0(b_1) = \widehat{b_1}, \dots, j_0(b_n) = \widehat{b_n}\}.$$

Note that (by Lemma 4.9) $a \in \mathrm{Cl}^{j_0}_{M_0.\alpha}$ when $\alpha = \rho(X, \widehat{b_1}, \dots, \widehat{b_n})$. Obviously, Γ is a Σ_{inac+}-operator in the parameters $\widehat{b_1}, \dots, \widehat{b_n}$.

Claim 1: If $\widehat{b} \in \boldsymbol{Pow}(S/F)$ and $b \notin \{b_1, \dots, b_n\}$, and if $\sigma \in \boldsymbol{Aut}(\boldsymbol{B}/F)$, then

$$\mathfrak{S}_{M.\gamma} \models x \mathring{\in} a[b_0 := b] \Leftrightarrow \sigma(x) \mathring{\in} a[b_0 := b].$$

Proof of claim 1. Let $M_0' := \{b, b_1, \dots, b_n\}$. Then $b, b_0 \notin \boldsymbol{supp_B}(f)$ and therefore $fb \simeq a[b_0 := b]$, consequently $\boldsymbol{supp_B}(a[b_0 := b]) \cap M \subseteq M_0'$, which by [8], Lemma 4.12 gives $a[b_0 := b] \in \boldsymbol{CL}_{M_0'.\gamma}$.

For $\sigma \in \boldsymbol{Aut}(\boldsymbol{B}/F)$ and $x \in S$ choose some $b' \in M$ such that $\widehat{b} = \widehat{b'}$ but $b' \notin \boldsymbol{supp_B}(x) \cup \boldsymbol{supp_B}(\sigma(x)) \cup \{b_1, \dots, b_n\}$. Use this to define $\sigma' \in \boldsymbol{Aut}(\boldsymbol{B}/\{b'\} \cup F)$ which agrees with σ on $\boldsymbol{supp_B}(x)$. Then we easily compute:

$$\begin{aligned}
\text{(3)} \qquad \mathfrak{S}_{M.\gamma} \models x \mathring{\in} a[b_0 := b] &\Leftrightarrow \mathfrak{S}_{M.\gamma} \models x \mathring{\in} a[b_0 := b'] \\
&\quad \text{because } fb' \simeq a[b_0 := b'] \text{ and} \\
&\quad f \text{ is extensional on } M. \\
&\Leftrightarrow \mathfrak{S}_{M.\gamma} \models \sigma'(x) \mathring{\in} a[b_0 := b'] \\
&\quad \text{by [8], Lemma 4.9} \\
&\Leftrightarrow \mathfrak{S}_{M.\gamma} \models \sigma(x) \mathring{\in} a[b_0 := b] \text{ by} \\
&\quad \text{extensionality of } f \text{ on } M.
\end{aligned}$$

Claim 2: For $\widehat{b} \in \boldsymbol{Pow}(S/F)$,

$$\mathfrak{S}_{M.\gamma} \models x \mathring{\in} fb \text{ if and only if } x \in \Gamma(\widehat{b}).$$

Proof of claim 2. First we consider the case that $b \notin \{b_1, \dots, b_n\}$. We commence by showing

$$\text{(4)} \qquad \mathfrak{S}_{M.\gamma} \models x \mathring{\in} a[b_0 := b] (= fb) \Leftrightarrow \widetilde{\mathfrak{S}_{M_0.\gamma}} \models x \mathring{\in} a$$

where the latter model is based on $\tilde{\ }: M_0 \to \boldsymbol{Pow}(S/F)$ with $\widetilde{b_0} = \widehat{b}$ and $\widetilde{b_i} = \widehat{b_i}$ for $i = 1, \ldots, n$. To this end, choose $\sigma \in \boldsymbol{Aut}(\boldsymbol{B}/F)$ with $\sigma(b) = b_0$ and let $M_0' = \{b, b_1, \ldots, b_n\}$ and $M_1 = \{b, b_0, b_1, \ldots, b_n\}$. We can extend $\hat{\ }$ from M_0' to a mapping $\tilde{\ } : M_1 \to \boldsymbol{Pow}(S/F)$ by additionally defining $\widetilde{b_0} := \widehat{b}$. Since $a[b_0 := b]$ in $\boldsymbol{CL}_{M_0',\gamma}$ by [8], Lemma 4.9 we have

$$\begin{aligned}\mathfrak{S}_{M,\gamma} \models x \mathrel{\mathring{\in}} a[b_0 := b] &\Leftrightarrow \mathfrak{S}_{M_0',\gamma} \models x \mathrel{\mathring{\in}} a[b_0 := b]\\ &\Leftrightarrow \mathfrak{S}_{M_0',\gamma} \models \sigma^{-1}(x) \mathrel{\mathring{\in}} a[b_0 := b] \text{ by claim 1.}\\ &\Leftrightarrow \widetilde{\mathfrak{S}}_{M_0,\gamma} \models x \mathrel{\mathring{\in}} \sigma(a[b_0 := b]) = a\end{aligned}$$

where in the final equivalence we used [8], Lemma 4.9 for $M_0' \subseteq M_1$ and the mapping $\tilde{\ } : M_1 \to \boldsymbol{Pow}(S/F)$.

This finishes the case when $b \notin \{b_1, \ldots, b_n\}$.

If $b \in \{b_1, \ldots, b_n\}$ holds, then choose $b' \notin \{b_1, \ldots, b_n\}$ such that $\widehat{b'} = \widehat{b}$. By extensionality of f on M, we conclude

$$\mathfrak{S}_{M,\gamma} \models x \mathrel{\mathring{\in}} fb \Leftrightarrow x \mathrel{\mathring{\in}} fb'.$$

Hence, using (3) and (4),

$$\begin{aligned}(5)\qquad \mathfrak{S}_{M,\gamma} \models x \mathrel{\mathring{\in}} a[b_0 := b](= fb) &\Leftrightarrow \mathfrak{S}_{M,\gamma} \models x \mathrel{\mathring{\in}} a[b_0 := b'](= fb)\\ &\Leftrightarrow \widetilde{\mathfrak{S}}_{M_0,\gamma} \models x \mathrel{\mathring{\in}} a\end{aligned}$$

where the latter model is also based on $\tilde{\ } : M_0 \to \boldsymbol{Pow}(S/F)$ with $\widetilde{b_0} = \widehat{b'} = \widehat{b}$ and $\widetilde{b_i} = \widehat{b_i}$ for $i = 1, \ldots, n$.

Note that (by Lemma 4.9)

$$\widetilde{\mathfrak{S}}_{M_0,\gamma} \models x \mathrel{\mathring{\in}} a \Leftrightarrow \mathfrak{S}^{j_0}_{M_0,\alpha} \models x \mathrel{\mathring{\in}} a$$

where $\alpha = \rho(\widehat{b}, \widehat{b_1}, \ldots, \widehat{b_n})$ and $j_0(b_0) = \widehat{b}$, $j_0(b_1) = \widehat{b_1}, \ldots, j_0(b_n) = \widehat{b_n}$. Claim 2 therefore follows from (4) and (5).

Claim 3: If $\widehat{b} \in \boldsymbol{Pow}(S/F)$, then $\Gamma(\widehat{b}) \in \boldsymbol{Pow}(S/F)$.

Proof of claim 3. Assume $\widehat{b} \in \boldsymbol{Pow}(S/F)$, $\sigma \in \boldsymbol{Aut}(\boldsymbol{B}/F)$, and choose $b' \notin \{b_1, \ldots, b_n\}$ such that $\widehat{b} = \widehat{b'}$. Then we have

$$\begin{aligned}x \in \Gamma(\widehat{b}) = \Gamma(\widehat{b'}) &\Leftrightarrow \mathfrak{S}_{M,\gamma} \models x \mathrel{\mathring{\in}} fb'\\ &\Leftrightarrow \mathfrak{S}_{M,\gamma} \models x \mathrel{\mathring{\in}} a[b_0 := b']\\ &\Leftrightarrow \mathfrak{S}_{M,\gamma} \models \sigma(x) \mathrel{\mathring{\in}} a[b_0 := b']\\ &\Leftrightarrow \mathfrak{S}_{M,\gamma} \models \sigma(x) \mathrel{\mathring{\in}} fb'\\ &\Leftrightarrow \sigma(x) \in \Gamma(\widehat{b'}) = \Gamma(\widehat{b}). \qquad \dashv\end{aligned}$$

LEMMA 4.11. *Let $F \subseteq \boldsymbol{B}$ be finite and Γ a Σ_{inac+}-operator as in the preceding lemma such that for all $b \in M$ with $\widehat{b} \in \boldsymbol{Pow}(S/F)$*

$$x \in \Gamma(\widehat{b}) \text{ and } \mathfrak{S}_{M.\gamma} \models x \mathrel{\mathring{\in}} fb \quad \Leftrightarrow \quad x \in \Gamma(\widehat{b}).$$

Define the operator $\Gamma' : \boldsymbol{Pow}(\omega) \cap \boldsymbol{L}_\pi \to \boldsymbol{Pow}(\omega) \cap \boldsymbol{L}$ by

$$\Gamma'(X) = \Gamma(\bigcup\{\mathrm{tr}_{\boldsymbol{B}/F}(x) : x \in X\}).$$

Then the following hold:

a) *Γ' is a Σ_{inac+}-operator.*
b) *If f is monotone, then Γ is monotone on $\mathcal{M}_F = \boldsymbol{L}_\gamma \cap \boldsymbol{Pow}(S/F)$ and Γ' is monotone on $\boldsymbol{L}_\gamma$.*
c) *Let f be monotone on M. Let $X' \subseteq \omega$ be minimal in $\boldsymbol{L}_\gamma$ such that $\Gamma'(X') \subseteq X'$, which exists by Corollary 4.8. Then $X = \bigcup\{\mathrm{tr}_{\boldsymbol{B}/F}(x) : x \in X'\} \in \boldsymbol{Pow}(S/F)$, therefore there is some $b \in M_F$ such that $X = \widehat{b}$. For this b*

$$\mathfrak{S}_{M.\gamma} \models fb \subseteq b.$$

Moreover, for all $a \in \mathrm{Cl}_{M.\gamma}$ we can conclude

$$\mathfrak{S}_{M.\gamma} \models fa \subseteq a \to b \subseteq a.$$

PROOF. a) follows from [8], Lemma 4.4(a).
b) The monotonicity of Γ follows from that of f using the equivalence characterizing Γ. From this, the monotonicity of Γ' is obvious since $\bigcup\{\mathrm{tr}_{\boldsymbol{B}/F}(x) : x \in X\} \in \boldsymbol{Pow}(S/F)$ by [8], Lemma 4.4(b).
c) Since $\mathrm{tr}_{\boldsymbol{B}/F}(x) \subseteq X'$ is arithmetical in x, X' by [8], Lemma 4.4(a), the set X is in $\boldsymbol{L}_\gamma$ if X' is and moreover it is in $\boldsymbol{Pow}(S/F)$.

Note that

$$\Gamma(X) = \Gamma'(X') \subseteq X' \subseteq X$$

from which $\mathfrak{S}_{M.\gamma} \models fb \subseteq b$ follows since $b \in M_F$.

Now assume $\mathfrak{S}_{M.\gamma} \models fa \subseteq a$. By [8], Lemma 4.15 we have

$$\mathfrak{S}_{M.\gamma} \models x \mathrel{\mathring{\in}} a \Leftrightarrow x \in Y$$

for some $Y \in \mathcal{M} = \{X \in \boldsymbol{L}_\gamma : X \in \boldsymbol{Pow}(S/F)$ for some finite $F \subseteq \boldsymbol{B}\}$. Then the set $Y' = \{x \in S : \mathrm{tr}_{\boldsymbol{B}/F}(x) \subseteq Y\}$ is in $\boldsymbol{L}_\gamma \cap \boldsymbol{Pow}(S/F)$ by [8], Lemma 4.4. So there is some $b' \in M_F$ such that $\widehat{b'} = Y'$. Obviously also $\mathfrak{S}_{M.\gamma} \models b' \subseteq a$ and so by monotonicity of f we have $\mathfrak{S}_{M.\gamma} \models fb' \subseteq fa \subseteq a$. Since $b' \in M_F$ this means $\Gamma(Y') \subseteq Y$ and $\Gamma(Y') \in \mathcal{M}_F$.

Therefore for all $x \in \Gamma(Y')$ and $\sigma \in \boldsymbol{Aut}(\boldsymbol{B}/F)$ we have $\sigma(x) \in \Gamma(Y') \subseteq Y$, which means $\mathrm{tr}_{\boldsymbol{B}/F}(x) \subseteq Y$, leading to $\Gamma(Y') \subseteq Y'$. Since $Y' \in \mathcal{M}_F$, we moreover have $\Gamma'(Y') = \Gamma(Y') \subseteq Y'$. The minimality of X' yields $X' \subseteq Y'$ and thus $X \subseteq X' \subseteq Y' \subseteq Y$. But this means $\mathfrak{S}_{M.\gamma} \models b \subseteq a$. ⊣

Theorem 4.1 is a consequence of the next result.

COROLLARY 4.12. $\mathfrak{S}_{M.\gamma} \models \boldsymbol{T_0} + \boldsymbol{MID}$.

PROOF. From Lemma 4.10 combined with Lemma 4.11 we obtain $\mathfrak{S}_{M.\gamma} \models$ ***MID***. That $\mathfrak{S}_{M.\gamma}$ is a model all axioms of $\boldsymbol{T_0}$ except for (Join) is immediate. To show that it models (Join) as well, assume $a \in \mathrm{Cl}_{M.\gamma}$ and $\mathfrak{S}_{M.\gamma} \models \forall x \mathrel{\mathring{\in}} a \, \exists Y \, (fx \simeq Y)$. Let $M_0 := (\boldsymbol{supp_B}(a) \cup \boldsymbol{supp_B}(f)) \cap M$. By Proposition 4.13 below we get $a \in \mathrm{Cl}_{M_0.\gamma}$ and

$$\mathfrak{S}_{M_0.\gamma} \models \forall x \mathrel{\mathring{\in}} a \, \exists Y \, (fx \simeq Y)$$

and thus $\boldsymbol{j}(a, f) \in \mathrm{Cl}_{M_0.\gamma}$ by Lemma 4.9. Hence $\boldsymbol{j}(a, f) \in \mathrm{Cl}_{M.\gamma}$, verifying that $\mathfrak{S}_{M.\gamma}$ is a model of (Join). ⊣

PROPOSITION 4.13. *Let M_0 be finite subsets of M. If $\widehat{\ } \restriction M_0 : M_0 \to \boldsymbol{Pow}(S/F)$ for a finite set F such that $F \cap M \subseteq M_0$, then we have for all $c \in \boldsymbol{CL}_{M.\alpha}$*

$$\boldsymbol{supp_B}(c) \cap M \subseteq M_0 \Rightarrow c \in \boldsymbol{CL}_{M_0.\alpha}.$$

PROOF. We use induction on α. In the most important case, that is $c = \boldsymbol{j}(a, f)$, we use Lemma 4.14 below. The induction hypothesis guarantees the assumption of that lemma. ⊣

LEMMA 4.14. *Let M_0, F be as in Proposition 4.13. If $a \in \boldsymbol{CL}_{M_0.\alpha}$,*

$$(\boldsymbol{supp_B}(a) \cup \boldsymbol{supp_B}(f)) \cap M \subseteq M_0 \text{ and } \forall x \mathrel{\mathring{\in}} a \exists y \mathrel{\mathring{\in}} \boldsymbol{CL}_{M.\alpha}[fx \simeq y]$$

and furthermore

$$\forall x \mathrel{\mathring{\in}} a \forall y \in \boldsymbol{CL}_{M.\alpha}[fx \simeq y \wedge \boldsymbol{supp_B}(y) \cap M \subseteq M_0 \Rightarrow y \in \boldsymbol{CL}_{M_0.\alpha}],$$

then, for $x \mathrel{\mathring{\in}} a$ and $fx \simeq y$ we have $y \in \boldsymbol{CL}_{M_0.\alpha}$.

PROOF. Put $G := \boldsymbol{supp_B}(a) \cup \boldsymbol{supp_B}(f)$ in [8], Corollary 4.11. Then the hypotheses of [8], Corollary 4.11 are satisfied and so there is an automorphism

$$\sigma \in \boldsymbol{Aut}(\boldsymbol{B}/M_0 \cup F \cup \boldsymbol{supp_B}(a) \cup \boldsymbol{supp_B}(f))$$

such that $\sigma(x) \in a$ and $\boldsymbol{supp_B}(\sigma(x)) \cap M \subseteq M_0$.

From $fx \simeq y$ we therefore conclude

$$\boldsymbol{supp_B}(\sigma(y)) \subseteq \boldsymbol{supp_B}(\sigma(f)) \cup \boldsymbol{supp_B}(\sigma(x)) \subseteq \boldsymbol{supp_B}(f) \cup \boldsymbol{supp_B}(\sigma(x))$$

since $\sigma(f) = f$. This gives $\boldsymbol{supp_B}(\sigma(y)) \cap M \subseteq M_0$. Using $\sigma(y)$ instead of y in the additional hypothesis of the lemma we get $\sigma(y) \in \boldsymbol{CL}_{M_0.\alpha}$ and from this $y \in \boldsymbol{CL}_{M_0.\alpha}$ applying [8], Corollary 4.10 to σ^{-1}. ⊣

REFERENCES

[1] W. Buchholz, S. Feferman, W. Pohlers, W. Sieg: *Iterated inductive definitions and subsystems of analysis*, Lecture Notes in Math. 897 (Springer, Berlin, 1981) 78–142.

[2] D. Cenzer: *Ordinal recursion and inductive definitions* in: E. Fenstad and P. Hinman: *Generalized Recursion Theory I* (North-Holland, Amsterdam, 1974) 221-264.

[3] S. Feferman: *A language and axioms for explicit mathematics*, in: J.N. Crossley (ed.): *Algebra and Logic*, Lecture Notes in Math. 450 (Springer, Berlin 1975) 87–139.

[4] S. Feferman: *Constructive theories of functions and classes* in: Boffa, M., van Dalen, D., McAloon, K. (eds.), *Logic Colloquium '78* (North-Holland, Amsterdam 1979) 159–224.

[5] S. Feferman: *Monotone inductive definitions* in: Troelstra, A. S., van Dalen, D. (eds), *The L.E.J. Brouwer Centenary Symposium* (North-Holland, Amsterdam, 1982) 77–89.

[6] S. Feferman and W. Sieg: *Proof-theoretic Equivalences between classical and constructive theories of analysis*. in: W. Buchholz, S. Feferman, W. Pohlers, W. Sieg: *Iterated inductive definitions and subsystems of analysis*, Lecture Notes in Math. 897 (Springer, Berlin, 1981) 78–142.

[7] T. Glaß: *Standardstrukturen für Systeme Expliziter Mathematik*, Inaugural-Dissertation (Münster, 1993).

[8] T. Glaß, M. Rathjen, A. Schlüter: *The strength of monotone inductive definitions in explicit mathematics*, Annals of Pure and Applied Logic 85 (1997) 1–46.

[9] L.A. Harrington: *Kolmogorov's R-operator and the first nonprojectible ordinal*, mimeographed notes (1975) 13 pages.

[10] L.A. Harrington and A.S. Kechris: *On monotone versus nonmonotone induction*, Bull. Am. Math. Soc. 82, 888–890 (1976).

[11] L.A. Harrington and A.S. Kechris: *Inductive definability*, unpublished, type-written manuscript, 8 pages.

[12] G. Jäger: *A well-ordering proof for Feferman's theory* T_0, Archiv f. Math. Logik 23 (1983) 65–77.

[13] G. Jäger and W. Pohlers: *Eine beweistheoretische Untersuchung von* Δ^1_2*- CA + BI und verwandter Systeme*, Sitzungsberichte der Bayerischen Akademie der Wissenschaften, Mathematisch-Naturwissenschaftliche Klasse (1982).

[14] A.S. Kechris: *Spector second order classes and reflection*, in: J.E. Fenstad, R.O. Gandy, G.E. Sacks: *Generalized recursion theory II* (North-Holland, Amsterdam, 1978) 147–183.

[15] G. Kreisel: *Generalized inductive definitions*, in: Stanford Report on the Foundations of Analysis (mimeographed), CH. III, Stanford 1963.

[16] T. John: *Recursion in Kolmogorov's R-operator and the ordinal* σ_3. Journal of Symbolic Logic 51 (1986) 1–11.

[17] M. Rathjen: *Untersuchungen zu Teilsystemen der Zahlentheorie zweiter Stufe und der Mengenlehre mit einer zwischen* $\Delta^1_2 - CA$ *und* $\Delta^1_2 - CA + BI$ *liegenden Beweisstärke* (Publication of the Institute for Mathematical Logic and Foundational Research of the University of Münster, 1989).

[18] M. Rathjen: *Monotone inductive definitions in explicit mathematics*. Journal of Symbolic Logic 61 (1996) 125–146.

[19] M. Rathjen: *Explicit mathematics with the monotone fixed point principle*. Journal of Symbolic Logic 63 (1998) 509–542.

[20] M. Rathjen: *Explicit mathematics with the monotone fixed point principle. II: Models*. Journal of Symbolic Logic 64 (1999) 517–550.

[21] S. Takahashi: *Monotone inductive definitions in a constructive theory of functions and classes*. Ann. Pure Appl. Logic 42 (1989) 255-279.

DEPARTMENT OF PURE MATHEMATICS
UNIVERSITY OF LEEDS
LEEDS LS2 9JT, ENGLAND
E-mail: rathjen@amsta.leeds.ac.uk

PART IV: PHILOSOPHY OF MODERN MATHEMATICAL AND LOGICAL THOUGHT

ON THE CONSTRUCTIVITY OF PROOFS. A DEBATE AMONG BEHMANN, BERNAYS, GÖDEL, AND KAUFMANN

PAOLO MANCOSU

To Sol, on the occasion of his 70th birthday[1]

§1. Introduction. The question of whether proofs by contradiction can be eliminated in favor of direct proofs has occupied the minds of many philosophers and mathematicians for centuries. Aristotle, Arnauld, Kant, Bolzano, and many others up to our day have tackled this topic.[2] In the present paper I present the details of a debate centered around the relationship between constructivity and indirect proofs which took place in 1930 and involved some of the most able logicians and philosophers of mathematics of the century, namely Heinrich Behmann, Paul Bernays, Rudolf Carnap, Kurt Gödel, and Felix Kaufmann. The debate never made its way into print but it is still accessible in its entirety in documents preserved in the Kaufmann archive in Konstanz, the Bernays archive at the ETH in Zurich, and the Behmann archive in Erlangen. Related materials are also found in the Carnap archive in Pittsburgh. The Kaufmann archive in Konstanz contains, among other things, a large folder entitled *Zur Frage der Konstruktivität von Beweisen* which is of great interest to the historian of logic and the foundations of mathematics. Felix Kaufmann was an associate of the Vienna circle and he was in constant contact with, among others, Carnap, Gödel, Hahn, Waismann, Hempel, and Behmann. Most of the material we will discuss is found in this folder but supplementary materials from the other archives are used to fill in the picture when needed.[3]

§2. A conjecture by Kaufmann. On October 24, 1930 Heinrich Behmann sent a letter to Paul Bernays, assistant editor of *Mathematische Annalen*, submitting a manuscript for publication entitled *Zur Frage der Konstruktivität von Beweisen*. He explained to Bernays the circumstances from which the paper originated:

I would like to thank the curator of the Felix Kaufmann archive in Konstanz, Dr. Martin Endress, the curators of the Behmann archive in Erlangen, Prof. C. Thiel and Dr V. Peckhaus, the curator of the Bernays Nachlaß in Zurich, Dr. Yvonne Vögeli, for their help and kindness. In addition, many thanks to John Dawson Jr., Johannes Hafner, Richard Zach, and an anonymous referee for several helpful suggestions. Finally, it is a pleasure to acknowledge that the research leading to this paper was carried out during my tenure as fellow of the Wissenschaftskolleg zu Berlin in 1997–1998.

Reflections on the foundations of mathematics
Essays in honor of Solomon Feferman
Edited by Wilfried Sieg, Richard Sommer, and Carolyn Talcott

> [The paper] concerns the solution of a problem Herr Kaufmann suggested to me during my stay in Vienna in September. Although I do not share the constructivist point of view I still deem such investigations important so that the practical import of the constructivist principles will not be overestimated and consequently great parts of mathematics be, unjustifiably, put into doubt.[4]

He also added "Herr Carnap has already checked the proof and he agrees with it in every respect."[5]

Behmann had visited Kaufmann for 11 days in September 1930 from September 9 to September 20.[6] Neurath had first suggested to Behmann in 1929 that he should deliver a talk for the Schlick circle.[7] Indeed, on the occasion of his visit to Vienna, Behmann delivered a lecture at the invitation of the Schlick circle entitled *Widersprüche in Mathematik und Logik* (September 15, 1930). This took place just after the famous Königsberg meeting.

During his visit Kaufmann had proposed a problem on the constructivity of existence proofs to which Behmann proposed a solution just a few days after leaving Vienna.[8] The problem raised by Kaufmann was, roughly, the following. Is it possible to prove that all proofs of existence claims, with the exception of those obtained by means of the axiom of choice, can be shown to rely on an implicit exhibition of an instance? Kaufmann's conjecture was that this would in fact be the case. The consequences for the debate on the foundations of mathematics would have been remarkable. A proof of the conjecture would have implied that most of classical mathematics – the only exception being those parts relying on the axiom of choice – is already "constructive." Before we look at Behmann's solution let us try to be more precise on the nature of Kaufmann's conjecture. Kaufmann's position is expressed in his *The Infinite in Mathematics*, which was published in 1930. Since the thesis is somewhat unclear, let me quote the entire relevant passage:

> Our train of thought is this: if we make perfectly clear to ourselves what the criteria are for a mathematical assumption being absurd, we recognize that they lie in the fact that two signs for natural numbers presupposed as synonymous turn out, at the end of a proof, not to be synonymous. Every contradiction therefore relates to certain natural numbers: that it arises at all means that it occurs at certain points and in exhibiting it we must therefore implicitly determine those points. That this determination does not always show in mathematical proofs themselves must be traced back to the abbreviations in the formation of mathematical concepts (symbolism); but a completely articulated theory of proof, as aimed at by Hilbert, would have to bring it into the open. If so, the illusion that there are pure existence proofs would vanish and Brouwer's demand for constructivity throughout would have been met; but it is precisely the fulfilment of this demand that will show that whenever the question is that of the existence of finite bounds, the only question that arises in the

building up of mathematics (apart from set theory, to be analysed later), a divergence between absurdity of absurdity and mathematical existence did not even 'originally' exist, so that the results of existence proofs in classical mathematics will be perfectly justified from Brouwer's point of view as well, leaving aside differences in terminology.

However, this does not hold for the pseudo-existence proofs in the doctrine of the non-denumerable infinite, which, as we shall see, do not have a constructive basis. ([Kaufmann, 1978 [1930]], pp. 57–8)

If we abstract for a moment from the appeal to Hilbert's proof theory in the motivation of Kaufmann's conjecture the overall thesis appears to consist of two main parts:

(a) There is no distinction between constructive existence proofs and general existence proofs in the domain of "countable" mathematics;

(b) Such a distinction arises as soon as the "uncountable" makes its appearance. It should be noted that the thesis displays a certain amount of ambiguity; the boundaries between countable and uncountable mathematics are not spelled out in formal terms; moreover, it is not clear how the proof of such a claim should be carried out. But the general nature of the claim is clear enough. It is a question of showing that all existence claims proved in those parts of classical mathematics that do not rely on non-constructive existence axioms, such as the axiom of choice, are constructive existence claims, i.e. can be shown to be grounded on the exhibition of an instance for the claim. Kaufmann did not even hint at an argument for establishing this conjecture but he spoke about it to an able logician, Heinrich Behmann, who shortly thereafter provided a "proof" of the conjecture.

Behmann sent a paper containing a proof of Kaufmann's conjecture to Kaufmann on October 6, 1930 requesting that the paper be shown to Carnap, Waismann and others. Moreover, in the same letter, Behmann discussed the opportunity ("wegen der Wichtigkeit der Sache") of publishing the result in a mathematical journal such as *Mathematische Annalen* or *Mathematische Zeitschrift*. Kaufmann's reply to Behmann showed complete agreement with the solution proposed by Behmann and contained two additions which Kaufmann proposed to add as an appendix to Behmann's paper.

§3. Behmann's typescript and its contents. Let us now consider Behmann's typescript. Although it was never published it survives in many archives (in Zurich, Konstanz, Erlangen, Pittsburgh). Kaufmann had made copies for Waismann, Carnap, Gödel, Wittgenstein, and Hahn. Moreover, a copy was sent to Bernays and it is still found among Bernays' papers at the ETH in Zurich. The latter version of the paper shows signs of reworking that are absent from the original copy sent to Kaufmann. Occasionally, I will remark on the difference between the different versions of the typescript in the notes.[9]

The first section of the paper is entitled "A conjecture [Vermutung] by Kaufmann"[10]. Behmann begins by discussing the contemporary constructive tendencies and their restriction of classical mathematics in methods of proof and, consequently, in their theorems. This takes place in the form of three restrictions:

(1) Elimination of non-constructive existence axioms (in particular the axiom of choice)
(2) Elimination of non-predicative existence proofs (in as far as they are not already excluded by (1))
(3) Elimination of concepts of higher type (that is, restriction to the so-called *engeren Funktionenkalkül*)

Behmann has no doubts that the first restriction is an actual restriction, for it was well known that there are statements of classical mathematics that would not hold without the axiom of choice.[11]

Concerning the restriction in (2) Behmann observes that existence proofs can be obtained either directly by exhibition [Aufweisung] or indirectly, by means of a proof by contradiction. (2) would thus constitute an essential restriction only if there are existence proofs proved by contradiction that cannot be made constructive. Equivalently, (2) represents a restriction if there are proofs of existence claims which, although not relying on non-constructive existence axioms, can only be proved by contradiction.[12]

Kaufmann's conjecture is exactly that an indirect existence proof, provided it satisfies (1), is also constructive:

> Herr Kaufmann has recently formulated an interesting conjecture by means of which the aforementioned question would be answered negatively. He is of the opinion that an indirect proof, in so far as it satisfies condition (1), is also in essence constructive; more precisely, that it relies on a hidden instance, which through an appropriate analysis of the demonstration can automatically [zwangsläufig] be made explicit.[13]

I hasten to warn the reader that the use of the word "non-predicative" in condition (2) was an oversight by Behmann, which in fact generated confusion in the exchange with Bernays. Behmann simply meant "non-constructive" rather than "non-predicative". Let me also add a word about condition (3). In other parts of the Behmann-Kaufmann correspondence we find many discussions as to whether quantification over properties can be eliminated without loss of any parts of classical mathematics. This was not a new problem as Weyl's system *Das Kontinuum* (1918) was in fact an attempt to develop mathematics without assuming quantification over subsets of the domain but only over individuals of the domain. One of the topics of discussion between Behmann and Kaufmann was whether the theory of real numbers and theorems such as Weierstrass' theorem on the convergence of bounded monotone sequences can be proved within a framework which does not assume quantification over classes or predicates. However, I will not enter into this issue as it is not relevant to our main concern.[14]

Behmann claimed to have proved Kaufmann's conjecture and the rest of the paper offers a proof of the claim.

§4. Behmann's theorem: every indirect proof can be transformed into a direct one. The second section of the paper is entitled *Die Struktur des mathematischen Beweises*. Despite the promising title Behmann's treatment of the topic is extremely sketchy and somewhat confusing. The source of the confusion lies in the fact that there is no specification of a language or of the rules of inference. Thus, the notion of proof is not given by a derivation system and it is left completely unspecified. Moreover, in the course of the presentation Behmann uses certain diagrammatic configurations to represent – at different times – proofs, sentences, and inferences. At the source of this lack of clarity is the fact, I claim, that Behmann tacitly identifies inference and logical implication. I will come back to this after we have gone through Behmann's presentation.

Behmann begins by representing [darstellen] direct and indirect proofs [Beweisführungen] through certain schematic diagrams [Schemata]. Let a, b, c, etc. be sentences used for the proof of a certain proposition d.
A direct proof is represented by

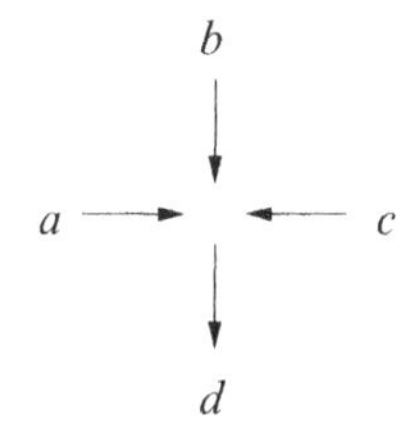

An indirect proof is represented by

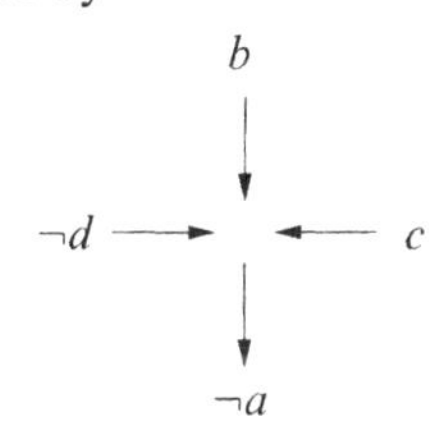

Behmann then remarks that in passing from a direct to an indirect proof the negation of one of the assumptions becomes the conclusion and the negation of the conclusion one of the assumptions.

The above schemata stand for proofs. However, Behmann goes on to claim that the same schemata hold for each single inferential step of the corresponding proof. In the case of a direct proof each inferential step can be seen as leading from a set of premises a, b, c, ... (taken to be true on account of the fact that they are obtained in the course of a direct proof) to a proposition d, considered as the conclusion of the inferential step. In the case of an indirect proof $\neg d$, c, b will stand for the assumptions and $\neg a$ for the conclusion of the inferential step. In short, Behmann is claiming that the global structure of direct and indirect proofs is replicated at the local level for each inferential step.[15]

Behmann now describes a procedure that transforms one diagram into another by using negations and reversing the direction of the arrows. In order to make this step more comprehensible he provides a new interpretation of the "arrow" symbolism, which is now interpreted as a representation of sentences, in particular of disjunctions. To see how this works let us begin with some easy configurations.

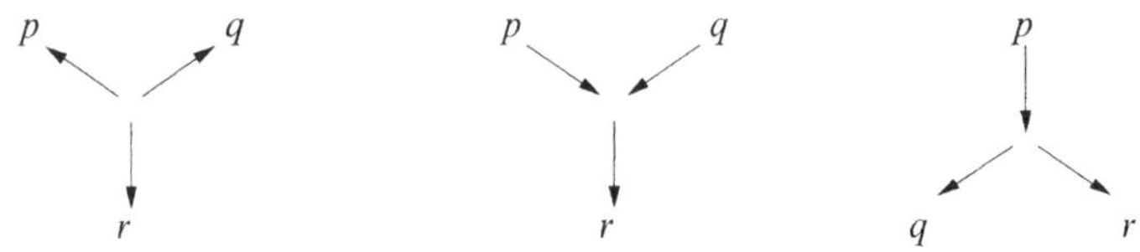

The first configuration stands for the disjunction $p \vee q \vee r$. But if we reverse the direction of the arrow, this means that the negation of the formula goes into the disjunction. Thus in the second configuration we have $\neg p \vee \neg q \vee r$, i.e. $(p\&q) \supset r$. In short, one can reverse the direction of the arrow when one wants to substitute the occurrence of a formula with its negation. The third configuration will be read as $p \supset (q \vee r)$. Behmann denotes the arrows pointing inside as "condition-arrows" and those pointed outwards "conclusion-arrows".

Any proof can now be seen, according to Behmann, as a system of such configurations [Sternfiguren]. Under the condition – commonly assumed in proof theory, says Behmann – that each single occurrence of the same sentence in the proof is considered independent of other occurrences (and thus is proved afresh every time it is introduced) we have the following characterization of a proof structure:

> At the bottom stands (with arrows pointing downwards) the configuration of the last inference of the proof. Insofar as the sentences related by the arrows are not the assumptions of the whole proof there is an associated configuration with its arrows. [...] Thus with the exception of the assumptions and of the conclusion every single occurrence of a proposition stands at the beginning and at the end of exactly one arrow.[16]

The scheme has the structure of a tree – the terminology is Behmann's – with the conclusion as the root, and the immediately connected configurations are the branches. At the end of the branching we find the assumptions of the proof, i.e. the leaves [Zweigende]. Behmann specifies that all arrows are necessarily open and that there cannot be a closed polygon of arrows. He then goes on to assert that if one abstracts from the direction of the arrows, the proof structure can be looked at as if any of the leaves of the proof were in fact the conclusion. He says:

> It is now intuitively evident [anschaulich evident] that within the described structure as such, if one abstracts from the direction of the arrows, the position of the conclusion is, from a topological point of view, equivalent to that of the assumptions, so that one can think of the [proof-]schema as if it were constructed in exactly the same way from any leaf [Zweigende]. If one turns the direction of the associated

configuration in such a way that the leaf in question is directed downwards (i.e. the arrow is directed upwards) then it is again true that to those configurations, which contain neither the assumptions nor the conclusion of the proof, is associated exactly one configuration whose arrows we can think as reversed etc.[17]

I find remarkable the idea of applying to proofs, even if at such an elementary level, graph-theoretic and topological considerations. This might very well be the first time such talk occurs.[18]

We have seen that the characterization of proofs given by Behmann uses a diagrammatic device that can be interpreted in three different ways, depending on whether we take the arrow configurations to stand for proofs, inferences, or disjunctive propositions. How can the three interpretations be reconciled? The only reading that makes sense of this is to interpret Behmann as claiming that to every inference there corresponds a disjunctive tautological sentence (or, equivalently, a tautological implication). Consequently, the whole proof corresponds to a tautological sentence (see section five for a concrete example). To see that this assumption cannot be granted to Behmann it is enough to think that while the claim is true for derivations in propositional logic, it is certainly not the case for inferences in first-order logic. Indeed, some of the inferential rules in first order logic, such as the rule of universal generalization, do not correspond to tautological implications.

We now move to section three of the paper where Behmann outlines his strategy for transforming an indirect proof into a direct one. Without loss of generality assume that the negation of the proposition to be proved appears only at one end of the proof structure (this can be obtained by making the relevant transformations, according to the preceding explanations). One then follows these steps: 1. Eliminate the negation sign of the conclusion sentence and reverse the direction of the arrow making it into a conclusion arrow; 2. Keep reversing all the arrows connecting the conclusion to the negation of the assumption which is the desired conclusion.

Example:

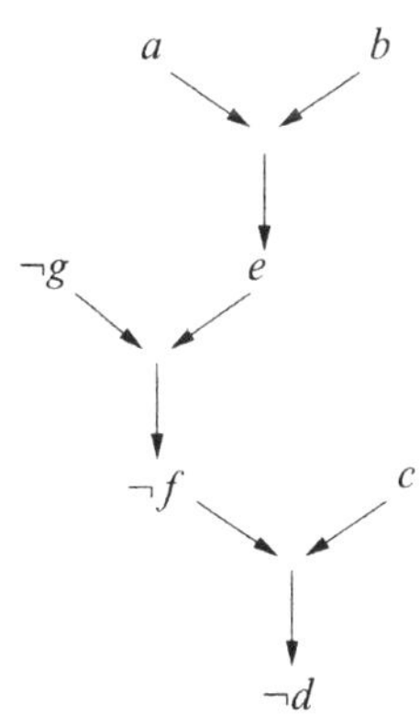

Assume that d is an established theorem and that $\neg d$ is the last step needed for inferring g by contradiction. Then:

Step 1.

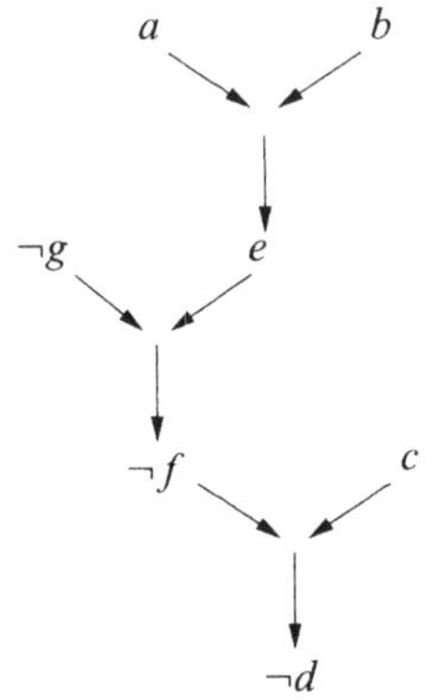

Step 2.

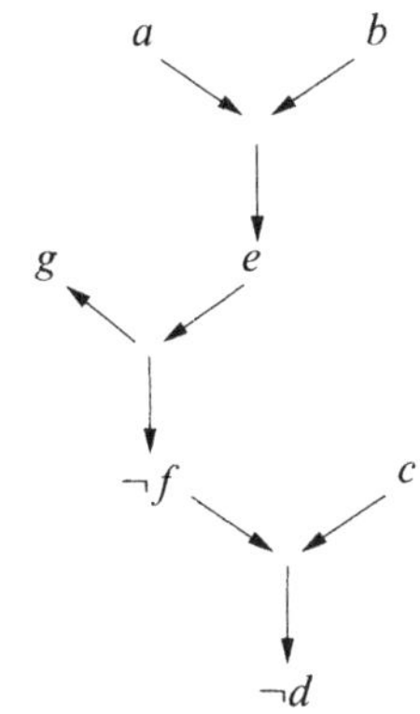

Step 3.

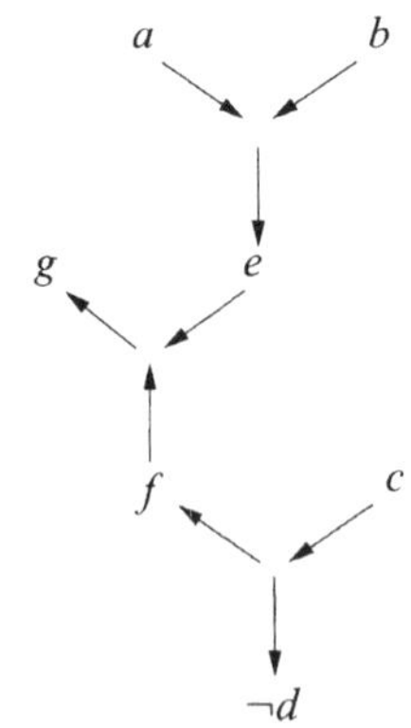

Step 4.

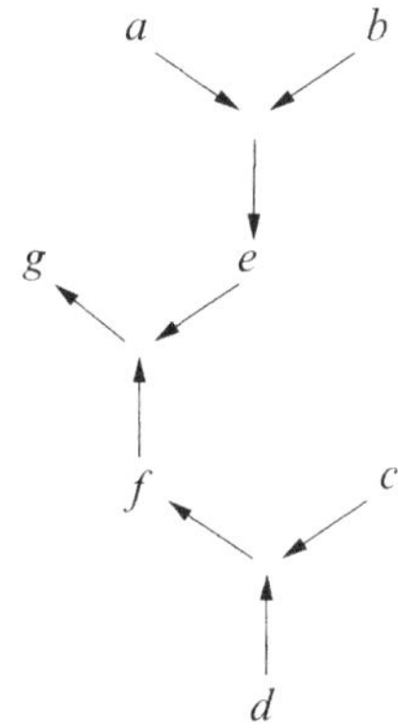

If the negation of the claim to be proved occurs several times in the proof as a leaf then one has to apply the transformation for all the leaves in question. The claim thus appears as end-inference in several places of the direct proof obtained by transforming the indirect proof. The whole procedure boils down to reversing arrows and taking the negation of the formulas whose arrows are being reversed. How is this going to establish Kaufmann's conjecture?

Since in the absence of a non-constructive existence axiom every existence proposition can only be proved directly by providing an instance – or so does Behmann claim – then at some stage of the proof there must be an occurrence of the form $fa \supset \exists x fx$. If the arrow in question goes in the reverse direction in the original proof then the inference has the form $(x)\neg fx \supset \neg fa$. According to Behmann this proves the claim. The rest of the paper goes on to show in the specific case of Euler's proof of the infinitude of primes how to transform the indirect proof into a direct one. We need not delve into the example since we will show that Behmann's argument is faulty. In order to do this we will attempt to clarify some of the hidden presuppositions of Behmann's argument by working out in detail an example of a formal derivation in the propositional system of *Principia Mathematica*.

§5. The propositional axioms of *Principia Mathematica* and Behmann's argument. In order to gain a better insight into Behmann's argument I am going to supplement his general exposition by looking at one specific system, the propositional system of *Principia Mathematica*, and a specific example of a derivation.[19] In order to keep the similarity with the exposition given by Behmann the axioms of the system in question and the rule of inference will be expressed using disjunction and negation as primitives. The axioms of the system are:

1. $\neg(p \vee p) \vee p$
2. $\neg p \vee (p \vee q)$
3. $\neg(p \vee q) \vee (q \vee p)$
4. $\neg(\neg p \vee q) \vee (\neg(r \vee p) \vee (r \vee q))$

Rules of inference:

(R1) Substitution

(R2) Modus ponens: From p and $\neg p \vee q$ infer q.

A proof of $p \vee \neg p$ can be given as follows (without explicit mention of the appropriate substitutions in the axioms):

$(a)\ \neg(\neg(p \vee p) \vee p) \vee [\neg(\neg p \vee (p \vee p)) \vee (\neg p \vee p)]$	(by axiom 4)
$(b)\ \neg(p \vee p) \vee p$	(by axiom 1)
$(c)\ \neg(\neg p \vee (p \vee p)) \vee (\neg p \vee p)$	(by R2)
$(d)\ \neg p \vee (p \vee p)$	(by axiom 2)
$(e)\ \neg p \vee p$	(by R2)
$(f)\ \neg(\neg p \vee p) \vee (p \vee \neg p)$	(by axiom 3)
$(g)\ p \vee \neg p$	(by R2)

In Behmann's setup this would have the following tree structure:

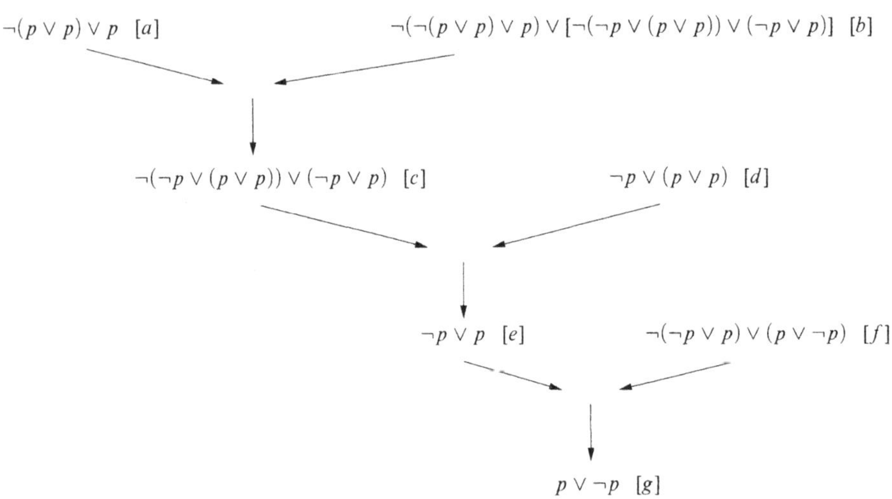

To this tree structure corresponds a propositional tautology, constructed as follows:

$$(([a]\&[b]) \supset [c])\&(([c]\&[d]) \supset [e])\&(([e]\&[f]) \supset [g])$$

Let us remain within the boundaries of this formalization of propositional logic. Notice that in this system we proceed from truths to truths. However, following Behmann's idea, we could now turn the arrows of the graph and obtain a scheme for an indirect proof, where $(*)$ denotes the negation of the result to be proved.

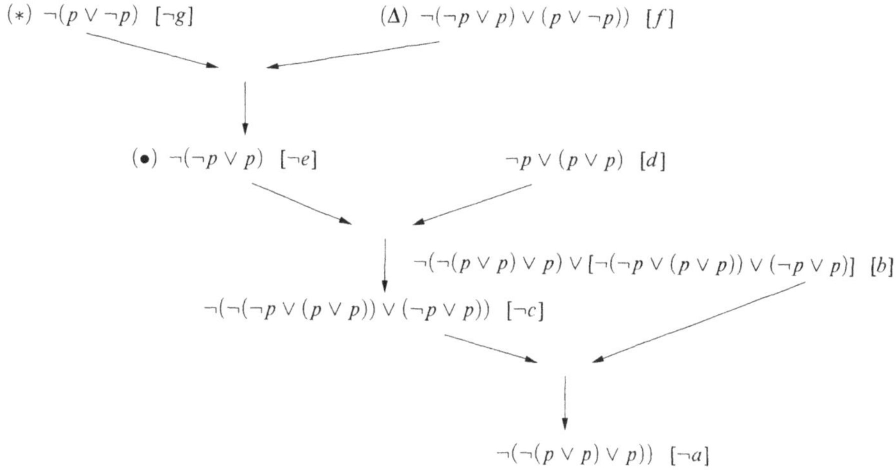

We see that the end-sentence is a negation of an axiom and thus would be the pivotal point of a proof by contradiction. The result of the transformation once again gives rise to a propositional tautology:

$$(([\neg g]\&[f]) \supset [\neg e])\&(([\neg e]\&[d]) \supset [\neg c])\&(([\neg c]\&[b]) \supset [\neg a])$$

However, it is also important to notice that although the transformation proposed by Behmann sends a graph into a graph which preserves logical consequence, we do not get a proof after the transformation has been carried out. For instance, in our system we have no rule that would allow us to go from $(*)$ and (Δ) to $(\bullet)$. However, under the assumption of completeness of the calculus, a proof can in principle be found for each of the steps. Likewise, if we start with a proof by contradiction the transformation will not in general lead us to a proof.

We thus see how in the propositional case Behmann could move between the several interpretations of his arrow diagrams without trouble. However, things get more complicated as soon as we move to the first-order calculus, where the Behmannian assumption that each inference rule corresponds to a true disjunction does not hold anymore. Consider the usual rule for universal generalization: from $A(x)$ infer $(x)A(x)$. Assume that the rule is applied as an inferential step in an indirect proof. Behmann's strategy for obtaining a direct proof would call for a reversal of the arrows of implication. But this amounts to confusing the rule of inference with the formula $A(x) \supset (x)A(x)$. In his argument Behmann made two assumptions, one implicit and the other explicit. The implicit assumption is that only the inferential step concerning the introduction of the existential quantifier is relevant for the success of the transformation of every indirect proof into a direct proof. But, as I pointed out above, there are other quantificational steps, which do not correspond to implications, involved in a first-order demonstration that cause problems.

The explicit assumption concerns the introduction of the existential quantifier. According to Behmann, in the absence of non-constructive existence axioms the only way to introduce $\exists x f(x)$ is by first obtaining $f(a)$ (or by the equivalent rule: from $(x)\neg fx$ infer $\neg fa$). From the reactions to Behmann's proof we will discover the roots of the explicit assumption and its problematic nature.

§6. **Kaufmann, Carnap, and Gödel on Behmann's proof.** The reception of the proof in Vienna was marked by acceptance of its validity. Kaufmann wrote back on October 9, 1930: "Thank you very much for your letter and the expected proof. It seems to me absolutely conclusive and of paradigmatic clarity"[20] The letter continues by saying that copies of the essay are being made for Hahn, and Waismann (intended for Wittgenstein). At the same time Kaufmann asks permission on behalf of Carnap to show the essay to Gödel. The rest of the letter discusses the appropriateness of trying to publish the essay either in *Mathematische Zeitschrift* or *Mathematische Annalen*. On account of the importance of the essay Kaufmann said he would not talk to Hahn about possible publication in the *Monatshefte für Mathematik und Physik* (which, ironically, will publish Gödel's paper on incompleteness just a few months later). Another point raised in the letter concerns the similarities between a constructivization of Euler's proof for the infinitude of primes recently given by Wittgenstein and the one contained in the essay. Finally, Kaufmann also adds two appendices that he would like to publish together with the essay by Behmann. In the first, Kaufmann tries to point out the philosophical importance of Behmann's result for the dispute over the foundations of mathematics. The second appendix concerns the issue of whether quantification over classes or predicates, can be eliminated and answers the question positively. Part of the later correspondence between Kaufmann and Behmann delves more into this issue but I will not follow this up. Let me just report the passages where Kaufmann spells out the philosophical importance of Behmann's result:

> In my opinion, the claim proved by Mr. Behmann and put forward in my book [*The Infinite in Mathematics and its Elimination*] on p. 66 f decides (as was set out there on pp. 59 ff) the methodological dispute between formalism and neo-intuitionism (Brouwer). In the first place this follows for those propositions which are proved without the help of the axiom of choice. For the mathematical core of Brouwer's teachings consists in posing the following trilemma:
>
> (1) A mathematical existence claim E, e.g., the occurrence of a sequence 0, 1, 2, 3, ..., 9, in the decimal expansion of π, is verified by direct exhibition.
>
> (2) The absurdity of E is proved, i.e., $\neg E$ is proved for all numbers.
>
> (3) The absurdity of (2), thus the absurdity of the absurdity of (1) is proved without it being possible to give a construction (1).
>
> The possibility of (3) would mean that one could not infer the validity of an existence claim from the absurdity of the absurdity of that existence claim. On the assumption of this possibility rests Brouwer's thesis that the law of the excluded middle is not valid for infinite domains. This thesis

> coincides, as Brouwer himself has stressed (see e.g., Intuitionistische Mengenlehre, Jahresber. d. Deutsch. Math. Ver., Bd 28, pp. 203–208, 1919), with his negation of the definiteness of decision of arithmetic and analysis.
>
> So the basis for this claim disappears for all propositions which are proved without the help of the axiom of choice.
>
> We obtain the following result:
>
> The intuitionistic requirement of universal constructivity is legitimate; but it appears to be implicitly satisfied in the so-called existence proofs of classical mathematics, and so its realization does not entail a reduction of the assets of mathematics.[21]

The latter claim is qualified later by Kaufmann by remarking that the claim holds with the exception of results depending on the axiom of choice: in particular, it does not hold for results not reducible to countable cardinalities. It is thus clear why Behmann's result generated such attention. If true, it would have shown that classical mathematics was, to an unsuspected extent, in essence intuitionistic. It should be remarked that Kaufmann's characterization of the philosophical nature of the project emphasizes the goal of providing a positive contribution to the debate between formalism and intuitionism. In this connection, then, it should be mentioned that Behmann's "theorem" can be seen as one of the many contributions to the clarification of the relationship between classical and intuitionistic mathematics, such as those of Kolmogorov, Glivenko, Gödel and others. I am referring to the negative translations (or double negation interpretations) aimed at interpreting fragments of classical mathematics into intuitionistic mathematics. Indeed, as presented in the above quote, Behmann's theorem would have provided a translation from classical (countable) mathematics into intuitionistic mathematics such that if in classical mathematics we have $\exists x A(x)$ then intuitionistic mathematics proves $A^*(t)$, where * stands for some suitable translation. As a consequence, if intuitionistic mathematics proves $\neg\neg\exists x A(x)$, it would also prove $A^*(t)$ for some term t.

In his reply of October 12, Behmann declines the suggestion to publish Kaufmann's remarks with his essay, motivating his choice on strategic grounds by a desire not to alienate acceptance of his result on account of ideological resistance, and states that he is submitting the essay to Bernays for publication in *Mathematische Annalen*. On the same day Kaufmann writes to Behmann to announce that Carnap and Waismann find the proof absolutely compelling:

> Today I had a longish, very nice discussion with Mr. Carnap; he is very happy about the successful proof, which he finds – just as I expected – completely transparent and he recognizes its immense importance. The same is true of Mr. Waismann with whom I spoke on the phone today.[22]

Let us now look at Gödel's contribution to this debate.

In a letter dated October 9, 1930 Kaufmann asks permission on Carnap's behalf to send a copy of the paper to Gödel.[23] A handwritten letter from Kaufmann to Gödel, which informs Gödel of some corrections to the proof made by Behmann (in a letter to Kaufmann dated October 11, 1930) is still to be found in Kaufmann's papers. The letter is a handwritten draft and it is dated October 14, 1930.[24] Carnap and Gödel probably met to discuss the issue on October 16, 1930 and Carnap made a record of the discussion in his diary.[25] Dawson ([1997], p. 73) mentions an entry in Carnap's diary (dated October 16, 1930) which "recorded Gödel's refutation of a claim by Heinrich Behmann that every existence proof could be made constructive". Dawson rightly relates Gödel's refutation to Brouwer's doubt as to "whether any formal system could comprise all of intuitionistic mathematics". A fuller account of Gödel's objections can be gathered from a letter from Kaufmann to Behmann. Indeed, just three days after meeting Carnap Gödel went to see Kaufmann and expressed his doubts concerning Behmann's result.

Vienna, October 19, 1930.

Dear Mr Behmann,

Dr Gödel is with me right now and he raises the following objections against my constructivity thesis and your proof of it:

He claims to be able to construct examples where clearly an existence claim is proved although one could not give a construction. The simplest example mentioned by Mr Gödel – which may serve in principle as representative for all the remaining examples – is the following:

Let a one-to-one mapping between the natural numbers and certain rational numbers from the interval (0, 1) be given. Then one can prove in the usual fashion that the given sequence of rational numbers has an accumulation point, although it would in general not be possible to specify it. Let us consider an example where one certainly cannot give an accumulation point! A number is called a Goldbach number if all even numbers smaller than it are sums of two prime numbers. The sequence of rational numbers is now defined as follows; the natural number n is mapped to $\frac{1}{n}$ if n is a Goldbach number, and to the number $1-\frac{1}{n}$ if n is not a Goldbach number. Then one can prove that this sequence has an accumulation point (and indeed that it is rational), namely either 0 or 1. However, without a solution of Goldbach's problem no accumulation point can be specified.

Mr Gödel has also talked with Mr Carnap about it who finds the objection very plausible.

I would be very grateful if you could let me know your reaction to this objection and in the meantime I will also rack my brains about it.

Mr Gödel also adds that although in this case it is a question of a disjunction between *two* possibilities, this is not essential; the disjunction between infinitely many cases could also remain undecided.

I am yours with warmest regards

Felix Kaufmann[26]

The letter was completed by a *Nachtrag* which gave the immediate reaction of Kaufmann to the result. Kaufmann's letter to Behmann was written together with Gödel and as such it represents an important testimony to Gödel's activities during this period. Indeed, the copy of the letter in Behmann's archive contains a handwritten note by Gödel on the side ("Die besten Grüße, Ihr erg. Kurt Gödel"). Gödel's objection can be restated as follows. Working within the background of classical mathematics let $E(x)$ stand for *x is even* and $P(x)$ for *x is prime*. Define $G(n)$ to be $(\forall x < n)(E(x) \supset \exists yz(P(y)\&P(z)\&x = y + z))$ Let

$$\alpha(n) = \begin{cases} \frac{1}{n}, & \text{if } G(n); \\ 1 - \frac{1}{n}, & \text{if } \neg G(n). \end{cases}$$

$\alpha(n)$ converges to 0 if Goldbach's conjecture is true; it converges to 1 if Goldbach's conjecture is false. We thus have a sentence of classical mathematics that says that there exists an x such that $\alpha(n)$ converges to x. The proof of the last claim does not rely on a proof by contradiction. The argument starts from the premise that either Goldbach's conjecture is true or it is false. In either case it concludes that $\alpha(n)$ converges. We thus have a direct proof for an existence claim, which however, does not rely on a specific instance of the claim, for as a matter of fact we have established neither that $\alpha(n)$ converges to 0 nor that $\alpha(n)$ converges to 1. This shows that there are ways to introduce an existence statement in a direct proof for which we have no instance. It should be pointed out that the counterexample provided by Gödel shows the influence of the Brouwerian counterexamples on Gödel's thinking.

In his answer to Kaufmann dated October 21, 1930, Behmann relativizes the impact of the objection by saying: "First of all, it must be said that the undoubtedly very interesting objection by Mr Gödel in any case does not affect my proof as it stands; it affects indeed the presupposition of my proof that every direct existence proof rests on the exhibition of an instance [auf einer Aufweisung]."[27] However, it was exactly that unquestioned assumption that made Behmann's result of interest. Moreover, as we have seen, it is not the case that even granting that assumption the proof goes through. In the same letter Behmann also gives examples similar to the one provided by Gödel. In subsequent correspondence between Behmann and Kaufmann the discussion about the proof is continued but nothing of great importance emerges, except the realization that the claim had not been proved conclusively. Kaufmann also suggested that a revised version of the paper might include a discussion of the recently published article by Hölder, who claimed to have shown that proofs by contradiction were not eliminable in favour of direct ones (see [Hölder, 1929, 1930]). However, the realization that much more work needed to be done if the claim were to be conclusively established came with the criticisms raised by Bernays to which we now turn.

§7. **Bernays' objections to Behmann's proof.** The correspondence between Bernays and Behmann on the subject of constructive proofs is preserved in the Bernays archives at the ETH. We saw that Behmann had submitted the paper to Bernays on October 24, 1930. Bernays replies on October 31 from Göttingen and raises a number of objections to the work. He begins by stating that he is not convinced that Behmann's result succeeds in showing the eliminability of non-predicative existence proofs. As mentioned earlier talk of non-predicativity was caused by a mistake in Behmann's orginal formulation of his second requirement. Bernays goes on to state that he has no qualms with the main lemma on the transformation of indirect into direct proofs. However, this goal had already been achieved by Löwenheim in a paper read to the Berlin Mathematical Society.[28] But this result does not allow one to conclude that anything follows for the elimination of non-predicative existence theorems. In particular, the opposition set up by Behmann between "direct through instantiation or indirect" is misleading in that indirect proof procedure is not what characterizes non-predicative proofs. Bernays then goes on to discuss the example of Euler's proof of the infinitude of primes. However, Bernays argues that also in this case the essential point of the transformation from a non-predicative to a predicative proof has nothing to do with the issue of direct vs. indirect proof. Many mathematicians, he remarks, have conjectured that it is in general true that every non-predicative proof can be transformed into a predicative one but, he adds, so far there is no proof of this claim. He diagnosed the difficulty of giving such a proof as follows: "The difficulty of the proof lies in the variety of forms in which the impredicative moment can be introduced."[29]
Suppose we infer, for instance,
$\exists x Bx$
from
$(x)\exists y Axy \supset \exists x Bx$,
$\exists x(y) \neg Axy \supset \exists x Bx$,
and
$(x)\exists y Axy \vee \exists x(y)\neg Axy$.
Another problem is definition by cases such as

$$\varphi(n) = \begin{cases} 0, & \text{if } (x)Axn; \\ 1, & \text{if } \exists x \neg Axn. \end{cases}$$

We see, as Gödel did in his counterexample, that Bernays makes use of definition by cases based on the law of the excluded middle.[30] Bernays also says that of course one could object that such definitions use the "incomplete symbol" "the x such that" (indicated by ιx) and that this symbol is in principle eliminable. But the proof, says Bernays, is not easy at all and it has given him quite a lot of trouble although he has just succeeded in the elimination of the ι-operator in the case of the first-order predicate calculus.[31] Problems only get worse when the ι-operator is used in connection with recursive definitions or with quantification over higher variables. The last part of the letter concerns the problem of the elimination of second-order variables and a few comments by Bernays to the effect that Kaufmann's results on elimination of second-order variables from theorems such as Weierstrass' theorem on monotone convergence are not original and are found in Weyl's *Das Kontinuum*. The letter ends with the advice to Behmann to avoid the use of imprecise terms such as "constructive" and "instance" and replace them with sharp logical notions. Finally, he asks Behmann why he keeps at a distance from Hilbert's proof theory, which can be the only framework adequate for treating with precision the problems Behmann is addressing.

Behmann's reply is dated November 1, 1930. First of all Behmann regrets having used the word non-predicative in that he realizes this led to misunderstandings. All he meant in point (2) of his essay was "non-constructive". He clarifies that by "instance" [Aufweisung] he means the replacement of an $\exists x f(x)$ by a suitable $f(a)$. By "constructive" he means that for every "there is" a corresponding "instance" is given. Behmann went on to say that it was far from his intentions to ignore the developments of proof theory but that he was not as familiar with the details of proof theory as Bernays was, as he never had occasion before to make use of proof theory.

The most important remark concerns the assumption that every direct existence proof, which used only constructive axioms, already contains an instance. He claims to have taken this from *Principia Mathematica* (p. 20) and adds that the other Viennese take this result for granted. The relevant passage in *Principia* reads:

> An asserted proposition of the form "$(Ex).fx$" expresses an "existence-theorem," namely "there exists an x for which $f(x)$ is true." The above proposition gives what is in practice the only way of proving existence-theorems: we always have to find some particular y for which fy holds, and thence to infer "$(Ex).fx$". If we were to assume what is called the multiplicative axiom, or the equivalent axiom enunciated by Zermelo, that would, in an important class of cases, give an existence-theorem where no particular instance of its truth can be found. ([Whitehead, Russell, 1910], p. 20)

Behmann recognized that, after Bernays' criticism, this claim could not be accepted as evident and that also the meaning of constructive axiom had to be clarified.

The final reply by Bernays came on November 11, 1930. Bernays acknowledges that if one reads "constructible" for "non-predicative" (in Behmann's essay) then Behmann's claim is more plausible. But still one must provide a proof along the following lines:

> It must be shown that in the case of the derivability of a formula E prefixed by unnegated universal and existential quantifiers, we can also derive a formula E^* which is obtained from E in that to each existential

> there is a corresponding instance; and indeed it must be shown that the derivation of E^* can be carried out in such a way that no existential quantifiers or negated universal quantifiers occur in it.[32]

Bernays then gave a precise axiom for the introduction of the ι-operator and advised Behmann to follow Hilbert-Ackermann's text [1928] for the details of formalization.

This concludes the exchange between Bernays and Behmann and the debate on the constructivity of proofs. Behmann saw the strength of Bernays' objections and wrote to Kaufmann:

> Following this it appears questionable whether publication of my article, in particular of the proof of replaceability of every indirect proof by a direct proof, is worthwhile at the moment. At the least it should not just be assumed that every direct existence proof contains an exhibition [of an instance], rather, this would have to be introduced explicitly as a mere conjecture. (November 2, 1930)[33]

The essay was never published and when asked in 1934 by a Japanese student, J. Hirano, whether the essay had been published, Behmann replied that it had not on account of the fact that he had never seen quite clearly into the matter:

> A few years ago I completed an article "On the question of constructivity of proofs," but never published it. For the most part, my aim was to show that indirect existence proofs, just like direct ones, are constructive, as long as they do not use the axiom of choice, i.e., they contain explicit exhibition of a thing of the kind in question. An important step to that conclusion was the proof that indirect proofs are nothing but transformations of direct ones.
>
> Prof. Bernays, to whom I had sent the manuscript for consideration, then wrote to me that this latter proof had been given some time before by Löwenheim, who presented it to the mathematicians in Berlin, but never published it. Furthermore, he gave a detailed critique of the article as a whole, which I did not follow in all the details and I therefore do not know in how far it was justified; it may be that he was somewhat misled by a mistake of mine ("not predicative" instead of "non-constructive"). Therefore I was in any case unsure whether it was reasonable to publish. (Behmann to Hirano, December 8, 1934)[34]

§8. Conclusion. We have seen that Behmann's paper has several shortcomings, both technical and conceptual, and that his principal claim is simply false. The aims and ideas underlying his paper, together with the discussions with Bernays and Gödel, however, provide interesting additions to the history of notions such as "constructive proof" and "proof transformation." Behmann's idea of a tree-like proof configuration and of a proof transformation proceeding by local transformations in the proof structure is, although vaguely formulated, one of the first times that the structure of a formal proof is made a subject of investigation. We find similar ideas already in Hilber's work on proof theory in the early 1920s, where derivations are transformed from linear into tree-like form, but it is only with Gentzen's work on proof theory in 1934 that the notion of tree-like derivation with branches and leaves is as central a topic as in Behmann's paper. The paper's title, "On the Constructivity of Proofs," emphasizes the notion of constructivity. Behmann's aim was to show how one could extract an instance from a proof – direct or indirect – of an existence claim. His argument rested in part on the false assumption – taken over from Russell – that any direct proof of an existential rests on the proof of an instance. Although the argument fails, Behmann's effort and the ensuing discussion is an interesting record of the state of thought about constructivity around 1930. The basic idea, i.e., that it should be possible to extract concrete information from a proof of an existence statement, and the conditions for when this is possible, is certainly the most important aspect of the debate. Again, Hilbert's work on proof theory bears some resemblance here – the epsilon-substitution method provided a way to replace quantified formulas by instances in a proof. The clearest early examples pointing in similar directions are, however, Herbrand's theorem and the closely related midsequent theorem by Gentzen. Both results establish that it is possible to extract from a proof of an existence statement in the predicate calculus a finite set of terms one of which is a witness to the existential quantifier. Whereas Behmann's result wanted to provide one instance from a proof of an existential in a classical theory, Herbrand's and Gentzen's results show that one can obtain at least a disjunction of instances in predicate logic, which is a propositional tautology, from the proof of an existential. Although more limited in scope than Behmann's original intentions, Herbrand's and Gentzen's results laid the foundation for more sophisticated results for arithmetical theories, such as the no-counterexample interpretation and witnessing theorems in bounded arithmetic.

NOTES

[1] In recent years I have been occupied, among other things, with the development of the foundations of mathematics in the twenties [Mancosu, 1998, 1999a, 1999b]. My decision to work in this area, and in particular on the development of Hilbert's program and Weyl's predicative approach to mathematics, is very much a consequence of Sol's influence on my interests. In addition to his technical engagement in proof theoretical work Sol has also displayed a refined interest in the historical development of the programs to which he so successfully contributed (see, for instance, the essays contained in [Feferman, 1998] and his general editorship of Gödel's *Collected Works*). My paper for this Festschrift is also a contribution to this part of my research. The debate I discuss centers around the topic of how "constructive" classical mathematics is – a theme dear to Sol – and contains, among other things, some hitherto unknown information concerning Gödel's contributions to the foundations of mathematics. Sol's influence on my work has been a decisive one and it extends to all of my research since my days as a graduate student (including my work on mathematical logic, on seventeenth century mathematics, and on mathematical explanation). And that this influence reaches far beyond Sol's specific expertise points to the fact that – in addition to what might be specific transmission of knowledge – Sol's greatest legacy is that of his style of thought. I dedicate this paper to him in gratitude.

[2] On the literature on proofs by contradiction see [Mancosu, 1996], chapter 4.

[3]Recent work on Behmann include [Mancosu, 1999b], [Thiel, 1999]. For a catalogue of the Behmann Nachlaß see [G. Haas, E. Stemmler, 1981].

[4]Es handelt sich um die Durchführung einer Aufgabe, zu der mich Herr Kaufmann während meines Aufenthalts in Wien im September angeregt hat. Obgleich ich ja nicht selbst die Auffassung des Konstruktivismus teile, halte ich derartige Untersuchungen doch für wichtig, damit die praktische Tragweite der konstruktivistischen Prinzipien nicht überschätzt und infolgedessen unberechtigerweise große Teile der Mathematik in Frage gestellt werden. [Bernays Nachlaß, ETH Zurich, Hs 975: 275]. For Behmann's attitude towards constructivism see also [Behmann to Kaufmann, October 22, 1930, Kaufmann archive, Konstanz, 006313].

[5]Herr Carnap hat den Aufsatz bereits geprüft und ist in jeder Hinsicht einverstanden. [Bernays Nachlaß, ETH Zurich, Hs 975: 275]

[6]See [Kaufmann to Hempel, November 30, 1930] and [Behmann to Kaufmann, Sept. 5, 1930]: both letters are in the Kaufmann archive in Konstanz.

[7]See [Behmann to Kaufmann, July 9, 1930, Kaufmann archive, Konstanz].

[8]Ich möchte Ihnen gern noch erzählen, daß ich, als ich am Montag Abend in Dresden wegen des im Speisewagen genossenen Kaffees – koffeinfreien gab es leider nicht – nicht sogleich einschlafen konnte, über Ihr Problem der Konstruktivität der Existenzbeweise nachgedacht habe und dabei zu dem Ergebnis gekommen bin, daß Sie mit Ihrer Vermutung nach meiner Beurteilung Recht haben. [Behmann to Kaufmann, September 26, 1930, Kaufmann archive, Konstanz, 006271]

[9]Copies of the essay by Behmann are found in the Bernays Nachlaß (ETH, Zurich, Hs 974:18) in the Rudolf Carnap Nachlaß (Pittsburgh/Konstanz), in the Behmann Nachlaß (Erlangen) and in the Felix Kaufmann Nachlaß (Konstanz). The correspondence between Behmann and Kaufmann can be reconstructed by using jointly the Behmann and the Kaufmann archives.

[10]In the original to Kaufmann it is "Der Satz von Kaufmann."

[11]See [Moore, 1982].

[12]"Die Anforderung (2) wird nur in dem Fall eine positive Beschränkung bedeuten, daß es Existenzbehauptungen gibt, die kein nicht-konstruktives Axiom voraussetzen und nur auf die zweite, indirekte Art bewiesen werden können." As Bernays will be quick to point out the opposition between constructive by instance vs proof by contradiction is misleading [Behmann, 1930]. See below.

[13]Hierzu hat neuerdings Herr Kaufmann eine interessante Vermutung asugesprochen durch welche die angedeutete Frage verneinend beantwortet würde. Er ist der Ansicht, daß auch ein indirekter Existenzbeweis, sofern er der Anforderung (1) genügt, seiner Natur nach konstruktiv sei, genauer gesagt, auf einer versteckten Aufweisung beruhe, die sich durch eine geeignete Analyse des Beweises zwangsläufig zur Evidenz bringen lassen müsse. [Behmann 1930]

[14]On the elimination of second order quantification see Kaufmann archive, 006243.

[15]Das gleiche Schema gilt für jeden einzelnen Schluß des Beweises, wenn im ersten Fall, d.h. falls der Schluß einer direkt Argumentation angehört, a, b, c, (usw.) als die (auf Grund der vorausgehenden direkten Beweisschritte durchweg wahren) Vordersätze und d als der Schlußsatz des Einzelschlusses [...] betrachtet werden. ([Behmann, 1930], p. 3)

[16]Ganz unten steht (mit abwärts gerichtetem Schlußpfeil) die Sternfigur des letzten Schlusses des Beweises. Soweit die an den Bedingungspfeilen stehenden Aussagen nicht Voraussetzungen des Gesamtbeweises sind, schließt sich an diese je eine weitere Sternfigur mit ihrem Schlußpfeil (der ungeachtet etwaiger Überdeckungen der Sternfiguren wiederum abwärts gerichtet vorgestellt werden möge), an jeden Bedingungspfeil dieser Sternfiguren, der nicht durch eine Voraussetzung des Gesamtbeweises besetzt ist, in genau der gleichen Weise je eine weitere Sternfigur, usf. Mit Ausnahme der Vordersätze und des Schlußsatzes des Gesamtbeweises steht also jede auftretende Aussage am Anfang und am Ende genau eines Pfeiles. ([Behmann, 1930], p. 3)

[17]Es ist nun anschaulich evident, daß innerhalb der beschriebenen Struktur als solcher, wenn man vom Richtungssinn der Pfeile absieht, die Stelle des Schlußsatzes mit denen der Vordersätze topologisch gleichberechtigt ist, daß man also das Schema in genau der gleichen Weise von irgend einem "Zweigende" her aufgebaut werden kann. Drehen wir nämlich die zugehörige Sternfigur so, daß das fragliche Zweigende abwärts (d.h. der Pfeil aufwärts) gerichtet ist, so gilt wiederum, daß an

diejenigen Sternzacken, an welchen weder Vordersätze noch der Schlußsatz des Beweises stehen, sich genau eine weitere Sternfigur anschließt, deren berührenden Pfeil wir wiederum senkrecht gestellt denken können, usf. ([Behmann, 1930], pp. 3–4)

[18]For the history of graph theory see [Biggs, 1976].

[19]I use the version due to Bernays, who had shown the redundancy of one of the original axioms in the propositional system of *Principia* and explicitly states the need for a substitution rule. For Bernays' work on propositional logic see [Zach, 1999].

[20]Herzlichen Dank für Ihren Brief und den erwarteten Beweis. Er erscheint mir als absolut schlüssig und von mustergültiger Klarheit. Kaufmann archive, Konstanz, 006231.

[21]Die von Herrn Behmann bewiesene, in meinem genannten Buche [*Das Unendliche in der Mathematik und seine Ausschaltung*] S. 66 f. aufgestellte Behauptung bringt (wie dortselbst S. 59 ff näher ausgeführt wurde) m.E. die Entscheidung in dem Methodenstreit zwischen Formalismus und Neo-Intuitionismus (Brouwer). Dies ergibt sich zunächst im Hinblick auf diejenigen Sätze, deren Beweis ohne Hilfe des Auswahlaxioms durchgeführt wird. Denn der mathematische Kern der Brouwerschen Lehre liegt in der Aufstellung des folgenden Trilemmas:

(1) Eine mathematische Existentialbehauptung E z.B. das Auftreten einer Sequenz 0, 1, 2, 3, ... 9, in der Dezimalbruchentwicklung von π ist durch direkte Aufweisung verifiziert.

(2) der Beweis der Absurdität von E ist geliefert, d.h. $\neg E$ ist für alle Zahlen bewiesen.

(3) Die Absurdität von (2) also die Absurdität der Absurdität von (1) wird bewiesen, aber ohne dass zugleich eine Konstruktion (1) angebbar wäre.

Die Möglichkeit von (3) würde bedeuten, dass aus der Absurdität der Absurdität einer Existenzialbehauptung nicht die Gültigkeit dieser Existenzialbehauptung gefolgert werden könnte und die Annahme dieser Möglichkeit ist es, die Brouwers These, dass in unendlichen Bereichen der Satz vom ausgeschlossenen Dritten keine Geltung habe, zugrundeliegt. Diese These deckt sich, wie Brouwer selbst mit Nachdruck hervorgehoben hat (z.B. Intuitionistische Mengenlehre, Jahresber. d. Deutsch. Math.-Ver., Bd 28, S 203–208, 1919) mit seiner Negierung der Entscheidungsdefinitheit der Arithmetik und Analysis.

Die Basis für diese Behauptung fällt also zunächst für alle Sätze, die ohne Hilfe des Auswahlaxioms bewiesen werden, weg.

Hier erhält man das folgende Ergebnis:

Die intuitionistische Forderung nach durchgängiger Konstruktivität ist berechtigt; aber sie erscheint auch in den sogenannten Existenzbeweisen der klassichen Mathematik implizit erfüllt, daher bringt ihre Verwirklichung keine Einbusse des Besitzstandes der Mathematik mit sich. [Kaufmann archive, Konstanz, 006228-006230]

[22]Heute habe ich eine längere, sehr schöne Diskussion mit Herrn Carnap gehabt; er ist äusserst erfreut über den gelungenen Beweis, den er – wie ich es nicht anders erwartet habe – für vollkommen greifbar hält und dessen gewaltige Tragweite er erkennt. Das Gleiche gilt von Herrn Waismann, mit dem ich heute telephonisch gesprochen habe. [Kaufmann archive, Konstanz, 006287]

[23]Carnap möchte auch Gödel gerne Ihren Beweis zeigen, wofür Ihre Genehmigung erbitte. [Kaufmann to Behmann, Vienna, October 9, 1930, Kaufmann archive, Konstanz, 006231]

[24]Kaufmann archive, Konstanz, 006218.

[25]According to [Dawson, 1997] the diary entry is to be found under RC 028-06-19.

[26][Kaufmann archive, Konstanz, 006304] and [Behmann archive, Erlangen]. The transcription is from the copy in the Behmann archive.

Wien, am 19. Oktober 1930.

Sehr geehrter Herr Behmann!

Eben ist Herr Dr. Gödel bei mir, der gegenüber meiner Konstruktivitätsbehauptung bzw. Ihrem Beweise derselben folgende Einwände erhebt:

Er behauptet sehr wohl Beispiele konstruieren zu können, wo offenkundig eine Existenzialbehauptung bewiesen ist, ohne daß doch eine Konstruktion angegeben werden könne. Das einfachste von Herrn Gödel genannte Beispiel, welches aber auch prinzipiell als Repräsentant für die übrigen Beispiele gelten darf, ist folgendes:

Gegeben sei eine eineindeutige Zuordnung zwischen den natürlichen Zahlen und gewissen Rationalzahlen des Intervalles (0, 1). Dann läßt sich auf die bekannte Art zeigen, daß die angegebene Folge von Rationalzahlen einen Häufungspunkt hat, ohne daß es jedoch allgemein möglich wäre, einen solche anzugeben. Betrachten wir ein Beispiel, wo man einen Häufungspunkt sicher nicht angeben kann! Eine Zahl heiße Goldbach-Zahl, wenn alle kleinern geraden Zahlen Summe zweier Primzahlen sind. Die Folge rationaler Zahlen werde nun so definiert: Der natürlichen Zahl n sei die Zahl $\frac{1}{n}$ zugeordnet, wenn n eine Goldbach-Zahl ist und die Zahl $1 - \frac{1}{n}$, wenn n keine Goldbach-Zahl ist. Dann kann man beweisen, daßdiese Folge einen Häufungspunkt hat (und zwar einen rationalen) nämlich entweder 0 oder 1. Es läßt sich aber ohne Lösung des Goldbach'schen Problems kein Häufungspunkt angeben.

Herr Gödel hat auch mit Herrn Carnap darüber gesprochen und diesem erscheint der gemachte Einwand sehr einleuchtend.

Ich wäre Ihnen sehr dankbar, wenn Sie mir mitteilen wollten, wie Sie sich zu diesem Einwand stellen, und will mir inzwischen auch selbst darüber der Kopf zerbrechen.

Herr Gödel fügt noch hinzu, daß es sich hier um eine Disjunktion zwischen *zwei* Möglichkeiten handle, daß dies aber nicht prinzipiell sei: es könne auch die Disjunktion zwischen unendlich vielen Fällen unentschieden bleiben.

Mit den herzlichsten Grüßen bin ich Ihr,

Felix Kaufmann

[27] Zunächst ist zu sagen, daß der zweifellos sehr interessante Einwand Herrn Gödels jedenfalls meinen Beweis als solchen nicht berührt; er betrifft ja nur die Voraussätzung meines Beweises, daß jeder direkte Existenzbeweise auf einer Aufweisung beruht. Auch Ihre Vermutung wird genau genommen nicht betroffen, wenn man sie dahin versteht, daß "in gleicher Weise wie der direkte Existenzbeweis auch der indirekte eine Aufweisung implizit enthält." [Kaufmann archive, Konstanz, 006306]

[28] Löwenheim presented the paper in Berlin in 1917. The German version was never published but a translation into English was published by Quine in 1946.

[29] Die Schwierigkeit des Nachweises liegt in der Mannigfaltigkeit der Formen, in denen das imprädikative Moment auftreten kann. [Bernays to Behmann, October 31, 1930, Bernays Nachlaß, ETH Zurich, Hs 975: 5768]

[30] Indeed, it should be pointed out that in intuitionistic logic it holds that if $\exists x A x$ is provable then $A(t)$ (for some t) is provable, which is exactly the property Behmann is after.

[31] This result appears in volume I of [Hilbert-Bernays, 1934].

[32] Es muss gezeigt werden, dass im Falle der Ableitbarkeit einer Formel E mit vorausstehenden unnegierten All- und Seinszeichen auch eine Formel E^* ableitbar ist, welche aus E hervorgeht, indem jedem Seinszeichen in E eine entsprechende "Aufweisung" zugeordnet ist; und zwar ist zu zeigen, dass die Ableitung von E^* so geführt werden kann, dass darin keine Seinszeichen und keine negierten Allzeichen vorkommen. [Bernays to Behmann, November 11, 1930, Bernays Nachlaß, ETH Zurich, Hs 975: 5769]

[33] Hiernach erscheint es doch fraglich, ob die Veröffentlichung meines Aufsatzes, insbesondere meines Beweises für die Ersetzbarkeit jedes indirekten Beweises durch einen direkten sich im Augenblick lohnt. Zum mindesten dürfte nicht einfach vorausgesetzt werden, daß jeder direkte Existenzbeweis eine Aufweisung enthält, sondern dies müßte ausdrücklich als bloße Vermutung eingeführt werden. (November 2, 1930) [Kaufmann archive, Konstanz, 006329]

[34] Einen Aufsatz "Zur Frage der Konstruktivität von Beweisen" habe ich vor einigen Jahren ausgearbeitet, aber nicht veröffentlicht. Ich wollte hauptsächlich darauf hinaus, daß der indirekte Existenzbeweis wie der direkte, sofern er nicht vom Auswahlaxiom Gebrauch macht, "konstruktiv" ist, d.h. die explizite Aufweisung eines Dinges der fraglichen Art enthält. Ein wesentlicher Schritt hierfür war der Nachweis, daß ein indirekter Beweis nichts weiter als eine Umwandlung eines direkten ist.

Prof. Bernays, dem ich die Ausarbeitung zur Beurteilung geschickt hatte, schrieb mir darauf, daß dieser letzte Nachweis schon vor längerer Zeit von Löwenheim gegeben worden sei, der ihn den Berliner Mathematikern vorgetragen, aber nicht veröffentlicht habe. Er gab weiterhin eine ausführliche Kritik des Aufsatzes im ganzen, die ich freilich in ihren Einzelheiten nicht klar durchschaut habe und von der ich daher nicht weiß, inwieweit sie berechtigt war; es mag vielleicht sein, daß er sich durch einen Schreibfehler von mir ("nicht-prädikativ" statt "nicht-konstruktiv") etwas irreführen ließ. Demzufolge war ich jedenfalls im Zweifel, ob die Veröffentlichung noch angebracht war. [Behmann to Hirano, December 8, 1934, Behmann archive, Erlangen]

REFERENCES

H. Behmann, 1930, "Zur Frage der Konstruktivität von Beweisen", unpublished typescript, Wissenschaftshistorische Sammlung, ETH, Zürich, Hs 974: 18.

N. L. Biggs et al., 1976, Graph Theory, 1736–1936, Oxford University Press.

J. W. Dawson, 1997, Logical Dilemmas, A.K. Peters, Wellesley, Massachusetts.

S. Feferman, 1998, In the Light of Logic, Oxford University Press, New York.

G. Haas, E. Stemmler, 1981, Der Nachlaß Heinrich Behmanns (1891–1970) - Gesamtverzeichnis-, in Aachener Schriften zur Wissenschaftstheorie, Logik und Logikgeschichte, Heft 1, pp. 1–39.

D. Hilbert and W. Ackermann, 1928, Grundzüge der theoretischen Logik, Springer, Berlin.

D. Hilbert and P. Bernays, 1934, Grundlagen der Mathematik, vol. 1, Springer.

O. Hölder, 1929, Der indirekte Beweis in der Mathematik, Berichte der Sächsischen Akademie der Wissenschaften zu Leipzig, 81, pp. 201–216.

O. Hölder, 1930, Nachtrag zu meinen Aufsatz über den indirekten Beweis, Berichte der Sächsischen Akademie der Wissenschaften zu Leipzig, 82, pp. 97–104.

F. Kaufmann, 1930, Das Unendliche in der Mathematik und seine Ausschaltung, Deuticke, Wien. English translation: The Infinite in Mathematics, Reidel, Dordrecht, 1978.

L. Löwenheim, 1946, On making indirect proofs direct, Scripta Mathematica XII, pp. 125–139.

P. Mancosu, 1996, Philosophy of Mathematics and Mathematical Practice in the Seventeenth Century, Oxford University Press.

P. Mancosu, 1998, From Brouwer to Hilbert. The debate on the foundations of Mathematics in the 1920s, Oxford University Press.

P. Mancosu, 1999a, Between Vienna and Berlin: the immediate reception of Gödel's incompleteness theorems, History and Philosophy of Logic, 20, 33–45

P. Mancosu, 1999b, Between Russell and Hilbert: Behmann on the foundations of mathematics, The Bulletin of Symbolic Logic, 5, no.3, 303–330

G. Moore, 1982, Zermelo's Axiom of Choice, Springer, Berlin.

C. Thiel, 1999, Gödels Anteil am Streit über Behmanns Behandlung der Antinomien, to appear in W. De Pauli-Schimanovich, P. Weibel, eds., Kurt Gödel–Ein mathematischer Mythos, Hölder-Pichler-Tempsky, Wien.

A. N. Whitehead, B. Russell, 1910, Principia Mathematica, vol. 1, Cambridge, Cambridge University Press. Second edition 1927. Quotes from the 1978 reprint of the second edition.

R. Zach, 1999, Completeness before Post: Bernays, Hilbert, and the development of propositional logic, The Bulletin of Symbolic Logic, 5, no.3, 331–366

DEPARTMENT OF PHILOSOPHY
UNIVERSITY OF CALIFORNIA, BERKELEY
BERKELEY, CA 94720-2390, USA
E-mail: mancosu@socrates.berkeley.edu

REALISM AND THE DEBATE ON IMPREDICATIVITY, 1917–1944*

CHARLES PARSONS

It is fair to say that the acceptability of impredicative definitions and reasoning in mathematics is not now, and hasn't been in recent years, a matter of major controversy. Solomon Feferman may regret that state of affairs, even though his own work contributed a great deal to bringing it about. I see the work of Feferman and Kurt Schütte on the analysis of predicative provability in the 1960's as bringing to closure one aspect of the discussion of predicativity that began with Poincaré 's protests against "non-predicative definitions" in the first decade of the twentieth century and with Russell's making a "vicious circle principle" a major principle by which constructions in logic should be assessed. Although in the 1950's Paul Lorenzen and Hao Wang had undertaken to reconstruct mathematics in such a way that impredicativity would be avoided, insistence on this (to which even Wang did not subscribe) was very much a minority view, and Feferman in particular sought principally to analyze what predicativity is, with the understanding that some aspects of this enterprise would require impredicative methods. The picture has changed since then by work to which he has also contributed, which has brought to light how much of classical analysis in particular can be done by methods that are logically very weak, in particular predicative.

The pioneer of this latter effort, as Feferman has analyzed in detail, was Hermann Weyl.[1] Weyl also brought about the most dramatic episode in the early history by claiming that there is a "vicious circle" of the kind pointed to by

*An earlier version of this paper was presented to the symposium *Reflections*, honoring Solomon Feferman on his 70th birthday, at Stanford University December 13, 1998. It was an honor and a pleasure to participate in this event honoring Sol, a valued friend whose leadership in the foundations of mathematics, both intellectual and institutional, has benefited all of us. In earlier years, when I worked in proof theory, I learned much from his writings, and he was a valuable source of information and advice for my own work. Since then, circumstances have brought us closer, and we have been collaborators on the long-running task of editing Gödel's works, where Sol's leadership and wide-ranging grasp of logic have been essential.

§6 of the paper is an inadequate response to a question asked by Sol after the lecture, which I continue to ponder. I am also indebted to Paolo Mancosu, Wilfried Sieg, W. W. Tait, and an anonymous referee.

Page references to works cited in an original and an English translation are to the original, with the page reference to the translation in brackets. I have, however, not necessarily followed the translations cited.

[1]Weyl 1918 and Feferman 1988

Reflections on the foundations of mathematics
Essays in honor of Solomon Feferman
Edited by Wilfried Sieg, Richard Sommer, and Carolyn Talcott

Poincaré and Russell in some basic reasonings in analysis. Curiously, his promising beginning for a predicative reconstruction of analysis was not pursued further at the time either by him or by others.[2] It may seem that Weyl's charge of a vicious circle found few adherents, but this appearance may be misleading because intuitionist analysis, which was just then being developed seriously by Brouwer, was not thought to be subject to the same difficulty. For some time thereafter, however, workers in foundations who accepted classical mathematics thought it necessary to reply to Weyl.

How did we get from the situation in, say, the time from 1906 to 1925, when impredicative methods seemed questionable to many of the best minds, to a later situation in which one could examine in a calm way the scope and limits of predicative methods, and in particular it was accepted that the tools of this examination might themselves be impredicative? Obviously, that would be a longer story than could be told in one lecture. Furthermore, much of it has already been well told by others. What I would like to do is briefly to sketch Weyl's criticism of analysis as he found it, which presents with exemplary clarity the assumptions behind the rejection of impredicative methods, and then say something about reactions to it by some leading figures: Hilbert, Bernays, Ramsey, Carnap, and Gödel. There is a conventional wisdom, to which I myself have subscribed in some published remarks, that the defense of impredicativity in classical mathematics[3] rests on a realist or platonist conception. Such a view is fostered by Gödel's famous discussion of Russell's vicious circle principle. Still, I want to argue that this conventional wisdom is to some degree oversimplified, both as a story about the history and as a substantive view. I don't think it even entirely does justice to Gödel. The history will end with Gödel's 1944 paper on Russell. Thus I will not deal with Feferman's work or with its very interesting background in the logical work of the 1950's; fortunately he has given a clear and informative presentation of that background himself.[4]

The debates about impredicativity, and the later work on the analysis of predicativity, both centered on analysis. Above and below that level there are issues left over, where there may still be room for controversy. One is the impredicativity involved in defining sets by quantification over the universe of sets. The other is the question whether there is a sense in which the concept of natural number is impredicative. I will be able to make only a brief remark about the first of these issues. To my regret, the second will have to be left for another occasion.[5]

§1. Weyl's charge of a vicious circle in analysis. It is well known that Weyl doubted on grounds of impredicativity the theorem that every bounded set of

[2]In the case of Weyl himself, this was in the first instance due to his temporary conversion to intuitionism; see Weyl 1921 and van Dalen 1995.

[3]As opposed to constructive, in particular intuitionist.

[4]Feferman 1964, part I.

[5]A particular reason for regret is that on this matter Feferman has taken issue with published views of mine; see Feferman and Hellman 2000.

real numbers has a least upper bound. What is less well known is the precise way in which Weyl developed this objection, why he thought it *was* an objection. Poincaré's and Russell's objections to impredicativity were much more theoretical, and in the case of Poincaré bound up with his objections to logicism and in the case of Russell insulated from actual mathematics by his adopting the axiom of reducibility.[6]

All three thinkers held views close to the view of Frege that what are called sets are extensions of concepts.[7] Weyl begins the exposition in *Das Kontinuum* with a discussion of judgments and predication. The particular formulations are visibly influenced by Husserl, but nothing derived from Husserl is essential except that Weyl puts things in terms of properties, relations, and judgments rather than beginning with language as most contemporary logicians would. However, he considers judgments formed from atomic ones by operations of first-order logic. Concerning existential judgments, however, he thinks of the range of the quantifier as being a category of objects (close to the range of significance of a propositional function for Russell), although apparently they might be further restricted; existential judgments presuppose that the range of the quantifier is "a closed system of definite objects existing in themselves" (p. 4 [8]). In §2 Weyl describes modes of construction of judgments that correspond to operations of first-order logic; such judgments with empty places he calls derived properties and relations.

This development is the background of what Weyl says about the concept of set. Although finite sets can be given by enumerating their elements, an infinite set can be given only as the extension of a property. Since one rule for forming properties is the substitution of an object in an argument place of a relation, the means of giving a finite set is itself reducible to characterizing it by a property.[8] Weyl calls the "mathematical process" the construction, beginning with a domain of objects and certain basic properties and relations, of the domain of sets and relations in extension, essentially those that are first-order definable with parameters. He states quite clearly that sets are extensions of properties so constructed.[9] He rejects in advance what Bernays later called a "quasi-combinatorial" conception:

---

[6] I will not discuss Russell's attempt, in the second edition of *Principia*, to dispense with the axiom of reducibility. But see Potter 2000, which also goes more deeply into Ramsey's position than I do here.

[7] Weyl mentions this view of Frege with approval (1918, p. 35 [47]).

[8] In the case of judgments with more than one open argument place, Weyl talks of "many-dimensional sets", what we would call relations (in extension). He also uses the term "functional connection", which is somewhat confusing because the relations involved need not be functions in the usual sense. But it is by way of such relations that he introduces functions, which I neglect in the text.

[9] If the domain is the natural numbers, some form of definition by induction, of which he admits he is not yet giving a precise formulation, is also available (p. 18 [26]). Such a formulation is given in §7; cf. note 11 below.

> The representation of an infinite set as a "collection" brought together by infinitely many individual arbitrary acts of choice, collected, and then overviewed as a whole by consciousness is nonsensical (p. 15 [23]).[10]

He opposes his own conception to the "totally vague concept of function, which has become canonical in analysis since Dirichlet" and to the similar concept of set.

So far we might think that the only sets that Weyl admits are those first-order definable from some basic relations, including perhaps relations or functions introduced by recursion. However, he is willing to view the sets obtained in this way as themselves a domain of objects and to continue the process, so that by iterating it one obtains what for us would be a model of a ramified higher-order theory. In his positive construction of analysis, as is well known, he does not use this possibility, so that what he develops is what we would call arithmetic analysis.[11] But he observes that if one construes real numbers as Dedekind cuts, if one has given a bounded set of real numbers of order 1, then the definition of the least upper bound as the union of the lower sections of the corresponding cuts is of order 2.[12]

In *Das Kontinuum* Weyl proceeds on the basis of a conception of sets as extensions of predicates that evidently must be antecedently understood, and although he does not make very clear why the usual definitions, such as the unramified version of the definition of the least upper bound, should involve a vicious circle, it is easy enough to see the conflict between them and the underlying understanding of the concept of set. What is less clear is why Weyl rejects the alternative. He expands on the matter in Weyl 1919. If a concept B is extensionally definite (*umfangsdefinit*), then for a clearly understood property E of the objects falling under B the question whether there is an object satisfying B with the property E has a definite sense, and the law of excluded middle applies.

On the basis of what he calls the "intuition of iteration," we are convinced that the concept of natural number is extensionally definite. Weyl offers the general concepts of object and property as non-extensionally definite concepts, and then claims that one can prove this even for "property of natural numbers." This is based on the intuition that if we have an extensionally definite concept of "κ-property" of properties of natural numbers, and A is a property of properties of natural numbers, then the property E_A that a number x has if and only if there is a κ-property E satisfying A such that x has E is "according to its sense different from any κ-property". The point is maybe clearer if put linguistically: Suppose $A(F)$ is a second-level predicate, whose arguments are filled by predicates

---

[10]Cf. also Weyl 1921, pp. 49–50 [94].

[11]Feferman shows (1988, pp. 264–65) that Weyl's recursion principle creates a difficulty for this. He is able to reconstruct Weyl's construction of the elements of analysis so as to get around it.

[12]This example has continued to be a problem for later work in predicative analysis allowing much more powerful methods. Thanks to Stephen Simpson and Solomon Feferman for setting me straight on this point.

of numbers, and $\boldsymbol{K}(F)$ is also a second-level predicate that is assumed to be extensionally definite. Then '$\exists F[\boldsymbol{K}(F) \wedge A(F) \wedge F(x)]$' has a definite sense, but this sense must be different from that of any predicate expressing a κ-property, i.e. satisfying $\boldsymbol{K}$.[13] Thus if we assume that properties are individuated by the senses of predicates, it follows that any extensionally definite property of properties of numbers cannot be satisfied by all such properties of numbers.

Weyl concludes that the property "property of rational numbers" is not extensionally definite, and therefore the attempt to define the least upper bound of a set of real numbers as the union, which on his view of sets involves a quantification over properties of rationals, is meaningless.

Recall that the κ-properties are assumed to be an extensionally definite domain. As Weyl notes, it does not follow from the above observations that this domain is not closed under existential quantification in the weaker sense that in the above case, the property E_A is *coextensive* with a κ-property; indeed such closure would follow from a translation into this situation of Russell's axiom of reducibility. About this he says

> It is, however, at the outset extraordinarily improbable that it is possible to set up an extensionally definite concept of "κ-property" in an exact way, so that every property E_A to be defined by the above schema from the *totality* of κ-properties is coextensive with a κ-property. In any case we have *not the shadow of a proof* of such a possibility (p. 87 [112]).

It seems from Weyl's examples that properties are given by predicates constructed by logical operations from the fundamental relations of given structures, in particular the natural numbers. On that assumption, it certainly is improbable that there is such an extensionally definite domain of properties that is closed under existential quantification as Weyl envisages here. Weyl's view on this point was close to that of a number of researchers of the time in considering the axiom of reducibility; it was widely thought that the existence of a property or propositional function of lower order coextensive with a given one of higher order is not a logical truth and is not evident.

One might, however, consider a reply suggested by Russell's often cited remark that the axiom of reducibility accomplishes what is accomplished by assuming the existence of classes. If we have among our objects arbitrary *sets* of natural numbers, and allow $\in$ as a fundamental relation, then by Weyl's own principles for a given set a the predicate $x \in a$ will be admissible. Then if we assume the second-level predicates involved extensional, and assume the usual comprehension schema for sets of natural numbers, then for Weyl's E_A we can obtain a coextensive property by replacing quantification over properties by quantification over sets. So the assumption that the property "a is a set of natural numbers" is extensionally definite would accomplish the end. Obviously Weyl would regard

[13] $\boldsymbol{K}$ need not be extensional. Weyl clearly does not assume that a property coextensive with a κ-property is necessarily a κ-property.

this assumption as hopelessly question-begging, and indeed on the conception of sets as extensions of predicates, there is no reason to think it true. But it is still well to keep that reply in mind.

§2. **Hilbert and Bernays.** A prime concern of Hilbert's foundational work was always to defend the practice of classical mathematics, so that one would expect him to respond to Weyl's critique of the elements of analysis. In fact, Hilbert and Bernays both commented on Weyl's views at the beginning of their publication on foundations in the 1920's. Since it derives from an earlier lecture than Hilbert's, I will comment first on Bernays 1922. He interprets Weyl as attempting to build up mathematics in a constructive way and as observing that in analysis and set theory "in the attempt to carry out consistently the replacement of existential axioms by methods of construction, one falls into logical circles at every step." This leads Weyl to limit the allowed inferences so that such circles are avoided.

Evidently Bernays saw Weyl's constructivism as consisting in his insistence that sets are the extensions of predicates constructed step by step, although he does not always distinguish the Weyl of 1918-19 from the intuitionism of Weyl 1921. Both positions, however, require giving up some of existing analysis. And Bernays does not see that either approach leads to a superior theory, granted that existing analysis lacks a relevant kind of intuitive evidence.

> Concerning the question of pure mathematics, what matters is only whether the usual, axiomatically characterized mathematical continuum is itself a possible, that is, a consistent, structure.[14]

The question of consistency leads him into a discussion of Hilbert's views.

Hilbert also attributes Weyl's criticism to a constructive standpoint (1922, p. 158 [199]). He says that the circle claimed by Weyl does not arise under two alternative standpoints commonly taken by mathematicians.

> First one says something like: A real number is a partition of the rational numbers that possesses the Dedekind cut property; there the concept of a partition of the rational numbers is sharp in its content and exactly limited in its extension.

Hilbert says the objection to this standpoint is that the concept of set has led to paradoxes. He replies that the fact that the general concept of set may lead to paradoxes in no way implies that the concept of set of natural numbers does so, and the known paradoxes offer no evidence of it. "On the contrary, all our mathematical experiences speak for the correctness and consistency of this concept."

The second standpoint is one alluded to briefly by Bernays: the axiomatic. Then the real numbers are simply a structure and are not defined by Dedekind cuts; instead one has the Archimedean axiom and the so-called completeness axiom. Hilbert says that "the standpoint presented is logically completely free of objections and it only remains undecided, whether a system of the required kind

[14] Bernays 1922, p. 14 [218].

is thinkable, that is whether the axioms don't perhaps lead to a contradiction." Then Hilbert remarks that real analysis has been pursued more thoroughly than almost any domain of mathematics, with complete agreement in the results. It is thus reasonable to accept the axioms on which the theory rests, although the problem remains to prove their consistency.

Paolo Mancosu has called attention to a manuscript in Bernays' hand in Hilbert's papers that discusses Weyl's criticism.[15] This manuscript contains the basic points of Hilbert's response; in places the latter seems to be following it almost verbatim. Both of the standpoints Hilbert mentions are set forth in similar terms in the manuscript, and in common with both published papers it attributes to Weyl a constructive standpoint, opposed to the axiomatic one:

> He would like pure mathematics to create for itself the objects with which it deals and not put forth statements about an unknown system of things.

In Hilbert's lectures *Probleme der mathematischen Logik* (SS 1920) some of the same issues come up.[16] Hilbert views both Russell and Weyl as seeking to reduce the concept of set to predication. He agrees with Weyl that a vicious circle arises if one attempts to define a predicate by quantification over all predicates of certain objects:

> Namely we must ask what "there is a predicate P" should mean. In axiomatic set theory "there is" always relates to the underlying domain B. In logic we can, to be sure, think of predicates as collected to a domain, but this domain of predicates cannot in this case be considered as something given at the outset, but the predicates must be constructed by logical operations, and the domain of predicates is determined only afterward by the rules of logical construction.
>
> From this it is clear, that in the rules of logical construction of predicates reference to the domain of predicates cannot be allowed. For otherwise a *circulus vitiosus* would arise (p. 31).

Russell's axiom of reducibility represents abandoning a constructive standpoint in favor of an axiomatic one.

I have gone on at some length describing these early replies of Hilbert and Bernays to Weyl. Now I wish to pose the question what it tells us about the conventional wisdom with which I began. It may seem supported by the fact that Weyl's position is described as constructive; constructivism vs. platonism or realism is a quite standard opposition in the philosophy of mathematics. On the other hand, we do not think of the Hilbert of the 1920's as a platonist, and there seems to be basic agreement in the responses of Bernays and Hilbert.

---

[15]Mancosu 1998, p. 75. I am indebted to Mancosu for providing me with a copy of this text. He thinks it likely that this was the text for a lecture Bernays gave in Zürich in 1917, thus responding to *Das Kontinuum* before its publication.

[16]I am indebted to Wilfried Sieg for pointing this out and providing copies.

The second of these contrary impressions is supported by the fact that what Weyl's constructive standpoint is primarily contrasted with is an axiomatic standpoint; in fact Hilbert's second alternative to Weyl's position could well be described as structuralist. The relation of such views to platonism or realism has been much discussed in our own day, but there is no evidence that Hilbert saw the issue in those terms. The axiomatic standpoint allows him to *assume* that the real numbers are, in Weyl's terms, an extensionally definite domain. The question how the facts about them relate to our knowledge, always the central concern of realism, is not raised. What matters is that the conception of such a domain, as described by the axioms, should be a coherent one, in particular that the resulting theory should be consistent.

It is true that Weyl holds, and Hilbert and Bernays deny, that sets are is some sense constructions. But one must exercise some care about this point. Weyl sees the positive integers as constructed by beginning with 1 and iterating the operation of passing to the successor. We can view the numbers as constructions because we obtain in this way *all* natural numbers; that is of course the idea that the principle of induction cashes in. Evidently for this reason, Weyl regards the natural numbers as extensionally definite and is ready to apply the law of excluded middle to statements involving quantification over them.

Weyl describes a similar way of constructing sets, by constructing properties and relations by the logical operations, reducing them by not distinguishing coextensive ones, and iterating what he calls the "mathematical process". But the resulting concept of set of natural numbers is what Michael Dummett calls an "indefinitely extensible" concept. In this case this has a pretty definite meaning: Our procedure of construction can always be continued, but we never have reason to believe that we have captured all sets of integers.[17] We don't have a conception of a procedure of construction that gives the conviction that it ultimately yields all sets of integers.

There may be reasons for doubting how fundamental the distinction of these two cases is; even Weyl himself came to doubt it when, under the influence of Brouwer, he questioned treating the natural numbers as extensionally definite. I should stress that the difference is not one of predicativity. Suppose one grants the thesis expressed by Dummett, myself, and Edward Nelson, that the concept of natural number is impredicative. I don't think that would have moved the Weyl of 1917-19. The "intuition of iteration" yields the extensional definiteness of the natural numbers, so that predicates formed by quantification over the natural numbers have a definite sense.

However, what is more relevant to my concern is the question of the extent to which Weyl's view reflects a form of anti-realism, and if so whether that is what separates him from Hilbert and Bernays. One might infer sympathy for such views from the evident influence of Husserl and the suggestion that Fichte could

[17]Putting Weyl aside, even if we introduced from outside arbitrary ordinals, so that the process would cease to yield new sets once we got beyond the countable ordinals, the conception we would get would be that of the constructible sets of integers.

offer illumination on questions of the philosophy of logic,[18] and perhaps also from Weyl's conversion to intuitionism shortly after the time that concerns us. But the difference that makes the difference between them concerns the relation of sets or functions to constructions of a particular kind, logical or linguistic.

There is no doubt that later, in Bernays 1935, Bernays did link the defense of impredicativity to platonism. An important feature of Bernays's discussion is that he distinguishes platonistic assumptions with respect to different domains of mathematics, beginning with what he calls the assumption of the totality of integers. In its consequences, this is close to what Weyl calls regarding the integers as extensionally definite. The next step would be to make a similar assumption about sets or sequences of integers; Bernays asserts that acceptance of impredicative definitions involving quantification over real numbers depends on an assumption of this kind (p. 55 [260]). Eliminating the second platonistic assumption would give rise to a program of a more thorough arithmetization of analysis than the usual one. "One emphasizes that an infinite sequence or a decimal fraction can be given only by an arithmetical law, and one regards the continuum as a set of elements defined by such laws" (p. 57 [261]). The reference is to the arithmetic analysis undertaken in Weyl 1918.

But what does Bernays mean by "platonism"? Before he introduces the term, he remarks that objections have been made to modes of reasoning in analysis and set theory:

> [According to these modes of reasoning] the objects of a theory are viewed as elements of a totality such that one can reason as follows: For each property expressible using the notions of the theory, it is [an] objectively determinate [fact] whether there is or there is not an element of the totality which possesses this property (pp. 52–3 [258]).

He then contrasts Euclid's language of construction with the existential character of the axioms in Hilbert's axiomatization of geometry. "This example shows already that the tendency of which we are speaking consists[19] in viewing the objects as cut off from all links with the reflecting subject" (p. 53 [259]).

However, in part because platonism so characterized is restricted in that platonistic assumptions are introduced stepwise, he says that platonism "does not claim to be more than, so to speak, an ideal projection of a domain of thought" (p. 56 [261]). He contrasts this with "absolute platonism" which he characterizes as "conceptual realism, postulating the existence of a world of ideal objects containing all the objects and relations of mathematics." This view, he thinks, is refuted by the paradoxes.

Bernays does not explain what he means by an "ideal projection of a domain of thought". However, it is clear from much of what he says that the difference between the platonism he defends and alternatives is a methodological issue.

[18] Weyl 1918, p. 2; cf. 1919, p. 86.

[19] The French is "consiste"; the later German version has "dahin geht", which might be rendered as "amounts to [viewing]" (Bernays 1976, p. 63). It's not clear that this difference makes a difference.

Platonistic assumptions are a certain way of setting up theories, and whether or not it is appropriate depends on the nature of the investigation. He contrasts an intuitive point of view (of which the paradigm is Hilbert's version of finitism) with platonism and considers the former more appropriate to number theory, even though the platonistic assumption of the totality of numbers is acceptable and necessary for analysis. Bernays's linking platonism with the acceptance of cases of the law of excluded middle is a step in the direction of Dummett's later logical characterizations of realism. But so far as the Bernays of 1935 is a realist with regard to mathematics, his realism is a methodological stance. In particular, I don't think it commits him to the view that on any question that might be posed in the language of a platonistic theory, there must be an objectively determinate answer. Some years ago Charles Chihara discussed a hypothetical view he called "mythological platonism". Bernays' view of 1935 might well be called "methodological platonism".

§3. **Ramsey.** The idea that the acceptance of impredicativity requires a platonistic point of view may have originally derived its currency from a remark of Rudolf Carnap about F. P. Ramsey's attempt to reconstruct the system of *Principia Mathematica* in a way that would make the axiom of reducibility unnecessary. Carnap interprets Ramsey as assuming "that the totality of properties already exists before their characterization by definition" (Carnap 1931, p. 102 [50]). He goes on to say, "Such a conception, I believe, is not far removed from belief in a platonic realm of ideas which exist in themselves, independently of *if* and *how* finite human beings are able to think them."

What was the view that prompted this comment? Ramsey 1926a, the paper to which Carnap refers, is a wide-ranging discussion of problems in foundations, attempting to reconstruct the position of *Principia* with the help of ideas from Wittgenstein's *Tractatus*.[20] Probably its most enduring contribution concerns the paradoxes: Ramsey worked out the distinction of mathematical and semantical paradoxes, pointed out that the simple theory of types was sufficient to circumvent the set-theoretical paradoxes, and showed how concepts of a different nature are required to generate the other paradoxes. But he tried to give a more fundamental defense of abandoning the position of the first edition of *Principia*, with the ramified theory and the axiom of reducibility. It is there that his defense of impredicativity comes in.

The guiding idea is that of Wittgenstein's view of logic (not his view of mathematics), that the "propositions" of logic are tautologies. Ramsey undertook to develop this view as a view of mathematics. It would follow that the propositions of mathematics "say nothing" and lack sense in the proper sense of that word, contrary to Carnap's reading and our conventional wisdom. Another Wittgensteinian idea, which Ramsey sees as the key to overcoming the difficulties

[20]It is remarkable that this paper was submitted for publication in 1925, when Ramsey was 22 years old.

of *Principia*, is that every proposition is a truth-function of elementary propositions. The decisive move is to admit such functions without regard to whether the number of arguments is finite:

> A *predicative function* of individuals is one which is any truth-function of arguments which, whether finite or infinite in number, are all either atomic functions of individuals or propositions. ... Admitting an infinite number involves that we do not define the range of functions as those which could be constructed in a certain way, but determine them by a description of their meanings (p. 39).

A little later, contrasting '$(\varphi).\varphi a$' with an ordinary finite truth-function, he remarks that the former cannot be elementarily expressed "owing to our inability to write propositions of infinite length, which is logically a mere accident" (p. 41).

Ramsey finds even this generous concept of predicative function not quite adequate, because of his view inherited from Wittgenstein that identities are not genuine propositions. So that at the level of individuals he introduces "functions in extension", which are hardly distinguishable from functions in the set-theoretic sense from individuals to propositions.[21]

Carnap's characterization of Ramsey's position seems to me on the whole just, and this case supports the conventional wisdom, even though Ramsey might have thought that the idea that logical truths are tautologies minimized the significance of his realistic commitment. But not long afterward, he saw difficulties about the axiom of infinity,[22] and before the end of his life changed his position more radically.

§4. **Carnap.** Carnap's critical discussion of Ramsey's view prefaces a proposal of his own, which is quite different from those we have considered so far.[23] Carnap asks how we verify a second-order universal statement. If, in impredicative logic, we had to run through all cases for this verification, we would indeed encounter a circle. Rather, we can verify it by logical derivation, where the universally quantified second-order variable will in general play the role of a parameter for an arbitrary property.

> If we reject the belief that it is necessary to run thorough individual cases and rather make it clear to ourselves that the complete verification of a statement about an arbitrary property means nothing more than its logical (more exactly, tautological) validity for an arbitrary property, we will come to the conclusion that impredicative definitions are logically admissible (1931, p. 104 [51]).

Carnap's suggestion is very sketchy, but it makes no noticeable appeal to any form of realism. Its focus on verification, in fact, makes it adaptable to intuitionism. It is hard to see that it says more than that we are able to obtain conclusions

[21] Ramsey argues that a corresponding move is not needed at higher levels.

[22] Ramsey 1926b, pp. 79–81.

[23] A similar suggestion is made in Langford 1927.

about arbitrary predicates by reasoning schematically, and that impredicativity is no obstacle to the verification of generalizations about them by such reasoning. The intuitionist will naturally ask whether Carnap's suggestion gives us any reason to believe that the law of excluded middle will hold where second-order quantifiers are involved or even tells us what meaning to assign to the negation of a second-order generalization.

Carnap returns to the matter in §44 of *Logical Syntax* but hardly gives more detail. However, the view of that work offers, rather than an answer to the questions just raised, a reason for brushing them off. Since the choice of predicative or impredicative logic is part of the setting up of a logical framework, and questions of meaning can only be answered in the context of such a framework, obviously the question of the meaning of negations is an improper question or is answered by giving the rules of inference involving negation. Since, by the principle of tolerance, "in logic there are no morals," neither the admission of impredicativity nor the use of classical logic with them can be rejected on the grounds of philosophical argument. The matter could be settled in a given context by working out the consequences of the different choices and deciding whether a given choice suits the purpose at hand. But on this specific issue, he doesn't do anything to elaborate the consequences.

Carnap's proposal is given a more direct reading by Gödel, who reads it and Langford's as interpreting "all" "as meaning analyticity or necessity or demonstrability" (1944, p. 136). The interpretation as necessity fits better Langford's proposal than Carnap's, since modalities are not mentioned in either of the Carnapian texts. Reflecting on the view of *Logical Syntax*, Carnap might well have said that $\forall F\ A(F)$ is true by virtue of the analyticity of $A(F)$, but not, I think, that $\forall F\ A(F)$ *means that* $A(F)$ is analytic. The reading in terms of demonstrability is let in because in 1931 Carnap is not explicit about what he means by logical validity. Indeed, if it is put in terms of a semantic definition of higher-order consequence (not that clear to anyone in 1931), then it is not so clear how it differs from what Gödel rightly sees Carnap as seeking to avoid, namely the reading of the quantifier as infinite conjunction.

§5. Gödel: Platonism and the vicious circle principle. Gödel already addresses the question of impredicativity in his Cambridge lecture *1933o, where he mentions it as one of the weak spots in justifying set theory. He has already said that in the application of excluded middle, the law is applied "just as if in some objective realm of ideas this question was settled quite independently of any human knowledge" (p. 49). In turning to the problem of predicativity, he says that the notion of class of integers is "essentially the same" as that of property of integers, and in giving an example of an impredicative definition formulates it as involving quantification over all properties (presumably of a given type).

> Again, as in the case of the law of excluded middle, this process of definition presupposes that the totality of all properties exists somehow independently of our knowledge and our definitions, and that our

> definitions merely serve to pick out certain of these previously existing properties (p. 50).

Gödel is prompted by these considerations to conclude that "our axioms, if interpreted as meaningful statements, necessarily presuppose a kind of Platonism, which cannot satisfy any critical mind and which does not even produce the conviction that they are consistent" (ibid.). It is well known that Gödel over the course of the next ten years did become satisfied.

It's not surprising that in the discussion of these issues in 1944, Gödel in effect defends the platonism about which he had made the above critical remark in 1933. He distinguishes three versions of the vicious circle principle, of which he says that only the first, "No totality can contain members definable only in terms of this totality," rules out impredicative definitions (1944, p. 135). Gödel observes, in agreement with both Weyl and the defenders of impredicativity, that classical mathematics does not satisfy the principle in this form; he considers this an objection to the principle rather than to classical mathematics; he concludes that the system of the first edition of *Principia* (with reducibility) also does not satisfy it if "'definable' means 'definable within the system'" (p. 135).

Gödel mentions the Carnap suggestion, but his main direct claim against the principle is that

> the vicious circle principle in its first form applies only if the entities involved are constructed by ourselves. In this case there must clearly exist a definition (namely the description of the construction) which does not refer to a totality to which the object defined belongs, because the construction of a thing can certainly not be based on a totality of things to which the construction itself belongs (p. 136).

A little later he describes the standpoint from which the principle applies as constructivistic or nominalistic. This would, I think, include the definitionism of Poincaré and Weyl as well as what he sees as Russell's own tendency to try to eliminate reference to classes and concepts by paraphrase. The term "nominalistic" is probably inserted so as to include reduction of sets or classes to linguistic entities, i.e. predicates.

At this point Gödel makes a strong affirmation of realism concerning classes and concepts, making one of his most quoted statements:

> It seems to me that the assumption of such objects is quite as legitimate as the assumption of physical bodies and there is quite as much reason to believe in their existence. They are in the same sense necessary to obtain a satisfactory system of mathematics as physical bodies are necessary for a satisfactory system of our sense perceptions (p. 137).

However, Gödel has clearly made some distinction, not made in *1933o, between independence of our constructions and definitions and independence of our knowledge. It's the former independence that is decisive for the defense of impredicativity. It seems possible to be quite realistic about what is constructible in the relevant sense and still reject impredicativity on the basis of this conception of set. Gödel does not quite say this, but I think it is implicit in the characterization of the constructivistic or nominalistic standpoint as one toward the objects of logic and mathematics. Intuitionism, by contrast, though it is often presented in that way, is first of all a standpoint about mathematical truth and its relation to proof.[24]

A distinctive feature of Gödel's realism is that it is not just realism about sets but also about "concepts". Although I can't defend this claim here, I don't think Gödel has as good an argument for the claim that the assumption of concepts on his conception of them is "necessary for a satisfactory system of mathematics" as he has for the corresponding more straightforward claim about sets. Moreover, his position on this point was a source of difficulty for him; he seems never to have arrived at a formal theory of concepts that satisfied him, even very late in life he was troubled by what he called the "intensional paradoxes", and issues about concepts troubled him in a directly philosophical way. It appears from *1933o and 1944 that he does not distinguish the issues about impredicativity with reference to properties (and in the terms of 1944, concepts) from those with reference to sets as clearly as most of us would today, even though he saw very clearly that the iterative conception was satisfactory as an account of the universe of sets while nothing corresponding satisfied him as an account of concepts.

Although it goes beyond our historical limit, I will make one further remark about Gödel. Gödel clearly paid attention to the fact that ZF licenses what one might call "super-impredicative" definitions, specifications of sets by conditions involving quantification over the universe of sets. He refers to this explicitly in a justification of the axiom of replacement published in Wang 1974 (p. 186).[25] Parallel to defenses of impredicativity at lower levels would be a defense of super-impredicativity by maintaining or assuming that the universe of sets is extensionally definite. Many of those who reflect on set theory reject this defense and hold that if that were right then the universe would be a set, which standard set theory denies. I think Gödel accepted it. It may have been connected in his mind with his realism about concepts; I don't now see clearly what the connection would

[24]This is not to say that intuitionistic mathematics does not differ in its ontology from classical; for example, as a referee suggests, Brouwer's argument for his Bar Theorem and the BHK interpretation of the logical constants make proofs fundamental mathematical objects as they are not in standard formulations of classical theories.

W. W. Tait has suggested to me that Gödel might nonetheless have thought impredicativity compatible with constructivity, in view of his introduction of what are often called the impredicative primitive recursive functionals of finite type, so called because such a functional Φ can take as arguments functionals defined in terms of Φ or functionals of higher type than Φ. Gödel points to this characteristic of these functionals in his first introduction of the concept, in the lecture *1938a, p. 94. He does not use the term "impredicative" there or elsewhere with reference to these functionals. He does, however, say that this situation does not give rise to circularity "because expressions can be ordered in a series" (ibid.). Whatever precisely this means, he seems to be saying that any violation of a Russellian or related vicious circle principle would be only apparent. But whatever he thought about this question, his silence on the subject in 1944 would be accounted for by the fact that he had still not published his work on these functionals.

[25]Cf. also Wang 1996, p. 259.

be. But that he held this view is indicated by the following tantalizing remark in Wang's reconstruction of his conversations with Gödel:

> 8.3.4 To say that the universe of all sets is unfinishable does not mean objective undeterminedness, but merely a subjective inability to finish it.[26]

§6. Conclusions. Feferman describes the position of Poincaré and of Weyl 1918, and by implication of other opponents of impredicativity, as "definitionism":[27] sets are definable sets. What we have surveyed in §§2–5 are criticisms of definitionism that were made after Weyl dramatized the effect of this view on mathematical practice. The common denominator of the defenses of impredicativity might be called "anti-definitionism". This observation brings out the fact that what Poincaré and Weyl advanced, and the writers we have surveyed criticized, was a view about the *objects* of analysis. What makes this view constructive, as Weyl and others already made clear, is that it offers a way of taking sets of integers, and other sets that arise in mathematics, as constructed in a step-by-step way.

Some philosophers of mathematics have distinguished "object platonism" from "truth platonism". Where the defense of impredicativity does rest on a form of platonism, the view presented might be described as a form of object platonism, asserting the independence of sets and real numbers from definitions and constructions. However, defenses of impredicativity along these lines are available from a broadly intuitionist point of view, so that it is not evident that the rejection of "definitionism" needs to commit us to any form of realism. Therefore the term "object platonism" may only be appropriate in the classical context.

"Truth platonism" would hold that there is an objective truth of the matter about the domains of mathematics at issue that is independent of our knowledge, perhaps even of our possibilities of knowledge, if the latter notion is not hopelessly idealized. Of course the existence of such objective truth implies something about the existence of the objects talked about, since among the truths will be some of the form 'there exist (sets, numbers) such that ... '. There might be such truth about objects that are in some sense "constructions of our own", since it might be determined quite independently of our knowledge what we can construct, or what can be constructed by quite specific means such as first-order operations beginning with the basic relations of some given structure.[28] Thus it seems possible to hold truth platonism while denying object platonism in the specific form at issue here.

To what extent did the defenders of impredicativity we have surveyed assume truth platonism in this sense? Ramsey and Gödel certainly did; Carnap clearly did not. Gödel's actual argument against the first form of the vicious circle

[26] Wang 1996, p. 260.

[27] Feferman 1988, p. 254.

[28] The distinction of object- and truth-platonism made here is essentially that of Parsons 1995, where the former is called objects platonism. Others (in particular Isaacson 1994) have made such a distinction, not necessarily coinciding with the present one.

principle, however, does not in any obvious way turn on it. In §2 I argued that even Bernays's methodological platonism of 1935 falls short of truth platonism, if the latter view goes beyond admitting the law of excluded middle for the relevant domain of objects.

One way of distinguishing Bernays's platonism as a methodological stance adopted in specific domains of mathematics and platonism as a more metaphysical view would be to make the distinction between an internal and external perspective on a given theory. If we ask whether the continuum hypothesis must be true or false, of course there is a sense in which the answer is trivially affirmative: In set theory we assume $CH \vee \neg CH$, but if CH, then CH is true, and if $\neg CH$, CH is false; therefore it is either true or false. So long as we are using the language of classical set theory and taking it at face value, there is no room for refusing to affirm that CH is either true or false. But it does not follow that, if one reflects on set theory and considers what grounds there might be for the truth or falsity of CH, one is forced to assume that such grounds for one or the other conclusion must exist. One relevant sense of "grounds" would be epistemic, possible reasons that an agent might have for affirming or denying CH. A view that one might call "holistic verificationism" would accept the law of the excluded middle, and thus the "internal" affirmation that CH is either true or false, but in the external stance would maintain that there is no sense to be made of the idea that CH is true independently of any possibility an epistemic agent might have of coming to know it. Such a view has had its defenders,[29] and it no doubt also has its difficulties. But there is no reason to think it requires a conception of set like the one underlying Weyl's arguments.

In the discussion of my lecture at the Reflections conference, Solomon Feferman asked whether the history I recounted shows that there is a nonplatonist defense of impredicativity. I assume he had in mind the context of classical mathematics. The history certainly offers suggestions in that direction, but it also shows that the answer depends on what one counts as "platonism." Holistic verificationism is a view that would be an example of what Feferman was asking for. Much recent work on the philosophy of mathematics seeks a middle ground between such a view and more full-blown truth platonism, what one might call transcendental realism. One might well shift Feferman's question and ask whether contemporary work, whatever its relation to the history, yields the kind of defense of impredicativity he is asking for. That is a larger question than can be attacked on the present occasion.

REFERENCES

Paul Benacerraf and Hilary Putnam (editors) [1983], ***Philosophy of mathematics: Selected readings, 2d ed.***, Cambridge University Press.

[29]As a general view of truth, it is certainly implicit in the writings of W. V. Quine. Remarks expressing this tendency occur in Tait 1986 and in lectures by him I have heard, but he might reject the view as here stated because of skepticism about ideal possibility applied to epistemic concepts.

Paul Bernays [1922], *Über Hilberts Gedanken zur Grundlegung der Arithmetik*, ***Jahresbericht der Deutschen Mathematiker-Vereinigung***, vol. 31, pp. 10–19, Translation in Mancosu 1998.

Paul Bernays [1935], *Sur le platonisme dans les mathématiques*, ***L'enseignement mathématique***, vol. 34, pp. 52–69, Translation in Benacerraf and Putnam 1983. German translation in Bernays 1976.

Paul Bernays [1976], ***Abhandlungen zur Philosophie der Mathematik***, Darmstadt: Wissenschaftliche Buchgesellschaft.

Rudolf Carnap [1931], *Die logizistische Grundlegung der Mathematik*, ***Erkenntnis***, vol. 2, pp. 91–105. Translation in Benacerraf and Putnam 1983.

Rudolf Carnap [1934], ***Logische Syntax der Sprache***, Vienna: Springer.

Rudolf Carnap [1937], ***Logical Syntax of Language***, London: Kegan Paul, Translation. with additions, of Carnap 1934.

Solomon Feferman [1964], *Systems of predicative analysis*, ***The Journal of Symbolic Logic***, vol. 30, pp. 1–30.

Solomon Feferman [1988], *Weyl vindicated:* Das Kontinuum *70 years later*, ***Temi e prospettive della logica e della scienza contemporanee*** (Carlo Cellucci and Giovanni Sambin, editors). Bologna: CLUEB, vol. 1, pp. 59–93. Reprinted with corrections and additions in Feferman 1998. Cited according to reprint.

Solomon Feferman [1998], ***In the Light of Logic***, New York and Oxford: Oxford University Press.

Solomon Feferman and Geofrey Hellman [2000], *Challenges to predicative foundations of arithmetic*, ***Between Logic and Intuition: Essays in honor of Charles Parsons*** (Gila Sher and Richard Tieszen, editors). Cambridge University Press, pp. 317–338.

Kurt Gödel [*1933o], *The present situation in the foundations of mathematics. (Lecture to the Mathematical Association of America, 1933)*, ***Collected Works, volume III: Unpublished Essays and Lectures*** (Solomon Feferman et al., editors), New York and Oxford: Oxford University Press. 1995, pp. 45–53.

Kurt Gödel [*1938a], *Vortrag bei Zilsel. (Transcription of shorthand notes for a lecture to an informal circle, Vienna, 1938, with an English translation)*, ***Collected Works, volume III: Unpublished Essays and Lectures*** (Solomon Feferman et al., editors), New York and Oxford: Oxford University Press. 1995, pp. 86–113.

Kurt Gödel [1944], *Russell's mathematical logic*, ***The Philosophy of Bertrand Russell*** (Paul Arthur Schilpp, editor), Evanston: Northwestern University, pp. 125–153, reprinted in ***Collected Works, volume II: Publications 1938–1974***, Solomon Feferman et al., editors, (New York and Oxford: Oxford University Press, 1990). Cited according to original.

David Hilbert [1922], *Neubegründung der Mathematik. Erste Mitteilung*, ***Abhandlungen aus dem mathematischen Seminar der Universität Hamburg***, vol. 1, pp. 157–177, Translation in Mancosu 1998.

Daniel Isaacson [1994], *Mathematical intuition and objectivity*, ***Mathematics and Mind*** (Alexander George, editor), New York and Oxford: Oxford University Press.

C. H. Langford [1927], *On inductive relations*, ***Bulletin of the American Mathematical Society***, vol. 33, pp. 599–607.

Paolo Mancosu (editor) [1998], ***From Brouwer to Hilbert. The Debate on the Foundations of Mathematics in the 1920s***, New York and Oxford: Oxford University Press.

Charles Parsons [1995], *A distinction in platonism*, ***Under Tall Trees: A Tribute to Dirk van de Kaa***, Wassenaar: Netherlands Institute for Advanced Study, pp. 9–10.

Michael Potter [2000], ***Reason's Nearest Kin. Philosophies of Arithmetic from Kant to Carnap***, Oxford and New York: Oxford University Press.

F. P. Ramsey [1926a], *The foundations of mathematics*, ***Proceedings of the London Mathematical Society***, vol. 25, no. 2, pp. 338–384, Cited according to reprint in Ramsey 1931.

F. P. Ramsey [1926b], *Mathematical logic*, ***The Mathematical Gazette***, vol. 13, no. 184, pp. 185–194, Cited according to reprint in Ramsey 1931.

F. P. Ramsey [1931], ***The Foundations of Mathematics and other Logical Essays***, London: Kegan Paul, Edited by R. B. Braithwaite, with a preface by G. E. Moore.

Dirk van Dalen [1995], *Hermann Weyl's intuitionistic mathematics*, ***The Bulletin of Symbolic Logic***, vol. 1, pp. 145–169.

Hao Wang [1974], ***From Mathematics to Philosophy***, London: Routledge and Kegan Paul.

Hao Wang [1996], ***A Logical Journey. From Gödel to Philosophy***, Cambridge, Mass.: MIT Press.

Hermann Weyl [1918], ***Das Kontinuum. Kritische Untersuchungen über die Grundlagen der Analysis***, Leipzig: Veit, Translation, Weyl 1987.

Hermann Weyl [1919], *Der circulus vitiosus in der heutigen Begründung der Analysis*, ***Jahresbericht der deutschen Mathematiker-Vereinigung***, vol. 28, pp. 85–92, Translation in Weyl 1987.

Hermann Weyl [1921], *Über die neue Grundlagenkrise der Mathematik*, ***Mathematische Zeitschrift***, vol. 10, pp. 39–79, Translation in Mancosu 1998.

Hermann Weyl [1987], ***The Continuum. A Critical Examination of the Foundations of Analysis***, Kirksville, Mo.: Thomas Jefferson University Press, Translated by Stephen Pollard and T. Bole.

DEPARTMENT OF PHILOSOPHY
HARVARD UNIVERSITY
CAMBRIDGE, MA 02138, USA
E-mail: parsons2@fas.harvard.edu

CALCULATIONS BY MAN AND MACHINE: CONCEPTUAL ANALYSIS*

WILFRIED SIEG

§1. **Analysis & history.** To investigate calculations is to analyze symbolic processes carried out by calculators; that is a lesson we owe to Turing. Taking the lesson seriously, I will formulate restrictive conditions and well-motivated axioms for two types of calculators, namely, for *human (computing) agents* and *mechanical (computing) devices*.[1] My objective is to resolve central foundational problems in logic and cognitive science that require a deeper understanding of the nature of calculations. Without such an understanding, neither the scope of undecidability and incompleteness results in logic nor the significance of computational models in cognitive science can be explored in their proper generality. The claim for logic is almost trivial and implies the claim for cognitive science; after all, the relevant logical notions have been used when striving to create artificial intelligence or to model mental processes in humans.

The foundational problems come to the fore in arguments for Church's or Turing's Thesis, asserting that an informal notion of effective calculability is captured fully by a particular precise mathematical concept. Church's Thesis, for example, claims in its original form that the effectively calculable number theoretic functions are exactly those functions whose values are computable in Gödel's equational calculus. My strategy, when arguing for the adequacy of a notion, is to bypass theses altogether and avoid the fruitless discussion of their (un-)provability. This can be achieved by *conceptual analysis,* i.e., by sharpening the informal notion, formulating its general features axiomatically, and investigating the axiomatic framework. Such an analysis will be provided for the two types of calculators I mentioned, examining closely and recasting thoroughly work of Turing and Gandy. My paper builds on systematic and historical work I have

*This paper is dedicated to my teacher, colleague, collaborator, and friend Solomon Feferman on the occasion of his 70th birthday.

[1] I presented versions of this paper at the *Feferfest* at Stanford University on December 13, 1998, at the *Séminaire Philosophie des mathématiques et de l'informatique* at the Sorbonne on June 7, 1999, at the *Workshop on Logic and Physics of Computation* in Erice (Sicily) on July 29, 1999, and at the *11th International Congress of Logic, Methodology and Philosophy of Science* in Cracow on August 21, 1999. Thanks to Guglielmo Tamburrini and two anonymous referees for helpful suggestions that clarified some issues, and also to Dawn McLaughlin for turning an awkward Word document into a beautifully LaTeXed paper.

Reflections on the foundations of mathematics
Essays in honor of Solomon Feferman
Edited by Wilfried Sieg, Richard Sommer, and Carolyn Talcott

pursued for more than a decade, much of it in collaboration with John Byrnes, Daniele Mundici, and Guglielmo Tamburrini. The considerations presented here reshape and extend the earlier systematic work in a novel and, for me, unexpected way; their aim is nevertheless extremely classical, namely, to provide what Hilbert called *eine Tieferlegung der Fundamente*. It will also become evident that they are embedded in an illuminating historical context.

There is general agreement that Turing, in 1936, gave the most convincing analysis of *effective calculability* in his paper "On computable numbers - with an application to the Entscheidungsproblem". It can be argued that he gave the only convincing analysis and, in addition, that the nature of his analysis is still not clearly recognized. Thus, it seems worthwhile to go back to the discussion of the mid-thirties of the last century. Section 2 sets the stage by dealing with effective calculability of number theoretic functions as investigated by Gödel, Church, and Hilbert & Bernays, whereas section 3 makes explicit Turing's truly distinctive contribution and focuses on calculations by *mechanical man* (or computors).[2] Evident limitations of the human sensory apparatus motivate boundedness and locality conditions for computors, and broad physical limitations lead to similar requirements for mechanical devices in section 4. There I adapt Gandy's work on machine computability and treat calculations by (parallel) *machines*.[3] Shifting the methodological perspective and drawing consequences from these investigations, section 5 gives axiomatic characterizations of *Turing Computors* and *Gandy Machines*. For both notions suitable *representation theorems* can be proved. The theorems guarantee that the computations of any model of the axioms can be simulated by a standard Turing machine over a two-letter alphabet.

The detailed conceptual analysis of effective calculabilty yields rigorous characterizations that dispense with theses, reveal human and machine calculability as axiomatically given mathematical concepts, and allow their systematic reduction to Turing computability.

§2. Effective calculations. Church reviewed Turing's "On computable numbers" a few months after its publication for the Journal of Symbolic Logic. He contrasted Turing's notion for effective calculability (via idealized machines) with his own (via λ -definability) and with Gödel's (via the equational calculus). "Of these [notions]," Church remarked, "the first has the advantage of making the identification with effectiveness in the ordinary (not explicitly defined) sense evident immediately ... " Neither in this review nor anywhere else did Church give reasons, why the identification *is* immediately evident for Turing's notion, and why it *is not* for the others. In contrast, Gödel seemed to capture essential aspects of Turing's considerations when making a brief and enigmatic remark in the 1964

[2]For a detailed historical and systematic examination with extensive references to the literature see my papers (1994) and (1997), but also (Davis 1982), (Gandy 1988), and (Soare 1996).

[3]Mathematical details for the characterization of machines are presented in a companion paper, (Sieg 2000).

postscript to the Princeton Lectures he had delivered thirty years earlier: "Turing's work gives an analysis of the concept of 'mechanical procedure' ... This concept is *shown* to be equivalent with that of a 'Turing machine'."[4] But neither in this postscript nor in other writings did Gödel indicate the nature of Turing's analysis or give a proof showing that the analyzed concept is indeed equivalent to that of a Turing machine.

Gödel underlined the significance of Turing's analysis, repeatedly and emphatically. He claimed in 1964, for example, that only Turing's work provided "a precise and unquestionably adequate definition of the general concept of formal system". A formal system is defined to be a mechanical procedure for producing theorems; consequently, the adequacy of this definition rests squarely on the correctness of Turing's analysis of mechanical procedures. Gödel had an abiding interest in the issue, as it was crucial for his goal to give the most general mathematical formulation and the broadest philosophical interpretation of his incompleteness theorems. A general concept of formal system was needed to achieve that goal. Gödel himself had tried to arrive at such a concept in a quite different way, namely, by a direct characterization of effectively calculable number theoretic functions. As a first step towards such a characterization, Gödel had introduced in his Princeton Lectures *general recursive functions* via his equational calculus.[5] I will review briefly the crucial features of Gödel's definition.

The general recursive functions are taken by Gödel to be those number theoretic functions whose values can be calculated from basic equations via elementary substitution rules. This is an extremely natural approach and generalizes properly the idea underlying the definition of primitive recursive functions: the new class of functions includes in addition to all primitive recursive functions also the non-primitive recursive Ackermann function. Assume, Gödel suggests, you are given a finite sequence $\psi_1, \ldots, \psi_k$ of "known" functions and a symbol ϕ for an "unknown" one. Then substitute these symbols "in one another in the most general fashions" and equate certain pairs of the resulting expressions. If the selected set of functional equations has exactly one solution, consider ϕ as (denoting) a general recursive function. Gödel attributes this proposal to Herbrand[6] and proceeds to make two restrictions: (1) the left-hand side of given equations must be of the form $\phi(\psi_{i1}(x_1, \ldots, x_n), \ldots, \psi_{il}(x_1, \ldots, x_n))$, and (2) for every l-tuple of natural numbers the value of ϕ must be "computable in a calculus". The second condition demands more precisely that for every l-tuple $k_1, \ldots, k_l$ there is exactly one m such that $\phi(k_1, \ldots, k_l) = m$ is a "derived equation". So it remains to

[4]Gödel's *Collected Works*, volume I, 369–70. The emphases are mine. In the context of this paper (and reflecting the above discussion of Church and Gödel), I consider effective and mechanical procedure as synonymous.

[5]Recall that in the contemporaneous discusssion the class of *recursive* functions consisted of those functions we now call *primitive recursive*.

[6]As we know now, Gödel misremembered: see section 2.2 and the Appendix of (Sieg 1994). The reader should compare the discussion there with that in (Wang 1974), pp. 87-89.

specify the set of *derived equations*. That is done inductively using elementary substitution rules; the basic ones are:

(**A.1**) Replacing all the variables of a given equation by numerals yields a derived equation;

(**A.2**) All true equalities $\psi_{ij}(x_1, \ldots, x_n) = m$ are derived equations.

The rules allowing steps from already obtained equations to additional ones are formulated as follows:

(**R.1**) Replace occurrences of $\psi_{ij}(x_1, \ldots, x_n)$ by m, if $\psi_{ij}(x_1, \ldots, x_n) = m$ is a derived equation;

(**R.2**) Replace occurrences of $\phi(x_1, \ldots, x_l)$ on the right-hand side of a derived equation by m, if $\phi(x_1, \ldots, x_l) = m$ is a derived equation.

Kleene analyzed, using Gödel's arithmetization technique for describing provability in the equational calculus, the general recursive functions in his (1936). He established a version of what is now called *Kleene's Normal Form Theorem*: every general recursive function can be expressed in the form $\psi(\varepsilon y.\rho(x_1, \ldots, x_n, y) = 0)$, where ψ and ρ are primitive recursive and for every n-tuple $x_1, \ldots, x_n$ there is a y such that $\rho(x_1, \ldots, x_n, y) = 0$; ε is the ε-symbol as introduced by Hilbert and used by Gödel in his 1934 Princeton Lectures. This theorem (or rather its proof) is quite remarkable: the ease with which it allows to establish equivalences of different formulations makes it plausible that *some* stable notion has been isolated; what is needed for the proof is only that the inference or computation steps are all primitive recursive. However, the question, whether *that* stable notion corresponds to the informal concept of effective calculability, has to be answered independently.

Any formal system S that is even weakly adequate for number theory will allow the computations carried out in Gödel's equational calculus. Gödel made an important observation in 1936 and tried to use it in his answer to the question just formulated: no extension of such an S by higher types, even transfinite ones, will increase the class of calculable or computable functions.[7] "Thus," he concluded, "the notion 'computable' is in a certain sense 'absolute', while almost all metamathematical notions otherwise known (for example, provable, definable, and so on) quite essentially depend upon the system adopted." At the Princeton Bicentennial Conference in 1946, Gödel stressed again the importance of the concept of general recursiveness or Turing computability and reemphasized:

> It seems to me that this importance is largely due to the fact that with this concept one has for the first time succeeded in giving an

[7]This observation made it for the first time plausible to Gödel that a stable notion had been characterized; cf. (Sieg 1994), p. 88. He pointed out, quite forcefully, in a letter to Martin Davis of February 15, 1965 that Note 3 of his 1934 Princeton Lectures did not give a formulation of Church's Thesis. Indeed, he emphasized in the letter that "at the time of these lectures, [I was] not at all convinced that my concept of recursion comprises all possible recursions; and in fact the equivalence between my definition and Kleene's ... is not quite trivial." The reference is to Kleene's 1936 paper. For further details, cf. (Davis 1982) and (Sieg 1997), pp. 159–160.

absolute definition of an interesting epistemological notion, i.e., one not depending on the formalism chosen.[8]

Indeed, the details of the formalisms (extending arithmetic) do not matter, but it is crucial that we are dealing with formalisms at all; in other words, a precise aspect of the (unexplicated) *formal* character of the extending theories has to come into play, when arguing for the absoluteness of the concept computability. Gödel did not prove that computability is an absolute concept, neither here nor in the earlier paper. I conjecture that he used considerations similar to those underlying the proof of Kleene's Normal Form Theorem in order to convince himself of the claim in 1936. If that conjecture is correct, then Church's contemporaneous step-by-step argument for the co-extensiveness of effective calculability and general recursiveness is completely parallel.[9] Church required, when explicating effective calculability as computability in logical calculi, the inferential steps in such calculi not only to be effective, but—without any supporting reason—to be general recursive.

So we find in both these cases a hidden and semi-circular condition on "steps", a condition that allows the parallel arguments to go through. This step-condition was subsequently moved into the foreground by Hilbert & Bernays' marvelous analysis of "computations in deductive formalisms". Their analysis, presented in a supplement to the second volume of *Grundlagen der Mathematik*, used the concept of reckonable function (regelrecht auswertbare Funktion). General deductive formalisms were appropriately restricted by "recursiveness conditions", the crucial one requiring the proof predicate of such formalisms to be primitive recursive. Hilbert & Bernays showed that all reckonable functions are computable in a restricted number theoretic formalism, and that the functions computable in that formalism coincide with the general recursive ones. In this way, proper mathematical underpinnings were provided for Gödel's absoluteness claim and Church's argument, but they were provided *only relative* to the recursiveness conditions.

The conceptual work of Gödel, Church, and Hilbert & Bernays had intimate historical connections and is still of genuine and deep interest.[10] It explicated effective calculability of functions by *one core notion,* namely, computability of their values in a calculus via restricted rules. But no one gave convincing reasons

---

[8](CW II, p. 150): my emphasis. In the footnote added to this remark in 1965 Gödel wrote: "To be more precise: a function of integers is computable in any formal system containing arithmetic if and only if it is computable in arithmetic, where a function f is called computable in S if there is in S a computable term representing f." — Tarski's remarks at this conference, only recently published in (Sinaceur 2000), make so vivid, how important the issue of the "intuitive adequacy" of general recursiveness was taken to be.

[9]The argument was given in Church's (1936); the Thesis had been formulated publicly by Church in 1935 using Gödel's notion. — How closely related Gödel's and Church's considerations were can be seen from the letter Church wrote to the Polish logician Pepis on June 8, 1937; cf. my (1997), pp. 168–9 and Appendix A.

[10]Their work was complemented by important mathematical work of Kleene and Rosser; cf. (Sieg 1997).

for the proposed restrictions on the steps permitted in computations, i.e., all the analyses ran up against the very same stumbling block. A dramatic shift of perspective made for real progress. Instead of considering particular schemes for computing the values of number theoretic functions, Turing and Post proposed to look at underlying symbolic or combinatory processes. The shift is also dramatic in a different sense: though contiguous with the other work, it overcomes—through Turing's reflections—the stumbling block for a fundamental conceptual analysis.

§3. **By mechanical man.** In his specification of *finite combinatory processes* Post used a human worker who operates in a "symbol space" and carries out, over a two-letter alphabet, exactly the kind of operations Turing machines can perform. Post thought that the support for his model would be provided inductively, by considering ever-wider formulations and reducing them to the restricted formulation I just hinted at.[11] In the argument for his model, Post did not use the fact that a human worker does the computing; Turing, in describing his model, focused so strongly on machines that some commentators have taken him to analyze machine computations. Turing examined however *human* mechanical computability and exploited, in sharp contrast to Post, limitations of the human computing agent to motivate restrictive conditions. The latter replace the unconvincing implicit or explicit recursiveness requirements in the earlier analyses. The essence of Turing's formulation is brought out by an aphoristic remark of Wittgenstein's on *Turing's 'Machines'*. "These machines," Wittgenstein said, "are humans who calculate." That is not only right, but Turing asked in the historical context in which he found himself *the* pertinent question, namely, what are the possible processes a human being can carry out (when computing a number or, equivalently, determining algorithmically the value of a number theoretic function)? The general problematic required an analysis of the idealized capacities of calculators, and precisely this focus makes the analysis epistemologically significant.[12]

In my further discussion I shall use a convention suggested by Gandy and refer by *computor* to a human computing agent who proceeds mechanically. A computor operates on certain symbolic configurations, and Turing demands *immediate recognizability* of these symbolic configurations so that the most basic computation steps need not be subdivided any further. This normative demand

[11]Cf. (Sieg 1994), pp. 91–2, and (Sieg & Byrnes 1996), section 2.

[12]In my presentation I am taking an anachronistic step by having Turing discuss "computable functions". This is mainly for expository expediency, but it is systematically justifiable as follows. Turing discusses "computable numbers", but emphasizes that he could have equally easily investigated "computable functions of an integral variable or a real or computable variable, computable predicates, and so forth". Indeed, he states emphatically: "The fundamental problems involved are, however, the same in each case, and I have chosen the computable numbers for explicit treatment as involving the least cumbrous technique." (Turing 1936), p. 116. — That is clearly in the broad tradition of Leibniz's "Calculemus!", and it is needed for the negative resolution of the Entscheidungsproblem (and other mathematical and metamathematical issues). That some such "machine simulation" should be considered is clear from other contemporaneous remarks, e.g., in (Gödel 1933).

and the evident limitation of a computor's sensory apparatus[13] lead to a number of restrictive conditions that can be extracted from Turing's discussion.[14] To formulate them in the most straightforward way, I am using two ideas, both endorsed by Turing: (i) the elimination of internal states by "more physical counterparts" (proposed in section 9 (III) of his 1936 paper), and (ii) the representation of Turing machines as Post production systems (used by Turing for obtaining new mathematical results in 1950, but also for a wonderful informal exposition of solvable and unsolvable problems in 1953).[15]

Of course, the elimination of internal states is realized in the Post representation, as the physical counterparts of such states are joined to the ordinary symbolic configurations to form complete *instantaneous descriptions* or *ids*. Any id contains exactly one such physical counterpart, and the[16] immediately recognizable sub-configuration of an id must contain it. Given this compact description, the restrictive conditions are as follows:

(B) (Boundedness) *There is a fixed bound on the number of configurations a computor can immediately recognize.*

(L) (Locality) *A computor can change only immediately recognizable (sub-) configurations.*

Computors proceed deterministically; consequently, the computing process has to satisfy:

---

[13]Turing goes beyond the appeal to sensory limitations and views memory limitations as the ultimate reason for the restrictive conditions: cf. (Sieg 1994), p. 96.

[14]For ease of comparison, I attach here Turing's conditions as analyzed, for example, in my (1997); a similar analysis is presented in (Wang 1974) on pp. 90–95. (However, the crucial point of grounding the boundedness and locality conditions in the limitations of the computing subject is not brought out by Wang.) A first boundedness condition can be formulated as follows:

(B.1) *There is a fixed bound on the number of symbolic configurations a computor can immediately recognize.*

A second boundedness condition concerns "internal states" reflecting the computor's experience:

(B.2) *There is a fixed bound on the number of internal states a computor can be in.*

For a given computor there are consequently only boundedly many different combinations of symbolic configurations and internal states. His behavior is taken to be deterministic, i.e., it has to satisfy a determinacy condition:

(D) *A computor's internal state together with the observed configuration fixes uniquely the next computation step and the next internal state.*

Thus, he can carry out at most finitely many different operations. The operations are restricted by locality conditions.

(L.1) *A computor can change only elements of an observed symbolic configuration.*

A second locality condition allows a computor to turn attention to a configuration "close" to the one he had been observing.

(L.2) *A computor can shift attention from one symbolic configuration to another one, but the new observed configuration must be within a bounded distance of the immediately previously observed configuration.*

[15]Turing machines were represented as production systems, of course, by Post in his superb paper "Recursive unsolvability of a problem of Thue" (1947). This representation is also used in Davis' classical textbook *Computability and Undecidability*.

[16]There is a certain conventionality to determine "the" sub-configuration at this point, as a number of different ones may be observable; cf. (Sieg & Byrnes 1996), pp. 103–6.

(D) (Determinacy) *The immediately recognizable (sub-)configuration determines uniquely the next computation step (and id).*

Turing argued that the number theoretic functions calculable by such a computor are also computable by a Turing machine. Thus, Turing's assertion that effective calculability can be identified with machine computability is the result of a two-part analysis: the first part yields the boundedness condition for symbolic configurations and the locality condition for mechanical operations; the second part argues for the *claim* that every number theoretic function calculable by a computor, satisfying these conditions, is computable by a Turing machine. As we shall see, this latter part has to be broken into two separate steps.

The above restrictive conditions on computors are given quite informally. In order to formulate them precisely and make their investigation possible, it seems that the symbolic configurations must be specified; they were taken by Turing to be linear. Here is the starting-point of Turing's considerations together with a dimension-lowering step:

> Computing is normally done by writing certain symbols on paper. We may suppose this paper is divided into squares like a child's arithmetic book. In elementary arithmetic, the two-dimensional character of the paper is sometimes used. But such a use is always avoidable, and I think that it will be agreed that the two-dimensional character of paper is no essential of computation. I assume then that the computation is carried out on one-dimensional paper, *i.e.* on a tape divided into squares.[17]

In his further reductive argument Turing constructed machines that mimic the work of computors on linear configurations. These more general machines are best seen as allowing the replacement of finite strings on the tape by finite strings. Calling such machines *string machines* and standard Turing machines *letter machines,* the assertion rigorously established by Turing is this: computations of string machines can be carried out by letter machines.[18] Turing's considerations make plausible the claim of his *Central Thesis*: computations carried out by a computor satisfying the boundedness and locality conditions can be *directly* simulated by a string machine. Computations of a string machine, in turn, can provably be mimicked by a letter machine. The diagram below represents these reflections graphically and relates them to the standard formulation of Turing's Thesis.

Step 1 in the diagram is given by conceptual analysis, whereas step 2 indicates the application of the Central Thesis. The equivalence proof justifies an extremely

[17](Turing 1936), p. 135.

[18]Turing observed: "The machines just described do not differ very essentially from computing machines as described in §2, and corresponding to any machine of this type a computing machine can be constructed to compute the same sequence, that is to say the sequence computed by the computer [in my terminology: computor]."

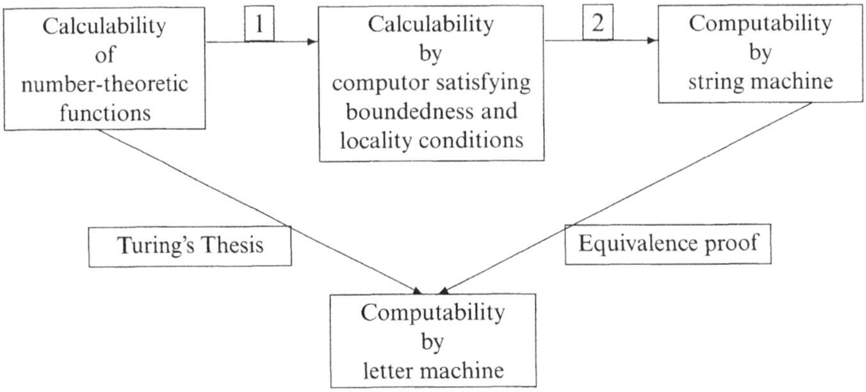

simple description of computations that is most useful for mathematical investigations, from the construction of a universal machine and the formulation of the halting problem to the proof of the undecidability of the Entscheidungsproblem. It should be underlined that step 2, not the equivalence proof, is for Turing the crucial one that goes beyond the conceptual analysis; for me it is the problematic one that requires further reflection.

In order to make Turing's Central Thesis, quite in Post's spirit, inductively more convincing, it seems sensible to allow larger classes of symbolic configurations and more general operations on them. Turing himself intended, as we saw, to give an analysis of mechanical procedures on two-dimensional configurations already in 1936. In 1953 he considered even three-dimensional configurations and mechanical operations on them, starting out with examples of puzzles: square piece puzzles, puzzles involving the separation of rigid bodies or the transformation of knots, i.e., puzzles in two and three dimensions. He viewed Post production systems as linear or *substitution puzzles*. As he considered them as puzzles in "normal form", he was able to formulate a suitable version of "Turing's Thesis":

> Given any puzzle we can find a corresponding substitution puzzle which is equivalent to it in the sense that given a solution of the one we can easily find a solution of the other ... A transformation can be carried out by the rules of the original puzzle if and only if it can be carried out by substitutions ... [19]

Turing admits that this formulation is "somewhat lacking in definiteness" and claims that it will remain so; he characterizes its status as lying between a theorem and a definition: "In so far as we know *a priori* what is a puzzle and what is not, the statement is a theorem. In so far as we do not know what puzzles are, the statement is a definition which tells us something about what they are." Of

[19](Turing 1953), p. 15.

course, Turing continues, one could define puzzle by a phrase beginning with "a set of definite rules", or one could reduce its definition to that of computable function or systematic procedure. A definition of any of these notions would provide one for puzzles. Neither in 1936 nor in 1953 did Turing characterize mathematically more general configurations and elementary operations on them. I want to describe briefly one particular attempt to do just that, by Byrnes and me in our (1996).

Our approach was influenced by Kolmogorov & Uspensky's work on algorithms and has three distinct components: the computor operates on certain finite connected and labeled graphs, we call *K(olmogorov)-graphs*; K-graphs contain a unique *distinguished element* that corresponds to the scanned square of a Turing machine tape; the operations *substitute neighborhoods* of the distinguished element by appropriate other neighborhoods and are given by a finite list of generalized Post production rules. Though broadening Turing's original considerations, we remain within his general analytic framework and establish that letter machines can mimic K-graph machines. Turing's Central Thesis is turned into the thesis that K-graph machines can do the work of computors directly. As a playful indication of how human algorithms can be carried out straightforwardly by K-graph machines, Byrnes and I programmed a K-graph machine to do ordinary, two-dimensional column addition. In sum, a much more general class of symbolic configurations and operations on them is considered, and the central thesis for K-graph machines seems even more plausible than the one for string machines.

The separation of informal conceptual analysis and mathematical equivalence proof is essential for recognizing that the correctness of *Turing's Thesis* (taken generically) rests on two pillars; namely, on the correctness of boundedness and locality conditions for computors, and on the correctness of the pertinent central thesis. The latter asserts explicitly that computations of a computor can be mimicked directly by a particular kind of machine. However satisfactory one may find this line of analytic argument, there are two weak spots: the looseness of the restrictive conditions (What are symbolic configurations? What changes can mechanical operations effect?) and the corresponding vagueness of the central thesis. We are, no matter how we turn ourselves, in a position that is methodologically still unsatisfactory. To make a step towards a more satisfactory stance, I shall first take a detour through Gandy's analysis of machine computations, abstracting further away from particular types of configurations and operations, and then return to this central methodological issue in section 5.

§4. By (parallel) machine. It has been claimed frequently that Turing analyzed computations of machines. That is historically and systematically inaccurate, as my exposition should have made quite clear. Only in 1980 did Turing's student Robin Gandy characterize machine computations. Gandy focused on "discrete mechanical devices", as opposed to possibly more general physical devices; he excluded, in particular, analogue machines from consideration. The two physical conditions characteristic for such devices are a lower bound on the size of their atomic parts (or simply "atoms") and an upper bound on the speed of signal propagation. This justifies Gandy's contention that states of such machines "can be adequately described in finite terms."[20] Calculations are taken to proceed in discrete and uniquely determined steps. Consequently, Gandy claims that these devices can be viewed, in a loose sense, as digital computers. The central and novel aspect of Gandy's analysis is the fact that it incorporates parallelism and covers cellular automata directly. This is of real interest, because cellular automata do not satisfy the locality condition (**L**); after all, the configurations affected in a single computation step are potentially unbounded.

Gandy's characterization—some details are presented in the Appendix—is given in terms of discrete dynamical systems $\langle \boldsymbol{S}, \boldsymbol{F} \rangle$, where $\boldsymbol{S}$ is the set of states and $\boldsymbol{F}$ governs the system's evolution. More precisely, $\boldsymbol{S}$ is a structural class, i.e., a subclass of the hereditarily finite sets $\boldsymbol{HF}$ over an infinite set $\boldsymbol{U}$ of atoms that is closed under $\in$-isomorphisms, and $\boldsymbol{F}$ is a structural operation from $\boldsymbol{S}$ to $\boldsymbol{S}$, i.e., a transformation that is, roughly speaking, invariant under permutations of atoms. These dynamical systems have to satisfy four restrictive principles. The first principle pertains to the *form of description* and states that any machine $\boldsymbol{M}$ can be presented by such a pair $\langle \boldsymbol{S}, \boldsymbol{F} \rangle$, and that $\boldsymbol{M}$'s computation, starting in an initial state $\boldsymbol{x}$, is given by the sequence $\boldsymbol{x}, \boldsymbol{F}(\boldsymbol{x}), \boldsymbol{F}(\boldsymbol{F}(\boldsymbol{x})), \ldots$. Gandy formulates three groups of substantive principles, the first of which—*The Principle of Limitation of Hierarchy*—requires that the set theoretic rank of the states is bounded, i.e., the structural class $\boldsymbol{S}$ is contained in a fixed initial segment of the $\boldsymbol{HF}$ hierarchy. Gandy argues that it is natural or convenient to think of a machine in hierarchical terms, and that "for a given machine the maximum height of its hierarchical structure must be bounded".[21] The second of the substantive principles—*The Principle of Unique Reassembly*—claims that any state can be "assembled" from "parts" of bounded size; its proper formulation requires care and a lengthy sequence of definitions. The informal idea, though, is wonderfully straightforward: any state of a concrete machine must be built up from (finitely many different types of) off-the-shelf components. Clearly, the components have a bound on their complexity. Both of these principles are concerned with the states in $\boldsymbol{S}$; the remaining third and central principle—*The Principle of Local Causality*—puts conditions on (the local determination of) the structural operation $\boldsymbol{F}$. It is formulated by Gandy in this preliminary way: "The next state, $\boldsymbol{Fx}$, of a machine can be reassembled from its restrictions to overlapping 'regions' s and these restrictions are locally caused." It requires that the parts from which $\boldsymbol{F}(\boldsymbol{x})$ can be reassembled depend only on bounded parts of $\boldsymbol{x}$.

Gandy's Central Thesis is naturally formulated as the claim that any mechanical device can be represented as a dynamical system satisfying the above principles. As to the basic set-up John Shepherdson remarked: "Although Gandy's principles

[20]Cf. (Gandy 1980) p. 126, but also pp. 135–6. For a more detailed argument see (Mundici & Sieg 1995), section 3.

[21](Gandy 1980), p. 131.

were obtained by a very natural analysis of Turing's argument they turned out to be rather complicated, involving many subsidiary definitions in their statement. In following Gandy's argument, however, one is led to the conclusion that that is in the nature of the situation."[22] — Nevertheless, in (Sieg & Byrnes 1999b) a greatly simplified presentation is achieved by choosing definitions appropriately, focusing sharply on the central informal ideas, and using one key suggestion made by Gandy in the Appendix to his paper. This simplification does not change at all the form of presentation. The subsidiary definitions are streamlined significantly, however, and of the four principles used by Gandy only a restricted version of the principle of local causality is explicitly retained. It is formulated in two separate parts, namely, as the principle of *Local Causation* and that of *Unique Assembly*. The separation reflects the distinction between the local determination of regions of the next state and their assembly into the next state. The resulting modified set-up is focused resolutely on the dynamics of processes: the static build-up of states from simple parts is no longer of concern; rather, the machines have to recognize patterns in a given state and act on them! That there are only a finite number of such patterns is justified *directly* by the two physical limitations appealed to above, namely, the lower bound on the size of atoms and the upper bound on the speed of signal propagation.

Before giving a synopsis of an even more simplified version of parallel computations, let me point to the crucial new methodological difference with Gandy's set-up. I no longer take a *Gandy machine* to be a dynamical system $\langle \boldsymbol{S}, \boldsymbol{F} \rangle$ (satisfying Gandy's principles), but rather a structure $\boldsymbol{M}$ consisting of a structural class $\boldsymbol{S}$ of states together with two kinds of patterns and operations on (instantiations of) the latter; these patterns and operations underlie also Gandy's principle of local causality. Given any dynamical system $\langle \boldsymbol{D}, \boldsymbol{F} \rangle$, where $\boldsymbol{D}$ need not be a structural class and $\boldsymbol{F}$ need not be a structural operation, $\boldsymbol{F}$ is now called *computable in parallel* if, and only if, there is a *Gandy machine* $\boldsymbol{M}$ on $\boldsymbol{S_D}$ that determines a sequence of states $\boldsymbol{z_0}, \boldsymbol{z_1}, \boldsymbol{z_2}, \ldots$ $\in$-isomorphic to the successive states $\boldsymbol{x}, \boldsymbol{F}(\boldsymbol{x}), \boldsymbol{F}(\boldsymbol{F}(\boldsymbol{x})), \ldots$ for each $\boldsymbol{x}$ in $\boldsymbol{D}$.[23] In general, a Gandy machine $\boldsymbol{M}$ on a structural class $\boldsymbol{S}$ includes then, first of all, a finite set $\boldsymbol{T_1}$ of stereotypes, i.e., isomorphism classes of *parts* of a state $\boldsymbol{x}$, and an associated structural function $\boldsymbol{G_1}$. $\boldsymbol{G_1}$ operates on "maximal" parts or *causal neighborhoods for* $\boldsymbol{x}$, abbreviated by $Cn_1(\boldsymbol{x})$, to yield *determined regions* of some state $\boldsymbol{z}$ out of which the next state can be assembled; the regions are said to be *locally caused.* $Dr_1(\boldsymbol{z}, \boldsymbol{x})$ denotes the set of determined regions.

The first principle (**LC.1**) states that every element of $Cn_1(\boldsymbol{x})$ yields a determined region or, to put it in Gandy's words, that every cause has an effect. Indeed, the principle guarantees also that the effects are unique by requiring that determined regions of $\boldsymbol{z}$ are identical, when they are $\in$-isomorphic over $\boldsymbol{x}$; otherwise there would not be, as Gandy put it, "any bounds on the number of distinct regions which arise from a given causal neighborhood".[24] A second finite set $\boldsymbol{T_2}$ of

[22](Shepherdson 1988), p. 586.

[23]$\boldsymbol{S_D}$ is obtained from $\boldsymbol{D}$ by closing under $\in$-isomorphisms.

[24](Gandy 1980), p. 139.

stereotypes together with its associated structural operation G_2 play a significant role in the assembly of the next state and have to satisfy the analogue of (**LC.1**): every element of $Cn_2(x)$ yields a determined region in $Dr_2(z, x)$, without requiring uniqueness; this is principle (**LC.2**). The determined regions in $Dr_1(z, x)$ are the building blocks of the next state and not only cover the next state, but fit together suitably, when they *overlap*, i.e., have new atoms in common. Here T_2 and G_2 come in, as they provide local information concerning overlaps. That is expressed by principle (**GA.1**): for any collection C of regions from $Dr_1(z, x)$ with common new atoms there must be a w in $Dr_2(z, x)$, such that all elements of C are parts of w. The final principle, (**GA.2**), requires that the next state be obtained as the union of the determined regions in $Dr_1(z, x)$. A quintuple $\langle S; T_1, G_1, T_2, G_2 \rangle$ is called a *Gandy Machine* M on S if and only if for every $x \in S$ there is a $z \in S$, such that the local causation conditions (**LC.1–2**) and the global assembly conditions (**GA.1–2**) are all satisfied; precise formulations are given in the Appendix.

Gandy established in his paper as the central mathematical fact that the sequence of states of such a system is Turing computable (to within $\in$-isomorphism). The proof relies on a lemma stating that, for every $x \in S$, $Dr_1(z, x)$ and z are unique up to $\in$-isomorphism over x.[25] The next state is thus computable by a Turing machine (via an exhaustive search). It may be that Gandy machines can be simplified further, for example, by a graph theoretic presentation or a category theoretic description.[26] But what is needed most, in my view, is their further mathematical investigation, e.g., for issues of complexity and speed-up, and their use in significant applications, e.g., for the analysis of DNA computations or of parallel distributed processes. The latter was done by De Pisapia in his (2000) for many important kinds of artificial neural networks.[27] The consequence is that artificial neural nets of these varieties can be simulated by Turing machines. Analogous results are obtained, using quite different techniques, by Siegelmann in (1998).

However, the most difficult and subtle aspect of Gandy machines, namely the addition of new atoms, is not used at all for these neural nets, as they have a fixed number of nodes. So there are two obvious questions: "Is there a natural subclass of Turing computable functions in which these neural nets lie?", and "Are there mental processes for whose representation this aspect of Gandy machines might

---

[25]The argument as given in Gandy is not quite correct: details can be found in (Sieg 2000).

[26]A graph theoretic presentation was proposed in (Byrnes & Sieg 1996). On the topic of a category theoretic definition Gandy wrote in his (1980), p. 147: "The heavy use made of restrictions ... suggests that a treatment using concepts analogous to those of sheaf theory or topos theory might be worth developing. However, it seems to me that the concepts from category theory which would be necessary would be too abstract to allow one to use them (as we have used the more concrete notions of set theory) as a justification for the main thesis of this paper." Perhaps this issue should be revisited twenty years after its original formulation. A starting-point can be found in (Herron 1995).

[27]De Pisapia considered in particular ANNs with Hebbian learning and the backpropagation algorithm. Here is a possibility for rich interaction with classical and contemporay work in the foundations of mathematics, namely, the formalization of analysis in very weak formal frameworks.

be crucial?". These are appealing and important questions, but let me turn back to the central methodological issue I set out to address.

§5. **Methodological point.** The analysis offered by Turing was radically new in 1936 and yet, as I argued, contiguous with the work of Gödel, Church, Hilbert & Bernays, and others. In particular, it was just a step away from Gödel's considerations in the spring of 1934: Turing analyzed in a novel and convincing way the processes underlying computations (say in Gödel's equational calculus), but—and that is crucial here—he provided also a basis for further reflections along Gödelian lines. What do I have in mind with this last remark? In a conversation with Church in early 1934, Gödel found Church's proposal to identify effective calculability with λ-definability "thoroughly unsatisfactory". As a counter-proposal he suggested "to state a set of axioms which would embody the generally accepted properties of this notion (i.e., effective calculability), and to do something on that basis".[28] Gödel did not articulate what the generally accepted properties of effective calculability might be, or what might be done on the basis of an appropriate set of axioms.

The sharpened version of Turing's work and a thorough-going re-interpretation of Gandy's approach allow us to fill in the blanks of Gödel's suggestion; this resolves, in my view, the methodological issue raised at the end of section 3. Let me bring out the central points for Gandy machines. First, the definition of a Gandy machine is an "abstract" mathematical definition that embodies generally accepted properties of parallel computations;[29] the axiomatic conditions enforce, for any Gandy machine, that the state immediately following a given state $\boldsymbol{x}$ can be obtained (uniquely up to $\in$-isomorphism over $\boldsymbol{x}$) from the determined regions. Second, Gandy machines share with groups and topological spaces the general feature of abstract axiomatic definitions, namely, that they admit a wide variety of different interpretations.[30] Third, the central fact, that the computations of any Gandy machine can be simulated by a letter machine, is best understood as a *representation theorem* for the axiomatic notion. With these broad observations in the background, I will recast now the earlier considerations for Turing computors in such a way that they parallel those for Gandy machines.

Let $\langle \boldsymbol{D}, \boldsymbol{F} \rangle$ be a discrete dynamical system; the operation $\boldsymbol{F}$ is called computable if, and only if, there is a *Turing computor* $\boldsymbol{M}$ *on* $\boldsymbol{S_D}$ that determines a sequence

[28] That is reported by Church in a letter to Kleene of November 29, 1935; cf. my (1997), pp. 159–160.

[29] In (Lamport and Lynch) it is asserted that "the theory of sequential computing rests upon fundamental concepts of computability that are independent of any particular computational model." In contrast, it is claimed, no such fundamental concepts underlying distributed computing have yet been developed. That is followed by some informal observations (p. 1166): "Underlying almost all models of concurrent systems is the assumption that an execution consists of a set of discrete events, each affecting only part of the system's state. Events are grouped into processes, each process being a more or less completely sequenced set of events sharing some common locality in terms of what part of the state they affect. For a collection of autonomous processes to act as a coherent system, the processes must be synchronized."

[30] In my paper (1996) I tried to argue for the distinctive character of "abstract definitions."

of states $z_0, z_1, z_2, \ldots$ $\in$-isomorphic to the successive states $\boldsymbol{x}, \boldsymbol{F}(\boldsymbol{x}), \boldsymbol{F}(\boldsymbol{F}(\boldsymbol{x})), \ldots$ for each $\boldsymbol{x}$ in $\boldsymbol{D}$. A Turing computor $\boldsymbol{M}$ on $\boldsymbol{S}$ is given by a triple $\langle \boldsymbol{S}; \boldsymbol{T}, \boldsymbol{G} \rangle$, where $\boldsymbol{S}$ is a structural class of states, $\boldsymbol{T}$ a finite set of patterns or stereotypes, and $\boldsymbol{G}$ the corresponding structural operation modifying instantiated patterns. Every state is required by (**C.0**) to contain exactly one observed configuration, its sole causal neighborhood, denoted by $\boldsymbol{cn}(\boldsymbol{x})$. The remaining axioms for Turing computors are similar to those for Gandy machines. First, the observed configuration $\boldsymbol{cn}(\boldsymbol{x})$ of state $\boldsymbol{x}$ yields a unique determined region of any possible next state $\boldsymbol{z}$; that is expressed by principle (**LC.1**). Second, denoting the uniquely fixed determined region of $\boldsymbol{z}$ by $\boldsymbol{dr}(\boldsymbol{z}, \boldsymbol{x})$, the next state is obtained as the union of $\boldsymbol{x} \setminus \boldsymbol{cn}(\boldsymbol{x})$ and $\boldsymbol{dr}(\boldsymbol{z}, \boldsymbol{x})$. This is the assembly condition (**GA.1**). The next state $\boldsymbol{z}$ is determined uniquely up to $\in$-isomorphism over $\boldsymbol{x}$. Looking back at the informal conditions on Turing computors, one notices readily that the boundedness and locality conditions are satisfied by $\boldsymbol{M}$. Note also, that the principles for a Turing computor are satisfied by string and K-graph machines. A suitable representation theorem can be established, showing that computations of any model of these principles can be simulated by a letter machine.

The axiomatic approach captures the essential nature of computation processes in an abstract way. The difference between the two types of calculators I have been describing is reduced to the fact that Turing computors modify *one* bounded part of a state, whereas Gandy machines operate in parallel on *arbitrarily many* bounded parts. The representation theorems guarantee that models of the axioms are computationally equivalent to Turing machines in their letter variety. In any event, these considerations remove any appeal to theses, whether central or not. They fit the bill of Turing's appealing intuitive puzzle-approach and satisfy Gödel's demand for an axiomatically characterized notion. Indeed, they give proper content to Gödel's enigmatic remark—discussed in section 2—stating: "Turing's work gives an *analysis* of the concept of 'mechanical procedure' .... This concept is *shown* to be equivalent with that of a 'Turing machine'." As to the correctness of the analysis, an appeal to intuition in Turing's sense can no more be avoided in this case than in any other case of an axiomatically characterized mathematical structure that is intended to model broad aspects of reality.[31]

In the case under discussion this is fraught with controversy and often misunderstanding. For example, Gödel spotted a "philosophical error" in Turing's work, *assuming* that Turing's argument in the 1936 paper was to show that "mental procedures cannot go beyond mechanical procedures."[32] Not surprisingly, he

[31]The argument in section 9 of (Turing 1936)—that was analyzed above—is characterized as "a direct appeal to intuition"; Turing discusses his concept of intuition at greater length in section 11 of his 1939 paper on ordinal logics. — A second significant example of such an axiomatic analysis, "modeling broad aspects of reality", was given by Dedekind for the continuum; cf. my (1997), pp. 173–4.

[32]In his Note from 1972; that is also reported with some additional comments in (Wang 1974), pp. 324–6.

considered the argument as inconclusive. Indeed, Turing does not give a conclusive argument for Gödel's claim, but it has to be added that he did not intend to argue for it. Even in his work of the late 1940's and early 1950's that deals explicitly with mental processes, Turing does not argue that "mental procedures cannot go beyond mechanical procedures." Mechanical processes are still made precise as Turing machine computations; in contrast, machines that might exhibit intelligence have a much more complex structure.[33] Conceptual idealization and empirical adequacy are being sought now for different purposes, and Turing is trying to capture quite clearly what Gödel found missing in the analysis of (a broader concept of humanly) effective calculability, namely, " ... that mind, in its use, is not static, but constantly developing." The real difference between Turing's and Gödel's views, it seems, is Gödel's belief that it is "a prejudice of our time" that "[t]here is no mind separate from matter." This is reported by Wang.[34] Gödel expected also, according to Wang, that this prejudice "will be disproved scientifically (perhaps by the fact that there aren't enough nerve cells to perform the observable operations of the mind)." Clearly, Turing did not share that expectation.

Appendix

This appendix provides some details for Gandy machines. Their states are non-empty hereditarily finite sets over an infinite set $\boldsymbol{U}$ of atoms. A class $\boldsymbol{S}$ of states is called *structural*, if $\boldsymbol{S}$ is closed under $\in$-isomorphisms. The lawlike connections between states are given by *structural operations* $\boldsymbol{F}$ *on* $\boldsymbol{S}$, i.e., from $\boldsymbol{S}$ to $\boldsymbol{S}$. Structural operations satisfy the condition:[35] for all permutations π on $\boldsymbol{U}$ and all $\boldsymbol{x} \in \boldsymbol{S}$, $\boldsymbol{F}(\boldsymbol{x}^\pi)$ is $\in$-isomorphic over $\boldsymbol{x}^\pi$ to $\boldsymbol{F}(\boldsymbol{x}^\pi)$. "$\boldsymbol{x}$ is $\in$-isomorphic over $\boldsymbol{z}$ to $\boldsymbol{y}$" means that the $\in$-isomorphism between $\boldsymbol{x}$ and $\boldsymbol{y}$ is the identity on $\sup(z)$, i.e., the atoms in the transitive closure of $\boldsymbol{z}$; I use the abbreviation "$\boldsymbol{x} \cong_z \boldsymbol{y}$".

If $\boldsymbol{x}$ is a given state, regions of the next state are *locally* determined. Thus it is important to describe suitable substructures of $\boldsymbol{x}$ on which operations can be performed. Proper subtrees $\boldsymbol{y}$ of the $\in$-tree for $\boldsymbol{x}$ are called *parts for* $\boldsymbol{x}$, briefly $\boldsymbol{y} <^* \boldsymbol{x}$, if they are specified as follows:[36] $\boldsymbol{y} \neq \boldsymbol{x}$ and $\boldsymbol{y}$ is a non-empty subset of

$$\{\boldsymbol{v} | (\exists \boldsymbol{z})(\boldsymbol{v} <^* \boldsymbol{z} \wedge \boldsymbol{z} \in \boldsymbol{x})\} \cup \{\boldsymbol{r} | \boldsymbol{r} \in \boldsymbol{x}\}$$

---

[33] Turing's speculations are described most carefully and compared thoroughly with early work by Newell and Simon in (Colvin 1997).

[34] (Wang 1974), p. 326. There one finds also this elaboration: "More generally, Gödel believes that mechanism in biology is a prejudice of our time which will be disproved. In this case, one disproval, in Gödel's opinion, will consist in a mathematical theorem to the effect that the formation within geological times of a human body by the laws of physics (and any other laws of a similar nature), starting from a random distribution of the elementary particles and the field, is about as unlikely as the separation by chance of the atmosphere into its components."

[35] For motivation of this particular condition see p. 154 of (Sieg & Byrnes 1999b).

[36] I am deviating quite consciously from Gandy's terminology; my "part" (is more general than, but) corresponds roughly to "located subassembly" in (Gandy 1980) and to "subassembly" in (Sieg & Byrnes 1999b). The reader should also compare it to Gandy's $\subseteq^*$, p. 136. Gandy remarks: "If one

If the non-empty subset in this $\in$-recursive definition consists at each stage of exactly one element, $\boldsymbol{y}$ is a *path through* $\boldsymbol{x}$. Paths through $\boldsymbol{x}$ are of the form $\{\boldsymbol{r}\}^n$, for a natural number n and some atom $\boldsymbol{r}$. A collection $\boldsymbol{C}$ of parts for $\boldsymbol{x}$ is a *cover for* $\boldsymbol{x}$ just in case for every path $\boldsymbol{y}$ through $\boldsymbol{x}$ there is a $\boldsymbol{z} \in \boldsymbol{C}$, such that $\boldsymbol{y}$ is a path through $\boldsymbol{z}$.

The local operations are given by a structural operation $\boldsymbol{G_1}$ that works on parts $\boldsymbol{y}$ for $\boldsymbol{x}$. Each $\boldsymbol{y}$ lies in one of a finite number of isomorphism classes (or stereotypes).[37] So let $\boldsymbol{T_1}$ be a fixed, finite class of stereotypes: a part for $\boldsymbol{x}$ that is a member of a stereotype of $\boldsymbol{T_1}$ is called, naturally enough, a $\boldsymbol{T_1}$ *-part for* $\boldsymbol{x}$. A $\boldsymbol{T_1}$ -part $\boldsymbol{y}$ for $\boldsymbol{x}$ is a *causal neighborhood for* $\boldsymbol{x}$ given by $\boldsymbol{T_1}$, briefly $\boldsymbol{y} \in Cn_1(\boldsymbol{x})$,[38] if there is no $\boldsymbol{T_1}$ -part $\boldsymbol{y}^*$ for $\boldsymbol{x}$ such that $\boldsymbol{y}$ is $\in$-embeddable into $\boldsymbol{y}^*$. $\boldsymbol{G_1}$ operates on such causal neighborhoods. The values of $\boldsymbol{G_1}$, however, are in general not exactly what is needed for the assembly of the next state. For that purpose, we introduce *determined regions* of a state $\boldsymbol{z}$ obtained from causal neighborhoods for $\boldsymbol{x}$: $\boldsymbol{v} \in Dr_1(\boldsymbol{z}, \boldsymbol{x})$ if and only if $\boldsymbol{v} <^* \boldsymbol{z}$ and there is a $\boldsymbol{y} \in Cn_1(\boldsymbol{x})$, such that $\boldsymbol{G_1}(\boldsymbol{y}) \cong_{\boldsymbol{y}} \boldsymbol{v}$ and $\sup(\boldsymbol{v}) \cap \sup(\boldsymbol{x}) \subseteq \sup(\boldsymbol{y})$. The last condition for Dr_1 guarantees that new atoms in $\boldsymbol{G_1}(\boldsymbol{y})$ correspond to new atoms in $\boldsymbol{v}$, and that the new atoms in $\boldsymbol{v}$ are new for $\boldsymbol{x}$. If one requires $\boldsymbol{G_1}$ to satisfy similarly $\sup(\boldsymbol{G_1}(\boldsymbol{y})) \cap \sup(\boldsymbol{x}) \subseteq \sup(\boldsymbol{y})$, then the condition "$\boldsymbol{G_1}(\boldsymbol{y}) \cong_{\boldsymbol{y}} \boldsymbol{v}$" can be strengthened to "$\boldsymbol{G_1}(\boldsymbol{y}) \cong_{\boldsymbol{x}} \boldsymbol{v}$". The new atoms are thus always taken from $\boldsymbol{U} \setminus \sup(\boldsymbol{x})$. Note that the number of new atoms introduced by $\boldsymbol{G_1}$ is bounded, i.e., $|\sup(\boldsymbol{G_1}(\boldsymbol{y})) \setminus \sup(\boldsymbol{x})| < n$ for some natural number n (any $\boldsymbol{x} \in \boldsymbol{S}$ and any causal neighborhood $\boldsymbol{y}$ for $\boldsymbol{x}$). The determined regions have to be assembled into the next state, and for that, a second structural operation $\boldsymbol{G_2}$ and a second set $\boldsymbol{T_2}$ of stereotypes are needed. Finally, we have all the ingredients of a Gandy Machine.

DEFNITION. $\boldsymbol{M} = \langle \boldsymbol{S}; \boldsymbol{T_1}, \boldsymbol{G_1}, \boldsymbol{T_2}, \boldsymbol{G_2} \rangle$ is a *Gandy Machine* on $\boldsymbol{S}$, where $\boldsymbol{S}$ is a structural class, $\boldsymbol{T_i}$ a finite set of stereotypes, $\boldsymbol{G_i}$ a structural operation on (the elements of) $\boldsymbol{T_i}$, if and only if, for every $\boldsymbol{x} \in \boldsymbol{S}$ there is a $\boldsymbol{z} \in \boldsymbol{S}$, such that

$$\textbf{(LC.1)}\quad (\forall \boldsymbol{y} \in Cn_1(\boldsymbol{x}))(\exists! \boldsymbol{v} \in Dr_1(\boldsymbol{z}, \boldsymbol{x}))\ \boldsymbol{v} \cong_{\boldsymbol{x}} \boldsymbol{G_1}(\boldsymbol{y})$$
$$\textbf{(LC.2)}\quad (\forall \boldsymbol{y} \in Cn_2(\boldsymbol{x}))(\exists \boldsymbol{v} \subset Dr_2(\boldsymbol{z}, \boldsymbol{x}))\ \boldsymbol{v} \cong_{\boldsymbol{x}} \boldsymbol{G_2}(\boldsymbol{y})$$
$$\textbf{(GA.1)}\quad (\forall \boldsymbol{C})[\boldsymbol{C} \subseteq Dr_1(\boldsymbol{z}, \boldsymbol{x}) \wedge \bigcap\{\sup(\boldsymbol{v}) \cap A(\boldsymbol{z}, \boldsymbol{x}) | \boldsymbol{v} \in \boldsymbol{C}\} \neq \emptyset \rightarrow (\exists \boldsymbol{w} \in Dr_2(\boldsymbol{z}, \boldsymbol{x}))(\forall \boldsymbol{v} \in \boldsymbol{C})\ \boldsymbol{v} <^* \boldsymbol{w}];^{39}$$
$$\textbf{(GA.2)}\quad \boldsymbol{z} = \bigcup Dr_1(\boldsymbol{z}, \boldsymbol{x}).$$

considers y as a tree of its $\in$-chains, then $u \subseteq^* y$ implies that u is a subtree with the same vertex as y." The relation is defined by the condition $(\exists s \subseteq Tc(y))u = y \uparrow s$.

[37]This operation is an operation on $\boldsymbol{x}$ and $\boldsymbol{y}$, as it introduces, possibly, new atoms—new for $\boldsymbol{x}$. It has in this very weak sense a "global" aspect: however, as it is a structural operation, the precise choice of the atoms does not matter at all.

[38]Causal neighborhoods are of course implicitly dependent on the set $\boldsymbol{T_1}$ of stereotypes.

[39]In Gandy's set-up the finiteness of the $\boldsymbol{C}$ has to be forced axiomatically: here it is a trivial consequence of the finiteness of Dr_1. Furthermore, principle (**GA.1**) forces a fixed upper bound on the number of determined regions that have new atoms in common.

LC stands for *Local Causation*, whereas **GA** abbreviates *Global Assembly*; $A(z, x)$ abbreviates $\sup(z) \setminus \sup(x)$. By a slight abuse of language a z satisfying the conditions is denoted by $M(x)$. That is justified by the central fact established by Gandy: $Dr_1(z, x)$ and z are unique up to $\in$-isomorphism over x.[40] This is the systematic background for the next definition.

DEFNITION. Let $\langle D, F \rangle$ be any discrete dynamical system; F is called *computable in parallel* if and only if there is a Gandy machine M on S_D, such that for each $x \in D : F(x) \cong_x M(x)$.

The second assembly condition implies that $Dr_1(z, x)$ is a cover for z. The central fact for Gandy's proof, establishing the Turing computability of the sequence of states, is formulated now in my setting as follows:

THEOREM. Let M be $\langle S; T_1, G_1, T_2, G_2 \rangle$ as above and $x \in S$; if there are z and z' in S satisfying principles (**LC.1**)–(**LC.2**), (**GA.1**), and such that $Dr_1(z, x)$ and $Dr_1(z', x)$ cover z and z', then $Dr_1(z, x) \cong_x Dr_1(z', x)$.

Now I formulate analogous conditions for Turing Computors.

DEFNITION. $M = \langle S; T, G \rangle$ is a *Turing Computor on* S, where S is a structural class, T a finite set of stereotypes, and G a structural operation on T, if and only if, for every $x \in S$ there is a $z \in S$, such that

(**LC.0**) $(\exists! y) y \in Cn(x)$
(**LC.1**) $(\exists! v <^* z)\ v \cong_x G(cn(x))$;
(**GA.1**) $z = (x \setminus cn(x)) \cup dr(z, x)$.

$cn(x)$ and $dr(z, x)$ denote the sole causal neighborhood of x, respectively the determined region of z.

REFERENCES

J. BYRNES AND W. SIEG [1996], *A graphical presentation of Gandy's parallel machines (abstract)*, ***The Bulletin of Symbolic Logic***, vol. 2, pp. 452–3.

A. CHURCH [1936], *An unsolvable problem of elementary number theory*, ***American Journal of Mathematics***, vol. 58, pp. 345–363, reprinted in (Davis 1965).

A. CHURCH [1937], *Review of (Turing 1936)*, ***The Journal of Symbolic Logic***, vol. 2, no. 1, pp. 40–41.

S. COLVIN [1997], *Intelligent machinery: Turing's ideas and their relation to the work of Newell and Simon*, ***Master's thesis***, Carnegie Mellon University, Department of Philosophy, 63pp.

M. DAVIS [1958], ***Computability and unsolvability***, McGraw-Hill, New York.

M. Davis (editor) [1965], ***The undecidable: Basic papers on undecidable propositions, unsolvable problems, and computable functions***, Raven Press, Hewlett, New York.

[40] In (Gandy 1980) this uniqueness up to $\in$-isomorphism over x is achieved in a much more complex way, mainly, because parts of a state are proper subtrees, in general non-located. Given an appropriate definition of cover, a collection C is called an *assembly for* x, if C is a cover for x and the elements of C are maximal. The fact that C is an assembly for exactly one x, if indeed it is, is expressed by saying that C *uniquely assembles to* x. Indeed, in my setting the axiom (**GA.2**) is equivalent to the claim that $Dr_1(z, x)$ uniquely assembles to z.

M. DAVIS [1982]. *Why Gödel didn't have Church's thesis*, ***Information and Control***, vol. 54, pp. 3–24.

R. GANDY [1980]. *Church's thesis and principles for mechanisms*, ***The Kleene symposium*** (J. Barwise, H.J. Keisler, and K. Kunen, editors), North-Holland, pp. 123–148.

R. GANDY [1988]. *The confluence of ideas in 1936*, ***The universal Turing machine—a half-century survey*** (R. Herken, editor), Oxford University Press, pp. 55–111.

K. GÖDEL [1933], *The present situation in the foundations of mathematics*, ***Collected works***, vol. III, Oxford University Press, pp. 45–53.

K. GÖDEL [1934], *On undecidable propositions of formal mathematical systems*, ***Collected works***, vol. I, Oxford University Press, pp. 356–371.

K. GÖDEL [1936], *Über die Länge von Beweisen*, ***Collected works***, vol. I, Oxford University Press, pp. 396–399.

K. GÖDEL [1946], *Remarks before the Princeton bicentennial conference on problems in mathematics*, ***Collected works***, vol. II, Oxford University Press, pp. 150–153.

K. GÖDEL [1972], *Some remarks on the undecidability results*, ***Collected works***, vol. II, Oxford University Press, pp. 305–306.

T. HERRON [1995], *An alternative definition of pushout diagrams and their use in characterizing K-graph machines*, Carnegie Mellon University.

D. HILBERT AND P. BERNAYS [1939], ***Grundlagen der Mathematik II***, Springer Verlag, Berlin.

S. C. KLEENE [1936], *General recursive functions of natural numbers*, ***Mathematische Annalen***, vol. 112, pp. 727–742, reprinted in (Davis 1965).

A. KOLMOGOROV AND V. USPENSKY [1963], *On the definition of an algorithm*, ***AMS Translations***, vol. 21, no. 2, pp. 217–245.

L. LAMPORT AND N. LYNCH [1990], *Distributed computing: Models and methods*, ***Handbook of theoretical computer science*** (J. van Leeuwen, editor), Elsevier Science Publisher B. V..

D. MUNDICI AND W. SIEG [1995], *Paper machines*, ***Philosophia Mathematica***, vol. 3, pp. 5–30.

N. DE PISAPIA [2000], *Gandy machines: an abstract model of parallel computation for Turing machines, the game of life, and artificial neural networks*, ***Master's thesis***, Carnegie Mellon Univeristy, Department of Philosophy, 75 pp.

E. POST [1947], *Recursive unsolvability of a problem of Thue*, ***The Journal of Symbolic Logic***, vol. 12, pp. 1–11.

J. SHEPHERDSON [1988], *Mechanisms for computing over arbitrary structures*, ***The universal Turing machine—a half-century survey*** (R. Herken, editor), Oxford University Press, pp. 581–601.

W. SIEG [1994], *Mechanical procedures and mathematical experience*, ***Mathematics and mind*** (A. George, editor), Oxford University Press, pp. 71–117.

W. SIEG [1996], *Aspects of mathematical experience*, ***Philosophy of mathematics today*** (E. Agazzi and G. Darvas, editors), Kluwer.

W. SIEG [1997], *Step by resursive step: Church's analysis of effective calculability*, ***The Bulletin of Symbolic Logic***, vol. 3, pp. 154–180.

W. SIEG [2000], *Calculations by man & machine: mathematical presentation*, Kluwer, to appear in: ***Proceedings of the 11th International Congress of Logic, Methodology and Philosophy of Science***.

W. SIEG AND J. BYRNES [1996], *K-graph machines: generalizing Turing's machines and arguments*, ***Gödel '96*** (P. Hajek, editor), Lecture Notes in Logic 6, Springer Verlag, pp. 98–119.

W. SIEG AND J. BYRNES [1999a], *Gödel, Turing, & K-graph machines*, ***Logic and the foundations of mathematics*** (A. Cantini, E. Casari, and P. Minari, editors), Synthese Library 280, Kluwer, (The paper was submitted for this volume already in January of 1997), pp. 57–66.

W. SIEG AND J. BYRNES [1999b], *An abstract model for parallel computations: Gandy's thesis*, ***The Monist***, vol. 82, no. 1, pp. 150–64.

H. T. SIEGELMANN [1998], ***Neural networks and analog computation—beyond the Turing limit***, Birkhäuser.

H. SINACEUR [2000], *Address at the Princeton University bicentennial conference on problems of mathematics (December 17–19, 1946), by Alfred Tarski*, ***The Bulletin of Symbolic Logic***, vol. 6, no. 1, pp. 1–44.

R. SOARE [1996], *Computability and recursion*, ***The Bulletin of Symbolic Logic***, vol. 2, no. 3, pp. 284–321.

G. TAMBURRINI [1987], *Reflections on mechanism*, ***Ph.D. thesis***, Columbia University, Department of Philosophy, New York.

G. TAMBURRINI [1997], *Mechanistic theories in cognitive science: The import of Turing's thesis*, ***Logic and scientific methods*** (M. L. Dalla Chiara, K. Doets, D. Mundici, and J. van Benthem, editors), Synthese Library 259, Kluwer, pp. 239–57.

A. TURING [1936], *On computable numbers with an application to the Entscheidungsproblem*, ***Proceedings of the London Mathematical Society***, 2, vol. 42, pp. 230–265.

A. TURING [1939], *Systems of logic based on ordinals*, ***Proceedings of the London Mathematical Society***, 2, vol. 45, pp. 161–228.

A. TURING [1950a], *Computing machinery and intelligence*, ***Mind***, vol. 59, pp. 433–460.

A. TURING [1950b], *The word problem in semi-groups with cancellation*, ***Annals of Mathematics***, vol. 52, pp. 491–505.

A. TURING [1953], *Solvable and unsolvable problems*, ***Science News***, vol. 31, pp. 7–23.

H. WANG [1974], ***From mathematics to philosophy***, Routledge & Kegan Paul, London.

DEPARTMENT OF PHILOSOPHY
CARNEGIE MELLON UNIVERSITY
PITTSBURGH, PENNSYLVANIA 15213, USA
E-mail: ws15+@andrew.cmu.edu

REMARKS ON FINITISM

W. W. TAIT†

The background of these remarks is that in 1967, in "Constructive reasoning" [27], I sketched an argument that finitist arithmetic coincides with primitive recursive arithmetic, *PRA*; and in 1981, in "Finitism" [28], I expanded on the argument. But some recent discussions and some of the more recent literature on the subject lead me to think that a few further remarks would be useful.

The question "What is finitism?" is really two questions, which I did separate in 1981, but perhaps not sufficiently clearly. First, and, in truth, for me most importantly, there is the *conceptual* problem of making sense of the idea of a 'finitist' function or 'finitist' proof of a finitist arithmetic proposition such as $\forall xy[x + y = y + x]$, which seems to refer to the infinite totality of numbers. And secondly, there is the *historical* question of what Hilbert—or perhaps better, Hilbert and Bernays[1]—meant by "finitism". The two questions are not entirely independent, of course, since it was Hilbert and Bernays who originally marked out the conception of finitist mathematics. But, if we take as the central issue the question of the 'finite' in finitism, we may be led to reject some aspects of the Hilbert-Bernays account. Indeed, my analysis rejects the Kantian element in their discussions. Their account of finitist mathematics begins with the restriction to objects representable in intuition or obtained by 'intuitive abstraction'—'formal objects' as Bernays calls them [12, 1]. For Bernays the finiteness of mathematical objects is a *consequence* of their representability in intuition. (See [2, p. 40].) But our problem is, of course, not the finiteness of a number, but the infinity of numbers. There is, I think, a difficulty with Bernays' notion of formal object, where this is intended to extend to numbers so large as, not only to be beyond processing by the human mind, but possibly to be beyond representablity in the physical world. [2, p. 39]. This difficulty ought to be discussed more adequately then I have sofar done; but I won't take it up here. The point I want to emphasize

†This paper is based on a talk that I was very pleased to give at the conference *Reflections*, December 13–15, 1998, in honor of Solomon Feferman on his seventieth birthday. The choice of topic is especially appropriate for the conference in view of recent discussions we had had about finitism. I profited from the discussion following my talk and, in particular, from the remarks of Richard Zach. I have since had the advantage of further discussions with Zach and of reading his paper 1998; and I use his scholarship here shamelessly for my own purposes. Finally, I want to thank the two anonymous referees of this paper, whose comments have led to several changes in the final version.

[1]See [22, §3.4] for a discussion of Bernays contribution to the conception of finitism.

Reflections on the foundations of mathematics
Essays in honor of Solomon Feferman
Edited by Wilfried Sieg, Richard Sommer, and Carolyn Talcott

now is that I don't see in this notion of formal object the means for reasoning about the *totality* of numbers. My argument, which I don't want to repeat or expand upon here, is that the idea of iteration, which is of the essence of the idea of number and, in particular, is the means by which numerical functions are defined and numerical equations are proved, is not found or represented in intuition but is a creature of reason. That we may define a function by primitive recursion on a numerical variable is not a consequence of the representability of numbers in intuition, but rather follows from what we mean by (finite) iteration.

I attempted to answer the first, conceptual, question by taking seriously the notion of an *arbitrary* or *generic* object X of a given finitist type, where a finitist type of the first kind is a product $\mathbb{N} \times \cdots \times \mathbb{N}$, $\mathbb{N}$ being the type of the natural numbers, and a finitist type of the second kind is a product whose factors are numerical equations $m = n$. An object of the latter type, if there is any, consists of a proof of each of the factors $m = n$ from the axiom $0 = 0$ using the inference $a = b \Rightarrow a' = b'$. My argument is that one can understand the idea of an arbitrary object of a given finitist type independently of that of the totality of objects of that type; and on its basis, we may proceed to construct objects of possibly other finitist types, which may depend on X. Thus, when X is of finitist type of the first kind, we may construct from it other objects $f(X)$ of types of the first kind. I claimed that, when we identify just what means of construction from X are implicit in the idea of such an arbitrary object, they turn out to be precisely those by means of which we define the primitive recursive functions. Likewise, when we identify, for given finitist functions $f(X)$ and $g(X)$, what constructions of a proof of $f(X) = g(X)$ are implicit in the idea of an arbitrary X, they turn out to yield proofs of exactly those equations deducible in *PRA*.

Concerning the conceptual question, some doubts have been raised about the identification of finitism with *PRA*. For example, Kreisel [20] takes finitism to include quantifier-free induction up to any ordinal below ε_0 and Ignjatović [18, p. 323] writes that my analysis of finitistic reasoning is not beyond any doubt. I have criticized Kreisel's conception in [28]. In footnote 5 (p. 323) Ignjatović writes: "Of course, one cannot rule out the possibility that any basis sufficient to justify what is formalized in (PRA) and which satisfies some necessary closure properties in order to be acceptable as an epistemologically distinguished system of methods, is also sufficient to justify ε_0-induction." But the basis on which [28] derived *PRA* is the finitist types; the claim is that, to go beyond *PRA* requires the introduction of higher types, e.g. of numerical functions and proofs. One would think that this restriction on the types of objects admitted would satisfy the condition of being "an epistemologically distinguished system of methods".

Regarding Gödel's position(s) on the conceptual question, the evidence is not entirely sraightforward. In [4, p. 198], he explicitly denies that his incompleteness theorems undermine Hilbert's attempt to obtain *finitary* consistency proofs, and it seems reasonably clear that he is referring to the conceptual rather than the historical question of the meaning of finitism.[2] In view of this, it would seem that he was open at that time to the possibility that finitism extends beyond *PRA*. On the other hand, in his notes for lectures in 1933 [5] and in 1938 [6], he explicitly attributes to Hilbert the aim of establishing consistency in *PRA*. Moreover, in the first of these he refers to *PRA* as the first in an ascending series of layers of intuitionistic or constructive mathematics [11, p. 51] and, in the second, he refers to it as finitary number theory, the lowest level of a hierarchy of finitist systems [11, p. 93]. In [7, p. 281], however, he refers to Hilbert's definition of finitism as "the mathematics in which evidence rests on what is *intuitive*" and so rejects as finitist in this sense the higher levels of "finitist systems", such as intuitionism and his own system T of primitive recursive functions of finite type.

Kreisel [21, p. 506] accepts the identification of finitism (at least as it is described in the beginning of [16]) with *PRA*, but doubts my argument for it. He writes of [27] that "its central point seems to be that the evidence of each proof has, in some essential way, a strictly *finite* character. As it stands the analysis ... is unconvincing since the understanding of any *one* rule goes beyond this." I would not like to talk about the evidence of a proof, but simply of the proof (i.e. evidence for the proposition). The 'some essential way' in which the proof $f(X)$ has a finite character is that it appeals only to what is implicit in the idea of an arbitrary object X of the given finitist type $\mathbb{N} \times \cdots \times \mathbb{N}$ and does not appeal to higher types of objects, such as numerical functions, proofs, etc. Whether or not this gives the proof a *strictly* finite character, it does seem to me to provide a sense in which we can say that certain numerical functions or proofs of equations, which apply to an infinite number of objects, are finite. It is certainly true that if we finitistically construct $f(X)$ from an arbitrary number X, then the construction applies to obtain $f(0)$, $f(1)$, etc.—and, if we understand the former, then presumably we understand $f(n)$ for each given n. But that is not to say that the validity of the construction $f(X)$ or our understanding of it depends on the validity or our understanding of each of the infinite number of instances of it.

Of course, an analysis of the notion of finitism cannot be presented as a theorem. It is, rather, analogous to Turing's analysis of the notion of a computable real: at one end is an intuitive notion and at the other, its explication in terms of a precise mathematical concept. There are however at least two crucial differences between Turing's analysis and the analysis of the notion of finiteness. The most prominent difference concerns the relative importance of the notions; the other difference is that Turing's analysis has stood up over a long period of time.[3] Far too little attention has been paid to the explication of the idea of finitism for one to feel

[2]He writes of "Hilbert's formalistic viewpoint": "For this viewpoint presupposes only the existence of a consistency proof in which nothing but finitary means of proof is used, and it is conceivable that there exist finitary proofs that *cannot* be expressed in the formalism of P (or M or A)."

[3]This is true both in the sense that his analysis of human computation and his argument that such computation can be carried out by a Turing machine has survived subsequent analysis and in the sense that the class of Turing-computable functions has turned out to contain all the functions that one might, from some other point of view, deem computable. For a discussion of the first sense, see [25]

that my analysis is beyond doubt. But sofar I have seen no serious discussion of it; and so I'm not going to discuss this question further now.

Concerning the second, historical, sense of the question "What is finitism?", there has been some serious discussion: I mention in particular Zach's paper [31]. Hilbert nowhere gives a precise characterization of what he means by this; and, indeed, in [16] the authors, in discussing the extent to which finitist methods include the principles formalized in first-order arithmetic (*PA*), write

> To be sure, even as we have formulated it, this question is not precise; for we have not introduced the expression "finit" as a sharply delimited term, but only as a designation of a methodological guideline which, to be sure, enables us to definitively recognize certain kinds of concept formations and inferences as finitist, certain others as not finitist, but which provides no precise dividing line between those which satisfy the demands of finitist methods and those which do not. [16, p. 361]

However, there is general agreement, well supported by Hilbert's own writings, that he regarded the kinds of concept formation and inferences formalized in primitive recursive arithmetic as finitist.[4] The substantive issue seems to be whether he was *committed* to there being concepts and modes of inference which are not formalizable in *PRA*. I will confine my attention to this question.

There are four sources of data relevant or at least thought to be relevant to the issue: various passages in the two volumes of *Grundlagen der Mathematik* [16, 17], the reference to Ackermann's function in "Über das Unendliche" [12] and Hilbert's so-called 'ω-rule' in "Grundlegung der elementaren Zahlenlehre" [14].

Grundlagen I

Zach mentions one passage in §7 of Volume I which he takes as evidence for the view that the authors took finitism to be more extensive than *PRA*.

> We may, however, admit certain *extensions of the schema of recursion* as well as of the induction schema, without taking away what is characteristic of the method of recursive number theory. [p. 325 (p. 330)][5]

[4]See in particular [16, p. 325 (330)]. Niebergal and Schirn [23] argue that the conception of finitism in [16] extends the original conception that Hilbert had in the 1920's. This view seems to be partly based on the view that finitism in the earlier period was concerned only with metamathematics and so not with numbers. But, of course, for Hilbert the numbers themselves were syntactic objects and so arithmetic was for him was a special case of metamathematics. The authors also seem to believe that Hilbert's finitism was restricted to making statements about particular syntactical objects and admitted no general statements (Π^0_1 statements, as they refer to them). Aside from the fact that they are then unable to give a convincing account of what the statement of consistency for a formal system such as *PA* would be (see pp. 297-302), [13] explicitly states the 'contentual' principle of mathematical induction as a finitist principle. One premise of this principle is surely a general proposition.

[5]Page references to both volumes of *Grundlagen der Mathematik* are to the first edition, with the corresponding page references to the second edition in parentheses.

The subsequent discussion is, first, of various forms of recursion and induction, e.g. simultaneous recursions, which reduce to the primitive forms. But they then go on to introduce examples of nested double recursive definition, namely the enumeration $f_x(y)$ of the primitive recursive functions of y and the Ackermann function, which they point out are unlike the previous examples in not being reducible to primitive recursion.[6] Did they at that time regard such nested double recursions, too, as partaking of what is characteristic of the method of recursive number theory? And, if so, does that imply that they thought at that time that such definitions are finitistically valid? Presumably Zach believes that "what is characteristic of the method of recursive number theory" refers to its finitistic character. But what immediately precedes the above passage is

> The distinction of recursive number theory from intuitive number theory consists in its formal constraints; its only method of concept formation, aside from explicit definition, is the schema of recursion, and also the methods of deduction are strictly circumscribed.

In view of this, I am inclined to take "what is characteristic of" its method to be the kind of definitions (recursion equations) and rules of inference (free-variable induction principles) that are involved. Just preceding the above passages, the authors state explicitly that recursive number theory, *as it had been developed up to that point*, is finitistically valid. Insofar as they can be reduced to *PRA*, this would then be so of what can be obtained by the various principles of definition and proof that they subsequently introduce in §7. However, nothing they write points clearly to an acceptance of the diagonal function or the Ackermann function as finitist. On the other hand, they don't explicitly reject them as finitist, either.

Grundlagen II

There is no doubt that in Volume II there is evidence that the authors regarded finitism as extending beyond *PRA*. Thus, Zach quotes a passage referring back to the passage just quoted from Volume I, in which they write

> Certain methods of finitist mathematics which go beyond recursive number theory (in the original sense) have been discussed already in §7 [of [16]], namely the introduction of functions by nested recursion and the more general induction schema. [17, p. 340 (354)]

But the question is whether or not this reflects their view of finitism in earlier writings, up to the publication of Volume I, which were unaffected either by Gödel's incompleteness theorems or by the Gentzen's success in proof theory using methods that go beyond *PRA*. Already in his introduction to [16], written in March 1934, after the completion of the text, in response to Gödel's incompleteness theorems, Hilbert spoke of the necessity for using the finitist standpoint in a sharper (*schärferen*) way than was required for the treatment of elementary formalisms. This would seem to indicate that he had in mind some resource that he regarded

[6] They prove that $f_x(x) + 1$ is not primitive recursive.

to be already contained in the finitist conception, rather than some more extensive conception, going beyond finitism. On the other hand, in the Introduction to Volume II, Bernays, in reference to the new methods needed to prove the consistency of less elementary formalisms such as PA, speaks of an *extension* of the finitist standpoint. I don't know whether this form of expression constitutes a recognition that these new methods go beyond finitism as originally conceived or whether, like Hilbert, he is refering to new resources contained within the original conception.

Further complicating the issue, on p. 224 the authors write about "contentual finite number theory, which", they say, "indeed is formalized through recursive number theory"; and here they refer back to Volume I. This would seem to contradict the later statement, quoted above, that finitism has no precise boundary; however they go on to write that

> The original narrow concept of a finitist proposition in the field of number theory admits as finitist number theoretic propositions only those which can be expressed in the formalism of recursive number theory [PRA], *possibly* including symbols for certain computable functions ... or which are capable of a stricter interpretation by a proposition of this form. (My emphasis) [p. 348 (362)]

This would seem to leave the boundary of finitism, as conceived in Volume I, indeterminate without commiting it to more than PRA.

But we also see here the expression "the original narrow concept of a finitist proposition", suggesting a wider conception that is still, on their view, finitist. Indeed, they go on to discuss certain propositions (e.g. implications with universal antecedents) the employment of which "appears not to be a violation of the basic ideas of the methodology (methodischen Grundgedanken) of finitist proof theory." In fact, they suggest that, having taken this step, one can extend the methods of finitist proof theory even further.

Certainly, from my own point of view, this amounts to a clear transgression of finitist mathematics, which admits as a limiting case, only, proofs of statements (i.e. functions of type) $\forall x[s(x) = t(x)]$. The authors are now contemplating functions of types $\forall x F(x) \rightarrow \forall x G(x)$ *which take functions (or proofs) as arguments*. At this stage, I don't know what the authors mean by "finitism": I certainly can't understand what it has to do with finite.

Zach does quote from a letter from Bernays to Gödel from 1970:

> These nested recursions ... appear to me to be finite in the same sense as the primitive recursions, i.e., if one regards them as statements of computation procedures where one can recognize that the function defined by the respective process satisfies the recursion equations (for every system of numerical values for the arguments). Indeed, the computation of the value of a function according to a nested recursion, when the numerical values of the arguments are given, comes down to

> the application of several primitive recursions, the number of which is determined by a numerical argument.

But one should note, certainly with admiration but also as a caution, that Bernays wrote this in a letter when he was 82 years old. And, in any case, his reasoning here is defective. To understand his point about the computation of the Ackerman function $f(x, y)$ amounting to the computation of x primitive recursive functions of y, let $f(x, y)$ instead be the enumeration $f_x(y)$ of the primitive recursive functions of y mentioned above. Now, it is true that, for any particular number m, we can construct $f_m[Y]$ from arbitrary Y; but how are we to construct $f_X[Y]$ from arbitrary X and Y? On my analysis, a *particular* primitive recursive function is finitist, but not an 'arbitrary' one f_X. The construction or computation of $f_X[Y]$ from X and Y requires, not the arbitrary iterations X and Y, but the transfinite iteration up to $\omega^X \times Y$, which can make no claim to finiteness.[7]

Über das Unendliche

My original foray into history concerned a reference to the Ackermann function by Hilbert in [12], which Kreisel cites in [21, fn 42, p. 514] as evidence against my thesis that finitism is PRA. I responded to this in [28]; but the matter seems not to be entirely laid to rest. So let me expand on it and convince you that *this* issue is not at all open!

In his paper, Hilbert wanted to sketch, as an application of his proof theory, a proof of the continuum hypothesis CH, and a key to understanding what he intends is to remember that his proof theory involved formalizationj in the ε-calculus and then proving the eliminability of all ε-terms. He assumed that a certain formal system Σ, in which CH can be formulated, is complete; and so it sufficed to prove that $\neg CH$ cannot be derived in Σ. Σ is a many-sorted theory containing the types N of the second number class and Z of the first number class. The negation of CH can be expressed by

$$\forall F : N \to Z^Z \exists g \forall \alpha [F(\alpha) \neq g]$$

(To improve readability, I am using arrows and exponentiation interchangeably.) Using ε-symbols, this takes the form

$$\forall F : N \to Z^Z \forall \alpha [F(\alpha) \neq t(F)]$$

where $t = t(F)$ is a term for a numerical function, built up by means of the ε-operator from F. He now invokes his Lemma I, which he states to be a conseqence of his general principle that every mathematical problem can be solved:

> If a proof of a proposition contradicting the continuum theorem is given in a formalized version with the aid of functions defined by means of the transfinite symbol ε ..., then in this proof these functions can

[7] See [26] for the analysis of nested double recursions (i.e. nested recursions on ω^2) or, more generally, nested α-recursions.

always be replaced by functions defined, without the use of the symbol ε, by means merely of ordinary and transfinite recursion .... [8]

By ordinary and transfinite recursions, he means recursions on Z and on N, respectively. So now we may assume t to be defined by recursions on Z and N and explicit definitions from F. The recursions involved cannot all be assumed to be of the form of primitive recursions; but Hilbert believes that all the recursions involved in the proof can be reduced to primitive recursions in the impredicative sense, if we admit the transfinite hierarchy $\langle N_\alpha \mid \alpha \in N \rangle$ of function types, where $N_0 = N$. *The Ackermann function is introduced at this point as a simple example of a numerical function 'defined by recursion' but not by predicative primitive recursion, and which can then be defined by impredicative primitive recursion on Z.* Call the functions obtained by impredicative primitive recursion relative to this hierarchy $N - PR$. So, $t(F)$ is an $N - PR$ function of F. Now we may define another hierarchy $\langle Z_\alpha \mid \alpha \in N \rangle$, where $Z_0 = Z$, obtaining also the class of $Z - PR$ functions.[9] The class of $Z - PR$ functions has the power of N and so we can take F to be an enumeration of all the $Z - PR$ numerical functions, itself defined by recursion on N. So $t = t(F)$ is a $N - PR$ numerical function. Hilbert gives an argument to show that any $N - PR$ numerical function and, in particular. t is $Z - PR$ and hence in the range of the enumeration F—a contradiction.

Hilbert goes on to say that, even aside from the holes in the argument that need filling, the proof would require "a recasting strictly faithful to the finitist attitude". Clearly what he has in mind is the formalization Σ of the theory of the second number class N, including the theories of the $N - PR$ and $Z - PR$ functions, and the translation of the proof of $\neg CH$ in the ε-calculus into a proof in Σ of a contradiction. Since he presumably believed that a finitist proof of the consistency of Σ could be obtained, he could conclude that there can be no proof of $\neg CH$ and, hence, by completeness, CH is true.

Hilbert's so-called ω-*Rule*

There is also a prevailing confusion concerning Hilbert's paper "Die Grundlegung der elementaren Zahlenlehre" [14], in which he proposes the following "Schlussregel":

If it is proved that the formula

$$A(z)$$

is a true numerical formula for each given numeral z, then the formula

$$\forall x A(x)$$

may be admitted as an initial formula.

[8]This assertion has of course never been proved. In [13] he asserts that it is dispensible for his proof of CH; but it isn't clear to me how he intended to establish this.

[9]This is Gödel's hierarchy extended into the transfinite.

Gödel reviewed this paper and, in the introductory note to the review in [9], Feferman discusses the question of whether or not Hilbert was here reacting to Gödel's incompleteness theorems. But, more to the point, he notes that the principle in question is mis-named the ω-rule. In the first place, the formula $A(x)$ is to be quantifier-free; in fact, it is to be the formalization of a finitistically meaningful property of numbers. But, secondly, in spite of Hilbert calling it a "Schlussregel", it is not even a restricted case of the ω-rule. Rather, it provides a criterion under which the universal formula $\forall x A(x)$ may be taken as an initial formula—i.e. an *axiom* in a deduction. Hilbert is describing a system Σ, which is obtained from the formal system of elementary number theory, with definition by primitive recursion included, by admitting such axioms. Of course, Σ is not itself a formal system—*unless we accept the thesis that finitism* $= PRA$ or some other formal system. But it is misleading to call it a *quasi-formal* system, at least as that term was introduced by Schütte for deductions admitting the ω-rule. One might note also that it is precisely this system Σ which Gentzen described in [3] and to which he applied his consistency proof.

Conclusion

Sofar, on the evidence I have seen, I remain inclined towards the view that there was no commitment to finitist principles that go beyond PRA in Volume I or in earlier writings. In Volume II, by their own admission, the authors are including more under the term "finitist" than was originally included. Perhaps further evidence will settle the historical question decisively.[10]

REFERENCES

[1] P. Bernays, *Die Philosophie der Mathematik und die Hilbertsche Beweistheorie*, ***Blätter für deutsche Philosophie***, vol. 4 (1930–31), pp. 326–367. Reprinted in [2]. A translation by P. Mancosu appears in[22], pp. 234–265.

[2] ———, ***Abhandlungen zur philosophie der mathematik***, Darmstadt: Wissenschaftliche Buchgesellschaft, 1976.

[3] G. Gentzen, *Die Widerspruchfreiheit der reinen Zahlentheorie*, ***Mathematische Annalen***, vol. 112 (1936), pp. 493–565.

[4] K. Gödel, *Über formal unentscheidbare Sätze der* Principia Mathematica *und verwandter Systeme I*, ***Monatshefte für Mathematik und Physik***, vol. 38 (1931), pp. 173–198.

[5] ———, *The present situation in the foundations of mathematics*, ***[11]***, *1933o.

[6] ———, *Lecture at Zilsel's*, ***[11]***, *1938a.

[10] I confess that, up to the time of my talk at Feferfest and aside from noting the passage from p. 224 quoted above and recalling in a general way that the authors had extended the notion of finitism, I h adn't really looked thoroughly at Volume II to see what light it might shed on the earlier conception of finitism; and, in particular, although it had been pointed out to me by Howard Stein shortly after the publication of [28], I had forgotten the reference to nested double recursive functions in the passage on p. 354 until it was pointed out again by Zach. I have since sought expiation by resigning all of my memberships in historical associations.

[7] ———, *Über eine bisher noch nicht benützte Erweiterung des finiten Standpunktes*, ***Dialectica***, vol. 12 (1958), pp. 280–287. Reprinted with an Englsh translation in [10]. [8] is a revised version.

[8] ———, *On an extension of finitary mathematics which has not yet been used*, ***[10]***, 1972. Revised version of [7].

[9] K. Gödel, ***Collected works, vol. I***, Oxford: Oxford University Press, 1986.

[10] ———, ***Collected works, vol. II***, Oxford: Oxford University Press, 1990.

[11] ———, ***Collected works, vol. III***, Oxford: Oxford University Press, 1995.

[12] D. Hilbert, *Über das Unendliche*, ***Mathematische Annalen***, vol. 95 (1926), pp. 161–90. Translated by Stefan Bauer-Mengelberg in [29, 367-92].

[13] ———, *Die Grundlagen der Mathematik*, ***Abhandlungen aus dem mathematischen Seminar der Hamburgischen Universität 6 (1928)***, (1927). Translated in [29] pp. 464–479.

[14] ———, *Die Grundlegung der elementaren Zahlenlehre*, ***Mathematische Annalen***, vol. 104 (1930), pp. 485–494. Reprinted in part in [15].

[15] ———, ***Gesammelte Abhandlungen***, vol. 3, Berlin: Springer, 1935.

[16] D. Hilbert and P. Bernays, ***Grundlagen der Mathematik I***, Berlin: Springer-Verlag, 1934, The second edition was published in 1968.

[17] ———, ***Grundlagen der Mathematik II***, Berlin: Springer-Verlag, 1939. The second edition was published in 1970.

[18] A. Ignjatović, *Hilbert's program and the omega rule*, ***The Journal of Symbolic Logic***, vol. 59 (1994), pp. 322–343.

[19] A. Kino, J. Myhill, and R. Vesley (editors), ***Intuitionism and proof theory***, Amsterdam: North-Holland, 1970.

[20] G. Kreisel, *Ordinal logics and the characterization of informal notions of proof*, ***Proceedings of the International Congress of Mathematicians, Edinburgh***, (1958), pp. 289–299.

[21] ———, *Principles of proof and ordinals implicit in given concepts*, ***[19]***, (1970), pp. 489–516.

[22] P. Mancosu (editor), ***From Brouwer to Hilbert: The debate on the foundations of mathematics in the 1920's***, Oxford: Oxford University Press, 1998.

[23] K-G. Niebergall and M. Schirn, *Hilbert's finitism and the notion of infinity*, ***[24]***, (1998), pp. 271–306.

[24] M. Schirn (editor), ***The philosophy of mathematics today***, Oxford: Clareendon Press, 1998.

[25] W. Sieg, *Calculations by man and machine*, ***Technical report***, Carnegie Mellon, 2000.

[26] W.W. Tait, *Nested recursion*, ***Mathematische Annalen***, vol. 143 (1961), pp. 236–250.

[27] ———, *Constructive reasoning*, ***[30]***, (1967), pp. 185–199.

[28] ———, *Finitism*, ***Journal of Philosophy***, vol. 78 (1981), pp. 524–556.

[29] J. van Heijenoort (editor), ***From Frege to Gödel: A source book in mathematical logic***, Cambridge:Harvard University Press, 1967.

[30] B. van Rootselaar and J.F. Staal (editors), ***Logic, methodology and philosophy of science III***, Amsterdam: North-Holland, 1968.

[31] R. Zach, *Numbers and functions in Hilbert's finitism*, ***Taiwanese Journal for Philosophy and History of Science***, vol. 10 (1998), pp. 33–60.

DEPARTMENT OF PHILOSOPHY
UNIVERSITY OF CHICAGO
CHICAGO, IL 60637, USA
E-mail: wwtx@uchicago.edu

LOGICAL CONSTANTS:
THE VARIABLE FORTUNES OF AN ELUSIVE NOTION

JOHAN VAN BENTHEM

To Sol Feferman, in friendship and admiration

§1. Logic: inferential and expressive power. Logic is usually taken to be the study of reasoning - and the highest rank of nobility for an assertion or an inference is 'logical validity'. There are proof-theoretic methods for describing all validities for a given language, and completeness theorems assure us that we have found 'all there is' to a particular style of reasoning. But in addition to this inferential power, logic is also about expressive power. Reasoning needs a language supplying 'logical forms', and what determines the choice of *these*? Logical languages contain various forms of expression: propositional connectives, quantifiers, or modalities, whose meanings are analyzed in addition to their inferential behavior. But what makes these particular notions 'logical', as opposed to others? Perhaps the usual expressive completeness argument reassures us that the Boolean connectives capture all there is to two-valued propositional reasoning. But no similar result is known for first-order predicate logic, the major working system of modern logic.

This concern is not part of the basic 'agenda' in textbooks, partly because it is seldom raised, and partly because there is no consensus on an answer. Sol Feferman is one of the small group of authors who do think about this basic issue (cf. [9]), and I am happy to contribute a little piece of my own thinking on these matters. But before doing so, let me note that not all great logicians would find the effort worthwhile. E.g., Bernard Bolzano's system, pioneering in so many ways, did not contain any privileged set of logical operations. For him, the distinction logical/non-logical is merely one of methodology. Forms of assertion for a style of reasoning arise by fixing the meanings of some expressions (these will have a 'logic' then), and letting others vary. But this distinction can go in more than one way. Granting this liberality, I still would like to find out what makes the usual 'logical constants' tick.

My aims in this are different from Feferman's excellent analysis, who mentions a 'demarcation' of logic as a major concern - and first-order logic as a preferred target. I myself feel no need for a principled separation of logic from other territories, mathematics, linguistics, computer science, psychology, or whatever. And fixing some target set of expressions, no matter how noble its ancestry, is

Reflections on the foundations of mathematics
Essays in honor of Solomon Feferman
Edited by Wilfried Sieg, Richard Sommer, and Carolyn Talcott

too conservative for me. I prefer accounts of logicality which bring to light interesting 'parameters' that can be set in different ways, giving us a handle on *analogies* with other fields. Note the phrasing here. 'Logical constants' suggests some expressions in general use among logicians, perhaps a finite set. What intrigues me is not the book-keeping job of capturing these - but the analysis of their appealing general features. Out of that we get notions of *logicality*, that many expressions might have to greater or smaller degrees. Now you may find this a slippery slope. Soon, someone who thinks like this will also start doubting everyone's dearest intuitions, and consider even logical *validity* a matter of degree. That is exactly what I do think - but enough of this confessional!

§2. Logicality: proposals from different stances. There are a few scattered attempts in the literature at elucidating intuitive notions of logicality. Not surprisingly, these follow the major stances found in logic. First, there are *semantic* accounts, in terms of *invariance for permutations* of objects in the domain of discourse. What this mathematical condition intends to capture is the intuitive 'topic-neutrality' of logical expressions. They do not depend on any specific individual features of objects or predicates. This proposal, made by many authors independently (including Tarski in [21]) seems the most successful account to date, and it will occupy us in most of this paper. But there are also other semantic aspects of logicality, as we shall see below (cf. [3]). Next, there have been *proof-theoretic* accounts of logicality, addressing different intuitions. E.g., one might think of a logical constant as one supporting a particularly rich set of inference patterns - or as one playing a crucial role in the 'organization of proof'. The latter is the thrust of the expressive completeness theorems in [26], [27]. Alternatively, [12] analyzes logical constants in terms of 'little content', being those expressions that are 'conservative': when added to a proof calculus, they do not imply any new consequences in which their inference rules do not occur.

A word on 'stances'. By this I mean general perspectives in a field, which support a certain methodology - say like 'algebra' and 'geometry' in mathematics. Most logicians recognize a semantic stance on validity ("that which transmits truth"), as well as a proof-theoretic one ("that which follows by natural proof steps"). Are there further natural stances in logic? I think there are. E.g., the *game-theoretic* stance goes back to the appealing intuition that valid arguments are the ones that you can *win* as a player in a debate. This stance, too, carries its view on logicality [13]. In a game-theoretic setting, basic operations have to do with various choices that players make, or role switches. ('Putting yourself in my position' is an eminently logical operation!) Thus, on this stance, logicality would have to do with the basic workings of 'control'. This paper takes a semantic stance. But we discuss connections with some other approaches at the end.

Finally, another plausible necessary condition for logicality is *algorithmic* in nature. Should logical expressions have meanings that are 'easiest to compute' in some way? Or should they rather be the process combinations that make for 'maximal effective computing power'? We will discuss this angle, too. I mention these different angles here in order to show that there are many intuitive aspects to 'logicality', and that we need neither expect, nor desire one simple formal characteristic exhausting the notion.

§3. **Permutation invariance.** In this section, we state the proposal, discuss its known properties, putting it in context. Our purpose here is not original. We just want to make some basic things common knowledge - as the subject is rife with re-inventions of wheels and epicycles.

3.1. The proposal. Logical structure is not about individual peculiarities. Intuitively, 'logical' operators do not care about specific objects inside their arguments. Thus, permuting these should not make a difference to the workings of the operator. Consider logical negations. A set complement $-A$ just returns the outside of A, without inspecting any particular feature of the objects in A. Likewise, a relational complement $-R$ works uniformly on just the structure of R, the 'arrow pattern' of its ordered pairs. Similar observations hold for predicate intersection (conjunctions) and projection (quantifiers). This is in contrast to non-logical predicate operations like 'Dutch' whose action depends on the content of its arguments. ("Dutch dining" makes one's guests pay for themselves, "Dutch climbers" overtake their companions just at the summit.) The common mathematical generalization behind these examples involves *invariance under permutations* π of the underlying universe of individual objects.

DEFNITION. *Let π be a permutation of the base universe U. This can be lifted to sets, relations, and all functions in a finite type hierarchy over U in a canonical way. An object F is permutation-invariant if $\pi(F) = F$ for all permutations π.*

One should not think of this proposal as falling out of the sky. There is an analogy here with something we all know, the semantic *Isomorphism Lemma* for standard logics. Think of operations defined by some formula of the logic with free variables for objects, predicates, etc. If π is an isomorphism between two models $\boldsymbol{M}$ and $\boldsymbol{N}$ of the appropriate similarity type, then one proves - by induction on formulas - that

$$\boldsymbol{M} \models \phi[x, y, \dots, X, Y, \dots] \text{ iff } \boldsymbol{N} \models \phi[\pi(x), \pi(y), \dots, \pi[X], \pi[Y], \dots]$$

Equivalently, this says that π maps the $\boldsymbol{M}$-denotation of ϕ to its $\boldsymbol{N}$-denotation:

$$\pi[[[\phi]]^{\boldsymbol{M}}] = [[\phi]]^{\boldsymbol{N}}$$

which is the obvious cross-model generalization of permutation invariance. This Isomorphism Invariance is a well-known defining constraint on 'logics' in Abstract Model Theory.

3.2. How it works. In concrete cases, this notion has useful practical equivalents. E.g., permutation invariance means either of the following for classical set operations:

$$\pi[O(X, Y, \dots)] = O(\pi[X], \pi[Y], \dots) \text{ ('}O\text{ commutes with }\pi\text{')}$$

$$x \in O(X, Y, \dots) \iff \pi(x) \in O(\pi[X], \pi[Y], \dots) \text{ ('}\pi\text{ respects }O\text{')}$$

Many expressions denote permutation-invariant objects. The account lifts from the standard logical operations in first-order types to other expressions, bringing out unexpected relationships. In particular, many second-order categories in natural language contain invariant items. Useful examples include (a) certain *generalized quantifiers*, such as "ten A are B", "most A are B", taking two predicates (viewed as sets) to an assertion, (b) the *reflexivizer* "self", taking binary verbs R to unary ones $\lambda x \bullet Rxx$. The viewpoint also carries over from natural to programming languages. When the objects are computational states, programs denote binary transition relations, and then the standard *program constructs* such as sequential composition, guarded choice, or iteration, are permutation invariant 'logical constants'.

Let us dispel any air of mystery here. These observations have a simple background in standard logic - since they follow from the above Isomorphism Lemma. Here is why. The Lemma holds for any well-behaved logical language, not just first-order, but also in higher types, or with infinitary expressions. Thus, any definable expression is permutation-invariant. But all examples mentioned so far - and all in the literature that I am aware of - *are* definable in such logical languages. E.g., the above program constructs are operations in Relational Algebra, which have a first-order definition. Likewise, the best accepted readings of linguistic generalized quantifiers like "most" have definitions in higher-order logic which are then invariant by their logical form.

3.3. Classification. Even more interestingly, one can also *reverse* the perspective, and classify all invariant items in simple types in some format of definition, as has been done by many authors. Here is a simple illustration [3]:

FACT. *All invariant set operations must have values $O(X, Y, \dots)$ made up out of exhaustive Boolean combinations of their arguments $X, Y, \dots$*

PROOF. If the value set $O(X, Y, \dots)$ properly overlapped any of the 'minimal zones' in the Venn Diagram for any arguments X, Y, one could then permute two objects in such a zone across the boundary (keeping all other objects fixed), leaving all images $\pi[X]$ equal to X, and yet changing the image set $\pi[O(X, Y, \dots)]$. Note: We get no *uniform* Boolean definition in this way: the value of O may depend on argument sizes. ⊣

More complex classifications are found in [4] and ([5], Chapter 10). These are the attractions of the approach when one comes to it first. Permutation invariance is simple, it covers all standard logical constants and relates them to others across categories - and complete classifications of invariants may be viewed as interesting 'expressive completeness theorems' for certain important types of expression.

Permutation invariance has been proposed in many places independently, in full type-theoretic generality by [21, 15, 3] and others. Papers are scattered across logic, linguistics, and computer science (cf. [22]). Philosophers have caught up a bit later, again in relative isolation: cf. [16, 19] and others. Since the ideas are so simple, one good logic textbook should do the job of making this 'universal knowledge' into 'common knowledge'.

3.4. Historical background. Dual viewpoints on definability of basic concepts and their invariance under suitable transformations date back to the mid-19th century. In the ground-breaking work by Helmholtz on the perceptual foundations of mathematics, invariants of spatial *movement* (moving in a straight line, turning around) correspond to natural primitives for geometrical theories. For instance, my observing that one point lies in between two others will not change, no matter how I view this constellation across my natural movements in space. Thus, primitive 'betweenness' emerges out of experience, not linguistic convention. Movements are given, their invariants force themselves upon us. Later mathematicians took these ideas further. Klein's 'Erlanger Programm' defines a mathematical structure by some family of transformations, which then generate the invariants to be dealt with in one's theories. A major habitat of invariance analysis in mathematics is the theory of Lie groups. Transformations and their invariants as the source of our primitives are also quite alive today in empirical fields like psychology and computational image processing.

Now think of logical constants as the *most robust invariants* of all, resistant to every permutation. The latter are the roughest transformation, leaving no structure intact except bare distinctions between objects. The dark side of this robustness, of course, is that logical constants have little to no content. Moreover, as with Klein's Erlanger turn, the 'transformations' are now in the mind, chosen by us, rather than by Nature. Again, we see the same move here that we noticed earlier. Mathematicians sometimes talk about invariants inside one single structure, and then again across structures. This seems no big issue, and the two viewpoints have been around from the start. Either way, this is the appropriate setting for invariance accounts of logicality, as an extreme case in a broader scientific tradition which is still very much alive.

3.5. General logical results. Permutation invariance has been around before the 1990s, and some of its basic properties deserve to be known more broadly. These include expressive completeness results, but also some types of result that are less obvious from the above. First, the Isomorphism Lemma implied that any operation defined in a logical formalism that does not refer to specific objects (such as first-order logic with identity, higher-order logic, or finite type theory) will be permutation invariant. And the observation is more general. By the same reasoning,

> Formulas in a language with a non-logical vocabulary L define type-theoretic objects that are *invariant for L-isomorphisms.*

Inside one model, these objects are invariant for *L-automorphisms* - i.e., those permutations that leave L-predicates and operations the same. As for converse results, the earliest reference I have found so far is Hermann Weyl's 1926 book "Philosophy of Mathematics and the Natural Sciences" [23], where he observes that all logically definable geometrical relations, starting from basic primitives such as 'betweenness', are invariant under the usual geometric transformations. Weyl then mentions the converse question, *whether all such invariants are logically* *definable*, and says that this is a difficult question which will probably never be solved.

Here are some well-known facts, showing that at least some things of interest have been discovered since. No general equivalence between L-automorphism invariance and L-definability can hold for finitary languages, for cardinality reasons. The only automorphism of $\mathbb{N}$ is the identity - and so there are uncountably many invariant subsets of natural numbers, which outrun the resources of standard logical languages. Things get better on special models, witness the following folklore result.

PROPOSITION (Invariance equals first-order definability on the finite models). *On finite models, the L-automorphism invariant predicates of individuals are exactly the first-order L-definable ones.*

Model theory has a more sophisticated result implying this:

THEOREM (Svenonius' Theorem). *If a predicate Q is invariant for all L-automorphisms of the models for some first-order theory T, then T implies some finite disjunction of explicit definitions for Q in terms of L.*

The Proposition follows by taking T to be the complete first-order theory of the finite model. If one wants to have the Proposition on all models without restrictions, the language must adapt, and become infinitary [17]:

THEOREM (Invariance equals infinitary definability on all models). *On any model, the L-automorphism invariant predicates are those definable in the infinitary language $L_{\infty\infty}(L, =)$.*

There are other ways of approaching this match: cf. the use of categories of Boolean-valued models in [8], which does get by with first-order logic. Going in another direction, one can also lift the analysis to more general linguistic categories of expressions using finite type theory [4, 5]:

THEOREM (Invariance for type-heoretic objects). *For all terms τ_b with n free variable occurrences of types $a_1, \ldots, a_n$, and all permutations π of the individuals, suitably lifted to higher types,*

$$\pi_b([[\tau(u_1, \ldots, u_n)]]) = [[\tau(\pi_{a_1}(u_1), \ldots, \pi_{a_n}(u_n)]].$$

THEOREM (From invariance to definability). *Over* finite *base domains, all permutation invariants in their finite type function hierarchy are explicitly definable in finite type theory.*

These results are not all there is to say in this area. Here is a beautiful classical result [15]:

THEOREM (Invariants need intuitionistic types). *The purely functional types that contain invariant items are just those whose implicational forms are provable in* intuitionistic logic.

The 'truth value' type t is read as "true" or "provable" in this setting. Thus, this explains why there are permutation invariants taking objects to predicates (e.g., the 'Quiner' $\lambda x \bullet \lambda y \bullet y = x$ lives in the functional type $e \to (e \to t))$), while there are no invariant 'choice functions' in the type $(e \to t) \to e$: the latter is not a valid formula.

Here is another type of result. Usually, one allows permutations of entities, while keeping the truth values fixed. As a result every object in the pure t-type hierarchy over the base set $\{0, 1\}$ counts as 'permutation invariant'. Now we know that the operations of first-order type here are all definable by standard Booleans $\neg$, &, $\vee$. Here is an expressive completeness result for the full hierarchy [5], extending the usual reasoning in propositional logic to higher types like $(t \to t) \to t$:

THEOREM. *All objects in the finite t-type hierarchy are definable in the t-typed* lambda calculus *with just Boolean* $\neg$, &, $\vee$.

Thus, at least for 'truth table objects', our usual languages do just the job required. Finally, here is yet another type of question concerning the extent of invariance, which has remained open so far:

> *Find an explicit* counting formula *for the number of invariant items contained in any given finite type.*

Though most of these results are elementary qua proofs, they do show that there is some technical substance to the notion of invariance, backing up its intuitive appeal.

3.6. Cross-model versions. Next, we consider some 'cross-model results' related to logicality. This time, our purpose is to show that some well-known model-theoretic results normally not related to invariance are highly related to the above intra-model versions. After that, we discuss the significance of this perspective. Again, we follow up on the standard Isomorphism Lemma with a simple folklore result.

PROPOSITION. *For finite models* $\boldsymbol{M}, \boldsymbol{N}$ *with tuples of objects* $\boldsymbol{d}$, $\boldsymbol{e}$, *TFAE*:

1. *there is an L-isomorphism between* $\boldsymbol{M}$, $\boldsymbol{N}$ *sending* $\boldsymbol{d}$ *to* $\boldsymbol{e}$
2. $\boldsymbol{d}$, $\boldsymbol{e}$ *satisfy the same first-order formulas in the vocabulary L.*

This result is often interpreted as a constraint on language design. A good language matching a structural similarity relation should be invariant, but it should also detect any pair of 'dissimilar' models by providing some specific definable difference. Thus, expressive power of a language can be measured, either by its being able to *define enough invariants* inside models (our focus in Section 3.4), or by *providing enough distinctions* between models. But the technical connection at this level is not tight yet.

REMARK. *Specialized to single models, the Proposition does not claim directly that all automorphism invariants in a model are definable. It only says that, if a predicate P is invariant, each tuple* $\boldsymbol{e}$ *outside of it will disagree on at least one first-order formula with each tuple* $\boldsymbol{d}$ *inside P. In finite models, these formulas can be 'patched' to provide a first-order definition for P: but in general we need infinitary combinations.*

REMARK. *As a constraint on language design, logicality becomes a 'holistic' notion concerning a whole language, rather than on its separate expressions. This shift would be in line with folklore views among logicians holding that truly 'logical' languages should satisfy desirable 'system properties', a favorite example being an* interpolation theorem. *We will not pursue this holistic take on logicality here.*

But then, something much stronger holds. Here is a generalization to arbitrary models and the infinitary logic $L_{\infty\omega}$ of *Scott's Theorem*, which says that each countable model has one formula in countably infinitary logic defining its isomorphism class [1]:

THEOREM. *For every model* $\boldsymbol{M}$ *and tuple of objects* $\boldsymbol{d}$ *there is a formula* $\phi^{\boldsymbol{M}}(\boldsymbol{d})$ *of the infinitary language* $L_{\infty\omega}$ *such that TFAE*:

1. $\boldsymbol{N}, \boldsymbol{e} \models \phi^{\boldsymbol{M}}(\boldsymbol{d})$
2. $\boldsymbol{N}, \boldsymbol{e}$ *is potentially isomorphic to* $\boldsymbol{M}, \boldsymbol{d}$.

Potential isomorphisms are non-empty families of finite partial isomorphisms which satisfy the usual back-and-forth extension properties. This notion is the set-theoretic 'absolute' analogue of isomorphism. It is easy to derive corresponding intra-model versions for this theorem in terms of invariance for 'potential automorphisms'.

At another extreme of syntactic resources, these results are related to another well-known topic in elementary logic, which is not usually considered as doing the job of analyzing 'logicality'. Consider the usual analysis of first-order logic in terms of *Ehrenfeucht-Fraïssé games*. At least with finite relational vocabularies, we can define models using single formulas up to Duplicator's finite winning levels. In a sense then, the usual Adequacy Theorem is also an invariance characterization of first-order logic!

THEOREM. *For any two models* $\boldsymbol{M}, \boldsymbol{N}$ *with object tuples* $\boldsymbol{d}, \boldsymbol{e}$ *TFAE*

1. $\boldsymbol{M}, \boldsymbol{d}, \boldsymbol{N}, \boldsymbol{e}$ *agree on all first-order formulas up to quantifier depth* k
2. *Duplicator has a winning strategy in the* k*-round comparison game between the two models* $\boldsymbol{M}, \boldsymbol{N}$ *starting from the match* $\boldsymbol{d} - \boldsymbol{e}$.

This is a combination of two intuitive requirements: (a) invariance, plus (b) finiteness. It might be worth the philosophers' while thinking about the plausibility of this line of analysis.

Finally, invariance for cross-model relations is also used in analyzing other languages. E.g., in the *lambda calculus*, Plotkin has partially captured the set of lambda-definable items via invariance for 'logical relations' that may also identify individuals. We shall return to this in Section 4.2.

3.7. Digression: invariance and definability in science. Logical theory does not *quite* fit invariance thinking in the sciences. Physicists and mathematicians work with 'symmetries', invariances inside one model, or 'transformations' across models. Consider the Galilei or Lorentz transformations in physics. These coordinate mappings from, say, $\Re^4$ to itself preserve relevant 'physical assertions', including fundamental laws. But precise languages are never specified, and key results often have a different flavor. E.g., consider the celebrated theorem that "Causality implies the Lorentz Group" [25] - a phrase which a well-bred logician might have trouble parsing ... In Minkowski Space, the Lorentz transformations preserve 'causal precedence': the relation which holds between a point and all points in its future light cone. Zeeman showed that conversely, each automorphism w.r.t. causal precedence is a composition of Lorentz transformations with a few minor domestic ones. Notice that there is no explicit role of any language here. Even so, the result is reminiscent of Robb's logical work on special relativity, showing that the second-order theory of causal precedence suffices for defining the full metric structure of Minkowski space. Here is another interesting gap to be bridged. Consider Beth's Theorem, a classic on logical definability. Its transition from implicit definability (an invariance property!) to explicit definability seems directly applicable to elementary geometrical reasoning. We often 'see' that some geometrical object is determined by others. E.g., the angles of a triangle are fixed once we have the sides. A geometrical rule of thumb then says one should be able to explicitly 'define' the former in terms of the latter, as is indeed the case for angles and sides. But Beth's Theorem only helps us here when the problem can be stated in *first-order* terms (a constraint quite alien to working scientists), and the 'determination' must hold in *all* models of the geometrical theory, not just 'ordinary space' ... It would be of interest to analyze such concrete examples, confronting logical theory with invariance practice.

§**4. Critique.** Let us now look at the merits of permutation invariance as an account of logicality. In this section, I will state what I dislike and what I like about it. Also, I will briefly discuss Feferman's verdict in [9].

4.1. Bad points. First, there is a problem of 'size': *too many items* for comfort *are permutation-invariant*. For instance, in the type of monadic generalized quantifiers (denoting properties of sets), already infinitely many items $\lambda X \bullet Q(X)$ are definable in higher order-logic. Related to this is a second problem. Permutation-invariance is local, and *allows wild variations* across models: say, a quantifier meaning "some" on domains of even size and "all" on domains of odd size. Isomorphism invariance does work across models, but still allows for these variations across different cardinalities. Third, the over-inclusiveness also involves the following. *Arbitrary* infinite set conjunctions and disjunctions, no matter how complex and non-constructive, of invariant objects are themselves invariant. Now these are all 'book-keeping objections': the set of items that passes is 'not quite to our taste'. But there are further problems. To begin with, permutation invariance ignores *further semantic aspects* of logicality, such as intuitive features of 'uniformity' or 'finiteness'. But our main objection is what follows.

Permutation invariance is *blatantly circular* as an account of logicality! And it shares this feature with any transformation/invariant account of a set of notions. For, to get going, we must make a number of prior decisions that manipulate the final outcome:

Which objects are taken? And then, which transformations?

E.g., in geometry, affine transformations are supposed to 'justify' "betweenness" as a geometrical primitive. But betweenness is needed to define these transformations in the first place, when all is said and done. The same holds for logicality. Do we want identity to come out as a logical notion? Then look at permutations, whose definition presupposes the notion of identity (surjective functions respecting non-identities). The case is even more blatant with Boolean truth value operations. These come out as invariant for the totally trivial reason that type-theoretic permutations leave the truth values the same. So, they do not effect truth value operations at all. Evidently, this is no real 'explanation', and permutation invariance is not even a genuine 'test' that the Boolean operations pass. Examples of this circularity are easily multiplied at many places throughout Section 3.

The prior choice of objects is even more insidious. For instance, geometrical invariance works on extensionless 'points', and therefore only derives primitive relations in dependence on some prior notion of basic object. But we might want to say, for instance, that 'betweenness' is a primitive notion in spatial reasoning even without making any commitment as to the underlying objects: extensionless points, primitive extended regions, etc. More generally, should truly logical notions not be *independent* from particular choices of objects over which they are supposed to work? Here is a linguistic example. Natural languages tend to have *two* quantifier systems. One system works on 'count nouns' ("every girl left", "many boys left"), the other on so-called 'mass nouns' ("all wine was spilled", "much wine was spilled"). The first requires what linguists call countable objects, the second so-called measurable ones. Now, the 'same' logical quantifiers seem involved over both these domains. But how to get this out of invariance for transformations? And though a unification may be feasible in this particular case, we will still be left with the worry that we are not really done. We may have to redo the exercise once we find yet other sorts of object. Thus,

permutation invariance cannot serve as an independent foundational account of logicality.

As an independent story of logicality, it is about as circular as - to mention one blatant hoax - the 'justification' of first-order axioms by informal soundness arguments which presuppose all the axioms they are justifying (plus usually a whole lot more).

There, we have said it! But please realize that this is only a knock-down objection if one had fundamentalist designs in the first place. I myself think that there cannot be absolute foundations, neither for logical laws nor for logical notions. Therefore the circularity does not bother me at all, and my story will continue.

4.2. Feferman's proposal. Here are some points from Sol Feferman's discussion in [9]. Its main critical points concerning permutation invariance are its (a) 'making logic dependent on set theory', (b) 'set-theoretic non-absoluteness' and

(c) 'non-uniformity across domains'. The former two are of a mathematical nature, more tangential to my main concerns here. But as for (c), he proposes a stronger version of isomorphism invariance, which is of immediate interest. It specializes a notion from the Lambda Calculus [18], viz. *invariance for logical relations*:

Not just bijections or isomorphisms, but *any* binary relation R between individuals in two models $\boldsymbol{M}$, $\boldsymbol{N}$ can be lifted canonically to a relation between objects in all types:

$$f\ R_{a\bullet b}\ g \text{ iff } \mathrm{proj}^L(f)\ R_a\ \mathrm{proj}^L(g) \textit{ and } \mathrm{proj}^R(f)\ R_b\ \mathrm{proj}^R(g)$$
$$f\ R_{a\to b}\ g \text{ iff } \forall x \in D_a^{\boldsymbol{M}}\ \forall y \in D_a^{\boldsymbol{N}},\ x\ R_a\ y \text{ only if } f(x)\ R_b\ g(y)$$

Now, Plotkin observed that all lambda-terms F denote definable functions $F^{\boldsymbol{M}}$, $F^{\boldsymbol{N}}$ which are *relation – invariant* in the following sense:

> if we feed R-related arguments to a function F, then their F-values are also R-related.

An object F in one model is called 'relation-invariant' if

$$F\ R\ F \text{ for all relations } R.$$

Conversely, Plotkin proved that all relation-invariant objects must be lambda-definable, at least *up to second-order types*. The general situation is discussed in [20]. Variations on this definition have been proposed in ([5], Chapter 12), which lifts Boolean monotonic functions and modal bisimulations in the same style, obtaining richer classes of invariant items. Now here is what Feferman proposes for 'logical invariance'. He takes the base relations to be 'homomorphisms', viz. relations $x\ R\ y$ of the special form

> $y = f(x)$ for some surjective function f from individuals in $\boldsymbol{M}$ to those in $\boldsymbol{N}$

This will make the standard operations of negation ($\neg$), conjunction ($\wedge$) and existential quantification ($\exists$) relation-invariant. Then he proves what amounts to an invariance characterization of first-order logic:

1. all items lambda-definable from $\neg, \wedge, \exists$ are homomorphism-invariant
2. all homomorphism-invariant *monadic quantifiers* are lambda-definable from $\neg, \wedge, \exists$ (in fact, they are definable in monadic first-order logic)

For an assessment of what this achieves, a little technical detail is needed. Here is a typical case. Consider a monadic quantifier taking sets to truth values. Feferman's criterion amounts to saying that the same truth value will be assigned in any 'contraction' to a single argument set. It follows immediately that the invariant items, across the board, are determined by two truth-value assignments: to the empty set, and a non-empty set. Here is a more general perspective on what is going on (not mentioned in [9]), which shows its relation to the above discussion in Sections 3.2, 3.3, 3.6. For operators of the type $F(x_1, \ldots, x_k, A_1, \ldots, A_m)$ with object and predicate arguments, the stated invariance criterion works out very simply to a much more familiar one:

Let f be any surjective strong homomorphism between models $\boldsymbol{M}$, $\boldsymbol{N}$ (i.e., this function respects all predicates both ways, except for identity). Then we must have the following equivalence:

$$F^{M}(d_1, \ldots, d_k, A_1, \ldots, A_m) \text{ iff } F^{N}(f(d_1), \ldots, f(d_k), f[A_1], \ldots, f[A_m]).$$

Thus, we restrict attention to objects that are *invariant for strong homomorphisms*, not just for isomorphisms. As we know from model theory, this holds for all notions definable in any logical language with the shown parameters *without using identity*. This fact explains why the above monadic definitions will do (the lambdas are not crucial), and it also explains why *well-foundedness* will be 'logical' on this account (as discussed in [9]). In this light, the monadic case is rather special, as we can classify all invariant items. But the same huge variety as before will return with operators of a binary relational type $F(R)$. For, there are many equivalence classes for binary relations under strong homomorphism: e.g., all ordinals fall into different ones. And on each of these, the criterion allows us a different identity-free definition.

I conclude that Feferman's account suffers from the same over-generation that permutation invariance has. And this is to be expected. Any reasonable invariance analysis must postulate some relevant equivalence relation on models. On each of its invariance classes, we are free to choose our denotation. Now the relation may be so coarse that only finitely many equivalence classes remain, and no explosion of variety occurs among denotations for 'logical constants'. But then we have given away the game in its definition. Or, like with most interesting semantic equivalence relations, we have infinitely many equivalence classes, and then the variety of denotations will explode.

Another critical point is related. Given the equivalence in the definition of invariance under model relations, the following is bound to happen:

all infinitary Boolean combinations of invariants will be automatically invariant.

This did not matter in the special monadic case, but it will be a side-effect for whatever more complex types we try to analyze. Thus, infinitary Booleans take a 'free ride' on such analyses - and cannot really be considered as having been 'explained'. Finally, in the earlier conceptual terms, Sol's proposal is as 'circular' as any. The choice of the 'homomorphisms' as the relevant equivalence relation invests precisely the amount of 'indifference' needed to engineer the landing in monadic first-order logic.

I conclude that Feferman's account has its merits, but changes nothing in the general limitations of invariance approaches. But to me, this is not a bug, but a feature!

4.3. Good points. As for permutation invariance in general, all the objections so far show that it is not a good *foundational* account of logicality. But then, I do not consider this a great tragedy. First, invariance accounts have the virtue of tying in logicality with ways of thinking in other sciences whose utility is beyond doubt - which is a benefit in itself. Next, even circular accounts of notions can be useful, when they couple 'free-floating' notions in *enlightening* ways to others. (As David Lewis once remarked about another 'logical explication', a fixed *connection* between two weak swaying reeds may be a hard fact.) Invariance can even help pinpoint useful connections. Suppose we find some 'non-logical' item F failing the permutation invariance test. The interesting thing is not this negative information by itself. Usually, analyzing the counter-example will point to 'hidden variables': additional structure that needs to be preserved by a transformation to get the expression as an invariant. For instance, *prepositions* ("in", "behind", "along") are ubiquitous linguistic expressions that are not quite 'logical', but not quite 'non-logical' either. Arbitrary permutations do not preserve them, but the failure suggests a better-tuned class of automorphisms respecting the right information, e.g., suitable *geometrical transformations* of visual scenes (cf. [24]). In other words, the invariance approach comes with a hierarchy of levels, and hence it is 'parametrized' in the sense we were looking for in Section 1.

§5. **A procedural viewpoint.** One attraction of the invariance approach are the analogies it offers across different areas where it is used. In this section, we explore one such analogy, which leads us to a more computational or *dynamic* view of 'logical constants'.

5.1. Bisimulation and modal process invariants. In computer science, notions of similarity between computation spaces (e.g., labeled transition systems) determine a notion of process. For instance, taking *bisimulation* between labeled transition systems for one's process equivalence, one gets the usual modal process logics.

DEFINITION. *A* bisimulation *is a binary relation* $\equiv$ *between states in models* $\boldsymbol{M}$, $\boldsymbol{N}$ *satisfying the following two conditions*:

1. **Atomic Harmony** *If* $x \equiv y$, *then* $\boldsymbol{M}, x \models p$ *iff* $\boldsymbol{N}, y \models p$, *for all proposition letters* p.
2. **Action Zigzag**
   (a) *If* $x \equiv y$ *and* $xR_a z$, *then there exists* u *such that* $yR_a u$ *and* $z \equiv u$,
   (b) *If* $x \equiv y$ *and* $yR_a u$, *then there exists* z *such that* $xR_a z$ *and* $z \equiv u$.

For a more systematic account, cf. ([6], *Chapters* 3-5).

Invariant items are then those properties P of processes that are mimicked by the simulation:

whenever $x \equiv y$, then $\boldsymbol{M}, x \models P$ iff $\boldsymbol{N}, y \models P$

These invariant properties can often be classified in some appropriate language. E.g., with bisimulation, the invariants are the items expressible in (infinitary) *modal logic*. In this setting, the role of 'good' process constructions is this:

they are 'safe,' in the sense of being automatically simulated.

More precisely, consider an operation $F(R, S, \dots)$ sending binary transition relations to a new transition relation. Examples are the usual program constructions of composition, guarded choice, or iteration. Assume we have any two models with a bisimulation $\equiv$ satisfying the back-and-forth clauses for the arguments $R, S, \ldots$

> F is *safe* if this same $\equiv$ automatically satisfies the back-and-forth clauses for the transition relation which is its value $F(R, S, \ldots)$.

Examples of safe operations are *relational composition* ; , *union* $\cup$ (finite or infinite), as well as *strong negation* $\sim$ (being the test that the relation has no successor). Typically non-safe are relational *intersection* and Boolean *complement*. Here is an expressive completeness result from ([6], Chapter 5):

THEOREM. *The safe first-order definable relational operations are precisely those defined by means of* $\{;, \cup, \sim\}$.

Note that these operations are a kind of dynamic counterparts to the usual Booleans in standard propositional logic.

5.2. Logical constants and simulation. Safety is also implicit for logical operations in the inductive proof for the Isomorphism Lemma between two models $\boldsymbol{M}, \boldsymbol{N}$. This proof shows how *evaluating* a formula (a process taking place inside one model) can be *simulated* step by step by passing to another model. This, in effect, is what is achieved by familiar subroutines like:

$$\boldsymbol{M}, \boldsymbol{d} \models \neg\phi \text{ iff } \textit{not } \boldsymbol{M}, \boldsymbol{d} \models \phi \text{ iff } \textit{not } \boldsymbol{N}, f(\boldsymbol{d}) \models \phi \text{ iff } \boldsymbol{N}, f(\boldsymbol{d}) \models \neg\phi$$

Also, when evaluating $\boldsymbol{M}, \boldsymbol{d} \models \exists x\phi$, two routes lead to the same match between object tuples. We can take a witness d for x in $\boldsymbol{M}$, and then pass from $\boldsymbol{M}, d, \boldsymbol{d} \models \phi$ to $\boldsymbol{N}, f(d), f(\boldsymbol{d}) \models \phi$. Or we first pass to the other side, observing that $\boldsymbol{N}, f(\boldsymbol{d}) \models \exists x\phi$, and then find the witness in $\boldsymbol{N}$. This yields exactly the kind of commutative diagram that we also had with bisimulation. Thus, in simple arguments like this, logical constants appear as natural moves in evaluation processes. In a slogan

> *logical constants are* process glue!

Let us apply these ideas to first-order logic. For a process equivalence, take *potential isomorphism* between models $\boldsymbol{M}, \boldsymbol{N}$, used in the theory of the infinitary logic $L_{\infty\omega}$. This is a bisimulation between transition systems whose 'states' are maps from finite sets of variables to objects in the models. The relevant transitions are of two kinds:

testing a fact	(while staying in the same state)
picking an object	(resetting the value of some existing variable, or assigning a value to some fresh variable)

Clearly, these are the only primitive actions in evaluation of first-order formulas, viewed dynamically. All the rest is indeed 'process glue'. Correspondingly, matched states in a potential isomorphism satisfy the same atomic statements, plus the back-and-forth condition of bisimulation with respect to object-picking moves.

Now, to define invariance as simulation 'safety', we need to view first-order formulas as evaluation processes. But what are the corresponding transition relations? Here we take a leaf from 'dynamic predicate logic' [11] which fits very closely with potential isomorphisms as bisimulations (cf. also [10]):

DEFNITION (Update Conditions for Dynamic Predicate Logic).
Atoms are tests

$$s_1[[Pt]]^{\boldsymbol{M}} s_2 \text{ iff } s_1 = s_2 \textit{ and } \boldsymbol{M}, s_1 \models Pt$$

Conjunctions are relational compositions

$$s_1[[\phi \& \psi]]^{\boldsymbol{M}} s_2 \text{ iff } \textit{for some } s_3, s_1[[\phi]]^{\boldsymbol{M}} s_3 \textit{ and } s_3[[\psi]]^{\boldsymbol{M}} s_2$$

Existential quantification are random assignments

$$s_1[[\exists x \phi]]^{\boldsymbol{M}} s_2 \text{ iff } \textit{for some } s_3, s_1 =_x s_3 \textit{ and } s_3[[\psi]]^{\boldsymbol{M}} s_2$$

Negations are strong failure tests

$$s_1[[\sim \phi]]^{\boldsymbol{M}} s_2 \text{ iff } s_1 = s_2 \textit{ and for no } s_3, s_1[[\phi]]^{\boldsymbol{M}} s_3$$

By the earlier modal analysis, clearly safe operations in this setting are (a) dynamic conjunction-as-composition, and (b) dynamic negation as an impossibility-test. Less clear is the role of existential quantification. The proper way of viewing its clause is as (c) a sequential composition of two actions: one of object picking for the prefix $\exists x$, and another of processing the remaining assertion ϕ. Thus, on this dynamic analysis,

an existential quantifier is not a logical constant!

It is rather an atomic semantic action, which needs to be 'glued' to what follows by a 'hidden logical constant', viz. composition. Thus, replacing invariance by safety as an account of logicality has surprising aspects, which look 'non-standard'. On the other hand, the reader may also find them innovative.

5.3. Discussion. Safety effects a drastic reduction in the space of 'logical items'. It is very easy to see that safe operations must also be isomorphism-invariant, but the converse holds by no means. But this 'counting' is not our main concern. Neither is the circularity, which will rear its head once more, as we only get going by settling on some prior notion of 'process equivalence'. In this sense, safety does not change anything fundamental from earlier sections. More interesting, however, is the question whether the above dynamic account has some intrinsic plausibility as a view of logicality. To pursue this, one would need to see whether it generalizes well. A good test case would be simulations for languages with generalized quantifiers.

But our main concern would be another one. Like permutation invariance, safety does not really provide an account of the standard Boolean operations! One reason is that the above transition relations are rather coarse. They only record input-output states of succesful verifications, without recording the fine-structure of evaluation processes 'along the way'. But if we look at the latter, Boolean operations do play a crucial role in the 'control' of where to go. This observation illustrates demonstrates a general point, and at the same time, a further 'parameter' for the present analysis. Not just process equivalences over the same class of structures are a choice point for our invariance analysis. Process representations *themselves* may come at various levels of detail. And the level we pick will determine what 'invariant' or 'safe' items we can discern, and hence, what sort of 'logical items' we will encounter. This point is related to our discussion in Section 4.1, and we think this open-endedness is inevitable. More concretely, we do think the step-by-step fine structure of evaluation is of logical relevance, and we expect to find a more genuine analysis of the Booleans at that level. But at the moment, we have nothing more to offer than a vague sense that this requires an analysis in terms of *evaluation games* [7, 13] of which the above transitions are just the 'externally visible' traces. Some logical constants only emerge at this finer level, as the crucial operations regulating control for the players involved.

The proposals in this section are certainly not the final word. But we hope to have shown at least how invariance thinking can still offer genuinely new perspectives on logicality.

§6. **Further semantic aspects.** Does logicality have intuitive semantic aspects that are different from invariance? There is some evidence for this in the linguistic literature on *Generalized Quantifier Theory* [14]. We review some relevant themes here, using notions and results from [3] to show concretely how one can broaden the analysis of logicality.

6.1. Determiners and quantifiers. Linguistic determiners are expressions like "all", "most", "three", "enough", that take common nouns and verb phrases to form sentences according to the basic pattern

$$Det\ N\ VP$$

Across languages, determiners satisfy some universal semantic constraints, such as *Conservativity*, restricting the statement made to just those objects which satisfy the first argument:

$$Det\ AB \Leftrightarrow Det\ A\ (B \cap A)$$

Next, imposing permutation invariance for special determiners of 'quantification' allows us to represent these linguistic 'quantifiers' Q *numerically*, as the set of all ordered pairs

> (a, b) for for which there exists some pair of sets A, B with $Q\ AB$ such that $|A - B| = a$, $|A \cap B| = b$

This representation has been used extensively for semantic analysis of quantifier classes, via the so-called 'Tree of Numbers'.

6.2. Smoothness and monotonicity. Basic quantifier expressions in natural languages are not just permutation invariant. A semantic feature with much extra bite is their

> *persistence across related situations*

which can be viewed either as a semantic property of 'smoothness', or as an inferential one of 'preservation'. More precisely, all basic quantifiers are *monotone*, either upward or downward, in both their arguments. E.g., "all" satisfies the two syllogistic inference patterns:

$$\frac{Q\,AB \quad B \subseteq B'}{Q\,AB'} \qquad \frac{Q\,AB \quad A' \subseteq A}{Q\,A'B}$$

We can use this property to characterize the four classical quantifiers "all", "some", "no", "not all". For this we need one extra semantic condition:

Variety If A is non-empty, there exist B, B' with $Q\,AB$, $\neg Q\,AB'$.

FACT. *The four quantifiers in the Square of Opposition are the only determiners that satisfy Conservativity, Double Monotonicity, and Variety.*

PROOF. Let $Q\,XY$ be any determiner satisfying these three conditions. We do one case, the others are similar. Suppose Q has monotonicity type 'left-down, right-up'.

CLAIM. *Q is "all", i.e., $Q\,XY$ iff $X \subseteq Y$*

First, by Variety, we have $Q\,AB$ for some A, B. Hence we have $Q\emptyset B$ ($Q\,AB$ plus left downward monotonicity) and hence $Q\,\emptyset\emptyset$ (conservativity) and so $Q\,\emptyset Y$ (right upward monotonicity).

From right to left. (a) If X is empty, then we have just shown $Q\,\emptyset Y$. (b) If X is non-empty, then we have $Q\,XA$ (variety), $Q\,X(A \cap X)$ (conservativity), and so $Q\,XY$ (by right upward monotonicity).

From left to right. Suppose $Q\,XY$, but not $X \subseteq Y$. By left downward monotonicity, $Q\,(X - Y)Y$, and by conservativity, $Q\,(X - Y)\emptyset$. By right upward monotonicity then $Q\,(X - Y)Z$ for all sets Z, which contradicts variety for the non-empty set $X - Y$. ⊣

Again, monotonicity is a semantic 'parameter' which expressions can have in degrees. E.g., the non-standard quantifier "most" is still right upward monotone, but it lacks monotonicity either way for its left argument. Without Variety, more quantifiers make the grade, including "at least three" or "at most five". Here is a more complicated expressive completeness result [3].

THEOREM. *The 'doubly monotone' quantifiers on finite universes are the four classical ones plus all numerical variants "at least n", "at most n", etc.*

A nice proof is by geometrical inspection of the number representation for quantifiers. Right upward monotonicity makes the set Q of ordered pairs (a, b) closed under rightward steps $(a + 1, b) \Rightarrow (a, b + 1)$, left upward monotonicity makes Q closed under steps $(a, b) \Rightarrow (a + 1, b)$ and $(a, b + 1)$ that 'shift levels'. The geometrical shapes for quantifiers Q allowed by these conditions are easily enumerated. Higher-order quantifiers such as "most" have a more complex, but still regular 'number geometry'.

Monotonicity is a ubiquitous phenomenon in natural reasoning and semantics. The basic lexical items in the determiner category are monotone across all languages. Surely, this is also a key feature of 'logical' constants: stable description of situations, and supporting corresponding inferences.

6.3. Simple computability and semantic automata. Another attractive feature of basic logical items is their *simple uniform computability*. For the basic quantifiers, this involves very simple procedures: so-called *semantic automata*. E.g., to compute whether "all AB" we just need a finite automaton with two states. This will survey the relevant set of objects A (which suffices by conservativity) in some order, reading in symbols a for each object in $A - B$, and b for each object in $A \cap B$: We start in a recognizing state, and stay there as long as we read symbols

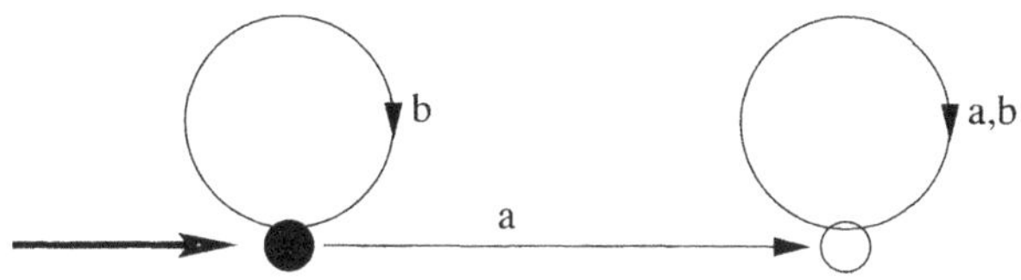

b. As soon as a symbol a is encountered, we move to a second, rejecting state and stay there, whatever is read afterwards. The general result here is this:

THEOREM. *The first-order definable quantifiers are precisely those computed by* acyclic finite automata.

Again we have a 'parameter' now, as we can vary the type of automaton to get further levels of semantic computability. Arbitrary finite automata add the power of counting (say, "an odd number of"). The next step would be adding memory storage. Indeed, the semantic test procedure for a quantifier like "most" needs the unbounded finite memory of a *push-down store automaton*. Push-down storage seems the most complex computational procedure needed for natural language quantifiers. [3] characterizes the corresponding numerical content as follows:

THEOREM. *The quantifiers computable by push-down automata are precisely those whose numerical conditions on the characteristic numbers* (a, b) *with* $a = |A - B|, b = |A \cap B|$ *are definable in additive 'Pressburger Arithmetic'.*

Pressburger Arithmetic is decidable, and hence many questions about natural language quantifiers will also be. Thus, the usual Automata Hierarchy now provides fine-structure levels for logical constants, via their corresponding computational requirements.

6.4. What this means. Generalized Quantifier Theory shows how we can zoom in on the standard quantifiers, amidst all permutation-invariant ones, through various natural additional semantic and computational properties. Now this harmony may be a lucky coincidence between a bunch of different notions. This suggests that the core group of 'logical constants' is best analyzed as an intersection of different natural kinds of expression: they are clever, attractive, and good cooks into the bargain ...

§7. Comparing stances. We have analyzed logicality in a semantic perspective here. But relations to other logical stances have occurred occasionally: computational (Sections 5, 6.3), game-theoretic (Section 5), and inferential (Section

6.2). There are certainly independent intuitions along the latter lines that should be explored. In particular, a sustained analysis of the proof-theoretic stance on logicality might be very illuminating, throwing also additional light on semantic invariance analyses. For instance,

why are the core invariant items so rich in valid inferences?

I know of no good reason; but one would want to understand the phenomenon. Likewise, I would want to couple the semantic and algorithmic aspects more closely. But to avoid a misunderstanding, on my philosophy, no 'completeness theorem' is needed stating that all these approaches 'produce the same yield'. Indeed, even in the analysis of logical consequence, completeness seems often somewhat engineered. And in any case, it does not show that the different logical stances 'reduce' to each other.

Of more interest to me would be *combinations* of perspectives. E.g., intuitively, reasoning seems to involve two kinds of step. Standard inferences are 'local'. We investigate some situation, observe A, and know for general reasons that B 'follows' in this same situation. But often we want to use A to derive a statement about some *other* situation, perhaps even couched in a different language, provided it is suitably related. The latter step is closer in spirit to invariance, which allows us to cross models. The two activities together seem basic to how we reason in practice. To account for this theoretically, a notion of 'entailment along a model relation' was proposed in [2], allowing inference from one model to its submodels, bisimilar models, homomorphic images, and whatever jump seems relevant.

Likewise, this combined view makes us think about designing inference systems and languages *at the same time*, and this joint task may put its own constraints on 'logicality'. One wants the right 'balance' between the vocabulary of a proof system, and its expressive power on corresponding models. E.g., *intuitionistic logic*, ignored in the classical setting of this paper, makes stronger demands on logical consequence, demanding *constructive* provability. What are its natural 'invariants'? (Bisimulations between Kripke models have been suggested in this connection.) Is there any reason to assume that the classical 'logical constants' are the right language from a constructivist point of view? Similar issues arise with *non-monotonic logics* in AI. Their designers question the laws of classical logic, replacing valid consequence by new schemes like "the conclusion only has to be true in all *most-preferred* models of the premises". But they do stick to the language in which the old dogmas were formulated. True radicals would seek a balance, designing a set of logical constants for non-monotonic reasoning that reflects the new preference structure up to some suitable invariance.

§8. Conclusions. Our discussion of logicality has not produced a demarkation of logic. This is good. The essence of 'logic' does not lie in canonical lists of 'logical laws' or 'logical expressions' anyway, but in its general properties and methods. What we did present were the basic facts about the semantic invariance aspect of logicality. This account is circular from a foundational point of view, but suggestive and useful all the same. Moreover, it still has the potential of suggesting new views of logical constants, witness our dynamic analysis of logicality as safety for semantic process operations. But there are certainly other aspects to logicality, coming from different stances: proof-theoretic, or game-theoretic, which eventually deserve an equal opportunity.

Ah, and yes, whatever reservations one may have, the basic invariance folklore definitely belongs in logic textbooks!

REFERENCES

[1] J. BARWISE, ***Admissible Sets and Structures***, Springer, Berlin, 1975.

[2] J. BARWISE and J. VAN BENTHEM, *Interpolation, Preservation, and Pebble Games*, ***Journal of Symbolic Logic***, vol. 64 (1999), no. 2, pp. 881–903.

[3] J. VAN BENTHEM, ***Essays in Logical Semantics***, Reidel, Dordrecht, 1986.

[4] ———, *Logical Constants across Varying Types*, ***Notre Dame Journal of Formal Logic***, vol. 30 (1989), no. 3, pp. 315–342.

[5] ———, ***Language in Action***, North Holland, Amsterdam, 1991 (new edition with addenda. MIT Press, Boston 1995).

[6] ———, ***Exploring Logical Dynamics***, CSLI Publications, Stanford, 1996.

[7] ———, *Logic and Games*, ***Technical report***, (ILLC-X-2000-03) ILLC, University of Amsterdam, 2000.

[8] C. BUTZ and I. MOERDIJK, *An Elementary Definability Theorem for First-Order Logic*, ***Journal of Symbolic Logic***, vol. 64 (1999), no. 3, pp. 1028–1036.

[9] S. FEFERMAN, *Logic, Logics, Logicism*, ***Notre Dame Journal of Formal Logic***, vol. 40 (1999), special issue: George Boolos.

[10] T. FERNANDO, *Bisimulations and Predicate Logic*, ***Journal of Symbolic Logic***, vol. 59 (1994), pp. 924–944.

[11] J. GROENENDIJK and M. STOKHOF, *Dynamic Predicate Logic*, ***Linguistics and Philosophy***, vol. 14 (1991), pp. 39–100.

[12] I. HACKING, *What is Logic?*, ***Journal of Philosophy***, vol. 76 (1979), pp. 285–319.

[13] J. HINTIKKA, ***Logic, Language, Games and Information***, Clarendon, Oxford, 1973.

[14] E. KEENAN and D. WESTERSTÅHL, *Quantifiers*, ***Handbook of Logic and Language*** (J. van Benthem and A. ter Meulen, editors), Elsevier, Amsterdam, 1997.

[15] A. LÄUCHLI, *An Abstract Notion of Realizability for which the Predicate Calculus is Complete*, ***Intuitionism and Proof Theory*** (J. Myhill, A. Kino, and A. Vesley, editors), North-Holland, Amsterdam, 1970, pp. 227–234.

[16] T. MCCARTHY, *The Idea of a Logical Constant*, ***Journal of Philosophy***, vol. 78 (1981), pp. 499–523.

[17] V. MCGEE, *Logical Operations*, ***Journal of Philosophical Logic***, vol. 25 (1996), pp. 567–580.

[18] G. PLOTKIN, *Lambda Definability in the the Full Type Hierarchy*, ***To H.B. Curry: Essays on Combinatory Logic, Lambda Calculus and Formalism*** (J. Seldin and J. Hindley, editors), Academic Press, London, 1980, pp. 363–373.

[19] G. SHER, ***The Bounds of Logic***, MIT Press, Cambridge Mass., 1991.

[20] R. STATMAN, *Completeness, Invariance, and λ-Definabilty*, ***Journal of Symbolic Logic***, vol. 47 (1982), pp. 17–26.

[21] A. TARSKI, *What are Logical Notions?*, ***History and Philosophy of Logic***, vol. 7 (1986), pp. 143–154.

[22] B. TRAKHTENBROT, *On 'Logical Relations' in Program Semantics*, ***Mathematical Logic and its Applications*** (D. Skordev, editor), Plenum Press, New York, 1987, pp. 213–229.

[23] H. Weyl, ***Philosophy of Mathematics and Natural Science***, Atheneum, New York, 1963 (translated reprint from 1926).

[24] Y. Winter and J. Zwarts, *A Semantic Characterization of Locative Prepositional Phrases*, ***Proceedings of Semantics and Linguistic Theory (SALT 97)***, 1997.

[25] E.C. Zeeman, *Causality implies the Lorentz Group*, ***Journal of Mathematical Physics***, vol. 5 (1964), pp. 490–493.

[26] J. Zucker, *The Adequacy Problem for Classical Logic*, ***Journal of Philosophical Logic***, vol. 7 (1978), pp. 517–535.

[27] J. Zucker and R. Tragesser, *The Adequacy Problem for Inferential Logic*, ***Journal of Philosophical Logic***, vol. 7 (1978), pp. 501–516.

INSTITUTE FOR LOGIC, LANGUAGE AND COMPUTATION (ILLC)
UNIVERSITY OF AMSTERDAM
PLANTAGE MUIDERGRACHT 24
1018 TV AMSTERDAM, THE NETHERLANDS
and
CENTER FOR THE STUDY OF LANGUAGE AND INFORMATION
STANFORD UNIVERSITY
VENTURA HALL, STANFORD 94305-4115, USA
E-mail: johan@science.uva.nl
URL: http://turing.wins.uva.nl/~johan

SYMPOSIUM PROGRAM

REFLECTIONS

A Symposium Honoring Solomon Feferman

on his 70th Birthday

December 11–13 1998, Stanford University

Organizing Committee:
Jon Barwise (co-chair), Wilfried Sieg (co-chair),
Rick Sommer, Carolyn Talcott

This symposium (a.k.a. the Feferfest) is centered around proof theoretically inspired foundational investigations. Solomon Feferman has been a contributor to these investigations during the last forty years in a most systematic and significant way, and the main themes of the Symposium are themes in his work. The Symposium is a tribute to him on the occasion of his 70-th birthday—a tribute both to his specific contributions and to his influence on the direction of current research.

Program

December 11; 100 Cordura Hall

Opening remarks. J. Barwise

Session I: Proof theoretic ordinals
(Chair: G. Mints).

W. Pohlers; Proof-theoretic ordinals for theories in the language of (second order) arithmetic and set theory
J. Avigad; Ordinal analysis without proofs
R. Sommer; Iterating reflection

Session II: Foundational reductions
(Chair: W. Sieg).

P. Martin-Löf; Modelling versus Tarski semantics.
J. Barwise; Symbolic and presymbolic logic
J. van Bentham; Logical constants: the variable fortunes of an elusive notion

December 12; Gates Building, B03

Session III: Formalizations in restricted systems (Chair: S. Buss).

G. Takeuti; Gödel sentences of bounded arithmetic

R. Constable; Admiring proof reflections and working with them

U. Kohlenbach; Classical analysis in weak systems of finite type

S. Simpson; Predicativity: the outer limits

Session IV: Applicative and self-applicative theories (Chair: M. Beeson).

D. Scott; Project update: logics of types and computations

M. Rathjen; Monotone inductive definitions in explicit mathematics

A. Cantini; On extensionality, uniformity and comprehension in explicit mathematics

I. Mason/ C. Talcott; Feferman-Landin logic

December 13; Gates Building, B03

Session V: Philosophy and history of modern mathematical thought (Chair: D. Follesdal).

C. Parsons; Reflections on predicativity

J. Dawson; The unity of mathematics—a foundational touchstone?

W. Sieg; Calculations by man and machine

W. Tait; Some remarks about finitism

Session VI: Generalized computation and reflective closure (Chair: J. Etchemendy).

S. Wainer; Accessible recursive functions

J. Fenstad; Computability theory: structure or algorithms?

T. Strahm; Reflective closures of formal systems

Closing remarks: W. Sieg

LECTURE NOTES IN LOGIC

General Remarks

This series is intended to serve researchers, teachers, and students in the field of symbolic logic, broadly interpreted. The aim of the series is to bring publications to the logic community with the least possible delay and to provide rapid dissemination of the latest research. Scientific quality is the overriding criterion by which submissions are evaluated.

Books in the Lecture Notes in Logic series are printed by photo-offset from master copy prepared using LaTeX or (preferably) AMS-LaTeX and the ASL style files. For this purpose the Association for Symbolic Logic provides technical instructions to authors. Careful preparation of manuscripts will help keep production time short, reduce costs, and ensure quality of appearance of the finished book. Authors receive 50 free copies of their book. No royalty is paid on LNL volumes.

Commitment to publish may be made by letter of intent rather than by signing a formal contract, at the discretion of the ASL Publisher. The Association for Symbolic Logic secures the copyright for each volume.

The editors prefer email contact and encourage electronic submissions.

Editorial Board

Editorial Policy

1. Submissions are invited in the following categories:

i) Research monographs
ii) Lecture and seminar notes
iii) Reports of meetings
iv) Texts which are out of print

Those considering a project which might be suitable for the series are strongly advised to contact the publisher or the series editors at an early stage.

2. Categories i) and ii). These categories will be emphasized by Lecture Notes in Logic and are normally reserved for works written by one or two authors. The goal is to report new developments quickly, informally, and in a way that will make them accessible to non-specialists. Books in these categories should include

– at least 100 pages of text;
– a table of contents;
– an informative introduction, perhaps with some historical remarks, which should be accessible to readers unfamiliar with the topic treated;
– a subject index.

In the evaluation of submissions, timeliness of the work is an important criterion. Texts should be well-rounded and reasonably self-contained. In most cases the work will contain results of others as well as those of the authors. In each case, the author(s) should provide sufficient motivation, examples, and applications. In this respect, Ph.D. theses will be suitable for this series only when they are of exceptional interest and of high expository quality.

Proposals for volumes in this category should be submitted (preferably in duplicate) to one of the series editors, and will be refereed. A provisional judgment on the acceptability of a project can be based on partial information about the work: a first draft, or a detailed outline describing the contents of each chapter, the estimated length, a bibliography, and one or two sample chapters. A final decision whether to accept will rest on an evaluation of the completed work.

3. Category iii). Reports of meetings will be considered for publication provided that they are of lasting interest. In exceptional cases, other multi-authored volumes may be considered in this category. One or more expert participant(s) will act as the scientific editor(s) of the volume. They select the papers which are suitable for inclusion and have them individually refereed as for a journal. Organizers should contact the Managing Editor of Lecture Notes in Logic in the early planning stages.

4. Category iv). This category provides an avenue whereby out-of-print books that are still in demand can be made available to a new generation of logicians.

5. Format. Works in English are preferred. After the manuscript is accepted in its final form, an electronic copy in LaTeX or (preferably) AMS-LaTeX format will be appreciated and will advance considerably the publication date of the book. Authors are strongly urged to seek typesetting instructions from the Association for Symbolic Logic at an early stage of their manuscript preparation.

Lecture Notes in Logic

From 1993 to 1999 this series was published under an agreement between the Association for Symbolic Logic and Springer-Verlag. Since 1999 the ASL is Publisher and A K Peters, Ltd. is Co-publisher. The ASL is committed to keeping all books in the series in print.

Current information may be found at `http://www.aslonline.org`, the ASL Web site. Editorial and submission policies and the list of Editors may also be found above.

Previously published books in the *Lecture Notes in Logic* are:

1. *Recursion theory.* J. R. Shoenfield. (1993, reprinted 2001; 84 pp.)

2. *Logic Colloquium '90; Proceedings of the Annual European Summer Meeting of the Association for Symbolic Logic, held in Helsinki, Finland, July 15–22, 1990.* Eds. J. Oikkonen and J. Väänänen. (1993, reprinted 2001; 305 pp.)

3. *Fine structure and iteration trees.* W. Mitchell and J. Steel. (1994; 130 pp.)

4. *Descriptive set theory and forcing: how to prove theorems about Borel sets the hard way.* A. W. Miller. (1995; 130 pp.)

5. *Model theory of fields.* D. Marker, M. Messmer, and A. Pillay. (1996; 154 pp.)

6. *Gödel '96; Logical foundations of mathematics, computer science and physics; Kurt Gödel's legacy. Brno, Czech Republic, August 1996, Proceedings.* Ed. P. Hajek. (1996, reprinted 2001; 322 pp.)

7. *A general algebraic semantics for sentential objects.* J. M. Font and R. Jansana. (1996; 135 pp.)

8. *The core model iterability problem.* J. Steel. (1997; 112 pp.)

9. *Bounded variable logics and counting.* M. Otto. (1997; 183 pp.)

10. *Aspects of incompleteness.* P. Lindstrom. (1997; 133 pp.)

11. *Logic Colloquium '95; Proceedings of the Annual European Summer Meeting of the Association for Symbolic Logic, held in Haifa, Israel, August 9–18, 1995.* Eds. J. A. Makowsky and E. V. Ravve. (1998; 364 pp.)

12. *Logic Colloquium '96; Proceedings of the Colloquium held in San Sebastian, Spain, July 9–15, 1996.* Eds. J. M. Larrazabal, D. Lascar, and G. Mints. (1998; 268 pp.)

13. *Logic Colloquium '98; Proceedings of the Annual European Summer Meeting of the Association for Symbolic Logic, held in Prague, Czech Republic, August 9–15, 1998.* Eds. S. R. Buss, P. Hájek, and P. Pudlák. (2000; 541 pp.)

14. *Model Theory of Stochastic Processes.* S. Fajardo and H. J. Keisler. (2002; 136 pp.)

15. *Reflections on the Foundations of Mathematics; Essays in honor of Solomon Feferman.* Eds. W. Seig, R. Sommer, and C. Talcott. (2002; 444 pp.)

For Product Safety Concerns and Information please contact our EU representative GPSR@taylorandfrancis.com Taylor & Francis Verlag GmbH, Kaufingerstraße 24, 80331 München, Germany

Printed and bound by CPI Group (UK) Ltd, Croydon, CR0 4YY
14/09/2026
02216240-0005